AF717637

Monographs in

Number Theory

Volume 12

Analytic and Combinatorial Number Theory: The Legacy of Ramanujan

Contributions in Honor of Bruce C. Berndt

Monographs in Number Theory ISSN 1793-8341

Series Editors: Bruce C. Berndt
(*University of Illinois at Urbana-Champaign, USA*)

Wadim Zudilin
(*Radboud University, The Netherlands*)

Editorial Board Members:

William Duke (*University of California at Los Angeles, USA*)
Wen-Ching Winnie Li (*The Pennsylvania State University, USA*)
Kannan Soundararajan (*Stanford University, USA*)

Published

Vol. 12 *Analytic and Combinatorial Number Theory: The Legacy of Ramanujan Contributions in Honor of Bruce C. Berndt*
edited by George E. Andrews, Michael Filaseta & Ae Ja Yee

Vol. 11 *Analytic Methods in Number Theory: When Complex Numbers Count*
by Wadim Zudilin

Vol. 10 *Recent Progress on Topics of Ramanujan Sums and Cotangent Sums Associated with the Riemann Hypothesis*
by Helmut Maier, Michael Th Rassias & Lázló Tóth

Vol. 9 *Modular and Automorphic Forms & Beyond*
by Hossein Movasati

Vol. 8 *Topics and Methods in q-Series*
by James Mc Laughlin

Vol. 7 *The Theory of Multiple Zeta Values with Applications in Combinatorics*
by Minking Eie

Vol. 6 *Development of Elliptic Functions According to Ramanujan*
by K. Venkatachaliengar, edited and revised by Shaun Cooper

Vol. 5 *Hecke's Theory of Modular Forms and Dirichlet Series (2nd Printing with Revisions)*
by Bruce C. Berndt & Marvin I. Knopp

Vol. 4 *Number Theory: An Elementary Introduction Through Diophantine Problems*
by Daniel Duverney

Vol. 3 *Analytic Number Theory for Undergraduates*
by Heng Huat Chan

Vol. 2 *Topics in Number Theory*
by Minking Eie

The complete list of the published volumes in the series can also be found at
https://www.worldscientific.com/series/mnt

Monographs in Number Theory

Volume 12

Analytic and Combinatorial Number Theory: The Legacy of Ramanujan

Contributions in Honor of Bruce C. Berndt

Editors

George E. Andrews
The Pennsylvania State University, USA

Michael Filaseta
University of South Carolina, USA

Ae Ja Yee
The Pennsylvania State University, USA

World Scientific

NEW JERSEY • LONDON • SINGAPORE • BEIJING • SHANGHAI • HONG KONG • TAIPEI • CHENNAI

Published by

World Scientific Publishing Co. Pte. Ltd.
5 Toh Tuck Link, Singapore 596224
USA office: 27 Warren Street, Suite 401-402, Hackensack, NJ 07601
UK office: 57 Shelton Street, Covent Garden, London WC2H 9HE

British Library Cataloguing-in-Publication Data
A catalogue record for this book is available from the British Library.

Monographs in Number Theory — Vol. 12
ANALYTIC AND COMBINATORIAL NUMBER THEORY: THE LEGACY OF RAMANUJAN
Contributions in Honor of Bruce C. Berndt

Copyright © 2024 by World Scientific Publishing Co. Pte. Ltd.

All rights reserved. This book, or parts thereof, may not be reproduced in any form or by any means, electronic or mechanical, including photocopying, recording or any information storage and retrieval system now known or to be invented, without written permission from the publisher.

For photocopying of material in this volume, please pay a copying fee through the Copyright Clearance Center, Inc., 222 Rosewood Drive, Danvers, MA 01923, USA. In this case permission to photocopy is not required from the publisher.

ISBN 978-981-12-7736-8 (hardcover)
ISBN 978-981-12-7737-5 (ebook for institutions)
ISBN 978-981-12-7738-2 (ebook for individuals)

For any available supplementary material, please visit
https://www.worldscienti ic.com/worldscibooks/10.1142/13440#t=suppl

© 2024 World Scientific Publishing Company
https://doi.org/10.1142/9789811277375_fm

Contents

© 2024 World Scientific Publishing Company
https://doi.org/10.1142/9789811277375_fm

Photo gallery

Bruce Berndt, February 7, 2021.

Bruce Berndt holding Ramanujan's school slate.

Bruce with a subset of his students and postdocs. Front row from left to right: Amita Malik, Bruce Berndt, Helen Berndt, Heekyoung Hahn, Soon-Yi Kang. Row 2: Byungchan Kim, Chadwick Gugg, Heng Huat Chan, Paul Bialek. Row 3: Song Heng Chan, Jaebum Sohn, Hannah Burson, Victoria Reuter, Atul Dixit, Doug Bowman, Ae Ja Yee. Row 4: Youn-Seo Choi, Tim Huber, Zhu Cao, Dennis Eichhorn, Armin Straub.

A coffee mug gifted by Bruce's family, one for each participant.

Group photo of the conference participants.

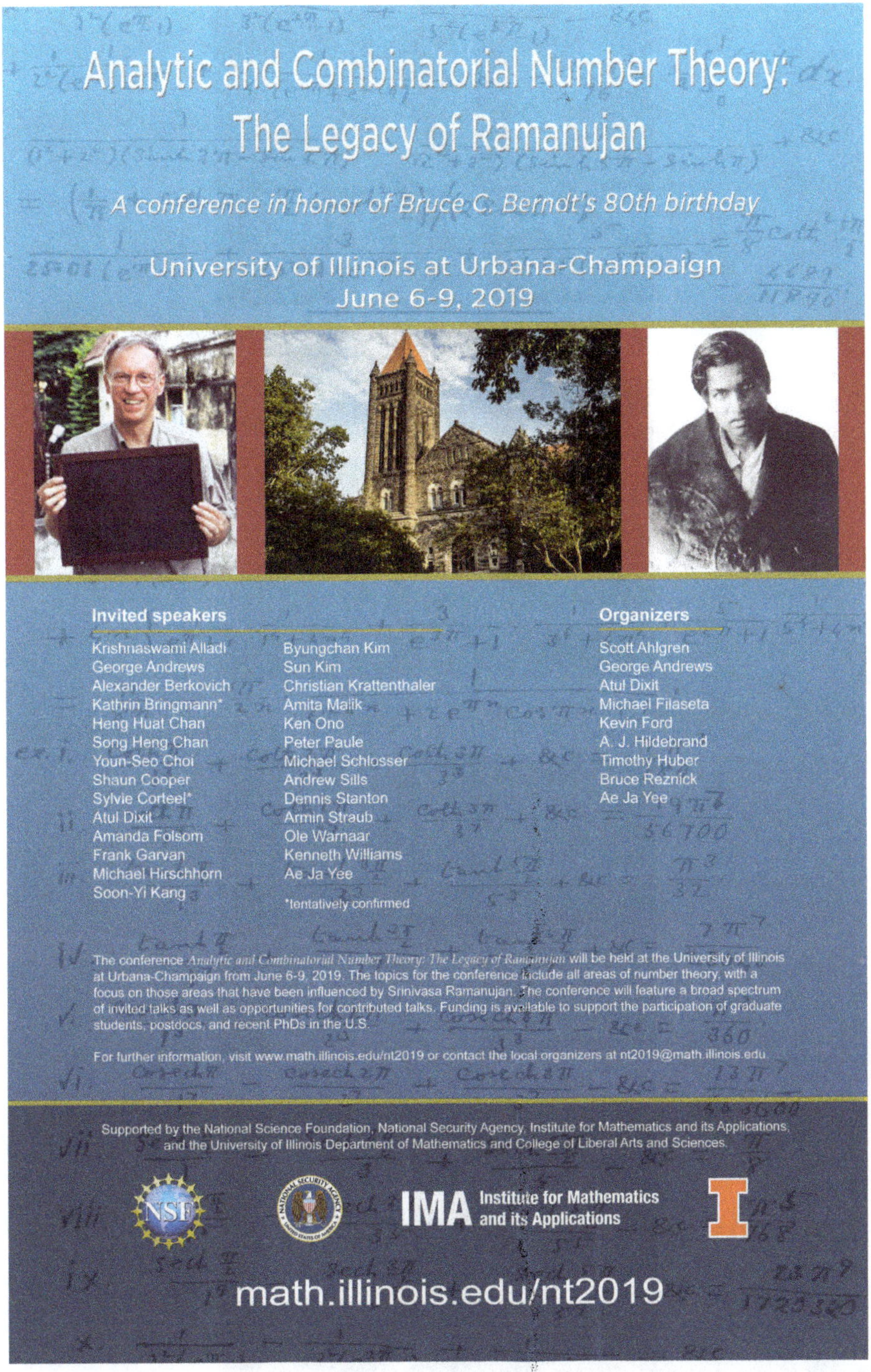

The conference poster. (We thank the Department of Mathematics of the University of Illinois at Urbana-Champaign for allowing us to include this poster, which was wonderfully designed by Tori Corkery for them.)

George Andrews at the banquet.

Harold Diamond at the banquet.

Heng Huat Chan at the banquet.

Krishna Alladi at the banquet.

Bruce Berndt at the banquet.

Helen Berndt at the banquet.

Brooks Berndt at the banquet.

Ken Stolarsky at the banquet.

Karen Mortensen at the banquet.

A singing quartet. From left to right: Matthew Ando, Paul Weichsel, Robin Sahner, and Doug West.

© 2024 World Scientific Publishing Company
https://doi.org/10.1142/9789811277375_fm

Banquet speeches

1. Richard Askey
Emeritus Professor of Mathematics, University of Wisconsin–Madison

Dear Bruce,

I'm very sorry to have missed this meeting. There are some very interesting sounding titles, and it is always pleasant to attend a meeting you are involved in.

Along with many others, I appreciate the incredible work you have done with Ramanujan's work. However, this time I would like to point out a different aspect of your work which has delighted me a few times. You have not been able to find proofs of some of Ramanujan's claims, and you wrote to people you thought might help. One question you asked me was where did Ramanujan get a few of his approximations to some classes of series which gave approximations using very few terms, including in some cases values not in the series. I was not able to tell you how Ramanujan came up with his claims, but was able to show how his claims could be found using Gaussian quadrature with respect to a discrete measure. Recall that Gauss's derivation of Gaussian quadrature with respect to the usual measure on a finite interval came from continued fractions, and the derivation using orthogonal polynomials was done later by Jacobi. People who are curious about these claims of Ramanujan should read in the fifth volume of your treatment of the Ramanujan Notebooks he did before going to England. Some of Ramanujan's work can be done using orthogonal polynomials, but this seems to be the closest Ramanujan came to work which we now think of exclusively in terms of orthogonal polynomials, and without your request for help in explaining Ramanujan's claims, I would not have known about this aspect of his work.

Best wishes and thanks once again for a very interesting problem.
Dick

Note: Late Professor Dick Askey could not attend the conference due to his personal reasons, so George Andrews read his letter on his behalf.

2. George Andrews
Evan Pugh Professor of Mathematics, The Pennsylvania State University

Bruce, one of the most significant and satisfying projects of my entire career was writing the five volumes of *Ramanujan's Lost Notebook* with you. I've said often before — and I will repeat here — I cannot imagine a better co-author than you. You are smart, tolerant, thoughtful, kind and humorous, a great partner to have on a decades-long adventure.

It was suggested that there should be some light stories. I think the best I can do here is to repeat the beginning of my talk on March 25, 2009 at the Illinois Number Theory Conference celebrating your 70th birthday and Paul Bateman's 90th. I began as follows, "Today is a day in which we celebrate the birthdays of two people who are very important to me. I'm not talking about Paul and Bruce. Bruce has already had his birthday and Paul's isn't until June. Instead, I'm referring to my daughter-in-law, Laurie, and my wife, Joy. Both were born on March 25th. Joy asked me, 'Are you really going to be away on my birthday?' And I replied, 'But honey, these guys are really important to me.' At least that's the case I'm putting to the marriage counselor next week."

That introduction was 10 years ago and my marriage is still intact, so I guess I was persuasive. All's well that ends well. Congratulations on reaching 80. Let's hope another lost notebook of Ramanujan turns up, we can write another five volumes.

3. Harold Diamond
Emeritus Professor, University of Illinois

Thank you to all who have come to the conference and banquet honoring Bruce on the occasions of his 80th birthday and retirement.

As Bruce reported in his lecture on Thursday, his contact with the University of Illinois started inauspiciously with the rejection of his grad school application. He went on to the University of Wisconsin, where he earned his Ph.D. After a postdoctoral year at the University of Glasgow, Bruce tried Illinois again, this time more successfully, and he came to the Math Department in 1967. We have been friends and colleagues in the number theory group since that time. Bruce has served as an active faculty member for 52 years, a record for our department.

He has spent this time well and has brought distinction to our group and to the Illinois Math Department through his outstanding teaching and research. To date, Bruce has written or co-authored some 250 papers and 14 served on innumerable national committees; he is currently an editor for a dozen journals; and he has held visiting positions and delivered invited addresses in 26 foreign countries.

Bruce has won many awards for his scholarship and for his teaching. To mention a few of these: he was appointed Michio Suzuki Distinguished Research Professor of Mathematics in our department, he won an expository Steele Prize from the

American Mathematical Society for his work editing Ramanujan's Notebooks, twice shared a Lester R. Ford Award for excellent exposition from the Mathematical Association of America, and his place on the University's list of Excellent Teachers could be chiseled in stone. Bruce is a prominent member of the UIUC Center for Advanced Study, received an honorary doctorate from SASTRA University in India, and he is a Fellow of the American Mathematical Society.

Early in his career, Bruce took up the challenge of analyzing Ramanujan's 3000+ assertions, and over the past forty years, he, along with George Andrews, a few other colleagues, and several students, have carried out this herculean task. This is a remarkable body of scholarship, achieved by enormous effort and insight, and the 10 volumes that Bruce and his collaborators prepared are of great value to the mathematical and scientific communities.

Bruce has tied our department to the world of Ramanujan through his work and activities. He has made several trips to India, including a visit to Ramanujan's widow. In 1987, to mark Ramanujan's 100th birthday, the University of Illinois sponsored a major conference, to which even the Indian consul came — along with his guards. For the occasion, Bruce's wife, Helen, and Shantha Rao, the wife of Ranga Rao, organized a genuine Indian dinner to feed the entire conference — a gigantic undertaking. This conference is the most recent event connecting the University with the Ramanujan universe.

The Berndts have been married over 50 years and raised three pleasant and successful children. Kristin has been a chaplain at Carle Hospital in Urbana for several years, and is now pursuing a Ph.D. in Community Health. Brooks is the Environmental Justice Minister of the United Church of Christ, educating about climate change, especially its effects on the poor. And Sonja is a senior investigator with scientific tenure at the National Cancer Institute, which is part of National Institutes of Health. (Incidentally, she has many more scientific papers to her name than does Bruce.) The Berndts have five grandchildren — with Bruce's retirement, more babysitting is on the horizon.

Bruce and Helen are most hospitable, and the community, the number theory group, and many individuals have benefited from their kindness and interest. They have provided innumerable meals and parties for guests and have opened their home to many Math Department visitors. Helen's cookies, containing dozens of digits of π, have been a treat for mind and body.

Bruce has been enthusiastic about sports his entire life. We went together to football games with the Berndts back in the day when Illinois was competitive, and the Berndt's loyalty continued in this for years afterward. Also, the Berndts were fixtures at Illinois basketball games for decades. And being athletic, Bruce came to the Department by bike for a good part of each year.

Bruce's plans for aftermath, beyond babysitting, are to continue the work that he has been doing for the past 52 years, except that he will stay out of classrooms on Mondays, Wednesdays, and Fridays. Bruce has already planned his next trip abroad: in October, he will go to India for three weeks to give lectures.

We congratulate you, Bruce, for all you have done for people, mathematics, and the University.

4. Heng Huat Chan
Professor of Mathematics, National University of Singapore

In 1987, when I was in the military, my commanding officer showed me a reprinted article on Ramanujan in *Straits Times* written by James Gleick of *The New York Times*. That was the first time that I saw (but did not register) the name of Bruce C. Berndt. In 1988, in an attempt to understand Rogers–Ramanujan continued fractions, I was led to *Ramanujan's Notebook Part I* and *Part II* written by Bruce C. Berndt and that was the second time I saw the name. Two years later in 1990, in the office of Professor Derek Robinson, who was a visiting professor to National University of Singapore and who was my algebra professor for the whole of my third year, Berndt's name surfaced again. Professor Robinson told me that he had a colleague who specialized in the work of Ramanujan and his name was Bruce C. Berndt. It was because of this incident that I felt that I was destined to apply to University of Illinois at Urbana-Champaign to pursue my Ph.D. under the supervision of Professor Berndt.

With formal recommendation letters from Professor Robinson, Professor Berrick and Dr. Cheng Kai Nah and an informal letter from Professor J. P. Serre, I succeeded in securing a place at UIUC, the only university that I applied to for the Ph.D. program.

Professor Berndt accepted me as his Ph.D. student right from the first day of my study at UIUC. I still remember him giving me his newly published *Ramanujan's Notebook: Part III* as a gift on the first day we met.

I stayed in UIUC for my Ph.D. program from 1991 to 1995. During the four years, Professor Berndt's office door was always opened during working hours and I had the opportunity to discuss mathematics with him regularly. He always greeted me with enthusiasm even when he was very busy. He would listen to me patiently and often give me fruitful suggestions. He is truly a dedicated Ph.D. advisor.

I was Professor Berndt's 8th Ph.D. student from 1974–1995. Within a few years after I graduated, Professor Berndt's number of Ph.D. students increased tremendously, with nine Ph.D. students graduating from 1997 to 2001. I wonder if I were responsible for the sudden increase in Ph.D. students. If that were the case, Professor Andrews might have found "The Lost Notebook", but I found "The Lost Super Ph.D. Advisor."

It is a fact that Professor Berndt is an excellent researcher and a hero in the field of number theory. He is also responsible for popularizing Ramanujan's mathematics. I hope that through sharing with you my personal experience as his student, I have succeeded in showing that Professor Berndt is also an excellent Ph.D. advisor and teacher. Many of his students have benefited from his guidance, care and generosity in sharing his work with them. I am truly very fortunate to be one of his many students.

5. Krishna Alladi
Professor of Mathematics, University of Florida

Bruce Berndt is the world's authority on the mathematics in Ramanujan's Notebooks. In the course of an intense investigation of the Notebooks, he also made a close study of Ramanujan's life and the many persons who had influence on Ramanujan's career and mathematics, and so apart from Robert Kanigel, Bruce is the leading authority on various aspects of Ramanujan's life. Over a period of about four decades, I have had close association with Bruce and here I shall allude to some major aspects of our interaction.

My first meeting with Bruce was in December 1976 during the West Coast Number Theory Conference that took place in San Diego. At that time, I was a second year graduate student at UCLA. We were all staying at the Torrey Pines Inn, close to the UCSD campus, and perched on a cliff overlooking the sea. One evening during the conference, Hugh Montgomery, at the request of Harold Diamond, was scheduled to give a talk on the Large Sieve at the fireside chat room of the Torrey Pines Inn. As we were waiting for the talk to begin, Bruce walked in with Harald Niederreiter, who that year was visiting UCLA. Niederreiter introduced me to Bruce and said that Bruce was keenly interested in working on Ramanujan's Notebooks. Niederreiter then asked if I had any suggestions as to how to get a copy of Ramanujan's Notebooks. I responded that the Tata Institute had brought out photostat copies of Ramanujan's Notebooks, and suggested to Bruce that he should write to the Tata Institute to get a copy. Bruce did write to the Tata Institute, and after receiving the Notebooks started working on them with unabated intensity.

After getting my Ph.D. at UCLA in the Spring of 1978, I went to the University of Michigan in Ann Arbor as a Hildebrandt Assistant Professor. Meanwhile, Bruce had started writing a few papers on certain chapters in Ramanujan's Notebooks in the process of "editing" the Notebooks. He kept me "in the loop" of the progress he was making by regularly sending me preprints of his papers on various chapters he had edited. During the year 1979–1980, the two senior number theorists in Ann Arbor, Don Lewis and Hugh Montgomery, were both on sabbatical. So I was left in charge of the Number Theory Seminar in Ann Arbor. Since I was well aware of the significant progress that Bruce was making on the Ramanujan Notebooks, I invited him to address the Number Theory Seminar, and also arranged for him to deliver a Colloquium talk at the University of Michigan. I think that Colloquium by Bruce must have been either his first, or among the first few, on the Notebooks. Our close friendship began with this visit of his to Ann Arbor.

In August 1980, the Summer Meeting of the AMS took place in Ann Arbor. Bruce, who was organizing a Special Session at that meeting, invited me to speak at that session. That was my first talk at an AMS Meeting, and I appreciate the opportunity given to me by Bruce.

After my three year term at the University of Michigan, and a year at the Institute for Advanced Study, Princeton, and the University of Texas, Austin, I was back in India in August 1983 at MATSCIENCE, the Institute of Mathematical Sciences, that my father Alladi Ramakrishnan had founded in 1962 in Madras. Upon return to MATSCIENCE, I organized an International Conference in Number Theory in January 1984 in honor of Paul Erdős for his 70th birthday, at the charming hill station of Ootacamund (Ooty for short). Bruce was one of the mathematicians I invited as one of the main speakers. That was Bruce's first trip to India, and he used the occasion to visit Ramanujan's home in Kumbakonam. Following this trip, Bruce wrote an article entitled "A Pilgrimage", which began with a reference to the Ooty conference and then dealt with his visit to Ramanujan's home. This appeared in *The Mathematics Intelligencer* in 1986.

In his fine speech at the Berndt 80th birthday conference banquet, Bruce's son referred to Bruce instructing everyone not to divulge the score of the UCLA–Illinois Rose Bowl game that took place on January 2, 1984, because he had taped it and wanted to see it with eagerness upon return to Illinois after the Indian trip. Bruce mentioned this tape to me in Ooty. He was rooting for Illinois and I was hoping that UCLA would win. Sadly for Bruce, but happily for me, the UCLA Bruins destroyed the Fighting Illini of Illinois by the score of 45-9!

After the conference at Ooty, most of the international delegates, including Bruce, visited MATSCIENCE for lectures. It was a pleasure for my father and our family to welcome Bruce and other delegates to our family home for a dinner party in honor of Erdős that was graced by the Governor of Madras.

The year of 1987 was the Ramanujan Centenary year. It was an occasion when mathematicians around the world had gathered in India to pay homage to the Indian mathematical genius, and assess the impact his work would have in the future. Naturally, Richard Askey, George Andrews, and Bruce, were three of the stars of the Ramanujan Centenary. In America, they are jocularly referred to as The Gang of Three. As a devout Hindu, I prefer to refer to them as The Trinity, after the three main gods of the Hindu pantheon — Brahma, Vishnu, and Shiva! The Trinity attended almost all the Ramanujan conferences that took place during December 87–January 88. In particular, I organized a one day number theory conference in Madras at Anna University on December 21, 1987, one day before Ramanujan's 100th birthday, and was pleased that The Trinity spoke at that conference. I must say that I was working on classical analytic number theory at that time, but after listening to the lectures of The Trinity during the Ramanujan Centenary, I decided to shift the focus of my research to the theory of partitions and q-series. Askey would say that I was inflicted with the q-disease; and indeed I was! It took a quarter century for me to recover from this (wonderful!) disease; I have just returned to analytic number theory and my talk at this conference was in this area.

After I joined the University of Florida in December 1986, we have had the pleasure of hosting Bruce several times, as a plenary speaker at conferences we have organized regularly, and for Colloquium talks and History Lectures. Our conferences were held usually in March, Bruce's birthday month, and so it was an occasion to say "Happy Birthday" to Bruce, sometimes during the conference party hosted by Mathura and me at our home.

Bruce has invited me on many occasions to the University of Illinois to speak at number theory conferences or to address the Colloquium and seminars. And more than once, I have enjoyed the gracious hospitality of Bruce and Helen at their home where I have stayed.

Since 1987, after my move into the World of Ramanujan, I have been involved in a variety of ways to foster the legacy of Ramanujan, and I have often counted on Bruce for his support. For instance, in launching *The Ramanujan Journal* in 1997, I valued his role as a Coordinating Editor of the Journal.

In 2003, I became involved with SASTRA University in India, and in 2005, I helped SASTRA launch the SASTRA Ramanujan Prize, a $10,000 annual prize to very young mathematicians. Owing to this, I have very close contact with SASTRA University and I organize annual conferences in December when the prize is given. In 2012, for the Ramanujan 125 celebrations in India, we felt that The Trinity had to be recognized in a special way with Honorary Doctorates at the SASTRA University campus in Kumbakonam, Ramanujan's hometown. So a special convocation ceremony was held and The Trinity were welcomed with traditional Indian music played on the Nadaswaram (a wind instrument like the clarinet, but bigger and louder) as they walked to the dais.

Bruce has visited India regularly since 1984, and has lectured at MATSCIENCE as well as at our family home Ekamra Nivas on several of those trips. After my father passed away in 2008, Bruce delivered the first Alladi Ramakrishnan Memorial Lecture in June 2009 at Ekamra Nivas. Our family appreciated this very much. Bruce's son, Brooks, was also visiting India that year and we were pleased that he came to our home in Madras in connection with this lecture.

Bruce has defied the passage of time by continuing to be productive all these years. His contributions, not just by himself, but also through the dozens of graduate students and postdocs he has groomed, are immense and everlasting. We wish him many more years of good health so that he can continue to inspire and influence us with his mathematical works.

6. Gratitude: On Retirement by Bruce Berndt

I would like to begin with a quote from a theologian:

Gratitude is one of the finest feelings in life... nothing beats gratitude for sheer joy; it is, perhaps, the genesis of all other really good feelings in the human repertoire.

Please permit me to express my gratitude to several people. It is, indeed, a joy for me to do so.

First, I am very grateful to the members of the organizing committee, in particular, A. J. Hildebrand, Scott Ahlgren, Ae Ja Yee, and Atul Dixit, for their dedication and work in making this conference a joy for all of us, and especially for me.

Second, I would like to thank my parents. My mother was one of 17 children, and so my family was rather large, but I am the first in our family to go to college. My father went to high school for only two years, but getting a strong education for my sister and me was a high priority. My work ethic was inherited from my father. For several years, he worked daily on our farm from about 7 a.m. to 3 p.m., when he would go to Whirlpool Corporation to work an 8-hour shift until midnight.

Third, I thank my wife Helen and our three children Kristin, Sonja, and Brooks for their loving support while I worked on Ramanujan, doubtless frequently thinking that maybe I inherited too much of a work ethic from my father.

Fourth, I have had inspiring teachers beginning with Keith Moore who taught me almost all of my undergraduate mathematics at Albion College, and Rod Smart, Marvin Knopp, and Dick Askey at Wisconsin.

Fifth, despite the unusual thrust of my research, I have always had the support of our mathematics department, especially from the number theory group, which has always been the most talented and cordial group of number theorists in the number theory world.

Sixth, our department has always been blessed with exceptional staff, who have aided me an infinite number of times, beginning several years ago with Pat Coombs and Hilda Britt, among others, and continuing to current times with Tori, Paula, Jane, Ann, Tony, and many others.

Seventh, although I have written some papers and books, and indeed it has been a pleasure to do so, my greatest joy as a mathematician has been to supervise 37 graduate students and about a dozen postdocs. Although I have lost contact with a few of them, and one died much too early, their continued friendship via mathematics or otherwise has been a source of great happiness for me.

Lastly, thanks to all of you for sharing these enjoyable moments with me.

7. Helen Berndt, Wife of Bruce Berndt

When we came to Urbana/Champaign, Bruce certainly did not plan to stay here for his entire career. It is because of you, the mathematics community, that he stayed 52 years. I am very sorry that Ron Evans was not able to be here tonight. When Ron asked Bruce to be his Ph.D. advisor Bruce did not know how to respond. Ron had shown himself to be such a mathematically talented person that Bruce thought he should probably study under someone more experienced. Bruce told Paul Bateman his concern and received an immediate response. "Take him on! Those are the best kinds of Ph.D. students. You don't have to do anything!" So he did and their friendship and collaborations have continued to this very day. It is because of Ron that Bruce looked forward to taking on subsequent students. Ron really paved the way!

In our early years the Batemans were very influential in establishing a very congenial atmosphere for new staff members and we were able to form many friendships. The Diamonds arrived when we did and we have enjoyed their friendship over the years.

Bruce went to Albion College, a small liberal arts school, as an undergraduate and greatly appreciated the experience. He liked the interaction that students at small colleges had with the faculty and he wanted to be part of that kind of nurturing environment. Listen up, Hannah! Bruce wanted to teach at St. Olaf. However, he discovered that he could provide that type of experience for his graduate students. He always enjoyed interacting with his students and considered them to be his second family.

Bruce and I owe you, the mathematics community, a debt of gratitude for making our life here such a rich and rewarding experience. Thank you for coming to celebrate with us tonight.

8. Brooks Berndt, Son of Bruce Berndt

Good evening! I want to turn back the clock tonight to a most remarkable and extraordinary time, a time very different from our own. To illustrate just how different it was, I want to give three examples. The first thing I should mention to demonstrate how incredibly different this time was is that Illinois had a winning football team. Yes, not only were they a winning team, but their record at the end of the regular season was 10 and 1. That's 10 wins, not 10 losses. I know it's hard to believe such a time existed, but that was the state of the world when my father embarked on his first trip to India in December of 1983.

My father was a big Illinois football fan at the time, and he was quite distressed that he would have to miss watching the Rose Bowl live on TV. It promised to be a great game for Illinois as they took on a UCLA team that had a record of 6-4-1. Fortunately, he had a friend who could help him out of this distress. And, here we come to the second demonstration of how incredibly different this time was. This friend of my father had what at the time was an advanced form of technology that my family did not have: a VHS video cassette player!

My father's family and friends were under strict instructions not to tell him the result of the Rose Bowl, so almost immediately after he got back to the United States, he went to his friend's house and proceeded to fast forward his way through a game in which the 4th ranked Illinois lost to an unranked UCLA by a score of 45 to 9, an omen of things to come. As depressing as that homecoming was for my father, his coming home was a big event for the family. I was seven years old at the time, and I can still recall how amazing the items and gifts were that he brought back. There were beautiful, ornate carvings, statues, and bronze plates. There were silk ties and clothes of vibrant artistry. And, then most importantly, there were Hindu comic books and wondrous toys bought from street vendors.

The third and final demonstration of just how different things were back in 1984 is this. When my father returned, we had not yet seen any pictures of his trip on Facebook or Instagram, things that would still cause my father in this very moment to whisper to my mother, "What's that?" Back then, there were no instant photos documenting one's life and travels. There weren't even DVD movies that one could quickly produce on one's computer. No, what one had was a long process of getting photos developed as slides and then putting those slides in a circular holder in the correct order facing the correct direction so that they could be projected onto a portable screen in one's living room.

It might take a week or more to get these slides ready, but eventually, my family would gather to watch them after dinner like a family gathering for movie night. My father would then proceed to show us his photos from India as he told the most remarkable stories and showed us pictures of Ramanujan's wife, Hindu temples, and more.

This evening I want to invite each of you to use your imaginations to create your own slideshow as we celebrate the life and work of my father, as we celebrate Ramanujan, and as we celebrate the extraordinary moments of life that transcend culture, time and place.

9. A Song Dedicated to Bruce by Kenneth Stolarsky Emeritus Professor of Mathematics, University of Illinois

Hymn to Clarity (The music for this song is taken from the civil war (US) song "Battle Hymn of the Republic".)

There's a notion of partition
that is really very nice,
In which there is no number
that occurs within it twice.
And the length of such partitions
has a special quality,
We call it parity.

Lengths are either odd or even,
Lengths are either odd or even,
Lengths are either odd or even,
We call it parity,

Could the odds compared to evens
have a commanding lead?
Or could the even ones
be those that do exceed?

These both seem unlikely
to our common sense indeed.
Some balance must exist.

How shall we establish here
a universal rule?
Could some clever person
Find a really useful tool?
Fabian Franklin was clearly
not a fool.
He chose a Ferrers graph.

Lengths are either odd or even,
Lengths are either odd or even,
Lengths are either odd or even,
Franklin chose a Ferrers graph.

His right-most diagonal
spoke to the bottom line.
His right-most diagonal
spoke to the bottom line.
Parity would alter if we
side by side combine.
Shall it be your place or mine?

This bijection seems to
make it all quite clear.
But now is not yet time
to give a rousing cheer;
We may have fallen into
a somewhat subtle trap.
These lines might overlap!

There are yet some details here
to be put in place.
In this remaining
but quite important case.
And surely we know
what wisdom does advise –
Leave it to the reader
as an exercise!

We shall leave it to the reader.
We shall leave it to the reader.

We shall leave it to the reader,
and Franklin's proof is clear!

Mathematics must
have a solid base.
And the exact truth
is its public face;
But it cannot
serve the human race,
unless it has clarity!

Solid clarity forever,
solid clarity forever,
solid clarity forever.
Franklin's proof made it clear.

Ramanujan's theorems
once were hard to see,
but Bruce stepped up and gave to them
a wondrous clarity;
he led us from an obscure thicket
to a sunny shore.
Bruce did what Franklin did
a hundred-fold or more!

Solid clarity forever,
solid clarity forever,
solid clarity forever.
Bruce made Ramanujan clear.

(The singers repeated the final four lines with the entire audience joining in.)

© 2024 World Scientific Publishing Company
https://doi.org/10.1142/9789811277375_fm

Talks at the conference

Krishnaswami Alladi ...... *Parity of the generalized divisor function with restrictions on the prime factors*

George Andrews .... *George Beck's conjectures on ranks of partitions*

Zafer Selcuk Aygin .... *Ramanujan's convolution sum twisted by Dirichlet characters*

Nilufar Mana Begum ...... *Generating functions and congruences for partition functions related to third order mock theta functions*

Paul Beirne .... *Knots, the colored Jones polynomial and stability*

Alexander Berkovich ...... *On a q-theoretic approach to Capparelli's partition theorems*

Bruce Berndt .... *Ramanujan — The Ultimate Superhero (public lecture)*

Gaurav Bhatnagar .... *On Entry II.16.12: A continued fraction of Ramanujan*

Arunabha Biswas .... *Search for higher Mahler measure of cyclotomic polynomials.*

Aubrey Blecher .... *Distinct tuples in integer partitions*

Douglas Bowman .... *Euler combinatorics for divergent continued fractions after Ramanujan*

Walter Bridges .... *Kadell-type partition inequalities and applications to sum-product conjectures of Kanade–Russell*

Hannah Burson .... *Simultaneous core partitions into k distinct parts*

Zhu Cao .... *Exact covering systems, quadratic forms, and product identities for theta functions*

Heng Huat Chan .... *Ramanujan and the partition function $p(n)$*

Song Heng Chan .... *Ranks of partitions and some identities related to mock theta functions*

Tsz Chan .... *On the congruence equation $\overline{a} + \overline{b} \equiv \overline{c} \pmod{p}$*

Sunil Chebolu . . . . *Mersenne primes and Fermat primes*

Shane Chern . . . . *Weighted partition rank and crank moments and Andrews–Beck type congruences*

Hyunsoo Cho . . . . *A bijection between self-conjugate and ordinary partitions and counting simultaneous cores as its application*

Youn-Seo Choi . . . . *Hypergeometric series in my papers*

Madeline Dawsey . . . . *Integer partitions and a conjecture of John Thompson*

Karl Dilcher . . . . *Infinite products involving Dirichlet characters and cyclotomic polynomials*

Atul Dixit . . . . *On values of the Riemann zeta function at odd positive integers*

Alexander Dunn . . . . *The twisted second moment of modular half integral weight L-functions*

Dennis Eichhorn . . . . *The differences between $p(n)$, $p(2n)$, and $p(2n+1)$*

Larry Ericksen . . . . *Properties and polynomials related to restricted binary overpartitions*

Susan Ficken . . . . *Asymptotics of the zeros of Fine's function*

Amanda Folsom . . . . *Quantum modular forms and applications*

Craig Franze . . . . *An asymptotic expansion related to the Dickman function*

Frank Garvan . . . . *Mock theta functions: yesterday and today*

Oscar Gonzales . . . . *An observation of Rankin on Hankel determinants*

Chadwick Gugg . . . . *Ramanujan's 40 identities for the Rogers–Ramanujan functions*

Lakshika Gunawardana . . . *A primitive 15–theorem?*

Rajat Gupta . . . . *A Ramanujan-type formula for $\zeta^2(2m+1)$ and its generalizations*

Heekyoung Hahn . . . . *Poles of triple product L-functions of monomial representations*

Bernhard Heim . . . . *Ramanujan's τ function and non-sign changes*

Wade Hindes . . . . *Dynamical and arithmetic degrees of random iterates*

Michael Hirschhorn . . . . . . *Some identities of Ramanujan, Andrews and Berndt*

Nankun Hong . . . . *Rank function and Lambert series*

Timothy Huber . . . . *A Ramanujan–Sato series primer*

Brad Isaacson .... *On some elementary character sum identities*

Uha Isnaini .... *On k-restricted overpartitions*

Paul Jenkins .... *The arithmetic of modular grids*

Lin Jiu .... *On harmonic sums: Integral and matrix representations with connections to partition-theoretic generalization of the Riemann zeta-function and random walks*

Yeongseong Jo .... *Rankin–Selberg L-functions via good sections*

Edna Jones .... *Representations by ternary quadratic forms*

Soon-Yi Kang .... *Generators of the space of harmonic weak Maass forms*

Pin Hung Kao .... *Almost-prime values of reducible polynomials at prime arguments*

William Keith .... *(k, j)-colored partitions: Congruences and other behaviors*

Byungchan Kim .... *Partial or false theta series in q-series*

Louis Kolitsch .... *A pair of congruences for $(3, 5)$-regular partitions modulo 5*

Shin-ya Koyama .... *Applications of Ramanujan's method on the behavior of Euler products to Selberg zeta functions*

Christian Krattenthaler ... *A joint central limit theorem for the sum-of-digits function, and asymptotic divisibility of Catalan-like sequences*

Brandt Kronholm .... *A technique in partitions*

Rahul Kumar .... *Zeros of combinations of the Riemann Ξ-function and the confluent hypergeometric function on bounded vertical shifts*

Debanjana Kundu .... *On the relation between the classical Iwasawa $\mu = 0$ conjecture and the Coates–Sujatha Conjecture A*

Kagan Kursungoz .... *Andrews–Gordon type series for Schur's partition identity*

Bibekananda Maji .... *Recent developments in the theory of the restricted partition function $p(n, N)$ and $\mathrm{spt}(n, N)$*

Amita Malik .... *Two stopovers in BCB's mathematical journey: Riemann zeta and partitions*

Nicholas Mayers .... *Integer partition statistics arising from seaweed algebras*

James McLaughlin .... *Some observations on Lambert series, vanishing coefficients and dissections of infinite products and series*

Shashika Petta Mestrige .. *Congruences for the partition function* $p_{[1^c \ell^d]}(n)$

Victor Moll .... *Valuations of integer sequences*

Mojtaba Moniri .... *Pairs of relatively long, mod 4 disjoint consecutive runs under Mahler's recurrence*

Seyyed Hamed Mousavi .. *Pentagonal number theorem and zeroes of the Riemann zeta function*

Hayan Nam .... *How to determine a partition up to conjugation using hook multisets*

Ken Ono .... *The Jensen–Polya program for the Riemann Hypothesis and related problems*

Jiakun Pan .... *Quantum unique equidistribution conjecture for Eisenstein series in the level aspect*

Peter Paule .... *Ramanujan, Weierstrass, and computer algebra*

Bogdan Petrenko .... *Some properties of nontrivial coefficients of cyclotomic polynomials*

Meenakshi Rana .... *Combinatorial interpretations of mock theta functions by attached weights*

Bruce Reznick .... *Equal sums of cubes of quadratic forms*

Herman Schaumburg ...... *Partition identities encoded in Ramanujan's three limit continued fraction*

Michael Schlosser .... *The bilateral afterlife of Srinivasa Ramanujan*

Robert Schneider .... *Eulerian series and the algebra of partitions*

Andrew Sills .... *Rogers–Ramanujan type identities, old and new*

Satyanand Singh .... *Generating fourth terms of Nathanson's lambda sequences*

Nicolas Smoot .... *Some new identities from the Ramanujan–Kolberg algorithm*

Fikreab Solomon .... *Zeta functions of nilpotent groups*

Garima Sood .... *A finite analogue of the Beck–Chern theorem and some results on self-conjugate S-partitions*

Jonathan Sorenson .... *Recent work on the Erdős–Selfridge function* $g(k)$

Emily Stamm .... *An investigation of (nearly) weakly primes*

Dennis Stanton .... *Comments on cranks and the rank*

Kenneth Stolarsky .... *Rogers–Szegő polynomials and products of algebraic numbers*

Armin Straub .... *On the Gaussian binomial coefficients, the simplest of q-series*

Karen Taylor .... *Dirichlet series and generalized Hecke groups*

Pee Choon Toh .... *On a symmetric identity of Ramanujan*

Thomas Tran .... *Secondary terms in asymptotics for the number of zeros of quadratic forms over integers*

Ian Wagner .... *Schwartz functions and modular forms*

Tanay Wakhare .... *Special classes of q-bracket operators*

Liuquan Wang .... *Representations of mock theta functions*

Ole Warnaar .... *Some (q-)integrals Ramanujan would have liked*

Sarah Wesley .... *Generalizations of Fine's iterative identities*

Kenneth Williams .... *Ramanujan and ternary quadratic forms*

Ae Ja Yee .... *Singular overpartitions and truncated partition theorems*

Wen-Bin Zhang .... *Exact Wiener–Ikehara theorems and Beurling's generalized numbers*

© 2024 World Scientific Publishing Company
https://doi.org/10.1142/9789811277375_fm

List of conference participants

Scott Ahlgren, University of Illinois
Abdulaziz Alanazi, University of Tabuk
Krishnaswami Alladi, University of Florida
Michael Allen, Oregon State University
Patrick Allen, University of Illinois
George Andrews, Pennsylvania State University
Emily Anible, Michigan Technological University
Zafer Selcuk Aygin, University of Lethbridge
Siegfred Baluyot, University of Illinois
Koustav Banerjee, RISC University of Linz
George Beck, Wolfram Research
Nilufar Mana Begum, Tezpur University
Paul Beirne, University College Dublin
Alexander Berkovich, University of Florida
Bruce Berndt, University of Illinois
Gaurav Bhatnagar, University of Vienna
Paul Bialek, Trinity International University
Arunabha Biswas, University of North Texas
Aubrey Blecher, University of Witwatersrand
Florin Boca, University of Illinois
Douglas Bowman, Northern Illinois University
Matthew Boylan, University of South Carolina
Stevo Bozinovski, South Carolina State University
Walter Bridges, Louisiana State University
Hannah Burson, University of Illinois
Zhu Cao, Kennesaw State University
Hei-Chi Chan, University of Illinois at Springfield
Heng Huat Chan, National University of Singapore
Song Heng Chan, Nanyang Technological University
Tsz Chan, University of Memphis
Sunil Chebolu, Illinois State University
Shane Chern, Penn State University
Hyunsoo Cho, Yonsei University
Youn-Seo Choi, Korea Institute for Advanced Study
Shashank Chorge, University of Rochester
Jack Dalton, University of South Carolina
Agniva Dasgupta, Texas A&M University
Madeline Dawsey, Emory University
Harold Diamond, University of Illinois
Robert Dicks, University of Illinois
Karl Dilcher, Dalhousie University
Atul Dixit, IIT Gandhinagar
Alexander Dunn, University of Illinois
Dennis Eichhorn, UC Irvine
Thomas Engelsma, None Given
Larry Ericksen, Cornell
Susan Ficken, Anne Arundel Community College
Michael Filaseta, University of South Carolina
Amanda Folsom, Amherst College
Kurt Foster, None Given
Craig Franze, Ohio State University
Frank Garvan, University of Florida
Cyndi Garvan, University of Florida
Nicholas Geis, Ohio State University
Kaare Gjaldbaek, CUNY-Lehman College
Oscar Gonzales, University of Illinois
Sid Graham, Central Michigan University
Chadwick Gugg, Georgia Southwestern State University
Lakshika Gunawardana, Southern Illinois University
Zhixing Guo, University of Illinois
Rajat Gupta, IIT Gandhinagar
Heekyoung Hahn, Duke University
Tanja Haxhoviq, CUNY-Lehman College
Bernhard Heim, German Univ. of Tech. in Oman
A.J. Hildebrand, University of Illinois
Wade Hindes, Texas State University
Aimo Hinkkanen, University of Illinois
Michael Hirschhorn, University of New South Wales
Nankun Hong, Nanyang Technological University
Timothy Huber, University of Texas Rio Grande Valley
Brad Isaacson, New York City College of Tech.
Uha Isnaini, National Institute of Education Singapore
Paul Jenkins, Brigham Young University
Lin Jiu, Dalhousie University
Yeongseong Jo, University of Iowa
Edna Jones, Rutgers University
Soon-Yi Kang, Kangwon National University
Pin Hung Kao, Flagler College
William Keith, Michigan Technological University
Chad Kelterborn, Univ. of North Carolina Chapel Hill
Mizan Khan, Eastern Connecticut State University
Byungchan Kim, SeoulTech
Eunmi Kim, Korea Institute for Advanced Study
Kwang-Seon Kim, Chosun University
Louis Kolitsch, University of Tennessee at Martin
Shin-ya Koyama, Toyo University
Stephen Kraemer, Saint Joseph's College
Christian Krattenthaler, Universität Wien
Brandt Kronholm, University of Texas Rio Grande Valley
Rochelle Kronzek, World Scientific Publishing Company
Rahul Kumar, IIT Gandhinagar
Debanjana Kundu, University of Toronto

Kagan Kursungoz, Sabanci University
Yeong-Wook Kwon, Sungkyunkwan University
Lauren LaRue, Flagler College
Chong Gyu Lee, Soongsil University
Wing Hong Leung, Ohio State University
Xiaomin Li, University of Illinois
Elizabeth Loew, Springer Verlag
Wenjun Ma, Tianjin University
Karl Mahlburg, Lousiana State University
Bibekananda Maji, IIT Gandhinagar
Amita Malik, Rutgers University
Nicholas Mayers, Lehigh University
James McLaughlin, West Chester University
Julian Mejia Cordero, Ohio State University
Shashika Petta Mestrige, Louisiana State University
Victor Moll, Tulane University
Mojtaba Moniri, Normandale Community College
Dania Morales, University of Illinois
Karen Mortensen, University of Illinois
Seyyed Hamed Mousavi, Georgia Institute of Tech.
Hayan Nam, UC Irvine
Evangelos Nastas, Florida Atlantic University
Ken Ono, Emory University
Jiakun Pan, Texas A&M University
Dhir Patel, Ohio State University
Peter Paule, RISC - Univ. of Linz
Bogdan Petrenko, Eastern Illinois University
Meenakshi Rana, Thapar Ins. of Eng. and Tech. Punjab
Vicki Reuter, University of Wisconsin - Platteville
Bruce Reznick, University of Illinois
Demmas Salim, University of South Carolina
Praneel Samanta, University of Iowa
Herman Schaumburg, Northern Illinois University
Michael Schlosser, University of Vienna
Robert Schneider, University of Georgia
Robert Sczech, Rutgers University Newark
Arseniy Sheydvasser, CUNY-Graduate Center
Yahui Shi, Tianjin University
Andrew Sills, Georgia Southern University
Diljit Singh, Stony Brook University
Satyanand Singh, New York City College of Tech.
Simone Sisneros-Thiry, University of Illinois
Nicolas Smoot, RISC - Univ. of Linz
Jaebum Sohn, Yonsei University
Fikreab Solomon, CUNY-Graduate Center
Garima Sood, IIT Gandhinagar
Brianna Sorenson, Butler University
Jonathan Sorenson, Butler University
Emily Stamm, Allstate
Dennis Stanton, University of Minnesota
Kenneth Stolarsky, University of Illinois
Armin Straub, University of South Alabama
Leah Sturman, Oregon State University
Karen Taylor, Bronx Community College-CUNY
Johann Thiel, New York City College of Tech.
Pee Choon Toh, Nanyang Technological University
Thomas Tran, Duke University
Minh Tran, University of Illinois
Zack Tripp, Vanderbilt University
Steve Ullom, University of Illinois
Fernando Venegas, Ohio State University
Ian Wagner, Emory University
Tim Wagner, Michigan Tech University
Samuel Wagstaff, Purdue University
Tanay Wakhare, University of Maryland
Mesa Walker, Oregon State University
Biao Wang, SUNY at Buffalo
Liuquan Wang, Wuhan University and NCTS
Ole Warnaar, University of Queensland
Bin Wei, Tianjin University
Sarah Wesley, inte Q
Kenneth Williams, Carleton University
Daodao Yang, Indiana University
Heesung Yang, Dalhousie University
Wijit Yangjit, University of Michigan
Ae Ja Yee, Penn State University
Wen-Bin Zhang, Univ. of West Indies and UIUC
Scott Zinzer, Aurora University

© 2024 World Scientific Publishing Company
https://doi.org/10.1142/9789811277375_fm

Bruce C. Berndt's biography*

Bruce C. Berndt was born on March 13, 1939, in St. Joseph, Michigan. The first six years of his life were turbulent. He almost died on three occasions. Between the ages of 1 and 2, he became very weak, and it was feared that he would die. He was in the hospital for two weeks. Doctors thought that he was losing his eyesight. When recovering, his eyesight was tested by doctors holding playing cards at various distances for him to identify. (Among the first words that he learned were the numbers on playing cards.) At the age of 4, he narrowly escaped from drowning in Lake Michigan. At the age of 6, he was badly injured from being hit by a car.

From kindergarten through 7th grade, Bruce went to a two-room country school near Stevensville (6 miles south of St. Joseph; abbrev. St. Joe), where the education was very poor. Bruce recalls that the school devoted more time to playing baseball during the three long recesses than to lessons in the classroom. He was more interested in sports at that time, but he had difficulty with them due to very little sight in his left eye. From ages 6–14, he worked every day in the summer picking fruit (mostly black raspberries, sour cherries, peaches, and grapes) on his family's small fruit farm. In 8th grade, his parents sent him to a Lutheran school in St. Joe. The education was much better. By that time, because of poor crop yields, his father had started to work daily from 3:30 to midnight at Whirlpool Corp. in St. Joe, although his father continued to farm. Bruce went to high school in St. Joe. When a senior, although his father was laid off, he was still eligible to apply for the annual Whirlpool scholarship for children of employees. Fortunately, Bruce received the Whirlpool scholarship, which paid $1,000 per year for four years, and he chose to go to a small liberal arts college, Albion College. Bruce says he could have had a partial track scholarship at a couple large universities, because as a senior at St. Joe he was the state high school 440 yard champion; however, he did not want to attend a large university.

Bruce entered college intending to major in physics. However, he felt that two of his close friends were much better at physics than he was, and he discovered that he liked math more than physics. Consequently, he majored in math, but continued to take all the physics courses. (Both friends later got Ph.D.'s, one of them at Penn State, and the other at Wisconsin.) He applied for graduate school at Michigan, Wisconsin, Illinois, and Purdue. He was accepted at Wisconsin and Purdue; he

*For some biographical details, we have consulted with Bruce Berndt and his webpage.

was rejected at Illinois and Michigan. He chose Wisconsin. He thought he would specialize in complex analysis. In the spring semester of his third year, he took a course in modular forms, because he was told that modular forms used a lot of complex analysis. He enjoyed the course so much that he asked the professor, Rod Smart, to be his advisor. In the following fall semester, he took a course in modular forms with applications to number theory, taught by Marvin Knopp. In that course, he learned about Ramanujan for the first time. In particular, he learned about Ramanujan's congruences for the partition function $p(n)$ and the circle method.

After receiving his Ph.D. in 1966, Bruce went to the University of Glasgow for one year, where his mentor was Robert Rankin, one of the world leaders in modular forms. Rankin was Hardy's last Ph.D. student and had a life-long interest in Ramanujan. One day in the office of Rankin, their conversation centered around Ramanujan, and Rankin told Bruce about Ramanujan's notebooks offering to loan his copy (published by the Tata Institute in 1957), but Bruce declined Rankin's offer.

In February, 1967, Bruce received an offer from Paul Bateman to be an Assistant Professor at Illinois. This was a huge surprise to him. He had actually forgotten that he had applied for a position at Illinois. He had only done so at the insistence of Marvin Knopp, who had received his Ph.D. at Illinois under Paul Bateman (one of George Andrews' mathematical brothers) and always had a fondness for Illinois. Bruce recalls he had planned to teach at a small liberal arts college and did have a couple offers, but he was not eager to accept either one. He therefore accepted Bateman's offer, thinking he would teach at Illinois for only a few years. The few turned out to be 52 years and he never got to a small college. Bruce says he really enjoyed teaching at Illinois, and his colleagues were very friendly, and so he stayed. A turning point came in his third year when Ron Evans asked him to be his thesis advisor. According to Bruce, Evans was an outstanding student, and it was a joy to work with him.

It was in his third year at Illinois that Bruce came across two papers of Emil Grosswald, in which some formulas from Chapter 14 of Ramanujan's second notebook were proved. However, Bruce did not think much more about them until his sabbatical year at the Institute for Advanced Study. In February, 1974, while working on transformation formulas for Eisenstein series, he discovered that he could use his work to prove the formulas from Ramanujan's notebooks that Grosswald had proved. Bruce remembers that it was a very exciting revelation. He checked out the notebooks from the Princeton University library and saw all of those many wonderful identities. When he returned to Illinois, the first thing that he did was to order a copy of the notebooks from the Tata Institute. The cost was $25, including shipping. At the end of the spring semester in 1976, as a personal challenge, he set a goal of trying to prove each of the 87 results in Chapter 14 of Ramanujan's second notebook, the chapter from which Grosswald had proved some formulas. He completed this task in a little more than one year. Then in a conversation with George

Andrews, Bruce was told that when Andrews discovered Ramanujan's lost notebook at Trinity College, Cambridge, he also discovered that the notes that B. M. Wilson and G. N. Watson had accumulated in their efforts to edit the notebooks were preserved there. So, Bruce asked for a copy of these notes. He then went to the beginning of the second notebook, and in 21 years finished the task of editing the notebooks in five volumes with the help of lots of people, especially his graduate students.

As Bruce was completing his work on the notebooks, he asked Andrews whether Andrews would be interested in working with him on a similar project with the lost notebook and the many unpublished manuscripts and fragments published with the lost notebook. According to Bruce, Andrews anticipated that maybe two volumes would be needed to complete the work. After about 20 years, they completed their work in five volumes.

Bruce retired in June, 2019, after serving at Illinois for 52 years. He says the most enjoyable part of his career was working with 37 graduate students. He also advised 10 postdocs.

Bruce devoted almost all of his academic career to Ramanujan and his mathematics. The ten volumes of Ramanujan's notebooks and lost notebooks are a culmination of his life-long work. He spent his earlier career working on the five volumes of Ramanujan's notebooks, which serve as excellent references to the work of Ramanujan for the mathematical community, and the other half of his career working on the five volumes on the lost notebook (with Andrews), which guide us to a better understanding about the mathematics Ramanujan had built during his last few years. Bruce edited two books on Ramanujan with Rankin: *Ramanujan: Letters and Commentary*, and *Ramanujan: Essays and Surveys*. Also, his book *Number Theory in the Spirit of Ramanujan* is an excellent introductory textbook on number theory for undergraduate and graduate students.

Bruce has been interested in Ramanujan as a human being as well. He has visited India 11 times (years 1984, 1987, 1999, 2000, 2002, 2009, 2011, 2012, 2012, 2016, 2019) to trace back the life of Ramanujan there and publicize the work of Ramanujan. Some of the memorable events to him are: two conversations with Ramanujan's widow, S. Janaki, in 1984 and 1987 (her foster son, W. Narayanan, served as their translator); a visit to Ramanujan's home in Kumbakonam, which was in the same condition as when Ramanujan lived in it and has now been turned into a museum; visits to Kumbakonam High School and the Government College of Kumbakonam, which Ramanujan attended; twice seeing and holding Ramanujan's slate, which is in the possession of the family of S. Narayana Aiyar, the Chief Accountant at the Madras Port Trust where Ramanujan worked; and visiting the University of Madras and viewing Ramanujan's notebooks.

Bruce was involved in various commemorative events for Ramanujan in India, including the Ramanujan Birth Centenary celebrations in Chennai in 1987 and the Ramanujan 125th birth celebrations in Tamil Nadu and Delhi in 2012. His dedication to Ramanujan has had strong influences on the Indian mathematical community and the younger generation there. As an acknowledgement of his dedication, he received an honorary doctorate from SASTRA University in 2012 along with George Andrews and the late Richard Askey.

Bruce also organized and co-organized many conferences and meetings on Ramanujan in the US. In particular, there was a large meeting at the University of Illinois in June, 1987, to celebrate the 100th anniversary of Ramanujan's birth. He recalls that of the meetings that he has co-organized during his career, this one involved the most time that he has ever devoted to such a task.

According to MathSciNet, as of September, 2022, Bruce has 282 publications with 86 coauthors. Even after his retirement in 2019, Bruce continues to do excellent research in number theory which is evident from a large number of research articles that he is continually producing.

In 1996, he was awarded the Leroy P. Steele prize of the American Mathematical Society for his monumental work on editing Ramanujan's notebooks. He is an inaugural fellow of the American Mathematical Society. He was a Guggenheim Fellow in 1998–1999, and the Mahler Lecturer of the Australian Mathematical Society in 2005. He holds editorships for various journals including the Ramanujan Journal and Journal of the Ramanujan Mathematical Society. He is the Founding Editor of the International Journal of Number Theory.

Atul Dixit
Indian Institute of Technology Gandhinagar

Ae Ja Yee
The Pennsylvania State University

© 2024 World Scientific Publishing Company
https://doi.org/10.1142/9789811277375_fm

Bruce C. Berndt's publications

Main Research Publications:

[1] Bruce C. Berndt, Atul Dixit, and Rajat Gupta, *Generalizations of the Andrews-Yee identities associated with the mock theta functions $\omega(q)$ and $\nu(q)$*, J. Algebraic Combin. **55** (2022), no. 4, 1031–1062.

[2] Bruce C. Berndt, Martino Fassina, Sun Kim, and Alexandru Zaharescu, *Balanced derivatives, identities, and bounds for trigonometric and Bessel series*, Adv. Math. **395** (2022), Paper No. 108085, 40.

[3] ———, *Two-parameter identities for divisor sums in algebraic number fields*, J. Math. Anal. Appl. **506** (2022), no. 2, Paper No. 125679, 25.

[4] Bruce C. Berndt and Örs Rebák, *Explicit values for Ramanujan's theta function $\varphi(q)$*, Hardy-Ramanujan J. **44** (2021), 41–50.

[5] Bruce C. Berndt and Atul Dixit, *Ramanujan's beautiful integrals*, Hardy-Ramanujan J. **43** (2020), 69–82.

[6] George E. Andrews, Bruce C. Berndt, Song Heng Chan, Sun Kim, and Amita Malik, *Four identities for third order mock theta functions*, Nagoya Math. J. **239** (2020), 173–204.

[7] Bruce C. Berndt, Junxian Li, and Alexandru Zaharescu, *The final problem: an identity from Ramanujan's lost notebook*, J. Lond. Math. Soc. (2) **100** (2019), no. 2, 568–591.

[8] Bruce C. Berndt, Atul Dixit, Sun Kim, and Alexandru Zaharescu, *Sums of squares and products of Bessel functions*, Adv. Math. **338** (2018), 305–338.

[9] Bruce C. Berndt, Amita Malik, and Alexandru Zaharescu, *Partitions into kth powers of terms in an arithmetic progression*, Math. Z. **290** (2018), no. 3-4, 1277–1307.

[10] Bruce C. Berndt, Sun Kim, and Alexandru Zaharescu, *The circle problem of Gauss and the divisor problem of Dirichlet—still unsolved*, Amer. Math. Monthly **125** (2018), no. 2, 99–114.

[11] Bruce C. Berndt, *The appearance of H. F. Baker and E. W. Hobson in "The man who knew infinity"*, Analytic number theory, modular forms and q-hypergeometric series, Springer Proc. Math. Stat., vol. 221, Springer, Cham, 2017, pp. 95–97.

[12] Bruce C. Berndt and Armin Straub, *Ramanujan's formula for $\zeta(2n+1)$*, Exploring the Riemann zeta function, Springer, Cham, 2017, pp. 13–34.

[13] Bruce C. Berndt, Atul Dixit, Sun Kim, and Alexandru Zaharescu, *On a theorem of A. I. Popov on sums of squares*, Proc. Amer. Math. Soc. **145** (2017), no. 9, 3795–3808.

[14] Bruce C. Berndt, Atul Dixit, Arindam Roy, and Alexandru Zaharescu, *New pathways and connections in number theory and analysis motivated by two incorrect claims of Ramanujan*, Adv. Math. **304** (2017), 809–929.

[15] Bruce C. Berndt, *Integrals associated with Ramanujan and elliptic functions*, Ramanujan J. **41** (2016), no. 1-3, 369–389.

[16] Bruce C. Berndt and Armin Straub, *On a secant Dirichlet series and Eichler integrals of Eisenstein series*, Math. Z. **284** (2016), no. 3-4, 827–852.

[17] ———, *Certain integrals arising from Ramanujan's notebooks*, SIGMA Symmetry Integrability Geom. Methods Appl. **11** (2015), Paper 083, 11.

[18] Bruce C. Berndt and Sun Kim, *Identities for logarithmic means: a survey*, Advances in the theory of numbers, Fields Inst. Commun., vol. 77, Fields Inst. Res. Math. Sci., Toronto, ON, 2015, pp. 1–10.

[19] Bruce C. Berndt and Roberta R. Zhou, *Identities for partitions with distinct colors*, Ann. Comb. **19** (2015), no. 3, 397–420.

[20] Bruce C. Berndt and Sun Kim, *Logarithmic means and double series of Bessel functions*, Int. J. Number Theory **11** (2015), no. 5, 1535–1556.

[21] Bruce C. Berndt, Sun Kim, and Alexandru Zaharescu, *Weighted divisor sums and Bessel function series, V*, J. Approx. Theory **197** (2015), 101–114.

[22] Bruce C. Berndt, Soon-Yi Kang, and Jaebum Sohn, *Finite and infinite Rogers–Ramanujan continued fractions in Ramanujan's lost notebook*, J. Number Theory **148** (2015), 112–120.

[23] Bruce C. Berndt, Sun Kim, and Alexandru Zaharescu, *Analogues of Koshliakov's formula*, Ramanujan 125, Contemp. Math., vol. 627, Amer. Math. Soc., Providence, RI, 2014, pp. 41–48.

[24] Bruce C. Berndt and Roberta R. Zhou, *Proofs of conjectures of Sandon and Zanello on colored partition identities*, J. Korean Math. Soc. **51** (2014), no. 5, 987–1028.

[25] Bruce C. Berndt, Sun Kim, and Alexandru Zaharescu, *The circle and divisor problems, and Ramanujan's contributions through Bessel function series*, The legacy of Srinivasa Ramanujan, Ramanujan Math. Soc. Lect. Notes Ser., vol. 20, Ramanujan Math. Soc., Mysore, 2013, pp. 111–127.

[26] ———, *Weighted divisor sums and Bessel function series, III*, J. Reine Angew. Math. **683** (2013), 67–96.

[27] ———, *Dirichlet L-functions, elliptic curves, hypergeometric functions, and rational approximation with partial sums of power series*, Math. Res. Lett. **20** (2013), no. 3, 429–448.

[28] Bruce C. Berndt, Sun Kim, M. Tip Phaovibul, and Alexandru Zaharescu, *Diophantine approximation with partial sums of power series*, Acta Arith. **161** (2013), no. 3, 249–266.

[29] Bruce C. Berndt, Sun Kim, and Alexandru Zaharescu, *Diophantine approximation of the exponential function and Sondow's Conjecture*, Adv. Math. **248** (2013), 1298–1331.

[30] Bruce C. Berndt, Byungchan Kim, and Kenneth S. Williams, *Euler products in Ramanujan's lost notebook*, Int. J. Number Theory **9** (2013), no. 5, 1313–1349.

[31] Bruce C. Berndt and Sun Kim, *A problem in Diophantine approximation found in Ramanujan's lost notebook*, Ramanujan J. **31** (2013), no. 1-2, 83–95.

[32] Bruce C. Berndt, Sun Kim, and Alexandru Zaharescu, *Circle and divisor problems, and double series of Bessel functions*, Adv. Math. **236** (2013), 24–59.

[33] Bruce C. Berndt, Chadwick Gugg, and Sun Kim, *Ramanujan's elementary method in partition congruences*, Partitions, q-series, and modular forms, Dev. Math., vol. 23, Springer, New York, 2012, pp. 13–21.

[34] Bruce C. Berndt and Prapanpong Pongsriiam, *Discarded fragments from Ramanujan's papers*, Analytic and probabilistic methods in number theory, TEV, Vilnius, 2012, pp. 49–59.

[35] Bruce C. Berndt, George Lamb, and Mathew Rogers, *Two-dimensional series evaluations via the elliptic functions of Ramanujan and Jacobi*, Ramanujan J. **29** (2012), no. 1-3, 185–198.

[36] Bruce C. Berndt, Sun Kim, and Alexandru Zaharescu, *Weighted divisor sums and Bessel function series, IV*, Ramanujan J. **29** (2012), no. 1-3, 79–102.

[37] Bruce C. Berndt, Heng Huat Chan, and Yoshio Tanigawa, *Two Dirichlet series evaluations found on page 196 of Ramanujan's lost notebook*, Math. Proc. Cambridge Philos. Soc. **153** (2012), no. 2, 341–360.

[38] Bruce C. Berndt, Sun Kim, and Alexandru Zaharescu, *Weighted divisor sums and Bessel function series, II*, Adv. Math. **229** (2012), no. 3, 2055–2097.

[39] Bruce C. Berndt and Daniel Schultz, *On sums of powers in Ramanujan's lost notebook*, Appl. Anal. **90** (2011), no. 3-4, 725–730.

[40] Bruce C. Berndt, *The chief accountant and mathematical friend of Ramanujan—S. Narayana Aiyar*, Amer. Math. Monthly **118** (2011), no. 9, 767–776.

[41] Bruce C. Berndt and Byungchan Kim, *Asymptotic expansions of certain partial theta functions*, Proc. Amer. Math. Soc. **139** (2011), no. 11, 3779–3788.

[42] Bruce C. Berndt, Atul Dixit, and Jaebum Sohn, *Character analogues of theorems of Ramanujan, Koshliakov and Guinand*, Adv. in Appl. Math. **46** (2011), no. 1-4, 54–70.

[43] Bruce C. Berndt, Chadwick Gugg, Sarachai Kongsiriwong, and Johann Thiel, *A proof of the general theta transformation formula*, Ramanujan rediscovered, Ramanujan Math. Soc. Lect. Notes Ser., vol. 14, Ramanujan Math. Soc., Mysore, 2010, pp. 53–62.

[44] Bruce C. Berndt, *What is a q-series?*, Ramanujan rediscovered, Ramanujan Math. Soc. Lect. Notes Ser., vol. 14, Ramanujan Math. Soc., Mysore, 2010, pp. 31–51.

[45] Bruce C. Berndt and Atul Dixit, *A transformation formula involving the gamma and Riemann zeta functions in Ramanujan's lost notebook*, The legacy of Alladi Ramakrishnan in the mathematical sciences, Springer, New York, 2010, pp. 199–210.

[46] Nayandeep Deka Baruah and Bruce C. Berndt, *Eisenstein series and Ramanujan-type series for* $1/\pi$, Ramanujan J. **23** (2010), no. 1-3, 17–44.

[47] Bruce C. Berndt, Heng Huat Chan, Song Heng Chan, and Wen-Chin Liaw, *Cranks—really, the final problem*, Ramanujan J. **23** (2010), no. 1-3, 3–15.

[48] Bruce C. Berndt, Byungchan Kim, and Ae Ja Yee, *Ramanujan's lost notebook: combinatorial proofs of identities associated with Heine's transformation or partial theta functions*, J. Combin. Theory Ser. A **117** (2010), no. 7, 857–973.

[49] Nayandeep Deka Baruah and Bruce C. Berndt, *Ramanujan's Eisenstein series and new hypergeometric-like series for* $1/\pi^2$, J. Approx. Theory **160** (2009), no. 1-2, 135–153.

[50] Bruce C. Berndt and Ping Xu, *An integral analogue of theta functions and Gauss sums in Ramanujan's lost notebook*, Math. Proc. Cambridge Philos. Soc. **147** (2009), no. 2, 257–265.

[51] Bruce C. Berndt and Dimitris Koukoulopoulos, *A reciprocity theorem for certain hypergeometric series*, Proc. Amer. Math. Soc. **137** (2009), no. 7, 2369–2373.

[52] Bruce C. Berndt, Yoonbok Lee, and Jaebum Sohn, *Koshliakov's formula and Guinand's formula in Ramanujan's lost notebook*, Surveys in number theory, Dev. Math., vol. 17, Springer, New York, 2008, pp. 21–42.

[53] Bruce C. Berndt and Tim Huber, *A fragment on Euler's constant in Ramanujan's lost notebook*, South East Asian J. Math. Math. Sci. **6** (2008), no. 2, 17–22.

[54] Nayandeep Deka Baruah and Bruce C. Berndt, *Partition identities arising from theta function identities*, Acta Math. Sin. (Engl. Ser.) **24** (2008), no. 6, 955–970.

[55] Bruce C. Berndt, O-Yeat Chan, Sung-Geun Lim, and Alexandru Zaharescu, *Questionable claims found in Ramanujan's lost notebook*, Tapas in experimental mathematics, Contemp. Math., vol. 457, Amer. Math. Soc., Providence, RI, 2008, pp. 69–98.

[56] Nayandeep Deka Baruah and Bruce C. Berndt, *Ramanujan's series for* $1/\pi$ *arising from his cubic and quartic theories of elliptic functions*, J. Math. Anal. Appl. **341** (2008), no. 1, 357–371.

[57] George E. Andrews and Bruce C. Berndt, *Your hit parade: the top ten most fascinating formulas in Ramanujan's lost notebook*, Notices Amer. Math. Soc. **55** (2008), no. 1, 18–30.

[58] Nayandeep Deka Baruah, Bruce C. Berndt, and Heng Huat Chan, *Ramanujan's series for* $1/\pi$*: a survey*, Math. Student, Special Centenary Volume (2007), 1–24 (2008).

[59] Bruce C. Berndt, *Ramanujan's congruences for the partition function modulo 5, 7, and 11*, Int. J. Number Theory **3** (2007), no. 3, 349–354.

[60] Bruce C. Berndt and Song Heng Chan, *Sixth order mock theta functions*, Adv. Math. **216** (2007), no. 2, 771–786.

[61] Nayandeep Deka Baruah and Bruce C. Berndt, *Partition identities and Ramanujan's modular equations*, J. Combin. Theory Ser. A **114** (2007), no. 6, 1024–1045.

[62] Bruce C. Berndt, *Partition-theoretic interpretations of certain modular equations of Schröter, Russell, and Ramanujan*, Ann. Comb. **11** (2007), no. 2, 115–125.

[63] Bruce C. Berndt, Geumlan Choi, Youn-Seo Choi, Heekyoung Hahn, Boon Pin Yeap, Ae Ja Yee, Hamza Yesilyurt, and Jinhee Yi, *Ramanujan's forty identities for the Rogers-Ramanujan functions*, Mem. Amer. Math. Soc. **188** (2007), no. 880, vi+96.

[64] Bruce C. Berndt and William B. Hart, *An identity for the Dedekind eta-function involving two independent complex variables*, Bull. Lond. Math. Soc. **39** (2007), no. 2, 345–347.

[65] James M. Hill, Bruce C. Berndt, and Tim Huber, *Solving Ramanujan's differential equations for Eisenstein series via a first order Riccati equation*, Acta Arith. **128** (2007), no. 3, 281–294.

[66] Bruce C. Berndt, Song Heng Chan, Boon Pin Yeap, and Ae Ja Yee, *A reciprocity theorem for certain q-series found in Ramanujan's lost notebook*, Ramanujan J. **13** (2007), no. 1-3, 27–37.

[67] Bruce C. Berndt and Wenchang Chu, *Two entries on bilateral hypergeometric series in Ramanujan's lost notebook*, Proc. Amer. Math. Soc. **135** (2007), no. 1, 129–134.

[68] Bruce C. Berndt and Alexandru Zaharescu, *Weighted divisor sums and Bessel function series*, Math. Ann. **335** (2006), no. 2, 249–283.

[69] Bruce C. Berndt and Hamza Yesilyurt, *New identities for the Rogers-Ramanujan functions*, Acta Arith. **120** (2005), no. 4, 395–413.

[70] Matthias Beck, Bruce C. Berndt, O-Yeat Chan, and Alexandru Zaharescu, *Determinations of analogues of Gauss sums and other trigonometric sums*, Int. J. Number Theory **1** (2005), no. 3, 333–356.

[71] Bruce C. Berndt and Paul R. Bialek, *On the power series coefficients of certain quotients of Eisenstein series*, Trans. Amer. Math. Soc. **357** (2005), no. 11, 4379–4412.

[72] Bruce C. Berndt and Geumlan Choi, *A continued fraction from Ramanujan's lost notebook*, Aequationes Math. **69** (2005), no. 3, 257–262.

[73] Bruce C. Berndt, Heng Huat Chan, Song Heng Chan, and Wen-Chin Liaw, *Ramanujan and cranks*, Theory and applications of special functions, Dev. Math., vol. 13, Springer, New York, 2005, pp. 77–98.

[74] George E. Andrews, Bruce C. Berndt, Jaebum Sohn, Ae Ja Yee, and Alexandru Zaharescu, *Continued fractions with three limit points*, Adv. Math. **192** (2005), no. 2, 231–258.

[75] Bruce C. Berndt, Heng Huat Chan, Song Heng Chan, and Wen-Chin Liaw, *Cranks and dissections in Ramanujan's lost notebook*, J. Combin. Theory Ser. A **109** (2005), no. 1, 91–120.

[76] Bruce C. Berndt and Ae Ja Yee, *q-Gauss summation via Ramanujan and combinatorics*, South East Asian J. Math. Math. Sci. **3** (2004), no. 1, 15–22.

[77] Bruce C. Berndt and Alexandru Zaharescu, *Finite trigonometric sums and class numbers*, Math. Ann. **330** (2004), no. 3, 551–575.

[78] Bruce C. Berndt, *Some integrals in Ramanujan's lost notebook*, Proc. Amer. Math. Soc. **132** (2004), no. 10, 2983–2988.

[79] ———, *An unpublished manuscript of Ramanujan on infinite series identities*, J. Ramanujan Math. Soc. **19** (2004), no. 1, 57–74.

[80] Bruce C. Berndt, Song Heng Chan, Zhi-Guo Liu, and Hamza Yesilyurt, *A new identity for* $(q;q)^{10}_\infty$ *with an application to Ramanujan's partition congruence modulo 11*, Q. J. Math. **55** (2004), no. 1, 13–30.

[81] Bruce C. Berndt, Ae Ja Yee, and Alexandru Zaharescu, *New theorems on the parity of partition functions*, J. Reine Angew. Math. **566** (2004), 91–109.

[82] Bruce C. Berndt and Ae Ja Yee, *On the generalized Rogers-Ramanujan continued fraction*, Ramanujan J. **7** (2003), no. 1-3, 321–331. Rankin memorial issues.

[83] Bruce C. Berndt, Winfried Kohnen, and Ken Ono, *The life and work of R. A. Rankin (1915–2001)*, Ramanujan J. **7** (2003), no. 1-3, 9–38. Rankin memorial issues.

[84] Bruce C. Berndt and Ae Ja Yee, *Combinatorial proofs of identities in Ramanujan's lost notebook associated with the Rogers-Fine identity and false theta functions*, Ann. Comb. **7** (2003), no. 4, 409–423.

[85] Bruce C. Berndt, Ae Ja Yee, and Jinhee Yi, *Theorems on partitions from a page in Ramanujan's lost notebook*, Proceedings of the International Conference on Special Functions and their Applications (Chennai, 2002), 2003, pp. 53–68.

[86] Bruce C. Berndt, Heng Huat Chan, and Alexandru Zaharescu, *A quasi-theta product in Ramanujan's lost notebook*, Math. Proc. Cambridge Philos. Soc. **135** (2003), no. 1, 11–18.

[87] Bruce C. Berndt, Ae Ja Yee, and Alexandru Zaharescu, *On the parity of partition functions*, Internat. J. Math. **14** (2003), no. 4, 437–459.

[88] George E. Andrews, Bruce C. Berndt, Jaebum Sohn, Ae Ja Yee, and Alexandru Zaharescu, *On Ramanujan's continued fraction for* $(q^2;q^3)_\infty/(q;q^3)_\infty$, Trans. Amer. Math. Soc. **355** (2003), no. 6, 2397–2411.

[89] Bruce C. Berndt and Ae Ja Yee, *A page on Eisenstein series in Ramanujan's Lost Notebook*, Glasg. Math. J. **45** (2003), no. 1, 123–129.

[90] Bruce C. Berndt and Alexandru Zaharescu, *A multi-variable theta product*, Indag. Math. (N.S.) **13** (2002), no. 1, 11–22.

[91] Bruce C. Berndt and Ae Ja Yee, *Ramanujan's contributions to Eisenstein series, especially in his lost notebook*, Number theoretic methods (Iizuka, 2001), Dev. Math., vol. 8, Kluwer Acad. Publ., Dordrecht, 2002, pp. 31–53.

[92] Bruce C. Berndt, *The problems solved by Ramanujan in the Journal of the Indian Mathematical Society*, Number theory and discrete mathematics (Chandigarh, 2000), Trends Math., Birkhäuser, Basel, 2002, pp. 189–200.

[93] ———, *The influence of Carr's synopsis on Ramanujan*, Number theory and discrete mathematics (Chandigarh, 2000), Trends Math., Birkhäuser, Basel, 2002, pp. 31–35. Reprinted from *Ramanujan: essays and surveys* [Amer. Math. Soc., Providence, RI, 2001; MR1862735].

[94] Bruce C. Berndt and Boon Pin Yeap, *Explicit evaluations and reciprocity theorems for finite trigonometric sums*, Adv. in Appl. Math. **29** (2002), no. 3, 358–385.

[95] Bruce C. Berndt and Alexandru Zaharescu, *An integral of Dedekind eta-functions in Ramanujan's lost notebook*, J. Reine Angew. Math. **551** (2002), 33–39.

[96] Bruce C. Berndt, Blair K. Spearman, and Kenneth S. Williams, *Commentary on an unpublished lecture by G. N. Watson on solving the quintic*, Math. Intelligencer **24** (2002), no. 4, 15–33.

[97] Scott Ahlgren, Bruce C. Berndt, Ae Ja Yee, and Alexandru Zaharescu, *Integrals of Eisenstein series and derivatives of L-functions*, Int. Math. Res. Not. **32** (2002), 1723–1738.

[98] Bruce C. Berndt and Ae Ja Yee, *Congruences for the coefficients of quotients of Eisenstein series*, Acta Arith. **104** (2002), no. 3, 297–308.

[99] Bruce C. Berndt, Paul R. Bialek, and Ae Ja Yee, *Formulas of Ramanujan for the power series coefficients of certain quotients of Eisenstein series*, Int. Math. Res. Not. **21** (2002), 1077–1109.

[100] Bruce C. Berndt, Heng Huat Chan, Soon-Yi Kang, and Liang-Cheng Zhang, *A certain quotient of eta-functions found in Ramanujan's lost notebook*, Pacific J. Math. **202** (2002), no. 2, 267–304.

[101] Bruce C. Berndt and Jaebum Sohn, *Asymptotic formulas for two continued fractions in Ramanujan's Lost Notebook*, J. London Math. Soc. (2) **65** (2002), no. 2, 271–284.

[102] Bruce C. Berndt, *Flowers which we cannot yet see growing in Ramanujan's garden of hypergeometric series, elliptic functions, and q's*, Special functions 2000: current perspective and future directions (Tempe, AZ), NATO Sci. Ser. II Math. Phys. Chem., vol. 30, Kluwer Acad. Publ., Dordrecht, 2001, pp. 61–85.

[103] Bruce C. Berndt and K. Venkatachaliengar, *On the transformation formula for the Dedekind eta-function*, Symbolic computation, number theory, special functions, physics and combinatorics (Gainesville, FL, 1999), Dev. Math., vol. 4, Kluwer Acad. Publ., Dordrecht, 2001, pp. 73–77.

[104] Bruce C. Berndt and Heng Huat Chan, *Eisenstein series and approximations to* π, Illinois J. Math. **45** (2001), no. 1, 75–90.

[105] Bruce C. Berndt, Heng Huat Chan, and Wen-Chin Liaw, *On Ramanujan's quartic theory of elliptic functions*, J. Number Theory **88** (2001), no. 1, 129–156.

[106] Bruce C. Berndt and Robert A. Rankin, *The books studied by Ramanujan in India*, Amer. Math. Monthly **107** (2000), no. 7, 595–601.

[107] Bruce C. Berndt and Douglas C. Bowman, *Ramanujan's short unpublished manuscript on integrals and series related to Euler's constant*, Constructive, experimental, and nonlinear analysis (Limoges, 1999), CRC Math. Model. Ser., vol. 27, CRC, Boca Raton, FL, 2000, pp. 19–27.

[108] Bruce C. Berndt, Heng Huat Chan, and Sen-Shan Huang, *Incomplete elliptic integrals in Ramanujan's lost notebook*, q-series from a contemporary perspective (South Hadley, MA, 1998), Contemp. Math., vol. 254, Amer. Math. Soc., Providence, RI, 2000, pp. 79–126.

[109] Bruce C. Berndt, *Modular equations in Ramanujan's lost notebook*, Number theory, Trends Math., Birkhäuser, Basel, 2000, pp. 55–74.

[110] Bruce C. Berndt, Heng Huat Chan, Jaebum Sohn, and Seung Hwan Son, *Eisenstein series in Ramanujan's lost notebook*, Ramanujan J. **4** (2000), no. 1, 81–114.

[111] Bruce C. Berndt and William F. Galway, *On the Brocard-Ramanujan Diophantine equation* $n!+1=m^2$, Ramanujan J. **4** (2000), no. 1, 41–42.

[112] Bruce C. Berndt, Sen-Shan Huang, Jaebum Sohn, and Seung Hwan Son, *Some theorems on the Rogers-Ramanujan continued fraction in Ramanujan's lost notebook*, Trans. Amer. Math. Soc. **352** (2000), no. 5, 2157–2177.

[113] Bruce C. Berndt, *Fragments by Ramanujan on Lambert series*, Number theory and its applications (Kyoto, 1997), Dev. Math., vol. 2, Kluwer Acad. Publ., Dordrecht, 1999, pp. 35–49.

[114] Bruce C. Berndt and Heng Huat Chan, *Ramanujan and the modular j-invariant*, Canad. Math. Bull. **42** (1999), no. 4, 427–440.

[115] Bruce C. Berndt and Ken Ono, *Ramanujan's unpublished manuscript on the partition and tau functions with proofs and commentary*, Sém. Lothar. Combin. **42** (1999), Art. B42c, 63. The Andrews Festschrift (Maratea, 1998).

[116] Bruce C. Berndt, Heng Huat Chan, Sen-Shan Huang, Soon-Yi Kang, Jaebum Sohn, and Seung Hwan Son, *The Rogers-Ramanujan continued fraction*, J. Comput. Appl. Math. **105** (1999), no. 1-2, 9–24. Continued fractions and geometric function theory (CONFUN) (Trondheim, 1997).

[117] Bruce C. Berndt and Heng Huat Chan, *Notes on Ramanujan's singular moduli*, Number theory (Ottawa, ON, 1996), CRM Proc. Lecture Notes, vol. 19, Amer. Math. Soc., Providence, RI, 1999, pp. 7–16.

[118] Bruce C. Berndt, Youn-Seo Choi, and Soon-Yi Kang, *The problems submitted by Ramanujan to the Journal of the Indian Mathematical Society*, Continued fractions: from analytic number theory to constructive approximation (Columbia, MO, 1998), Contemp. Math., vol. 236, Amer. Math. Soc., Providence, RI, 1999, pp. 15–56.

[119] Bruce C. Berndt and Fritz Gesztesy, *L. J. Lange on his 70th birthday*, Continued fractions: from analytic number theory to constructive approximation (Columbia, MO, 1998), Contemp. Math., vol. 236, Amer. Math. Soc., Providence, RI, 1999, pp. xi–xv.

[120] Bruce C. Berndt, *The remaining* 40% *of Ramanujan's lost notebook*, Sūrikaisekikenkyūsho Kōkyūroku **1060** (1998), 111–118. Number theory and its applications (Japanese) (Kyoto, 1997).

[121] ———, *An overview of Ramanujan's notebooks*, Charlemagne and his heritage. 1200 years of civilization and science in Europe, Vol. 2 (Aachen, 1995), Brepols, Turnhout, 1998, pp. 119–146.

[122] Bruce C. Berndt, Heng Huat Chan, and Liang-Cheng Zhang, *Radicals and units in Ramanujan's work*, Acta Arith. **87** (1998), no. 2, 145–158.

[123] ———, *Ramanujan's singular moduli*, Ramanujan J. **1** (1997), no. 1, 53–74.

[124] ———, *Ramanujan's association with radicals in India*, Amer. Math. Monthly **104** (1997), no. 10, 905–911.

[125] ———, *Ramanujan's remarkable product of theta-functions*, Proc. Edinburgh Math. Soc. (2) **40** (1997), no. 3, 583–612.

[126] ———, *Ramanujan's class invariants with applications to the values of q-continued fractions and theta functions*, Special functions, q-series and related topics (Toronto, ON, 1995), Fields Inst. Commun., vol. 14, Amer. Math. Soc., Providence, RI, 1997, pp. 37–53.

[127] ———, *Ramanujan's class invariants, Kronecker's limit formula, and modular equations*, Trans. Amer. Math. Soc. **349** (1997), no. 6, 2125–2173.

[128] ———, *Explicit evaluations of the Rogers-Ramanujan continued fraction*, J. Reine Angew. Math. **480** (1996), 141–159.

[129] Bruce C. Berndt and Heng Huat Chan, *Ramanujan's explicit values for the classical theta-function*, Mathematika **42** (1995), no. 2, 278–294.

[130] Bruce C. Berndt, Heng Huat Chan, and Liang Cheng Zhang, *Ramanujan's class invariants and cubic continued fraction*, Acta Arith. **73** (1995), no. 1, 67–85.

[131] Bruce C. Berndt and Paul Bialek, *Five formulas of Ramanujan arising from Eisenstein series*, Number theory (Halifax, NS, 1994), CMS Conf. Proc., vol. 15, Amer. Math. Soc., Providence, RI, 1995, pp. 67–86.

[132] Bruce C. Berndt and Heng Huat Chan, *Some values for the Rogers-Ramanujan continued fraction*, Canad. J. Math. **47** (1995), no. 5, 897–914.

[133] Bruce C. Berndt, S. Bhargava, and Frank G. Garvan, *Ramanujan's theories of elliptic functions to alternative bases*, Trans. Amer. Math. Soc. **347** (1995), no. 11, 4163–4244.

[134] Bruce C. Berndt and Liang Cheng Zhang, *A new class of theta-function identities originating in Ramanujan's notebooks*, J. Number Theory **48** (1994), no. 2, 224–242.

[135] Bruce C. Berndt and Ronald J. Evans, *On Rademacher's multiplier system for the classical theta-function*, The Rademacher legacy to mathematics (University Park, PA, 1992), Contemp. Math., vol. 166, Amer. Math. Soc., Providence, RI, 1994, pp. 1–7.

[136] Bruce C. Berndt, *Hans Rademacher (1892–1969)*, The Rademacher legacy to mathematics (University Park, PA, 1992), Contemp. Math., vol. 166, Amer. Math. Soc., Providence, RI, 1994, pp. xiii–xxxvi.

[137] Bruce C. Berndt and James Lee Hafner, *Two remarkable doubly exponential series transformations of Ramanujan*, Proc. Indian Acad. Sci. Math. Sci. **104** (1994), no. 1, 245–252. K. G. Ramanathan memorial issue.

[138] Bruce C. Berndt and S. Bhargava, *Ramanujan—for lowbrows*, Amer. Math. Monthly **100** (1993), no. 7, 644–656.

[139] Bruce C. Berndt, *Ramanujan's theory of theta-functions*, Theta functions: from the classical to the modern, CRM Proc. Lecture Notes, vol. 1, Amer. Math. Soc., Providence, RI, 1993, pp. 1–63.

[140] Bruce C. Berndt and James Lee Hafner, *A theorem of Ramanujan on certain alternating series*, A tribute to Emil Grosswald: number theory and related analysis, Contemp. Math., vol. 143, Amer. Math. Soc., Providence, RI, 1993, pp. 47–57.

[141] Bruce C. Berndt, *On a certain theta-function in a letter of Ramanujan from Fitzroy House*, Ganita **43** (1992), no. 1-2, 33–43.

[142] Bruce C. Berndt and S. Bhargava, *A remarkable identity found in Ramanujan's third notebook*, Glasgow Math. J. **34** (1992), no. 3, 341–345.

[143] Bruce C. Berndt and Ronald J. Evans, *Asymptotic expansion of a series of Ramanujan*, Proc. Edinburgh Math. Soc. (2) **35** (1992), no. 2, 189–199.

[144] Bruce C. Berndt, *Hans Rademacher (1892–1969)*, Acta Arith. **61** (1992), no. 3, 209–225.

[145] Bruce C. Berndt and Liang Cheng Zhang, *Ramanujan's identities for eta-functions*, Math. Ann. **292** (1992), no. 3, 561–573.

[146] G. E. Andrews, Bruce C. Berndt, Lisa Jacobsen, and Robert L. Lamphere, *The continued fractions found in the unorganized portions of Ramanujan's notebooks*, Mem. Amer. Math. Soc. **99** (1992), no. 477, vi+71.

[147] Bruce C. Berndt and Ronald J. Evans, *Some elegant approximations and asymptotic formulas of Ramanujan*, J. Comput. Appl. Math. **37** (1991), no. 1-3, 35–41.

[148] Bruce C. Berndt and S. Bhargava, *Ramanujan's inversion formulas for the lemniscate and allied functions*, J. Math. Anal. Appl. **160** (1991), no. 2, 504–524.

[149] Bruce C. Berndt and Ronald J. Evans, *An integral functional equation of Ramanujan related to the dilogarithm*, Number theory (Banff, AB, 1988), de Gruyter, Berlin, 1990, pp. 1–5.

[150] Bruce C. Berndt, *Some asymptotic formulas for q-series found in Ramanujan's third notebook and "lost notebook"*, Indian J. Math. **32** (1990), no. 2, 179–185.

[151] ———, *Ramanujan's formulas for Eisenstein series*, Number theory and related topics (Bombay, 1988), Tata Inst. Fund. Res. Stud. Math., vol. 12, Tata Inst. Fund. Res., Bombay, 1989, pp. 23–29.

[152] ———, *Ramanujan and the theory of prime numbers*, Number theory, Madras 1987, Lecture Notes in Math., vol. 1395, Springer, Berlin, 1989, pp. 122–139.

[153] George E. Andrews, Bruce C. Berndt, Lisa Jacobsen, and Robert L. Lamphere, *Variations on the Rogers-Ramanujan continued fraction in Ramanujan's notebooks*, Number theory, Madras 1987, Lecture Notes in Math., vol. 1395, Springer, Berlin, 1989, pp. 73–83.

[154] Bruce C. Berndt, *Srinivasa Ramanujan*, Amer. Scholar **58** (1989), no. 2, 234–244.

[155] ———, *Introduction to Ramanujan's modular equations*, Proceedings of the Ramanujan Centennial International Conference (Annamalainagar, 1987), RMS Publ., vol. 1, Ramanujan Math. Soc., Annamalainagar, 1988, pp. 15–20.

[156] Gert Almkvist and Bruce Berndt, *Gauss, Landen, Ramanujan, the arithmetic-geometric mean, ellipses, π, and the Ladies diary*, Amer. Math. Monthly **95** (1988), no. 7, 585–608.

[157] Bruce C. Berndt, *Ramanujan—100 years old (fashioned) or 100 years new (fangled)?*, Math. Intelligencer **10** (1988), no. 3, 24–29.

[158] ———, *Ramanujan's modular equations*, Ramanujan revisited (Urbana-Champaign, Ill., 1987), Academic Press, Boston, MA, 1988, pp. 313–333.

[159] Bruce C. Berndt, Anthony J. Biagioli, and James M. Purtilo, *Ramanujan's modular equations of "large" prime degree*, J. Indian Math. Soc. (N.S.) **51** (1987), 75–110 (1988).

[160] ———, *Ramanujan's modular equations of degrees* 7 *and* 11, Indian J. Math. **29** (1987), no. 3, 215–228 (1988).

[161] ———, *Ramanujan's "mixed" modular equations*, J. Ramanujan Math. Soc. **1** (1986), no. 1-2, 46–70.

[162] Bruce C. Berndt and Ronald J. Evans, *Chapter 15 of Ramanujan's second notebook. II. Modular forms*, Acta Arith. **47** (1986), no. 2, 123–142.

[163] Bruce C. Berndt, *A pilgrimage*, Math. Intelligencer **8** (1986), no. 1, 25–30.

[164] Bruce C. Berndt, Robert L. Lamphere, and B. M. Wilson, *Chapter 12 of Ramanujan's second notebook: continued fractions*, Rocky Mountain J. Math. **15** (1985), no. 2, 235–310. Number theory (Winnipeg, Man., 1983).

[165] Bruce C. Berndt and Ronald J. Evans, *Extensions of asymptotic expansions from Chapter 15 of Ramanujan's second notebook*, J. Reine Angew. Math. **361** (1985), 118–134.

[166] Bruce C. Berndt, *Remarks on some of Ramanujan's number theoretical discoveries found in his second notebook*, Number theory (Ootacamund, 1984), Lecture Notes in Math., vol. 1122, Springer, Berlin, 1985, pp. 47–55.

[167] ———, *The gamma function and the Hurwitz zeta-function*, Amer. Math. Monthly **92** (1985), no. 2, 126–130.

[168] Chandrashekar Adiga, Bruce C. Berndt, Srinivasamurthy Bhargava, and George N. Watson, *Chapter* 16 *of Ramanujan's second notebook: theta-functions and q-series*, Mem. Amer. Math. Soc. **53** (1985), no. 315, v+85.

[169] Bruce C. Berndt and Ronald J. Evans, *Chapter 13 of Ramanujan's second notebook: integrals and asymptotic expansions*, Exposition. Math. **2** (1984), no. 4, 289–347.

[170] Bruce C. Berndt, *Ramanujan's quarterly reports*, Bull. London Math. Soc. **16** (1984), no. 5, 449–489.

[171] Bruce C. Berndt and Larry A. Goldberg, *Analytic properties of arithmetic sums arising in the theory of the classical theta functions*, SIAM J. Math. Anal. **15** (1984), no. 1, 143–150.

[172] Bruce C. Berndt and Ronald J. Evans, *Chapter* 7 *of Ramanujan's second notebook*, Proc. Indian Acad. Sci. Math. Sci. **92** (1983), no. 2, 67–96.

[173] Bruce C. Berndt, *The quarterly reports of S. Ramanujan*, Amer. Math. Monthly **90** (1983), no. 8, 505–516.

[174] ———, *Chapter 6 of Ramanujan's second notebook*, Results Math. **6** (1983), no. 1, 7–26.

[175] Bruce C. Berndt, Ronald J. Evans, and B. M. Wilson, *Chapter 3 of Ramanujan's second notebook*, Adv. in Math. **49** (1983), no. 2, 123–169.

[176] Bruce C. Berndt, *Chapter 11 of Ramanujan's second notebook*, Bull. London Math. Soc. **15** (1983), no. 4, 273–320.

[177] ———, *Chapter 8 of Ramanujan's second notebook*, J. Reine Angew. Math. **338** (1983), 1–55.

[178] ———, *Chapter 10 of Ramanujan's second notebook*, J. Indian Math. Soc. (N.S.) **46** (1982), no. 1-4, 31–76 (1985).

[179] ———, *An arithmetic Poisson formula*, Pacific J. Math. **103** (1982), no. 2, 295–299.

[180] Bruce C. Berndt and Ulrich Dieter, *Sums involving the greatest integer function and Riemann-Stieltjes integration*, J. Reine Angew. Math. **337** (1982), 208–220.

[181] Bruce C. Berndt and Ronald J. Evans, *Corrigendum to: "The determination of Gauss sums" [Bull. Amer. Math. Soc. (N.S.) **5** (1981), no. 2, 107–129; MR 82h:10051]*, Bull. Amer. Math. Soc. (N.S.) **7** (1982), no. 2, 441.

[182] Bruce C. Berndt and B. M. Wilson, *Chapter 5 of Ramanujan's second notebook*, Analytic number theory (Philadelphia, Pa., 1980), Lecture Notes in Math., vol. 899, Springer, Berlin-New York, 1981, pp. 49–78.

[183] ———, *Chapter 4 of Ramanujan's second notebook*, Proc. Roy. Soc. Edinburgh Sect. A **89** (1981), no. 1-2, 87–109.

[184] Bruce C. Berndt, Padmini T. Joshi, and B. M. Wilson, *Chapter 2 of Ramanujan's second notebook*, Glasgow Math. J. **22** (1981), no. 2, 199–216.

[185] Bruce C. Berndt and Ronald J. Evans, *The determination of Gauss sums*, Bull. Amer. Math. Soc. (N.S.) **5** (1981), no. 2, 107–129.

[186] Bruce C. Berndt and Lowell Schoenfeld, *Corrigendum to the paper: "Periodic analogues of the Euler-Maclaurin and Poisson summation formulas with applications to number theory" [Acta Arith. 28 (1975/76), no. 1, 23–68; MR 52 #5586]*, Acta Arith. **38** (1980/81), no. 3, 323.

[187] Bruce C. Berndt, *Chapter 14 of Ramanujan's second notebook*, Enseign. Math. (2) **26** (1980), no. 1-2, 1–65.

[188] Bruce C. Berndt and Ronald J. Evans, *Half Gauss sums*, Math. Ann. **249** (1980), no. 2, 115–125.

[189] ———, *Sums of Gauss, Jacobi, and Jacobsthal*, J. Number Theory **11** (1979), no. 3, S. Chowla Anniversary Issue, 349–398.

[190] ———, *Sums of Gauss, Eisenstein, Jacobi, Jacobsthal, and Brewer*, Illinois J. Math. **23** (1979), no. 3, 374–437.

[191] Bruce C. Berndt, *Analytic Eisenstein series, theta-functions, and series relations in the spirit of Ramanujan*, J. Reine Angew. Math. **303(304)** (1978), 332–365.

[192] ———, *Ramanujan's notebooks*, Math. Mag. **51** (1978), no. 3, 147–164.

[193] Bruce C. Berndt and Ronald J. Evans, *Dedekind sums and class numbers*, Monatsh. Math. **84** (1977), no. 4, 265–273.

[194] Bruce C. Berndt and M. L. Glasser, *A new class of Bessel function integrals*, Aequationes Math. **16** (1977), no. 1-2, 183–186.

[195] Bruce C. Berndt and Sarvadaman Chowla, *The reckoning of certain quartic and octic Gauss sums*, Glasgow Math. J. **18** (1977), no. 2, 153–155.

[196] Bruce C. Berndt, *Reciprocity theorems for Dedekind sums and generalizations*, Advances in Math. **23** (1977), no. 3, 285–316.

[197] ———, *Modular transformations and generalizations of several formulae of Ramanujan*, Rocky Mountain J. Math. **7** (1977), no. 1, 147–189.

[198] ———, *Classical theorems on quadratic residues*, Enseign. Math. (2) **22** (1976), no. 3-4, 261–304.

[199] ———, *Dedekind sums and a paper of G. H. Hardy*, J. London Math. Soc. (2) **13** (1976), no. 1, 129–137.

[200] ———, *On Eisenstein series with characters and the values of Dirichlet L-functions*, Acta Arith. **28** (1975/76), no. 3, 299–320.

[201] ———, *Periodic Bernoulli numbers, summation formulas and applications*, Theory and application of special functions (Proc. Advanced Sem., Math. Res. Center, Univ. Wisconsin, Madison, Wis., 1975), Math. Res. Center, Univ. Wisconsin, Publ. No. 35, Academic Press, New York, 1975, pp. 143–189.

[202] Bruce C. Berndt and Lowell Schoenfeld, *Periodic analogues of the Euler-Maclaurin and Poisson summation formulas with applications to number theory*, Acta Arith. **28** (1975/76), no. 1, 23–68.

[203] Bruce C. Berndt, *Character analogues of the Poisson and Euler-MacLaurin summation formulas with applications*, J. Number Theory **7** (1975), no. 4, 413–445.

[204] ———, *The evaluation of certain classes of nonabsolutely convergent double series*, SIAM J. Math. Anal. **6** (1975), no. 6, 966–977.

[205] ———, *Elementary evaluation of $\zeta(2n)$*, Math. Mag. **48** (1975), 148–154.

[206] ———, *Identities involving the coefficients of a class of Dirichlet series. VII*, Trans. Amer. Math. Soc. **201** (1975), 247–261.

[207] ———, *A generalization of a theorem of Gauss on sums involving $[x]$*, Amer. Math. Monthly **82** (1975), 44–51.

[208] ———, *Problem 73-26, "A double summation"*, SIAM Rev. **16** (1974), 555–557.

[209] ———, *Generalized Eisenstein series and modified Dedekind sums*, J. Reine Angew. Math. **272** (1974), 182–193.

[210] Bruce C. Berndt and Ronald J. Evans, *Least positive residues and the law of quadratic reciprocity*, Delta (Waukesha) **4** (1974/75), 65–69.

[211] Bruce C. Berndt, *A new proof of the reciprocity theorem for Dedekind sums*, Elem. Math. **29** (1974), 93–94.

[212] Bruce C. Berndt and Sarvadaman Chowla, *Zero sums of the Legendre symbol*, Nordisk Mat. Tidskr. **22** (1974), 5–8.

[213] Bruce C. Berndt, *Generalized Dedekind eta-functions and generalized Dedekind sums*, Trans. Amer. Math. Soc. **178** (1973), 495–508.

[214] ———, *A new method in arithmetical functions and contour integration*, Canad. Math. Bull. **16** (1973), 381–387.

[215] ———, *Character transformation formulae similar to those for the Dedekind eta-function*, Analytic number theory (Proc. Sympos. Pure Math., Vol. XXIV, St. Louis Univ., St. Louis, Mo., 1972), Amer. Math. Soc., Providence, R.I., 1973, pp. 9–30.

[216] ———, *The evaluation of infinite series by contour integration*, Univ. Beograd. Publ. Elektrotehn. Fak. Ser. Mat. Fiz. **412–460** (1973), 119–122.

[217] ———, *An elementary proof of some character sum identities of Apostol*, Glasgow Math. J. **14** (1973), 50–53.

[218] ———, *On Gaussian sums and other exponential sums with periodic coefficients*, Duke Math. J. **40** (1973), 145–156.

[219] ———, *A new proof of the functional equation of Dirichlet L-functions*, Proc. Amer. Math. Soc. **37** (1973), 355–357.

[220] ———, *On Gaussian sums*, Proceedings of the Number Theory Conference (Univ. Colorado, Boulder, Colo., 1972), Univ. Colorado, Boulder, Colo., 1972, pp. 14–18.

[221] ———, *The Voronoï summation formula*, The theory of arithmetic functions (Proc. Conf., Western Michigan Univ., Kalamazoo, Mich., 1971), Lecture Notes in Math., Vol. 251, Springer, Berlin, 1972, pp. 21–36.

[222] ———, *The evaluation of character series by contour integration*, Univ. Beograd. Publ. Elektrotehn. Fak. Ser. Mat. Fiz. **381-409** (1972), 25–29.

[223] ———, *Two new proofs of Lerch's functional equation*, Proc. Amer. Math. Soc. **32** (1972), 403–408.

[224] ———, *On the Hurwitz zeta-function*, Rocky Mountain J. Math. **2** (1972), no. 1, 151–157.

[225] ———, *The functional equation of some Dirichlet series. II*, Proc. Amer. Math. Soc. **31** (1972), 24–26.

[226] ———, *On the average order of a class of arithmetical functions. II*, J. Number Theory **3** (1971), 288–305.

[227] ———, *On the average order of a class of arithmetical functions. I*, J. Number Theory **3** (1971), 184–203.

[228] ———, *Identities involving the coefficients of a class of Dirichlet series. V, VI*, Trans. Amer. Math. Soc. 160 (1971), 139-156; ibid. **160** (1971), 157–167.

[229] ———, *The functional equation of some Dirichlet series*, Proc. Amer. Math. Soc. **29** (1971), 457–460.

[230] ———, *The number of zeros of the Dedekind zeta-function on the critical line*, J. Number Theory **3** (1971), 1–6.

[231] ———, *On the zeros of a class of Dirichlet series. I*, Illinois J. Math. **14** (1970), 244–258.

[232] ———, *On the zeros of a class of Dirichlet series. II*, Illinois J. Math. **14** (1970), 678–691.

[233] ———, *The number of zeros for* $\zeta^{(k)}(s)$, J. London Math. Soc. (2) **2** (1970), 577–580.

[234] ———, *On the average order of ideal functions and other arithmetical functions*, Bull. Amer. Math. Soc. **76** (1970), 1270–1274.

[235] ———, *On the average order of some arithmetical functions*, Bull. Amer. Math. Soc. **76** (1970), 856–859.

[236] ———, *Identities involving the coefficients of a class of Dirichlet series. IV*, Trans. Amer. Math. Soc. **149** (1970), 179–185.

[237] ———, *Erratum: "On the zeros of the Riemann zeta-function"*, Proc. Amer. Math. Soc. **24** (1970), 839.

[238] ———, *Identities involving the coefficients of a class of Dirichlet series. III*, Trans. Amer. Math. Soc. **146** (1969), 323–348.

[239] ———, *A note on the number of integral ideals of bounded norm in a quadratic number field*, Bull. Amer. Math. Soc. **75** (1969), 1283–1285.

[240] ———, *On the zeros of the Riemann zeta-function*, Proc. Amer. Math. Soc. **22** (1969), 183–188.

[241] ———, *Identities involving the coefficients of a class of Dirichlet series. I, II*, Trans. Amer. Math. Soc. **137** (1969), 345–359, 361–374.

[242] ———, *Arithmetical identities and Hecke's functional equation*, Proc. Edinburgh Math. Soc. (2) **16** (1968/69), 221–226.

[243] ———, *An identity for certain Dirichlet series*, Glasgow Math. J. **9** (1968), 79–82.

[244] ———, *Generalised Dirichlet series and Hecke's functional equation*, Proc. Edinburgh Math. Soc. (2) **15** (1966/67), 309–313.

Books:

[245] George E. Andrews and Bruce C. Berndt, *Ramanujan's lost notebook. Part V*, Springer, Cham, 2018.

[246] ———, *Ramanujan's lost notebook. Part IV*, Springer, New York, 2013.

[247] ———, *Ramanujan's lost notebook. Part III*, Springer, New York, 2012.

[248] ———, *Ramanujan's lost notebook. Part II*, Springer, New York, 2009.

[249] Bruce C. Berndt and Marvin I. Knopp, *Hecke's theory of modular forms and Dirichlet series*, Monographs in Number Theory, vol. 5, World Scientific Publishing Co. Pte. Ltd., Hackensack, NJ, 2008.

[250] Bruce C. Berndt, *Number theory in the spirit of Ramanujan*, Student Mathematical Library, vol. 34, American Mathematical Society, Providence, RI, 2006.

[251] George E. Andrews and Bruce C. Berndt, *Ramanujan's lost notebook. Part I*, Springer, New York, 2005.

[252] Bruce C. Berndt, Ronald J. Evans, and Kenneth S. Williams, *Gauss and Jacobi sums*, Canadian Mathematical Society Series of Monographs and Advanced Texts, John Wiley & Sons, Inc., New York, 1998. A Wiley-Interscience Publication.

[253] Bruce C. Berndt, *Ramanujan's notebooks. Part V*, Springer-Verlag, New York, 1998.

[254] Bruce C. Berndt and Robert A. Rankin, *Ramanujan*, History of Mathematics, vol. 9, American Mathematical Society, Providence, RI; London Mathematical Society, London, 1995. Letters and commentary.

[255] Bruce C. Berndt, *Ramanujan's notebooks. Part IV*, Springer-Verlag, New York, 1994.

[256] ———, *Ramanujan's notebooks. Part III*, Springer-Verlag, New York, 1991.

[257] ———, *Ramanujan's notebooks. Part II*, Springer-Verlag, New York, 1989.

[258] ———, *Ramanujan's notebooks. Part I*, Springer-Verlag, New York, 1985. With a foreword by Subrahmanyan Chandrasekhar.

[259] Bruce C. Berndt and Padmini T. Joshi, *Chapter 9 of Ramanujan's second notebook*, Contemporary Mathematics, vol. 23, American Mathematical Society, Providence, RI, 1983. Infinite series identities, transformations, and evaluations.

Other Publications:

[260] Bruce C. Berndt, *Ramanujan, his lost notebook, its importance*, George E. Andrews—80 years of combinatory analysis, Trends Math., Birkhäuser/Springer, Cham, 2021, pp. 33–52.

[261] ———, *Living with Ramanujan for 40 years*, Philos. Trans. Roy. Soc. A **378** (2020), no. 2163, 20180437, 15.

[262] Krishna Alladi, Bruce Berndt, Peter Paule, James Sellers, and Ae Ja Yee, *Combinatory analysis 2018: in honor of George Andrews' 80th birthday [Editorial]*, Ann. Comb. **23** (2019), no. 3-4, 425–426. Held at Pennsylvania State University, PA, June 21–24, 2018.

[263] Bruce Berndt, *The power of q: a personal journey [book review of MR3699428]*, Amer. Math. Monthly **126** (2019), no. 2, 189–192.

[264] Krishnaswami Alladi, Bruce C. Berndt, YoungJu Choie, and Wladimir Pribitkin, *Preface [special issue in memory of Marvin Knopp]*, Ramanujan J. **41** (2016), no. 1-3, 1–2.

[265] Bruce C. Berndt and Harold G. Diamond, *Preface [Special issue: In memory of Paul T. Bateman and Heini Halberstam]*, Int. J. Number Theory **11** (2015), no. 5, v–xiii.

[266] Bruce C. Berndt, *Lecture given by B. M. Wilson, University of Leeds, May, 1927*, Math. Newsl. **24** (2013), no. 1, 1–6.

[267] ———, *Ramanujan reaches his hand from his grave to snatch your theorems from you*, Asia Pac. Math. Newsl. **1** (2011), no. 2, 8–13.

[268] ———, *K. Venkatachaliengar*, Ramanujan rediscovered, Ramanujan Math. Soc. Lect. Notes Ser., vol. 14, Ramanujan Math. Soc., Mysore, 2010, pp. xiii–xvi.

[269] Nayandeep Deka Baruah, Bruce C. Berndt, and Heng Huat Chan, *Ramanujan's series for $1/\pi$: a survey*, Amer. Math. Monthly **116** (2009), no. 7, 567–587.

[270] Bruce C. Berndt and YoungJu Choie, *Preface [A special issue in celebration of Marvin Knopp's 73rd birthday]*, Int. J. Number Theory **3** (2007), no. 3, v–viii.

[271] Bruce C. Berndt and C. A. Reddi, *Two exams taken by Ramanujan in India*, Amer. Math. Monthly **111** (2004), no. 4, 330–339.

[272] Bruce C. Berndt and Ae Ja Yee, *A survey on Eisenstein series in Ramanujan's lost notebook*, Sūrikaisekikenkyūsho Kōkyūroku **1274** (2002), 130–141. New aspects of analytic number theory (Japanese) (Kyoto, 2001).

[273] Bruce C. Berndt, Youn-Seo Choi, and Soon-Yi Kang, *The problems submitted by Ramanujan to the Journal of the Indian Mathematical Society [MR1665361 (2000i:11003)]*, Ramanujan: essays and surveys, Hist. Math., vol. 22, Amer. Math. Soc., Providence, RI, 2001, pp. 215–258.

[274] Bruce C. Berndt, *π related developments since 1988*, Ramanujan: essays and surveys, Hist. Math., vol. 22, Amer. Math. Soc., Providence, RI, 2001, pp. 201.

[275] ———, *An overview of Ramanujan's notebooks [MR1672414 (99k:01032)]*, Ramanujan: essays and surveys, Hist. Math., vol. 22, Amer. Math. Soc., Providence, RI, 2001, pp. 143–164.

[276] ———, *A short biography of S. Narayana Iyer*, Ramanujan: essays and surveys, Hist. Math., vol. 22, Amer. Math. Soc., Providence, RI, 2001, pp. 97–98.

[277] ———, *S. Janaki Ammal (Mrs. Ramanujan)*, Ramanujan: essays and surveys, Hist. Math., vol. 22, Amer. Math. Soc., Providence, RI, 2001, pp. 83–87.

[278] ———, *The influence of Carr's synopsis on Ramanujan*, Ramanujan: essays and surveys, Hist. Math., vol. 22, Amer. Math. Soc., Providence, RI, 2001, pp. 17–20.

[279] Bruce C. Berndt and Robert A. Rankin, *The books studied by Ramanujan in India [MR1786233 (2002b:01036)]*, Ramanujan: essays and surveys, Hist. Math., vol. 22, Amer. Math. Soc., Providence, RI, 2001, pp. 9–15.

[280] Bruce C. Berndt, *The four photographs of Ramanujan*, Ramanujan: essays and surveys, Hist. Math., vol. 22, Amer. Math. Soc., Providence, RI, 2001, pp. 3–8.

[281] ———, *The notebooks of Srinivasa Ramanujan*, Srinivasa Ramanujan (1887–1920), Macmillan India, Madras, 1988, pp. 42–51.

[282] Bruce Carl Berndt, *Identities involving the coefficients of a class of Dirichlet series*, ProQuest LLC, Ann Arbor, MI, 1966. Thesis (Ph.D.), The University of Wisconsin-Madison.

© 2024 World Scientific Publishing Company
https://doi.org/10.1142/9789811277375_fm

Preface

The volume you hold in your hands constitutes research articles stemming from the conference, Analytic and Combinatorial Number Theory: The Legacy of Ramanujan, held at the University of Illinois from June 6–9, 2019, honoring Bruce Berndt on his 80th birthday.

This preface is somewhat unusual. Often the preface of a volume like this provides a summary of the talks, thanks the contributors and others responsible for the meeting and concludes with a few nice words about the honoree.

Obviously, it is necessary to follow this precedent to some degree. I thank the other organizers, Scott Ahlgren, Atul Dixit, Michael Filaseta, Kevin Ford, A. J. Hildebrand, Tim Huber, Bruce Reznick and Ae Ja Yee for doing a great job in putting this grand event together. In addition, deep thanks should go to the sponsoring organizations: the National Science Foundation, the National Security Agency, the Institute for Mathematics and its Applications, and the University of Illinois Department of Mathematics and College of Liberal Arts and Sciences.

It was fortuitous that this conference occurred shortly after the appearance of the final volume of the series, *Ramanujan's Lost Notebook, Parts I–V*, which Bruce and I co-authored. It seems appropriate here to describe a bit of the decades-long adventure that Bruce and I undertook.

In the spring of 1976, I stumbled upon the Lost Notebook in the Wren Library of Trinity College Cambridge. The discovery is a long story that has been told elsewhere. I started giving talks about it the next year. In the fall of 1977, Bruce and I were both speaking at the conference, Number Theory Pure and Simple, at the Miami University of Ohio in Oxford, Ohio. The title of Bruce's talk was "Ramanujan's Notebooks." By then, Bruce was already engaged in his decades-long effort to provide proofs of everything in Ramanujan's original notebooks, also a five-volume effort which ended in 1998. While I did not speak on Ramanujan at this conference, Bruce had heard about the Lost Notebook discovery (indeed, I had given an invited address on it at the recent AMS/MAA summer meetings in Seattle). Bruce asked if I might be interested in a joint project, perhaps a couple of books, proving everything in the Lost Notebook. I was delighted with the idea. There was much in the Lost Notebook that was likely to defeat me completely, and most of that material was in Bruce's wheelhouse. So, the collaboration was set up, but not set in motion until around 1995. At that time, we started serious discussions about the "one or two" volumes in the Lost Notebook. However, it was not until 2005 that the first of five volumes appeared. In 2018, the fifth volume was published. Throughout it all, Bruce was a steady, organizing force who contributed most to the success of the venture.

I have known Bruce since we met at the AMS/MAA summer meeting in 1969, held at Penn State. This gives me 50-odd years of perspective on his career. As we all are aware, Bruce is best known for his contribution to the understanding of Ramanujan. Thirteen of Bruce's 15 books have "Ramanujan" in the title. Academic Influence recently listed Bruce as one of the five most influential mathematicians of the last decade. However, it would be a gross mistake to think that his research fully describes Bruce's impact on the world.

In 1986, the late Richard Guy published an article entitled, "How Good a Mathematician Are You?" He began by noting that most of us would read the title and reflect on our research output, but that was not what Guy was after. The point of the article is to reflect on your contributions to the community and to the world. From this standpoint, Bruce would again rank very high. Bruce has had 36 Ph.D. students, and they worship him for his thoughtfulness, kindness, and good advice. He has served on numerous committees for the AMS and MAA, and has given an immense amount of effort (both time and money) to numerous charities.

It has been a privilege to collaborate with this grand man and grand mathematician. It is great that 170 mathematicians gathered in Urbana in the spring of 2019 to celebrate Bruce's 80th birthday. This magnificent volume is a fitting tribute to the life and accomplishments of Bruce Berndt.

George E. Andrews
The Pennsylvania State University, USA

© 2024 World Scientific Publishing Company
https://doi.org/10.1142/9789811277375_0001

A triple integral analog of a multiple zeta value*

Tewodros Amdeberhan

Department of Mathematics, Tulane University
6823 St Charles Ave, New Orleans, LA 70118, USA
tamdeber@tulane.edu

Victor H. Moll

Department of Mathematics, Tulane University
6823 St Charles Ave, New Orleans, LA 70118, USA
vhm@tulane.edu

Armin Straub[†]

Department of Mathematics and Statistics, University of South Alabama
411 University Blvd N, Mobile, AL 36688, USA
straub@southalabama.edu

Christophe Vignat

Department of Mathematics, Tulane University
6823 St Charles Ave, New Orleans, LA 70118, USA
cvignat@tulane.edu

Dedicated to Bruce Berndt on the occasion of his 80th birthday

We establish the triple integral evaluation

$$\int_1^\infty \int_0^1 \int_0^1 \frac{dz\,dy\,dx}{x(x+y)(x+y+z)} = \frac{5}{24}\zeta(3),$$

as well as the equivalent polylogarithmic double sum

$$\sum_{k=1}^\infty \sum_{j=k}^\infty \frac{(-1)^{k-1}}{k^2} \frac{1}{j\,2^j} = \frac{13}{24}\zeta(3).$$

This double sum is related to, but less approachable than, similar sums studied by Ramanujan. It is also reminiscent of Euler's formula $\zeta(2,1) = \zeta(3)$, which is the simplest instance of duality of multiple polylogarithms. We review this duality and apply it to derive a companion identity. We also discuss approaches based on computer algebra. All of our approaches ultimately require the introduction of polylogarithms and nontrivial

*This article originally appeared in *International Journal of Number Theory*, **17**(2), pp. 223–237 © 2021 World Scientific Publishing Company. DOI: 10.1142/S1793042120400102
[†]Corresponding author.

relations between them. It remains an open challenge to relate the triple integral or the double sum to $\zeta(3)$ directly.

Keywords: Multiple polylogarithms; integral representation; multiple zeta values; Riemann zeta function.

Mathematics Subject Classification 2020: 11M32

1. Introduction

The evaluation

$$\zeta(3) = \int_0^1 \int_x^1 \int_y^1 \frac{dz\,dy\,dx}{(1-x)yz} \tag{1.1}$$

is given by Kontsevich and Zagier [11] as an illustration that $\zeta(3)$ is a *period*, in the sense that it is the value of an absolutely convergent integral of a rational function with rational coefficients, over a domain in $\mathbb{R}^3$ given by polynomial inequalities $(0 < x < y < z < 1)$ with rational coefficients. The goal of this work is to prove and discuss the following, much less obvious, variation of a triple integral evaluation.

Theorem 1.1. *We have*

$$Z_3 := \int_1^\infty \int_0^1 \int_0^1 \frac{dz\,dy\,dx}{x(x+y)(x+y+z)} = \frac{5}{24}\zeta(3). \tag{1.2}$$

In contrast to (1.1), it appears to be a rather tricky problem to relate the triple integral (1.2) to $\zeta(3)$ directly. Indeed, all of our approaches to this integral have ultimately required the introduction of polylogarithms and nontrivial relations between them. We give such a proof in Sec. 3.

The integral (1.2) might be seen as a continuous analog of multiple zeta values defined by

$$\zeta(s_1, \ldots, s_k) = \sum_{n_1 > n_2 > \cdots > n_k \geq 1} \frac{1}{n_1^{s_1} n_2^{s_2} \cdots n_k^{s_k}}. \tag{1.3}$$

These sums were introduced by Euler. The reader is referred to the site [10] maintained by Hoffman for a large collection of papers related to these series. For instance, when $k = 3$, the multiple zeta values can be written as

$$\zeta(s_1, s_2, s_3) = \sum_{p=1}^\infty \sum_{q=1}^\infty \sum_{r=1}^\infty \frac{1}{p^{s_3}(p+q)^{s_2}(p+q+r)^{s_1}}, \tag{1.4}$$

where the similarity with Z_3 becomes apparent.

Remark 1.2. Algorithmic approaches to computing (period) integrals such as the one in (1.2) are described in [5, 16]. In particular, Panzer implemented his symbolic integration approach [16] using hyperlogarithms in a Maple package called `HyperInt`. Using this package, the integral (1.2) is automatically evaluated as

$$Z_3 = \frac{19}{8}\zeta(3) - 2\log(2)\zeta(2) - \mathrm{Li}_{2,1}\left(\frac{1}{2}, 2\right) - \mathrm{Li}_{1,1,1}\left(\frac{1}{3}, \frac{3}{2}, 2\right)$$

featuring the multiple polylogarithms reviewed in Sec. 2. Simplifying the right-hand side to a multiple of $\zeta(3)$, however, is not straightforward. Further comments on evaluating the integral Z_3 with the help of computer algebra are included in Sec. 7.

In Sec. 4, we show that the triple integral evaluation of Theorem 1.1 is equivalent to the following relation between polylogarithms.

Theorem 1.3. *We have*

$$S_3 := \sum_{k=1}^{\infty}\sum_{j=k}^{\infty} \frac{(-1)^{k-1}}{k^2} \frac{1}{j\,2^j} = \frac{13}{24}\zeta(3). \tag{1.5}$$

As in the case of the triple integral, it is not clear how to relate this double sum directly to $\zeta(3)$. In the notation for multiple polylogarithms reviewed in Sec. 2, the double sum in (1.5) can be expressed as

$$S_3 = -\operatorname{Li}_{1,2}\left(\frac{1}{2}, -1\right) - \operatorname{Li}_3\left(-\frac{1}{2}\right). \tag{1.6}$$

Readers familiar with multiple polylogarithms might therefore wonder whether the identity (1.5) is an instance of duality [7, Sec. 6.1]. However, it appears that duality does not help in evaluating (1.6). We review duality in Sec. 5 and show that it instead naturally provides the companion identity

$$\sum_{\substack{k,m,n\geq 1\\ k \text{ odd}}} \frac{1}{3^n} \frac{1}{k(k+m)(k+m+n)} = \frac{13}{48}\zeta(3). \tag{1.7}$$

We note that (1.5) is reminiscent of Ramanujan's identity [2, p. 259] (see also [7, (7.4)])

$$\sum_{k=1}^{\infty}\sum_{j=k}^{\infty} \frac{1}{k} \frac{1}{j^2\,2^j} = \zeta(3) - \frac{\pi^2}{12}\log(2). \tag{1.8}$$

Indeed, we show in Sec. 6 that Ramanujan's approach can be applied to evaluate the non-alternating version of (1.5) as

$$\sum_{k=1}^{\infty}\sum_{j=k}^{\infty} \frac{1}{k^2} \frac{1}{j\,2^j} = \frac{5}{8}\zeta(3). \tag{1.9}$$

We further indicate that (1.5) does not succumb readily to the same approach.

Remark 1.4. We conclude this introduction by observing that the 2-dimensional analog of (1.2), namely,

$$Z_2 := \int_1^{\infty}\int_0^1 \frac{dy\,dx}{x(x+y)} \tag{1.10}$$

is simple to evaluate. For instance, we can transform the domain of integration to the unit square via the change of variables $x \mapsto 1/x$ to obtain

$$Z_2 = \int_0^1\int_0^1 \frac{dydx}{1+xy} = \sum_{n=0}^{\infty}\int_0^1\int_0^1 (-xy)^n dydx = \sum_{n=0}^{\infty} \frac{(-1)^n}{(n+1)^2} = \frac{1}{2}\zeta(2).$$

2. Polylogarithms

The polylogarithm function is defined, for $|z| < 1$, by the power series

$$\operatorname{Li}_s(z) = \sum_{k=1}^{\infty} \frac{z^k}{k^s}. \tag{2.1}$$

The function Li_2 is also called the *dilogarithm* and Li_3 the *trilogarithm*. For other values of z, the polylogarithms are defined by analytic continuation, with the principal branch obtained from a cut along the positive real axis from $z = 1$ to ∞. We record that the analytic continuation of Li_2 is provided by the integral representation

$$\operatorname{Li}_2(z) = \int_1^{1-z} \frac{\log(t)}{1-t}\,dt. \tag{2.2}$$

These functions satisfy a large collection of identities. Among these are the duplication formula

$$\operatorname{Li}_s(-z) = -\operatorname{Li}_s(z) + 2^{1-s}\operatorname{Li}_s(z^2), \tag{2.3}$$

as well as the inversion formula

$$\operatorname{Li}_n(z) + (-1)^n \operatorname{Li}_n(1/z) = -\frac{(2\pi i)^n}{n!} B_n\left(\frac{1}{2} + \frac{\log(-z)}{2\pi i}\right), \tag{2.4}$$

which holds for positive integers n and $z \notin [0,1]$. Here, the B_n are the Bernoulli polynomials, of which we will only use $B_2(x) = x^2 - x + \frac{1}{6}$ and $B_3(x) = x^3 - \frac{3}{2}x^2 + \frac{1}{2}x$. In the cases $n = 2$ and $n = 3$, the inversion formula thus becomes

$$\operatorname{Li}_2(z) = -\operatorname{Li}_2(1/z) - \frac{1}{2}\log^2(-z) - \frac{\pi^2}{6}, \tag{2.5}$$

$$\operatorname{Li}_3(z) = \operatorname{Li}_3(1/z) - \frac{1}{6}\log^3(-z) - \frac{\pi^2}{6}\log(-z). \tag{2.6}$$

The dilogarithm and the trilogarithm satisfy a number of additional relations. A small selection of formulæ is recorded in [15, Sec. 25.12] (see also [2, Chap. 9]). A more complete collection appears in [12, 13]. For our purposes, we record the reflection formulæ

$$\operatorname{Li}_2(z) + \operatorname{Li}_2(1-z) = \frac{\pi^2}{6} - \log(z)\log(1-z) \tag{2.7}$$

and

$$\begin{aligned}\operatorname{Li}_3(z) &+ \operatorname{Li}_3(1-z) + \operatorname{Li}_3(1-1/z)\\ &= \zeta(3) + \frac{1}{6}\log^3(z) + \frac{\pi^2}{6}\log(z) - \frac{1}{2}\log^2(z)\log(1-z),\end{aligned} \tag{2.8}$$

as well as the identity [12, (6.34)]

$$\begin{aligned}\operatorname{Li}_3\left(\frac{1-z}{1+z}\right) - \operatorname{Li}_3\left(-\frac{1-z}{1+z}\right) &= \frac{1}{2}\operatorname{Li}_3\left(\frac{-z^2}{1-z^2}\right) - 2\operatorname{Li}_3\left(\frac{-z}{1-z}\right) - 2\operatorname{Li}_3\left(\frac{z}{1+z}\right)\\ &\quad + \frac{7}{4}\zeta(3) + \frac{\pi^2}{4}\log\left(\frac{1-z}{1+z}\right)\\ &\quad + \frac{1}{4}\log^2\left(\frac{1+z}{1-z}\right)\log\left(\frac{1-z^2}{z^2}\right).\end{aligned} \tag{2.9}$$

It follows from (2.7) and (2.8), together with $\mathrm{Li}_3(-1) = -\frac{3}{4}\zeta(3)$, that

$$\mathrm{Li}_2\left(\frac{1}{2}\right) = \frac{\pi^2}{12} - \frac{1}{2}\log^2(2), \tag{2.10}$$

$$\mathrm{Li}_3\left(\frac{1}{2}\right) = \frac{7}{8}\zeta(3) - \frac{\pi^2}{12}\log(2) + \frac{1}{6}\log^3(2). \tag{2.11}$$

Combining the mentioned identities as done, for instance, in [3, (39.1), (39.4); p. 324] where Ramanujan considers such combinations, we can further derive the relation

$$2\,\mathrm{Li}_2\left(\frac{1}{3}\right) - \mathrm{Li}_2\left(-\frac{1}{3}\right) = \frac{\pi^2}{6} - \frac{1}{2}\log^2(3). \tag{2.12}$$

Similarly, by setting $z = 1/2$ in (2.10), we find

$$2\,\mathrm{Li}_3\left(\frac{1}{3}\right) - \mathrm{Li}_3\left(-\frac{1}{3}\right) = \frac{13}{6}\zeta(3) - \frac{\pi^2}{6}\log(3) + \frac{1}{6}\log^3(3). \tag{2.13}$$

Generalizing (2.1), the *multiple polylogarithms* are the sums

$$\mathrm{Li}_{s_1,\ldots,s_k}(z_1,\ldots,z_k) = \sum_{n_1>n_2>\cdots>n_k\geq 1} \frac{z_1^{n_1} z_2^{n_2} \cdots z_k^{n_k}}{n_1^{s_1} n_2^{s_2} \cdots n_k^{s_k}}, \tag{2.14}$$

though notation varies throughout the literature (in particular, the order of the arguments is sometimes reversed; the choice here is consistent with multiple zeta values as defined in (1.3), so that (2.14) becomes $\zeta\ (s_1,\ldots,s_k)$ when $z_1 = z_2 = \cdots = z_k = 1$). A vast literature, see, for instance, [7] and the references therein, exists concerning relations between values of multiple polylogarithms.

3. A Proof of the Triple Integral Evaluation

In this section, we evaluate the triple integral Z_3 as claimed in Theorem 1.1. We begin by integrating with respect to x and, then, with respect to y to obtain

$$\begin{aligned} Z_3 &= \int_0^1 \int_0^1 \left[\frac{\log(y+1)}{yz} - \frac{\log(y+z+1)}{z(y+z)}\right] dy\,dz \\ &= \int_0^1 [\mathrm{Li}_2(-z-1) - \mathrm{Li}_2(-1) - \mathrm{Li}_2(-z)]\,\frac{dz}{z} \\ &= \int_0^1 [\mathrm{Li}_2(-z-1) - \mathrm{Li}_2(-1)]\,\frac{dz}{z} + \frac{3}{4}\zeta(3). \end{aligned} \tag{3.1}$$

For the second equality, we used the fact that the derivative of $\mathrm{Li}_2(-x)$ is $-\log(x+1)/x$, while, for the third, we integrated term by term to find

$$\int_0^1 \mathrm{Li}_2(-z)\frac{dz}{z} = \int_0^1 \sum_{n=1}^{\infty} \frac{(-z)^n}{n^2}\frac{dz}{z} = \sum_{n=1}^{\infty} \frac{(-1)^n}{n^3} = -\frac{3}{4}\zeta(3). \tag{3.2}$$

We now use the integral representation (2.2) of the dilogarithm to conclude

$$Z_3 = \int_0^1 \int_2^{2+z} \frac{\log(t)\,dt}{1-t}\frac{dz}{z} + \frac{3}{4}\zeta(3).$$

Exchanging the order of integration and evaluating the inner integral, we obtain

$$Z_3 = \int_0^1 \frac{\log(t)\log(t+2)}{1+t}\,dt + \frac{3}{4}\zeta(3). \tag{3.3}$$

We now replace

$$2\log(t)\log(t+2) = \log^2(t) + \log^2(t+2) - \log^2\left(\frac{t}{t+2}\right) \tag{3.4}$$

in the integral (3.3), and make the change of variables $t/(t+2) \to t$ in the third of the three resulting integrals, to find

$$\begin{aligned} Z_3 &= \frac{1}{2}\int_0^1 \frac{\log^2(t)}{1+t}\,dt + \frac{1}{2}\int_0^1 \frac{\log^2(t+2)}{1+t}\,dt - \int_0^{1/3} \frac{\log^2(t)}{1-t^2}\,dt + \frac{3}{4}\zeta(3) \\ &= \frac{1}{2}\int_0^1 \frac{\log^2(t+2)}{1+t}\,dt - \int_0^{1/3} \frac{\log^2(t)}{1-t^2}\,dt + \frac{3}{2}\zeta(3). \end{aligned} \tag{3.5}$$

Here, we evaluated the first integral by expanding $1/(1+t)$ as a geometric series, integrating term by term and using

$$\int_0^x t^{n-1}\log^2(t)\,dt = \frac{x^n}{n^3}\left(2 - 2n\log(x) + n^2\log^2(x)\right), \tag{3.6}$$

(which is readily verified by differentiating both sides) as well as the final sum in (3.2). Summing (3.6), we find

$$\begin{aligned} \int_0^x \frac{\log^2(t)}{1-t^2}\,dt = {} & \mathrm{Li}_3(x) - \mathrm{Li}_3(-x) - \log(x)(\mathrm{Li}_2(x) - \mathrm{Li}_2(-x)) \\ & + \frac{1}{2}\log^2(x)\log\left(\frac{1+x}{1-x}\right). \end{aligned} \tag{3.7}$$

In light of the relations (2.12) and (2.13), this shows that

$$\int_0^{1/3} \frac{\log^2(t)}{1-t^2}\,dt = \frac{13}{6}\zeta(3) - \mathrm{Li}_3\left(\frac{1}{3}\right) - \mathrm{Li}_2\left(\frac{1}{3}\right)\log(3) - \frac{1}{6}\log\left(\frac{9}{8}\right)\log^2(3). \tag{3.8}$$

By appealing to analytic continuation, we could similarly approach the remaining integral in (3.5). Alternatively, we use the formula [13, (6.28)]

$$\begin{aligned} \int_0^t \frac{\log^2(u+1)}{u}\,du = {} & \log t\log^2(t+1) - \frac{2}{3}\log^3(t+1) \\ & - 2\log(t+1)\,\mathrm{Li}_2\left(\frac{1}{t+1}\right) - 2\,\mathrm{Li}_3\left(\frac{1}{t+1}\right) + 2\zeta(3) \end{aligned} \tag{3.9}$$

to evaluate

$$\begin{aligned} \frac{1}{2}\int_0^1 \frac{\log^2(t+2)}{1+t}\,dt &= \frac{1}{2}\int_0^2 \frac{\log^2(t+1)}{t}\,dt - \frac{1}{2}\int_0^1 \frac{\log^2(t+1)}{t}\,dt \\ &= \frac{7}{8}\zeta(3) - \mathrm{Li}_3\left(\frac{1}{3}\right) - \mathrm{Li}_2\left(\frac{1}{3}\right)\log(3) \\ &\quad - \frac{1}{6}\log\left(\frac{9}{8}\right)\log^2(3). \end{aligned} \tag{3.10}$$

Here, we used (2.10) and (2.11) to reduce the polylogarithms at 1/2. The claimed evaluation, $Z_3 = \frac{7}{8}\zeta(3) - \frac{13}{6}\zeta(3) + \frac{3}{2}\zeta(3) = \frac{5}{24}\zeta(3)$, now follows from using (3.8) and (3.10) in (3.5).

4. An Equivalent Double Sum Evaluation

In this section, we relate the triple integral Z_3 of Theorem 1.1 to the double sum in Theorem 1.3 in the following way.

Lemma 4.1. *We have*

$$Z_3 = \frac{3}{4}\zeta(3) - \sum_{k=1}^{\infty}\sum_{j=k}^{\infty} \frac{(-1)^{k-1}}{k^2} \frac{1}{j\,2^j}. \tag{4.1}$$

Proof. Expanding the integrand of (1.2) as a geometric series, we obtain

$$\begin{aligned}\int_0^1 \frac{dz}{x(x+y)(x+y+z)} &= \frac{1}{x(x+y)^2}\int_0^1 \frac{dz}{1+\frac{z}{x+y}} \\ &= \frac{1}{x}\sum_{k=0}^{\infty}\int_0^1 \frac{(-1)^k z^k\,dz}{(x+y)^{k+2}} \\ &= \frac{1}{x}\sum_{k=0}^{\infty} \frac{(-1)^k}{(k+1)(x+y)^{k+2}}.\end{aligned}$$

We then integrate with respect to y to find

$$\begin{aligned}\int_0^1\int_0^1 \frac{dz\,dy}{x(x+y)(x+y+z)} &= \frac{1}{x}\int_0^1 \sum_{k=0}^{\infty} \frac{(-1)^k\,dy}{(k+1)(x+y)^{k+2}} \\ &= \frac{1}{x}\sum_{k=0}^{\infty}\frac{(-1)^k}{(k+1)^2}\left[\frac{1}{x^{k+1}} - \frac{1}{(x+1)^{k+1}}\right] \\ &= \sum_{k=0}^{\infty}\frac{(-1)^k}{(k+1)^2 x^{k+2}} - \frac{1}{x}\sum_{k=0}^{\infty}\frac{(-1)^k}{(k+1)^2}\frac{1}{(x+1)^{k+1}}.\end{aligned}$$

Integrating the first term with respect to x results in

$$\sum_{k=0}^{\infty}\frac{(-1)^k}{(k+1)^2}\int_1^{\infty}\frac{dx}{x^{k+2}} = \sum_{k=0}^{\infty}\frac{(-1)^k}{(k+1)^3} = \frac{3}{4}\zeta(3). \tag{4.2}$$

On the other hand, the second term contributes

$$\sum_{k=0}^{\infty}\frac{(-1)^k}{(k+1)^2}\int_1^{\infty}\frac{dx}{x\,(x+1)^{k+1}}. \tag{4.3}$$

Finally, we note that

$$\int_1^{\infty}\frac{dx}{x(x+1)^{k+1}} = \sum_{j=k}^{\infty}\frac{1}{j\,2^j}, \tag{4.4}$$

which follows from the change of variables $w = 1/(x+1)$ and expanding a geometric series. Combining terms, we arrive at (4.1). □

Observe that Lemma 4.1 and Theorem 1.1 immediately imply Theorem 1.3.

Let us conclude this section by deriving an alternative integral representation for the double sum

$$S_3 = \sum_{k=1}^{\infty} \frac{(-1)^{k-1}}{k^2} \sum_{j=k}^{\infty} \frac{1}{j\,2^j} \tag{4.5}$$

of Theorem 1.3.

Theorem 4.2. *We have*

$$S_3 = -\int_0^{1/2} \frac{\mathrm{Li}_2(-t)}{t(1-t)}\,dt = -4\int_0^{\pi/4} \frac{\mathrm{Li}_2(-\sin^2\varphi)}{\sin(2\varphi)}\,d\varphi. \tag{4.6}$$

Proof. It is not hard to show (and can even be done automatically, for instance, using creative telescoping) that, for positive integers k, the inner sum can be expressed in hypergeometric terms as

$$\sum_{j=k}^{\infty} \frac{1}{j\,2^j} = \log 2 - \sum_{j=1}^{k-1} \frac{1}{j\,2^j} = \frac{1}{k}\,{}_2F_1\left(\left.\begin{matrix} k \quad k \\ k+1 \end{matrix}\right| -1\right). \tag{4.7}$$

The double sum therefore equals

$$S_3 = \sum_{k=1}^{\infty} \frac{(-1)^{k-1}}{k^3}\,{}_2F_1\left(\left.\begin{matrix} k \quad k \\ k+1 \end{matrix}\right| -1\right). \tag{4.8}$$

The classical integral representation for the hypergeometric function

$${}_2F_1\left(\left.\begin{matrix} a \quad b \\ c \end{matrix}\right| z\right) = \frac{1}{B(b,c-b)} \int_0^1 x^{b-1}(1-x)^{c-b-1}(1-xz)^{-a}\,dx, \tag{4.9}$$

see, for instance, [1, p. 65], gives

$$S_3 = \int_0^1 \left[\sum_{k=1}^{\infty} \frac{(-1)^{k-1}}{k^2} \left(\frac{x}{x+1}\right)^k\right] \frac{dx}{x}. \tag{4.10}$$

The series definition (2.1) for the dilogarithm, followed by the change of variables $t = \frac{x}{x+1}$, therefore implies the claimed integral representation. □

We note that, alternatively, the series S_3 can be obtained starting from the second integral in (4.6) by expanding the dilogarithm and using [9, Entry 2.516.2]

$$\int \frac{\sin^{2k-1}\varphi}{\cos\varphi}\,d\varphi = -\sum_{j=1}^{k-1} \frac{\sin^{2j}\varphi}{2j} - \log(\cos\varphi). \tag{4.11}$$

5. A Companion Identity via Duality

Recall that Theorem 1.3 is equivalent to

$$-\operatorname{Li}_{2,1}\left(-1, \frac{1}{2}\right) - \operatorname{Li}_3\left(-\frac{1}{2}\right) = \frac{13}{24}\zeta(3). \tag{5.1}$$

In this section, we review duality [7, Sec. 6.1] and apply it to the identity (5.1) to deduce the following companion identity.

Theorem 5.1. *We have*

$$\sum_{\substack{k,m,n\geq 1\\ k\ odd}} \frac{1}{3^n} \frac{1}{k(k+m)(k+m+n)} = \frac{13}{48}\zeta(3). \tag{5.2}$$

We use the notation of [7] and write

$$l\begin{pmatrix} s_1,\ldots,s_k \\ y_1,\ldots,y_k \end{pmatrix} := \operatorname{Li}_{s_1,\ldots,s_k}\left(\frac{1}{y_1}, \frac{y_1}{y_2}, \frac{y_2}{y_3}, \ldots, \frac{y_{k-1}}{y_k}\right)$$

for multiple polylogarithms as well as

$$\int_0^1 \omega(a_1)\omega(a_2)\cdots\omega(a_n) := \int_0^1 \int_0^{x_1} \cdots \int_0^{x_{n-1}} \frac{dx_n \cdots dx_2 dx_1}{(x_n - a_n)\cdots(x_2-a_2)(x_1-a_1)}$$

for iterated integrals. These integrals provide a natural way to express multiple polylogarithms. Indeed, we have the weight-dimensional iterated integral representation [7, (4.9)]

$$l\begin{pmatrix} s_1,\ldots,s_k \\ y_1,\ldots,y_k \end{pmatrix} = (-1)^k \int_0^1 \prod_{j=1}^k \omega(0)^{s_j-1}\omega(y_j). \tag{5.3}$$

Note that the weight of the multiple polylogarithm is defined to be $s_1 + s_2 + \cdots + s_k$ (and that this matches the number of integrations in the iterated integral representation (5.3)).

Reversing the order of integration in (5.3) and replacing each integration variable x by $1-x$ results in the dual iterated integral representation (5.4) of the multiple polylogarithm.

Theorem 5.2 ([7, (6.1)]). *We have*

$$l\begin{pmatrix} s_1,\ldots,s_k \\ y_1,\ldots,y_k \end{pmatrix} = (-1)^{s_1+s_2+\cdots+s_k+k} \int_0^1 \prod_{j=0}^{k-1} \omega(1-y_{k-j})\omega(1)^{s_{k-j}-1}. \tag{5.4}$$

A famous instance of duality is Euler's formula (see, for instance, [6])

$$\begin{aligned} \zeta(2,1) = l\begin{pmatrix} 2 & 1 \\ 1 & 1 \end{pmatrix} &= \int_0^1 \omega(0)\omega(1)\omega(1) \\ &= -\int_0^1 \omega(0)\omega(0)\omega(1) = l\begin{pmatrix} 3 \\ 1 \end{pmatrix} = \zeta(3), \end{aligned} \tag{5.5}$$

where both (5.3) and (5.4) have been used to relate one polylogarithm to its dual.

Proof of Theorem 5.1. Similar to (5.5), for the polylogarithms involved in (5.1), we find that

$$\begin{aligned}
\mathrm{Li}_{1,2}\left(\frac{1}{2},-1\right) &= l\begin{pmatrix}1 & 2\\ 2 & -2\end{pmatrix} = \int_0^1 \omega(2)\omega(0)\omega(-2)\\
&= -\int_0^1 \omega(3)\omega(1)\omega(-1) = l\begin{pmatrix}1 & 1 & 1\\ 3 & 1 & -1\end{pmatrix} = \mathrm{Li}_{1,1,1}\left(\frac{1}{3},3,-1\right)\\
&= \sum_{n_1>n_2>n_3\geq 1} \frac{(-1)^{n_3}}{3^{n_1-n_2}}\frac{1}{n_1n_2n_3},
\end{aligned}$$

as well as

$$\begin{aligned}
\mathrm{Li}_3\left(-\frac{1}{2}\right) &= l\begin{pmatrix}3\\ -2\end{pmatrix} = -\int_0^1 \omega(0)^2\omega(-2)\\
&= \int_0^1 \omega(3)\omega(1)^2 = -l\begin{pmatrix}1 & 1 & 1\\ 3 & 1 & 1\end{pmatrix} = -\,\mathrm{Li}_{1,1,1}\left(\frac{1}{3},3,1\right)\\
&= -\sum_{n_1>n_2>n_3\geq 1} \frac{1}{3^{n_1-n_2}}\frac{1}{n_1n_2n_3}.
\end{aligned}$$

Taken together, we obtain

$$\sum_{n_1>n_2>n_3\geq 1} \frac{1}{3^{n_1-n_2}}\frac{1}{n_1n_2n_3} - \sum_{n_1>n_2>n_3\geq 1} \frac{(-1)^{n_3}}{3^{n_1-n_2}}\frac{1}{n_1n_2n_3} = \frac{13}{24}\zeta(3)$$

or, equivalently, (5.2). □

6. A Related Simpler Double Sum

Ramanujan derived (1.8) by evaluating [2, Entry 9; p. 251] the function

$$g(z) = \sum_{k=1}^{\infty} H_k \frac{z^{k+1}}{(k+1)^2}, \quad \text{where } H_k = \sum_{j=1}^{k}\frac{1}{j},$$

in polylogarithmic terms as

$$g(1-z) = \frac{1}{2}\log^2(z)\log(1-z) + \mathrm{Li}_2(z)\log(z) - \mathrm{Li}_3(z) + \zeta(3), \tag{6.1}$$

and observing that the double sum (1.8) is given by $g(1/2) + \mathrm{Li}_3(1/2)$. We next show that (1.9) can be derived in an analogous manner.

Lemma 6.1. *For $|z| < 1$, define*

$$h(z) = \sum_{k=1}^{\infty} H_k^{(2)} \frac{z^{k+1}}{k+1}, \quad \text{where } H_k^{(2)} = \sum_{j=1}^{k}\frac{1}{j^2}.$$

Then,

$$h(1-z) = 2\,\mathrm{Li}_3(z) - \mathrm{Li}_2(z)\log(z) - \frac{\pi^2}{6}\log(z) - 2\zeta(3). \tag{6.2}$$

Proof. We note that

$$h'(z) = \sum_{k=1}^{\infty} H_k^{(2)} z^k = \frac{\mathrm{Li}_2(z)}{1-z}.$$

Integrating by parts, we find

$$h(z) = -\log(1-z)\,\mathrm{Li}_2(z) - \int_0^z \frac{\log^2(1-t)}{t}dt = -\log(1-z)\,\mathrm{Li}_2(z) - 2g(z),$$

so that the claim follows from (6.1) combined with (2.7). □

It is now straightforward to deduce (1.9).

Corollary 6.2. *We have*

$$\sum_{k=1}^{\infty}\sum_{j=k}^{\infty} \frac{1}{k^2}\frac{1}{j\,2^j} = \frac{5}{8}\zeta(3).$$

Proof. Observe that

$$\sum_{k=1}^{\infty}\sum_{j=k}^{\infty} \frac{1}{k^2}\frac{1}{j\,2^j} = h\left(\frac{1}{2}\right) + \mathrm{Li}_3\left(\frac{1}{2}\right),$$

so that the claimed evaluation follows from (6.2) together with (2.10) and (2.11). □

Let us indicate that similarly approaching the double sum S_3 in Theorem 1.3 is not sufficient to evaluate it. Define, for $|z|<1$,

$$\tilde{h}(z) = \sum_{k=1}^{\infty} \tilde{H}_k^{(2)} \frac{z^{k+1}}{k+1}, \quad \text{where } \tilde{H}_k^{(2)} = \sum_{j=1}^{k} \frac{(-1)^{j-1}}{j^2},$$

so that

$$S_3 = \tilde{h}\left(\frac{1}{2}\right) - \mathrm{Li}_3\left(-\frac{1}{2}\right).$$

Observing that

$$\tilde{h}'(z) = \sum_{k=1}^{\infty} \tilde{H}_k^{(2)} z^k = -\frac{\mathrm{Li}_2(-z)}{1-z},$$

we then find

$$\tilde{h}(z) = \log(1-z)\,\mathrm{Li}_2(-z) + \int_0^z \frac{\log(1-t)\log(1+t)}{t}dt. \tag{6.3}$$

In particular,

$$S_3 = \int_0^{1/2} \frac{\log(1-t)\log(1+t)}{t}dt - \mathrm{Li}_3\left(-\frac{1}{2}\right) - \log(2)\,\mathrm{Li}_2\left(-\frac{1}{2}\right). \tag{6.4}$$

The integral in (6.3) can be expressed as a sum of polylogarithms at various arguments, though the result takes a rather more complicated form than (6.2). As a consequence, it still requires a considerable amount of polylogarithmic relations to deduce the simple evaluation $S_3 = \frac{13}{24}\zeta(3)$ from (6.4).

7. Computer Algebraic Approaches

We indicated in Remark 1.2 that (period) integrals such as

$$Z_3 = \int_1^\infty \int_0^1 \int_0^1 \frac{dz\,dy\,dx}{x(x+y)(x+y+z)} \tag{7.1}$$

can be algorithmically evaluated in terms of multiple polylogarithms. These evaluations, however, are not typically in simplified form and establishing the necessary relations between special values of polylogarithms can be (sometimes prohibitively) difficult. For instance, Theorem 1.3 is equivalent to the polylogarithmic relation

$$-\operatorname{Li}_{1,2}\left(\frac{1}{2},-1\right) - \operatorname{Li}_3\left(-\frac{1}{2}\right) = \frac{13}{24}\zeta(3). \tag{7.2}$$

We note that $-\operatorname{Li}_{1,2}(x,-1) = H(1,-2;x)$ is an instance of the harmonic polylogarithm introduced in [18]. A Mathematica implementation of algorithms for working with and simplifying harmonic polylogarithms is provided by Maitre [14]. For instance, it is possible to algorithmically convert the harmonic polylogarithm $H(1,-2;x)$ to trilogarithms $\operatorname{Li}_3(x)$ and lower order terms:

$$\begin{aligned} H(1,-2;x) &= \operatorname{Li}_3(-x) - \operatorname{Li}_3(1-x) + \operatorname{Li}_3\left(\frac{x}{1+x}\right) + \operatorname{Li}_3\left(\frac{1-x}{2}\right) \\ &\quad + \operatorname{Li}_3\left(\frac{1+x}{2}\right) - \operatorname{Li}_3\left(\frac{2x}{x-1}\right) - \operatorname{Li}_3\left(\frac{2x}{1+x}\right) - \operatorname{Li}_2(x)\log(x+1) \\ &\quad + \frac{1}{6}\log\left(\frac{1-x}{8}\right)\log^2(1-x) - \frac{1}{2}\log(x)\log^2(1-x) \\ &\quad + \frac{\log^2(2)}{2}\log(1-x^2) - \frac{\log(2)}{2}\log^2(1+x) + \frac{\pi^2}{12}\log\left(\frac{1-x}{1+x}\right) \\ &\quad - \frac{3}{4}\zeta(3) - \frac{\log^3(2)}{3} + \frac{\pi^2}{6}\log(2). \end{aligned}$$

With a bit of human effort, we can then prove (7.2) by employing relations for the tri- and dilogarithm including those mentioned in Sec. 2.

The question of whether an integral like (7.1) can be automatically evaluated by a general purpose computer algebra system evolves in time. For instance, given (7.1), *Maple* 18 evaluates two of three integrals and returns a single integral similar to (3.1). On the other hand, *Mathematica* 9 returns the integral (7.1) unevaluated, while *Mathematica* 10 produces

$$\begin{aligned} Z_3 &= \operatorname{Li}_3\left(-\frac{1}{3}\right) - 2\operatorname{Li}_3\left(\frac{1}{3}\right) + \frac{19}{8}\zeta(3) + \frac{1}{2}\log(3)\operatorname{Li}_2\left(\frac{1}{9}\right) - 3\log(3)\operatorname{Li}_2\left(\frac{1}{3}\right) \\ &\quad - \frac{1}{3}\log^3(3) + \pi i\left[\frac{1}{2}\operatorname{Li}_2\left(\frac{1}{9}\right) - 3\operatorname{Li}_2\left(\frac{1}{3}\right) - \frac{1}{2}\log^2(3) + \frac{\pi^2}{6}\right]. \end{aligned}$$

Assuming that this calculation is correct, and observing that Z_3 is real, we can conclude the relation

$$\operatorname{Li}_2\left(\frac{1}{9}\right) = 6\operatorname{Li}_2\left(\frac{1}{3}\right) + \log^2(3) - \frac{\pi^2}{3}, \tag{7.3}$$

which follows from (2.3) combined with (2.12) (see also [3, (39.4); p. 324]). Using (7.3), the expression for the triple integral reduces to

$$Z_3 = \mathrm{Li}_3\left(-\frac{1}{3}\right) - 2\,\mathrm{Li}_3\left(\frac{1}{3}\right) + \frac{19}{8}\zeta(3) - \frac{\pi^2}{6}\log(3) + \frac{1}{6}\log^3(3). \tag{7.4}$$

The remaining polylogarithms can be simplified using (2.13), which results in the desired evaluation

$$Z_3 = \frac{5}{24}\zeta(3), \tag{7.5}$$

which we established in Theorem 1.1.

It is an interesting phenomenon that symmetrizing a problem can occasionally make it considerably more tractable by computer algebra. An impressive instance is [17], where Paule significantly reduces the order of recurrences for certain q-sums by "creative symmetrizing". In the remainder of this section, we indicate that a symmetrizing transformation also makes the integral Z_3 more palatable for symbolic evaluation. To begin with, the change of variables $x \mapsto 1/x$ transforms the integral (7.1) to the unit cube:

$$Z_3 = \int_0^1 \int_0^1 \int_0^1 \frac{x\,dx\,dy\,dz}{(1+xy)(1+x(y+z))}. \tag{7.6}$$

Denote the integrand of (7.6) by $u(x,y,z)$. The symmetrization of the integral is defined in terms of

$$u^{\mathrm{sym}}(x,y,z) = \frac{1}{3}\left[u(x,y,z) + u(y,z,x) + u(z,x,y)\right] \tag{7.7}$$

as

$$Z_3 = \int_0^1 \int_0^1 \int_0^1 u^{\mathrm{sym}}(x,y,z)\,dx\,dy\,dz. \tag{7.8}$$

This form seems to be more favorable to a symbolic calculation. Indeed, *Mathematica* is able to directly evaluate the symmetrized integral (7.8) as

$$Z_3 = \frac{5}{24}\zeta(3) - \frac{\pi i}{6}\left[\mathrm{Li}_2\left(\frac{1}{9}\right) - 6\,\mathrm{Li}_2\left(\frac{1}{3}\right) - \log^2(3) + \frac{\pi^2}{3}\right]. \tag{7.9}$$

Identity (7.3), or simply observing that (7.8) is real, then gives the value $Z_3 = \frac{5}{24}\zeta(3)$ of Theorem 1.1.

8. Conclusions

We discussed several approaches to the triple integral (1.2) as well as the equivalent double sum (1.5). However, in each case, we eventually required the introduction of polylogarithms and nontrivial relations between them. It remains an interesting challenge to relate either of (1.2) and (1.5) to $\zeta(3)$ directly.

In another direction, let us note that the unit square integral for $\frac{1}{2}\zeta(2)$ in Remark 1.4, a 2-dimensional analog of the integral (1.2), has other simpler higher-dimensional generalizations. Indeed, for any $m \geq 2$, we have the following integrals evaluating in terms of $\zeta(m)$:

$$\int_0^1 \cdots \int_0^1 \frac{dx_1 \cdots dx_m}{1 - x_1 \cdots x_m} = \zeta(m), \quad \int_0^1 \cdots \int_0^1 \frac{dx_1 \cdots dx_m}{1 + x_1 \cdots x_m} = (1 - 2^{1-m})\zeta(m).$$

In particular,

$$\int_0^1 \int_0^1 \int_0^1 \frac{dxdydz}{1 - xyz} = \zeta(3), \quad \int_0^1 \int_0^1 \int_0^1 \frac{dxdydz}{1 + xyz} = \frac{3}{4}\zeta(3).$$

Other interesting triple integrals involving $\zeta(3)$ have been considered in the literature. For instance, in his proof, inspired by Apéry, of the irrationality of $\zeta(3)$, Beukers [4] considers the integrals

$$J_n = \frac{1}{2} \int_0^1 \int_0^1 \int_0^1 \frac{x^n(1-x)^n y^n (1-y)^n z^n (1-z)^n}{(1-(1-xy)z)^{n+1}} dxdydz$$

and shows that

$$J_n = A(n)\zeta(3) + B(n),$$

where

$$A(n) = \sum_{k=0}^{n} \binom{n}{k}^2 \binom{n+k}{k}^2$$

are the Apéry numbers and $B(n)$ are certain rational numbers (satisfying the same three-term recurrence as the Apéry numbers). For instance

$$\begin{aligned} J_0 &= \frac{1}{2} \int_0^1 \int_0^1 \int_0^1 \frac{1}{1-(1-xy)z} dxdydz = \zeta(3), \\ J_1 &= \frac{1}{2} \int_0^1 \int_0^1 \int_0^1 \frac{x(1-x)y(1-y)z(1-z)}{(1-(1-xy)z)^2} dxdydz = 5\zeta(3) - 6. \end{aligned}$$

Beukers [4] shows that $d_n^3 J_n = a_n \zeta(3) + b_n$, with $d_n = \mathrm{lcm}(1, 2, \ldots, n)$, are integer linear combinations of $\zeta(3)$ and 1 (that is, $a_n, b_n \in \mathbb{Z}$). He then deduces the irrationality of $\zeta(3)$ from the bounds $0 < |a_n\zeta(3) + b_n| < (4/5)^n$, which hold for large enough n.

Recently, Brown [8] introduced cellular integrals, generalizing Beukers' integrals J_n, and showed that these cellular integrals are special linear forms in multiple zeta values, which reproduce (and vastly generalize) many of the known constructions related to irrationality questions for zeta values. We close by wondering whether the triple integral (1.2) can be similarly embedded in an infinite family of linear forms in $\zeta(3)$ and 1.

References

[1] G. E. Andrews, R. Askey and R. Roy, *Special Functions*, Encyclopedia of Mathematics and its Applications, Vol. 71 (Cambridge University Press, New York, 1999).

[2] B. Berndt, *Ramanujan's Notebooks, Part I* (Springer-Verlag, New York, 1985).

[3] B. Berndt, *Ramanujan's Notebooks, Part IV* (Springer-Verlag, New York, 1994).

[4] F. Beukers, A note on the irrationality of $\zeta(2)$ and $\zeta(3)$, *Bull. London Math. Soc.* **11** (1979) 268–272.

[5] C. Bogner and F. Brown, Feynman integrals and iterated integrals on moduli spaces of curves of genus zero, *Comm. Number Theory Phys.* **9** (2015) 189–238.

[6] J. M. Borwein and D. Bradley, Thirty two Goldbach variations, *Int. J. Number Theory* **2** (2006) 65–103.

[7] J. M. Borwein, D. M. Bradley, D. J. Broadhurst and P. Lisonek, Special values of multiple polylogarithms, *Trans. Amer. Math. Soc.* **353** (2001) 907–941.

[8] F. Brown, Irrationality proofs for zeta values, moduli spaces and dinner parties, *Mosc. J. Comb. Number Theory* **6**(2–3) (2016) 102–165.

[9] I. S. Gradshteyn and I. M. Ryzhik, *Table of Integrals, Series, and Products*, eds. D. Zwillinger and V. Moll, 8th edn. (Academic Press, New York, 2015).

[10] M. Hoffman, *References on Multiple Zeta Values and Euler Sums*, https://www.usna.edu/Users/math/meh/biblio.html.

[11] M. Kontsevich and D. Zagier, Periods, in *Mathematics Unlimited*–2001 *and Beyond*, (Springer-Berlin, 2000), pp. 771–808.

[12] L. Lewin, *Dilogarithms and Associated Functions*, 2nd edn. (Elsevier, North Holland, 1981).

[13] L. Lewin, *Polylogarithms and Associated Functions* (North Holland, New York, Oxford, 1981).

[14] D. Maitre, HPL, a mathematica implementation of the harmonic polylogarithms, *Comput. Phys. Comm.* **174** (2006) 222–240.

[15] F. W. J. Olver, D. W. Lozier, R. F. Boisvert and C. W. Clark (eds.), *NIST Handbook of Mathematical Functions* (Cambridge University Press, 2010).

[16] E. Panzer, Algorithms for the symbolic integration of hyperlogarithms with applications to Feynman integrals, *Comput. Phys. Comm.* **188** (2015) 148–166.

[17] P. Paule, Short and easy computer proofs of the Rogers-Ramanujan identities and of identities of similar type, *Electron. J. Combin.* **1** (1994) 1–9.

[18] E. Remiddi and J. A. M. Vermaseren, Harmonic polylogarithms, *Int. J. Mod. Phys. A* **15** (2000) 725–754.

© 2024 World Scientific Publishing Company
https://doi.org/10.1142/9789811277375_0002

The Ramanujan–Dyson identities and George Beck's congruence conjectures*

George E. Andrews
The Pennsylvania State University
University Park, PA 16802, USA
gea1@psu.edu

Dedicated to a grand mathematician, and my good friend, Bruce Berndt

Dyson's famous conjectures (proved by Atkin and Swinnerton-Dyer) gave a combinatorial interpretation of Ramanujan's congruences for the partition function. The proofs of these results center on one of the universal mock theta functions that generate partitions according to Dyson's rank. George Beck has generalized the study of partition function congruences related to rank by considering the total number of parts in the partitions of n. The related generating functions are no longer part of the world of mock theta functions. However, George Beck has conjectured that certain linear combinations of the related enumeration functions do satisfy congruences modulo 5 and 7. The conjectures are proved here.

Keywords: Dyson's rank; George Beck's conjectures; partitions; Ramanujan congruences.

Mathematics Subject Classification 2020: 11P84, 11P83, 11P81, 05A19, 33P15

1. Introduction

Among the celebrated achievements of Ramanujan, we find [1, p. 159]

$$p(5n+4) \equiv 0 \pmod{5}, \tag{1.1}$$

where $p(n)$ is the integer partition function [4], seeking to provide a combinatorial setting for (1.1), defined the *rank* of a partition to be the largest part minus the number of parts. For example, the rank of $5+5+3+1$ is $5-4=1$.

Let us define (cf. [2]) $N(m,k,n)$ to be the number of partitions of n with rank congruent to m modulo k. Dyson conjectured [4], and Atkin and Swinnerton-Dyer proved [2] that for $0 \leq i \leq 4$

$$N(i,5,5n+4) = \frac{1}{5}p(5n+4), \tag{1.2}$$

*This article originally appeared in *International Journal of Number Theory*, **17**(2), pp. 239–249 © 2021 World Scientific Publishing Company. DOI: 10.1142/S1793042120400060

thus revealing five disjoint, equinumerous subsets of the partitions of $5n+4$. Also treated in this way [2, 4] is the congruence

$$N(i,7,7n+5) = \frac{1}{7}p(7n+5), \tag{1.3}$$

where $0 \le i \le 6$, and (1.4), of course, implies the Ramanujan congruence

$$p(7n+5) \equiv 0 \pmod{7}. \tag{1.4}$$

Beck [3] has made surprising new conjectures along these lines. Now we define $NT(m,k,n)$ to be the total number of parts in the partitions of n with rank congruent to m modulo k. Beck's conjectures are formulated in the following two theorems.

Theorem 1. *If* $i = 1$ *or* 4

$$\begin{aligned} &NT(1,5,5n+i) - NT(4,5,5n+i) \\ &\qquad + 2NT(2,5,5n+i) - 2NT(3,5,5n+i) \\ &\equiv 0 \pmod{5} \end{aligned} \tag{1.5}$$

Theorem 2. *If* $i = 1$ *or* 5,

$$\begin{aligned} &NT(1,7,7n+i) - NT(6,7,7n+i) \\ &\qquad + NT(2,7,7n+i) - NT(5,7,7n+i) \\ &\qquad - NT(3,7,7n+i) + NT(4,7,7n+i) \\ &\equiv 0 \pmod{7}. \end{aligned} \tag{1.6}$$

Note that if the above assertions concerned N instead of NT, they would trivially be implied by the fact that

$$N(m,k,n) = N(k-m,k,n) \tag{1.7}$$

which follows by using partition conjugation to transform the partition enumerated by $N(m,k,n)$ into those enumerated by $N(k-m,k,n)$. However, (1.7) is generally false if N is replaced by NT.

As an example of Theorem 1, take $n=0$, $i=4$. The partitions of 4 are 4, $3+1$, $2+2$, $2+1+1$, $1+1+1+1$, with ranks 3, 1, 0, -1, -3. Consequently $NT(1,5,4)=2$, $NT(2,5,4)=4$, $NT(3,5,4)=1$, $NT(4,5,4)=3$, and

$$\begin{aligned} &NT(1,5,4) - NT(4,5,4) + 2NT(2,5,4) - 2NT(3,5,4) \\ &= 2 - 3 + 2\cdot 4 - 2\cdot 1 = 10 - 5 = 5 \equiv 0 \pmod{5}. \end{aligned}$$

For Theorem 2, take $n=0$, $i=5$, the partitions of 5 are 5, $4+1$, $3+2$, $3+1+1$, $2+2+1$, $2+1+1+1$, $1+1+1+1+1$, with ranks 4, 2, 1, 0, -1,

-2, -4. Consequently $N(1,7,5)=2$, $N(2,7,5)=2$, $N(3,7,5)=5$, $N(4,7,5)=1$, $N(5,7,5)=4$, $N(6,7,5)=3$, and

$$\begin{aligned} &NT(1,7,5)-NT(6,7,5)+NT(2,7,5)-NT(5,7,5)-NT(3,7,5)+NT(4,7,5)\\ &\quad =2-3+2-4-5+1=-7\equiv 0 \pmod 7. \end{aligned}$$

The obstacles to proving Theorems 1 and 2 lie in the nature of the generating function. Recall that the universal mock theta function

$$\sum_{n\geq 0}\frac{q^{n^2}}{(zq;q)_n(q/z;q)_n}, \tag{1.8}$$

where $(A;q)_n=(1-A)(1-Aq)\cdots(1-Aq^{n-1})$ is the generating function in which q marks the number being partitioned and z marks the rank [5, p. 66].

To keep track of the total number of parts, we must add an extra variable x; and the generating function which weights partitions by the number of parts while keeping track of the rank is (following Garvan's argument [5, p. 66])

$$\frac{\partial}{\partial x}\Big|_{x=1}\sum_{n\geq 0}\frac{x^n q^{n^2}}{(zq;q)_n(\frac{xq}{z};q)_n}. \tag{1.9}$$

This is a decidedly more complicated function than that appearing in (1.8).

In Sec. 2, we shall collect the necessary results and notations from the paper of Atkin and Swinnerton-Dyer [2]. Section 3 is devoted to obtaining a tractable representation of the series appearing in (1.9). Section 4 provides a proof of Theorem 1, and Sec. 5 treats Theorem 2. We conclude with a discussion of further problems.

2. Background

In this section, we shall collect the necessary results from the paper of Atkin and Swinnerton-Dyer [2] and a result from classical q-hypergeometric series.

We shall stick closely to the notation in [2, p. 106]. The main change consists of replacing their q by k. Also we shall use the subscript k explicitly to avoid confusion of the cases $k=5$ and $k=7$.

$$P_k(a)=(q^{ka};q^{k^2})_\infty(q^{k(k-a)};q^{k^2})_\infty \tag{2.1}$$

$$P_k(0)=(q^{k^2};q^{k^2})_\infty. \tag{2.2}$$

$$\Sigma_k(a,b)=\sum_{n=-\infty}^{\infty}\frac{(-1)^n q^{kbn+3kn(n+1)/2}}{1-q^{k(kn+a)}},\qquad a\not\equiv 0 \pmod k \tag{2.3}$$

$$S_k(b) = \sum_{\substack{n=-\infty \\ n\neq 0}}^{\infty} \frac{(-1)^n q^{n(3n+1)/2+bn}}{1-q^{kn}} \tag{2.4}$$

$$S_k(b) = -S_k(k-1-b) \tag{2.5}$$

$$S_k(b) - S_k(b+k) = \sum_{\substack{n=-\infty \\ n\neq 0}}^{\infty} (-1)^n q^{bn+n(3n+1)/2} \tag{2.6}$$

$$S_k((k-1)/2) = 0 \tag{2.7}$$

Each of (2.5)–(2.7) appears on page 99 of [2, Eqs. (6.2)–(6.4)].

$$r_b(d,k) = \sum_{n=0}^{\infty} N(b,k,kn+d)q^{kn}, \tag{2.8}$$

$$r_{bc}(d,k) = r_b(d,k) - r_c(d,k). \tag{2.9}$$

Beyond definitions, we require several identities from [2].

$$\begin{aligned} &2S_5(1) + S_5(4), \\ &\quad = \left\{ q^5 \frac{\Sigma_5(1,0)}{P_5(0)} + q^2 \frac{P_5(0)}{P_5(2)} - 2q^7 \frac{\Sigma_5(2,0)}{P_5(0)}, \right. \\ &\qquad \left. - q^3 \frac{P_5(0)P_5(1)}{P_5^2(2)} \right\} (q;q)_\infty \qquad \text{([2, mid page 102])} \end{aligned} \tag{2.10}$$

$$r_{13}(1,7) = 0 \qquad \text{([2, eq. (6.27)])} \tag{2.11}$$

$$r_{13}(5,7) = 0 \qquad \text{([2, eq. (6.39)])} \tag{2.12}$$

$$\begin{aligned} &-S_7(1) - 2S_7(4) \\ &\quad = \left\{ -q^{16} \frac{\Sigma_7(3,0)}{P_7(0)} + q^2 \frac{P_7(0)}{P_7(2)} - q^4 \frac{P_7(0)}{P_7(3)} \right. \\ &\qquad \left. + 2q^{13} \frac{\Sigma_7(2,0)}{P_7(0)} + q^6 \frac{P_7(0)P_7(1)}{P_7(2)P_7(3)} \right\} (q;q)_\infty \qquad \text{([2, eq. (6.49), p. 104])} \end{aligned} \tag{2.13}$$

$$\sum_{n\geq 0} N(1,7,n)q^n = \frac{1}{(q;q)_\infty} (S_7(1) - S_1(0)) \qquad \text{(by [2, p. 103])} \tag{2.14}$$

$$\sum_{n\geq 0} N(3,7,n)q^n \frac{1}{(q;q)_\infty} S_7(4). \qquad \text{(by [2, p. 103])} \tag{2.15}$$

Finally we require the following identity which follows from [6, Eq. (III.18),

p. 242] by letting $n \to \infty$:

$$\sum_{j=0}^{\infty} \frac{(a;q)_j(1-aq^{2j})(b;q)_j(c;q)_j(d;q)_j(e;q)_j(-1)^j\left(\frac{a^2q^2}{bcde}\right)^j q^{\binom{j}{2}}}{(q;q)_j(1-a)\left(\frac{aq}{b};q\right)_j\left(\frac{aq}{c};q\right)_j\left(\frac{aq}{d};q\right)_j\left(\frac{aq}{e};q\right)_j}$$

$$= \frac{(aq;q)_\infty\left(\frac{aq}{de};q\right)_\infty}{\left(\frac{aq}{d};q\right)_\infty\left(\frac{aq}{e};q\right)_\infty} \sum_{j=0}^{\infty} \frac{\left(\frac{aq}{bc};q\right)_j(d;q)_j(e;q)_j\left(\frac{aq}{de}\right)^j}{(q;q)_j\left(\frac{aq}{b};q\right)_j\left(\frac{aq}{e};q\right)_j}. \tag{2.16}$$

3. New Representation of (1.9)

Actually we shall not differentiate the series in (1.9) until late in the proofs of Theorems 1 and 2.

Theorem 3.

$$\sum_{n=0}^{\infty} \frac{x^n q^{n^2}}{(zq;q)_n\left(\frac{xq}{z};q\right)_n}$$

$$= 1 + \frac{1}{(xq;q)_\infty} \sum_{n=1}^{\infty} \frac{(xq;q)_n}{(q;q)_{n-1}}(-1)^{n-1}q^{n(3n+1)/2}x^n$$

$$\times \left(\frac{1}{q^n(1-zq^n)} + \frac{x}{z\,(1-xq^n/z)}\right). \tag{3.1}$$

Remark. As noted in Sec. 1, the series on the left in (3.1) generates the partitions of n where q marks the number being partitioned, z marks the rank, and x marks the number of parts.

Proof of Theorem 3. We apply (2.14) with $d = e = \frac{1}{\tau}$, $a = x$, $b = \frac{x}{z}$ $c = z$ and $\tau \to \infty$. Hence

$$\sum_{n=0}^{\infty} \frac{x^n q^{n^2}}{(zq;q)_n\left(\frac{xq}{z};q\right)_n}$$

$$= \lim_{\tau\to 0} \sum_{n=0}^{\infty} \frac{(q;q)_n\left(\frac{1}{\tau};q\right)_n^2 (x\tau^2 q)^n}{(q;q)_n(zq;q)_n\left(\frac{xq}{z};q\right)_n}$$

$$= \frac{1}{(xq;q)_\infty} \lim_{t\to 0} \sum_{n=0}^{\infty} \frac{(x;q)_n(1-xq^{2n})\left(\frac{x}{z};q\right)_n(z;q)_n\left(\frac{1}{\tau};q\right)_n^2}{(q;q)_n(1-x)(zq;q)_n\left(\frac{xq}{z};q\right)_n(xq\tau;q)_n^2}$$

$$\times \left(\frac{x^2q^2\tau^2}{\left(\frac{x}{z}\right)z}\right)(-1)^n q^{\binom{n}{2}}$$

$$= \frac{1}{(xq;q)_\infty}\left(1+\sum_{n\geq 1}\frac{(xq;q)_{n-1}(1-xq^{2n})}{(q;q)_n}\frac{(1-z)\left(1-\frac{x}{z}\right)}{(1-zq^n)\left(1-\frac{x}{z}q^n\right)}\right.$$
$$\left.\times x^n(-1)^n q^{n(3n+1)/2}\right) \tag{3.2}$$

Now note that

$$\frac{(1-z)(1-\frac{x}{z})(1-xq^{2n})}{(1-zq^n)(1-\frac{x}{z}q^n)}$$
$$= -(1-q^n)(1-xq^n)\left(\frac{1}{q^n(1-zq^n)}+\frac{x}{z(1-\frac{q^n x}{z})}\right)$$
$$+\frac{(1-xq^{2n})}{q^n}. \tag{3.3}$$

Also we require Sylvester's identity [1, p. 140]

$$1+\sum_{n=1}^{\infty}\frac{(xq;q)_{n-1}(1-xq^{2n})(-1)^n q^{n(3n-1)/2}}{(q;q)_n}$$
$$= (xq;q)_\infty. \tag{3.4}$$

Therefore, substituting (3.3) into (3.2), we find

$$\sum_{n=0}^{\infty}\frac{x^n q^{n^2}}{(zq;q)_n(\frac{xq}{z};q)_n}$$
$$= \frac{1}{(xq;q)_\infty}\left\{1+\sum_{n=1}^{\infty}\frac{(xq;q)_n}{(q;q)_{n-1}}(-1)^{n-1}q^{n(3n+1)/2}x^n\right.$$
$$\times\left(\frac{1}{q^n(1-zq^n)}+\frac{x}{z(1-\frac{xq^n}{z})}\right)$$
$$\left.+\sum_{n\geq 1}\frac{(-1)^n(xq;q)_{n-1}(1-xq^{2n})x^n q^{n(3n-1)/2}}{(q;q)_n}\right\}$$
$$= 1+\frac{1}{(xq;q)_\infty}\sum_{n=1}^{\infty}\frac{(xq;q)_n}{(q;q)_{n-1}}(-1)^{n-1}q^{n(3n+1)/2}x^n$$
$$\left(\frac{1}{q^n(1-zq^n)}+\frac{x}{z(1-\frac{xq^n}{z})}\right)\quad \text{(by (3.4))}. \tag{3.5}$$

□

It is now an easy matter to expand the terms involving z into a geometric series which leads directly to the generating functions for $NT(b,k,n)$.

Corollary 4. *If* $0 \le b \le k$,

$$\sum_{n\ge 0} NT(b,k,n)q^n$$
$$= \frac{\partial}{\partial x}\Big|_{x=1} \frac{1}{(xq;q)_\infty} \sum_{n\ge 1} \frac{(xq;q)_n}{(q;q)_{n-1}} (-1)^{n-1} x^n q^{n(3n+1)/2}$$
$$\times \left(\frac{q^{n(b-1)}}{1-q^{nk}} + \frac{x^{k-b} q^{(k-1-b)n}}{1-x^k q^{kn}} \right) \tag{3.6}$$

Proof. We write

$$\frac{1}{1-zq^n} = \sum_{j=0}^{\infty} z^j q^{nj}, \tag{3.7}$$

and

$$\frac{1}{z(1-\frac{xq^n}{z})} = \sum_{j=0}^{\infty} z^{-j-1} x^j q^{nj}. \tag{3.8}$$

We want only those partitions with ranks $\equiv b \pmod{k}$. So we take exponents in (3.7) of the form $kj+b$, and in (3.8) of the form $kj+(k-1-b)$.

Thus in the right-side of (3.1),

$$\frac{1}{q^n(1-zq^n)} \text{ is replaced by } \frac{q^{n(b-1)}}{1-q^{nk}},$$

and

$$\frac{1}{z(1-\frac{xq^n}{z})} \text{ is replaced by } \frac{x^{k-b} q^{(k-1-b)}}{1-x^k q^{kn}},$$

and this yields (3.6) once we differentiate by x to count each partition with a weight equal to its number of parts. □

4. Proof of Theorem 1

To prove Theorem 1, it is sufficient to show that in

$$\sum_{n\ge 0} \Big(NT(1,5,n) - NT(4,5,n)$$
$$+ 2NT(2,5,n) - 2NT(3,5,n) \Big) q^n \tag{4.1}$$

the coefficients of q^{5n+1} and q^{5n+4} are congruent to 0 modulo 5.

By Corollary 4, the expression in (4.1) is equal to

$$\begin{aligned}
&\frac{\partial}{\partial x}\Big|_{x=1}\Bigg(\frac{1}{(xq;q)_\infty}\sum_{n\geq 1}\frac{(xq;q)_n}{(q;q)_{n-1}}(-1)^{n-1}x^n q^{n(3n+1)/2}\\
&\quad\times\Bigg\{\left(\frac{1}{1-q^{5n}}+\frac{x^4q^{3n}}{1-x^5q^{5n}}\right)\\
&\quad-\left(\frac{q^{3n}}{1-q^{5n}}+\frac{x}{1-x^5q^{5n}}\right)\\
&\quad+2\left(\frac{q^{n}}{1-q^{5n}}+\frac{x^3q^{3n}}{1-x^5q^{5n}}\right)\\
&\quad-2\left(\frac{q^{2n}}{1-q^{5n}}+\frac{x^2q^{n}}{1-x^5q^{5n}}\right)\Bigg\}\Bigg)\\
&=\frac{\partial}{\partial x}\Big|_{x=1}+\frac{1}{(xq;q)_\infty}\sum_{n\geq 1}\frac{(xq;q)_n}{(q;q)_{n-1}}(-1)^{n-1}q^{n(3n+1)/2}\\
&\quad\times\frac{(1-x)(1-xq^{2n})(1-q^n)(1-xq^n)}{(1-q^{5n})(1-xq^{5n})}\\
&\quad\times\{(q^{4n}+3q^{3n}+q^{2n})x^2+(3q^{3n}+4q^{2n}+3q^n)x\\
&\quad+q^{2n}+3q^n+1\}.
\end{aligned}\tag{4.2}$$

Now we note the simple fact that

$$\frac{\partial}{\partial x}\Big|_{x=1}(1-x)F(x,q,z)=-F(1,q,z).\tag{4.3}$$

Consequently the expression in (4.1) is equal to

$$\begin{aligned}
&-\frac{1}{(q;q)_\infty}\sum_{n\geq 1}(1-q^n)(-1)^{n-1}q^{n(3n+1)/2}\frac{(1-q^{2n})(1-q^n)^2}{(1-q^{5n})^2}\\
&\quad\times(q^{4n}+6q^{3n}+6q^{2n}+6q^n+1)\\
&\equiv-\frac{1}{(q;q)_\infty}\sum_{n\geq 1}\frac{(1-q^n)^2(1-q^{2n})(-1)^{n-1}q^{n(3n+1)/2}}{1-q^{5n}}\pmod 5\\
&=-\frac{1}{(q;q)_\infty}(S_5(0)-2S_5(1))\\
&=-\frac{1}{(q;q)_\infty}(-S_5(4)-2S_5(1))\qquad\text{(by (2.5))}\\
&=\Bigg(q^5\frac{\Sigma_5(1,0)}{P_5(0)}+q^2\frac{P_5(0)}{P_5(2)}-2q^7\frac{\Sigma_5(2,0)}{P_5(0)}\\
&\quad-q^3\frac{P_5(0)P_5(1)}{P_5^2(2)}\Bigg)\qquad\text{(by (2.10))}.
\end{aligned}\tag{4.4}$$

We conclude by noting that there are no exponents on $q \equiv 1$ or $4 \pmod 5$ in this last expression which, in turn, is congruent to the expression in (4.1) modulo 5. Thus Theorem 1 is proved.

5. Proof of Theorem 2

To prove Theorem 2, it is sufficient to show that in

$$\sum_{n\geq 0}(NT(1,7,n)-NT(6,7,n) \\ +NT(2,7,n)-NT(5,7,n) \\ -NT(3,7,n)+NT(4,7,n))q^n \tag{5.1}$$

the coefficients of q^{7n+1} and q^{7n+5} are congruent to 0 modulo 7.

By Corollary 4, the expression in (4.1) is equal to

$$\frac{\partial}{\partial x}\Big|_{x=1}\left(\frac{1}{(xq;q)_\infty}\sum_{n\geq 1}\frac{(xq;q)_n}{(q;q)_{n-1}}(-1)^{n-1}x^n q^{n(3n+1)/2}\sum_{b=1}^{7}\alpha(b)F(b)\right), \tag{5.2}$$

where

$$\alpha(b)=\begin{cases}1 & \text{if } b=1,2,4\\ -1 & \text{if } b=3,5,6\end{cases} \tag{5.3}$$

and

$$F(b)=\frac{q^{n(b-1)}}{1-q^{7n}}+\frac{x^{7-b}q^{(6-b)n}}{1-x^7q^{7n}}. \tag{5.4}$$

Now

$$\sum_{b=1}^{7}\alpha(b)F(b) \\ =-\frac{(1-x)(1-xq^{2n})(1-q^n)(1-xq^n)}{(1-xq^{7n})(1-q^{7n})} \\ \times\{x^4(q^{8n}+2q^{7n}+q^{6n}+2q^{5n}+q^{4n}) \\ +x^3(2q^{7n}+3q^{6n}+4q^{5n}+3q^{4n}+2q^{3n}) \\ \times x^2(q^{6n}+4q^{5n}+4q^{4n}+4q^{3n}+q^{2n}) \\ \times x(2q^{5n}+3q^{4n}+4q^{3n}+3q^{2n}+2q^n) \\ \times q^{4n}+2q^{3n}+q^{2n}+2q^n+1\} \tag{5.5}$$

So if we set $x=1$ in the { } portion of (5.5), the result is

$$\frac{(1+q^n)^3(1-q^{7n})}{(1-q^n)}+7(q^{5n}+q^{4n}+q^{3n}).$$

Thus if we invoke (4.3), we find that the expression in (5.1) is congruent modulo 7 to

$$\frac{1}{(q;q)_\infty}\sum_{n\geq 1}\frac{(1-q^n)^2(1-q^{2n})(1+3q^n+q^{2n})(-1)^{n-1}q^{n(3n+1)/2}}{1-q^{7n}}$$

$$=\frac{1}{(q;q)_\infty}(S_7(0,7)+S_7(1,7)-5S_7(2,7))$$

$$=\frac{1}{(q;q)_\infty}(S_7(0)+2(S_7(1,7)+2S_7(4,7))$$

$$-S_7(1,7)+S_7(4,7)) \qquad \text{(by (2.5))}$$

$$=\frac{2}{(q;q)_\infty}(S_7(1,7)+2S_7(4,7))$$

$$-\sum_{n=0}^{\infty}N(1,7,n)q^n+\sum_{n=0}^{\infty}N(3,7,n)q^n) \qquad \text{(by (2.14) and (2.15))}$$

$$=2\Bigg\{-q^{16}\frac{\Sigma_7(3,0)}{P_7(0)}+q^2\frac{P_7(0)}{P_7(2)}-q^4\frac{P_7(0)}{P_7(3)}$$

$$+2q^{13}\frac{\Sigma_7(2,0)}{P_7(0)}+q^6\frac{P_7(0)P_7(1)}{P_7(2)P_7(3)}\Bigg\}$$

$$\times\sum_{n=0}^{\infty}(N(3,7,n)-N(1,7,n))q^n. \tag{5.6}$$

Now the expression inside { } has no powers of q congruent to 1 or 5 mod 7. And by (2.11) and (2.12) we see that

$$\sum_{n=0}^{\infty}(N(3,7,n)-N(1,7,n))q^n$$

also contains no powers of q congruent to 1 or 5 modulo 7. Thus the assertion about (5.1) is proved, and Theorem 2 is established.

6. Conclusion

The real surprise in this paper is that anyone would have guessed Theorem 1 and 2 in the first place. For this fact, we are indebted to George Beck's diligence and ingenuity.

George Beck has similar conjectures related to the crank which have been treated by Shane Chern.

References

[1] G. E. Andrews, *The Theory of Partition*, Encyclopedia of Mathemaics and Its Applications, Vol. 2 (Addison-Wesley, Reading, 1976) (Reissued: Cambridge University Press, 1998).

[2] A. O. L. Atkin and H. P. F. Swinnerton-Dyers, Some properties of partitions, *Proc. London Math. Soc. (3)* **4** (1954) 84–106.
[3] G. Beck, *private communication* (2017).
[4] F. J. Dyson, Some guesses in the theory of partitions, *Eureka (Cambridge)* **8** (1944) 10–15.
[5] F. G. Garvan, New combinatorial interpretations of Ramanujan's partition congruences mod 5, 7, and 11, *Trans. Amer. Math. Soc.* **305** (1988) 47–77.
[6] G. Gasper and M. Rahman, *Basic Hypergeometric Series*, Encyclopedia of Mathematics and Its Applications, Vol. 35 (Cambridge University Press, Cambridge, 1990).

© 2024 World Scientific Publishing Company
https://doi.org/10.1142/9789811277375_0003

On Entry II.16.12: A continued fraction of Ramanujan*

Gaurav Bhatnagar†
School of Physical Sciences, Jawaharlal Nehru University, Delhi, India
bhatnagarg@gmail.com

Mourad E. H. Ismail
Department of Mathematics, University of Central Florida, Orlando, FL 32816, USA
ismail@math.ucf.edu

Dedicated to Bruce Berndt on the occasion of his 80th birthday

We study a continued fraction due to Ramanujan, that he recorded as Entry 12 in Chapter 16 of his second notebook. It is presented in Part III of Berndt's volumes on Ramanujan's notebooks. We give two alternate approaches to proving Ramanujan's Entry 12, one using a method of Euler, and another using the theory of orthogonal polynomials. We consider a natural generalization of Entry 12 suggested by the theory of orthogonal polynomials.

Keywords: Continued fractions; orthogonal polynomials; Entry 12 of Chapter 16; Ramanujan's second notebook.

Mathematics Subject Classification 2020: 33D45, 30B70

1. Introduction

Ramanujan's Entry 12 in Chapter 16, Notebook II is stated as follows. Let $|q| < 1$, and $|ab| < 1$. Then we have (see Berndt [3, p. 24])

$$\frac{(a^2q^3;q^4)_\infty(b^2q^3;q^4)_\infty}{(a^2q;q^4)_\infty(b^2q;q^4)_\infty} = \frac{1}{1-ab}+\frac{(a-bq)(b-aq)}{(1-ab)(1+q^2)}+\frac{(a-bq^3)(b-aq^3)}{(1-ab)(1+q^4)}+\frac{(a-bq^5)(b-aq^5)}{(1-ab)(1+q^6)}+\cdots, \tag{1.1}$$

where we have used the notation for q-rising factorials

$$(a;q)_\infty = (1-a)(1-aq)(1-aq^2)\cdots.$$

*This article originally appeared in *International Journal of Number Theory*, **17**(2), pp. 251–266 © 2021 World Scientific Publishing Company. DOI: 10.1142/S1793042120400072
†Corresponding author.

Hardy [14, p. XXV] remarked

> There is always more in one of Ramanujan's formulae than meets the eye, as anyone who gets to work to verify those which look the easiest will soon discover.

Indeed, the verification of this beautiful continued fraction evaluation has proved to be most troublesome, and is due to the efforts of many mathematicians. The first proof was given by Adiga, Berndt, Bhargava and Watson [1], who further acknowledged the help they received from Askey and Bressoud. The proof is rather complicated. A few years later, Jacobsen [10] clarified and simplified this proof. Ramanathan [13] gave another formal proof, via contiguous relations. Our own proofs, and in this paper we give two, are also not without complication.

The difficulties arise because the first term of Ramanujan's continued fraction is a bit off; it doesn't fit the pattern of the rest of the terms. Perhaps the continued fraction should have been

$$\frac{1}{2(1-ab)} + \frac{(a-bq)(b-aq)}{(1-ab)(1+q^2)} + \frac{(a-bq^3)(b-aq^3)}{(1-ab)(1+q^4)} + \cdots.$$

Indeed, from the point of view of orthogonal polynomials, the continued fraction to consider is

$$\begin{aligned} H(x) := \frac{1-ab}{x(1-ab)+(1-ab)} &+ \frac{(a-bq)(b-aq)}{x(1-ab)+(1-ab)q^2} \\ + \frac{(a-bq^3)(b-aq^3)}{x(1-ab)+(1-ab)q^4} &+ \frac{(a-bq^5)(b-aq^5)}{x(1-ab)+(1-ab)q^6} + \cdots. \end{aligned} \tag{1.2}$$

This is an example of J-fraction, that is, a continued fraction of the form

$$\frac{A_0}{A_0x+B_0} - \frac{C_1}{A_1x+B_1} - \frac{C_2}{A_2x+B_2} - \cdots.$$

The kth convergent of this continued fraction is given by

$$\frac{N_k(x)}{D_k(x)} := \frac{A_0}{A_0x+B_0} - \frac{C_1}{A_1x+B_1} - \cdots - \frac{C_{k-1}}{A_{k-1}x+B_{k-1}}.$$

The numerators and denominators of the convergents are polynomials and satisfy a three-term recurrence relation.

Using [9, Theorem 2.6.1, p. 35], we see that the recurrence relation corresponding to (1.2) is given by

$$\begin{aligned} y_{k+1}(x) &= ((1-ab)x + (1-ab)q^{2k})y_k(x) \\ &\quad + ab(1-bq^{2k-1}/a)(1-aq^{2k-1}/b)y_{k-1}(x), \quad \text{for } k>0. \end{aligned} \tag{1.3}$$

The numerator and denominator satisfy (1.3), with the initial values

$$N_0(x)=0, \quad N_1(x)=1-ab; \quad D_0(x)=1, \quad D_1(x)=(1-ab)(x+1).$$

It is clear that $N_k(x)$ and $D_k(x)$ are polynomials, of degree $k-1$ and k, respectively. From the theory of orthogonal polynomials, it follows that the denominator polynomials are orthogonal with respect to a positive measure, provided we assume a and b are real numbers with different signs.

In a previous study, the first author [4] showed that an elementary method due to Euler can be used to derive all of Ramanujan's q-continued fractions, except for one notable exception. The exception was Entry 12. In Sec. 2, we remedy this omission. The primary benefit of this approach is that it only relies on the q-binomial theorem. Further, the algebraic manipulations appear motivated. However, this approach only gives modified convergence (see Lorentzen and Waadeland [11]) of the continued fraction. One needs some further restrictions on the parameters to obtain ordinary convergence.

Next, in Sec. 3, we provide a proof of Entry 12 using what we may call the standard orthogonal polynomial approach. The proof requires asymptotic techniques and is obtained from the generating function of the numerator and denominator polynomials. Again, the proof is motivated, but rather surprisingly, one still needs the algebraic manipulations from Euler's method. This is because of the missing 2 in the first denominator of (1.1); this is what causes the complications. In both these proofs we can spot elements of the aforementioned proofs due to Adiga *et al.*, Ramanathan and Jacobsen.

Finally, in Sec. 4, we compute the value of the more general continued fraction $H(x)$, and comment briefly on the orthogonality measure of the associated denominator polynomials.

We have mentioned that Ramanujan's Entry 12 is slightly off. On examining our proof in Sec. 3, it appears that Ramanujan would have been unable to write the continued fraction in terms of infinite products if the first denominator followed the pattern. Moreover, our proof of Sec. 2 begins with the products and shows how one can find the continued fraction expansion. If indeed this was similar to Ramanujan's own approach, it will explain why the first term of many of Ramanujan's continued fractions do not quite fit the pattern of the rest of the terms.

We require the following definitions and notations from Lorentzen and Waadeland [11] and Gasper and Rahman [8].

(1) **Continued fractions:** Let $\{a_n\}_{n=1}^{\infty}$ and $\{b_n\}_{n=0}^{\infty}$ be sequences of complex numbers, $a_n \neq 0$. Consider the sequence of mappings

$$s_0(w) = b_0 + w; \quad s_n(w) = \frac{a_n}{b_n + w}, \qquad \text{for } n = 1, 2, \ldots.$$

These mappings are from $\mathbb{C}^*$ to $\mathbb{C}^*$, where $\mathbb{C}^*$ denotes the extended complex numbers $\mathbb{C} \cup \{\infty\}$. Let $S_n(w)$ be defined as follows:

$$S_0(w) = s_0(w) = b_0 + w; \quad S_n(w) = S_{n-1}(s_n(w)).$$

Then $S_n(w)$ can be written as

$$S_n(w) = b_0 + \frac{a_1}{b_1} + \frac{a_2}{b_2} + \cdots + \frac{a_n}{b_n + w}.$$

Note that since $a_n \neq 0$, the $s_k(w)$ are non-singular linear fractional transformations. The S_n are compositions of these, and are thus also non-singular linear fractional transformations. A continued fraction is an ordered pair $((\{a_n\}, \{b_n\}), S_n(0))$, where a_n, b_n and S_n are as above.

(2) **Convergence of continued fractions:** When $S_n(0)$ converges to an extended complex number S, we say that the continued fraction converges, and we write

$$S = b_0 + \frac{a_1}{b_1} + \frac{a_2}{b_2} + \cdots.$$

The convergence of $S_n(w)$ (or even $S_n(w_n)$, for suitably chosen w_n) is called *modified convergence.*

(3) **Factorial notation:** The *q-rising factorial* $(a;q)_n$ is defined as

$$(a;q)_n := \begin{cases} 1, & \text{for } n = 0, \\ (1-a)(1-aq)\cdots(1-aq^{n-1}), & \text{for } n = 1, 2, \ldots. \end{cases}$$

In addition

$$(a;q)_\infty := \prod_{k=0}^{\infty}(1-aq^k), \quad \text{for } |q| < 1.$$

We use the short-hand notation

$$(a_1, a_2, \ldots, a_r; q)_k := (a_1;q)_k (a_2;q)_k \cdots (a_r;q)_k.$$

(4) **Basic hypergeometric series (or ${}_r\phi_s$ series):** This series is of the form

$${}_r\phi_s \left[\begin{matrix} a_1, a_2, \ldots, a_r \\ b_1, b_2, \ldots, b_s \end{matrix}; q, z \right] := \sum_{k=0}^{\infty} \frac{(a_1, a_2, \ldots, a_r; q)_k}{(q, b_1, b_2, \ldots, b_s; q)_k} \left((-1)^k q^{\binom{k}{2}}\right)^{1+s-r} z^k.$$

When $r = s+1$, the series converges for $|z| < 1$.

2. Entry 12 by Euler's Method

In this section, we give a formal proof of (1.1) using the approach Euler [7] used to find the continued fraction of a divergent series. We use the elementary identity

$$\frac{U}{V} = 1 + \frac{U-V}{V} \tag{2.1}$$

to 'divide' two series of the form $1 + a_1 z + a_2 z^2 + \cdots$. A repeated application leads to a continued fraction, much like Euclid's algorithm is used to find a continued fraction of a ratio of two numbers. We call this Euler's method. In [4], all but one of Ramanujan's q-continued fractions from Chapter 16 of Ramanujan's second notebook [3] and from the Lost Notebook [2, Chap. 6] were proved using this method, or obtained as special cases. The only exception was Entry 12 of [3, Chap. 16]. The purpose of this section is to rectify this omission.

Define, for $s = 0, 1, 2, 3, \ldots$

$$D(s) := \sum_{k=0}^{\infty} \frac{(bq^{2s-1}/a, -bq/a; q^2)_k}{(q^2, -q^{2s}; q^2)_k} (a^2 q)^k. \tag{2.2}$$

Theorem 2.1. *Let $D(s)$ be as above. Then, for $|q| < 1$ and $|a| < 1$, and $s = 0, 1, 2, 3, \dots$, we have*

$$\begin{aligned}&\frac{(a^2q^3, b^2q^3; q^4)_\infty}{(a^2q, b^2q; q^4)_\infty} \\ &= \frac{1}{1-ab} + \frac{(a-bq)(b-aq)}{(1-ab)(1+q^2)} + \frac{(a-bq^3)(b-aq^3)}{(1-ab)(1+q^4)} + \cdots \\ &\quad + \frac{(a-bq^{2s-1})(b-aq^{2s-1})}{(1-ab)(1+q^{2s})} + \frac{(a-bq^{2s+1})(b-aq^{2s+1})}{(1+q^{2s+2})\dfrac{D(s+1)}{D(s+2)}}.\end{aligned} \tag{2.3}$$

Remark 2.2. From Theorem 2.1, we immediately get modified convergence of the infinite continued fraction in Entry 12. For ordinary convergence, we need a further condition and a further argument. See [4] for an example in the context of Ramanujan's continued fractions.

Proof. We prove (2.3) in three steps. First observe that

$$\frac{(a^2q^3, b^2q^3; q^4)_\infty}{(a^2q, b^2q; q^4)_\infty} = \frac{\sum_{k=0}^\infty \frac{(bq/a, -bq/a; q^2)_k}{(q^2, -q^2; q^2)_k}(a^2q)^k}{\sum_{k=0}^\infty \frac{(b/aq, -b/aq; q^2)_k}{(q^2, -q^2; q^2)_k}(a^2q^3)^k}. \tag{2.4}$$

To show this we use the q-binomial theorem [8, Eq. (I.3)]. We can see that

$$\begin{aligned}\text{Left-hand side of (2.4)} &= \frac{(b^2q^3; q^4)_\infty / (a^2q; q^4)_\infty}{(b^2q; q^4)_\infty / (a^2q^3; q^4)_\infty} \\ &= \frac{\sum_{k=0}^\infty \frac{((bq/a)^2; q^4)_k}{(q^4; q^4)_k}(a^2q)^k}{\sum_{k=0}^\infty \frac{((b/aq)^2; q^4)_k}{(q^4; q^4)_k}(a^2q^3)^k}\end{aligned}$$

which can be written as the right-hand side of (2.4) using some elementary identities involving q-rising factorials.

Second, we show

$$\frac{{}_2\phi_1\left[\begin{matrix} b/aq, \ -b/aq \\ -q^2 \end{matrix}; q^2, \ a^2q^3\right]}{{}_2\phi_1\left[\begin{matrix} bq/a, \ -bq/a \\ -q^2 \end{matrix}; q^2, \ a^2q\right]} = 1 - ab + \frac{(a-bq)(b-aq)}{(1+q^2)\dfrac{D(1)}{D(2)}}. \tag{2.5}$$

To prove this we will use (2.1). Let U (respectively, V) denote the ${}_2\phi_1$ sum in the numerator (respectively, denominator) above. Observe that

$$\begin{aligned}\frac{U}{V} &= \frac{\sum_{k=0}^{\infty}\frac{(b/aq,-b/aq;q^2)_k}{(q^2,-q^2;q^2)_k}(a^2q^3)^k}{\sum_{k=0}^{\infty}\frac{(bq/a,-bq/a;q^2)_k}{(q^2,-q^2;q^2)_k}(a^2q)^k}\\ &= 1+\frac{1}{V}\left(\sum_{k=0}^{\infty}\frac{(b/aq,-b/aq;q^2)_k}{(q^2,-q^2;q^2)_k}(a^2q^3)^k-\sum_{k=0}^{\infty}\frac{(bq/a,-bq/a;q^2)_k}{(q^2,-q^2;q^2)_k}(a^2q)^k\right)\\ &= 1+\frac{1}{V}\left(\sum_{k=0}^{\infty}\frac{(b/aq,-b/aq;q^2)_k}{(q^2,-q^2;q^2)_k}(a^2q)^k\left((-1)\frac{(1-q^{2k})\left(1+b^2q^{2k}/a^2q^2\right)}{(1-b/aq)(1+b/aq)}\right)\right).\end{aligned}$$

Note that the first term of the sum corresponding to $k=0$ is 0. We cancel $(1-q^{2k})$, and then shift the index to obtain

$$\begin{aligned}\frac{U}{V} &= 1+\frac{(-1)}{V}\left(\sum_{k=1}^{\infty}\frac{(bq/a,-bq/a;q^2)_{k-1}}{(q^2;q^2)_{k-1}(-q^2;q^2)_k}(a^2q)^k\left(1+b^2q^{2k}/a^2q^2\right)\right)\\ &= 1+\frac{(-1)}{V}\left(\sum_{k=0}^{\infty}\frac{(bq/a,-bq/a;q^2)_k}{(q^2,-q^2;q^2)_k}(a^2q)^k\left[\frac{a^2q\left(1+b^2q^{2k}/a^2\right)}{1+q^{2k+2}}\right]\right).\end{aligned}$$

Next, we use

$$\frac{a^2q\left(1+b^2q^{2k}/a^2\right)}{1+q^{2k+2}} = ab+\frac{a\,(aq-b)\left(1-bq^{2k+1}/a\right)}{1+q^{2k+2}}, \tag{2.6}$$

to obtain

$$\begin{aligned}\frac{U}{V} &= 1+\frac{(-ab)}{V}\left(\sum_{k=0}^{\infty}\frac{(bq/a,-bq/a;q^2)_k}{(q^2,-q^2;q^2)_k}(a^2q)^k\right)\\ &\quad -\frac{a(aq-b)(1-bq/a)}{(1+q^2)\,V}\left(\sum_{k=0}^{\infty}\frac{(bq^3/a,-bq/a;q^2)_k}{(q^2,-q^4;q^2)_k}(a^2q)^k\right)\\ &= 1-ab+\cfrac{(a-bq)(b-aq)}{\cfrac{(1+q^2)\left(\sum_{k=0}^{\infty}\frac{(bq/a,-bq/a;q^2)_k}{(q^2,-q^2;q^2)_k}(a^2q)^k\right)}{\left(\sum_{k=0}^{\infty}\frac{(bq^3/a,-bq/a;q^2)_k}{(q^2,-q^4;q^2)_k}(a^2q)^k\right)}}.\end{aligned}$$

In this manner, we obtain (2.5).

Note that (2.4) and (2.5) give us (2.3) for $s=0$. The third fact we require is the recursion, for $s=0,1,2,\ldots$,

$$(1+q^{2s})\frac{D(s)}{D(s+1)} = (1-ab)(1+q^{2s})+\cfrac{(a-bq^{2s+1})(b-aq^{2s+1})}{(1+q^{2s+2})\cfrac{D(s+1)}{D(s+2)}}. \tag{2.7}$$

To prove (2.7), we again apply (2.1). The steps are similar to those above. We have

$$\frac{D(s)}{D(s+1)} = \frac{\sum_{k=0}^{\infty} \frac{(bq^{2s-1}/a,-bq/a;q^2)_k}{(q^2,-q^{2s};q^2)_k}(a^2q)^k}{\sum_{k=0}^{\infty} \frac{(bq^{2s+1}/a,-bq/a;q^2)_k}{(q^2,-q^{2s+2};q^2)_k}(a^2q)^k}$$
$$= 1 + \frac{1}{D(s+1)} \sum_{k=0}^{\infty} \left(\frac{(bq^{2s-1}/a, -bq/a; q^2)_k}{(q^2, -q^{2s}; q^2)_k} (a^2q)^k \right.$$
$$\left. \times \left((-1) \frac{(1-q^{2k})(q^{2s} + bq^{2s-1}/a)}{(1-bq^{2s-1}/a)(1+q^{2k+2s})} \right) \right).$$

Once again, we realize the first term is 0, and after canceling $(1-q^{2k})$, we shift the index so that it again begins from 0.

$$\frac{D(s)}{D(s+1)} = 1 + \frac{(-1)}{D(s+1)} \sum_{k=0}^{\infty} \frac{(bq^{2s+1}/a, -bq/a; q^2)_k}{(q^2, -q^{2s+2}; q^2)_k} (a^2q)^k$$
$$\times \left(\left[\frac{(1+bq^{2k+1}/a)(a^2q^{2s+1} + abq^{2s})}{(1+q^{2s})(1+q^{2k+2s+2})} \right] \right).$$

We use

$$\frac{(1+bq^{2k+1}/a)(a^2q^{2s+1} + abq^{2s})}{(1+q^{2s})(1+q^{2k+2s+2})} = ab + \frac{a(aq^{2s+1} - b)(1-bq^{2k+2s+1}/a)}{(1+q^{2s})(1+q^{2k+2s+2})}, \quad (2.8)$$

to obtain, after multiplying by $(1+q^2)$,

$$(1+q^2)\frac{D(s)}{D(s+1)} = (1-ab)(1+q^{2s})$$
$$- \frac{a(aq^{2s+1} - b)(1-bq^{2s+1}/a)}{(1+q^{2s+2})D(s+1)}$$
$$\times \left(\sum_{k=0}^{\infty} \frac{(bq^{2s+3}/a, -bq/a; q^2)_k}{(q^2, -q^{2s+4}; q^2)_k} (a^2q)^k \right).$$

This reduces to the right hand side of (2.7).

Using (2.4) and (2.5), we obtain (2.3) for $s = 0$. We iterate using (2.7) to complete the derivation. □

Remark 2.3. On the surface, it appears that the algebraic simplifications (2.6) and (2.8) are unmotivated, and the difficult parts of the proof. However, the key difficulty in applying Euler's method is to write the continued fraction as a ratio of two series. There are multiple ways to do so, and not all are amenable to Euler's approach. Once this is done in (2.4), the rest follows by applying Euler's method. Note that if we ignore the quantity in square brackets in the sum in the expression before (2.8), the sum is $D(s+1)$. A similar remark applies to the sum above (2.6). This observation motivates the algebraic manipulations required in the proof. Even so, the algebraic manipulations required for obtaining other continued fractions due to Ramanujan by this method (see [4]) are, comparatively, relatively straightforward.

3. A Proof of Entry 12 using Darboux's Method

The objective of this section to provide another proof of Ramanujan's Entry 12, by what may be called the standard orthogonal polynomial method. Let C denote Ramanujan's continued fraction.

$$C = \frac{1}{1-ab} + \frac{(a-bq)(b-aq)}{(1-ab)(1+q^2)} + \frac{(a-bq^3)(b-aq^3)}{(1-ab)(1+q^4)} + \cdots.$$

On the other hand, the continued fraction $K = H(1)/(1-ab)$ is

$$K = \frac{1}{2(1-ab)} + \frac{(a-bq)(b-aq)}{(1-ab)(1+q^2)} + \frac{(a-bq^3)(b-aq^3)}{(1-ab)(1+q^4)} + \cdots.$$

Evidently

$$\frac{1}{K} - (1-ab) = \frac{1}{C}. \tag{3.1}$$

Our approach is to first find an expression for $H(1)$ and then use (3.1) to find the value of C.

Theorem 3.1. *Let $|q| < 1$, $|ab| < 1$ and $|a^2q| < 1$ and let $H(x)$ be the continued fraction (1.2). Then*

$$H(1) = \frac{(1-ab)}{2} \cdot \frac{{}_2\phi_1\left[\begin{matrix} -bq/a, \; bq/a \\ -q^2 \end{matrix}; q^2, \; a^2q\right]}{{}_2\phi_1\left[\begin{matrix} -bq/a, \; b/aq \\ -1 \end{matrix}; q^2, \; a^2q\right]}. \tag{3.2}$$

Proof. Consider once again the recurrence relation (1.3), and the numerator and denominator polynomials, respectively, $N_k(x)$ and $D_k(x)$. Let

$$\widehat{N}_k(x) := \frac{N_k(x)}{(bq/a;q^2)_k},$$

and

$$\widehat{D}_k(x) := \frac{D_k(x)}{(bq/a;q^2)_k}.$$

We see that these polynomials satisfy the recurrence equation

$$\begin{aligned}(1-bq^{2k+1}/a)y_{k+1}(x) &= ((1-ab)x + (1-ab)q^{2k})y_k(x) \\ &\quad + ab(1-aq^{2k-1}/b)y_{k-1}(x),\end{aligned}$$

for $k > 0$; with initial values

$$\widehat{N}_0(x) = 0, \quad \widehat{N}_1(x) = \frac{1-ab}{1-bq/a}; \quad \widehat{D}_0(x) = 1, \quad \widehat{D}_1(x) = \frac{(1-ab)(x+1)}{1-bq/a}.$$

Let $\widehat{N}(t)$ and $\widehat{D}(t)$ denote the respective generating functions. We find that

$$\widehat{N}(t)=\frac{(1-ab)t}{(1-\delta_1 t)(1-\delta_2 t)}\sum_{m=0}^{\infty}\frac{\left(-aqt/b,a^2qt;q^2\right)_m}{(\delta_1 q^2t,\delta_2 q^2 t;q^2)_m}\left(\frac{bq}{a}\right)^m,$$

and

$$\widehat{D}(t)=\frac{(1-b/aq)}{(1-\delta_1 t)(1-\delta_2 t)}\sum_{m=0}^{\infty}\frac{\left(-aqt/b,a^2qt;q^2\right)_m}{(\delta_1 q^2t,\delta_2 q^2 t;q^2)_m}\left(\frac{b}{aq}\right)^m,$$

where δ_1 and δ_2 are such that

$$1-(1-ab)xt-abt^2=(1-\delta_1 t)(1-\delta_2 t).$$

We are interested in the value of the continued fraction when $x=1$. But when $x=1$, we have

$$\delta_1=1,\quad \delta_2=-ab.$$

If we assume $|ab|<1$, then $t=1$ is the singularity closest to the origin.

To find the asymptotic behavior of $\widehat{N}_k(1)$, we apply Darboux's method [9, Theorem 1.2.4] with the comparison function

$$N^*(t)=\sum_{k=0}^{\infty}N_k^* t^k=\frac{(1-ab)}{(1-t)(1+ab)}\sum_{m=0}^{\infty}\frac{\left(-aq/b,a^2q;q^2\right)_m}{(q^2,-abq^2;q^2)_m}\left(\frac{bq}{a}\right)^m.$$

Note that $\widehat{N}(t)-N^*(t)$ has a removable singularity at $t=1$, and is thus continuous in the disk $|t|\le r$, for some $r>1$. By Darboux's method, we obtain

$$\widehat{N}(1)=N_k^*+o(r^{-k}).$$

Similarly, we can find an asymptotic formula for $\widehat{D}(1)$.

In this manner, when $|ab|<1$, we obtain

$$\widehat{N}_k(1)\sim\frac{(1-ab)}{1+ab}{}_2\phi_1\left[\begin{matrix}-aq/b,\ a^2q\\ -abq^2\end{matrix};q^2,\ \frac{bq}{a}\right],$$

and

$$\widehat{D}_k(1)\sim\frac{(1-b/aq)}{1+ab}{}_2\phi_1\left[\begin{matrix}-aq/b,\ a^2q\\ -abq^2\end{matrix};q^2,\ \frac{b}{aq}\right].$$

The continued fraction $H(1)$ is the ratio

$$H(1)=\frac{N_k(1)}{D_k(1)}=\frac{\widehat{N}_k(1)}{\widehat{D}_k(1)},$$

which is the ratio of the above basic hypergeometric series. Finally, we transform both the ${}_2\phi_1$ series using Heine's transformation formula [8, Eq. (III.1)] to complete the proof. □

Corollary 3.2 (Ramanujan's Entry II.16.12). *Let* $|q| < 1$ *and* $|ab| < 1$. *Then, we have*

$$\frac{(a^2q^3, b^2q^3; q^4)_\infty}{(a^2q, b^2q; q^4)_\infty} = \frac{1}{1-ab} + \frac{(a-bq)(b-aq)}{(1-ab)(1+q^2)} + \frac{(a-bq^3)(b-aq^3)}{(1-ab)(1+q^4)} + \cdots .$$

Proof. We recognize that the ${}_2\phi_1$ series in (3.2) are $D(0)$ and $D(1)$, where $D(s)$ is defined in (2.2). So we have

$$K = H(1)/(1-ab) = \frac{D(1)}{2D(0)}.$$

Now from (3.2) and using (2.7) (with $s = 0$), we have

$$\begin{aligned} \frac{1}{C} = \frac{1}{K} - (1-ab) &= 2\frac{D(0)}{D(1)} - (1-ab) \\ &= (1-ab) + \frac{(a-bq)(b-aq)}{(1+q^2)\dfrac{D(1)}{D(2)}} \\ &= \frac{{}_2\phi_1\left[\begin{matrix} b/aq, & -b/aq \\ & -q^2 \end{matrix}; q^2, \; a^2q^3\right]}{{}_2\phi_1\left[\begin{matrix} bq/a, & -bq/a \\ & -q^2 \end{matrix}; q^2, \; a^2q\right]}, \end{aligned}$$

where we used (2.5) to obtain the last equality. Finally, using (2.4), we complete the proof. □

It is rather remarkable that we require the same algebraic manipulations in both our proofs of Entry 12.

Remarks

(1) If $|ab| > 1$, then we see that $t = -1/ab$ is the singularity closest to the origin. Again we can find the value of the continued fraction by using Darboux's theorem. Alternatively, perhaps it is easier to see that the continued fraction is equivalent to

$$\frac{-1/ab}{1-(ab)^{-1}} + \frac{(a^{-1}-b^{-1}q)(b^{-1}-a^{-1}q)}{(1-(ab)^{-1})(1+q^2)} + \frac{(a^{-1}-b^{-1}q^3)(b^{-1}-a^{-1}q^3)}{(1-(ab)^{-1})(1+q^4)} + \cdots .$$

The value of this (for $|q| < 1$) is obtained from Corollary 3.2 by replacing a by $1/a$, b by $1/b$ and dividing the products obtained by $-ab$.

(2) When $|q| > 1$, we see that the original continued fraction of Ramanujan is equivalent to

$$\frac{1}{1-ab} + \frac{(a-bp)(b-ap)}{(1-ab)(1+p^2)} + \frac{(a-bp^3)(b-ap^3)}{(1-ab)(1+p^4)} + \cdots ,$$

where now $p = 1/q$ and $|p| < 1$. We now use Entry 12, with q replaced by p, to find the value of the continued fraction.

4. Generalization of Ramanujan's Entry 12

We now evaluate the generalization of Ramanujan's Entry 12, the J-fraction (1.2). We will again use generating functions and Darboux's method. We will find it convenient to scale the polynomials suitably before proceeding.

We consider the set of polynomials given by

$$P_k(x) = \frac{D_k(\eta x)}{\eta^k(1-ab)^k},$$

where η is to be prescribed shortly, and $D_k(x)$ are the denominator polynomials satisfying (1.3) and the initial values mentioned above. We see that $P_k(x)$ satisfies the recurrence relation

$$\begin{aligned} xP_k(x) &= P_{k+1}(x) - (q^{2k}/\eta)P_k(x) \\ &\quad -\frac{ab}{\eta^2(1-ab)^2}(1-bq^{2k-1}/a)(1-aq^{2k-1}/b)P_{k-1}(x). \end{aligned}$$

We choose

$$\eta^2 = \frac{-4ab}{(1-ab)^2}$$

to obtain a recurrence relation of the form

$$\begin{aligned} xP_k(x) &= P_{k+1}(x) + cq^{2k}P_k(x) \\ &\quad +\frac{1}{4}(1-bq^{2k-1}/a)(1-aq^{2k-1}/b)P_{k-1}(x), \qquad \text{for } k>0, \end{aligned} \tag{4.1}$$

where

$$c = -\frac{(1-ab)}{2\sqrt{-ab}}.$$

We assume that $P_k(x)$ satisfies the initial conditions

$$P_0(x) = 1, \quad P_1(x) = x - c.$$

We require a and b to be of different signs, so that c is real.

This is now in the form of the standard three-term recurrence relation

$$xy_k(x) = y_{k+1}(x) + \alpha_k y_k(x) + \beta_k y_{k-1}(x), \qquad \text{for } k>0, \tag{4.2}$$

where α_k and β_k are real. In our case, we have

$$\alpha_k = cq^{2k} \quad \text{and} \quad \beta_k = \frac{1}{4}(1-bq^{2k-1}/a)(1-aq^{2k-1}/b).$$

Since a and b have opposite signs, $\beta_k > 0$ for $k>0$. So in view of [9, Theorem 2.5.2] we know that the $P_k(x)$ are orthogonal with respect to a positive measure. Further ahead in this section, we will comment further on the measure.

Let $P_k^*(x)$ be the associated polynomials that satisfy (4.1) with $P_0^*(x)=0$ and $P_1^*(x) = 1$. The $P_k^*(x)$ are polynomials of degree $k-1$, and correspond to the numerator polynomials of the associated continued fraction. Let $X(x)$ denote the continued fraction associated with (4.1), that is, let

$$X(x) = \lim_{k\to\infty} \frac{P_k^*(x)}{P_k(x)}. \tag{4.3}$$

Our objective in this section is to compute $X(x)$.

We will take $x = \cos\vartheta$, so that

$$e^{\pm i\vartheta} = x \pm \sqrt{x^2-1}.$$

We choose a branch of $\sqrt{x^2-1}$ in such a way that

$$\sqrt{x^2-1} \sim x, \quad \text{as } x \to \infty,$$

so that $|e^{-i\vartheta}| < |e^{i\vartheta}|$ in the upper half plane, and $|e^{i\vartheta}| < |e^{-i\vartheta}|$ in the lower half plane. We use the notation $\rho_1 = e^{-i\vartheta}$ and $\rho_2 = e^{i\vartheta}$.

Theorem 4.1. *Let $X(x)$ be the continued fraction in (4.3). Let ρ_1 and ρ_2 be as above. Let γ_1 and γ_2 given by*

$$\gamma_1, \gamma_2 = \frac{aq}{2b}(c \pm \sqrt{c^2-1}), \tag{4.4}$$

Let F and G be defined as follows:

$$F(\rho) = {}_2\phi_1\left[\begin{matrix} 2\gamma_1\rho,\ 2\gamma_2\rho \\ q^2\rho^2 \end{matrix}; q^2, \frac{bq}{a}\right],$$

and

$$G(\rho) = (1 - b/aq)\ {}_2\phi_1\left[\begin{matrix} 2\gamma_1\rho,\ 2\gamma_2\rho \\ q^2\rho^2 \end{matrix}; q^2, \frac{b}{aq}\right].$$

Then $X(x)$ converges for all complex numbers $x \notin (-1,1)$, except possibly a finite set of points, and is given by

$$X(x) = 2\rho\frac{F(\rho)}{G(\rho)},$$

where ρ is given by

$$\rho = \begin{cases} \rho_1, & \text{if } \operatorname{Im}(x) > 0, \text{ or } x > 1 \ (x \text{ real}) \\ \rho_2, & \text{if } \operatorname{Im}(x) < 0, \text{ or } x < -1 \ (x \text{ real}) \\ 1, & \text{if } x = 1, \\ -1, & \text{if } x = -1. \end{cases}$$

Proof. We will comment on the convergence of $X(x)$ shortly. We first compute asymptotic formulas for $P_k^*(x)$ and $P_k(x)$ in the upper half plane.

Consider

$$Q_k(x) = \frac{P_k(x)}{(bq/a;q^2)_k}.$$

Then $Q_k(x)$ satisfies

$$\begin{aligned} xQ_k(x) &= (1 - bq^{2k+1}/a)Q_{k+1}(x) + cq^{2k}Q_k(x) \\ &\quad + \frac{1}{4}(1 - aq^{2k-1}/b)Q_{k-1}(x), \qquad \text{for } k > 0, \end{aligned} \tag{4.5}$$

where

$$Q_0(x) = 1, \quad Q_1(x) = \frac{(x-c)}{(1-bq/a)}.$$

We will apply Darboux's method to find an asymptotic formula for $Q_k(x)$, from which we will get the required formula for $P_k(x)$. Let $Q(t)$ denote the generating function of $Q_k(x)$, that is, let

$$Q(t) := \sum_{k=0}^{\infty} Q_k(x)t^k.$$

Multiply (4.5) by t^{k+1} and sum over $k \geq 0$ to find that

$$Q(t) = \frac{(1-b/aq)}{1-xt+t^2/4} + \frac{(1-acqt/b + a^2q^2t^2/4b^2)}{1-xt+t^2/4}\left(\frac{b}{aq}\right)Q(tq^2).$$

We change the variable by taking

$$x = \frac{e^{i\vartheta} + e^{-i\vartheta}}{2} \quad (= \cos\vartheta)$$

so,

$$\begin{aligned} 1 - xt + t^2/4 &= (1 - e^{-i\vartheta}t/2)(1 - e^{i\vartheta}t/2) \\ &= (1 - \rho_1 t/2)(1 - \rho_2 t/2) \text{ (say)}. \end{aligned}$$

Let γ_1, γ_2 be such that

$$(1 - acqt/b + a^2q^2t^2/4b^2) = (1 - \gamma_1 t)(1 - \gamma_2 t),$$

so

$$\gamma_1, \gamma_2 = \frac{aq}{2b}\left(c \pm \sqrt{c^2 - 1}\right).$$

Using ρ_1, ρ_2, γ_1 and γ_2, we can write the q-difference equation for $Q(t)$ in the form

$$Q(t) = \frac{(1-b/aq)}{(1-\rho_1 t/2)(1-\rho_2 t/2)} + \frac{(1-\gamma_1 t)(1-\gamma_2 t)}{(1-\rho_1 t/2)(1-\rho_2 t/2)}\left(\frac{b}{aq}\right)Q(tq^2).$$

On iteration, we obtain

$$Q(t) = \frac{(1-b/aq)}{(1-\rho_1 t/2)(1-\rho_2 t/2)} \sum_{k=0}^{\infty} \frac{(\gamma_1 t, \gamma_2 t; q^2)_k}{(\rho_1 q^2 t/2, \rho_2 q^2 t/2; q^2)_k}\left(\frac{b}{aq}\right)^k.$$

Note that the poles of $Q(t)$ are at

$$t = 2\rho_1, 2\rho_1/q^2, 2\rho_1/q^4, \ldots; \qquad \text{and} \qquad t = 2\rho_2, 2\rho_2/q^2, 2\rho_2/q^4 \ldots.$$

Since $0 < |q| < 1$, the pole nearest to $t = 0$ is at $t = 2\rho_1$ or $t = 2\rho_2$.

Next let

$$Q_k^*(x) = \frac{P_k^*(x)}{(bq/a;q^2)_k},$$

so $Q_k^*(x)$ satisfies (4.5) with the initial conditions

$$Q_0^*(x) = 0 \quad \text{and} \quad Q_1^*(x) = 1/(1 - bq/a).$$

As before, let γ_1 and γ_2 be given by (4.4). The generating function of Q_k^* is given by

$$Q^*(t) = \frac{t}{(1-\rho_1 t/2)(1-\rho_2 t/2)} \sum_{k=0}^{\infty} \frac{(\gamma_1 t, \gamma_2 t; q^2)_k}{(\rho_1 q^2 t/2, \rho_2 q^2 t/2; q^2)_k} \left(\frac{bq}{a}\right)^k.$$

In the upper half-plane, the singularity nearest the origin is at $t = 2\rho_1$. Let $S^*(t)$ be the series

$$S^*(t) = \frac{2\rho_1}{(1-\rho_1^2)(1-\rho_2 t/2)} \sum_{m=0}^{\infty} \frac{(2\gamma_1\rho_1, 2\gamma_2\rho_1; q^2)_m}{(q^2, q^2\rho_1^2; q^2)_m} \left(\frac{bq}{a}\right)^m.$$

and note that $Q^*(t) - S^*(t)$ has a removable singularity at $t = 2\rho_1$. By Darboux's method, we have

$$Q_k^*(x) \sim \frac{2\rho_1\rho_2^k}{2^k(1-\rho_1^2)} \sum_{m=0}^{\infty} \frac{(2\gamma_1\rho_1, 2\gamma_2\rho_1; q^2)_m}{(q^2, q^2\rho_1^2; q^2)_m} \left(\frac{bq}{a}\right)^m = \frac{2\rho_1\rho_2^k}{2^k(1-\rho_1^2)} F(\rho_1) \text{ (say)}.$$

Similarly, considering the generating function of $Q_k(x)$ in the upper half plane, we find that

$$Q_k(x) \sim \frac{\rho_2^k(1-b/aq)}{2^k(1-\rho_1^2)} \sum_{m=0}^{\infty} \frac{(2\gamma_1\rho_1, 2\gamma_2\rho_1; q^2)_m}{(q^2, q^2\rho_1^2; q^2)_m} \left(\frac{b}{aq}\right)^m = \frac{\rho_2^k}{2^k(1-\rho_1^2)} G(\rho_1) \text{ (say)}.$$

Thus in the upper half-plane, we find that

$$X(x) = \lim_{k\to\infty} \frac{P_k^*(x)}{P_k(x)} = \lim_{k\to\infty} \frac{Q_k^*(x)(bq/a;q^2)_k}{Q_k(x)(bq/a;q^2)_k} = 2\rho_1 \frac{F(\rho_1)}{G(\rho_1)}.$$

The same calculation works for $x > 1$, when x is real. In the statement of our theorem, we write F and G in terms of the ${}_r\phi_s$ notation.

In the lower half-plane, since $t = 2\rho_2$ is the singularity nearest to the origin, a similar calculation yields

$$X(x) = \lim_{k\to\infty} \frac{P_k^*(x)}{P_k(x)} = 2\rho_2 \frac{F(\rho_2)}{G(\rho_2)}.$$

This is also valid for $x < -1$, for x real.

For $x = 1$, we find that the dominating term of the comparison function for the generating function of $P_k^*(1)$ is given by

$$S^*(t) = \frac{2}{(1-t/2)^2} \sum_{m=0}^{\infty} \frac{(2\gamma_1\rho_1, 2\gamma_2\rho_1; q^2)_m}{(q^2, q^2\rho_1^2; q^2)_m} \left(\frac{bq}{a}\right)^m.$$

(There is another term of the form $A/(1-t/2)$, but that does not have any contribution.) By Darboux's theorem, we find that

$$Q_k^*(1) \sim \frac{2(k+1)}{2^k} F(1).$$

Similarly, we have

$$Q_k(1) \sim \frac{(k+1)}{2^k} G(1).$$

Thus

$$X(1) = 2\frac{F(1)}{G(1)}.$$

A similar calculation is required for $x = -1$. $\square$

Recall that the $P_k(x)$ satisfy the three term recurrence relation (4.2), with α_k and β_k real, bounded, and $\beta_k > 0$ for $k > 0$. This implies that the polynomials $P_k(x)$ are orthogonal with respect to a positive measure μ. Further, the support of μ is bounded. From Blumenthal's theorem [6, Theorem IV-3.5, p. 117] it follows that μ has an absolutely continuous component μ'. We close this section with some further remarks on the orthogonality measure.

Remarks

(1) We can apply Nevai's theorem [12, Theorem 40, p. 143] to obtain an expression for μ' supported on $(-1, 1)$. This requires the use of Darboux's theorem as above. See [5, §3] for an example of a very similar calculation, in the context of another continued fraction of Ramanujan.

(2) Nevai's theorem further states that if μ has a discrete part, it lies outside $(-1, 1)$. Since the measure is bounded, there are only a finite set of isolated mass points of μ outside $(-1, 1)$.

(3) Markov's theorem [9, Theorem 2.6.2] implies that the value of the continued fraction $X(x)$ is the Stieltjes transform of the measure for $x \notin (a, b)$, where (a, b) is the true interval of orthogonality. This implies that the continued fraction $X(x)$ converges for all complex numbers $x \notin (-1, 1)$, except possibly a finite set of points, as asserted by Theorem 4.1.

(4) As done in [5, §4], we can obtain another expression for μ by inverting the Stieltjes transform, which is equal to $X(x)$ in Theorem 4.1.

(5) See also the remarks after Theorem 4.1 in [5].

The details of the analysis mentioned above are very similar to the analogous calculations in [5], and are omitted.

Acknowledgments

This work was done at the sidelines of conferences and summer schools organized by the members of the Orthogonal Polynomials and Special Functions (OPSF) group of SIAM. We thank the organizers of the following: Summer research institute on q-series, (July–Aug 2018), Chern Institute of Mathematics, Nankai University, Tianjin, P. R. China and OPSFA2019, Hagenberg, Austria. The research of the first author was supported by a grant of the Austrian Science Fund (FWF): F 50-N15.

References

[1] C. Adiga, B. C. Berndt, S. Bhargava and G. N. Watson, *Chapter 16 of Ramanujan's Second Notebook: Theta-Functions and q-Series*, Mem. Amer. Math. Soc. vol. 315, (AMS, 1985), v+85.

[2] G. E. Andrews and B. C. Berndt, *Ramanujan's Lost Notebook. Part II* (Springer, New York, 2009).

[3] B. C. Berndt, *Ramanujan's Notebooks. Part III* (Springer-Verlag, New York, 1991).

[4] G. Bhatnagar, How to prove Ramanujan's q-continued fractions, in *Ramanujan 125, Contemporary Mathematics* (American Mathematical Society, Providence, RI, 2014), pp. 49–68.

[5] G. Bhatnagar and M. E. H. Ismail, Orthogonal polynomials associated with a continued fraction of Hirschhorn, preprint (2019) arXiv:1901.09985, 22 pp.

[6] T. S. Chihara, *An Introduction to Orthogonal Polynomials* (Gordon and Breach Science Publishers, New York-London-Paris, 1978).

[7] L. Euler, De transformatione seriei divergentis $1 - mx + m(m+n)x^2 - m(m+n)(m+2n)x^3 + m(m+n)(m+2n)(m+3n)x^4 +$ etc. in fractionem continuam, *Nova Acta Academiae Scientarum Imperialis Petropolitinae*, vol. 2, (AMS, 1788), pp. 36–45. Reprinted in Opera Omnia: Series 1, Vol. 16, pp. 34–46; English translation by C. E. Sandifer available at http://www.math.dartmouth.edu/~euler/pages/E616.html.

[8] G. Gasper and M. Rahman, *Basic Hypergeometric Series*, 2nd edn. Encyclopedia of Mathematics and its Applications, vol. 96, (Cambridge University Press, Cambridge, 2004).

[9] M. E. H. Ismail, *Classical and Quantum Orthogonal Polynomials in One Variable*, Encyclopedia of Mathematics and its Applications, vol. 98, (Cambridge University Press, Cambridge, 2009).

[10] L. Jacobsen, Domains of validity for some of Ramanujan's continued fraction formulas, *J. Math. Anal. Appl.* **143**(2) (1989) 412–437.

[11] L. Lorentzen and H. Waadeland, *Continued Fractions with Applications*, Studies in Computational Mathematics, vol. 3 (North-Holland Publishing Co., Amsterdam, 1992).

[12] P. G. Nevai. *Orthogonal Polynomials*, Mem. Amer. Math. Soc. vol. 213 (AMS, 1979), v+185.

[13] K. G. Ramanathan, Hypergeometric series and continued fractions, *Proc. Indian Acad. Sci. Math. Sci.* **97**(1–3) (1988) 277–296.

[14] S. Ramanujan, *Collected papers of Srinivasa Ramanujan*, eds. G. H. Hardy, P. V. Seshu Aiyar and B. M. Wilson, (AMS Chelsea Publishing, Providence, RI, 2000), Third printing of the 1927 original, With a new preface and commentary by Bruce C. Berndt.

© 2024 World Scientific Publishing Company
https://doi.org/10.1142/9789811277375_0004

General Fine transformations I*

Douglas Bowman[†] and Sarah Wesley[‡]

Department of Mathematical Sciences
Northern Illinois University, 1425 Lincoln Hwy
DeKalb, IL 60115, USA
[†]*dbowman@niu.edu*
[‡]*wesleys0409@gmail.com*

We would like to dedicate this paper to Bruce Berndt
on the occasion of his eightieth birthday.

We describe an extension of Fine's method of deriving basic hypergeometric series transformations and derive new transformations from the method. Combinatorial proofs of two of the examples are provided.

Keywords: q-series; basic hypergeometric functions; integer partitions.

Mathematics Subject Classification 2020: 11P81, 33D15, 05A17

1. Introduction and Results

1.1. *Background*

In 1988, Fine published the monograph, "Basic Hypergeometric Series and Applications" [6]. The first chapter of [6] focuses primarily on establishing transformations for the function F defined by the series

$$F(a,b;t) = \sum_{n\geq 0} \frac{(a)_n}{(bq)_n} t^n, \tag{1}$$

where q is a fixed complex number with $|q| < 1$, $|t| < 1$, and as usual

$$(z)_n = (z;q)_n = \begin{cases} (1-z)(1-zq)(1-zq^2)\cdots(1-zq^{n-1}) & n > 0, \\ 1 & n = 0, \end{cases}$$

for $z \in \mathbb{C}$. Finally, $(z)_\infty = \prod_{i=0}^{\infty}(1-zq^i)$.

*This article originally appeared in *International Journal of Number Theory*, **17**(2), pp. 267–283 © 2021 World Scientific Publishing Company. DOI: 10.1142/S1793042120400163

Our notation for F differs slightly from that used by Fine, his $F(a,b;t)$ being our $F(aq,b;t)$. This change in notation was found to simplify the statements of most transformations. Throughout we consistently use our notation when quoting results from Fine. The parameters a and b are taken to be fixed complex numbers, $b \neq q^{-n}$, $n > 0$.

It is clear that this series for F converges absolutely for $|t| < 1$, and thus is analytic in the unit disk. As a function of a, F is entire away from singular values of b and t. As a function of b, F is regular except for the simple poles of $b = q^{-n}$, $n \geq 1$.

In his monograph, Fine gave numerous applications of his transformations to number theory. In the notes on [6, Chap. 1] Andrews also describes further applications. He also remarks on results of Fine that do not seem to fit in with known q-series transformations.

An example of this is Fine's transformation (13.1) ([6]):

$$\frac{1-t}{1-b}F(a,b;t) = \sum_{n\geq 0} \frac{(a)_n(at/b)_{2n}}{(tq)_n(a/b)_n} b^n - a\sum_{n\geq 0} \frac{(a)_n(at/b)_{2n+1}}{(tq)_n(a/b)_{n+1}}(bq)^n. \tag{2}$$

George Andrews writes [6]:

> *"This section seems more than any in Chap.* 1 *to be begging for further study. Identities* (13.1), (13.2), *and* (13.5) *do not seem to fit into the known q-hypergeometric hierarchy of results."*

This quote on its own begs for further study into Fine's method and suggests that extensions of his method may yield new transformations. Here, we show that this is indeed the case; for example, similar to the above identity of Fine, we have,

$$\begin{aligned}\frac{1-t}{1-b}F(a,b;t) &= \sum_{n=0}^{\infty} \frac{(a)_{2n}(at/b)_{3n}}{(tq)_n(a/b)_{2n}} b^n - a\sum_{n=0}^{\infty} \frac{(a)_{2n}(at/b)_{3n+1}}{(tq)_n(a/b)_{2n+1}}(bq^2)^n \\ &\quad - aq\sum_{n=0}^{\infty} \frac{(a)_{2n+1}(at/b)_{3n+2}}{(tq)_n(a/b)_{2n+2}}(bq^2)^n.\end{aligned}$$

This is (9) of Theorem 2.1 below, which consists of nine new transformations.

1.2. *Outline*

Section 2 gives the new transformations, a summary of Fine's method that is developed in [6, Chap. 1], and a proof of Theorem 2.1. Section 2.1 gives a statement of the results. Section 2.2 outlines Fine's method of iteration and also lists Fine's transformations. Section 2.3 gives a generalization of Fine's method and shows how we obtain new "seed" identities. Section 2.5 gives the proof of Theorem 2.1.

Section 3 gives a few applications of the new transformations to generating functions for distinct integer partitions and the divisor function. In Sec. 3.1, we give combinatorial proofs of two of these examples.

Section 4 concludes with a discussion of future work.

2. New Results and Fine's Method

For convenience, the next section begins with Theorem 2.1 which lists the new transformations derived in this paper.

2.1. *New identities*

The identities in the following theorem hold whenever both sides converge.

Theorem 2.1.

$$F(a,b;t) = \sum_{n=0}^{\infty} \frac{(a)_n (bq/a)_n}{(t)_{n+1}(bq)_{2n}} (1 - atq^{2n})(-at^2)^n q^{\frac{n(3n-1)}{2}} + bt \sum_{n=0}^{\infty} \frac{(a)_{n+1}(bq/a)_n}{(t)_{n+1}(bq)_{2n+1}} (-at^2)^n q^{\frac{(n+1)(3n+2)}{2}}; \tag{3}$$

$$F(a,b;t) = \sum_{n=0}^{\infty} \frac{(a)_{2n}(at/b)_n}{(bq)_{2n}(t)_{n+1}} (1 - atq^{3n})(bt^2)^n q^{2n^2} + bt \sum_{n=0}^{\infty} \frac{(a)_{2n+1}(at/b)_{n+1}}{(bq)_{2n+1}(t)_{n+1}} (bt^2)^n q^{(n+1)(2n+1)}; \tag{4}$$

$$F(a,b;t) = \sum_{n=0}^{\infty} \frac{(a)_{2n}(at/b)_{2n}}{(t)_{n+1}(bq)_n (a/b)_n} (1 - atq^{3n})(bt)^n q^{n^2} - atq \sum_{n=0}^{\infty} \frac{(a)_{2n+1}(at/b)_{2n+1}}{(t)_{n+1}(bq)_n (a/b)_{n+1}} (bt)^n q^{n(n+3)}; \tag{5}$$

$$F(a,b;t) = \sum_{n=0}^{\infty} \frac{(a)_n (at/b)_n (bq/a)_n}{(t)_{2n+1}(bq)_{2n}} (1 - atq^{3n})(-abt^2)^n q^{\frac{n(7n-1)}{2}} + btq \sum_{n=0}^{\infty} \frac{(a)_{n+1}(at/b)_{n+1}(bq/a)_n}{(t)_{2n+2}(bq)_{2n+1}} (-abt^2)^n q^{\frac{7n(n+1)}{2}}; \tag{6}$$

$$F(a,b;t)=\sum_{n=0}^{\infty}\frac{(a)_{2n}(at/b)_{3n}}{(t)_{2n+1}(a/b)_n(bq)_n}(1-atq^{4n})(b^2t)^nq^{2n^2}$$
$$+bt\sum_{n=0}^{\infty}\frac{(a)_{2n+1}(at/b)_{3n+1}}{(t)_{2n+2}(a/b)_n(bq)_n}(b^2t)^nq^{(n+1)(2n+1)}$$
$$-abt\sum_{n=0}^{\infty}\frac{(a)_{2n+1}(at/b)_{3n+2}}{(t)_{2n+2}(a/b)_{n+1}(bq)_n}(b^2t)^nq^{(n+2)(2n+1)};\tag{7}$$

$$F(a,b;t)=\sum_{n=0}^{\infty}\frac{(bq/a)_{2n}}{(t)_{n+1}(bq)_{2n}(bq/at)_n}(-at)^nq^{\binom{n}{2}}$$
$$+b\sum_{n=0}^{\infty}\frac{(bq/a)_{2n+1}}{(t)_{n+1}(bq)_{2n+1}(bq/at)_{n+1}}(-at)^nq^{\binom{n+2}{2}}.\tag{8}$$

For $|b|<1$,

$$F(a,b;t)=(1-b)\sum_{n=0}^{\infty}\frac{(a)_{2n}(at/b)_{3n}}{(t)_{n+1}(a/b)_{2n}}b^n$$
$$-a(1-b)\sum_{n=0}^{\infty}\frac{(a)_{2n}(at/b)_{3n+1}}{(t)_{n+1}(a/b)_{2n+1}}(bq^2)^n$$
$$-a(1-b)q\sum_{n=0}^{\infty}\frac{(a)_{2n+1}(at/b)_{3n+2}}{(t)_{n+1}(a/b)_{2n+2}}(bq^2)^n;\tag{9}$$

$$F(a,b;t)=(1-b)\sum_{n=0}^{\infty}\frac{(a)_n(at/b)_{3n}}{(t)_{2n+1}(a/b)_n}b^{2n}$$
$$-a(1-b)\sum_{n=0}^{\infty}\frac{(a)_n(at/b)_{3n+1}}{(t)_{2n+1}(a/b)_{n+1}}(b^2q)^n$$
$$+b(1-b)\sum_{n=0}^{\infty}\frac{(a)_{n+1}(at/b)_{3n+2}}{(t)_{2n+2}(a/b)_{n+1}}b^{2n}.\tag{10}$$

Finally, for $|t|<1$,

$$F(a,b;t)=\sum_{n=0}^{\infty}\frac{(a)_n(bq/a)_n}{(bq)_{2n}(bq/at)_n}t^n-\frac{bq}{a}\sum_{n=0}^{\infty}\frac{(a)_{n+1}(bq/a)_n}{(bq)_{2n+1}(bq/at)_{n+1}}(tq)^n.\tag{11}$$

The identity (4) is a special case of a generalization of (30), the Rogers–Fine identity. Our generalization, not given here, will be explored in a future paper. It is interesting to compare (9) and (10) to (2).

Seven "conjugate identities" are given in the following corollary. Conjugate identities are defined at the end of Sec. 2.2. The identities in the corollary are respectively conjugates of identities (3), (4), (7)–(11).

Corollary 2.2. *When both sides converge,*

$$F(a,b;t) = \sum_{n=0}^{\infty} \frac{(at/b)_n(bq/a)_n}{(t)_{2n+1}(bq)_n}(-abt)^n q^{\frac{n(3n-1)}{2}}$$

$$- at \sum_{n=0}^{\infty} \frac{(at/b)_n(bq/a)_{n+1}}{(t)_{2n+2}(bq)_n}(-abt)^n q^{3\binom{n+1}{2}}; \tag{12}$$

$$F(a,b;t) = \sum_{n=0}^{\infty} \frac{(a)_n(at/b)_{2n}}{(t)_{2n+1}(bq)_n}(1 - atq^{3n})(b^2t)^n q^{2n^2}$$

$$+ bt \sum_{n=0}^{\infty} \frac{(a)_{n+1}(at/b)_{2n+1}}{(t)_{2n+2}(bq)_n}(b^2t)^n q^{(n+1)(2n+1)}; \tag{13}$$

$$F(a,b;t) = \sum_{n=0}^{\infty} \frac{(at/b)_{2n}(a)_{3n}}{(t)_{n+1}(bq)_{2n}(a/b)_n}(1 - atq^{4n})(bt^2)^n q^{2n^2}$$

$$+ bt \sum_{n=0}^{\infty} \frac{(at/b)_{2n+1}(a)_{3n+1}}{(t)_{n+1}(bq)_{2n+1}(a/b)_n}(bt^2)^n q^{(n+1)(2n+1)}$$

$$- at^2 \sum_{n=0}^{\infty} \frac{(at/b)_{2n+1}(a)_{3n+2}}{(t)_{n+1}(bq)_{2n+1}(a/b)_{n+1}}(bt^2)^n q^{(n+2)(2n+1)}; \tag{14}$$

$$F(a,b;t) = \sum_{n=0}^{\infty} \frac{(bq/a)_{2n}}{(t)_{2n+1}(q/a)_n(bq)_n}(-at)^n q^{\binom{n}{2}}$$

$$+ b \sum_{n=0}^{\infty} \frac{(bq/a)_{2n+1}}{(t)_{2n+2}(q/a)_{n+1}(bq)_n}(-at)^n q^{\binom{n+2}{2}}. \tag{15}$$

For $|t| < 1$,

$$F(a,b;t) = \sum_{n=0}^{\infty} \frac{(at/b)_{2n}(a)_{3n}}{(a/b)_{2n}(bq)_n} t^n - \frac{at}{b} \sum_{n=0}^{\infty} \frac{(at/b)_{2n}(a)_{3n+1}}{(a/b)_{2n+1}(bq)_n} t^n q^{2n}$$

$$- \frac{atq}{b} \sum_{n=0}^{\infty} \frac{(at/b)_{2n+1}(a)_{3n+2}}{(a/b)_{2n+2}(bq)_n} t^n q^{2n}; \tag{16}$$

$$F(a,b;t) = \sum_{n=0}^{\infty} \frac{(at/b)_n(a)_{3n}}{(bq)_{2n}(a/b)_n} t^{2n} - \frac{at}{b} \sum_{n=0}^{\infty} \frac{(at/b)_n(a)_{3n+1}}{(bq)_{2n}(a/b)_{n+1}} t^{2n} q^n$$

$$+ t \sum_{n=0}^{\infty} \frac{(at/b)_{n+1}(a)_{3n+2}}{(bq)_{2n+1}(a/b)_{n+1}} t^{2n}. \tag{17}$$

Finally, for $|b| < 1$,

$$F(a,b;t) = (1-b)\sum_{n=0}^{\infty} \frac{(at/b)_n (bq/a)_n}{(t)_{2n+1}(q/a)_n} b^n$$
$$- \frac{b(1-b)q}{a}\sum_{n=0}^{\infty} \frac{(at/b)_{n+1}(bq/a)_n}{(t)_{2n}(q/a)_{n+1}}(bq)^n. \tag{18}$$

2.2. *Fine's method of iteration*

To explain Fine's method, it is helpful to start with an example. Such a transformation is (7.2) of [6]:

$$F(a,b;t) = -\frac{b}{at}\sum_{n\geq 0} \frac{(bq/a)_n}{(bq)_n (bq/at)_{n+1}} q^{n+1} + \frac{(bq/a)_\infty}{(bq)_\infty (bq/at)_\infty}\sum_{n\geq 0} (a)_n t^n. \tag{19}$$

Fine obtains this transformation by iterating what we refer to as a "seed" identity. The seed from which (19) is derived is the following:

$$F(a,b;t) = -\frac{bq}{(1-bq/at)at} + \frac{(1-bq/a)}{(1-bq)(1-bq/at)}F(a,bq;t). \tag{20}$$

More generally, Fine's seeds are identities of the form

$$F = B + AF(aq^{n_1}, bq^{n_2}; tq^{n_3}),$$

where $n_i = \{0,1\}$ and $A, B \in \mathbb{Q}(a,b,t,q)$. Fine denotes a transformation of the above form by the symbols $(a,b,t) \mapsto (aq^{n_1}, bq^{n_2}, tq^{n_3})$. In this paper, we employ this notation, but do not restrict the n_i to the set $\{0,1\}$. Equations involving only a single parameter being q-shifted are written without parentheses; thus (20) is denoted by $b \mapsto bq$. Also, parameters that are not shifted are not listed in the parentheses. The transformation $b \mapsto bq^2$ can be found by replacing b with bq in (20), and then substituting back into (20). Thus,

$$F(a,b,t) = -\frac{bq}{(1-bq/at)at} - \frac{(1-bq/a)bq^2}{(1-bq)(1-bq/at)(1-bq^2/at)at}$$
$$+ \frac{(1-bq/a)(1-bq^2/a)}{(1-bq)(1-bq^2)(1-bq/at)(1-bq^2/at)}F(a,bq^2;t)$$
$$= -\frac{b}{at}\sum_{i=0}^{1} \frac{(bq/a)_i}{(bq)_i (bq/at)_{i+1}} q^{i+1} + \frac{(bq/a)_2}{(bq)_2 (bq/at)_2}F(a,bq^2;t). \tag{21}$$

Continuing in this way, one can compute $b \mapsto bq^n$:

$$F(a,b;t) = -\frac{b}{at}\sum_{i=0}^{n-1} \frac{(bq/a)_i}{(bq)_i (bq/at)_{i+1}} q^{i+1} + \frac{(bq/a)_n}{(bq)_n (bq/at)_n}F(a,bq^n;t). \tag{22}$$

Letting $n \to \infty$, re-indexing, as well as noting that since $|q| < 1$, then $bq^n \to 0$ as $n \to \infty$, gives (19).

To illustrate the power of this method, Fine shows that special cases of (19) yield famous results of Gauss and Jacobi, such as

$$\sum_{n\geq 0} q^{(n^2+n)/2} = \prod_{n\geq 1} \frac{1-q^{2n}}{1-q^{2n-1}}, \tag{23}$$

$$\sum_{n=-\infty}^{+\infty} (-1)^n q^{n^2} = \prod_{n\geq 1} \frac{1-q^n}{1+q^n} = \prod_{n\geq 1} (1-q^{2n})(1-q^{2n-1})^2, \tag{24}$$

and

$$\sum_{n=-\infty}^{+\infty} q^{n^2} = \prod_{n\geq 1} (1-q^{2n})(1+q^{2n-1})^2. \tag{25}$$

Fine obtains six non-trivial identities by iterating seeds such as (20). We list Fine's transformations here for convenience.

By iterating the transformation $a \mapsto aq$, [6, Eq. (11.1)], Fine obtained:

$$F(a,b;t) = -\frac{a(1-b)}{b} \sum_{n\geq 0} \frac{(a)_n (at/b)_n}{(a/b)_{n+1}} q^n + \frac{(a)_\infty (at/b)_\infty}{(a/b)_\infty} \sum_{n\geq 0} \frac{t^n}{(bq)_n}. \tag{26}$$

The iterations $t \mapsto tq$ for $|b| < 1$, and $(b,t) \mapsto (bq,tq)$ [6, Eqs. (6.3) and (12.1)] give, respectively:

$$F(a,b;t) = (1-b) \sum_{n=0}^{\infty} \frac{(at/b)_n}{(t)_{n+1}} b^n, \tag{27}$$

and

$$(1-t)F(a,b;t) = \sum_{n\geq 0} \frac{(bq/a)_n}{(bq)_n (tq)_n} (-at)^n q^{(n^2-n)/2}. \tag{28}$$

Fine's iteration $(a,t) \mapsto (aq,tq)$ for $|b| < 1$ results in [6, 13.1]:

$$\frac{1-t}{1-b} F(a,b;t) = \sum_{n\geq 0} \frac{(aq)_n (at/b)_{2n}}{(tq)_n (aq/b)_n} b^n - aq \sum_{n\geq 0} \frac{(aq)_n (at/b)_{2n+1}}{(tq)_n (aq/b)_{n+1}} (bq)^n. \tag{29}$$

Finally, the iteration $(a,b,t) \mapsto (aq,bq,tq)$ yields the "Rogers-Fine" identity [6, (14.1)]:

$$(1-t)F(a,b;t) = \sum_{n\geq 0} \frac{(a)_n (at/b)_n}{(bq)_n (tq)_n} (1-atq^{2n})(bt)^n q^{n^2}. \tag{30}$$

Identities (19) and (26) are special in that infinite products occur on the right sides. None of Fine's other identities, nor the identities proved here, have this characteristic.

Identity (27) shows that for $0 < |b| < 1$, $(1-t)F(a,b;t)$ is invariant under the change of variables $a \mapsto at/b$, $b \mapsto t$ and $t \mapsto b$. Throughout this paper, we refer to (27) as *the involution* since applying the change of variables another time recovers $(1-t)F(a,b;t)$. One common application of (27), which is used systematically by Fine, is to apply the involution to identities derived using Fine's method to obtain new transformations. We call these new identities generated by the involution *conjugate* identities. This terminology is motivated by the fact that the resulting identities come from conjugation in associated Ferrers diagrams of integer partitions. We apply this same technique to identify throughout this paper. Indeed, Alladi [1] showed that the involution (27) arises combinatorially by applying conjugation to Ferrers diagrams of integer partitions.

2.3. *Generalizing Fine's method*

In this section, we consider periodic iterations of transformations of the form $(a,b,t)\mapsto(aq^{n_1},bq^{n_2},tq^{n_3})$ with $n_i\in\mathbb{Z}_{\geq 0}$, $\max(n_i)\leq 2$, and $\gcd(n_1,n_2,n_3)=1$. From these periodic transformations, taking the limit $n\to\infty$ yields new identities.

The condition $\gcd(n_1,n_2,n_3)=1$ is necessary to avoid redundant equations. For example, the iteration $b\mapsto bq^2$, or (21), generates the same transformation as $b\mapsto bq$ (20). More precisely, the identity arising from iterating (Eq. (21), for example) is equivalent to simply grouping pairs of consecutive terms on the right side of (19).

In addition to new identities obtained by extending Fine's method, we obtain the corollary using the involution (27). Setting $a'=at/b$, $b'=t$, and $t'=b$ in $(a,b,t)\mapsto(aq^{n_1},bq^{n_2},tq^{n_3})$, notice that $(a',b',t')\mapsto(a'q^{n_1+n_3-n_2},b'q^{n_3},t'q^{n_2})$. This defines a function ι on $\mathbb{Z}^3$ given by $\iota:(n_1,n_2,n_3)\mapsto(n_1+n_3-n_2,n_3,n_2)$. Notice that (n_1,n_2,n_3) is fixed under the involution if and only if $n_2=n_3$.

Fine's seed identities are the seven non-trivial transformations $(a,b,t)\mapsto(aq^{n_1},bq^{n_2},tq^{n_3})$, with $n_i\in\{0,1\}$. They are given in Table 1. From these, it is clear that one can obtain any transformation with $n_i\in\mathbb{Z}$. Multiplying through such a relation to make the affine term 1, and subtracting any two such normalized transformations gives that the vector space $\mathcal{F}$ of functions spanned by $F(aq^{n_1},bq^{n_2};tq^{n_3})$ over the field $\mathbb{Q}(a,b,t,q)$ has dimension 2. We reformulate Fine's method in terms of q-difference operators; after this we give a generalization.

Consider the group $\mathcal{S}$ (isomorphic to $\mathbb{Z}^3$) of q-shift operators generated by $\alpha: a\mapsto aq$, $\beta: b\mapsto bq$, and $\tau: t\mapsto tq$. General elements of this group are of the form $\sigma=\alpha^{n_1}\beta^{n_2}\tau^{n_3}$. This group acts on $\mathcal{F}$ in the following way: given $\sigma\in\mathcal{S}$, Fine's transformations $(a,b,t)\mapsto(aq^{n_1},bq^{n_2},tq^{n_3})$ with $n_i\in\{0,1\}$ imply that for general n_i,

$$F=B_\sigma+A_\sigma\sigma F,$$

for some $A_\sigma,B_\sigma\in\mathbb{Q}(a,b,t,q)$. The next proposition shows how to compute the action of $\mathcal{S}$ on the rational functions A_σ and B_σ.

Proposition 2.3. *For $\sigma,\sigma_i\in\mathcal{S}$, $1\leq i\leq n$,*

$$A_{\sigma^{-1}}=\frac{1}{\sigma^{-1}A_\sigma}, \tag{31}$$

$$B_{\sigma^{-1}} = -A_{\sigma^{-1}}\sigma^{-1}B_{\sigma}, \tag{32}$$

$$A_{\sigma_1\sigma_2} = A_{\sigma_1}\sigma_1 A_{\sigma_2}, \tag{33}$$

$$B_{\sigma_1\sigma_2} = B_{\sigma_1} + A_{\sigma_1}\sigma_1 B_{\sigma_2}. \tag{34}$$

By induction,

$$A_{\sigma_1\cdots\sigma_n} = A_{\sigma_1}(\sigma_1 A_{\sigma_2})(\sigma_1\sigma_2 A_{\sigma_3})\cdots(\sigma_1\cdots\sigma_{n-1}A_{\sigma_n}), \tag{35}$$

and

$$B_{\sigma_1\cdots\sigma_n} = B_{\sigma_1} + A_{\sigma_1}\sigma_1 B_{\sigma_2} + A_{\sigma_1\sigma_2}\sigma_1\sigma_2 B_{\sigma_3} + \cdots + A_{\sigma_1\cdots\sigma_{n-1}}\sigma_1\cdots\sigma_{n-1}B_{\sigma_n}. \tag{36}$$

Proof of Proposition 2.3. By definition, $F = B_{\sigma_{-1}} + A_{\sigma^{-1}}\sigma^{-1}F$. Applying σ^{-1} to $F = B_\sigma + A_\sigma\sigma F$, yields

$$\sigma^{-1}F = \sigma^{-1}B_\sigma + \sigma^{-1}(A_\sigma\sigma F).$$

Substituting this expression for $\sigma^{-1}F$ into the first equation, and matching terms on both sides yields (31) and (32).

To prove (33) and (34) use the equations

$$F = B_{\sigma_1} + A_{\sigma_1}\sigma_1 F,$$

and

$$F = B_{\sigma_2} + A_{\sigma_2}\sigma_2 F.$$

Substituting F from the second equation for F on the right side of the first equation and matching terms with

$$F = B_{\sigma_1\sigma_2} + A_{\sigma_1\sigma_2}\sigma_1\sigma_2 F$$

gives (33) and (34).

The proof of (35) and (36) follows by induction on (33) and (34), respectively. □

Thus, given an infinite sequence $\sigma_i \in \mathcal{S}$, we have a method to compute F in terms of $\sigma_1\sigma_2\cdots\sigma_n F$. Then, letting $n \to \infty$ potentially gives a new transformation of F.

We only consider periodic products of α, β, and τ. To this end, let $\rho_i \in \{\alpha, \beta, \tau\}$, for $1 \le i \le k$. Put $\sigma := \rho_1 \cdots \rho_k$. Then, we can compute F in terms of σ^n by using (35) and (36) with $\sigma_1 = \sigma_2 = \cdots = \sigma_n$. This yields

$$A_{\sigma^n} = A_\sigma(\sigma A_\sigma)(\sigma^2 A_\sigma)\cdots(\sigma^{n-1}A_\sigma) = \prod_{i=0}^{n-1} \sigma^i A_\sigma, \tag{37}$$

and

$$B_{\sigma^n} = B_\sigma + A_\sigma \sigma B_\sigma + A_{\sigma^2}\sigma^2 B_\sigma + \cdots + A_{\sigma^{n-1}}\sigma^{n-1}B_\sigma$$
$$= \sum_{i=0}^{n-1} A_{\sigma^i}\sigma^i B_\sigma \tag{38}$$
$$= \sum_{i=0}^{n-1}\left(\prod_{j=0}^{i-1}\sigma^j A_\sigma\right)\sigma^i B_\sigma. \tag{39}$$

Note that while (37) and (38) are actually equivalent to Fine's method of iteration when applied to the seed relation $F = B_\sigma + A_\sigma \sigma F$, Eqs. (35) and (36) are more general. Since Fine's results are confined to seeds derived from $(a, b, t) \mapsto (aq^{n_1}, bq^{n_2}; tq^{n_3})$ with $n_i \in \{0, 1\}$, in this paper we take the next step and consider seeds of the form $n_i \in \{0, 1, 2\}$ with $\gcd(n_1, n_2, n_3) = 1$ and some $n_i = 2$.

We conclude this section a table of Fine's seeds which are used in the proof of Theorem 2.1. They are Eqs. (2.2), (2.4), (4.1), (4.3)–(4.6) of [6] where they are proved.

2.4. *Proof of Theorem 2.1*

Because of the number of identities to be proved and the similarity of the proofs, we only give one proof in detail. However, we indicate the changes required to prove the remainder.

The identities in Theorem 2.1 arise from different cases of $\sigma = \alpha^{n_1}\beta^{n_2}\tau^{n_3}$, where $n_i \in \{0, 1, 2\}$. A_{σ^n} and B_{σ^n} are computed using (37) and (38) to get a new

Table 1. Fine's seeds.

σ	A_σ		B_σ	
α	$\frac{(1-a)(1-at/b)}{1-a/b}$	(40)	$-\frac{(1-b)a}{(1-a/b)b}$	(41)
β	$\frac{1-bq/a}{(1-bq)(1-bq/at)}$	(42)	$\frac{-bq}{(1-bq/at)at}$	(43)
τ	$\frac{1-at/b}{1-t}b$	(44)	$\frac{1-b}{1-t}$	(45)
$\beta\tau$	$-\frac{1-bq/a}{(1-bq)(1-t)}at$	(46)	$\frac{1}{1-t}$	(47)
$\alpha\beta$	$\frac{1-a}{1-bq}t$	(48)	1	(49)
$\alpha\tau$	$\frac{(1-a)(1-at/b)(1-atq/b)}{(1-t)(1-a/b)}b$	(50)	$\frac{1-b}{1-t} - \frac{(1-b)(1-at/b)a}{(1-t)(1-a/b)}$	(51)
$\alpha\beta\tau$	$\frac{(1-a)(1-at/b)}{(1-bq)(1-t)}btq$	(52)	$\frac{1-at}{1-t}$	(53)

transformation of F of the form $F = B_{\sigma^n} + A_{\sigma^n}\sigma^n F$. Finally, sending $n \to \infty$ yields the identities. Applying the involution (27) gives the corollary.

First, we list the values of σ used for each identity: for (3), $\sigma = \alpha\beta^2\tau$; for (4), $\sigma = \alpha^2\beta^2\tau$; for (5), $\sigma = \alpha^2\beta\tau$; for (6), $\sigma = \alpha\beta^2\tau^2$; for (7), $\sigma = \alpha^2\beta\tau^2$; for (8), $\sigma = \beta^2\tau$; for (9), $\sigma = \alpha^2\tau$; for (10), $\sigma = \alpha\tau^2$; for (11), $\sigma = \alpha\beta^2$.

We give the details of the proof of (3) and describe the changes needed for the other identities.

Proof of (3). Using (37), (33), (42), and (52),

$$\begin{aligned}
A_{(\alpha\beta^2\tau)^n} &= \prod_{i=0}^{n-1}(\alpha\beta^2\tau)^i A_{\alpha\beta^2\tau} \\
&= \prod_{i=0}^{n-1}(\alpha\beta^2\tau)^i A_{\alpha\beta\tau}(\alpha\beta\tau)A_\beta \\
&= \prod_{i=0}^{n-1}(\alpha\beta^2\tau)^i \left(\frac{(1-a)(1-at/b)}{(1-bq)(1-t)}btq \cdot \frac{(1-bq/a)}{(1-bq^2)(1-b/at)}\right) \\
&= \prod_{i=0}^{n-1}(\alpha\beta^2\tau)^i \left(-\frac{(1-a)(1-bq/a)}{(1-bq)(1-bq^2)(1-t)}at^2q\right) \\
&= \prod_{i=0}^{n-1} -\frac{(1-aq^i)(1-bq^{i+1}/a)}{(1-bq^{2i+1})(1-bq^{2i+2})(1-tq^i)}at^2q^{3i+1} \\
&= \frac{(a)_n(bq/a)_n}{(bq)_{2n}(t)_n}(-at^2)^n q^{(3n-1)n/2}.
\end{aligned}$$

Above we employed (33) with $\sigma_1 = \alpha\beta\tau$ and $\sigma_2 = \beta$. Identities (4)–(6), and (7) also use (33) with this choice of σ_1. Identities (9) and (10) have $\sigma_1 = \alpha\tau$, while identities (8) and (11) use $\sigma_1 = \beta\tau$ and $\sigma_1 = \alpha\beta$, respectively.

Using (34), (43) and (53),

$$\begin{aligned}
B_{\alpha\beta^2\tau} &= B_{\alpha\beta\tau} + A_{\alpha\beta\tau}(\alpha\beta\tau)B_\beta \\
&= \frac{1-at}{1-t} + \frac{(1-a)(1-at/b)}{(1-bq)(1-t)}btq(\alpha\beta\tau)\left(-\frac{bq}{(1-bq/at)at}\right) \\
&= \frac{1-at}{1-t} - \frac{(1-a)(1-at/b)}{(1-bq)(1-t)}btq \cdot \left(\frac{b}{(1-b/at)at}\right) \\
&= \frac{1-at}{1-t} + \frac{(1-a)}{(1-bq)(1-t)}btq.
\end{aligned}$$

The selections for σ_1 in (34) used for the proofs of the other identities are the same as the choices used in the last step when we employed (33).

By (38),

$$\begin{aligned}
B_{(\alpha\beta^2\tau)^n} &= \sum_{i=0}^{n-1} A_{(\alpha\beta^2\tau)^i}(\alpha\beta^2\tau)^i B_{\alpha\beta^2\tau} \\
&= \sum_{i=0}^{n-1} \frac{(a)_i(bq/a)_i}{(bq)_{2i}(t)_i}(-at^2)^i q^{i(3i-1)/2}(\alpha\beta^2\tau)^i \\
&\quad \times \left(\frac{1-at}{1-t} + \frac{(1-a)}{(1-bq)(1-t)}btq\right) \\
&= \sum_{i=0}^{n-1} \frac{(a)_i(bq/a)_i}{(bq)_{2i}(t)_i}(-at^2)^i q^{i(3i-1)/2} \\
&\quad \cdot \left(\frac{1-atq^{2i}}{1-tq^i} + \frac{(1-aq^i)}{(1-bq^{2i+1})(1-tq^i)}btq^{3i+1}\right) \\
&= \sum_{i=0}^{n-1} \frac{(a)_i(bq/a)_i}{(bq)_{2i}(t)_{i+1}}(-at^2)^i q^{i(3i-1)/2}(1-atq^{2i}) \\
&\quad + bt\sum_{i=0}^{n-1} \frac{(a)_{i+1}(bq/a)_i}{(bq)_{2i+1}(t)_{i+1}}(-at^2)^i q^{(i+1)(3i+2)/2}.
\end{aligned}$$

Thus, for $\sigma = \alpha\beta^2\tau$, the nth iteration is

$$\begin{aligned}
F(a,b;t) &= B_{\sigma^n} + A_{\sigma^n}\sigma^n F(a,b;t) \\
&= B_{\sigma^n} + A_{\sigma^n} F(aq^n, bq^{2n}; tq^n) \\
&= \sum_{i=0}^{n-1} \frac{(a)_i(bq/a)_i}{(bq)_{2i}(t)_{i+1}}(-at^2)^i q^{i(3i-1)/2}(1-atq^{2i}) \\
&\quad + bt\sum_{i=0}^{n-1} \frac{(a)_{i+1}(bq/a)_i}{(bq)_{2i+1}(t)_{i+1}}(-at^2)^i q^{(i+1)(3i+2)/2} \\
&\quad + \frac{(a)_n(bq/a)_n}{(bq)_{2n}(t)_n}(-at^2)^n q^{n(3n-1)/2} F(aq^n, bq^{2n}; tq^n).
\end{aligned}$$

Letting $n \to \infty$ and re-indexing gives (3). □

3. Some Special Cases

In this section, we consider some elementary applications of identities presented in Theorem 2.1. Certain special cases of identities in Theorem 2.1 recover well-known results while others yield new identities. For example, the case $t \to 1$, $a = 0$, and

$b = 1$ in (3), (8), (11), and other identities of Theorem 2.1 recover

$$\frac{1}{(q)_\infty} = \sum_{n=0}^{\infty} \frac{q^{n^2}}{(q)_n^2},$$

a well-known result of Euler. Also, the case $t \to 1$, $a = q$, and $b = 0$ in (3), (5), and (10) recovers Euler's pentagonal number theorem:

$$(q)_\infty = \sum_{n=0}^{\infty} (-1)^n q^{n(3n+1)/2}(1 - q^{2n+1}).$$

Before continuing, there is an interesting special case of (2), (3), and (30) which is a special case of an identity of Sylvester discussed by Andrews [3]. In [4, Theorem 9.2], let $x = -1$ and multiply both sides by $(-q)_\infty$ to obtain (54) below. Alternatively, in (2) or (30), let $t \to 1$, and then $b = 0$ and $a = -q$ to get

$$(-q)_\infty = \sum_{n=0}^{\infty} \frac{(-q)_n}{(q)_n} q^{n(3n+1)/2}(1 + q^{2n+1}). \tag{54}$$

This is a kind of analogue of Euler's pentagonal number theorem above. After multiplying by $1 - t$, the same substitutions in (3) also give this identity.

The next two identities are special cases of [2, Eq. (3.7)]. Multiply both sides of (4) by $1 - t$, and then let $b = 0$, $a = -q$, and $t \to 1$ to obtain

$$(-q)_\infty = \sum_{n=0}^{\infty} \frac{(-q)_{2n}}{(q)_n} q^{n(5n+1)/2}(1 + q^{3n+1}) + \sum_{n=0}^{\infty} \frac{(-q)_{2n+1}}{(q)_n} q^{(n+1)(5n+4)/2}. \tag{55}$$

This is the special case $z = 1$ of Eq. (3.9) of Andrews [2]. Andrews gives a combinatorial proof of the more general case, his Eq. (3.7). Making the same simplification of (6) instead of (4) yields

$$(-q)_\infty = \sum_{n=0}^{\infty} \frac{(-q)_n}{(q)_{2n}} q^{n(4n+1)}(1 + q^{3n+1}) + \sum_{n=0}^{\infty} \frac{(-q)_{n+1}}{(q)_{2n+1}} q^{2(n+1)(2n+1)}. \tag{56}$$

Equations (55) and (56) are similar to (54) and indeed a combinatorial proof in the spirit of Sylvester's school, discussed by Andrews [2, 3], can be given for a common generalization of these identities. It should be noted that (56) was given previously in [8], but no combinatorial proof was provided. We provide the combinatorial proofs (following Andrews) after listing a few more special cases of Theorem 2.1.

In Theorem 2.1, take $a = b = t = q$ and multiply both sides by $q/(1 - q)$ to get the generating function for the divisor function on the left side and new expressions for it on the right. (Below $d(n)$ denotes the divisor function, the number of positive integer divisors of n.) For example, this case in (3) yields

$$\sum_{n=1}^{\infty} d(n)q^n = \sum_{n=0}^{\infty} \frac{(1 + q^{n+1})}{(q^{n+1})_{n+1}} (-1)^n q^{\frac{(n+1)(3n+2)}{2}} + \sum_{n=0}^{\infty} \frac{1}{(q^{n+1})_{n+2}} (-1)^n q^{\frac{(n+1)(3n+8)}{2}}. \tag{57}$$

The same case of (4) yields

$$\sum_{n=1}^{\infty} d(n)q^n = \sum_{n=0}^{\infty} \frac{(1-q^{3n+2})}{(1-q^{2n+1})(1-q^{n+1})} q^{(n+1)(2n+1)} + \sum_{n=0}^{\infty} \frac{1}{(1-q^{2n+2})} q^{(n+2)(2n+2)}, \tag{58}$$

which looks like it should have an easy proof using partial fractions and simple series manipulations. Perhaps more interesting is the same case of (6),

$$\sum_{n=1}^{\infty} d(n)q^n = q\sum_{n=0}^{\infty} \frac{(q)_n}{(q^{n+1})_{n+1}^2}(-1)^n q^{\frac{7n(n+1)}{2}}(1-q^{3n+2}) + \sum_{n=0}^{\infty} \frac{(q)_n}{(q^{n+2})_{n+1}^2}(-1)^n q^{\frac{(7n+8)(n+1)}{2}}. \tag{59}$$

Letting $b = 1$ in (3), then let $q = q^2$ and $a = t = q$ along with Fine's (7.321), a well-known result of Gauss, gives

$$\frac{(q^2;q^2)_\infty}{(q;q^2)_\infty} = \sum_{n=0}^{\infty} q^{n(n+1)/2} = \sum_{n=0}^{\infty} \frac{(-1)^n q^{n(3n+2)}}{(-q)_{2n}(q^2;q^2)_n}(1+q^{2n+1}) + \sum_{n=0}^{\infty} \frac{(-1)^n q^{3n^2+8n+3}}{(1-q^{2n+1})(-q)_{2n+1}(q^2;q^2)_n}. \tag{60}$$

The same case in (7) yields

$$\frac{(q^2;q^2)_\infty}{(q;q^2)_\infty} = \sum_{n=0}^{\infty} q^{n(n+1)/2} = \sum_{n=0}^{\infty} (-q)_{3n}(q^{2n+1})_n(1+q^{4n+1})q^{n(4n+1)} + \sum_{n=0}^{\infty} \frac{(-q)_{3n+1}(q^{2n+1})_{n+1}}{(1-q^{4n+3})} q^{(n+1)(4n+3)} - \sum_{n=0}^{\infty} \frac{(-q)_{3n+2}(q^{2n+2})_{n+1}}{(1-q^{4n+3})} q^{(n+2)(4n+3)}. \tag{61}$$

3.1. *Combinatorial proofs of* (55) *and* (56)

Following Andrews [2], we now give combinatorial proofs of (55) and (56) in the spirit of Sylvester's school of partition theory.

We begin with (56). The left side is the well-known generating function for partitions into distinct parts, while the right side can be interpreted in terms of generating functions for classes of restricted partitions into distinct parts.

Rewrite (56) in the form:

$$(-q)_\infty = \sum_{i=0}^{\infty} \frac{(-q)_i}{(q)_{2i}} q^{2i^2+\frac{2i(2i+1)}{2}} + \sum_{i=0}^{\infty} \frac{(-q)_i}{(q)_{2i}} q^{i(2i+1)+\frac{(2i+1)(2i+2)}{2}}$$

$$+ \sum_{i=0}^{\infty} \frac{(-q)_{i+1}}{(q)_{2i+1}} q^{(i+1)(2i+1)+\frac{(2i+1)(2i+2)}{2}}. \tag{62}$$

The distinct partitions enumerated on the left are now divided into three classes corresponding to the sums on the right side of (62). Begin with a Ferrers diagram for a partition of n into distinct parts. Consider the largest integer i such that the first $2i$ parts are greater than i. If the remaining parts are less than or equal to i, put the partition into our first class, which we denote by $A_1(n)$. Otherwise, if the remaining parts are greater than i, our next case is that the $2i+1$st part is exactly equal to $i+1$, with the remaining parts less than or equal to i; we put these partitions of n into the class $A_2(n)$. The remaining case occurs if the $2i+1$st part is greater than $i+1$ and the remaining parts are less than or equal to $i+1$. The partitions in this last case are collected into the set $A_3(n)$.

Now, consider the ith term of the first sum on the right side of (62),

$$\frac{(-q)_i}{(q)_{2i}} q^{2i^2+\frac{2i(2i+1)}{2}}.$$

Construct a Ferrers diagram using the factor q^{2i^2} to generate a rectangle of dots representing a partition into $2i$ parts of size i. The factor $q^{\frac{2i(2i+1)}{2}}$ generates partitions into parts 1 to $2i$. We place this on the right side of the rectangle giving partitions into $2i$ distinct parts from size $i+1$ to $3i$. The factor $1/(q)_{2i}$ generates partitions into parts of size at most $2i$. We conjugate these partitions to get partitions into at most $2i$ parts and place them on the right side of the rectangle, giving partitions into $2i$ distinct parts greater than i. The factor $(-q)_i$ generates partitions into distinct parts of size ranging from 1 to i. These parts are adjoined below the above partition. Thus, this first factor generates partitions into distinct parts from the set $A_1(n)$.

Next, consider the ith term of the second sum on the right side of (62),

$$\frac{(-q)_i}{(q)_{2i}} q^{i(2i+1)+\frac{(2i+1)(2i+2)}{2}}.$$

The factor $q^{i(2i+1)}$ generates a rectangle as a partition into $2i+1$ parts of size i. The factor $q^{\frac{(2i+1)(2i+2)}{2}}$ generates a partition $1+2+\cdots+2i+1$. We place this on the right side of the rectangle to give partitions with $2i+1$ distinct parts of size between $i+1$ and $3i+1$. The factor $1/(q)_{2i}$ generates partitions into parts of size at most $2i$. We conjugate these partitions and place them on the right side of the rectangle, giving partitions into $2i+1$ distinct parts greater than or equal to $i+1$. The factor $(-q)_i$ generates partitions into distinct parts of size ranging from 1 to i. These parts are adjoined below parts constructed above. Thus, we have another partitions into distinct parts which satisfy the conditions of $A_2(n)$.

Finally, consider the ith term of the third sum on the right side of (62),

$$\frac{(-q)_{i+1}}{(q)_{2i+1}} q^{(2i+1)(i+1)+\frac{(2i+1)(2i+2)}{2}}.$$

The factor $q^{(2i+1)(i+1)}$ generates a rectangle as a partition into $2i+1$ parts of size $i+1$. The factor $q^{\frac{(2i+1)(2i+2)}{2}}$ generates a partition $1+2+\cdots+2i+1$. We place this on the right side of the rectangle to give partitions with $2i+1$ distinct parts of size between $i+2$ and $3i+2$. The factor $1/(q)_{2i+1}$ generates partitions into parts of size at most $2i+1$. We conjugate these partitions and place them on the right side of the rectangle, which gives partitions into $2i+1$ distinct parts larger than $i+2$. The factor $(-q)_{i+1}$ generates partitions into distinct parts of size ranging from 1 to $i+1$. These partitions are adjoined below the above parts. Thus, we have a partition satisfying the conditions of set $A_3(n)$.

A similar combinatorial proof of (55) is now outlined. First write (55) in the form,

$$(-q)_\infty = \sum_{i=0}^{\infty} \frac{(-q)_{2i}}{(q)_i} q^{2i^2+\frac{i(i+1)}{2}} + \sum_{i=0}^{\infty} \frac{(-q)_{2i}}{(q)_i} q^{(i+1)(2i+1)+\frac{i(i+1)}{2}} + \sum_{i=0}^{\infty} \frac{(-q)_{2i+1}}{(q)_i} q^{(i+1)(2i+2)+\frac{i(i+1)}{2}}. \tag{63}$$

Building classes of partitions using Durfee rectangles similar to the last proof we find that the three sums on the right side of (63) partitions the set of integer partitions with distinct parts into three sets $B_1(n)$, $B_2(n)$, and $B_3(n)$ defined as follows. $B_1(n)$ is the set of partitions of n for which there exists an i such that the first i parts are distinct and greater than or equal to $2i+1$ and the remaining parts are distinct and less than or equal to $2i$. $B_2(n)$ is the set of partitions of n for which there exists an i such that the first i parts are distinct and greater than or equal to $2i+2$, the $i+1$st part is exactly equal to $2i+1$, and the remaining parts distinct and less than or equal to $2i$. Finally, $B_3(n)$ is the set of partitions of n for which there exists an i such that the first i parts are distinct and greater than or equal to $2i+3$, the $i+1$st part is equal to $2i+2$, and all remaining parts are distinct and less than or equal to $2i+1$.

4. Future Work

Several papers are in progress building on the ideas presented here. First of all, a two-parameter generalization of the Rogers–Fine identity exists which we can prove both combinatorially as well as with Fine's method of iteration. This general identity also contains (4), but its combinatorial proof makes use of overpartitions and is more complicated than the proof of (55) (a special case of (4)) given here. We also plan to extend this approach as well. There is another general identity, a generalization of (12), which can be given a similar combinatorial treatment. Next, it is possible to iterate the general case $\sigma = \alpha^{n_1}\beta^{n_2}\tau^{n_3}$ and obtain a very general identity. This will also be the subject of a future paper.

Finally, we plan to investigate connections between our method and that of Kadell [7]. His method relies on applying the partial q-difference operators that act on the two parameters in the q-binomial theorem (equivalently, parameters t and a in Fine's function F, since when $b = 1$, F reduces to products by the q-binomial theorem). The interesting question is what results can be obtained by applying an extension of Kadell's technique to the partial q-difference opertors α, β, and τ working in a three-dimensional integer lattice, instead of the two-dimensional lattice studied by Kadell.

Acknowledgments

The authors would like to thank the conference organizers for their generous support of the second author's attendance of the conference.

References

[1] K. Alladi, *A Fundamental Invariant in the Theory of Partitions* (Springer, 1999), pp. 101–106.
[2] G. E. Andrews, Generalizations of the Durfee Square, *J. London Math. Soc. (2)* **3** (1971) 563–570.
[3] G. E. Andrews, *Partitions: Yesterday and Today* (The New Zealand Mathematical Society, 1979), pp. 10–13.
[4] G. E. Andrews, *The Theory of Partitions* (Cambridge University Press, 1998), pp. 140–142.
[5] S. Corteel and J. Lovejoy, Overpartitions, *Trans. Amer. Math. Soc.* **356** (2004) 1623–1635.
[6] N. J. Fine, *Basic Hypergeometric Series and Applications* (American Mathematical Society, 1988).
[7] K. W. J. Kadell, *Path Functions and Generalized Basic Hypergeometric Functions*, Vol. 360 (American Mathematical Society, 1987).
[8] S. Wesley, Variants of Fine transformations, Doctoral dissertation, Northern Illinois University, DeKalb, Illinois (2018).

© 2024 World Scientific Publishing Company
https://doi.org/10.1142/9789811277375_0005

Combinatorics of two second-order mock theta functions*

Hannah Burson
School of Mathematics, University of Minnesota, Vincent Hall
206 Church Street SE, Minneapolis, MN 55455, USA
hburson@umn.edu

We introduce combinatorial interpretations of the coefficients of two second-order mock theta functions. Then, we provide a bijection that relates the two combinatorial interpretations for each function. By studying other special cases of the multivariate identity proved by the bijection, we obtain new combinatorial interpretations for the coefficients of Watson's third-order mock theta function $\omega(q)$ and Ramanujan's third-order mock theta function $f(q)$.

Keywords: Partitions; mock theta functions; q-series.

Mathematics Subject Classification 2020: 05A17, 11P84

1. Introduction

In January 1920, Ramanujan wrote his final letter to Hardy [3], where he introduced a class of functions that he called mock theta functions due to how they mimic certain behaviors of classical modular forms, especially Jacobi theta functions. In this paper, Ramanujan listed properties of 17 such functions. Based on some of these properties, he assigned an order to each mock theta function, classifying his examples as third-, fifth-, or seventh-order mock theta functions.

In the century, since Ramanujan's death, number theorists have investigated the properties of these functions and worked to classify exactly how "almost modular" these functions are. This rigorous study started with Watson [12], who proved many of Ramanujan's initial claims and introduced some new mock theta functions. However, it was not until Zwegers's 2002 thesis [13] that we began to understand how mock theta functions fit in the general area of non-holomorphic modular forms. Zwegers showed that mock theta functions are, in fact, not modular, but can be completed to become a function that displays modular behavior by adding a certain non-holomorphic part. This completed function is called a weak Maass form. For

*This article originally appeared in *International Journal of Number Theory*, **17**(2), pp. 285–295 © 2021 World Scientific Publishing Company. DOI: 10.1142/S1793042120400229

more information on the history and definition of mock theta functions, see [2, 7, 9], or [11].

Zwegers's work allowed for the discovery of infinite families of mock theta functions. In particular, McIntosh studied some second-order mock theta functions in [10]. In this study, McIntosh stated the following second-order mock theta function identities:

$$A(q) = \sum_{n=0}^{\infty} \frac{q^{(n+1)^2}(-q;q^2)_n}{(q;q^2)_{n+1}^2} = \sum_{n=0}^{\infty} \frac{q^{n+1}(-q^2;q^2)_n}{(q;q^2)_{n+1}}, \tag{1.1}$$

$$B(q) = \sum_{n=0}^{\infty} \frac{q^{n(n+1)}(-q^2;q^2)_n}{(q;q^2)_{n+1}^2} = \sum_{n=0}^{\infty} \frac{q^{n}(-q;q^2)_n}{(q;q^2)_{n+1}}. \tag{1.2}$$

We use the q-shifted factorial

$$(a;q)_n = \prod_{k=0}^{n-1}(1-aq^k).$$

In this paper, we assume $|q| < 1$. Then, we can write

$$(a;q)_\infty = \lim_{n\to\infty}(a;q)_n = \prod_{k=0}^{\infty}(1-aq^k).$$

The main purpose of this paper is to provide combinatorial proofs of (1.1) and (1.2).

In Sec. 2, we collect some necessary definitions. In Sec. 3, we prove a general identity that contains (1.1) and (1.2) as special cases. In Sec. 4, we discuss other special cases of the combinatorial interpretation from Sec. 3, including a more direct combinatorial interpretation of (1.1). Finally, in Sec. 5, we generalize our interpretation of (1.2) to give a combinatorial interpretation of a q-hypergeometric series transformation identity.

2. Background

A partition λ of a non-negative integer n is a multiset of positive integers $(\lambda_1, \lambda_2, \ldots, \lambda_k)$ such that $\lambda_1 + \lambda_2 + \cdots + \lambda_k = n$. We use the convention that $\lambda_1 \geq \lambda_2 \geq \cdots \geq \lambda_k$ and let $p(n)$ denote the number of partitions of n. The generating function for $p(n)$ is given by

$$\sum_{n=0}^{\infty} p(n)q^n = \frac{1}{(q;q)_\infty}.$$

Partitions have a graphical representation called a Ferrers diagram (or Young diagram), which is a left-justified array of cells where the jth row has λ_j cells.

In this paper, we use two generalizations of partitions. A k-colored partition is a partition in which each part can appear in one of k colors. Let $p_k(n)$ denote

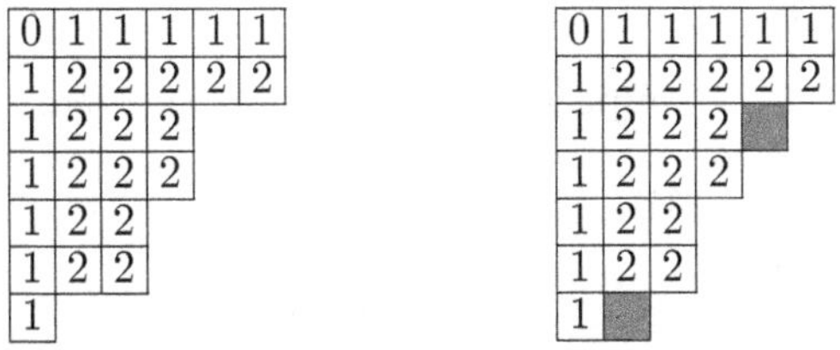

(a) an odd Ferrers diagram (b) a shaded odd Ferrers diagram

Fig. 1.

the number of k-colored partitions of n. Then, we have the following generating function:

$$\sum_{n=0}^{\infty} p_k(n)q^n = \frac{1}{(q;q)_\infty^k}.$$

In [1], George Andrews introduced odd Ferrers diagrams, which are obtained by taking a Ferrers diagram, placing a 0 in the top-left corner, filling the top row and left column with 1's, and filling the remainder of the diagram with 2's. We can think of these diagrams as representing a pair (k, λ) where k is a non-negative integer and λ is a partition into odd parts of size at most $2k+1$.

Overpartitions [5] are partitions where the first appearance of any size part may be overlined. If we let $\overline{p}(n)$ be the function that counts the number of overpartitions of n, we can see that the generating function for $\overline{p}(n)$ is given by

$$\frac{(-q;q)_\infty}{(q;q)_\infty}.$$

Overpartitions are a logical generalization of partitions and can be extended to provide a generalization of many partition-theoretic objects. In this paper, we work with an overpartition generalization of odd Ferrers diagrams called shaded odd Ferrers diagrams. In this generalization, we may add a shaded box to any interior corner of an odd Ferrers diagram. This is equivalent to considering pairs $(k, \overline{\lambda})$ where k is a non-negative integer and λ is an overpartition into odd parts of size at most $2k+1$ where all overlined parts must be of size at most $2k-1$. Shaded odd Ferrers diagrams are equivalent to the boxed 2-modular diagrams introduced by the author in [4]. We define the size of a shaded odd Ferrers diagram to be the sum of the entries in all the cells. If we let $\text{so}(n)$ denote the number of shaded odd Ferrers diagrams of size n, we have the following generating function:

$$\sum_{n=0}^{\infty} \text{so}(n)q^n = \sum_{k=0}^{\infty} \frac{(-q;q^2)_k}{(q;q^2)_{k+1}} q^k.$$

Note that the sum on the right-hand side is indexed over the number of ones in the top row.

3. A General Identity

In this section, we will focus on proving the following theorem.

Theorem 3.1. *For* $|q| < 1, |atz| \leq 1$ *and* $z \neq 0$,

$$\sum_{n=0}^{\infty} \frac{(\frac{-q^2}{z}; q^2)_n (a^2tz)^n q^{n(n+1)}}{(atq; q^2)_{n+1}(aq; q^2)_{n+1}} = \sum_{n=0}^{\infty} \frac{(at)^n q^n (-azq; q^2)_n}{(aq; q^2)_{n+1}}.$$

Proof. This identity can be obtained by a ${}_3\phi_2$ transformation (c.f. [6, Eq. 17.9.6]), which states that

$$\frac{(e/a, de/(bc); q)_\infty}{(e, de/(abc); q)_\infty} {}_3\phi_2\left(\begin{matrix} a, d/b, d/c \\ d, de/(bc) \end{matrix}; q, e/a\right) = {}_3\phi_2\left(\begin{matrix} a, b, c \\ d, e \end{matrix}; q, de/(abc)\right), \tag{3.1}$$

where

$$(a, b; q)_\infty = (a; q)_\infty (b; q)_\infty$$

and

$${}_3\phi_2\left(\begin{matrix} a, b, c \\ d, e \end{matrix}; q, z\right) = \sum_{n=0}^{\infty} \frac{(a; q)_n (b; q)_n (c; q)_n}{(d; q)_n (e; q)_n (q; q)_n} z^n, \quad |z| < 1.$$

In (3.1), replacing q by q^2 and taking $a = q^2$, $b = \frac{-q^2}{\tau}$, $c = \frac{-q^2}{z}$, $d = aq^3$ and $e = atq^3$ gives

$$\sum_{n=0}^{\infty} \frac{(-a\tau q; q^2)_n (-azq; q^2)_n}{(aq; q^2)_{n+1}(a^2t\tau z; q^2)_{n+1}} a^n t^n q^n = \sum_{n=0}^{\infty} \frac{(-q^2/\tau; q^2)_n (-q^2/z; q^2)_n}{(aq; q^2)_{n+1}(atq; q^2)_{n+1}} a^{2n} t^n \tau^n z^n.$$

Then, by letting $\tau \to 0$, we obtain

$$\begin{aligned} \sum_{n=0}^{\infty} \frac{(-azq; q^2)_n}{(aq; q^2)_{n+1}} a^n t^n q^n &= \lim_{\tau \to 0} \frac{1}{(1 - atq)(1 - aq)} \\ &\quad \times \sum_{n=0}^{\infty} \frac{(-q^2/\tau; q^2)_n (-q^2/z; q^2)_n}{(atq^3; q^2)_n (aq^3; q^2)_n} a^{2n} t^n \tau^n z^n \\ &= \sum_{n=0}^{\infty} \frac{(-q^2/z; q^2)_n a^{2n} t^n z^n q^{n(n+1)}}{(atq; q^2)_{n+1}(aq; q^2)_{n+1}}. \end{aligned} \tag{3.2}$$

□

In order to give a combinatorial proof of Theorem 3.1, we introduce some notation. Let $\mathcal{P}_{B,k}$ be the set of two-colored partitions into parts of size $\leq 2k + 1$, such that all even part sizes $\leq 2k$ are color 2 and appear exactly one or two times. Because all even parts are the same color, we will not label the color of the even parts. We let $\mathcal{P}_B = \cup_{k \geq 0} \mathcal{P}_{B,k}$. Let $\mathcal{SF}$ be the set of shaded odd Ferrers diagrams.

Given any partition λ, we let $\nu(\lambda)$ denote the number of parts and $\nu_o(\lambda)$ denote the number of odd parts. Analogously, for a shaded odd Ferrers diagram π, $\nu(\pi)$ is the number of rows in the diagram. For a partition $\lambda \in \mathcal{P}_B$, we define $\nu_{d,e}$ to be

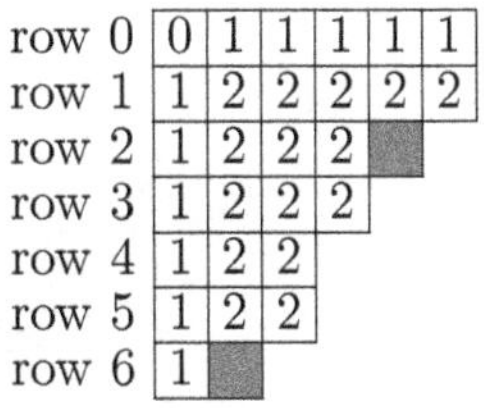

Fig. 2. A shaded odd Ferrers diagram π with $\nu(\pi) = 7$, $\overline{\nu}(\pi) = 2$, and $\text{top}(\pi) = 5$.

the number of even part sizes that appear, $\nu_{u,e}(\lambda)$ to be the number of even parts that appear exactly once and $\nu_1(\pi)$ to be the number of odd parts of color 1. For example, if we take $\lambda = 7_1+7_2+6+6+5_2+5_2+4+3_1+2+2+1_1+1_1+1_2 \in \mathcal{P}_{B,3}$, we have $\nu(\lambda) = 13$, $\nu_o(\lambda) = 8$, $\nu_{d,e}(\lambda) = 3$, $\nu_{u,e}(\lambda) = 1$, and $\nu_1(\pi) = 4$. Note that if $\lambda \in \mathcal{P}_{B,k}$, then $\nu_{d,e} = k$. For a shaded odd Ferrers diagram π, we define $\overline{\nu}(\pi)$ to be the number of shaded cells in π and $\text{top}(\pi)$ to be the sum of the entries in the top row of π. Additionally, when we talk about row numbers of a shaded odd Ferrers diagram, we label the rows starting with the top row as row 0. Figure 2 illustrates these statistics for a shaded odd Ferrers diagram.

Combinatorial proof of Theorem 3.1. First, notice that

$$\sum_{k=0}^{\infty} \frac{(\frac{-q^2}{z};q^2)_k (a^2tz)^k q^{k(k+1)}}{(atq;q^2)_{k+1}(aq;q^2)_{k+1}} = \sum_{\lambda \in \mathcal{P}_B} a^{\nu_o(\lambda)+2\nu_{d,e}(\lambda)} t^{\nu_{d,e}(\lambda)+\nu_1(\lambda)} z^{\nu_{u,e}(\lambda)} q^{|\lambda|}$$

and

$$\sum_{n=0}^{\infty} \frac{(at)^n q^n (-azq;q^2)_n}{(aq;q^2)_{n+1}} = \sum_{\pi \in \mathcal{SF}} a^{\text{top}(\pi)+\nu(\pi)-1} t^{\text{top}(\pi)} z^{\overline{\nu}(\pi)} q^{|\pi|}.$$

To prove Theorem 3.1 combinatorially, we will construct a bijection ϕ between $\mathcal{P}_B$ and $\mathcal{SF}$. Let $k \geq 0$ and let $\lambda \in \mathcal{P}_{B,k}$. We define $\phi(\lambda)$ by creating a shaded odd Ferrers diagram according to the following procedure (we illustrate the procedure on the partition $9_1+9_1+8+7_1+7_2+7_2+6+6+4+3_1+2+2+1_2 \in \mathcal{P}_{B,4}$):

Step 1. For each $1 \leq j \leq k$, turn the first part of size $2j$ into one shaded row of size $2j-1$ and add a single 1 to the top row of the diagram. After this step, there will be k 1s in the top row and k shaded rows (sizes $\{1, 3, \ldots, 2k-1\}$). Represent the result as a shaded odd Ferrers diagram.

$8+6+4+2 \rightarrow$

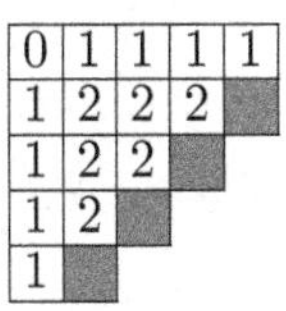

Step 2. Take each part of color 1, and add it as a column to the shaded odd Ferrers diagram.

$$+9_1+9_1 \quad +7_1+3_1 \quad \rightarrow$$

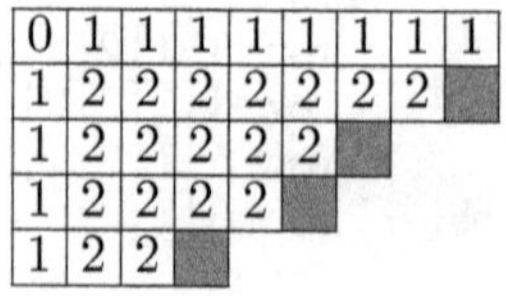

Step 3. Let $\{\tau_1, \tau_2, \ldots, \tau_m\}$ be the remaining even parts, listed in decreasing order. For each $1 \leq j \leq m$ define $g(j)$ to be the maximum integer n such that there are at least n parts of color 1 and size $\geq \tau_j + (2(j+n)-1)$. Then, remove the shaded box from row $j + g(j)$ and append τ_j as a 2-modular column starting at row $j + g(j)$.

$$+6+2 \rightarrow$$

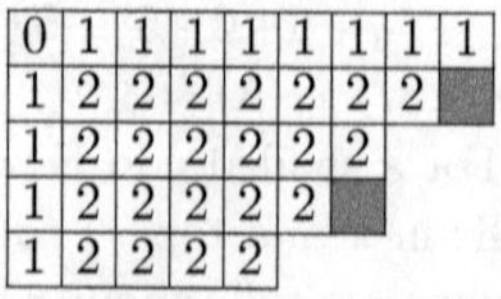

Step 4. Add odd parts of color 2 as unshaded 2-modular rows.

$$+7_2+7_2+1_2 \rightarrow$$

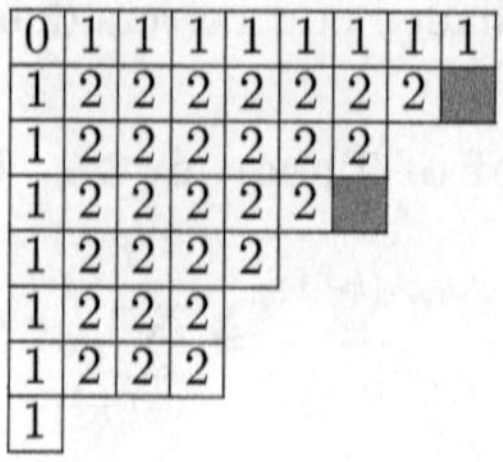

Note that, by the definitions in Step 3, for $1 \leq j < m$, λ has at least $g(j)$ parts of color 1 and size at least

$$\begin{aligned}\tau_j + 2(j+g(j)) - 1 &\geq \tau_{j+1} + 2 + 2(j+g(j)) - 1 \quad (\text{because } \tau_{j+1} \leq \tau_j - 2)\\ &= \tau_{j+1} + 2(j+1+g(j)) - 1.\end{aligned}$$

Therefore, $g(j+1) \geq g(j)$, so $j+1+g(j+1) > j+g(j)$. This inequality shows that Step 3 is well defined.

To help us define the inverse map, we prove the following claim.

Claim. Using the definitions in Step 3, for $1 \leq j < m$, we have

$$\tau_{j+1} + 2g(j+1) + 2j + 2 \leq \tau_j + 2g(j) + 2j. \tag{3.3}$$

Proof of Claim. Assume that the claim is false. Since $\tau_{j+1} < \tau_j$, we can define $x > 0$ such that $\tau_{j+1} = \tau_j - 2x$. Then, by our assumption, we have

$$2g(j+1) - 2x + 2 > 2g(j).$$

Then, by definition, λ has at least $g(j+1)$ parts of color 1 and size at least

$$\tau_j - 2x + 2(j+1+g(j+1)) - 1 \geq \tau_j + 2j + 2 + 2g(j) - 1.$$

Note that our assumption implies that $g(j+1) > g(j)$, so we now can say that there are at least $g(j)+1$ parts of color 1 and size at least $\tau_j + 2(j+g(j)+1) - 1$. This contradicts the assumption that $g(j)$ is the maximum integer n such that there are at least n parts of color 1 and size $\geq \tau_j + (2(j+n)-1)$. Therefore, our assumption must be false and our claim must be true. □

To prove that ϕ is a bijection, we give the inverse map ϕ^{-1}. We create a partition in $\bigcup_{k\geq 0} \mathcal{P}_{B,k}$ from each shaded odd Ferrers diagram according to the following procedure:

Step 1. Let $\pi \in \mathcal{SF}$. For $j \geq 1$, let n_j be the size of the jth largest non-shaded row (not including the top row). Create λ_2, a partition into odd parts by taking all n_j such that $n_j < 2\overline{\nu}(\pi) + 2j + 1$. Let π^* be the rows of π that were not used to create λ_2.

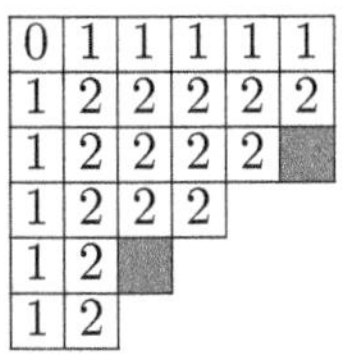

$\rightarrow \overline{\nu}(\pi) = 2,$
$n_1 = 11 \not< 7,$
$n_2 = 7 < 9,$
$n_3 = 3 < 11,$
$\lambda_2 = 7 + 3$

Step 2. For each $j \geq 1$, starting at 1, if row j of π^* is not shaded, create an even part by removing one cell containing a 2 from π^*_j and one cell containing a 2 from each row π^*_{j+m} with $\text{size}(\pi^*_{j+m}) > 2(\nu - m)$, where $\nu = \nu(\pi^*)$ is the number of rows in π^*. Then, add a shaded box to the end of row j. Let λ^*_e be the collection of the even parts obtained in this step, noting that λ^*_e will have distinct parts.

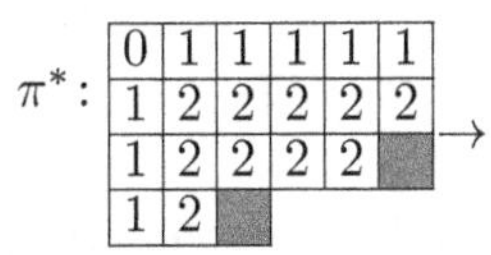

$\nu(\pi^*) = 3$
$\text{size}(\pi^*_2) = 9 >$
$4 = 2(3-1)$
$\text{size}(\pi^*_3) = 3 >$
$2 = 2(3-2)$
$\lambda^*_e = 6$

Step 3. Note that, after Step 2, the remaining diagram will have distinct parts. Thus, we can separate it into an $\nu \times \nu$ Durfee triangle odd Ferrers diagram (called τ) and the conjugate of a 2-modular diagram for a partition into odd parts of size $\leq 2\nu + 1$ (called λ_1).

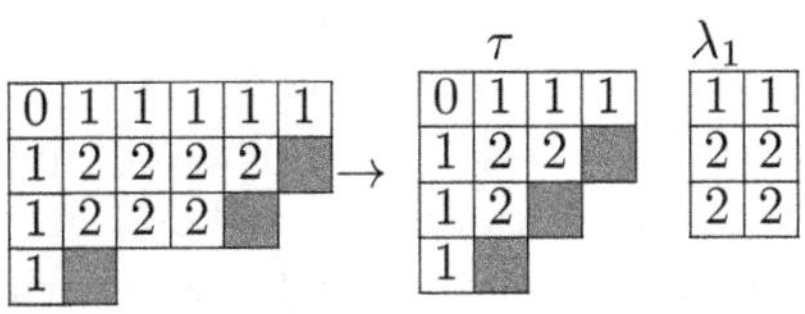

Step 4. From the triangle, create the even parts $2+4+\cdots+2\nu$. Inserting the parts of λ_e^*, the parts of λ_1 as parts of color 1, and the parts of λ_2 as parts of color 2, we obtain a partition in $\mathcal{P}_B$.

$$\begin{array}{l} \tau \to 6+4+2 \\ \lambda_1 \to 5_1+5_1 \\ \lambda_e^* = 6 \\ \lambda_2 \to 7_2+3_2 \end{array} \quad \to \quad \begin{array}{c} \phi^{-1}(\pi): \\ 7_2{+}6{+}6{+}5_1{+} \\ 5_1+4+3_2+2 \end{array}$$

To confirm that ϕ^{-1} is, in fact, the inverse of ϕ, we show how the threshold conditions are related. Let $\lambda \in \mathcal{P}_B$ and define $\pi = \phi(\lambda) \in \mathcal{SF}$. Note that any unshaded row in π that comes from an odd part of color 2 in λ must have size at most $2\nu_{d,e}(\lambda)+1$. Then, using (3.3), we can see that, once we remove the parts that came from an odd part of color 2, any remaining unshaded rows must satisfy $|\pi_{j+g(j)}| \geq 2(\nu_{d,e}(\lambda) - (j+g(j))) + 2g(j) + 2j + 1 = 2\nu_{d,e}(\lambda)+1$. Now, if we define $j_{\max}$ to be the number of even parts appearing twice in λ and if the jth unshaded part of π comes from a part of color 2 in λ, $j > j_{\max}$. Moreover, note that $\nu_{d,e}(\lambda) = \overline{\nu}(\pi) + j_{\max}$, where $j_{\max}$ is the number of even parts appearing twice in λ. In the language of Step 1 of ϕ^{-1}, we have $n_j \leq 2\overline{\nu}(\pi) + 2j_{\max} + 1 < 2\overline{\nu}(\pi) + 2j + 1$, as desired.

Next, we show how we obtain the threshold condition in Step 2 of ϕ^{-1}. Again, using (3.3) and the definition of $g(j)$, we can see that, for $1 \leq m \leq \frac{\tau_j}{2}$, $|\pi_{j+g(j)+m}| \geq 2(\nu_{d,e}(\lambda) - j - g(j) - m) + 1 + 2g(j) + 2j = 2\nu_{d,e}(\lambda) - 2m + 1$.

Note that our bijection takes a partition $\lambda \in \mathcal{P}_B$ with ν parts, ν_o odd parts, $\nu_{u,e}$ even parts appearing exactly once, $\nu_{d,e}$ different even part sizes, and ν_1 parts of color 1 to a shaded odd Ferrers diagram, π where π has $\nu_{u,e}$ shaded rows, $2k+\nu_o(\pi)$ 1's, and a top row of size $\nu_{d,e}+\nu_1$ Therefore, we have proved that

$$\sum_{\lambda \in \mathcal{P}_B} a^{\nu_o(\lambda)+2\nu_{d,e}(\lambda)} t^{\nu_{d,e}(\lambda)+\nu_1(\lambda)} z^{\nu_{u,e}(\lambda)} q^{|\lambda|} = \sum_{\pi \in \mathcal{SF}} a^{\mathrm{top}(\pi)+\nu(\pi)-1} t^{\mathrm{top}(\pi)} z^{\overline{\nu}(\pi)} q^{|\overline{\pi}|}.$$

Note that if we set $a = t = 1$ and $z = q$ in Theorem 3.1 and multiply both sides by q, we obtain (1.1). Furthermore, setting $a = t = z = 1$ in Theorem 3.1 results in (1.2).

4. Combinatorial Interpretations of Other Mock Theta Functions

In this section, we explore special cases of Theorem 3.1. We begin with a combinatorial interpretation of (1.1). Let $\mathcal{P}_{A,k}$ be the set of 3-colored partitions into odd parts, where the largest part is $2k+1$, all parts of color 3 and size $< 2k+1$ appear once or twice, and the part $(2k+1)_3$ appears exactly once. Let $\mathcal{B}$ be a set that generalizes odd Ferrers diagrams by taking shaded odd Ferrers diagrams, replacing the zero by a 1 and replacing the 1 in any shaded row with a 2.

Then, we can rewrite (1.1) as

$$\sum_{k\geq 0}\sum_{\lambda\in\mathcal{P}_{A,k}} q^{|\lambda|} = \sum_{\pi\in\mathcal{B}_2} q^{|\pi|}. \tag{4.1}$$

Note that, for $\lambda \in \mathcal{P}_{A,k}$, subtracting 1 from every part of color 3 and changing it to color 2 gives a partition $\lambda' \in \mathcal{P}_{B,k}$. Additionally, for $\pi \in \mathcal{B}$, subtracting 1 from every even part and replacing the 1 in the top left corner with a 0 results in a shaded odd Ferrers diagram $\pi' \in \mathcal{B}$. Therefore, the bijection described in Sec. also proves (4.1).

Next, we consider two third-order mock theta functions that appear as special cases of Theorem 3.1. As mentioned in [10], if we let $a = t = 1$ and let $z \to 0$, we obtain

$$\sum_{n=0}^{\infty} \frac{q^{2n(n+1)}}{(q;q^2)_{n+1}^2} = \sum_{n=0}^{\infty} \frac{q^n}{(q;q^2)_{n+1}}. \tag{4.2}$$

Note that this function is Watson's third-order mock theta function $\omega(q)$.

If, instead, we let $t = 1$, $a = -q$, and $z \to 0$, we obtain

$$\sum_{n=0}^{\infty} \frac{q^{2n(n+2)}}{(-q^2;q^2)_{n+1}^2} = \sum_{n=0}^{\infty} \frac{(-1)^n q^{2n}}{(-q^2;q^2)_{n+1}}.$$

If we multiply both sides by q^2 then take $q \to q^{1/2}$, we get

$$\sum_{n=1}^{\infty} \frac{q^{n^2}}{(-q;q)_n^2} = \sum_{n=1}^{\infty} \frac{(-1)^{n-1} q^n}{(-q;q)_n}. \tag{4.3}$$

Note that the left-hand side of (4.3) can be expressed in terms of Ramanujan's third-order mock theta function

$$f(q) := \sum_{n=0}^{\infty} \frac{q^{n^2}}{(-q;q)_n^2}.$$

Using these identities and the ideas presented in Sec. 3, we can obtain combinatorial interpretations of the coefficients of $\omega(q)$ and $f(q)$. From (4.2), we obtain the following interpretations for the coefficients of $\omega(q)$:

- The number of partitions where there are no skipped even parts, all even parts appear twice, and all odd parts are two colored.
- The number of odd Ferrers diagrams.

The second interpretation can also be found in [1]. Similarly, from (4.3), we note the following interpretations for the coefficients of $f(q)$:

- The number of two-colored partitions into odd parts, where all part sizes appear at least once with color 2, counted with weight $(-1)^{\text{\# parts}-\text{\# distinct part sizes}}$.
- The number of partitions counted with weight $(-1)^{\text{largest part}-\text{\# parts}}$.

A common interpretation of the left side of (4.3) is that it counts the number of partitions with even rank minus the number of partitions with odd rank, where the rank of a partition (as defined by Dyson in [8]) is the size of the largest part minus the number of parts. A fascinating result of our combinatorial interpretation is that, when traced through to this identity for $f(q)$, we get that the right side of (4.3) can also be interpreted as counting the number of partitions with even rank minus those with odd rank.

5. A Combinatorial Hypergeometric Series Transformation

We begin this section by noting that, in our proof of Theorem 3.1, we equated the coefficients of τ^0 on both sides of the identity

$$\sum_{n=0}^{\infty} \frac{(-a\tau q;q^2)_n(-azq;q^2)_n}{(aq;q^2)_{n+1}(a^2t\tau z;q^2)_{n+1}} a^n t^n q^n = \sum_{n=0}^{\infty} \frac{(-q^2/\tau;q^2)_n(-q^2/z;q^2)_n}{(aq;q^2)_{n+1}(atq;q^2)_{n+1}} a^{2n}t^n\tau^n z^n. \tag{5.1}$$

We can extend our combinatorial interpretation of Theorem 3.1 to give a new combinatorial interpretation of (5.1). Let $\mathcal{T}_1$ be the set of triples $(k, \overline{\pi}_1, \pi_2)$ where $k \in \mathbb{Z}_{\geq 0}$, $\overline{\pi}_1$ is an overpartition into odd parts of size $\leq 2k+1$ with all overlined parts $\leq 2k-1$, and π_2 is a partition into parts of size $\leq 2k$ with all odd parts distinct and parts of size 0 allowed. Let $\mathcal{T}_2$ be the set of triples $(m, \lambda_1, \lambda_2)$ where $m \in \mathbb{Z}_{\geq 0}$, λ_1 is a 2-colored partition into even parts of size $\leq 2m$ such that parts within each color are distinct, and λ_2 is a 2-colored partition into odd parts of size $\leq 2m+1$. Then (5.1) is equivalent to the following combinatorial identity.

Theorem 5.1.

$$\sum_{(k,\overline{\pi}_1,\pi_2)\in\mathcal{T}_1} a^{k+\nu(\overline{\pi}_1)+2\nu_e(\pi_2)+\nu_o(\pi_2)} t^{k+\nu_e(\pi_2)} z^{\overline{\nu}(\overline{\pi}_1)+\nu_e(\pi_2)} q^{k+|\overline{\pi}_1|+|\pi_2|} \tau^{\nu(\pi_2)}$$

$$= \sum_{(m,\lambda_1,\lambda_2)\in\mathcal{T}_2} a^{2m+\nu(\lambda_2)} t^{m+\nu_1(\lambda_2)} z^{m-\nu_2(\lambda_1)} \tau^{m-\nu_1(\lambda_1)} q^{m+|\lambda_1|+|\lambda_2|}.$$

A direct bijective proof of this identity would be a welcome addition to the q-hypergeometric series literature

References

[1] G. E. Andrews, Integer partitions with even parts below odd parts and the mock theta functions, *Ann. Comb.* **22**(3) (2018) 433–445.

[2] G. E. Andrews and B. C. Berndt, *Ramanujan's Lost Notebook, Part V.* (Springer, New York, 2018).

[3] B. C. Berndt and R. A. Rankin, *Ramanujan: Letters and Commentary* (American Mathematical Society, Providence, RI, 1995).

[4] H. E. Burson, A bijective proof of a false theta function identity from Ramanujan's lost notebook, *Ann. Comb.* **23**(3) (2019) 579–588.

[5] S. Corteel and J. Lovejoy, Overpartitions, *Trans. Amer. Math. Soc.* **356** (2004) 1623–1635.

[6] NIST Digital Library of Mathematical Functions, eds. F. W. J. Olver, A. B. Olde Daalhuis, D. W. Lozier, B. I. Schneider, R. F. Boisvert, C. W. Clark, B. R. Miller and B. V. Saunders, http://dlmf.nist.gov/, Release 1.0.19 of 2018-06-22.

[7] W. Duke, Almost a century of answering the question: What is a mock theta function? *Not. Amer. Math. Soc.* **61**(11) (2014) 1314–1320.

[8] F. Dyson, Some guesses in the theory of partitions, *Eureka* (*Cambridge*) **8** (1944) 10–15.

[9] A. Folsom, What is a mock modular form? *Not. Amer. Math. Soc.* **57**(11) (2010) 1441–1443.

[10] R. J. McIntosh, Second order mock theta functions, *Canad. Math. Bull.* **50**(2) (2007) 284–290.

[11] K. Ono, The last words of a genius, *Not. Amer. Math. Soc.* **57**(11) (2010) 1410–1419.

[12] G. N. Watson, The final problem: An account of the mock theta functions, *J. London Math. Soc.* **s1-11**(1) (1936) 55–80.

[13] S. Zwegers, *Mock theta functions*, Ph.D. thesis, Universiteit Utrecht (2002).

© 2024 World Scientific Publishing Company
https://doi.org/10.1142/9789811277375_0006

Chasing after cancellations: Revisiting a classic identity that implies the Rogers–Ramanujan identities*

Hei-Chi Chan

Mathematical Sciences Department
University of Illinois at Springfield, WUIS Building
Springfield, Illinois 62703-5407, USA
chan.hei-chi@uis.edu

In this paper, we revisit an identity which was proven by Ramanujan and from which he deduced the famous identities that are named after him and L. J. Rogers. Unlike Ramanujan's proof (which uses the method of q-difference equations), we examine directly the q-coefficients involved. We isolate and identify terms that cancel each other. Once these terms are paired up and canceled, we only need the geometric series to complete the proof.

Keywords: q-series; Rogers–Ramanujan identities.

Mathematics Subject Classification 2020: 05A15, 05A30

1. Introduction and the Main Result

The celebrated Rogers–Ramanujan identities are given by

$$\sum_{k=0}^{\infty} \frac{q^{k^2}}{(q)_k} = \frac{1}{(q;q^5)_\infty (q^4;q^5)_\infty}, \tag{1.1}$$

$$\sum_{k=0}^{\infty} \frac{q^{k^2+k}}{(q)_k} = \frac{1}{(q^2;q^5)_\infty (q^3;q^5)_\infty}. \tag{1.2}$$

Here, we have used standard notation for various symbols (such as $(q)_r$) — see Sec. 2 for details. The first published proof goes back to Rogers in 1894 [12, pp. 328–329, Formulas (1) and (2)]. Ramanujan discovered it without proof some time before 1913. See, for example, *Ramanujan's Notebooks*, Part III [3], Entries 38 (i) and (ii), for an excellent account of the history behind them. The research inspired by these identities is still going strong today. For more information, see [14], a recent and excellent monograph on the subject. It is a great time to study these great identities!

*This article originally appeared in *International Journal of Number Theory*, **17**(2), pp. 297–310 © 2021 World Scientific Publishing Company. DOI: 10.1142/S1793042120400266

In this paper, we revisit the following identity: For a being a variable and $|q| < 1$, the following is true:

$$\sum_{k=0}^{\infty} \frac{q^{k^2}}{(q)_k} a^k = \sum_{r=0}^{\infty} (-1)^r a^{2r} q^{r(5r-1)/2} (1 - aq^{2r}) \frac{1}{(q)_r (aq^r; q)_\infty}. \tag{1.3}$$

There is no doubt this is a classic result: It was proven by the Maestro himself in [13].[a] From (1.3) Ramanujan deduced (1.1) and (1.2). For an elegant account of the proof, see [10, Sec. 15.2].

Ramanujan's proof of (1.3) involves establishing and solving a q-difference equation, and he accomplishes all these in less than three pages. His proof (and that of Rogers) is a superb example of elegance and brilliance in the field of q-series.

Ramanujan's original proof, or versions of it (e.g., the one in Hardy and Wright [9]; see also [1, 2, 5, 10, 11, 14]), is very likely the first proof which beginning students encounter when they study the Rogers–Ramanujan identities.

The left-hand side of (1.3) is already a power series in a. So, it is natural to ask, what is the coefficient of a^k on the right-hand side?

Let us extract the coefficients of a^k from both sides of (1.3) — we outline the details in Appendix A. For $k \geq 1$, this process gives a polynomial identity:

$$\begin{aligned} q^{k^2} &= 1 + \sum_{r=1}^{\lfloor k/2 \rfloor} \begin{bmatrix} k-r \\ r \end{bmatrix} (-1)^r q^{r(r-1)/2+kr} (q^{k+1-r}; q)_r \\ &\quad + \sum_{r=1}^{\lfloor (k+1)/2 \rfloor} \begin{bmatrix} k-r \\ r-1 \end{bmatrix} (-1)^r q^{r(r-1)/2+k(r-1)} (q^{k+1-r}; q)_r, \end{aligned} \tag{1.4}$$

(note that we have multiplied $(q)_k$ to both sides.) This identity is not obvious at all. This should come as no surprise: This is the essence of (1.3) which implies the Rogers–Ramanujan identities, and none of the proofs of the Rogers–Ramanujan identities can be described as obvious.

When this identity is shown to beginning students in mathematics or even researchers in other math-intensive disciplines (such as physics or engineering), it is not uncommon for them to have questions along the following line of thought:

> Is there a way to prove (1.4) directly, which is independent of how Ramanujan proved it (i.e. by proving (1.3)), and, at the same time, elementary?

Usually audiences with such questions may be unaware of the more advanced and yet "elementary"[b] machineries (such as Bailey's Lemma; see [1, 2, 7, 11, 14] for a sample of such tools) that are designed to prove and generalize similar identities.

[a] Arranged by Hardy, [13] consists of two proofs of (1.1) and (1.2), one from Rogers and one from Ramanujan. Their proofs are essentially the same. Hardy included a note describing how the paper came about and the early proofs of the identities up to 1918.

[b] In the mathematical sense, which is not synonymous to *simple*.

Their questions are understandable: For many non-specialists, proving algebraic identities in an elementary way involves processes that are similar to proving, say, $4xy = (x+y)^2 - (x-y)^2$, in which one multiplies out the more complicated side and, after some manipulation, the other side emerges at the end.

While the above point of view and its reasoning may sound elementary (in the sense of beginner level), it does point to a relevant issue for this identity: The right-hand side of (1.4) is far more complicated than the left-hand side (which is just q^{k^2}). In order to reduce to the left, the expression on the right must have most of its terms canceled. In addition, the cancellations involved must be systematic, so that they work for all $k \geq 1$.

The purpose of this paper is to report a proof of (1.4) that underscores the cancellations involved: We start from its right-hand side (the more complicated side), then by rearranging terms and identifying terms that cancel each other, we are able to reduce it to the left-hand side (without using any difference equations). This is our key result.

To this end, let us first move the constant term 1 on the right-hand side of (1.4) to the left and divide both sides by $q^k - 1 = (-1)(1-q^k)$:

$$\begin{aligned} \frac{q^{k^2}-1}{q^k-1} &= (-1)\left\{\sum_{r=1}^{\lfloor k/2 \rfloor} \begin{bmatrix} k-r \\ r \end{bmatrix} (-1)^r q^{r(r-1)/2+kr}(q^{k+1-r};q)_{r-1} \right. \\ &\quad \left. + \sum_{r=1}^{\lfloor (k+1)/2 \rfloor} \begin{bmatrix} k-r \\ r-1 \end{bmatrix} (-1)^r q^{r(r-1)/2+k(r-1)}(q^{k+1-r};q)_{r-1} \right\} \\ &:= (-1)\left(\sum_{r=1}^{\lfloor k/2 \rfloor} \alpha(k)_r + \sum_{r=1}^{\lfloor (k+1)/2 \rfloor} \beta(k)_r\right). \end{aligned} \tag{1.5}$$

Note that $\alpha(k)_r$ (respectively, $\beta(k)_r$) is defined to be the rth term in the first sum (respectively, second sum) in the first equality. Their definitions are also stated in Sec. 2 — see (2.4) and (2.8).

The left-hand side of (1.5) suggests a connection to the geometric series. Indeed this is precisely what happens: We will see that the right-hand side of (1.5) — despite its complicated form — can be decomposed into terms in a systematic way, and many of these terms cancel each other. The remaining terms will give $\sum_m q^{mk}$, the geometric series that gives the left-hand side.

Let us focus on the nontrivial cases with $k \geq 2$. To move on, we need the following results (they will be proven in Sec. 3). First, a way to decompose $\alpha(k)_r$ and $\beta(k)_r$.

Theorem 1.1 (The Decomposition Theorem). *For $k \geq 2$ and $1 \leq r \leq \lfloor k/2 \rfloor$, the following is true:*

$$\alpha(k)_r = \alpha^+(k)_r + \alpha^-(k)_r - q^{(2r-1)k}. \tag{1.6}$$

For $k \geq 2$ and $1 \leq r \leq \lfloor (k+1)/2 \rfloor$, the following is true:

$$\beta(k)_r = \beta^+(k)_r + \beta^-(k)_r - q^{(2r-2)k}. \tag{1.7}$$

The definitions of $\alpha^\pm(k)_r$ and $\beta^\pm(k)_r$ can be found in Sec. 2; see (2.5) and (2.9).

Note that the last terms on the right-hand side in both equations (i.e. the power of q) are exactly the terms which will be combined to give the left-hand side of (1.5) through the geometric series. This also means, the rest of the terms ($\alpha^\pm(k)_r$ and $\beta^\pm(k)_r$ for each r) will be canceled. Indeed, we have the following theorem.

Theorem 1.2 (The Cancellation Theorem). *For $k \geq 2$ and $r \geq 1$, the followings are true:*

$$\beta^-(k)_r + \alpha^-(k)_r = 0, \tag{1.8}$$

$$\alpha^+(k)_r + \beta^+(k)_{r+1} = 0. \tag{1.9}$$

We also need the following to tie up the loose ends.

Theorem 1.3 (The Head & Tail Property). *For $k \geq 2$ the followings are true:*

(A) $\beta^+(k)_1 = 0$.
(B) *If k is odd, let $l = (k+1)/2$, then $\beta^-(k)_l = 0$.*
(C) *If k is even, let $l = k/2$, then $\alpha^+(k)_l = 0$.*

Assuming these theorems for the time being, let us prove (1.5) — and we can do it in several lines.

Starting with the right-hand side of (1.5), we have

$$(-1)\left(\sum_{r\geq 1}{}' \alpha(k)_r + \sum_{r\geq 1}{}'' \beta(k)_r\right)$$

$$\overset{(E1)}{=} (-1)\left(\sum_{r\geq 1}{}' \alpha^+(k)_r + \alpha^-(k)_r - q^{(2r-1)k}\right.$$

$$\left. + \sum_{r\geq 1}{}'' \beta^+(k)_r + \beta^-(k)_r - q^{(2r-2)k}\right)$$

$$= (-1)^2\left(\sum_{r\geq 1}{}' q^{(2r-1)k} + \sum_{r\geq 1}{}'' q^{(2r-2)k}\right)$$

$$-\left(\sum_{r\geq 1}{}' \alpha^+(k)_r + \alpha^-(k)_r + \sum_{r\geq 1}{}'' \beta^+(k)_r + \beta^-(k)_r\right)$$

$$= \sum_{m=0}^{k-1} q^{mk} - \{\beta^+(k)_1 + (\beta^-(k)_1 + \alpha^-(k)_1) + (\alpha^+(k)_1 + \beta^+(k)_2)$$

$$+ (\beta^-(k)_2 + \alpha^-(k)_2) + \cdots\} \overset{(E2)}{=} \frac{q^{k^2} - 1}{q^k - 1},$$

which is the left-hand side of (1.5). Here are the explanations for equalities:

(E1) On both sides of the equality, $\sum'$ denote the sum over r from 1 to $\lfloor k/2 \rfloor$. $\sum''$ denote the sum over r from 1 to $\lfloor (k+1)/2 \rfloor$. To obtain the right-hand side, we have used Theorem 1.1.

(E2) We have used Theorem 1.3(A) to set $\beta^+(k)_1$ to zero, and Theorem 1.2 to cancel all pairs. Regarding the ending ("tail") of the sum:

- ◇ If k is odd, write $k = 2l - 1$ and the tail is given by

$$\cdots + \underbrace{(\alpha^+(k)_{l-1} + \beta^+(k)_l)}_{\text{the last pair (= 0 by Theorem 1.2)}} + \underbrace{\beta^-(k)_l}_{= 0 \text{ by Theorem 1.3(B)}}$$

- ◇ If k is even, write $k = 2l$ and the tail is given by

$$\cdots + \underbrace{(\beta^-(k)_l + \alpha^-(k)_l)}_{\text{the last pair (= 0 by Theorem 1.2)}} + \underbrace{\alpha^+(k)_l}_{= 0 \text{ by Theorem 1.3(C)}} .$$

What remains is to complete the proof of the three theorems — see Sec. 3. Before that we list all major definitions in Sec. 2. Conclusion is in Sec. 4.

2. Notations and Definitions

First, we have several standard notations. The q-Pochhammer symbol is defined by

$$(a;q)_n := \prod_{k=0}^{n-1}(1 - aq^k) = (1-a)(1-aq)\cdots(1-aq^{n-1}) \quad \text{if } n \geq 1$$

and $(a;q)_0 := 1$, with a frequently used shorthand notation $(q)_n := (q;q)_n$.

The Gaussian polynomials (or the q-binomial numbers) are defined as

$$\begin{bmatrix} A \\ B \end{bmatrix} = \begin{cases} \dfrac{(1-q^A)(1-q^{A-1})\cdots(1-q^{A-B+1})}{(1-q^B)(1-q^{B-1})\cdots(1-q)} = \dfrac{(q)_A}{(q)_B\,(q)_{A-B}} & \text{if } 0 \leq B \leq A, \\ 0 & \text{otherwise.} \end{cases}$$

A useful property of the Gaussian polynomials is the following: For $n \geq 1$,

$$\begin{bmatrix} n \\ m \end{bmatrix} = \begin{bmatrix} n-1 \\ m \end{bmatrix} q^m + \begin{bmatrix} n-1 \\ m-1 \end{bmatrix}. \tag{2.1}$$

See, e.g., [1, 2, 4, 5, 7, 8, 10, 11, 14]. An equivalent version can be obtained by applying the property

$$\begin{bmatrix} n \\ m \end{bmatrix} = \begin{bmatrix} n \\ n-m \end{bmatrix}$$

to the right-hand side of (2.1).

Next, we define some functions that are used extensively in this paper. They may look complicated. The fact is, these definitions emerge naturally in the course of the proofs.

First, we have the following multiplicative factors (used mainly to mask out the loaded exponents):

$$\gamma(k)_{rs} := (-1)^{r+s} q^{r(r-1)/2+kr+s(s-1)/2+(k+1-r)s}, \tag{2.2}$$

$$\delta(k)_{rs} := \gamma(k)_{rs} q^{-k}. \tag{2.3}$$

For each of the functions defined below, its *admissible domain* for its arguments is stated on the right. Whenever one or more arguments lie outside the stated admissible domain, the function is defined to be zero. Also, if one or more intervals stated in an admissible domain become inconsistent (e.g., for $k = 2$, the interval $1 \leq r \leq \lfloor (k-1)/2 \rfloor$ becomes $1 \leq r \leq 0$), the function (associated to the domain) is defined to be zero.

We will define the α-related functions first, then the β-related ones. All these functions are defined for $k \geq 1$.

$$\alpha(k)_r := (-1)^r q^{r(r-1)/2+kr} \begin{bmatrix} k-r \\ r \end{bmatrix} (q^{k+1-r}; q)_{r-1} \quad 1 \leq r \leq \lfloor k/2 \rfloor. \tag{2.4}$$

This function first appeared in (1.5): It is the rth term in the first sum in the first equality.

The functions $\alpha^{\pm}(k)_r$ in Theorem 1.2 are defined by the following double sum:

$$\alpha^{\pm}(k)_r := \sum_{s=0}^{r-1} \sum_{j=1}^{s+1} \alpha^{\pm}(k)_{rsj} \quad 1 \leq r \leq \lfloor k/2 \rfloor, \tag{2.5}$$

where

$$\alpha^{+}(k)_{rsj} := \gamma(k)_{rs} \begin{bmatrix} k-r-j \\ r-j+1 \end{bmatrix} \begin{bmatrix} r-j \\ s-j+1 \end{bmatrix} q^{r-j+1} \quad \begin{array}{l} 1 \leq r \leq \lfloor (k-1)/2 \rfloor, \\ 0 \leq s \leq r-1, \\ 1 \leq j \leq s+1, \end{array} \tag{2.6}$$

$$\alpha^{-}(k)_{rsj} := \gamma(k)_{rs} \begin{bmatrix} k-r-j \\ r-j \end{bmatrix} \begin{bmatrix} r-j-1 \\ s-j+1 \end{bmatrix} q^{s-j+1} \quad \begin{array}{l} 1 \leq r \leq \lfloor k/2 \rfloor, \\ 0 \leq s \leq r-2, \\ 1 \leq j \leq s+1. \end{array} \tag{2.7}$$

Now the β-related functions:

$$\beta(k)_r := (-1)^r q^{r(r-1)/2+k(r-1)} \begin{bmatrix} k-r \\ r-1 \end{bmatrix} (q^{k+1-r}; q)_{r-1} \quad 1 \leq r \leq \lfloor (k+1)/2 \rfloor. \tag{2.8}$$

This function first appeared in (1.5): It is the rth term in the second sum in the first equality.

The functions $\beta^{\pm}(k)_r$ in Theorem 1.2 are defined by the following double sum:

$$\beta^{\pm}(k)_r := \sum_{s=0}^{r-1}\sum_{j=1}^{s+1} \beta^{\pm}(k)_{rsj} \quad 1 \le r \le \lfloor (k+1)/2 \rfloor, \tag{2.9}$$

where

$$\beta^{+}(k)_{rsj} := \delta(k)_{rs} \begin{bmatrix} k-r-j+1 \\ r-j \end{bmatrix} \begin{bmatrix} r-j-1 \\ s-j+1 \end{bmatrix} q^{s-j+1} \quad \begin{array}{l} 2 \le r \le \lfloor (k+1)/2 \rfloor, \\ 0 \le s \le r-2, \\ 1 \le j \le s+1, \end{array} \tag{2.10}$$

$$\beta^{-}(k)_{rsj} := \delta(k)_{rs} \begin{bmatrix} k-r-j \\ r-j \end{bmatrix} \begin{bmatrix} r-j-1 \\ s-j \end{bmatrix} q^{r-j} \quad \begin{array}{l} 1 \le r \le \lfloor k/2 \rfloor, \\ 0 \le s \le r-1, \\ 1 \le j \le s. \end{array} \tag{2.11}$$

With the above understood, we are ready for the proof of the three main theorems.

3. Proof of the Main Theorems

3.1. *Proof of Theorem* 1.1

We need the following lemma.

Lemma 3.1. *For* $n, r \ge 1$, *we have*

$$\begin{aligned} \begin{bmatrix} n \\ m \end{bmatrix} \begin{bmatrix} r \\ s \end{bmatrix} &\overset{(A)}{=} \begin{bmatrix} n-1 \\ m \end{bmatrix} \begin{bmatrix} r \\ s \end{bmatrix} q^m + \begin{bmatrix} n-1 \\ m-1 \end{bmatrix} \begin{bmatrix} r-1 \\ s \end{bmatrix} q^s + \begin{bmatrix} n-1 \\ m-1 \end{bmatrix} \begin{bmatrix} r-1 \\ s-1 \end{bmatrix}, \\ &\overset{(B)}{=} \begin{bmatrix} n \\ m \end{bmatrix} \begin{bmatrix} r-1 \\ s \end{bmatrix} q^s + \begin{bmatrix} n-1 \\ m \end{bmatrix} \begin{bmatrix} r-1 \\ s-1 \end{bmatrix} q^m + \begin{bmatrix} n-1 \\ m-1 \end{bmatrix} \begin{bmatrix} r-1 \\ s-1 \end{bmatrix}. \end{aligned}$$

Note that the last term on the right is the same bi-Gaussian polynomial on the left, with each of its arguments decreases by 1 unit (e.g., $n \to n-1$). As we will see, this feature enables us to derive recursive algorithms to prove Theorem 1.1.

Proof of Lemma 3.1. To prove (A), we apply (2.1) to the bi-Gaussian polynomial on the left with a small twist.

Let G and H be two Gaussian polynomials. Let us write $G = g_1 + g_2$ in accordance to (2.1), where g_1 (respectively, g_2) represents the first (respectively, second) term on the right-hand side of (2.1). Similarly, write $H = h_1 + h_2$ by using the same convention. We have

$$GH = (g_1 + g_2)(h_1 + h_2) = g_1h_1 + g_1h_2 + g_2h_1 + g_2h_2.$$

Note that this has four terms (but (A) only have three terms). Instead, we combine the first two terms (using $H = h_1 + h_2$) and obtain

$$GH = g_1H + g_2h_1 + g_2h_2.$$

Now, we have three terms. Applying this to the left-hand side of (A) gives the desired result. Note that the condition $n, r \geq 1$ allows us to apply (2.1).

For (B), we use (A) but interchange n with r and m with s. □

With specific choices for arguments, Lemma 3.1 implies the following identities (denoted as (A′) and (B′)) which will be used repeatedly in the rest of this section: with $k - r - i \geq 1$ and $r - i - 1 \geq 1$, we have

$$\gamma(k)_{rs}\begin{bmatrix} k-r-i \\ r-i \end{bmatrix}\begin{bmatrix} r-i-1 \\ s-i \end{bmatrix}$$
$$\overset{(A')}{=} \alpha^+(k)_{r,s,i+1} + \alpha^-(k)_{r,s,i+1} + \gamma(k)_{rs}\begin{bmatrix} k-r-i-1 \\ r-i-1 \end{bmatrix}\begin{bmatrix} r-i-2 \\ s-i-1 \end{bmatrix},$$
$$\delta(k)_{rs}\begin{bmatrix} k-r-i \\ r-i-1 \end{bmatrix}\begin{bmatrix} r-i-1 \\ s-i \end{bmatrix}$$
$$\overset{(B')}{=} \beta^+(k)_{r,s,i+1} + \beta^-(k)_{r,s,i+1} + \delta(k)_{rs}\begin{bmatrix} k-r-i-1 \\ r-i-2 \end{bmatrix}\begin{bmatrix} r-i-2 \\ s-i-1 \end{bmatrix},$$

(note that (A′) follows from Lemma 3.1(A) and (B′) from Lemma 3.1(B)).

Now, the proof of Theorem 1.1.

Proof of (1.6). First, we apply the q-binomial theorem [1, 2, 4, 5, 7, 8, 10, 11, 14] to the factor $(q^{k+1-r}; q)_{r-1}$ in $\alpha(k)_r$. This gives

$$\alpha(k)_r = \sum_{s=0}^{r-1} \gamma(k)_{rs}\begin{bmatrix} k-r \\ r \end{bmatrix}\begin{bmatrix} r-1 \\ s \end{bmatrix}. \tag{3.1}$$

Next, we prove the following identity — it shows how the sth term in the sum above decomposes into $\alpha^{\pm}(k)_{rsj}$:

$$\gamma(k)_{rs}\begin{bmatrix} k-r \\ r \end{bmatrix}\begin{bmatrix} r-1 \\ s \end{bmatrix}$$
$$= \sum_{j=1}^{s+1} \alpha^+(k)_{rsj} + \sum_{j=1}^{s+1} \alpha^-(k)_{rsj} - \delta_{s,r-1}q^{(2r-1)k}, \tag{3.2}$$

where $\delta_{\cdot,\cdot}$ in the last term is the Kronecker delta function.

The proof of (3.2) goes as follows. Take the sth term in (3.1). Apply (A′) to the bi-Gaussian polynomial recursively for a total of s times:

$$\begin{aligned}
&\gamma(k)_{rs} \underbrace{\begin{bmatrix} k-r \\ r \end{bmatrix} \begin{bmatrix} r-1 \\ s \end{bmatrix}}_{\text{use (A}') \text{ with } i=0} \\
&= (\alpha^+(k)_{rs1} + \alpha^-(k)_{rs1}) + \gamma(k)_{rs} \underbrace{\begin{bmatrix} k-r-1 \\ r-1 \end{bmatrix} \begin{bmatrix} r-2 \\ s-1 \end{bmatrix}}_{\text{use (A}') \text{ with } i=1} \\
&= (\alpha^+(k)_{rs1} + \alpha^-(k)_{rs1}) + (\alpha^+(k)_{rs2} + \alpha^-(k)_{rs2}) \\
&\quad + \gamma(k)_{rs} \underbrace{\begin{bmatrix} k-r-2 \\ r-2 \end{bmatrix} \begin{bmatrix} r-3 \\ s-2 \end{bmatrix}}_{\text{use (A}') \text{ with } i=2} \\
&\quad \vdots \qquad \vdots \\
&= \sum_{j=1}^{s} \alpha^+(k)_{rsj} + \sum_{j=1}^{s} \alpha^-(k)_{rsj} + \gamma(k)_{rs} \begin{bmatrix} k-r-s \\ r-s \end{bmatrix} \underbrace{\begin{bmatrix} r-s-1 \\ 0 \end{bmatrix}}_{=1}. \qquad (3.3)
\end{aligned}$$

To analyze the last term in (3.3), we consider two cases: $s < r-1$ and $s = r-1$.

Case 1: $s < r-1$.

Apply (2.1) to the remaining Gaussian polynomial in the last term in (3.3). This gives

$$\gamma(k)_{rs} \begin{bmatrix} k-r-s \\ r-s \end{bmatrix} = \alpha^+(k)_{r,s,s+1} + \alpha^-(k)_{r,s,s+1}.$$

This identity and (3.3) give (3.2) for the case of $s < r-1$.

Case 2: $s = r-1$.

Like the previous case, we apply (2.1) to the remaining Gaussian polynomial in the last term in (3.3). Using the fact that $s = r-1$, we have

$$\begin{aligned}
\gamma(k)_{r,r-1} \begin{bmatrix} k-2r+1 \\ 1 \end{bmatrix} &= \underbrace{\gamma(k)_{r,r-1} \begin{bmatrix} k-2r \\ 1 \end{bmatrix} q^1}_{=\alpha^+(k)_{r,r-1,r}} + \gamma(k)_{r,r-1} \underbrace{\begin{bmatrix} k-2r \\ 0 \end{bmatrix}}_{=1} \\
&= \alpha^+(k)_{r,r-1,r} - q^{(2r-1)k} \\
&= \alpha^+(k)_{r,r-1,r} + \alpha^-(k)_{r,r-1,r} - q^{(2r-1)k} \qquad (3.4)
\end{aligned}$$

(note: $\alpha^-(k)_{r,r-1,r} = 0$). Therefore, (3.3) and (3.4) imply (3.2) for the case of $s = r-1$. This completes the proof of (3.2).

Now, we can finish the proof of (1.6):

$$\begin{aligned}\alpha(k)_r &= \sum_{s=0}^{r-1} \gamma(k)_{rs} \begin{bmatrix} k-r \\ r \end{bmatrix} \begin{bmatrix} r-1 \\ s \end{bmatrix} \\ &= \sum_{s=0}^{r-1}\sum_{j=1}^{s+1} \alpha^+(k)_{rsj} + \sum_{s=0}^{r-1}\sum_{j=1}^{s+1} \alpha^-(k)_{rsj} - \sum_{s=0}^{r-1} \delta_{s,r-1} q^{(2r-1)k} \\ &= \alpha^+(k)_r + \alpha^-(k)_r - q^{(2r-1)k}.\end{aligned}$$

The first equality is (3.1). To obtain the second equality, we have used (3.2). For the last equality, we note that the two double sums are $\alpha^{\pm}(k)_r$, respectively. □

Proof of (1.7). The proof is similar to that of (1.6) and we will be brief.

First, we use the q-binomial theorem to expand $\beta(k)_r$.

$$\beta(k)_r = \sum_{s=0}^{r-1} \delta(k)_{rs} \begin{bmatrix} k-r \\ r-1 \end{bmatrix} \begin{bmatrix} r-1 \\ s \end{bmatrix}. \tag{3.5}$$

Next, we need the following identity:

$$\begin{aligned}&\delta(k)_{rs} \begin{bmatrix} k-r \\ r-1 \end{bmatrix} \begin{bmatrix} r-1 \\ s \end{bmatrix} \\ &\quad = \sum_{j=1}^{s+1} \beta^+(k)_{rsj} + \sum_{j=1}^{s+1} \beta^-(k)_{rsj} - \delta_{s,r-1}\, q^{2(r-1)k}.\end{aligned} \tag{3.6}$$

The proof is similar to that of (3.2). We apply (B′) (not (A′)) recursively for s times and obtain

$$\begin{aligned}\delta(k)_{rs} \begin{bmatrix} k-r \\ r-1 \end{bmatrix} \begin{bmatrix} r-1 \\ s \end{bmatrix} &= \sum_{j=1}^{s} \beta^+(k)_{rsj} + \sum_{j=1}^{s} \beta^-(k)_{rsj} \\ &\quad + \delta(k)_{rs} \begin{bmatrix} k-r-s \\ r-s-1 \end{bmatrix} \underbrace{\begin{bmatrix} r-s-1 \\ 0 \end{bmatrix}}_{=1}.\end{aligned} \tag{3.7}$$

To analyze the last term, again, we consider two cases:

Case 1: $s < r-1$.

Here, we have

$$\begin{aligned}\delta(k)_{rs} \begin{bmatrix} k-r-s \\ r-s-1 \end{bmatrix} &= \beta^+(k)_{r,s,s+1} \\ &= \beta^+(k)_{r,s,s+1} + \beta^-(k)_{r,s,s+1}\end{aligned} \tag{3.8}$$

(note: $\beta^-(k)_{r,s,s+1} = 0$). Hence (3.7) and (3.8) imply (3.6) with $s < r-1$.

Case 2: $s = r - 1$.

In this case, the last term in (3.7) reads

$$\delta(k)_{r,r-1} \underbrace{\begin{bmatrix} k - 2r + 1 \\ 0 \end{bmatrix}}_{=1} = -q^{2(r-1)k}$$

$$= \beta^+(k)_{r,r-1,r} + \beta^-(k)_{r,r-1,r} - q^{2(r-1)k} \tag{3.9}$$

(note: $\beta^{\pm}(k)_{r,r-1,r} = 0$). Hence (3.7) and (3.9) imply (3.6) with $s = r - 1$. This completes the proof of (3.6).

Finally, the proof of (1.7) goes exactly like the proof of (1.6): We substitute (3.6) into (3.5) and perform the sum. □

3.2. *Proof of Theorem* 1.2

We first prove a more refined cancellation result.

Lemma 3.2.

$$\alpha^+(k)_{rsj} + \beta^+(k)_{r+1,s,j} = 0, \tag{3.10}$$

$$\alpha^-(k)_{rsj} + \beta^-(k)_{r,s+1,j} = 0. \tag{3.11}$$

Proof of Lemma 3.2. For (3.10): We first note that both functions in (3.10) have the same admissible domain. Hence if one function is zero due to its arguments lying outside the admissible domain, the other function will also be zero for the same reason. The identity is then trivially satisfied.

Otherwise both functions are given by their definitions (see (2.6) and (2.10)), which immediately imply (3.10) — one simply needs to check the definitions involved.

Equation (3.11) can be proven by similar arguments. □

We now turn to the proof of Theorem 1.2. Carrying out the double sum on (3.11) gives

$$\sum_{s=0}^{r-1}\sum_{j=1}^{s+1} \alpha^-(k)_{rsj} + \underbrace{\sum_{s=0}^{r-1}\sum_{j=1}^{s+1} \beta^-(k)_{r,s+1,j}}_{:=Z^-(k)_r} = 0. \tag{3.12}$$

The first double sum is $\alpha^-(k)_r$. The second double sum can be written as

$$Z^-(k)_r = \beta^-(k)_r + \sum_{j=1}^{r}(\beta^-(k)_{rrj} - \beta^-(k)_{r,j-1,j}) = \beta^-(k)_r. \tag{3.13}$$

For the second equality, we note that all terms in the sum are zero (their arguments lie outside the admissible domain): $\beta^-(k)_{rrj} = 0$ because $s = r > r - 1$;

$\beta^-(k)_{r,j-1,j} = 0$ because $j > s = j-1$. Hence (3.12) and (3.13) imply (1.8), the first identity of Theorem 1.2.

The other identity, (1.9), can be proven by the same method. This completes the proof of Theorem 1.2.

3.3. *Proof of Theorem* 1.3

- For Part (A), we note that

$$\beta^+(k)_1 = \beta^+(k)_{101} = 0.$$

The first equality follows from the Definition (2.9). But $\beta^+(k)_{101} = 0$ because its arguments lie outside the admissible domain.

- For Part (B), since k is odd, write $l = (k+1)/2$, we have

$$\beta^-(k)_l = \sum_{s=0}^{l-1}\sum_{j=1}^{s+1} \beta^-(k)_{lsj} = 0.$$

Again, the first equality follows from the definition (2.9). And $\beta^-(k)_{lsj} = 0$ because $l = (k+1)/2 > \lfloor k/2 \rfloor$.

- For Part (C), since k is even, write $l = k/2$, we have

$$\alpha^+(k)_l = \sum_{s=0}^{l-1}\sum_{j=1}^{s+1} \alpha^+(k)_{lsj} = 0.$$

The first equality follows from the definition (2.5). And $\alpha^+(k)_{lsj} = 0$ because $l = k/2 > \lfloor (k-1)/2 \rfloor$.

This completes the proof of Theorem 1.3.

4. Conclusion

In this paper, we prove the polynomial identity (1.4). This, in turns, gives a new proof of (1.3) (by reversing the process described in Appendix A), from which Ramanujan derived the famous identities that named after him and Rogers. The aim of this work is neither to look for short or elegant proofs, nor to develop flexible methods that can produce board generalizations. This project has been guided by the questions and context discussed in Sec. 1.

The key take-away of this paper is the derivation at the end of Sec. 1. We are able to decompose various terms and identify how they cancel each others. At the end of the day, we pull a rabbit out of a hat through the geometric series.

Given the complexity of $\alpha(k)_r$ and $\beta(k)_r$, it is a bit surprising to see that only the powers of q (the last terms in (1.6) and (1.7)) survive the cancellation at the end–in fact, only one from each $\alpha(k)_r$ and one from each $\beta(k)_r$. All other parts ($\alpha^\pm(k)_r$ and $\beta^\pm(k)_r$) cancel each other (see Lemma 3.2 and Theorem 1.2).

In order to isolate the needed powers of q, we use recursively algorithms to decompose the bi-Gaussian polynomials involved (as in the proof of (3.2) and (3.5)).

The process is like peeling off "layers" of terms until we reach the specific powers of q.

We have mentioned several times that it is the geometric series that brings the proof across the finishing line. But this is not the same as saying that (1.4) (or (1.5)) is as simple as the geometric series. In fact, the essential features of (1.4) (or (1.5)) are hidden in the parts that are being canceled (i.e. $\alpha^{\pm}(k)_r$ and $\beta^{\pm}(k)_r$).

To explore the significance of these "hidden gems", so to speak, one would have to examine the combinatorial version of the proof of (1.4).[c] Such a combinatorial proof will shed light on the Decomposition Theorem (Theorem 1.1) and the Cancellation Theorem (Theorem 1.2).

One should not stop there: A combinatorial proof of (1.4) should lead to a combinatorial proof of (1.3). This may give a new combinatorial proof of the Rogers–Ramanujan identities (or shed light on existing ones).

Acknowledgments

We are deeply indebted to Professor Berndt. One of the lessons he has taught us through his prolific research output is that, it is absolutely worthwhile to study and understand the remarkable work of Ramanujan by as many different ways as one could find or discover. This is a main driving force behind this project. The reviewer's comments and suggestions are very helpful. They are greatly appreciated.

Happy Birthday Professor Berndt! Wishing you health, strength and many more fun-filled years!

Appendix A. Extracting the Coefficients of a^k of (1.3)

The main issue is the right-hand side of (1.3). To this end, we need an identity due to Euler:

$$\frac{1}{(t;q)_\infty} = \sum_{l\geq 0} \frac{t^l}{(q)_l}.$$

Substitute this (with $t = aq^r$) into the right-hand side of (1.3) and obtain (after distributing $(1 - aq^{2r})$)

$$\sum_{l,r\geq 0} (-1)^r \frac{a^{2r+l} q^{r(5r-1)/2+lr}}{(q)_r (q)_l} - \sum_{l,r\geq 0} (-1)^r \frac{a^{2r+l+1} q^{r(5r-1)/2+lr+2r}}{(q)_r (q)_l}. \tag{A.1}$$

For the first sum in (A.1), we define a new variable $k = 2r + l$. It becomes

$$\sum_{k\geq 0} \frac{a^k}{(q)_k} \sum_{r=0}^{\lfloor k/2\rfloor} \begin{bmatrix} k-r \\ r \end{bmatrix} (-1)^r q^{r(r-1)/2+kr} (q^{k+1-r};q)_r. \tag{A.2}$$

[c]The method of combinatorial telescoping, invented and investigated in [6], would be a promising approach.

Note that for $k = 0$, only (A.2) contributes (the following Eq. (A.3), does not contribute).

For the second sum in (A.1), we define a new variable $k = 2r + l + 1$. It becomes

$$\sum_{k\geq 1} \frac{a^k}{(q)_k} \sum_{r=1}^{\lfloor (k+1)/2 \rfloor} \begin{bmatrix} k-r \\ r-1 \end{bmatrix} (-1)^{r-1} q^{r(r-1)/2+k(r-1)} (q^{k+1-r}; q)_r, \tag{A.3}$$

(note that we have $(-1)^{r-1}$ instead of $(-1)^r$.) Hence

$$\text{The right-hand side of (1.3)} \; = \text{(A.2)} - \text{(A.3)},$$

from which extracting the coefficient of $a^k/(q)_k$ gives the right-hand side of (1.4).

References

[1] G. E. Andrews, *The Theory of Partitions* (Addison-Wesley, 1976).
[2] G. E. Andrews, R. Askey and R. Roy, *Special Functions* (Cambridge University Press, 1999).
[3] B. C. Berndt, *Ramanujan's Notebooks Part III* (Springer-Verlag, 1991).
[4] B. C. Berndt, *Number Theory in the Spirit of Ramanujan* (American Mathematical Society, 2004).
[5] H.-C. Chan, *An Invitation to q-Series* (World Scientific, 2011).
[6] W. Y. C. Chen, Q.-H. Hou and L. H. Sun, The method of combinatorial telescoping, *J. Combin. Theory Ser. A* **118** (2011) 899–907.
[7] W. Chu and L. Di Claudio, *Classical Partition Identities and Basic Hypergeometric Series* (Università deg;o Stido di Lecce, 2004).
[8] S. Cooper, *Ramanujan's Theta Functions* (Springer-Verlag, 2017).
[9] G. H. Hardy and E. M. Wright, *An Introduction to the Theory of Numbers*, 6th edn., eds. R. Heath-Brown, J. Silverman and A. Wiles (Oxford University Press, 2008).
[10] M. D. Hirschhorn, *The Power of q: A Personal Journey* (Springer-Verlag, 2017).
[11] J. McLaughlin, *Topics and Methods in q-Series* (World Scientific, 2017).
[12] L. J. Rogers, Second memoir on the expansion of certain infinite products, *Proc. London Math. Soc. Ser. 1* **25** (1894) 318–343.
[13] L. J. Rogers and S. Ramanujan, Proof of certain identities in combinatory analysis, *Cambridge Philos. Soc. Proc.* **19** (1919) 211–216.
[14] A. V. Sills, *An Invitation to the Rogers–Ramanujan Identities* (CRC Press, 2017).

© 2024 World Scientific Publishing Company
https://doi.org/10.1142/9789811277375_0007

A mock theta function identity related to the partition rank modulo 3 and 9*

Song Heng Chan†, Nankun Hong‡ and Jerry§

Division of Mathematical Sciences
School of Physical and Mathematical Sciences
Nanyang Technological University, 21 Nanyang Link
Singapore 637371, Republic of Singapore
†*chansh@ntu.edu.sg*
‡*s160001@e.ntu.edu.sg*
§*je0001ry@e.ntu.edu.sg*

Jeremy Lovejoy

CNRS, Université de Paris, Bâtiment Sophie Germain
Case courier 7014, 8 Place Aurélie Nemours
75205 Paris Cedex 13, France
lovejoy@math.cnrs.fr

For Bruce Berndt on the occasion of his 80th birthday

We prove a new mock theta function identity related to the partition rank modulo 3 and 9. As a consequence, we obtain the 3-dissection of the rank generating function modulo 9. We also evaluate all of the components of the rank–crank differences modulo 9. These are analogous to conjectures of Lewis [The generating functions of the rank and crank modulo 8, *Ramanujan J.* **18** (2009) 121–146] on rank–crank differences modulo 8, first proved by Mortenson [On ranks and cranks of partitions modulo 4 and 8, *J. Combin. Theory Ser. A* **161** (2019) 51–80].

Keywords: Mock theta functions; ranks of partition; ranks modulo 3, ranks modulo 9.

Mathematics Subject Classification 2020: 11P81, 11P83

1. Introduction

In a recent paper, Andrews *et al.* [2] gave new proofs of four identities for Ramanujan's third-order mock theta functions. These four identities were first proved by

*This article originally appeared in *International Journal of Number Theory*, **17**(2), pp. 311–327 © 2021 World Scientific Publishing Company. DOI: 10.1142/S1793042120400254
†Corresponding author.

Yesilyurt [24]. The most difficult identity to prove among the four is

$$\frac{1}{2}(1+e^{\pi i/4})\tilde{\phi}(iq)+\frac{1}{2}(1+e^{-\pi i/4})\tilde{\phi}(-iq)$$
$$= f_{\sqrt{2}}(q)+\frac{1}{\sqrt{2}}\psi(-q)(-q^2;q^4)_\infty\prod_{n=1}^{\infty}\frac{1}{1+\sqrt{2}q^n+q^{2n}}, \tag{1.1}$$

where

$$\tilde{\phi}(q):=\sum_{n=0}^{\infty}\frac{q^{n^2}}{(iq,q/i)_n},\quad \psi(q):=\sum_{n=0}^{\infty}q^{n(n+1)/2}=(-q;q)_\infty^2(q;q)_\infty$$

and for $a\in\mathbb{R}$

$$f_a(q):=\sum_{n=0}^{\infty}\frac{q^{n^2}}{(1+aq+q^2)(1+aq^2+q^4)\cdots(1+aq^n+q^{2n})}. \tag{1.2}$$

Here and in the rest of the paper, we use the standard notations

$$(x)_0:=(x;q)_0:=1,$$
$$(x)_n:=(x;q)_n:=\prod_{k=0}^{n-1}(1-xq^k),\quad (x)_\infty:=(x;q)_\infty:=\prod_{k=0}^{\infty}(1-xq^k),$$
$$(x_1,\ldots,x_m)_\infty:=(x_1,\ldots,x_m;q)_\infty:=(x_1;q)_\infty\cdots(x_m;q)_\infty,$$
$$J_a:=(q^a;q^a)_\infty,\quad J_{a,b}:=(q^a,q^{b-a},q^b;q^b)_\infty,\quad j(x;q):=(x,q/x,q;q)_\infty,$$

and we assume $|q|<1$ for convergence.

The q-series occurring in (1.1) and (1.2) are related to the generating function for partition ranks. Recall that Dyson [10] introduced and defined the *rank* of a partition as the largest part minus the number of parts. Let $N(m,n)$ denote the number of partitions of n with rank m. Then Dyson showed that the generating function for $N(m,n)$ is

$$G(z,q)=\sum_{m=-\infty}^{\infty}\sum_{n=0}^{\infty}N(m,n)q^nz^m=\sum_{n=0}^{\infty}\frac{q^{n^2}}{(zq)_n(q/z)_n},$$

and we see that f_a is a specialization of this generating function,

$$f_a(q)=G\left(\frac{-a\pm\sqrt{a^2-4}}{2},q\right).$$

If $|a|\le 2$ then $(-a\pm\sqrt{a^2-4})/2$ is a root of unity, and so $f_a(q)$ is a mock theta function for $-2<a\le 2$ by [25, Theorem 7.1]. Several of Ramanujan's third-order mock theta functions are instances of $f_a(q)$.

The new proof of (1.1) in [2] uses the 2-dissections of the rank generating function modulo 4 and 8, which are proved for the first time in [2]. A natural question that arises is whether there are other dissections of $G(z;q)$ that lead to analogous identities involving mock theta functions. This brings us to our first main result.

Theorem 1.1. *Let ζ be a primitive ninth root of unity and let $\omega = \zeta^3$. Define*

$$C(z,q) := \frac{(q)_\infty}{(zq, q/z)_\infty}, \tag{1.3}$$

and recall that by (1.2),

$$f_1(q) = \sum_{n=0}^{\infty} \frac{q^{n^2}}{(\omega q, q/\omega)_n} \quad \text{and} \quad f_{-2\Re(\zeta)}(q) = \sum_{n=0}^{\infty} \frac{q^{n^2}}{(\zeta q, q/\zeta)_n}.$$

Then

$$\begin{aligned} &\frac{1}{3}(1-\zeta)(f_1(\omega q) - \omega^2 C(\omega, \omega q)) + \frac{1}{3}(1-\zeta^{-1})(f_1(\omega^2 q) - \omega C(\omega, \omega^2 q)) \\ &\quad = f_{-2\Re(\zeta)}(q) + \frac{1}{3}(\zeta^2 + \zeta^{-2} - \zeta - \zeta^{-1}) \frac{J_3^3}{J_1 J_9}. \end{aligned} \tag{1.4}$$

The proof of identity (1.4) relies on properties of Appell–Lerch series [14, Eq. (16)] along with the 3-dissection of the rank modulo 3, which first appeared in [9], (see (2.1) and (2.2)).

Observe that (1.4), unlike (1.1), contains extra infinite products on the left-hand side. These products are specializations of $C(z,q)$, which is the generating function for the partition crank. The crank is a partition statistic that was predicted by Dyson [10] and finally discovered by Andrews and Garvan [3] in 1988. To define it, let $w(\lambda)$ denote the number of ones in a partition λ and let $v(\lambda)$ denote the number of parts greater than $w(\lambda)$. Then the crank is equal to the largest part, if $w(\lambda) = 0$, and $v(\lambda) - w(\lambda)$ otherwise. Andrews and Garvan showed that if $M(m,n)$ is defined via

$$C(z,q) = \sum_{n=0}^{\infty} M(m,n) z^m q^n,$$

then for $n \geq 2$, $M(m,n)$ is the number of partitions of n with crank equal to m.

A consequence of (1.4) is the 3-dissection of the rank generating function modulo 9, which is presented here for the first time. This is the second objective of this paper.

Theorem 1.2. *Let ζ be a primitive ninth root of unity. Then*

$$\begin{aligned} &\sum_{n=0}^{\infty} \frac{q^{n^2}}{(\zeta q, q/\zeta)_n} \\ &\quad = (\zeta^2 + \zeta^{-2}) \frac{q^3 J_{27}^3}{J_9 J_{12,27}} + \frac{J_{12,27} J_{27}^3}{J_9 J_{3,27} J_{6,27}} + \frac{\zeta + \zeta^{-1} - 2}{J_9} \\ &\quad \times \sum_{n=-\infty}^{\infty} \frac{(-1)^n q^{(27n^2+45n)/2+9}}{1 - q^{27n+21}} + \frac{q J_{27}^3}{J_9 J_{3,27}} + (\zeta + \zeta^{-1} - 1) \frac{q^4 J_{3,27} J_{27}^3}{J_9 J_{6,27} J_{12,27}} \end{aligned}$$

$$+\frac{\zeta+\zeta^{-1}+\zeta^{2}+\zeta^{-2}-1}{J_9}\sum_{n=-\infty}^{\infty}\frac{(-1)^n q^{(27n^2+45n)/2+7}}{1-q^{27n+12}}$$

$$+(\zeta+\zeta^{-1}-1)\frac{q^2J_{27}^3}{J_9J_{6,27}}-(\zeta^2+\zeta^{-2})\frac{q^2J_{6,27}J_{27}^3}{J_9J_{3,27}J_{12,27}}$$

$$+\frac{1+\zeta^{2}+\zeta^{-2}}{J_9}\sum_{n=-\infty}^{\infty}\frac{(-1)^n q^{(27n^2+45n)/2+2}}{1-q^{27n+3}}.$$

Dissections like this can be used to obtain results on $N(k,m,n)$, the number of partitions of n where the rank is congruent to k modulo m. Dyson [10] was the first to observe identities like

$$N(1,5,5n+1)=N(2,5,5n+1),$$
$$N(i,5,5n+4)=N(j,5,5n+4),$$
$$N(i,7,7n+5)=N(j,7,7n+5),$$
$$N(0,7,7n+6)+N(1,7,7n+6)=N(2,7,7n+6)+N(3,7,7n+6),$$

the middle two being true for all i,j. Dyson's observations were confirmed by Atkin and Swinnerton-Dyer [5], and since then there have been many studies of identities of this type — see [4, 8, 14, 15, 18, 20, 21], for example.

By comparing Theorem 1.2 with the 3-dissection of the rank modulo 3 given in (2.1), we see that the same series and products appear in the dissections. This leads to the following rank identities, which were conjectured by Lewis [16] and first proved by Santa-Gadea in his thesis [23]. This phenomenon is also present in [2], where dissections of the rank modulo 4 and 8 imply relations on ranks modulo 8.

Corollary 1.3. *For all $n\geq 0$, we have*

$$N(3,9,3n)=N(4,9,3n), \tag{1.5}$$

$$N(1,9,3n+1)+N(2,9,3n+1)=N(3,9,3n+1)+N(4,9,3n+1), \tag{1.6}$$

$$N(0,9,3n+2)=N(4,9,3n+2). \tag{1.7}$$

There is also an extensive literature on similar identities for the crank as well as mixed rank–crank identities (e.g., [11, 12, 16, 19, 20, 22]). For example, letting $M(k,m,n)$ denote the number of partitions of n whose crank is congruent to k modulo m, work of Lewis and Santa-Gadea [17, 20] contains identities such as

$$M(4,9,3n+3)=N(4,9,3n+3), \tag{1.8}$$

$$M(0,9,3n+1)+M(1,9,3n+1)=N(1,9,3n+1)+N(2,9,3n+1), \tag{1.9}$$

$$M(1,4,2n+2)=N(2,4,2n+2), \tag{1.10}$$

$$M(0,8,4n+3)+M(1,8,4n+3)=N(2,8,4n+3)+N(3,8,4n+3), \tag{1.11}$$

all valid for $n\geq 0$.

This brings us to the third aim of this paper. In 2009, Lewis [19] proposed three conjectures involving the difference between ranks and cranks modulo 8. Define

$$R_8(i,j,k) := \sum_{n\geq 0}(N(i,8;4n+k) - M(j,8;4n+k))q^n.$$

In his first conjecture, Lewis gave explicit formulas for $R_8(i,j,k)$ for all possible values of (i,j,k). Every $R_8(i,j,k)$ is expressed as a combination of at most one infinite sum and at most two infinite products. For several triples (i,j,k), we have $R_8(i,j,k) = 0$. A proof of this conjecture was recently given by Mortenson [22].

Here, we show that Theorem 1.2 and Mortenson's method can be used to find similar expressions for

$$R_9(i,j,k) := \sum_{n\geq 0}(N(i,9;3n+k) - M(j,9;3n+k))q^n$$

for all possible values of (i,j,k). See Theorem 3.6 for a list of all 75 identities.

The paper is organized as follows. In Sec. 2, we prove Theorems 1.1 and 1.2 as well as Corollary 1.3. In Sec. 3, we list the results on $R_9(i,j,k)$ for all combinations of (i,j,k) in Theorem 3.6 and provide a proof.

2. Proofs of Theorems 1.1 and 1.2

First, we record some results from [9]. Let ω be a primitive third root of unity. Theorem 1.1 and Corollary 1.2 of [9] give two versions of the 3-dissections of the rank modulo 3,

$$\begin{aligned}
&\sum_{n=0}^{\infty}\frac{q^{n^2}}{(\omega q, q/\omega)_n}\\
&= -\frac{q^3J_{27}^3}{J_9J_{12,27}} + \frac{J_{12,27}J_{27}^3}{J_9J_{3,27}J_{6,27}} - \frac{3}{J_9}\sum_{n=-\infty}^{\infty}\frac{(-1)^nq^{(27n^2+45n)/2+9}}{1-q^{27n+21}}\\
&\quad + \frac{qJ_{27}^3}{J_9J_{3,27}} + \frac{q^4J_{3,27}J_{27}^3}{J_9J_{6,27}J_{12,27}} + \frac{3}{J_9}\sum_{n=-\infty}^{\infty}\frac{(-1)^nq^{(27n^2+45n)/2+7}}{1-q^{27n+12}}\\
&\quad + \frac{q^2J_{27}^3}{J_9J_{6,27}} + \frac{q^2J_{6,27}J_{27}^3}{J_9J_{3,27}J_{12,27}} - \frac{3}{J_9}\sum_{n=-\infty}^{\infty}\frac{(-1)^nq^{(27n^2+45n)/2+2}}{1-q^{27n+3}} \qquad (2.1)\\
&= \frac{1}{J_9}\sum_{n=-\infty}^{\infty}\left\{\frac{(-1)^nq^{(27n^2+9n)/2}}{1-q^{27n+3}} - \frac{(-1)^nq^{(27n^2+27n)/2+3}}{1-q^{27n+12}}\right.\\
&\quad \left. - 2\frac{(-1)^nq^{(27n^2+45n)/2+9}}{1-q^{27n+21}}\right\} + \frac{q}{J_9}\sum_{n=-\infty}^{\infty}\left\{\frac{(-1)^nq^{(27n^2+27n)/2}}{1-q^{27n+3}}\right.
\end{aligned}$$

$$+2\frac{(-1)^n q^{(27n^2+45n)/2+6}}{1-q^{27n+12}}+\frac{(-1)^n q^{(27n^2+63n)/2+15}}{1-q^{27n+21}}\Bigg\}$$

$$+\frac{q^2}{J_9}\sum_{n=-\infty}^{\infty}\Bigg\{-2\frac{(-1)^n q^{(27n^2+45n)/2}}{1-q^{27n+3}}-\frac{(-1)^n q^{(27n^2+63n)/2+9}}{1-q^{27n+12}}$$

$$+\frac{(-1)^n q^{(27n^2+81n)/2+21}}{1-q^{27n+21}}\Bigg\}. \tag{2.2}$$

Similarly, from [9, Theorem 1.3 and Corollary 1.4], we have respectively,

$$\frac{(q)_\infty}{(\omega q, q/\omega)_\infty}=2\frac{q^3 J_{27}^3}{J_9 J_{12,27}}+\frac{J_{12,27}J_{27}^3}{J_9 J_{3,27}J_{6,27}}-2\frac{qJ_{27}^3}{J_9 J_{3,27}}$$

$$+\frac{q^4 J_{3,27}J_{27}^3}{J_9 J_{6,27}J_{12,27}}-2\frac{q^2 J_{27}^3}{J_9 J_{6,27}}+\frac{q^2 J_{6,27}J_{27}^3}{J_9 J_{3,27}J_{12,27}} \tag{2.3}$$

$$=\frac{1}{J_9}\sum_{n=-\infty}^{\infty}\Bigg\{\frac{(-1)^n q^{(27n^2+9n)/2}}{1-q^{27n+3}}+2\frac{(-1)^n q^{(27n^2+27n)/2+3}}{1-q^{27n+12}}$$

$$+\frac{(-1)^n q^{(27n^2+45n)/2+9}}{1-q^{27n+21}}\Bigg\}+\frac{q}{J_9}\sum_{n=-\infty}^{\infty}\Bigg\{-2\frac{(-1)^n q^{(27n^2+27n)/2}}{1-q^{27n+3}}$$

$$-\frac{(-1)^n q^{(27n^2+45n)/2+6}}{1-q^{27n+12}}+\frac{(-1)^n q^{(27n^2+63n)/2+15}}{1-q^{27n+21}}\Bigg\}$$

$$+\frac{q^2}{J_9}\sum_{n=-\infty}^{\infty}\Bigg\{\frac{(-1)^n q^{(27n^2+45n)/2}}{1-q^{27n+3}}-\frac{(-1)^n q^{(27n^2+63n)/2+9}}{1-q^{27n+12}}$$

$$-2\frac{(-1)^n q^{(27n^2+81n)/2+21}}{1-q^{27n+21}}\Bigg\}. \tag{2.4}$$

Next, by [9, Lemma 2.2], we see that

$$m(q,q^3,\omega)=\frac{1}{(1-\omega)J_9}\sum_{n=-\infty}^{\infty}\frac{(-1)^n q^{(27n^2+9n)/2}}{1-q^{27n+3}}(-\omega-\omega^2 q^{9n+1}-q^{18n+2})$$

$$+\frac{1}{(1-\omega)J_9}\sum_{n=-\infty}^{\infty}\frac{(-1)^n q^{(27n^2+27n)/2+3}}{1-q^{27n+12}}(\omega^2+q^{9n+4}+\omega q^{18n+8})$$

$$+\frac{1}{(1-\omega)J_9}\sum_{n=-\infty}^{\infty}\frac{(-1)^n q^{(27n^2+45n)/2+9}}{1-q^{27n+21}}(-1-\omega q^{9n+7}-\omega^2 q^{18n+14}) \tag{2.5}$$

and

$$\omega^2 m(q,q^3,\omega^2)$$

$$= \frac{1}{(1-\omega^2)J_9}\sum_{n=-\infty}^{\infty}\frac{(-1)^n q^{(27n^2+9n)/2}}{1-q^{27n+3}}\{-\omega - q^{9n+1} - \omega^2 q^{18n+2}\}$$

$$+ \frac{1}{(1-\omega^2)J_9}\sum_{n=-\infty}^{\infty}\frac{(-1)^n q^{(27n^2+27n)/2+3}}{1-q^{27n+12}}\{1 + \omega^2 q^{9n+4} + \omega q^{18n+8}\}$$

$$+ \frac{1}{(1-\omega^2)J_9}\sum_{n=-\infty}^{\infty}\frac{(-1)^n q^{(27n^2+45n)/2+9}}{1-q^{27n+21}}\{-\omega^2 - \omega q^{9n+7} - q^{18n+14}\}, \tag{2.6}$$

where

$$m(x,q,z) := \frac{1}{(z,q/z,q)_\infty}\sum_{n=-\infty}^{\infty}\frac{(-1)^n q^{n(n-1)/2} z^n}{1-q^{n-1}xz}$$

is the Appell–Lerch series.

Using the above results, we derive the following lemma.

Lemma 2.1.

$$m(q,q^3,\omega) = \frac{1}{3}(f_1(q) - \omega C(\omega,q)), \tag{2.7}$$

$$m(q,q^3,\omega^2) = \frac{1}{3}(f_1(q) - \omega^2 C(\omega,q)). \tag{2.8}$$

Proof. Identity (2.7) follows from applying (2.2) and (2.4) to the right-hand side of (2.5). Similarly, (2.8) follows from applying (2.2) and (2.4)–(2.6). □

Next, we need an expression for the rank generating function in terms of Appell–Lerch series.

Lemma 2.2.

$$\sum_{n=0}^{\infty}\frac{q^{n^2}}{(xq,q/x)_n} = (1-x)m(x^3q,q^3,x^{-3}z^{-1}) + (1-x^{-1})m(x^{-3}q,q^3,x^3z) + \frac{x(1-x)J_1^2 j(xz;q)j(z;q^3)}{j(x;q)j(z;q)j(x^3z;q^3)}.$$

Proof. From [14, Eq. (16)], we have

$$\sum_{n=0}^{\infty}\frac{q^{n^2}}{(xq,q/x)_n} = x(1-x)\left(-\frac{1}{x}m(x^{-3}q^2,q^3,x^3z) - \frac{1}{x^2}m(x^{-3}q,q^3,x^3z) + \frac{J_1^2 j(xz;q)j(z;q^3)}{j(x;q)j(z;q)j(x^3z;q^3)}\right) + (1-x)$$

$$= (1-x)(1-m(x^{-3}q^2,q^3,x^3z)) + (1-1/x)m(x^{-3}q,q^3,x^3z) + \frac{x(1-x)J_1^2 j(xz;q)j(z;q^3)}{j(x;q)j(z;q)j(x^3z;q^3)}. \tag{2.9}$$

Using [14, Proposition 3.1, Eqs. (21b) and (21c)], we get

$$\begin{aligned} 1-m(x^{-3}q^2,q^3,x^3z) &= 1-x^3q^{-2}m(x^3q^{-2},q^3,x^{-3}z^{-1}) \\ &= m(x^3q,q^3,x^{-3}z^{-1}). \end{aligned} \tag{2.10}$$

Substituting identity (2.10) into (2.9), we obtain

$$\begin{aligned} \sum_{n=0}^{\infty} \frac{q^{n^2}}{(xq,q/x)_n} &= (1-x)m(x^3q,q^3,x^{-3}z^{-1}) + (1-x^{-1})m(x^{-3}q,q^3,x^3z) \\ &\quad + \frac{x(1-x)J_1^2 j(xz;q)j(z;q^3)}{j(x;q)j(z;q)j(x^3z;q^3)}, \end{aligned} \tag{2.11}$$

and this completes the proof of the lemma. □

Set $x=\zeta$, and let z tend to 1 in Lemma 2.2. Noting that $j(\omega,q^3)=(1-\omega)J_9$, we obtain the following corollary.

Corollary 2.3.

$$\sum_{n=0}^{\infty} \frac{q^{n^2}}{(\zeta q,q/\zeta)_n} = (1-\zeta)m(\omega q,q^3,\omega^2) + (1-1/\zeta)m(\omega^2q,q^3,\omega) + \frac{\zeta(1-\zeta)}{1-\omega}\frac{J_3^3}{J_1J_9}. \tag{2.12}$$

We need one more lemma for our proof of Theorems 1.1 and 1.2.

Lemma 2.4.

$$\begin{aligned} \frac{1}{J_1} &= \frac{J_9}{J_3^4}\{(J_{12,27}^2 - q^3J_{3,27}J_{6,27}) + (qJ_{6,27}J_{12,27} + q^4J_{3,27}^2) \\ &\quad + (q^2J_{6,27}^2 + q^2J_{3,27}J_{12,27})\}. \end{aligned} \tag{2.13}$$

Proof. First note that a simple consequence of Jacobi triple product identity [6, p. 3 and 48, Entries 19 and 31] is

$$J_1 = J_{12,27} - qJ_{6,27} - q^2J_{3,27}. \tag{2.14}$$

Substituting (2.14) into (2.13), cross multiplying and simplifying, we see that it suffices to prove

$$\frac{J_9}{J_3^4}\{J_{12,27}^3 - q^3J_{6,27}^3 - q^6J_{3,27}^3 - 3q^3J_{3,27}J_{6,27}J_{12,27}\} = 1. \tag{2.15}$$

By applying the second and third identity from the bottom of [6, p. 351], we see that (2.15) is equivalent to

$$\frac{J_9}{J_3^4}\{J_1^3 + 3qJ_9^3 - 9q^3 J_{3,27}J_{6,27}J_{12,27}\} = 1. \tag{2.16}$$

Multiplying both sides by J_3^4 and rearranging, we see that (2.16) is equivalent to

$$J_9\{J_1^3 + 3qJ_9^3\} = J_3\{J_3^3 + 9q^3 J_{27}^3\}.$$

This is in turn equivalent to

$$qJ_9^4\left\{3 + \frac{J_1^3}{qJ_9^3}\right\} = J_3^4\left\{1 + 9q^3\frac{J_{27}^3}{J_3^3}\right\}. \tag{2.17}$$

By applying the first equality [6, Entry 1(iv), p. 345] on the left-hand side of (2.17) and the last equality of the same entry on the right-hand side, we see that Eq. (2.17) follows. This completes the proof of Lemma 2.4. □

Theorems 1.1 and 1.2 now follow readily from the work above.

Proof of Theorem 1.1. First, using

$$1 + \zeta^3 + \zeta^6 = 0, \tag{2.18}$$

it is easy to verify that $\zeta(1-\zeta)/(1-\omega) = \frac{1}{3}(\zeta + \zeta^{-1} - \zeta^2 - \zeta^{-2})$. To complete the proof, it suffices to apply Lemma 2.1 to Corollary 2.3 and rearrange terms. □

Proof of Theorem 1.2. Apply (2.1) and (2.3) on the left-hand side of (1.4) and Lemma 2.4 on the right-hand side and rearrange terms. □

We end this section with a proof of Corollary 1.3.

Proof of Corollary 1.3. We use the notation

$$N_a = \sum_{n\geq 0} N(a,9,n)q^n, \quad N_a(d) = \sum_{n\geq 0} N(a,9,3n+d)q^{3n+d},$$

and note that since the rank generating function is invariant under $z \leftrightarrow 1/z$, we have $N_a = N_{9-a}$ and $N_a(d) = N_{9-a}(d)$. Together with (2.18) and the fact that

$$\zeta + \zeta^2 + \zeta^4 + \zeta^5 + \zeta^7 + \zeta^8 = 0, \tag{2.19}$$

we find that

$$\sum_{n=0}^{\infty}\frac{q^{n^2}}{(\zeta q, q/\zeta)_n} = N_0 - N_3 + (\zeta + \zeta^{-1})(N_1 - N_4) + (\zeta^2 + \zeta^{-2})(N_2 - N_4) \tag{2.20}$$

and

$$\sum_{n=0}^{\infty} \frac{q^{n^2}}{(\omega q, q/\omega)_n} = N_0 - N_1 - N_2 + 2N_3 - N_4. \tag{2.21}$$

Comparing coefficients of $1, \zeta + \zeta^{-1}$, and $\zeta^2 + \zeta^{-2}$ in (2.20) and the first line of Theorem 1.2, we find

$$N_0(0) - N_3(0) = \frac{J_{12,27}J_{27}^3}{J_9 J_{3,27} J_{6,27}} - \frac{2}{J_9} \sum_{n=-\infty}^{\infty} \frac{(-1)^n q^{(27n^2+45n)/2+9}}{1 - q^{27n+21}},$$

$$N_1(0) - N_4(0) = \frac{1}{J_9} \sum_{n=-\infty}^{\infty} \frac{(-1)^n q^{(27n^2+45n)/2+9}}{1 - q^{27n+21}},$$

$$N_2(0) - N_4(0) = \frac{q^3 J_{27}^3}{J_9 J_{12,27}}.$$

Comparing this with (2.21) and (2.1), we deduce that

$$\begin{aligned} &N_0(0) - N_3(0) - N_1(0) + N_4(0) - N_2(0) + N_4(0) \\ &= N_0(0) + 2N_3(0) - N_1(0) - N_2(0) - N_4(0). \end{aligned}$$

Simplifying, we obtain (1.5). The proofs of equations (1.6) and (1.7) are similar. □

3. Differences of Ranks and Cranks

Recall that the 3-dissection of the crank modulo 3 is given in (2.3). We record the 3-dissection of the crank modulo 9 given in [12, Eq. (2.15)] as the following theorem.

Theorem 3.1. *Let ζ be a primitive ninth root of unity. Then*

$$\frac{(q)_\infty}{(\zeta q)_\infty (\zeta^{-1} q)_\infty} = \frac{J_3 J_{27}^2}{J_9} \left(\frac{1}{J_{3,27}} + (\zeta + \zeta^{-1} - 1) \frac{q}{J_{6,27}} + (\zeta^2 + \zeta^{-2}) \frac{q^2}{J_{12,27}} \right). \tag{3.1}$$

Next, we reproduce the definition of rank deviation and crank deviation as given by Hickerson and Mortenson [14, 22].

Definition 3.2. The rank and crank deviations are given, respectively, by

$$D(a, m) = D(a, m; q) = \sum_{n=0}^{\infty} \left(N(a, m; n) - \frac{p(n)}{m} \right) q^n$$

and

$$D_C(a, m) = D_C(a, m; q) = \sum_{n=0}^{\infty} \left(M(a, m; n) - \frac{p(n)}{m} \right) q^n.$$

We do not compute rank and crank deviations from the definitions. Instead, we use the next two formulas. The first is [14, Eq. (25)], given with a slight variation. The next is the analogous result for cranks which is given in [22, Eq. (2.12)].

Theorem 3.3. *Let ζ_m be a primitive mth root of unity. Then*

$$D(a,m) = \frac{1}{m}\sum_{j=1}^{m-1} \zeta_m^{-aj} \sum_{n=0}^{\infty} \frac{q^{n^2}}{(\zeta_m^j q)_n(\zeta_m^{-j}q)_n} \tag{3.2}$$

and

$$D_C(a,m) = \frac{1}{m}\sum_{j=1}^{m-1} \zeta_m^{-aj} \frac{(q)_\infty}{(\zeta_m^j q)_\infty(\zeta_m^{-j}q)_\infty}. \tag{3.3}$$

Since the generating functions for $N(m,n)$ and $M(m,n)$ are invariant under $z \leftrightarrow 1/z$, one easily sees that $N(a,m;n) = N(m-a,m;n)$ and $M(a,m;n) = M(m-a,m;n)$. Therefore, it suffices to examine the values of $D(i,9)$ and $D_C(j,9)$ for $i,j = 0,1,2,3,4$.

Theorem 3.4. *Let us denote*

$$A := \frac{J_3 J_{27}^2}{J_9 J_{3,27}}, \quad B := q\frac{J_3 J_{27}^2}{J_9 J_{6,27}}, \quad C := q^2\frac{J_3 J_{27}^2}{J_9 J_{12,27}},$$

$$D := 2q^3\frac{J_{27}^3}{J_9 J_{12,27}} + \frac{J_{12,27}J_{27}^3}{J_9 J_{3,27}J_{6,27}} - 2q\frac{J_{27}^3}{J_9 J_{3,27}} + q^4\frac{J_{3,27}J_{27}^3}{J_9 J_{6,27}J_{12,27}}$$
$$- 2q^2\frac{J_{27}^3}{J_9 J_{6,27}} + q^2\frac{J_{6,27}J_{27}^3}{J_9 J_{3,27}J_{12,27}}.$$

Then, we have

$$D_C(0,9) = \frac{1}{9}(6A - 6B + 2D),$$

$$D_C(1,9) = \frac{1}{9}(6B - 3C - D),$$

$$D_C(2,9) = \frac{1}{9}(-3B + 6C - D),$$

$$D_C(3,9) = \frac{1}{9}(-3A + 3B + 2D),$$

$$D_C(4,9) = \frac{1}{9}(-3B - 3C - D).$$

Proof. Using the above notation, (3.1) and (2.3) simplify, respectively, to

$$\frac{(q)_\infty}{(\zeta q)_\infty(\zeta^{-1}q)_\infty} = A + (\zeta + \zeta^{-1} - 1)B + (\zeta^2 + \zeta^{-2})C \tag{3.4}$$

and

$$\frac{(q)_\infty}{(\omega q)_\infty(\omega^{-1}q)_\infty} = D. \tag{3.5}$$

Computing directly using (3.3) and applying (3.4) and (3.5), we find that

$$
\begin{aligned}
D_C(0,9) &= \frac{1}{9}\sum_{j=1}^{8}\frac{(q)_\infty}{(\zeta^j q)_\infty(\zeta^{-j}q)_\infty} \\
&= \frac{1}{9}\left(2\frac{(q)_\infty}{(\zeta q, q/\zeta)_\infty} + 2\frac{(q)_\infty}{(\zeta^2 q, q/\zeta^2)_\infty} + 2\frac{(q)_\infty}{(\omega q, q/\omega)_\infty} + 2\frac{(q)_\infty}{(\zeta^4 q, q/\zeta^4)_\infty}\right) \\
&= \frac{2}{9}(A + (\zeta+\zeta^{-1}-1)B + (\zeta^2+\zeta^{-2})C + A + (\zeta^2+\zeta^{-2}-1)B \\
&\quad + (\zeta^4+\zeta^{-4})C + D + A + (\zeta^4+\zeta^{-4}-1)B + (\zeta^8+\zeta^{-8})C) \\
&= \frac{2}{9}(3A - 3B + D + (\zeta+\zeta^{-1}+\zeta^2+\zeta^{-2}+\zeta^4+\zeta^{-4})B \\
&\quad + (\zeta^2+\zeta^{-2}+\zeta^4+\zeta^{-4}+\zeta^8+\zeta^{-8})C) \\
&= \frac{1}{9}(6A - 6B + 2D),
\end{aligned}
$$

by (2.19). The other cases are handled similarly and so we omit the proofs. □

Theorem 3.5. *Denote by*

$$
A' = q^3\frac{J_{27}^3}{J_9 J_{12,27}} - q^2\frac{J_{6,27}J_{27}^3}{J_9 J_{3,27}J_{12,27}}, \quad D' = \frac{1}{J_9}\sum_{n=-\infty}^{\infty}\frac{(-1)^n q^{(27n^2+45n)/2+9}}{1-q^{27n+21}},
$$

$$
B' = \frac{J_{12,27}J_{27}^3}{J_9 J_{3,27}J_{6,27}}, +q\frac{J_{27}^3}{J_9 J_{3,27}}, \quad E' = \frac{1}{J_9}\sum_{n=-\infty}^{\infty}\frac{(-1)^n q^{(27n^2+45n)/2+7}}{1-q^{27n+12}},
$$

$$
C' = q^4\frac{J_{3,27}J_{27}^3}{J_9 J_{6,27}J_{12,27}} + q^2\frac{J_{27}^3}{J_9 J_{6,27}}, \quad F' = \frac{1}{J_9}\sum_{n=-\infty}^{\infty}\frac{(-1)^n q^{(27n^2+45n)/2+2}}{1-q^{27n+3}}.
$$

Then, we have

$$
\begin{aligned}
D(0,9) &= \frac{-2A' + 8B' - 4C' - 18D'}{9}, \\
D(1,9) &= \frac{-2A' - B' + 5C' + 9D'}{9}, \\
D(2,9) &= \frac{7A' - B' - 4C' + 9F'}{9}, \\
D(3,9) &= \frac{-2A' - B' + 5C' + 9E' - 9F'}{9}, \\
D(4,9) &= \frac{-2A' - B' - 4C' - 9E'}{9}.
\end{aligned}
$$

Proof. With the above notation, Theorem 1.2 and Eq. (2.1) can be rewritten as, respectively,

$$\sum_{n=0}^{\infty} \frac{q^{n^2}}{(\zeta q, q/\zeta)_n} = (\zeta^2+\zeta^{-2})A' + B' + (\zeta+\zeta^{-1}-1)C' + (\zeta+\zeta^{-1}-2)D' \\ + (\zeta+\zeta^{-1}+\zeta^2+\zeta^{-2}-1)E' + (1+\zeta^2+\zeta^{-2})F' \tag{3.6}$$

$$\sum_{n=0}^{\infty} \frac{q^{n^2}}{(\omega q, q/\omega)_n} = -A' + B' + C' - 3D' + 3E' - 3F'. \tag{3.7}$$

To prove the theorem, we compute each of $D(a, 9)$ using (3.2) by applying (3.6) and (3.7). The computations are straightforward and similar to those in the proof of Theorem 3.4 and so we omit them. □

Theorems 3.4 and 3.5 provide us the 3-dissections of $D(i, 9)$ and $D_C(j, 9)$, which in turn leads us to $R_9(i, j, k)$. To display all the results in a neat way, here we define some notation. First, let

$$f_n(i, j, k) := \frac{J_9^2}{J_3 J_{2^n,9}} (i \cdot J_{12,27} + j \cdot qJ_{6,27} + k \cdot q^2 J_{3,27}).$$

Next, for $k = 0, 1, 2$ we define h_k as follows:

$$h_0 := \frac{1}{J_3} \sum_{n=-\infty}^{\infty} \frac{(-1)^n q^{(9n^2+15n)/2+3}}{1-q^{9n+7}},$$

$$h_1 := \frac{1}{J_3} \sum_{n=-\infty}^{\infty} \frac{(-1)^n q^{(9n^2+15n)/2+2}}{1-q^{9n+4}},$$

$$h_2 := \frac{1}{J_3} \sum_{n=-\infty}^{\infty} \frac{(-1)^n q^{(9n^2+15n)/2}}{1-q^{9n+1}}.$$

We are now ready to list the expressions for $R_9(i, j, k)$.

Theorem 3.6. *For $R_9(i, j, k)$, we have*

$R_9(0,0,0) = f_0(0,0,2) - 2h_0,$
$R_9(0,0,1) = f_1(2,0,0),$
$R_9(0,0,2) = 0,$
$R_9(0,1,0) = f_0(1,0,1) - 2h_0,$
$R_9(0,1,1) = f_1(0,1,1),$
$R_9(0,1,2) = f_2(0,0,-1),$
$R_9(0,2,0) = f_0(1,0,1) - 2h_0,$
$R_9(0,2,1) = f_1(1,0,0),$
$R_9(0,2,2) = f_2(-1,1,0),$
$R_9(0,3,0) = f_0(1,-1,1) - 2h_0,$
$R_9(0,3,1) = f_1(1,1,1),$
$R_9(0,3,2) = 0,$

$$
\begin{aligned}
&R_9(0,4,0) = f_0(1,0,1) - 2h_0, && R_9(2,4,2) = f_2(-1,-1,-1) + h_2,\\
&R_9(0,4,1) = f_1(1,0,0) && R_9(3,0,0) = f_0(-1,0,1),\\
&R_9(0,4,2) = f_2(0,0,-1), && R_9(3,0,1) = f_1(1,0,-1) + h_1,\\
&R_9(1,0,0) = f_0(-1,0,1) + h_0, && R_9(3,0,2) = f_2(1,0,1) - h_2,\\
&R_9(1,0,1) = f_1(1,0,-1), && R_9(3,1,0) = 0,\\
&R_9(1,0,2) = f_2(1,0,1), && R_9(3,1,1) = f_1(-1,1,0) + h_1,\\
&R_9(1,1,0) = h_0, && R_9(3,1,2) = f_2(1,0,0) - h_2,\\
&R_9(1,1,1) = f_1(-1,1,0), && R_9(3,2,0) = 0,\\
&R_9(1,1,2) = f_2(1,0,0), && R_9(3,2,1) = f_1(0,0,-1) + h_1,\\
&R_9(1,2,0) = h_0, && R_9(3,2,2) = f_2(0,1,1) - h_2,\\
&R_9(1,2,1) = f_1(0,0,-1), && R_9(3,3,0) = f_0(0,-1,0),\\
&R_9(1,2,2) = f_2(0,1,1), && R_9(3,3,1) = f_1(0,1,0) + h_1,\\
&R_9(1,3,0) = f_0(0,-1,0) + h_0, && R_9(3,3,2) = f_2(1,0,1) - h_2,\\
&R_9(1,3,1) = f_1(0,1,0), && R_9(3,4,0) = 0,\\
&R_9(1,3,2) = f_2(1,0,1), && R_9(3,4,1) = f_1(0,0,-1) + h_1,\\
&R_9(1,4,0) = h_0, && R_9(3,4,2) = f_2(1,0,0) - h_2,\\
&R_9(1,4,1) = f_1(0,0,-1), && R_9(4,0,0) = f_0(-1,0,1),\\
&R_9(1,4,2) = f_2(1,0,0), && R_9(4,0,1) = f_1(1,-1,0) - h_1,\\
&R_9(2,0,0) = f_0(-1,1,0), && R_9(4,0,2) = 0,\\
&R_9(2,0,1) = f_1(1,-1,0), && R_9(4,1,0) = 0,\\
&R_9(2,0,2) = f_2(-1,-1,0) + h_2, && R_9(4,1,1) = f_1(-1,0,1) - h_1\\
&R_9(2,1,0) = f_0(0,1,-1), && R_9(4,1,2) = f_2(0,0,-1)\\
&R_9(2,1,1) = f_1(-1,0,1), && R_9(4,2,0) = 0\\
&R_9(2,1,2) = f_2(-1,-1,-1) + h_2, && R_9(4,2,1) = f_1(0,-1,0) - h_1,\\
&R_9(2,2,0) = f_0(0,1,-1), && R_9(4,2,2) = f_2(-1,1,0),\\
&R_9(2,2,1) = f_1(0,-1,0), && R_9(4,3,0) = f_0(0,-1,0),\\
&R_9(2,2,2) = f_2(-2,0,0) + h_2, && R_9(4,3,1) = f_1(0,0,1) - h_1,\\
&R_9(2,3,0) = f_0(0,0,-1), && R_9(4,3,2) = 0,\\
&R_9(2,3,1) = f_1(0,0,1), && R_9(4,4,0) = 0,\\
&R_9(2,3,2) = f_2(-1,-1,0) + h_2, && R_9(4,4,1) = f_1(0,-1,0) - h_1,\\
&R_9(2,4,0) = f_0(0,1,-1), && R_9(4,4,2) = f_2(0,0,-1).\\
&R_9(2,4,1) = f_1(0,-1,0), &&
\end{aligned}
$$

Proof. All of these components $R_9(i,j,k)$ can be computed directly from Theorems 3.4 and 3.5. We illustrate the method of proof for the cases $(i,j,k) = (0,0,1)$ and $(0,0,2)$. The other cases are proved similarly and we omit the details.

One can easily check that

$$
\begin{aligned}
D(0,9) - D_C(0,9) &= \frac{-2A' + 8B' - 4C' - 18D' - 6A + 6B - 2D}{9}\\
&= -\frac{6}{9}\frac{J_3 J_{27}^2}{J_9 J_{3,27}} + \frac{6}{9}q\frac{J_3 J_{27}^2}{J_9 J_{6,27}} - \frac{6}{9}q^3\frac{J_{27}^3}{J_9 J_{12,27}} + \frac{6}{9}\frac{J_{12,27} J_{27}^3}{J_9 J_{3,27} J_{6,27}}
\end{aligned}
$$

$$+\frac{12}{9}q\frac{J_{27}^3}{J_9J_{3,27}}-\frac{6}{9}q^4\frac{J_{3,27}J_{27}^3}{J_9J_{6,27}J_{12,27}}-\frac{18}{9}\frac{1}{J_9}$$

$$\times\sum_{n=-\infty}^{\infty}\frac{(-1)^nq^{(27n^2+45n)/2+9}}{1-q^{27n+21}}.$$

Note that all the coefficients of q^{3n+2} are 0 for all integers n. Therefore, $R_9(0,0,2)=0$. Collecting only terms with q^{3n+1} for all integers n, we have

$$\frac{6}{9}\frac{J_3J_{27}^2}{J_9J_{6,27}}+\frac{12}{9}\frac{J_{27}^3}{J_9J_{3,27}}-\frac{6}{9}q^3\frac{J_{3,27}J_{27}^3}{J_9J_{6,27}J_{12,27}}.$$

Therefore,

$$\begin{aligned}R_9(0,0,1)&=\frac{6}{9}\frac{J_1J_9^2}{J_3J_{2,9}}+\frac{12}{9}\frac{J_9^3}{J_3J_{1,9}}-\frac{6}{9}q\frac{J_{1,9}J_9^3}{J_3J_{2,9}J_{4,9}}\\&=\frac{2}{3}\frac{J_9^2}{J_3J_{2,9}}\left(J_1+2\frac{J_{2,9}J_9}{J_{1,9}}-q\frac{J_{1,9}J_9}{J_{4,9}}\right)\\&=\frac{2}{3}\frac{J_9^2}{J_3J_{2,9}}((J_{12,27}-qJ_{6,27}-q^2J_{3,27})\\&\quad+(2J_{12,27}+2qJ_{6,27})-(qJ_{6,27}-q^2J_{3,27}))\\&=2\frac{J_9^2J_{12,27}}{J_3J_{2,9}},\end{aligned}$$

where in the penultimate equality, we use (2.14) and [7, Eqs. (4.2)–(4.4)] (which were reproduced in [1, Eqs. (3.4.2)–(3.4.4)]). □

We close with a few remarks on Theorem 3.6. First, we have used the notation $f_n(i,j,k)$ for consistency, but some of these sums of infinite products can be simplified to a single infinite product. For example, using [7, Eq. (4.2)], we have

$$f_0(0,1,-1)=\frac{qJ_9^3}{J_3J_{4,9}}.$$

Similarly, by using [7, Eq. (4.3)], we have

$$f_2(1,0,1)=\frac{J_9^3}{J_3J_{2,9}}.$$

These appear in several places in Theorem 3.6.

Second, Theorem 3.6 immediately gives rank–crank identities like (1.8) and (1.9) and other results of Lewis [17]. For example, by inspection we see that $R_9(1,0,1)=-R_9(2,1,1)$, which implies (1.9). All of the rank–crank identities of Lewis in [17] follow in the same way.

Finally, the 75 results in Theorem 3.6 can also be used to deduce rank–crank inequalities. To give just one example out of many, we have

$$R_9(1,1,2) = f_2(1,0,0) = \frac{(q^9;q^9)_\infty^3}{(q^3;q^3)_\infty(1-q^4)(1-q^5)}\left(1+\sum_{n=1}^{\infty} a_n q^n\right),$$

where $a_n \geq 0$. Since

$$\frac{(q^9;q^9)_\infty^3}{(q^3;q^3)_\infty} = 1 + q^3 + 2q^6 + \cdots$$

is the generating function for the number of 3-core partitions of $n/3$, which follows from [13, Eq. (3)] and every natural number at least 12 can be written as $4x+5y$ with $x, y \geq 0$, we deduce that $R_9(1,1,2) > 0$ for all $n \geq 12$. Checking the expansion for $n < 12$, we find that

$$N(1,9,3n+2) > M(1,9,3n+2)$$

except for $n = 8$. We leave the pursuit of other inequalities to the interested reader.

References

[1] G. E. Andrews and B. C. Berndt, *Ramanujan's Lost Notebook. Part III* (Springer, New York, 2012).

[2] G. E. Andrews, B. C. Berndt, S. H. Chan, S. Kim and A. Malik, Four identities for third order mock theta functions, *Nagoya Math. J.* **239** (2020) 173–204.

[3] G. E. Andrews and F. G. Garvan, Dyson's crank of a partition, *Bull. Amer. Math. Soc. (N.S.)* **18**(2) (1988) 167–171.

[4] A. O. L. Atkin and S. M. Hussain, Some properties of partitions. II, *Trans. Amer. Math. Soc.* **89** (1958) 184–200.

[5] A. O. L. Atkin and H. P. F. Swinnerton–Dyer, Some properties of partitions, *Proc. London Math. Soc. (3)* **4** (1954) 84–106.

[6] B. C. Berndt, *Ramanujan's Notebooks, Part III* (Springer, New York, 1991).

[7] B. C. Berndt, H. H. Chan, S. H. Chan and W.-C. Liaw, Cranks and dissections in Ramanujan's lost notebook, *J. Combin. Theory Ser. A.* **109** (2005) 77–98.

[8] K. Bringmann, K. Ono and R. C. Rhoades, Eulerian series as modular forms, *J. Amer. Math. Soc.* **21** (2008) 1085–1104.

[9] S. H. Chan and R. Mao, The rank and crank of partitions modulo 3, *Int. J. Number Theory* **12** (2016) 1027–1053.

[10] F. J. Dyson, Some guesses in the theory of partitions, *Eureka Cambridge* **8** (1944) 10–15.

[11] F. G. Garvan, New combinatorial interpretations of Ramanujan's partition congruences mod 5, 7 and 11, *Trans. Amer. Math. Soc.* **305** (1988) 47–77.

[12] F. G. Garvan, The crank of partitions mod 8, 9 and 10, *Trans. Amer. Math. Soc.* **322** (1990) 79–94.

[13] A. Granville and K. Ono, Defect zero p-blocks for finite simple groups, *Trans. Amer. Math. Soc.* **348** (1996) 331–347.

[14] D. R. Hickerson and E. T. Mortenson, Dyson's ranks and Appell–Lerch sums, *Math. Ann.* **367** (2017) 373–395.

[15] R. Lewis, On the ranks of partitions modulo 9, *Bull. London Math. Soc.* **23** (1991) 417–421.

[16] R. Lewis, On some relations between the rank and the crank, *J. Combin. Theory Ser. A* **59** (1992) 104–110.
[17] R. Lewis, Relations between the rank and the crank modulo 9, *J. London Math. Soc. (2)* **45** (1992) 222–231.
[18] R. Lewis, The ranks of partitions modulo 2, *Discrete Math.* **167/168** (1997) 445–449.
[19] R. Lewis, The generating functions of the rank and crank modulo 8, *Ramanujan J.* **18** (2009) 121–146.
[20] R. Lews and N. Santa-Gadea, On the rank and crank modulo 4 and 8, *Trans. Amer. Math. Soc.* **341** (1994) 449–465.
[21] R. Mao, Ranks of partitions modulo 10, *J. Number Theory* **133** (2013) 3678–3702.
[22] E. T. Mortenson, On ranks and cranks of partitions modulo 4 and 8, *J. Combin. Theory Ser. A* **161** (2019) 51–80.
[23] N. A. Santa-Gadea, On the rank and the crank moduli 8, 9 and 12, Ph.D. thesis, The Pennsylvania State University (1990).
[24] H. Yesilyurt, Four identities related to third order mock theta functions in Ramanujan's lost notebook, *Adv. Math.* **190** (2005) 278–299.
[25] D. Zagier, Ramanujan's mock theta functions and their applications (after Zwegers and Ono-Bringmann), *Astérisque* **326** (2009) 143–164.

© 2024 World Scientific Publishing Company
https://doi.org/10.1142/9789811277375_0008

The congruence equation $\overline{a}+\overline{b} \equiv \overline{c} \pmod p$*

Tsz Ho Chan
Department of Mathematical Sciences, University of Memphis
Memphis, TN 38152, USA
thchan6174@gmail.com

In this paper, we consider the congruence equation $\overline{a}+\overline{b} \equiv \overline{c} \pmod p$ with $1 \le a,b,c \le H$ where $\overline{x}$ denotes the multiplicative inverse of $x \pmod p$. We prove that its number of solutions is asymptotic to H^3/p when $H > p^{2/3+o(1)}$ by estimating a certain average of Kloosterman sums via Gauss sums. On the other hand, when $H < p^{1/2}\sqrt{\log p}$, the number of solutions has order of magnitude at least $H \log H$. It would be interesting to better understand its transition of behavior. By transforming the question slightly, one can relate the problem to a certain first moment of Dirichlet L-functions at $s = 1$.

Keywords: Congruence equation; multiplicative inverses; Egyptian fractions; Kloosterman sums; Gauss sums; Dirichlet L-functions.

Mathematics Subject Classification 2020: 11A07, 11D45, 11L05, 11M06

1. Introduction and Main Results

In this paper, we are interested in the congruence equation

$$\overline{a}+\overline{b} \equiv \overline{c} \pmod p, \tag{1.1}$$

where $\overline{x}$ denotes the multiplicative inverse of $x \pmod p$. We want to count its number of solutions when $1 \le a,b,c \le H$. Let

$$\mathcal{N}(H;p) := \{1 \le a,b,c \le H : \overline{a}+\overline{b} \equiv \overline{c} \pmod p\}.$$

Heuristically, for each pair of a and b, $\overline{a}+\overline{b}$ is congruent to $\overline{c} \pmod p$ with probability H/p since there are H choices for c. Thus, we expect to have $H^2 \cdot H/p = H^3/p$ solutions to (1.1). Using exponential sums, Kloosterman sums and Gauss sums, we obtain the following theorem.

Theorem 1.1. *For* $1 \le H \le p$,

$$\mathcal{N}(H;p) = \frac{H^3}{p} + O(p \log^3 p).$$

*This article originally appeared in *International Journal of Number Theory*, **17**(2), pp. 329–340
© 2021 World Scientific Publishing Company. DOI: 10.1142/S1793042120400011

The above equation gives an asymptotic formula when $H \gg p^{2/3} \log p$. To prove Theorem 1.1, we make use of the following proposition which may be of independent interest.

Proposition 1.2. *Recall the Kloosterman sum*

$$S(a, b; p) := \sum_{n=1}^{p-1} e\left(\frac{an + b\overline{n}}{p}\right).$$

Then

$$\begin{aligned} S(l_1, l_2, l_3) &:= \sum_{k=1}^{p-1} S(l_1, k; p) S(l_2, k; p) S(l_3, k; p) \\ &= p^3 \mathcal{I}_{l_1} \mathcal{I}_{l_2} \mathcal{I}_{l_3} + p^2 \left(\frac{l_1^2 + l_2^2 + l_3^2 - 2l_1 l_2 - 2l_2 l_3 - 2l_3 l_1}{p}\right) \\ &\quad - p^2 \mathcal{I}_{l_1} - p^2 \mathcal{I}_{l_2} - p^2 \mathcal{I}_{l_3} + 2p - (p\mathcal{I}_{l_1} - 1)(p\mathcal{I}_{l_2} - 1)(p\mathcal{I}_{l_3} - 1), \end{aligned}$$

where

$$\mathcal{I}_n = \begin{cases} 1 & \text{if } n \equiv 0 \pmod{p}, \\ 0 & \text{otherwise.} \end{cases}$$

As an immediate consequence, we have the following corollary.

Corollary 1.3. *For any integers l_1, l_2, l_3 and prime p, we have $S(l_1, l_2, l_3) \ll p^2$.*

Also, observe that (1.1) contains "trivial" solutions when we have equality instead of congruence equivalence. Hence, we define

$$\mathcal{T}(H) := \left\{1 \le a, b, c \le H : \frac{1}{a} + \frac{1}{b} = \frac{1}{c}\right\}.$$

This seems to be a new counting problem related to Egyptian fractions and we have the following theorem.

Theorem 1.4. *For $H \ge 1$,*

$$\mathcal{T}(H) = \frac{6 \log 2}{\pi^2} H \log H + O(H).$$

Therefore, the expected number of solutions H^3/p is valid only when $H \gg p^{1/2} \log^{1/2} p$. Interestingly, the number of solutions of (1.1) is related to certain first moment of Dirichlet L-functions.

Theorem 1.5. *For $2\sqrt{p} \le H \le p$,*

$$\mathcal{N}(H; p) - \mathcal{T}(H) \gg H \sum_{H^2/(5p) < k \le H^2/(4p)} L\left(1, \left(\frac{kp}{\cdot}\right)\right) + O\left(\frac{H^{59/20} \log^2 H}{p}\right),$$

where $L(s, \chi) = \sum_{n=1}^{\infty} \frac{\chi(n)}{n^s}$ for any Dirichlet character χ.

Theorem 1.6. *For $2\sqrt{p}\le H\le p$ and $3\le x<H^2/p$,*

$$\sum_{k\le x} L\left(1,\left(\frac{kp}{\cdot}\right)\right)=\frac{6}{\pi^2}x+O((\log H)^{10})+O\left(x\exp\left(-\frac{\log x}{5\log\log x}\right)\right).$$

Combining Theorems 1.4–1.6, we have, for $H\ge p^{1/2}(\log p)^6$,

$$\mathcal{N}(H;p)\gg\frac{H^3}{p}.$$

Thus, although we cannot prove the asymptotic formula in Theorem 1.1 for H less than $p^{2/3}$, we have a lower bound of the correct order of magnitude when H is down to $p^{1/2}(\log p)^6$. On the other hand, when $2p^{1/2}\le H\le\sqrt{5}p^{1/2}$, it is possible that $L(1,(\frac{p}{\cdot}))$ can be as big as $e^{\gamma}\log\log p$ where $\gamma=0.5772\ldots$ is Euler–Mascheroni constant (see, for example, [1, Theorem 5a]). Therefore, apart from the "trivial" solutions, there can be other contributions that exceed the expected main term H^3/p when H is near $p^{1/2}$. So, the behavior of $\mathcal{N}(H;p)$ can be very delicate when H is close to $p^{1/2}$.

One possible future work is to obtain the second-order main term for $\mathcal{T}(H)$ in Theorem 1.4. Another direction is to obtain an upper bound for $\mathcal{N}(H;p)-\mathcal{T}(H)$ similar to the lower bound in Theorem 1.5.

This paper is organized as follows. First, we prove Theorem 1.4 concerning the "trivial" solutions. Then, we prove Theorem 1.1 assuming Proposition 1.2. After that, we establish Proposition 1.2. At the end, we discuss the connection with first moment of L-functions and prove Theorems 1.5 and 1.6.

Some Notations. Throughout the paper, p stands for a prime. We use the Legendre symbol $(\frac{\cdot}{p})$ and its generalization, the Jacobi symbol, $(\frac{\cdot}{\cdot})$. Also, $e(x)=e^{2\pi ix}$ and $||x||$ stands for the distance between a real number x and the nearest integer. The notations $f(x)\ll g(x)$, $g(x)\gg f(x)$ and $f(x)=O(g(x))$ are equivalent to $|f(x)|\le Cg(x)$ for some constant $C>0$.

2. Some Lemmas on Gauss Sums and Exponential Sum

For any non-principal Dirichlet character $\chi \pmod p$, let

$$\tau(\chi)=\sum_{n=1}^{p}\chi(n)e\left(\frac{n}{p}\right).$$

Lemma 2.1.

$$\tau(\chi)=\epsilon_p\sqrt{p},\quad \text{where } \epsilon_p=\begin{cases}1 & \text{if } p\equiv 1 \pmod 4,\\ i & \text{if } p\equiv 3 \pmod 4.\end{cases}$$

Proof. See, for example, [3, Theorem 3.3]. □

Lemma 2.2. *For any integer a,*

$$\sum_{n=1}^{p} \chi(n) e\left(\frac{an}{p}\right) = \overline{\chi}(a)\tau(\chi).$$

Proof. If $a \equiv 0 \pmod p$, then the above sum is simply zero by orthogonality of character sums. This matches with the right-hand side as $\overline{\chi}(a) = 0$. Otherwise, as $\overline{\chi}(a)\chi(a) = 1$, the above sum is equal to

$$\overline{\chi}(a) \sum_{n=1}^{p} \chi(an) e\left(\frac{an}{p}\right) = \overline{\chi}(a) \sum_{m=1}^{p} \chi(m) e\left(\frac{m}{p}\right) = \overline{\chi}(a)\tau(\chi)$$

by a change of variable $m = an$. □

Recall the quadratic Gauss sum

$$G\left(\frac{a}{p}\right) = \sum_{n=1}^{p} e\left(\frac{an^2}{p}\right).$$

Then, we have the following lemma.

Lemma 2.3. *For any integer $a \not\equiv 0 \pmod p$,*

$$G\left(\frac{a}{p}\right) = \left(\frac{a}{p}\right) \epsilon_p \sqrt{p},$$

where $(\frac{\cdot}{p})$ is the Legendre symbol.

Proof. See, for example, [3, Eq. (3.29)]. □

Finally, we have the following lemma.

Lemma 2.4. *For any integers $1 \le H \le k$,*

$$\sum_{l=1}^{k} \left| \sum_{a \le H} e\left(\frac{la}{k}\right) \right| \ll k \log k.$$

Proof. For $l \ne k$, by summing geometric series,

$$\sum_{a \le H} e\left(\frac{la}{k}\right) = \frac{1 - e\left(\dfrac{l(H+1)}{k}\right)}{1 - e\left(\dfrac{l}{k}\right)} = \frac{e\left(\dfrac{l(H+1)}{2k}\right)}{e\left(\dfrac{l}{2k}\right)} \frac{e\left(\dfrac{l(H+1)}{2k}\right) - e\left(-\dfrac{l(H+1)}{2k}\right)}{e\left(\dfrac{l}{2k}\right) - e\left(-\dfrac{l}{2k}\right)}.$$

Hence, for $1 \le l \le k/2$,

$$\left| \sum_{a \le H} e\left(\frac{la}{k}\right) \right| \ll \min\left(H, \frac{1}{\left|\sin \dfrac{\pi l}{k}\right|} \right) \ll \min\left(H, \frac{1}{\left\|\dfrac{l}{k}\right\|} \right)$$

as $\sin\theta \ge \frac{2}{\pi}\theta$ for $0\le\theta\le\pi/2$. Therefore, as $\sin(\pi-\theta)=\sin\theta$,

$$\sum_{l=1}^{k}\left|\sum_{a\le H} e\left(\frac{la}{k}\right)\right| \ll 2\left[\sum_{0\le l\le k/H} H + \sum_{k/H<l\le k/2}\frac{k}{l}\right] \ll k\log k. \qquad \square$$

3. Proof of Theorem 1.4

Recall the equation

$$\frac{1}{a}+\frac{1}{b}=\frac{1}{c} \tag{3.1}$$

has $\mathcal{T}(H)$ solutions with $1\le a,b,c\le H$. Let $d=(a,b)$ with $a=da'$, $b=db'$ and $(a',b')=1$. By collecting common denominator, we have

$$\frac{a'+b'}{da'b'}=\frac{1}{c} \quad \text{or} \quad c(a'+b')=da'b'.$$

Since $(a',b')=1$, $(a'+b',a'b')=1$, then

$$c=a'b't \quad \text{and} \quad d=(a'+b')t$$

for some integer t. One possibility is $a'=b'=1$ which gives $a=b=d=2t\le H$ and $c=t$. This contributes $H/2+O(1)$ solutions to (3.1). Otherwise, without loss of generality, we may assume $1\le a'<b'$ with $(a',b')=1$ and

$$a=(a'+b')a't, \quad b=(a'+b')b't \quad \text{and} \quad c=a'b't.$$

Therefore,

$$\mathcal{T}(H)=\frac{H}{2}+O(1)+2\sum_{t\le\frac{H}{6}}\sum_{1<b'\le\sqrt{\frac{H}{t}+\frac{1}{4}}-\frac{1}{2}}\sum_{\substack{1\le a'<\min(b',\frac{H}{b't}-b')\\(a',b')=1}}1=:\frac{H}{2}+O(1)+2\Sigma$$

since $(1+b')b'\le\frac{H}{t}$ and $(a'+b')b't\le H$. We have

$$\Sigma=\sum_{t\le\frac{H}{6}}\sum_{1<b'\le\sqrt{\frac{H}{2t}}}\sum_{\substack{1\le a'<b'\\(a',b')=1}}1+\sum_{t\le\frac{H}{6}}\sum_{\sqrt{\frac{H}{2t}}<b'\le\sqrt{\frac{H}{t}+\frac{1}{4}}-\frac{1}{2}}\sum_{\substack{1\le a'<\frac{H}{b't}-b'\\(a',b')=1}}1=:\Sigma_1+\Sigma_2.$$

Firstly, as $\sum_{n\le x}\phi(n)=3x^2/\pi^2+O(x\log x)$ (see, for example, [6, Theorem 4 p. 40]),

$$\Sigma_1=\sum_{t\le\frac{H}{6}}\sum_{1<b'\le\sqrt{\frac{H}{2t}}}\phi(b')=\sum_{t\le\frac{H}{6}}\left[\frac{3}{\pi^2}\frac{H}{2t}+O\left(\sqrt{\frac{H}{t}}\log\frac{H}{t}\right)\right]$$
$$=\frac{3}{2\pi^2}H\log H+O(H).$$

Secondly, as

$$\sum_{\substack{n\le x\\(n,k)=1}} 1=\sum_{n\le x}\sum_{d|n,d|k}\mu(d)=\sum_{\substack{d|k\\d\le x}}\mu(d)\left[\frac{x}{d}\right]=x\sum_{d|k}\frac{\mu(d)}{d}$$

$$+O(d(k))=\frac{\phi(k)}{k}x+O(d(k)),$$

we have

$$\Sigma_2=\sum_{t\le\frac{H}{6}}\sum_{\sqrt{\frac{H}{2t}}<b'\le\sqrt{\frac{H}{t}+\frac{1}{4}}-\frac{1}{2}}\left\{\frac{\phi(b')}{b'}\left(\frac{H}{b't}-b'\right)+O(d(b'))\right\}$$

$$=\sum_{t\le\frac{H}{6}}\left[\frac{3\log 2}{\pi^2}\frac{H}{t}-\frac{3}{\pi^2}\frac{H}{2t}\right]+O\left(\sum_{t\le\frac{H}{6}}\sqrt{\frac{H}{t}}\log\frac{H}{t}\right)$$

$$=\left[\frac{3\log 2}{\pi^2}-\frac{3}{2\pi^2}\right]H\log H+O(H).$$

Consequently, combining all the above,

$$\mathcal{T}(H)=\frac{6\log 2}{\pi^2}H\log H+O(H)$$

and we have Theorem 1.4.

4. Proof of Theorem 1.1

By orthogonality of additive characters,

$$\mathcal{N}(H;p)=\frac{1}{p}\sum_{a\le H}\sum_{b\le H}\sum_{c\le H}\sum_{k=1}^{p}e\left(\frac{k(\bar a+\bar b-\bar c)}{p}\right)$$

$$=\frac{H^3}{p}+\frac{1}{p}\sum_{k=1}^{p-1}\sum_{a\le H}e\left(\frac{k\bar a}{p}\right)\sum_{b\le H}e\left(\frac{k\bar b}{p}\right)\sum_{c\le H}e\left(\frac{-k\bar c}{p}\right).$$

Using standard completing the sum method,

$$\sum_{a\le H}e\left(\frac{k\bar a}{p}\right)=\frac{1}{p}\sum_{a\le H}\sum_{n=1}^{p-1}\sum_{l=1}^{p}e\left(\frac{k\bar n}{p}\right)e\left(\frac{l(n-a)}{p}\right)$$

$$=\frac{1}{p}\sum_{l=1}^{p}\sum_{n=1}^{p}e\left(\frac{ln+k\bar n}{p}\right)\sum_{a\le H}e\left(-\frac{la}{p}\right)$$

$$=\frac{1}{p}\sum_{l=1}^{p}S(l,k;p)F(l,H;p),$$

where

$$S(l,k;p)=\sum_{n=1}^{p-1}e\left(\frac{ln+k\bar n}{p}\right)$$

is the usual Kloosterman sum, and

$$F(l,H;p) := \sum_{a\le H} e\left(-\frac{la}{p}\right).$$

Hence,

$$\mathcal{N}(H;p) = \frac{H^3}{p} + \frac{1}{p^4}\sum_{k=1}^{p-1}\sum_{l_1=1}^{p} S(l_1,k;p)F(l_1,H;p)\sum_{l_2=1}^{p} S(l_2,k;p)F(l_2,H;p)$$

$$\times\sum_{l_3=1}^{p} S(l_3,-k;p)F(l_3,H;p)$$

$$= \frac{H^3}{p} + \frac{1}{p^4}\sum_{l_1=1}^{p}\sum_{l_2=1}^{p}\sum_{l_3=1}^{p} F(l_1,H;p)F(l_2,H;p)F(l_3,H;p)$$

$$\times\sum_{k=1}^{p-1} S(l_1,k;p)S(l_2,k;p)S(l_3,-k;p).$$

$$= \frac{H^3}{p} + \frac{1}{p^4}\sum_{l_1=1}^{p}\sum_{l_2=1}^{p}\sum_{l_3=1}^{p} F(l_1,H;p)F(l_2,H;p)F(l_3,H;p)S(l_1,l_2,-l_3)$$

as $S(a,-b;p)=S(-a,b;p)$. By Corollary 1.3 and Lemma 2.4,

$$\mathcal{N}(H;p) = \frac{H^3}{p} + O\left(\frac{1}{p^2}\sum_{l_1=1}^{p}\sum_{l_2=1}^{p}\sum_{l_3=1}^{p} |F(l_1,H;p)||F(l_2,H;p)||F(l_3,H;p)|\right)$$

$$= \frac{H^3}{p} + O(p\log^3 p),$$

which gives Theorem 1.1.

5. Proof of Proposition 1.2

In this section, we consider

$$S := S(l_1,l_2,l_3) = \sum_{k=1}^{p-1} S(l_1,k;p)S(l_2,k;p)S(l_3,k;p).$$

By definition of Kloosterman sum and orthogonality of additive characters,

$$S = \sum_{k=1}^{p-1}\sideset{}{'}\sum_{a \pmod p} e\left(\frac{l_1a+k\bar{a}}{p}\right)\sideset{}{'}\sum_{b \pmod p} e\left(\frac{l_2b+k\bar{b}}{p}\right)\sideset{}{'}\sum_{c \pmod p} e\left(\frac{l_3c+k\bar{c}}{p}\right)$$

$$= \sideset{}{'}\sum_{a \pmod p}\sideset{}{'}\sum_{b \pmod p}\sideset{}{'}\sum_{c \pmod p} e\left(\frac{l_1\bar{a}+l_2\bar{b}+l_3\bar{c}}{p}\right)\left[\sum_{k=1}^{p} e\left(\frac{k(a+b+c)}{p}\right)-1\right]$$

$$= p \sideset{}{'}\sum_{\substack{a \pmod p \\ a+b\not\equiv 0 \pmod p}} \sideset{}{'}\sum_{b \pmod p} e\left(\frac{l_1\overline{a} + l_2\overline{b} - l_3\overline{a+b}}{p}\right)$$

$$- \sideset{}{'}\sum_{a \pmod p} \sideset{}{'}\sum_{b \pmod p} \sideset{}{'}\sum_{c \pmod p} e\left(\frac{l_1\overline{a} + l_2\overline{b} + l_3\overline{c}}{p}\right)$$

$$=: pS_1 - (p\mathcal{I}_{l_1} - 1)(p\mathcal{I}_{l_2} - 1)(p\mathcal{I}_{l_3} - 1),$$

where $\mathcal{I}_k$ is the indicator function of whether $k \equiv 0 \pmod p$. Here after, $\sum'$ stands for a sum over reduced residues $(\bmod\ p)$. By a change of variables $u = \overline{a}$ and $v = \overline{v}$ and noting that $a \not\equiv -b \pmod p$ is equivalent to $u \not\equiv -v \pmod p$, the above first sum becomes

$$S_1 = \sideset{}{'}\sum_{\substack{u \pmod p \\ u+v\not\equiv 0 \pmod p}} \sideset{}{'}\sum_{v \pmod p} e\left(\frac{l_1 u + l_2 v - l_3 uv\overline{u+v}}{p}\right)$$

Let $w = u + v$. Then, as $u \not\equiv 0 \pmod p$, we have $v \not\equiv w \pmod p$ and

$$S_1 = \sideset{}{'}\sum_{\substack{w \pmod p \\ v\not\equiv w \pmod p}} \sideset{}{'}\sum_{v \pmod p} e\left(\frac{l_1 w + (l_2 - l_1)v - l_3(w - v)v\overline{w}}{p}\right)$$

$$= \sideset{}{'}\sum_{w \pmod p} \sideset{}{'}\sum_{v \pmod p} e\left(\frac{l_1 w + (l_2 - l_3 - l_1)v + l_3 v^2\overline{w}}{p}\right) - \sideset{}{'}\sum_{w \pmod p} e\left(\frac{l_2 w}{p}\right)$$

$$= \sideset{}{'}\sum_{w \pmod p} e\left(\frac{l_1 w}{p}\right) \sideset{}{'}\sum_{v \pmod p} e\left(\frac{l_3\overline{w}v^2 + (l_2 - l_3 - l_1)v}{p}\right) - \sideset{}{'}\sum_{w \pmod p} e\left(\frac{l_2 w}{p}\right)$$

$$= \sideset{}{'}\sum_{w \pmod p} e\left(\frac{l_1 w}{p}\right) \sum_{v \pmod p} e\left(\frac{l_3\overline{w}v^2 + (l_2 - l_3 - l_1)v}{p}\right)$$

$$- \sideset{}{'}\sum_{w \pmod p} e\left(\frac{l_1 w}{p}\right) - \sideset{}{'}\sum_{w \pmod p} e\left(\frac{l_2 w}{p}\right)$$

$$= \sideset{}{'}\sum_{w \pmod p} e\left(\frac{l_1 w}{p}\right) \sum_{v \pmod p} e\left(\frac{l_3\overline{w}v^2 + (l_2 - l_3 - l_1)v}{p}\right) - p\mathcal{I}_{l_1} - p\mathcal{I}_{l_2} + 2$$

by orthogonality of additive characters. For $l_3 \not\equiv 0 \pmod p$, we complete the square in v, make a change of variable and obtain

$$S_1 = \sideset{}{'}\sum_{w \pmod p} e\left(\frac{[l_1 - \overline{4}l_3(l_2 - l_3 - l_1)^2]w}{p}\right) \sum_{v' \pmod p} e\left(\frac{l_3\overline{w}v'^2}{p}\right) - p\mathcal{I}_{l_1} - p\mathcal{I}_{l_2} + 2$$

$$= \epsilon_p\sqrt{p} \sideset{}{'}\sum_{w \pmod p} \left(\frac{l_3\overline{w}}{p}\right) e\left(\frac{[l_1 + \overline{4}l_3(l_2 - l_3 - l_1)^2]w}{p}\right) - p\mathcal{I}_{l_1} - p\mathcal{I}_{l_2} + 2$$

$$= \left(\frac{l_3}{p}\right)\epsilon_p\sqrt{p}\left(\frac{l_1-\overline{4l_3}(l_2-l_3-l_1)^2}{p}\right)\epsilon_p\sqrt{p} - p\mathcal{I}_{l_1} - p\mathcal{I}_{l_2} + 2$$
$$= p\left(\frac{-4l_1l_3+(l_2-l_3-l_1)^2}{p}\right) - p\mathcal{I}_{l_1} - p\mathcal{I}_{l_2} + 2$$

by Lemmas 2.1–2.3 and the well-known fact that $(\frac{-1}{p}) = \epsilon_p^2$.

For $l_3 \equiv 0 \pmod p$, by orthogonality of additive characters,

$$S_1 = p\mathcal{I}_{l_1-l_2}(p\mathcal{I}_{l_1}-1) - p\mathcal{I}_{l_1} - p\mathcal{I}_{l_2} + 2 = p^2\mathcal{I}_{l_1-l_2}\mathcal{I}_{l_1} - p\mathcal{I}_{l_1-l_2} - p\mathcal{I}_{l_1} - p\mathcal{I}_{l_2} + 2.$$

Therefore,

$$S = \begin{cases} p^2\left(\frac{(l_2-l_3-l_1)^2-4l_1l_3}{p}\right) - p^2\mathcal{I}_{l_1} - p^2\mathcal{I}_{l_2} + 2p & \\ \qquad - (p\mathcal{I}_{l_1}-1)(p\mathcal{I}_{l_2}-1)(p\mathcal{I}_{l_3}-1) & \text{if } l_3 \not\equiv 0 \pmod p, \\ p^3\mathcal{I}_{l_1}\mathcal{I}_{l_2} - p^2\mathcal{I}_{l_1-l_2} - p^2\mathcal{I}_{l_1} - p^2\mathcal{I}_{l_2} + 2p & \\ \qquad - (p\mathcal{I}_{l_1}-1)(p\mathcal{I}_{l_2}-1)(p\mathcal{I}_{l_3}-1) & \text{if } l_3 \equiv 0 \pmod p. \end{cases}$$

One can check that the above equation is the same as

$$S = p^3\mathcal{I}_{l_1}\mathcal{I}_{l_2}\mathcal{I}_{l_3} + p^2\left(\frac{l_1^2+l_2^2+l_3^2-2l_1l_2-2l_2l_3-2l_3l_1}{p}\right)$$
$$- p^2\mathcal{I}_{l_1} - p^2\mathcal{I}_{l_2} - p^2\mathcal{I}_{l_3} + 2p - (p\mathcal{I}_{l_1}-1)(p\mathcal{I}_{l_2}-1)(p\mathcal{I}_{l_3}-1)$$

since the Legendre symbol is 1 when $l_3 \equiv 0 \pmod p$ unless $l_1 \equiv l_2 \pmod p$. This gives Proposition 1.2.

6. Proof of Theorems 1.5 and 1.6

Multiplying both sides of equation (1.1) by abc, we have

$$bc + ac \equiv ab \pmod p$$

or

$$c^2 \equiv (a-c)(b-c) \pmod p. \tag{6.1}$$

From (6.1), we have

$$c^2 = de + kp \tag{6.2}$$

for some integer k with $d = a-c$ and $e = b-c$. Rephrase the above equation as

$$c^2 \equiv kp \pmod e. \tag{6.3}$$

Assume $1 \le c \le H/2$ and $H/8 < e \le H/4$ is odd. For fixed e and $H^2/(5p) < k \le H^2/(4p)$ (which guarantees $d \le H/2$ and hence $a, b \le H$), the quadratic congruence equation (6.3) has at least

$$\prod_{q|e}\left(1+\left(\frac{kp}{q}\right)\right)\left\lfloor\frac{H}{2e}\right\rfloor \ge \prod_{q|e}\left(1+\left(\frac{kp}{q}\right)\right)$$

solutions in c as $e \le H/4$. Hereafter, q stands for an odd prime number. For fixed k, Eq. (6.3) has at least

$$\sum_{\substack{H/8<e\le H/4\\ e\text{ odd}}} \prod_{q|e}\left(1+\left(\frac{kp}{q}\right)\right) = \sum_{\substack{H/8<e\le H/4\\ e\text{ odd}}} \sum_{t|e}{}^{*}\left(\frac{kp}{t}\right) \tag{6.4}$$

solutions where $\left(\frac{\cdot}{\cdot}\right)$ stands for the Jacobi symbol, a generalization of the Legendre symbol. The $*$ in the summation means that we are only summing over squarefree integers. Borrowing from Soundararajan [5],

$$\frac{1}{2\pi i}\int_{c-i\infty}^{c+i\infty} \xi^s \frac{ds}{s^2} = \begin{cases} \log\xi & \text{if } \xi \ge 1,\\ 0 & \text{if } 0<\xi\le 1,\end{cases}$$

and

$$\frac{1}{2\pi i}\int_{c-i\infty}^{c+i\infty} \xi^s \frac{(e^{\delta s}-1)^2}{s^2}ds = \begin{cases} \min(\log(e^{2\delta}\xi), \log(1/\xi)) & \text{if } e^{-2\delta}\le\xi\le 1,\\ 0 & \text{otherwise.}\end{cases}$$

With $\sigma_0 = 1 + 1/\log H$ and

$$L\left(s,\left(\frac{D}{\cdot}\right)\right) = \sum_{n=1}^{\infty}\frac{\left(\frac{D}{n}\right)}{n^s}, \quad\text{and}\quad \frac{L\left(s,\left(\frac{D}{\cdot}\right)\right)}{Ls\left(2s,\left(\frac{D}{\cdot}\right)\right)} = \sum_{n=1}^{\infty}{}^{*}\frac{\left(\frac{D}{n}\right)}{n^s},$$

the quantity (6.4) is at least

$$\frac{1}{\sqrt{2}}\frac{1}{2\pi i}\int_{\sigma_0-i\infty}^{\sigma_0+i\infty}\frac{L\left(s,\left(\frac{kp}{\cdot}\right)\right)}{L\left(2s,\left(\frac{kp}{\cdot}\right)\right)}\zeta(s)\left(1-\frac{1}{2^s}\right)\left(\frac{H}{8}\right)^s\frac{(e^{s\log 2/2}-1)^2}{s^2}ds.$$

By moving the contour to the line $\text{Re}(s) = 3/4$, the above equals

$$\frac{(\sqrt{2}-1)^2 H}{16\sqrt{2}}\frac{L\left(1,\left(\frac{kp}{\cdot}\right)\right)}{L\left(2,\left(\frac{kp}{\cdot}\right)\right)}$$

$$+\frac{1}{\sqrt{2}}\frac{1}{2\pi i}\int_{3/4-i\infty}^{3/4+i\infty}\frac{L\left(s,\left(\frac{kp}{\cdot}\right)\right)}{L\left(2s,\left(\frac{kp}{\cdot}\right)\right)}\zeta(s)\left(1-\frac{1}{2^s}\right)\left(\frac{H}{8}\right)^s\frac{(e^{s\log 2/2}-1)^2}{s^2}ds.$$

Recall the subconvexity bounds for Riemann zeta-function (by [3, Eq. (8.22)]) and Dirichlet L-functions (by [2, Corollary 1]),

$$\zeta(\sigma+it) \ll t^{(1-\sigma)/3}\log^2 t \quad\text{and}\quad L\left(\sigma+it,\left(\frac{kp}{\cdot}\right)\right) \ll (kpt)^{2(1-\sigma)/5}\log^2(kp).$$

Hence, the above integral is

$$\ll (kp)^{1/10}H^{3/4}\log^2(kp) \ll H^{19/20}\log^2 H$$

as $kp \le H^2$. Therefore, the number of solutions to Eq. (6.3) subjected to $H^2/(5p) < k \le H^2/(4p)$ is much greater than

$$H \sum_{H^2/(5p)<k\le H^2/(4p)} L\left(1,\left(\frac{kp}{\cdot}\right)\right) + O\left(\frac{H^{59/20}\log^2 H}{p}\right).$$

This gives Theorem 1.5 as $k=0$ corresponds to "trivial" solutions.

For $1 \le x < H^2/p \le p$, one can study the first moment

$$\sum_{k\le x} L\left(1,\left(\frac{kp}{\cdot}\right)\right),$$

by the method in Granville and Soundararajan [1]. By [1, Lemma 2.3], with $Z = \exp((2\log H)^{10})$,

$$L\left(1,\left(\frac{kp}{\cdot}\right)\right) = \sum_{n=1}^{\infty} \frac{\left(\frac{kp}{n}\right)}{n} e^{-n/Z} + O\left(\frac{1}{kp}\right)$$

if the L-function $L(s,(\frac{kp}{\cdot}))$ has no (Landau–Siegel) zero on the interval $[1-\frac{c}{\log(ekp)},1)$ with some small constant $c>0$. Since there is at most one such L-function with Landau–Siegel zero for $x<k\le 2x$, by following the proof of [1, Theorem 2], we have

$$\sum_{k\le x} L\left(1,\left(\frac{kp}{\cdot}\right)\right) = \frac{6}{\pi^2}x + O((\log H)^{10}) + O\left(x\exp\left(-\frac{\log x}{5\log\log x}\right)\right),$$

which gives Theorem 1.6. Note that the error term $O((\log H)^{10})$ comes from the estimate

$$\sum_{n=1}^{\infty} \frac{1}{n}e^{-n/Z} \ll \log Z,$$

which arises in the proof.

Note to the proof: After finishing the paper, the author came across a paper by D. H. Lehmer and E. Lehmer [4]. Its Eq. (3.20) is essentially the same as Proposition 1.2 but our result exhibits symmetry in all three variables.

References

[1] A. Granville and K. Soundararajan, The distribution of values of $L(1,\chi_d)$, *Geom. Funct. Anal.* **13** (2003) 992–1028.

[2] D. R. Heath-Brown, Hybrid bounds for Dirichlet L-functions, *Invent. Math.* **47** (1978) 149–170.

[3] H. Iwaniec and E. Kowalski, *Analytic Number Theory* (American Mathematical Society, Providence, RI, 2004).

[4] D. H. Lehmer and E. Lehmer, The cyclotomy of Kloosterman sums, *Acta Arith.* **12**(4) (1967) 385–407.
[5] K. Soundararajan, Smooth numbers in short intervals, preprint (2010); arXiv:1009.1591v1 [math.NT].
[6] G. Tenenbaum, *Introduction to Analytic and Probabilistic Number Theory* (Cambridge University Press, 1995).

© 2024 World Scientific Publishing Company
https://doi.org/10.1142/9789811277375_0009

Weighted partition rank and crank moments. I. Andrews–Beck type congruences

Shane Chern

Department of Mathematics, Penn State University, University Park, PA 16802, USA
shanechern@psu.edu, chenxiaohang92@gmail.com

Dedicated to Professor Bruce Berndt on the occasion of his 80th birthday

Recently, George Beck conjectured and George Andrews proved a handful of novel congruences concerning the rank statistic of Dyson and the crank statistic of Andrews and Garvan. In this paper, we shall prove two identities concerning the weighted rank and crank moments, from which more congruences of the Andrews–Beck type may be deduced.

Keywords: Partition; Andrews–Beck type congruence; rank; crank; weighted moment.

Mathematics Subject Classification 2020: 11P83, 11P81, 05A19

1. Introduction

1.1. *Background*

As usual, a *partition* of a positive integer n is a weakly decreasing sequence of positive integers whose sum equals n. For example, 4 has five partitions: 4, $3+1$, $2+2$, $2+1+1$ and $1+1+1+1$. If we denote by $p(n)$ the number of partitions of n, then $p(4)=5$.

In the theory of partitions, one of the most fascinating results is due to Ramanujan, who discovered that $p(n)$ satisfies the following congruences:

$$p(5n+4) \equiv 0 \pmod{5}, \tag{1.1}$$

$$p(7n+5) \equiv 0 \pmod{7}, \tag{1.2}$$

$$p(11n+6) \equiv 0 \pmod{11}. \tag{1.3}$$

See [13–15] or [7] or [1, Chap. 10].

In his famous 1944 paper [8], Dyson defined the *rank* of a partition as the largest part minus the number of parts. Dyson then conjectured that the rank statistic may provide combinatorial interpretations of (1.1) and (1.2); this assertion was later confirmed by Atkin and Swinnerton-Dyer [6]. Dyson also conjectured the existence of a *crank* statistic, which is able to combinatorially explain all the three congruences of Ramanujan. Such a statistic was not discovered until over four decades later by Andrews and Garvan [4] based on Garvan's study of the vector crank [10, 11].

As in [6], we denote by $N(m,k,n)$ the number of partitions of n with rank congruent to m modulo k. Atkin and Swinnerton-Dyer proved (1.1) and (1.2) by showing that for $0 \leq i \leq 4$,

$$N(i,5,5n+4) = \frac{1}{5}p(5n+4)$$

and that for $0 \leq i \leq 6$,

$$N(i,7,7n+5) = \frac{1}{7}p(7n+5).$$

In a recent paper [3], Andrews recorded that George Beck has conjectured a number of new congruences along a somewhat different road. Instead of considering the $N(m,k,n)$ function, Beck studied the total number of parts in the partitions of n with rank congruent to m modulo k, which is defined by $NT(m,k,n)$. One of the results proved by Andrews reads as follows.

Theorem ([3, Theorem 1.1]). If $i = 1$ or 4, then for $n \geq 0$,

$$\begin{aligned} &NT(1,5,5n+i) + 2NT(2,5,5n+i) \\ &\quad - 2NT(3,5,5n+i) - NT(4,5,5n+i) \equiv 0 \pmod 5. \end{aligned} \tag{1.4}$$

Andrews also remarked that (1.4) is trivial if one replaces the $NT(m,k,n)$ function by the rank function $N(m,k,n)$. This is simply due to the symmetry

$$N(m,k,n) = N(k-m,k,n).$$

However, the above symmetry is generally false for the $NT(m,k,n)$ function. Therefore, the validness of (1.4) is in some sense exciting.

In his proof of (1.4), Andrews did not apply the differentiation technique directly to the trivariate generating function

$$\sum_{n\geq 0}\sum_{\lambda\vdash n} x^{\sharp(\lambda)} z^{\mathrm{rank}(\lambda)} q^n = \sum_{n\geq 0} \frac{x^n q^{n^2}}{(zq;q)_n (xq/z;q)_n}.$$

Instead, he transformed the above generating function as

$$\begin{aligned} \sum_{n\geq 0}\sum_{\lambda\vdash n} x^{\sharp(\lambda)} z^{\mathrm{rank}(\lambda)} q^n &= 1 + \frac{1}{(xq;q)_\infty} \sum_{n\geq 1} (-1)^{n-1} q^{n(3n+1)/2} x^n \frac{(xq;q)_n}{(q;q)_{n-1}} \\ &\quad \times \left(\frac{1}{q^n(1-zq^n)} + \frac{x}{z(1-\frac{xq^n}{z})} \right), \end{aligned}$$

so he may take advantage of several results proved already by Atkin and Swinnerton-Dyer [6]; see [3, Theorem 3.1].

However, we discover that if we consider a weighted rank moment, then it is possible to arrive at a connection with the second Atkin–Garvan rank moment defined in [5] so that, surprisingly, (1.4) follows as an immediate consequence. In analogy, a weighted crank moment will lead to proofs of several conjectures of Beck concerning the crank function.

1.2. *Notation and terminology*

Let $\mathcal{P}$ be the set of partitions. As usual, the notation $\lambda \vdash n$ means that λ is a partition of n. Below, let λ always be a partition. Let $|\lambda|$ be the size of λ. We use $\sharp(\lambda)$ and $\omega(\lambda)$ to denote the number of parts in λ and the number of ones in λ, respectively. Further, $\mathrm{rank}(\lambda)$ and $\mathrm{crank}(\lambda)$ denote the rank and crank of λ.

As already defined, $NT(m,k,n)$ equals the total number of parts in the partitions of n with rank congruent to m modulo k. We also denote by $N(m,n)$ the number of partitions of n whose rank is m. Then the second Atkin–Garvan rank moment $N_2(n)$ is defined by

$$N_2(n) := \sum_{m=-\infty}^{\infty} m^2 N(m,n) = \sum_{\lambda \vdash n} \mathrm{rank}^2(\lambda).$$

On the other hand, let $M_\omega(m,k,n)$ count the total number of ones in the partitions of n with crank congruent to m modulo k.

Finally, we adopt the standard q-Pochhammer symbol for $n \in \mathbb{N} \cup \{\infty\}$:

$$(A;q)_n := \prod_{k=0}^{n-1} (1 - Aq^k).$$

1.3. *Main results*

Our first result treats the following weighted rank moment.

Theorem 1.1. *We have*

$$\sum_{\lambda \in \mathcal{P}} \sharp(\lambda)\, \mathrm{rank}(\lambda) q^{|\lambda|} = -\sum_{n \geq 1} \frac{q^{n^2}}{(q;q)_n^2} \sum_{m=1}^{n} \frac{q^m}{(1-q^m)^2} \tag{1.5}$$

and

$$\sum_{\lambda \vdash n} \sharp(\lambda)\, \mathrm{rank}(\lambda) = -\frac{1}{2} N_2(n). \tag{1.6}$$

Remark 1.1. It is worth pointing out that the following generating function identity for $N_2(n)$ is used most frequently:

$$\sum_{n \geq 0} N_2(n) q^n = -\frac{2}{(q;q)_\infty} \sum_{n \geq 1} \frac{(-1)^n q^{n(3n+1)/2}(1+q^n)}{(1-q^n)^2}.$$

See [2, Eq. (3.4)].

Likewise, for the weighted crank moment, our result reads as follows.

Theorem 1.2. *We have*

$$\sum_{\lambda\in\mathcal{P}} \omega(\lambda)\,\mathrm{crank}(\lambda)q^{|\lambda|} = -\frac{1}{(q;q)_\infty}\sum_{n\geq 1}\frac{q^n}{(1-q^n)^2} \tag{1.7}$$

and

$$\sum_{\lambda\vdash n} \omega(\lambda)\,\mathrm{crank}(\lambda) = -np(n). \tag{1.8}$$

Remark 1.2. Let $M(m,n)$ denote the number of partitions of n with crank m. Atkin and Garvan [5] defined the k-th crank moment $M_k(n)$ by

$$M_k(n) = \sum_{m=-\infty}^{\infty} m^k M(m,n) = \sum_{\lambda\vdash n}\mathrm{crank}^k(\lambda).$$

It can be shown by means of a relation due to Dyson [9] that

$$M_2(n) = 2np(n).$$

It turns out that

$$\sum_{\lambda\vdash n} \omega(\lambda)\,\mathrm{crank}(\lambda) = -\frac{1}{2}M_2(n). \tag{1.9}$$

This is an analog of (1.6).

As consequences of Theorems 1.1 and 1.2, we arrive at a number of Andrews–Beck type congruences with (1.4) included.

Corollary 1.3. *If* $i=1$ *or* 4, *then for* $n\geq 0$,

$$\begin{aligned}&NT(1,5,5n+i)+2NT(2,5,5n+i)\\&\quad -2NT(3,5,5n+i)-NT(4,5,5n+i)\equiv 0 \pmod 5.\end{aligned} \tag{1.10}$$

Corollary 1.4. *If* $i=1$ *or* 5, *then for* $n\geq 0$,

$$\begin{aligned}&NT(1,7,7n+i)+2NT(2,7,7n+i)\\&\quad +3NT(3,7,7n+i)-3NT(4,7,7n+i)\\&\quad -2NT(5,7,7n+i)-NT(6,7,7n+i)\equiv 0 \pmod 7.\end{aligned} \tag{1.11}$$

Corollary 1.5. *If* $i=0$ *or* 4, *then for* $n\geq 0$,

$$\begin{aligned}&M_\omega(1,5,5n+i)+2M_\omega(2,5,5n+i)\\&\quad -2M_\omega(3,5,5n+i)-M_\omega(4,5,5n+i)\equiv 0 \pmod 5.\end{aligned} \tag{1.12}$$

Corollary 1.6. *If $i = 0$ or 5, then for $n \geq 0$,*

$$\begin{aligned}
&M_\omega(1,7,7n+i) + 2M_\omega(2,7,7n+i)\\
&\quad + 3M_\omega(3,7,7n+i) - 3M_\omega(4,7,7n+i)\\
&\quad - 2M_\omega(5,7,7n+i) - M_\omega(6,7,7n+i) \equiv 0 \pmod 7.
\end{aligned} \tag{1.13}$$

Corollary 1.7. *If $i = 0$ or 6, then for $n \geq 0$,*

$$\begin{aligned}
&M_\omega(1,11,11n+i) + 2M_\omega(2,11,11n+i) + 3M_\omega(3,11,11n+i)\\
&\quad + 4M_\omega(4,11,11n+i) + 5M_\omega(5,11,11n+i) - 5M_\omega(6,11,11n+i)\\
&\quad - 4M_\omega(7,11,11n+i) - 3M_\omega(8,11,11n+i) - 2M_\omega(9,11,11n+i)\\
&\quad - M_\omega(10,11,11n+i) \equiv 0 \pmod{11}.
\end{aligned} \tag{1.14}$$

Finally, let $\mathrm{spt}(n)$ denote the total number of appearances of the smallest parts in all partitions of n. It was shown by Andrews [2] that

$$\mathrm{spt}(n) = np(n) - \frac{1}{2}N_2(n).$$

In view of (1.6) and (1.8), we have the following intriguing relation.

Corollary 1.8. *For $n \geq 0$,*

$$\mathrm{spt}(n) = \sum_{\lambda \vdash n} \sharp(\lambda)\,\mathrm{rank}(\lambda) - \sum_{\lambda \vdash n} \omega(\lambda)\,\mathrm{crank}(\lambda). \tag{1.15}$$

2. The Weighted Rank Moment

We first study the weighted rank moment. It was shown in [3] that

$$\sum_{n\geq 0}\sum_{\lambda\vdash n} x^{\sharp(\lambda)} z^{\mathrm{rank}(\lambda)} q^n = \sum_{n\geq 0} \frac{x^n q^{n^2}}{(zq;q)_n (xq/z;q)_n}. \tag{2.1}$$

Proof of Theorem 1.1. We first apply the operator $[\partial/\partial x]_{x=1}$ to (2.1).

$$\begin{aligned}
&\sum_{n\geq 0}\sum_{\lambda\vdash n} \sharp(\lambda) z^{\mathrm{rank}(\lambda)} q^n\\
&\quad = \sum_{n\geq 0}\left[\frac{\partial}{\partial x}\frac{x^n q^{n^2}}{(zq;q)_n (xq/z;q)_n}\right]_{x=1}\\
&\quad = \sum_{n\geq 0}\left[\frac{x^n q^{n^2}}{(zq;q)_n (xq/z;q)_n}\frac{\partial}{\partial x}\log\left(\frac{x^n}{(xq/z;q)_n}\right)\right]_{x=1}
\end{aligned}$$

$$= \sum_{n\geq 0} \frac{q^{n^2}}{(zq;q)_n(q/z;q)_n} \left[\frac{\partial}{\partial x} \left(n\log x - \sum_{m=1}^{n} \log(1-xq^m/z) \right) \right]_{x=1}$$

$$= \sum_{n\geq 0} \frac{q^{n^2}}{(zq;q)_n(q/z;q)_n} \left[\frac{n}{x} + \sum_{m=1}^{n} \frac{q^m}{z-xq^m} \right]_{x=1}$$

$$= \sum_{n\geq 1} \frac{q^{n^2}}{(zq;q)_n(q/z;q)_n} \left(n + \sum_{m=1}^{n} \frac{q^m}{z-q^m} \right). \tag{2.2}$$

Next, we make an easy observation: for any $n \in \mathbb{N} \cup \{\infty\}$,

$$\left[\frac{\partial}{\partial z} \log \left(\frac{1}{(zq;q)_n(q/z;q)_n} \right) \right]_{z=1} = \left[\sum_{m=1}^{n} \left(\frac{q^m}{1-zq^m} + \frac{q^m}{zq^m - z^2} \right) \right]_{z=1} = 0. \tag{2.3}$$

Applying the operator $[\partial/\partial z]_{z=1}$ to (2.2) and making use of (2.3), we have

$$\sum_{n\geq 0} \sum_{\lambda \vdash n} \sharp(\lambda)\, \mathrm{rank}(\lambda) q^n$$

$$= \sum_{n\geq 1} \left[\frac{\partial}{\partial z} \frac{q^{n^2}}{(zq;q)_n(q/z;q)_n} \left(n + \sum_{m=1}^{n} \frac{q^m}{z-q^m} \right) \right]_{z=1}$$

$$= \sum_{n\geq 1} \left[\frac{nq^{n^2}}{(zq;q)_n(q/z;q)_n} \frac{\partial}{\partial z} \log \left(\frac{1}{(zq;q)_n(q/z;q)_n} \right) \right]_{z=1}$$

$$+ \sum_{n\geq 1} \sum_{m=1}^{n} \left[\frac{q^{n^2}}{(zq;q)_n(q/z;q)_n} \frac{q^m}{z-q^m} \frac{\partial}{\partial z} \log \left(\frac{1}{(zq;q)_n(q/z;q)_n(z-q^m)} \right) \right]_{z=1}$$

$$= -\sum_{n\geq 1} \frac{q^{n^2}}{(q;q)_n^2} \sum_{m=1}^{n} \frac{q^m}{(1-q^m)^2}. \tag{2.4}$$

This is the first part of Theorem 1.1.

On the other hand, if one applies the operator $\left[\frac{\partial}{\partial z}\left(z\frac{\partial}{\partial z}\right)\right]_{z=1}$ to

$$\sum_{n\geq 0} \sum_{\lambda \vdash n} z^{\mathrm{rank}(\lambda)} q^n = \sum_{n\geq 0} \frac{q^{n^2}}{(zq;q)_n(q/z;q)_n},$$

one shall find that

$$\sum_{n\geq 0} N_2(n) q^n = \sum_{n\geq 0} \sum_{\lambda \vdash n} \mathrm{rank}(\lambda)^2 q^n$$

$$= \sum_{n\geq 0} \left[\frac{\partial}{\partial z} \left(z \frac{\partial}{\partial z} \frac{q^{n^2}}{(zq;q)_n(q/z;q)_n} \right) \right]_{z=1}$$

$$= \sum_{n\geq 0} \left[\frac{\partial}{\partial z} \frac{zq^{n^2}}{(zq;q)_n(q/z;q)_n} \sum_{m=1}^{n} \left(\frac{q^m}{1-zq^m} + \frac{q^m}{zq^m - z^2} \right) \right]_{z=1}$$

$$= 2\sum_{n\geq 1} \frac{q^{n^2}}{(q;q)_n^2} \sum_{m=1}^{n} \frac{q^m}{(1-q^m)^2}. \tag{2.5}$$

This combining with (2.4) gives the second part of Theorem 1.1. □

The arithmetic properties satisfied by $N_2(n)$ have been well studied. For example, it was indicated in [12, p. 285] that

$$N_2(5n+1) \equiv N_2(5n+4) \equiv 0 \pmod 5$$

and

$$N_2(7n+1) \equiv N_2(7n+5) \equiv 0 \pmod 7.$$

On the other hand, it is trivial to see that

$$\sum_{\lambda\in\mathcal{P}} \sharp(\lambda)\,\mathrm{rank}(\lambda) q^{|\lambda|} \equiv \sum_{n\geq 0} (NT(1,5,n) + 2NT(2,5,n) - 2NT(3,5,n) - NT(4,5,n))q^n \pmod 5$$

and

$$\sum_{\lambda\in\mathcal{P}} \sharp(\lambda)\,\mathrm{rank}(\lambda) q^{|\lambda|} \equiv \sum_{n\geq 0} (NT(1,7,n) + 2NT(2,7,n) + 3NT(3,7,n) - 3NT(4,7,n) - 2NT(5,7,n) - NT(6,7,n))q^n \pmod 7.$$

Therefore, Corollaries 1.3 and 1.4 follow from the above.

3. The Weighted Crank Moment

Recall that the crank of a partition λ is defined by

$$\mathrm{crank}(\lambda) = \begin{cases} \ell(\lambda) & \text{if } \omega(\lambda) = 0, \\ \mu(\lambda) - \omega(\lambda) & \text{if } \omega(\lambda) > 0, \end{cases}$$

where $\omega(\lambda)$ denotes the number of ones in λ as before, $\ell(\lambda)$ denotes the largest part in λ and $\mu(\lambda)$ denotes the number of parts in λ that are larger than $\omega(\lambda)$.

As in [4], we have

$$\sum_{n\geq 0}\sum_{\lambda\vdash n} x^{\omega(\lambda)} z^{\mathrm{crank}(\lambda)} q^n = \frac{1-q}{(zq;q)_\infty} + \sum_{j\geq 1} \frac{x^j q^j z^{-j}}{(q^2;q)_{j-1}(zq^{j+1};q)_\infty}$$

$$= \frac{1-q}{(zq;q)_\infty} \sum_{j\geq 0} \frac{(zq;q)_j}{(q;q)_j} \left(\frac{xq}{z}\right)^j$$

$$= \frac{(1-q)(xq^2;q)_\infty}{(zq;q)_\infty (xq/z;q)_\infty}. \tag{3.1}$$

Here in the last equality we make use of the q-binomial theorem [1, Theorem 2.1]:

$$\sum_{n\geq 0} \frac{(a;q)_n t^n}{(q;q)_n} = \frac{(at;q)_\infty}{(t;q)_\infty}.$$

If we take $x = 1$ in (3.1), we recover the bivariate generating function in [4].

Proof of Theorem 1.2. This time we apply the operator $[\partial/\partial x]_{x=1}$ to (3.1).

$$\sum_{n\geq 0}\sum_{\lambda\vdash n} \omega(\lambda) z^{\mathrm{crank}(\lambda)} q^n$$

$$= \left[\frac{\partial}{\partial x} \frac{(1-q)(xq^2;q)_\infty}{(zq;q)_\infty (xq/z;q)_\infty}\right]_{x=1}$$

$$= \left[\frac{(1-q)(xq^2;q)_\infty}{(zq;q)_\infty (xq/z;q)_\infty} \frac{\partial}{\partial x} \log\left(\frac{(xq^2;q)_\infty}{(xq/z;q)_\infty}\right)\right]_{x=1}$$

$$= \frac{(q;q)_\infty}{(zq;q)_\infty (q/z;q)_\infty} \left[\frac{\partial}{\partial x} \sum_{n\geq 1} \left(\log(1-xq^{n+1}) - \log(1-xq^n/z)\right)\right]_{x=1}$$

$$= \frac{(q;q)_\infty}{(zq;q)_\infty (q/z;q)_\infty} \sum_{n\geq 1} \left(-\frac{q^{n+1}}{1-q^{n+1}} + \frac{q^n/z}{1-q^n/z}\right). \tag{3.2}$$

We then apply the operator $[\partial/\partial z]_{z=1}$ to (3.2) and use (2.3) to deduce

$$\sum_{n\geq 0}\sum_{\lambda\vdash n} \omega(\lambda)\,\mathrm{crank}(\lambda) q^n$$

$$= \left[\frac{\partial}{\partial z} \frac{(q;q)_\infty}{(zq;q)_\infty (q/z;q)_\infty} \sum_{n\geq 1} \left(-\frac{q^{n+1}}{1-q^{n+1}} + \frac{q^n/z}{1-q^n/z}\right)\right]_{z=1}$$

$$= \sum_{n\geq 1} \left[\frac{(q;q)_\infty}{(zq;q)_\infty (q/z;q)_\infty} \cdot \frac{q^n/z}{1-q^n/z} \cdot \frac{1}{q^n - z}\right]_{z=1}$$

$$= -\frac{1}{(q;q)_\infty} \sum_{n\geq 1} \frac{q^n}{(1-q^n)^2}. \tag{3.3}$$

To prove the second part of Theorem 1.2, we simply observe that

$$\sum_{n\geq 0} np(n) q^n = \left[\frac{\partial}{\partial z} \frac{1}{(zq;zq)_\infty}\right]_{z=1}$$

$$= \frac{1}{(q;q)_\infty} \sum_{n\geq 1} \frac{nq^n}{1-q^n}$$

$$= \frac{1}{(q;q)_\infty} \sum_{n\geq 1} \frac{q^n}{(1-q^n)^2}.$$

In view of (3.3), we arrive at the desired result. □

Once again, we notice that

$$\sum_{\lambda\in\mathcal{P}} \omega(\lambda)\operatorname{crank}(\lambda)q^{|\lambda|} \equiv \sum_{n\geq 0} \big(M_\omega(1,5,n) + 2M_\omega(2,5,n) - 2M_\omega(3,5,n) - M_\omega(4,5,n)\big)q^n \pmod 5,$$

$$\sum_{\lambda\in\mathcal{P}} \omega(\lambda)\operatorname{crank}(\lambda)q^{|\lambda|} \equiv \sum_{n\geq 0} \big(M_\omega(1,7,n) + 2M_\omega(2,7,n) + 3M_\omega(3,7,n) - 3M_\omega(4,7,n) - 2M_\omega(5,7,n) - M_\omega(6,7,n)\big)q^n \pmod 7$$

and

$$\sum_{\lambda\in\mathcal{P}} \omega(\lambda)\operatorname{crank}(\lambda)q^{|\lambda|} \equiv \sum_{n\geq 0} \big(M_\omega(1,11,n) + 2M_\omega(2,11,n) + 3M_\omega(3,11,n) + 4M_\omega(4,11,n) + 5M_\omega(5,11,n) - 5M_\omega(6,11,n) - 4M_\omega(7,11,n) - 3M_\omega(8,11,n) - 2M_\omega(9,11,n) - M_\omega(10,11,n)\big)q^n \pmod{11}.$$

Thanks to Ramanujan's celebrated congruences (1.1)–(1.3), we complete the proof of Corollaries 1.5–1.7 by recalling (1.8).

Acknowledgements

I would like to thank George Andrews for introducing this problem to me and sharing his draft of [3] in early 2019. Part of this work was finished during my attendance at the Joint Mathematics Meetings 2019 in Baltimore.

References

[1] G. E. Andrews, *The Theory of Partitions*, Reprint of the 1976 original, Cambridge Mathematical Library (Cambridge University Press, Cambridge, 1998).

[2] G. E. Andrews, The number of smallest parts in the partitions of n, *J. Reine Angew. Math.* **624** (2008) 133–142.

[3] G. E. Andrews, The Ramanujan–Dyson identities and George Beck's congruence conjectures, *Int. J. Number Theory* **17**(2) (2021) 239–249.

[4] G. E. Andrews and F. G. Garvan, Dyson's crank of a partition, *Bull. Amer. Math. Soc. (N.S.)* **18**(2) (1988) 167–171.

[5] A. O. L. Atkin and F. G. Garvan, Relations between the ranks and cranks of partitions, *Ramanujan J.* **7**(1–3) (2003) 343–366.

[6] A. O. L. Atkin and P. Swinnerton-Dyer, Some properties of partitions, *Proc. London Math. Soc. (3)* **4** (1954) 84–106.

[7] B. C. Berndt and K. Ono, Ramanujan's unpublished manuscript on the partition and tau functions with proofs and commentary, *Sém. Lothar. Combin.* **42** (1999), Art. B42c, 63 pp.; also in *The Andrews Festschrift*, eds. D. Foata and G.-N. Han (Springer–Verlag, Berlin, 2001), pp. 39–110.

[8] F. J. Dyson, Some guesses in the theory of partitions, *Eureka* **8** (1944) 10–15.

[9] F. J. Dyson, Mappings and symmetries of partitions, *J. Combin. Theory Ser. A* **51**(2) (1989) 169–180.

[10] F. G. Garvan, Generalizations of Dyson's rank, Ph.D. thesis, Pennsylvania State University (1986).

[11] F. G. Garvan, New combinatorial interpretations of Ramanujan's partition congruences mod 5, 7 and 11, *Trans. Amer. Math. Soc.* **305**(1) (1988) 47–77.

[12] F. G. Garvan, Congruences for Andrews' smallest parts partition function and new congruences for Dyson's rank, *Int. J. Number Theory* **6**(2) (2010) 281–309.

[13] S. Ramanujan, Some properties of $p(n)$, the number of partitions of n, *Proc. Cambridge Philos. Soc.* **19** (1919) 207–210.

[14] S. Ramanujan, Congruence properties of partitions, *Proc. London Math. Soc. (2)* **18** (1920), Records for 13 March 1919.

[15] S. Ramanujan, Congruence properties of partitions, *Math. Z.* **9**(1–2) (1921) 147–153.

© 2024 World Scientific Publishing Company
https://doi.org/10.1142/9789811277375_0010

Hankel determinants of linear combinations of moments of orthogonal polynomials*

J. Cigler and C. Krattenthaler†

Fakultät für Mathematik, Universität Wien
Oskar-Morgenstern-Platz 1, A-1090 Vienna, Austria
†Christian.Krattenthaler@univie.ac.at

We prove evaluations of Hankel determinants of linear combinations of moments of orthogonal polynomials (or, equivalently, of generating functions for Motzkin paths), thus generalizing known results for Catalan numbers.

Keywords: Hankel determinants; moments of orthogonal polynomials; Catalan numbers; Motzkin numbers; Fibonacci numbers; Lucas numbers; Schröder numbers; Riordan numbers; Fine numbers; central binomial coefficients; central trinomial numbers; Chebyshev polynomials; Rogers–Szegő polynomials; Dyck paths; Motzkin paths; non-intersecting lattice paths.

Mathematics Subject Classification 2020: 11C20, 05A10, 05A15, 05A19, 15A15, 33C45, 42C05

1. Introduction

Let $C_n = \frac{1}{n+1}\binom{2n}{n}$ denote the nth *Catalan number*, and let $(F_n)_{n\geq 0}$ be the sequence of *Fibonacci numbers* defined by $F_n = F_{n-1} + F_{n-2}$ with initial values $F_0 = 0$ and $F_1 = 1$.

In [6], Cvetković *et al.* proved the Hankel determinant evaluations

$$\det(C_{i+j} + C_{i+j+1})_{i,j=0}^{n-1} = F_{2n+1} \tag{1.1}$$

and

$$\det(C_{i+j+1} + C_{i+j+2})_{i,j=0}^{n-1} = F_{2n+2}. \tag{1.2}$$

In [7], Dougherty *et al.* proved a result of similar flavor, namely

$$\det(C_{i+j} + 2C_{i+j+1} + C_{i+j+2})_{i,j=0}^{n-1} = \sum_{j=0}^{n} F_{2j+1}^2. \tag{1.3}$$

*This article originally appeared in *International Journal of Number Theory*, **17**(2), pp. 341–369
© 2021 World Scientific Publishing Company. DOI: 10.1142/S1793042120400321
†Corresponding author.

One finds variations and generalizations of these identities at numerous places in the literature, for example in [2, 3, 5–9, 18, 21].

Very recently, Cigler became aware of

$$\frac{\det\left(\binom{2i+2j+2}{i+j+1}+2\binom{2i+2j+4}{i+j+2}+\binom{2i+2j+6}{i+j+3}\right)_{i,j=0}^{n-1}}{2^n}=\sum_{j=0}^{n}L_{2j+1}^2, \tag{1.4}$$

where L_n denotes the nth *Lucas number*, defined by the Fibonacci recurrence $L_n = L_{n-1} + L_{n-2}$ with initial values $L_0 = 2$ and $L_1 = 1$. This formula was posted without proof by a mathematical amateur, Tony Foster, in a Facebook group about Pascal's triangle.

We decided to search for the general background of this kind of Hankel determinant evaluations. Our main result in Theorem 1 in the next section provides a formula for the evaluation of Hankel determinants of a linear combination of three generating functions for Motzkin paths. In these generating functions, the weight of a path depends on its steps. In that section, in Corollary 2, we then record separately the corresponding result for Dyck paths. Indeed, by appropriate specializations, all the above identities as well as many more, in the literature or not, can be obtained, see Secs. 3 and 4. One result that is of particular note is Corollary 3, which gives a closed form evaluation, in terms of Chebyshev polynomials, of Hankel determinants as in Theorem 1 for the special case where weights of steps do not depend on their height, except if they are at height zero.

We provide two different proofs for Theorem 1 and Corollary 2, both of which are of interest as we believe. The proofs in Sec. 5 are based on the decomposition of the matrices of which we want to compute the determinant into the product of "easier to handle" matrices. In particular, the proof of Corollary 2 in that section shows that it indeed follows from our main theorem, Theorem 1, although this is not completely obvious from the outset. On the other hand, in Sec. 6, we present combinatorial proofs that make use of the combinatorics of non-intersecting lattice paths. Once the setup is explained, the proofs themselves then consist of just one picture in each case. We point out that the basic idea has appeared before in [17, paragraph after Theorem 29 and paragraph above Theorem 30] and [2] but has not been fully exploited there.

Finally, in Sec. 7 we consider the Hankel determinants of linear combinations of *four* path generating functions. Indeed, the combinatorial model from Sec. 6 can be readily extended to yield a corresponding formula, see Theorem 5. We also derive the extension of Corollary 3 for the specialization where the step weights do not depend on the height of the steps (except if they are at height zero), see Corollary 6. By a perusal of Corollaries 3 and 6, a pattern emerges. On this basis, we make a conjecture on the evaluation of the Hankel determinant of an *arbitrarily long* linear combination of path generating functions in the case of the aforementioned specialization, see Conjecture 8.

2. The Main Theorem and a Corollary

Let $(s_n)_{n\geq 0}$ and $(t_n)_{n\geq 0}$ be sequences of real numbers with $t_n \neq 0$ for all n, and define polynomials (in α and the s_i's and t_i's) $f_n(\alpha)$ by the recursion

$$f_n(\alpha) = (\alpha + s_{n-1})f_{n-1}(\alpha) - t_{n-2}f_{n-2}(\alpha), \tag{2.1}$$

with $f_0(\alpha) = 1$ and $f_{-1}(\alpha) = 0$.

Consider now *Motzkin paths*, i.e. lattice paths on the two-dimensional (2D) integer lattice $\mathbb{Z}^2$ with up-steps $(x, y) \to (x+1, y+1)$, horizontal steps $(x, y) \to (x+1, y)$, and down-steps $(x, y) \to (x + 1, y - 1)$, which start at $(0, 0)$ and never run below the x-axis.[a] The weight of such a path is the product of the weights of its steps, where the weight of an up-step is 1, the weight of a horizontal step on height h is s_h, and the weight of a down-step which ends on height h is t_h.

Let $m(n, k)$ denote the sum of the weights of all such paths with end point (n, k). Then, we have

$$m(n, k) = m(n - 1, k - 1) + s_k m(n - 1, k) + t_k m(n - 1, k + 1), \tag{2.2}$$

with $m(0, k) = [k = 0]$ and $m(n, k) = 0$ for $k < 0$, where the Iverson bracket $[\mathcal{A}]$ is defined by $[\mathcal{A}] = 1$ if $\mathcal{A}$ is true and $[\mathcal{A}] = 0$ otherwise. In order to explain the title of the paper: it is well known (see e.g., [16, Theorems 11–13]) that the polynomials $p_n(x)$ defined recursively by

$$p_n(x) = (x - s_{n-1})p_{n-1}(x) - t_{n-2}p_{n-2}(x), \tag{2.3}$$

with initial values $p_{-1}(x) = 0$ and $p_0(x) = 1$ are orthogonal with respect to the linear functional L defined by $L(p_n(x)) = [n = 0]$, and that their moments are $L(x^n) = m(n, 0)$. (In other words: the polynomials $f_n(\alpha)$ are related to the polynomials $p_n(\alpha)$ by the replacement of s_i by $-s_i$ for all i.) So let us write m_n for $m(n, 0)$. The reader should be aware that, if we set $s_i = t_i = 1$ for all i, then m_n reduces to the classical *Motzkin number* M_n given by $M_n = \sum_{k=0}^{\lfloor n/2\rfloor} \binom{n}{2k}\frac{1}{k+1}\binom{2k}{k}$, and $f_n(\alpha)$ becomes $U_n((\alpha + 1)/2)$, where $U_n(x)$ is the nth *Chebyshev polynomial of the second kind*

$$U_n(x) = \sum_{k\geq 0}(-1)^k \binom{n-k}{k} (2x)^{n-2k}. \tag{2.4}$$

With these notations, we are ready to state our main result about Hankel determinants of linear combinations of the path generating functions $m_n = m(n, 0)$.

Theorem 1. *Let α and β be indeterminates. Then, for all positive integers n, we have*

$$\frac{\det(\alpha\beta m_{i+j} + (\alpha+\beta)m_{i+j+1} + m_{i+j+2})_{i,j=0}^{n-1}}{\det(m_{i+j})_{i,j=0}^{n-1}} = \sum_{j=0}^{n} f_j(\alpha)f_j(\beta)\prod_{\ell=j}^{n-1} t_\ell. \tag{2.5}$$

[a]We should point out that we use a slightly extended version of the notion of "Motzkin path" here. Usually, a Motzkin path is required to return to the x-axis at the end. We do not make this requirement.

Remark. We should point out that it is well known (see e.g., [24, Chap. IV, Corollary 6]) that

$$\det(m_{i+j})_{i,j=0}^{n-1} = \prod_{i=0}^{n-1} t_i^{n-i-1}. \tag{2.6}$$

It is worthwhile to state the analogous result for *Dyck paths* separately.

Let $(T_n)_{n\ge 0}$ be a sequence of real numbers with $T_n \neq 0$ for all n. Define polynomials (in α and the T_i's) $g_n(\alpha)$ by

$$g_{2n}(\alpha) = \alpha g_{2n-1}(\alpha) + T_{2n-2} g_{2n-2}(\alpha),$$

$$g_{2n+1}(\alpha) = g_{2n}(\alpha) + T_{2n-1} g_{2n-1}(\alpha), \tag{2.7}$$

with $g_{-1}(\alpha) = 0$ and $g_0(\alpha) = 1$.

Consider now Dyck paths, i.e. lattice paths on the 2D integer lattice $\mathbb{Z}^2$ with up-steps $(x, y) \to (x+1, y+1)$ and down-steps $(x, y) \to (x+1, y-1)$, which start at $(0, 0)$ and never run below the x-axis.[b] As before, the weight of such a path is the product of the weights of its steps. Here, the weight of an up-step is 1, and the weight of a down-step which ends on height h is T_h.

Let $c(n, k)$ denote the sum of the weights of all such paths with end point (n, k). Then, we have

$$c(n, k) = c(n-1, k-1) + T_k c(n-1, k+1), \tag{2.8}$$

with $c(0, k) = [k = 0]$ and $c(n, -1) = 0$. Finally, set $c_n = c(2n, 0)$. Here, the reader should be aware that, if we set $T_i = 1$ for all i, then c_n reduces to the Catalan number C_n. Furthermore, it can be shown that $g_{2n}(\alpha) = U_n((\alpha+2)/2) - U_{n-1}((\alpha+2)/2)$ and $g_{2n+1}(\alpha) = U_n((\alpha+2)/2)$, where $U_n(x)$ denotes as before the nth Chebyshev polynomial (2.4).

Corollary 2. *Let α and β be indeterminates. Then, for all positive integers n, we have*

$$\frac{\det(\alpha\beta c_{i+j} + (\alpha+\beta) c_{i+j+1} + c_{i+j+2})_{i,j=0}^{n-1}}{\det(c_{i+j})_{i,j=0}^{n-1}}$$

$$= \sum_{j=0}^{n} T_{2j} T_{2j+1} \cdots T_{2n-1} g_{2j}(\alpha) g_{2j}(\beta), \tag{2.9}$$

$$\frac{\det(\alpha\beta c_{i+j+1} + (\alpha+\beta) c_{i+j+2} + c_{i+j+3})_{i,j=0}^{n-1}}{\det(c_{i+j+1})_{i,j=0}^{n-1}}$$

$$= \sum_{j=0}^{n} T_{2j+1} T_{2j+2} \cdots T_{2n} g_{2j+1}(\alpha) g_{2j+1}(\beta). \tag{2.10}$$

[b] Again, we are using a slightly extended version of the notion of "Dyck path" here, by *not* requiring a Dyck path to return to the x-axis at the end.

Remark. Again, we must point out that it is well known (see e.g., [17, Eqs. (5.44) and (5.45)]) that

$$\det(c_{i+j})_{i,j=0}^{n-1} = \prod_{i=0}^{n-1} (T_{2i}T_{2i+1})^{n-i-1}, \tag{2.11}$$

and

$$\det(c_{i+j+1})_{i,j=0}^{n-1} = \prod_{i=0}^{n-1} T_0(T_{2i+1}T_{2i+2})^{n-i-1}. \tag{2.12}$$

3. Immediate Consequences

In this section, we state some direct implications of Theorem 1 and of Corollary 2 explicitly which result by specializing β to 0, respectively, to ∞.

First, if we let $\beta = 0$ in (2.5), then we obtain

$$\frac{\det(\alpha m_{i+j+1} + m_{i+j+2})_{i,j=0}^{n-1}}{\det(m_{i+j})_{i,j=0}^{n-1}} = \sum_{j=0}^{n} f_j(\alpha) f_j(0) \prod_{\ell=j}^{n-1} t_\ell. \tag{3.1}$$

Next we read coefficients of β^n on both sides of (2.5). Since each element of the matrix in the numerator on the left-hand side of which the determinant is taken is linear in β and the matrix itself is an $n \times n$ matrix, reading the coefficient of β^n on the left-hand side amounts to reading the coefficient of β^1 in each element of the matrix. On the other hand, due to the recurrence (2.1), the coefficient of β^n in $f_j(\beta)$ is zero as long as $j < n$, whereas it is 1 for $f_n(\beta)$. Hence, we obtain

$$\frac{\det(\alpha m_{i+j} + m_{i+j+1})_{i,j=0}^{n-1}}{\det(m_{i+j})_{i,j=0}^{n-1}} = f_n(\alpha). \tag{3.2}$$

Remark. The identities (3.1) and (3.2) have been given earlier by Mu *et al.* in [21, Theorem 1.3] in a different but equivalent form.

In the same way, from Corollary 2 we obtain

$$\frac{\det(\alpha c_{i+j+1} + c_{i+j+2})_{i,j=0}^{n-1}}{\det(c_{i+j})_{i,j=0}^{n-1}} = \sum_{j=0}^{n} g_{2j}(\alpha) g_{2j}(0) \prod_{\ell=2j}^{2n-1} T_\ell, \tag{3.3}$$

$$\frac{\det(\alpha c_{i+j+2} + c_{i+j+3})_{i,j=0}^{n-1}}{\det(c_{i+j+1})_{i,j=0}^{n-1}} = \sum_{j=0}^{n} g_{2j+1}(\alpha) g_{2j+1}(0) \prod_{\ell=2j+1}^{2n} T_\ell, \tag{3.4}$$

$$\frac{\det(\alpha c_{i+j} + c_{i+j+1})_{i,j=0}^{n-1}}{\det(c_{i+j})_{i,j=0}^{n-1}} = g_{2n}(\alpha), \tag{3.5}$$

$$\frac{\det(\alpha c_{i+j+1} + c_{i+j+2})_{i,j=0}^{n-1}}{\det(c_{i+j+1})_{i,j=0}^{n-1}} = g_{2n+1}(\alpha). \tag{3.6}$$

4. Specializations

By (further) specializing α, β, the s_i's and t_i's, we obtain numerous Hankel determinant evaluations, covering many of those that one finds in the literature. In particular, they cover all of the evaluations that we mentioned in the introduction.

In fact, Mu *et al.* [21] provide an extensive list of specializations of the path generating functions $m_n = m(n,0)$ that give familiar combinatorial sequences. For the convenience of the reader, we reproduce it here (the codes in parentheses refer to the sequence number in The On-Line Encyclopedia of Integer Sequences [22]):

(i) *Motzkin numbers* M_n: $s_i = t_i = 1$ for all i (A001006).
(ii) *Catalan numbers* C_n: $s_0 = 1$, $s_i = 2$ for $i \geq 1$, and $t_i = 1$ for all i (A000108).
(iii) *Shifted Catalan numbers* C_{n+1}: $s_i = 2$ and $t_i = 1$ for all i.
(iv) *Central binomial coefficients* $\binom{2n}{n}$: $s_i = 2$ for all i, $t_0 = 2$, and $t_i = 1$ for $i \geq 1$ (A000984).
(v) *Central trinomial coefficients* T_n: $s_i = 1$ for all i, $t_0 = 2$, and $t_i = 1$ for $i \geq 1$ (A002426).
(vi) *Central Delannoy numbers* D_n: $s_i = 3$ for all i, $t_0 = 4$, and $t_i = 2$ for $i \geq 1$ (A001850).
(vii) *Large Schröder numbers* r_n: $s_0 = 2$, $s_i = 3$ for $i \geq 1$, and $t_i = 2$ for all i (A006318).
(viii) *Little Schröder numbers* S_n: $s_0 = 1$, $s_i = 3$ for $i \geq 1$, and $t_i = 2$ for all i (A001003).
(ix) *Fine numbers* Fine_n: $s_0 = 0$, $s_i = 2$ for $i \geq 1$, and $t_i = 1$ for all i (A000957).
(x) *Riordan numbers* R_n: $s_0 = 0$, $s_i = 1$ for $i \geq 1$, and $t_i = 1$ for all i (A005043).
(xi) *numbers of restricted hexagonal polyominoes* H_n: $s_i = 3$ and $t_i = 1$ for all i (A002212).
(xii) *Bell numbers* B_n: $s_i = t_i = i + 1$ for all i (A000110).
(xiii) *Factorials* $n!$: $s_i = 2i + 1$ and $t_i = (i + 1)^2$ for all i.

We refer the reader to [1, 23] for definitions and discussion of most of these numbers, and to [13] for the hexagonal polyominoes (called *polyhexes* in [13]).

We add a few more specializations:

(xiv) *Shifted and scaled central binomial coefficients* $\frac{1}{2}\binom{2n+2}{n+1}$: $s_0 = 3$, $s_i = 2$ for $i \geq 1$, and $t_i = 1$ for all i.
(xv) *Odd and even central binomial coefficients* $\binom{n}{\lfloor n/2 \rfloor}$: $s_0 = 1$, $s_i = 0$ for $i \geq 1$, and $t_i = 1$ for all i.
(xvi) *Catalan numbers interleaved with zeroes* $[n \text{ even}]C_{n/2}$: $s_i = 0$ and $t_i = 1$ for all i.
(xvii) *Central binomial coefficients interleaved with zeroes* $[n \text{ even}]\binom{n}{n/2}$: $s_i = 0$ for all i, $t_0 = 2$, and $t_i = 1$ for $i \geq 1$.
(xviii) *Fine numbers interleaved with zeroes* $[n = 0] + [n \text{ even}]\mathrm{Fine}_{n/2}$: $s_i = 0$ for all i, $t_0 = 1$, and $t_i = -1$ for $i \geq 1$.

Mu *et al.* note that in most of these specializations, s_i and t_i are constant for $i \geq 1$ (namely, in (i)–(xi)), and the same is true for (xiv)–(xviii). They then discuss that special case separately, find generating functions for the values of m_n and $f_n(\alpha)$ provided that $s_i \equiv s$ and $t_i \equiv t$ for $i \geq 1$, but — at the time — missed to notice that the specialized polynomials $f_n(\alpha)$ can be expressed in terms of Chebyshev polynomials, and consequently did not realize that, for this specialization, the sum on the right-hand side of (3.1) can actually be evaluated. The corresponding (missed) results are the contents of the following corollary.

Corollary 3. *With the setup in Sec. 2, let $s_i \equiv s$ and $t_i \equiv t$ for $i \geq 1$. Then $f_0(\alpha) = 1$ and*

$$f_n(\alpha) = t_0 t^{(n-2)/2} U_n\left(\frac{\alpha+s}{2\sqrt{t}}\right) + ((t-t_0)(\alpha+s) - t(s-s_0))t^{(n-3)/2} U_{n-1}\left(\frac{\alpha+s}{2\sqrt{t}}\right) \tag{4.1}$$

$$= t^{n/2} U_n\left(\frac{\alpha+s}{2\sqrt{t}}\right) - t^{(n-1)/2}(s-s_0) U_{n-1}\left(\frac{\alpha+s}{2\sqrt{t}}\right) + t^{(n-2)/2}(t-t_0) U_{n-2}\left(\frac{\alpha+s}{2\sqrt{t}}\right), \quad \text{for } n \geq 1. \tag{4.2}$$

Moreover, in that case we have

$$\frac{\det(\alpha\beta m_{i+j} + (\alpha+\beta) m_{i+j+1} + m_{i+j+2})_{i,j=0}^{n-1}}{\det(m_{i+j})_{i,j=0}^{n-1}} = \frac{Num(\alpha,\beta)}{\alpha-\beta}, \tag{4.3}$$

where

$$\begin{aligned} Num(\alpha,\beta) &= t^{n+1/2}(U_{n+1}(x)U_n(y) - U_n(x)U_{n+1}(y)) \\ &\quad - t^n(s-s_0)(U_{n+1}(x)U_{n-1}(y) - U_{n-1}(x)U_{n+1}(y)) \\ &\quad + t^{n-1/2}((s-s_0)^2 - (t-t_0))(U_n(x)U_{n-1}(y) - U_{n-1}(x)U_n(y)) \\ &\quad + t^{n-1/2}(t-t_0)(U_{n+1}(x)U_{n-2}(y) - U_{n-2}(x)U_{n+1}(y)) \\ &\quad - t^{n-1}(s-s_0)(t-t_0)(U_n(x)U_{n-2}(y) - U_{n-2}(x)U_n(y)) \\ &\quad + t^{n-3/2}(t-t_0)^2(U_{n-1}(x)U_{n-2}(y) - U_{n-2}(x)U_{n-1}(y)), \end{aligned}$$

with $x = (\alpha+s)/(2\sqrt{t})$ and $y = (\beta+s)/(2\sqrt{t})$.

Proof. The Chebyshev polynomials $U_n(x)$ are defined by

$$U_j(\cos\theta) = \frac{\sin((j+1)\theta)}{\sin\theta} = \frac{e^{\mathrm{i}(j+1)\theta} - e^{-\mathrm{i}(j+1)\theta}}{e^{\mathrm{i}\theta} - e^{-\mathrm{i}\theta}}. \tag{4.4}$$

In particular, they satisfy the recurrence

$$U_{n+1}(x) = 2xU_n(x) - U_{n-1}(x), \quad \text{for } n \geq 1, \tag{4.5}$$

with initial values $U_0(x) = 1$ and $U_1(x) = 2x$. Hence, the polynomials $t^{n/2}U_n((\alpha + s)/(2\sqrt{t}))$ satisfy the recurrence (2.1) under the given specializations of the s_i's and t_i's. Thus, also the right-hand side of (4.1) satisfies (2.1). It is then routine to check that it also has the same initial values as the $f_n(\alpha)$'s. This proves (4.1). The alternative expression (4.2) (to which we shall come back in Sec. 7) follows from this by an application of (4.5).

For the second claim, we use the right-most expression in (4.4) to rewrite $U_j(x)$ as $\frac{z^{j+1}-z^{-j-1}}{z-z^{-1}}$, where $z = e^{\mathbf{i}\theta}$ and $x = \cos\theta$. Then the evaluation of the sum on the right-hand side of (2.5) amounts to an exercise in computing geometric series and rearranging terms (which one obviously performs with the help of a computer algebra programme). In the end, one arrives at the right-hand side of (4.3). □

Obviously, we can now take any of the specializations (i)–(xi) and (xiv)–(xviii), substitute the corresponding values of s_0, s, t_0, t in (4.3), and get a Hankel determinant evaluation in which the entries are linear combinations of three consecutive members of a well-known sequence. Of course, of special interest are those cases where the evaluations of the Chebyshev polynomials in (4.3) have a closed form. In this regard, we have

$$U_n(0) = [n \text{ even}](-1)^{n/2}, \tag{4.6}$$

$$U_n(1/2) = \begin{cases} 1, & \text{if } n \equiv 0,1 \mod 6, \\ 0, & \text{if } n \equiv 2 \mod 3, \\ -1, & \text{if } n \equiv 3,4 \mod 6, \end{cases} \tag{4.7}$$

$$U_n(1) = n + 1, \tag{4.8}$$

$$U_n(3/2) = F_{2n+2}, \tag{4.9}$$

$$U_n(\mathbf{i}/2) = \mathbf{i}^n F_{n+1}, \tag{4.10}$$

$$U_n(\mathbf{i}) = \mathbf{i}^n P_{n+1}, \tag{4.11}$$

where P_n is the nth *Pell number* defined by $P_n = 2P_{n-1} + P_{n-2}$ with initial values $P_0 = 0$ and $P_1 = 1$.

(I) In order to obtain (1.1), we have to apply the specialization (ii) in (3.2) with $\alpha = 1$. In that case, we have $m_n = C_n$ and $\frac{\alpha+s}{2\sqrt{t}} = \frac{3}{2}$, and therefore $f_n(1) = F_{2n+2} - F_{2n} = F_{2n+1}$ by (4.1) and (4.9).

In the same vein, to obtain (1.2) we need to apply the specialization (iii) in (3.2) with $\alpha = 1$. Here we have $m_n = C_{n+1}$ and $\frac{\alpha+s}{2\sqrt{t}} = \frac{3}{2}$, and therefore $f_n(\alpha) = F_{2n+2}$ by (4.1) and (4.9).

Next, in view of the above, we see that application of specialization (ii) in (2.5) with $\alpha = \beta = 1$ leads directly to (1.3), while specialization (iii) produces

$$\det(C_{i+j+1} + 2C_{i+j+2} + C_{i+j+3})_{i,j=0}^{n-1} = \sum_{j=0}^{n} F_{2j+2}^2. \tag{4.12}$$

On the other hand, if instead we set $\alpha = \beta = -1$ in (2.5) then, by (4.7), we obtain

$$\det(C_{i+j} - 2C_{i+j+1} + C_{i+j+2})_{i,j=0}^{n-1} = \sum_{j=0}^{n} (U_j(1/2) - U_{j-1}(1/2))^2 = \left\lfloor \frac{2n+3}{3} \right\rfloor \tag{4.13}$$

and

$$\det(C_{i+j+1} - 2C_{i+j+2} + C_{i+j+3})_{i,j=0}^{n-1} = \sum_{j=0}^{n} U_j^2(1/2) = \left\lfloor \frac{2n+4}{3} \right\rfloor. \tag{4.14}$$

More generally, without specializing α and β in (2.5), we get

$$\det(\alpha\beta C_{i+j} + (\alpha+\beta)C_{i+j+1} + C_{i+j+2})_{i,j=0}^{n-1} = \frac{\begin{array}{c}(U_{n+1}(x)U_n(y) - U_n(x)U_{n+1}(y) + U_{n-1}(x)U_{n+1}(y) \\ - U_{n+1}(x)U_{n-1}(y) + U_n(x)U_{n-1}(y) - U_{n-1}(x)U_n(y))\end{array}}{\alpha - \beta}, \tag{4.15}$$

with $x = (\alpha+2)/2$ and $y = (\beta+2)/2$, and

$$\det(\alpha\beta C_{i+j+1} + (\alpha+\beta)C_{i+j+2} + C_{i+j+3})_{i,j=0}^{n-1} = \frac{U_{n+1}((\alpha+2)/2)U_n((\beta+2)/2) - U_n((\alpha+2)/2)U_{n+1}((\beta+2)/2)}{\alpha - \beta}. \tag{4.16}$$

(II) The most straightforward specialization, $s_i = t_i = 1$ for all i, summarized above in Case (i), when applied in (2.5), leads to

$$\det(\alpha\beta M_{i+j} + (\alpha+\beta)M_{i+j+1} + M_{i+j+2})_{i,j=0}^{n-1} = \sum_{j=0}^{n} U_j\left(\frac{\alpha+1}{2}\right) U_j\left(\frac{\beta+1}{2}\right). \tag{4.17}$$

In particular, for $\alpha = \beta = 1$, by (4.8) this reduces to

$$\det(M_{i+j} + 2M_{i+j+1} + M_{i+j+2})_{i,j=0}^{n-1} = \sum_{j=0}^{n} (j+1)^2 = \frac{(n+1)(n+2)(2n+3)}{6}. \tag{4.18}$$

On the other hand, if instead we set $\alpha = \beta = -1$, then by (4.6) we get

$$\det(M_{i+j} - 2M_{i+j+1} + M_{i+j+2})_{i,j=0}^{n-1} = \left\lfloor \frac{n+2}{2} \right\rfloor. \tag{4.19}$$

Another interesting specialization is $\alpha = \beta = 2$. In this case, Identity (3.2) becomes

$$\det(2M_{i+j} + M_{i+j+1})_{i,j=0}^{n-1} = F_{2n+2}, \tag{4.20}$$

and (2.5) becomes

$$\det(4M_{i+j}+4M_{i+j+1}+M_{i+j+2})_{i,j=0}^{n-1}=\sum_{j=0}^{n}F_{2j+2}^2. \tag{4.21}$$

(III) We come to determinants involving central binomial coefficients. In Case (iv) above, we have $m_n=\binom{2n}{n}$ and $\frac{\alpha+s}{2\sqrt{t}}=\frac{\alpha+2}{2}$, and therefore $f_0(1)=1$ and

$$f_n(1)=2U_n(3/2)-3U_{n-1}(3/2)=2F_{2n+2}-3F_{2n}=L_{2n},\quad \text{for } n\ge 1,$$

by (4.1) and (4.9). Hence, from (4.3) we obtain

$$\frac{\det\left(\alpha\beta\binom{2i+2j}{i+j}+(\alpha+\beta)\binom{2i+2j+2}{i+j+1}+\binom{2i+2j+4}{i+j+2}\right)_{i,j=0}^{n-1}}{2^{n-1}}$$
$$=\frac{\begin{aligned}&(U_{n+1}(x)U_n(y)-U_n(x)U_{n+1}(y)+U_n(x)U_{n-1}(y)\\&-U_{n-1}(x)U_n(y)-U_{n+1}(x)U_{n-2}(y)+U_{n-2}(x)U_{n+1}(y)\\&+U_{n-1}(x)U_{n-2}(y)-U_{n-2}(x)U_{n-1}(y))\end{aligned}}{\alpha-\beta}, \tag{4.22}$$

with $x=(\alpha+2)/2$ and $y=(\beta+2)/2$. For $\alpha=\beta=1$, Identity (2.5) tells us that the above reduces to

$$\frac{\det\left(\binom{2i+2j}{i+j}+2\binom{2i+2j+2}{i+j+1}+\binom{2i+2j+4}{i+j+2}\right)_{i,j=0}^{n-1}}{2^{n-1}}=-2+\sum_{j=0}^{n}L_{2j}^2. \tag{4.23}$$

(The "correction" of -2 on the right-hand side above is caused by the "discrepancy" between $f_0(\alpha)=1$ and $L_0=2$.)

In the same vein, in Case (xiv) we have $m_n=\frac{1}{2}\binom{2n+2}{n+1}$ and $\frac{\alpha+s}{2\sqrt{t}}=\frac{\alpha+2}{2}$, and therefore $f_n(1)=U_n(3/2)+U_{n-1}(3/2)=F_{2n+2}+F_{2n}=L_{2n+1}$ by (4.1) and (4.9). Hence, from (4.3) we obtain

$$\frac{\det\left(\alpha\beta\binom{2i+2j+2}{i+j+1}+(\alpha+\beta)\binom{2i+2j+4}{i+j+2}+\binom{2i+2j+6}{i+j+3}\right)_{i,j=0}^{n-1}}{2^{n}}$$
$$=\frac{\begin{aligned}&(U_{n+1}(x)U_n(y)-U_n(x)U_{n+1}(y)+U_{n+1}(x)U_{n-1}(y)\\&-U_{n-1}(x)U_{n+1}(y)+U_n(x)U_{n-1}(y)-U_{n-1}(x)U_n(y))\end{aligned}}{\alpha-\beta}, \tag{4.24}$$

with $x=(\alpha+2)/2$ and $y=(\beta+2)/2$. For $\alpha=\beta=1$, Identity (2.5) tells us that the above reduces to (1.4).

(IV) We continue with central binomial coefficients, but with even *and* odd central binomial coefficients, which are obtained by the choice described in Case (xv).

In this case, we have $m_n = \binom{n}{\lfloor n/2\rfloor}$ and $\frac{\alpha+s}{2\sqrt{t}} = \frac{\alpha}{2}$, and therefore

$$f_n(1) = U_n(1/2) + U_{n-1}(1/2) = \begin{cases} 1, & \text{if } n \equiv 0,2 \pmod 6, \\ -1, & \text{if } n \equiv 3,5 \pmod 6, \\ 2, & \text{if } n \equiv 1 \pmod 6, \\ -2, & \text{if } n \equiv 4 \pmod 6, \end{cases}$$

and

$$f_n(-1) = U_n(-1/2) + U_{n-1}(-1/2) = \begin{cases} 1, & \text{if } n \equiv 0 \pmod 3, \\ -1, & \text{if } n \equiv 2 \pmod 3, \\ 0, & \text{if } n \equiv 1 \pmod 3, \end{cases}$$

by (4.1) and (4.7). Further special evaluations are $f_n(2) = (n+1)+n = 2n+1$ and $f_n(3/2) = F_{2n+2} + F_{2n} = L_{2n+1}$. Hence, from (4.3) we obtain

$$\det\left(\alpha\beta\binom{i+j}{\lfloor (i+j)/2\rfloor} + (\alpha+\beta)\binom{i+j+1}{\lfloor (i+j+1)/2\rfloor} + \binom{i+j+2}{\lfloor (i+j+2)/2\rfloor}\right)_{i,j=0}^{n-1}$$
$$= \frac{\begin{array}{c}(U_{n+1}(x)U_n(y) - U_n(x)U_{n+1}(y) + U_{n+1}(x)U_{n-1}(y) \\ - U_{n-1}(x)U_{n+1}(y) + U_n(x)U_{n-1}(y) - U_{n-1}(x)U_n(y))\end{array}}{\alpha-\beta}, \tag{4.25}$$

with $x = \alpha/2$ and $y = \beta/2$. In particular, for $\alpha = \beta = 1$, this reduces to

$$\det\left(\binom{i+j}{\lfloor (i+j)/2\rfloor} + 2\binom{i+j+1}{\lfloor (i+j+1)/2\rfloor} + \binom{i+j+2}{\lfloor (i+j+2)/2\rfloor}\right)_{i,j=0}^{n-1}$$
$$= \begin{cases} 2n+1 & \text{if } n \equiv 0 \pmod 3, \\ 2n+3 & \text{if } n \equiv 1 \pmod 3, \\ 2n+4 & \text{if } n \equiv 2 \pmod 3. \end{cases} \tag{4.26}$$

On the other hand, for $\alpha = \beta = -1$ we get

$$\det\left(\binom{i+j}{\lfloor (i+j)/2\rfloor} - 2\binom{i+j+1}{\lfloor (i+j+1)/2\rfloor} + \binom{i+j+2}{\lfloor (i+j+2)/2\rfloor}\right)_{i,j=0}^{n-1}$$
$$= \left\lfloor \frac{2n+3}{3} \right\rfloor. \tag{4.27}$$

For $\alpha = 2$ and $\alpha = 3$, curious evaluations can be obtained from (3.2), namely

$$\det\left(2\binom{i+j}{\lfloor (i+j)/2\rfloor} + \binom{i+j+1}{\lfloor (i+j+1)/2\rfloor}\right)_{i,j=0}^{n-1} = 2n+1 \tag{4.28}$$

and

$$\det\left(3\binom{i+j}{\lfloor (i+j)/2\rfloor} + \binom{i+j+1}{\lfloor (i+j+1)/2\rfloor}\right)_{i,j=0}^{n-1} = L_{2n+1}. \tag{4.29}$$

(V) Another curious special case arises if we choose $s_i = 0$ and $t_i = -1$ for all i. Then $m_n = [n \text{ even}](-1)^{n/2}C_{n/2}$ and $f_n(1) = (-\mathbf{i})^n U_n(\mathbf{i}/2) = F_{n+1}$. (The important point here is to choose $\sqrt{t} = \sqrt{-1}$ *consistently*; our choice here is $\sqrt{-1} = -\mathbf{i}$.) From (3.2) we therefore get

$$\begin{aligned}\det&\left((-1)^{\lfloor (i+j+1)/2\rfloor} C_{\lfloor (i+j+1)/2\rfloor}\right)_{i,j=0}^{n-1}\\ &= \det(m_{i+j} + m_{i+j+1})_{i,j=0}^{n-1}\\ &= \det(m_{i+j})_{i,j=0}^{n-1} \cdot F_{n+1} = (-1)^{\binom{n}{2}} F_{n+1}. \end{aligned} \tag{4.30}$$

For example, for $n = 5$, we get

$$\det\begin{pmatrix} 1 & -1 & -1 & 2 & 2\\ -1 & -1 & 2 & 2 & -5\\ -1 & 2 & 2 & -5 & -5\\ 2 & 2 & -5 & -5 & 14\\ 2 & -5 & -5 & 14 & 14 \end{pmatrix} = 8 = F_6.$$

Formula (2.5) with the same choices of the s_i's and t_i's gives

$$\det(m_{i+j} + 2m_{i+j+1} + m_{i+j+2})_{i,j=0}^{n-1} = (-1)^{\binom{n+1}{2}} \sum_{j=0}^{n} (-1)^j F_{j+1}^2. \tag{4.31}$$

(VI) We point out that our formula (4.3) allows us to vastly generalize the second main result in [7, Theorem 3]. Namely, there Dougherty *et al.* establish a linear recurrence of order 4 for the Hankel determinant

$$\det(\lambda C_{i+j} + \mu C_{i+j+1} + C_{i+j+2})_{i,j=0}^{n-1},$$

where λ and μ are some constants. Our formula in (4.3) provides even a *closed form expression* for a determinant that has much more general matrix entries (and the reparametrization $\lambda = \alpha\beta$ and $\mu = \alpha + \beta$). Using this, we may deduce the following corollary.

Corollary 4. *With the setup in Sec. 2, let $s_i \equiv s$ and $t_i \equiv t$ for $i \geq 1$. Then the scaled Hankel determinants*

$$t_0^{-(n-1)} t^{-\binom{n-1}{2}} \det(\alpha\beta m_{i+j} + (\alpha+\beta) m_{i+j+1} + m_{i+j+2})_{i,j=0}^{n-1} \tag{4.32}$$

satisfy a linear recurrence of order 4 with constant coefficients.

Remark. The reader should note that the scaling in (4.32) is exactly the value of the determinant in the denominator on the left-hand side of (4.3).

Proof of Corollary 4. The reader should note that the expression on the right-hand side of (4.3) is a linear combination of products of the form $t^n U_{n+\delta}(x) U_{n+\varepsilon}(y)$ with constant coefficients, where $\delta, \varepsilon \in \{1, 0, -1, -2\}$. (We recall that the arguments of the Chebyshev polynomials are $x = (\alpha + s)/(2\sqrt{t})$ and $y = (\beta + s)/(2\sqrt{t})$.) Now, by (4.5), for each fixed δ the sequence $(t^{n/2}U_{n+\delta}(x))_{n\geq 0}$ satisfies the recurrence relation

$$p_{n+1} = (\alpha + s)p_n - tp_{n-1}. \tag{4.33}$$

Similarly, for each fixed ε the sequence $(t^{n/2}U_{n+\varepsilon}(y))_{n\geq 0}$ satisfies the same recurrence relation with α replaced by β. Hence, each product sequence $(t^n U_{n+\delta}(x)U_{n+\varepsilon}(y))_{n\geq 0}$ satisfies the same recurrence relation, namely, the one resulting from the "product" of (4.33) with (4.33) where α has been replaced by β. It follows from the proof of [23, Theorem 6.4.9] (which is actually a much more general theorem) that the order of this "product" recurrence is $2 \cdot 2 = 4$. □

(VII) We turn to q-analogues. We use the familiar notations $(a;q)_0 := 1$, $(a;q)_n := (1-a)(1-aq)\cdots(1-aq^{n-1})$, $n \geq 1$, for the *q-shifted factorials*, and

$$\begin{bmatrix} n \\ k \end{bmatrix}_q := \frac{(q;q)_n}{(q;q)_k\,(q;q)_{n-k}}$$

for the *q-binomial coefficients*. The *Rogers–Szegő polynomials* $r_n(t)$, defined by

$$r_n(t) := \sum_{k=0}^{n} \begin{bmatrix} n \\ k \end{bmatrix}_q t^k,$$

are the moments for the orthogonal polynomials (2.3) with $s_i = q^i(t+1)$ and $t_i = q^i t(q^{i+1}-1)$. The corresponding polynomials $f_n(\alpha)$ are given by

$$f_n(\alpha) = \sum_{k=0}^{n} \begin{bmatrix} n \\ k \end{bmatrix}_q \sum_{j=0}^{k} q^{\binom{j}{2}+\binom{k-j}{2}} \begin{bmatrix} k \\ j \end{bmatrix}_q t^j \alpha^{n-k}.$$

Thus, from (3.2) we obtain

$$\frac{\det(\alpha r_{i+j}(t) + r_{i+j+1}(t))_{i,j=0}^{n-1}}{(-t)^{\binom{n}{2}} q^{\binom{n}{3}} \prod_{i=1}^{n-1} (q,q)_i} = \sum_{k=0}^{n} \begin{bmatrix} n \\ k \end{bmatrix}_q \sum_{j=0}^{k} q^{\binom{j}{2}+\binom{k-j}{2}} \begin{bmatrix} k \\ j \end{bmatrix}_q t^j \alpha^{n-k}, \tag{4.34}$$

and from (2.5) we get

$$\frac{\det(\alpha\beta r_{i+j}(t) + (\alpha+\beta)t_{i+j+1}(t) + r_{i+j+2}(t))_{i,j=0}^{n-1}}{(-t)^{\binom{n}{2}} q^{\binom{n}{3}} \prod_{i=1}^{n-1} (q,q)_i}$$

$$= \sum_{j=0}^{n} f_j(\alpha) f_j(\beta) (-t)^{n-j} q^{\binom{n}{2}-\binom{j}{2}} (q^{j+1};q)_{n-j}, \tag{4.35}$$

with $f_j(\alpha)$ as above. We point out that the right-hand sides in (4.34) and (4.35) simplify for $t=-1$ since, in basic hypergeometric notation (cf. [10]), we have

$$\sum_{j=0}^{k} q^{\binom{j}{2}+\binom{k-j}{2}} \begin{bmatrix} k \\ j \end{bmatrix}_q t^j = q^{\binom{k}{2}} {}_1\phi_1\left[\begin{matrix} q^{-k} \\ 0 \end{matrix}; q, qt\right],$$

and this ${}_1\phi_1$-series can be evaluated by means of [10, (1.8.1); Appendix (II.9); $b\to\infty$]. To be precise, for $t=-1$ we have

$$\sum_{j=0}^{k} q^{\binom{j}{2}+\binom{k-j}{2}} \begin{bmatrix} k\\ j\end{bmatrix}_q (-1)^j = \begin{cases} (-1)^{k/2} q^{\frac14(k^2-2k)}(q;q^2)_{k/2} & \text{if } k \text{ is even,}\\ 0 & \text{if } k \text{ is odd}\end{cases}$$

and thus for example

$$\frac{\det(\alpha r_{i+j}(-1)+r_{i+j+1}(-1))_{i,j=0}^{n-1}}{q^{\binom{n}{3}}\prod_{i=1}^{n-1}(q,q)_i} = \sum_{k=0}^{n} \begin{bmatrix} n\\ 2k\end{bmatrix}_q (-1)^k q^{k^2-k}(q;q^2)_k \alpha^{n-2k}, \tag{4.36}$$

by (4.34). For $\alpha=1$, this reduces even further, namely, to

$$\frac{\det(r_{i+j}(-1)+r_{i+j+1}(-1))_{i,j=0}^{n-1}}{q^{\binom{n}{3}}\prod_{i=1}^{n-1}(q,q)_i} = q^{\binom{n}{2}}, \tag{4.37}$$

which is most easily seen by looking at the recursion (2.1) for $\alpha=1$, $s_i=0$, and $t_i=t^i(t^{i+1}-1)$ for all i.

(VIII) The powers $q^{\binom{n}{2}}$ are the moments for the orthogonal polynomials (2.3) with $s_i=q^{i-1}(q^{i+1}+q^i-1)$ and $t_i=q^{3i}(q^{i+1}-1)$. The corresponding polynomials $f_n(\alpha)$ are given by

$$f_n(\alpha)=\sum_{k=0}^{n}\begin{bmatrix} n\\ k\end{bmatrix}_q q^{(n-1)k}\alpha^{n-k}.$$

Thus, from (2.5) we obtain

$$\begin{aligned}&\frac{\det\left(\alpha\beta q^{\binom{i+j}{2}}+(\alpha+\beta)q^{\binom{i+j+1}{2}}+q^{\binom{i+j+2}{2}}\right)_{i,j=0}^{n-1}}{(-1)^{\binom{n}{2}}q^{3\binom{n}{3}}\prod_{i=1}^{n-1}(q;q)_i}\\ &\quad=\sum_{j=0}^{n}(-1)^{n-j}q^{3\binom{n}{2}-3\binom{j}{2}}(q^{j+1};q)_{n-j}\sum_{k_1=0}^{j}\begin{bmatrix} j\\ k_1\end{bmatrix}_q q^{(j-1)k_1}\alpha^{j-k_1}\\ &\quad\quad\times\sum_{k_2=0}^{j}\begin{bmatrix} j\\ k_2\end{bmatrix}_q q^{(j-1)k_2}\beta^{j-k_2}.\end{aligned}\tag{4.38}$$

(IX) For another example, we choose $T_i=q^{\lfloor (i+1)/2\rfloor}(1-q^{\lfloor (i+2)/2\rfloor})$ in the Dyck path setting in Sec. 2. Then, we get $c_n=(q;q)_{n/2}$ for even n and $c_n=0$ for odd n. The corresponding polynomials are

$$g_{2n}(\alpha)=\sum_{k=0}^{n} q^{\binom{k}{2}}(q;q)_k \begin{bmatrix} n\\ k\end{bmatrix}_q^2 \alpha^{n-k}$$

and

$$g_{2n+1}(\alpha) = \sum_{k=0}^{n} q^{\binom{k}{2}} (q;q)_k \begin{bmatrix} n+1 \\ k \end{bmatrix}_q \begin{bmatrix} n \\ k \end{bmatrix}_q \alpha^{n-k}.$$

Hence, from (3.5) with $\alpha = 1$ we obtain

$$\frac{\det\big((q;q)_{\lfloor (i+j+1)/2 \rfloor}\big)_{i,j=0}^{n-1}}{q^{\frac{1}{6}n(2n^2-3n+1)} \prod_{i=1}^{n-1} (q;q)_i^2} = \sum_{k=0}^{n} q^{\binom{k}{2}} (q;q)_k \begin{bmatrix} n \\ k \end{bmatrix}_q^2, \tag{4.39}$$

and similarly, from (3.6),

$$\frac{\det\big((q;q)_{\lfloor (i+j+2)/2 \rfloor}\big)_{i,j=0}^{n-1}}{q^{2\binom{n+1}{3}} (q;q)_n \prod_{i=1}^{n-1} (q;q)_i^2} = \sum_{k=0}^{n} q^{\binom{k}{2}} (q;q)_k \begin{bmatrix} n+1 \\ k \end{bmatrix}_q \begin{bmatrix} n \\ k \end{bmatrix}_q. \tag{4.40}$$

(X) Obviously, our identities in Secs. 2 and 3 lead to many more Hankel determinant evaluations involving well-known functions when applied to the *"orthogonal polynomials as moments"* from [14, 15].

5. Computational Proofs Using Matrix Algebra

Proof of Theorem 1. Define the matrices $G_n(\alpha)$ by

$$G_n(\alpha) := \begin{pmatrix} \alpha+s_0 & 1 & & & & \\ t_0 & \alpha+s_1 & 1 & & & \\ & t_1 & \alpha+s_2 & 1 & & \\ \cdots & \cdots & \cdots & \cdots & \cdots & \cdots \\ & & & t_{n-3} & \alpha+s_{n-2} & 1 \\ & & & & t_{n-2} & \alpha+s_{n-1} \end{pmatrix}$$

(we show only nonzero entries in the matrix). By expanding the determinant along the last row of the matrix, we obtain

$$\det G_n(\alpha) = (\alpha + s_{n-1}) \det G_{n-1}(\alpha) - t_{n-2} \det G_{n-2}(\alpha).$$

Since we have in addition

$$\det G_1(\alpha) = \alpha + s_0 = f_1(\alpha) \quad \text{and} \quad \det G_2(\alpha) = (\alpha+s_0)(\alpha+s_1) - t_0 = f_2(\alpha),$$

we see that

$$f_n(\alpha) = \det G_n(\alpha). \tag{5.1}$$

Let A_n be the matrix of path generating functions $(m(i,j))_{i,j=0}^{n-1}$, and let D_n be the diagonal matrix with entries $d(i,i) = \prod_{\ell=0}^{i-1} t_\ell$. By (2.2) we get

$$(m(i+1,j))_{i,j=0}^{n-1} = A_n R_n, \tag{5.2}$$

where

$$R_n := \begin{pmatrix} s_0 & 1 & & & & \\ t_0 & s_1 & 1 & & & \\ & t_1 & s_2 & 1 & & \\ \cdots & \cdots & \cdots & \cdots & \cdots & \cdots \\ & & t_{n-3} & s_{n-2} & 1 \\ & & & t_{n-2} & s_{n-1} \end{pmatrix}.$$

It is well known (cf. e.g., [4]) and easy to see by cutting a Motzkin path from $(0,0)$ to $(i+j,0)$ after i steps that

$$\sum_k m(i,k)m(j,k)\prod_{\ell=0}^{k-1} t_\ell = m(i+j,0). \tag{5.3}$$

This means that

$$(m_{i+j})_{i,j=0}^{n-1} = A_n D_n A_n^t,$$

and it implies that

$$(\alpha m_{i+j} + m_{i+j+1})_{i,j=0}^{n-1} = \alpha A_n D_n A_n^t + A_n R_n D_n A_n^t = A_n G_n(\alpha) D_n A_n^t.$$

Since A_n is a lower diagonal matrix and $m(i,i)=1$ for all i, we get

$$\det(A_n G_n(\alpha) D_n A_n^t) = \det D_n \cdot \det G_n(\alpha),$$

which, recalling (2.6) and (5.1), implies (3.2).

In the same way as before (5.2), we get

$$(m(i+2,j))_{i,j=0}^{n-1} = A_n(R_n^2 + t_{n-1}E_n), \tag{5.4}$$

where $E_n = (u_n(i,j))_{i,j=0}^{n-1}$ with $u_n(n-1,n-1)=1$ and $u_n(i,j)=0$ otherwise. The term $t_{n-1}E_n$ arises to account for the Motzkin paths that are included in the weighted enumeration $m(n+1,n-1)$ which end with an up-step followed by a down-step. Then, using (5.3) with i replaced by $i+2$, we infer

$$\begin{aligned}(\alpha\beta m(i+j,0) &+ (\alpha+\beta)m(i+j+1,0) + m(i+j+2,0))_{i,j=0}^{n-1} \\ &= A_n(\alpha\beta I_n + (\alpha+\beta)R_n + R_n^2 + t_{n-1}E_n)D_n A_n^t \\ &= A_n(G_n(\alpha)G_n(\beta) + t_{n-1}E_n)D_n A_n^t, \end{aligned} \tag{5.5}$$

where I_n denotes the $n\times n$ identity matrix.

If $[A]_k$ denotes the restriction of the matrix A to the first k rows and columns, then it is a simple exercise to verify that

$$[G_n(\alpha)G_n(\beta)]_{n-1} = G_{n-1}(\alpha)G_{n-1}(\beta) + t_{n-2}E_{n-1}.$$

If we write H_n for the matrix $G_n(\alpha)G_n(\beta)+t_{n-1}E_n$, then the above relation entails a recursion for the determinant of this matrix, namely

$$\begin{aligned}\det H_n &= \det(G_n(\alpha)G_n(\beta)) + t_{n-1}\det[G_n(\alpha)G_n(\beta)]_{n-1}\\ &= f_n(\alpha)f_n(\beta) + t_{n-1}\det H_{n-1}.\end{aligned}$$

Consequently, by induction, we get $\det H_n = \sum_{j=0}^{n} f_j(\alpha)f_j(\beta)\prod_{\ell=j}^{n-1} t_\ell$. If this is substituted in (5.5) and the determinant is taken on both sides of that identity, then we obtain (2.5) if we again recall (2.6). □

Proof of Corollary 2. There is a well-known correspondence between Dyck paths and Motzkin paths which arises by arranging the steps of a Dyck path in groups of two, and then interpreting a group of two up-steps as an up-step in the corresponding Motzkin path, a group of two down-steps as a down-step in the corresponding Motzkin path, and a group consisting of an up-step and a down-step as a level step in the corresponding Motzkin path. This idea stands behind all the computations below.

With the $c(n,k)$'s defined by (2.8), we let $a(n,k) = c(2n,2k)$ and $b(n,k) = c(2n+1,2k+1)$. Then, by iterating (2.8), we get

$$a(n,k) = a(n-1,k-1) + s_k a(n-1,k) + t_k a(n-1,k+1) \tag{5.6}$$

with $s_k = T_{2k-1} + T_{2k}$ and $t_k = T_{2k}T_{2k+1}$, where, by convention, $T_{-1} = 0$, and

$$b(n,k) = b(n-1,k-1) + s_k b(n-1,k) + t_k b(n-1,k+1) \tag{5.7}$$

with $s_k = T_{2k} + T_{2k+1}$ and $t_k = T_{2k+1}T_{2k+2}$.

Then, with the $g_n(\alpha)$'s as defined by (2.7), we get

$$\begin{aligned}g_{2n}(\alpha) &= \alpha g_{2n-1}(\alpha) + T_{2n-2}g_{2n-2}(\alpha)\\ &= \alpha g_{2n-2}(\alpha) + T_{2n-2}g_{2n-2}(\alpha) + \alpha T_{2n-3}g_{2n-3}(\alpha)\\ &= \alpha g_{2n-2}(\alpha) + T_{2n-2}g_{2n-2}(\alpha) + T_{2n-3}(g_{2n-2}(\alpha) - T_{2n-4}g_{2n-4}(\alpha))\\ &= (\alpha + T_{2n-2} + T_{2n-3})g_{2n-2}(\alpha) - T_{2n-3}T_{2n-4}g_{2n-4}(\alpha).\end{aligned}$$

This implies that for $a(n,k)$ the associated polynomials $f_n(\alpha)$ (in the sense of the setup in Sec. 2, where the s_k's and t_k's are as given below (5.6)) are given by $f_n(\alpha) = g_{2n}(\alpha)$.

Similarly, we have

$$\begin{aligned}g_{2n+1}(\alpha) &= g_{2n}(\alpha) + T_{2n-1}g_{2n-1}(\alpha)\\ &= \alpha g_{2n-1}(\alpha) + T_{2n-1}g_{2n-1}(\alpha) + T_{2n-2}g_{2n-2}(\alpha)\\ &= \alpha g_{2n-1}(\alpha) + T_{2n-1}g_{2n-1}(\alpha) + T_{2n-2}(g_{2n-1}(\alpha) - T_{2n-3}g_{2n-3}(\alpha))\\ &= (\alpha + T_{2n-1} + T_{2n-2})g_{2n-1}(\alpha) - T_{2n-2}T_{2n-3}g_{2n-3}(\alpha).\end{aligned}$$

This implies that for $b(n,k)$ the associated polynomials $f_n(\alpha)$ (defined by the s_k's and t_k's as given below (5.7)) are given by $f_n(\alpha) = g_{2n+1}(\alpha)$.

Therefore, Identity (2.5) with $s_k = T_{2k-1} + T_{2k}$ and $t_k = T_{2k}T_{2k+1}$ directly gives (2.9) since in that case we have $m_n = a(n,0) = c(2n,0) = c_{2n}$. Similarly, by observing that $c(2n+2,0) = T_0 c(2n+1,1)$ for $n \geq 0$, we see that Identity (2.5) with $s_k = T_{2k} + T_{2k+1}$ and $t_k = T_{2k+1}T_{2k+2}$ gives (2.10) since in that case we have $m_n = b(n,0) = c(2n+1,1) = T_0^{-1}c(2n+2,0) = T_0^{-1}c_{2n+2}$. □

6. Combinatorial Proofs Using Non-Intersecting Lattice Paths

Proof of Theorem 1. In order to give a combinatorial proof of (2.5), we need a combinatorial model for the polynomials $f_n(\alpha)$, and we need (another) combinatorial model for the Hankel determinant

$$\det(\alpha\beta m_{i+j} + (\alpha+\beta)m_{i+j+1} + m_{i+j+2})_{i,j=0}^{n-1}. \tag{6.1}$$

Our model for $f_n(\alpha)$ consists of $n \times 1$ "pillars" which are decomposed into "squares" (1×1 bricks) — which come in two kinds — and "dominoes" (2×1 bricks). An example for $n = 8$ can be found in Fig. 1. There, squares of the second kind are indicated by thick framing, whereas squares of the first kind are shown "normally". (The labeling should be ignored at this point.) This pillar consists — from bottom to top — of a square of the second kind, followed by two dominoes, followed by a square of the first kind, a square of the second kind, and a square of the first kind.

We define the weight of a square of the first kind whose bottom is at height h to be s_h, we define the weight of a square of the second kind to be α, and we define the weight of a domino whose bottom is at height h to be $-t_h$. The weight of a pillar p is by definition the product of the weights of its squares and dominoes and is denoted by $w(p)$. Thus, the weight of the pillar in Fig. 1 is

$$\alpha(-t_1)(-t_3)s_5\alpha s_7$$

(which now also explains the labels in Fig. 1).

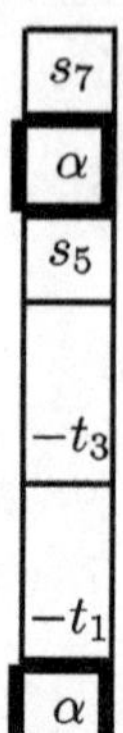

Fig. 1. A pillar.

Let $\mathcal{P}_n$ denote the set of all $n \times 1$ pillars. We claim that

$$\sum_{p \in \mathcal{P}_n} w(p) = f_n(\alpha). \tag{6.2}$$

This is indeed easy to see. For, on top of such a pillar we may either have a square of either kind, where the sum of the weights of these two possible squares is $\alpha + s_{n-1}$, or a domino, which has weight $-t_{n-2}$. If we remove the top brick, then there remains an $(n-1) \times 1$, respectively, an $(n-2) \times 1$ pillar. Consequently, the generating function for $n \times 1$ pillars, $\sum_{p \in \mathcal{P}_n} w(p)$, satisfies as well the recurrence (2.1), and the initial conditions are obvious. This establishes the claim.

We turn to our model for the Hankel determinant (6.1). Not very surprisingly, it is a model of non-intersecting lattice paths and is based on the corresponding enumeration theorem due to Lindström [20, Lemma 1] and Gessel and Viennot [11; 12, Theorem 1]. We recall that a family of paths in a (directed) graph is called *non-intersecting* if no two paths of the family have a *graph vertex* in common.

The underlying directed graph is the following: vertices are the points (x, y) in the 2D integer lattice $\mathbb{Z} \times \mathbb{Z}$ with non-negative ordinate (i.e. with $y \geq 0$). The edges are *up-steps* $(x, y) \to (x+1, y+1)$ with weight 1, *down-steps* $(x, y) \to (x+1, y-1)$ with weight t_{y-1} if $y \geq 1$, and *horizontal steps* $(x, y) \to (x+1, y)$ with weight s_y; in a strip of width 2 centred at the y-axis, we introduce further horizontal steps of a "second kind": in our directed graph we have *additional* horizontal steps $(-1, y) \to (0, y)$ with weight α, and *additional* horizontal steps $(0, y) \to (1, y)$ with weight β. As before, the weight $w(P)$ of a path P is the product of the weights of all its edges.

Let $e(A_i, E_j) := \sum_{P: A_i \to E_j} w(P)$, where the sum is over all paths from $A_i = (-i-1, 0)$ to $E_j = (j+1, 0)$, for $i, j \geq 0$. Then, we have

$$e(A_i, E_j) = \alpha\beta m_{i+j} + (\alpha + \beta) m_{i+j+1} + m_{i+j+2}.$$

For, the set of those paths which do not contain any of the additional horizontal steps contributes m_{i+j+2} to the total weight, the set of those paths which use one of the additional horizontal steps of the type $(-1, y) \to (0, y)$ contributes αm_{i+j+1}, the set of those paths which use one of the additional horizontal steps of the type $(0, y) \to (1, y)$ contributes βm_{i+j+1}, and the set of those paths which use both types of additional horizontal steps contributes $\alpha\beta m_{i+j}$.

The Lindström–Gessel–Viennot theorem then gives

$$\det\left(\alpha\beta m_{i+j} + (\alpha + \beta) m_{i+j+1} + m_{i+j+2}\right)_{i,j=0}^{n-1} = \det\left(e(A_i, E_j)\right)_{i,j=0}^{n-1}$$
$$= \sum_{\mathcal{P}} \operatorname{sgn}(\sigma(\mathcal{P})) \prod_{i=0}^{n-1} w(P_i). \tag{6.3}$$

Here, $\mathcal{P} = (P_0, P_1, \ldots, P_{n-1})$ runs over all families of non-intersecting paths, where P_i runs from A_i to $E_{\sigma(i)}$, $i = 0, 1, \ldots, n-1$, for some permutation σ of $\{0, 1, \ldots, n-1\}$. An example of such a family of non-intersecting lattice paths for $n = 9$

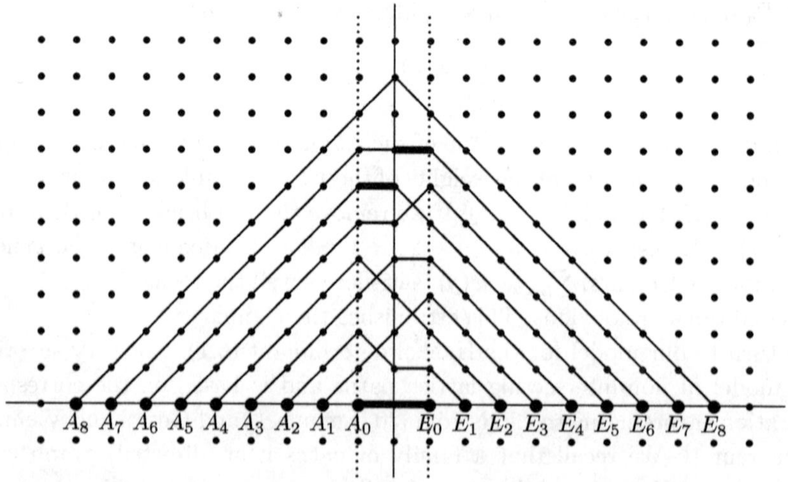

Fig. 2. A family of non-intersecting Motzkin paths.

is shown in Fig. 2. There, the additional horizontal steps between the abscissa -1 and $+1$ are indicated by thick segments. For example, path P_0 consists of two of these additional steps (hence its weight is $\alpha\beta$), while path P_6 contains an additional horizontal step with weight α, and path P_7 contains an additional horizontal step with weight β. The permutation σ associated with the path family in Fig. 2 is 031426578 (meaning that P_0 ends in E_0, P_1 ends in E_3, etc.).

A reader who is experienced in the combinatorics of non-intersecting lattice paths does not have to go through the following explanations in detail; it will suffice to look at Figs. 2 and 3, since they contain the essence of the argument. Everything else is — so-to-speak — just "book-keeping".

However, in order to properly explain, we start with the question of how such an above family of non-intersecting paths looks like?

In the figure, we have marked the vertical lines $x = -1$ and $x = +1$ by dotted lines. To the left of $x = -1$, the paths are uniquely determined by the condition of being non-intersecting, and the same is true to the right of $x = +1$; see the figure. Thus, it is only in the strip between the vertical lines $x = -1$ and $x = +1$ where something "interesting" may happen.

In that strip, there may be some paths on the top, say $P_{n-1}, \dots, P_j$, which, in that strip, have an up-step followed by a down-step. (In the figure, this is only the case for the top-most path P_8.) All other paths do not exceed height $j-1$, and they form patterns in that strip that can be described as follows. We first look into the left substrip between the vertical lines $x = -1$ and $x = 0$. There, a path P_i may do three different things (see Fig. 2):

(1) a horizontal step, which may be of two different kinds — "ordinary" or one of the "additional" steps;

(2) a down-step, but then there must be another path doing an up-step in the substrip crossing P_i;
(3) an up-step, but then there must be another path doing a down-step in the substrip crossing P_i.

The same is true in the substrip between the vertical lines $x = 0$ and $x = +1$, and what happens there is *independent* from what is happening in the other substrip.

In order to determine the contribution of all this to the overall generating function (6.3), we observe that the weight contribution of the path parts to the left of the vertical line $x = -1$ and to the right of the vertical line $x = +1$ is in total the right-hand side of (2.6), or, in other words

$$\det(m_{i+j})_{i,j=0}^{n-1},$$

the denominator in (2.5). Hence, the contribution of the various configurations between the vertical lines $x = -1$ and $x = +1$ must give the right-hand side of (2.5).

This contribution of the "free" part in the strip between the vertical lines $x = -1$ and $x = +1$ is composed of the following constituents:

(1) The top-most paths $P_{n-1}, \dots, P_j$, each of which have an up-step followed by a down-step in this strip, contribute a multiplicative factor of

$$t_j t_{j+1} \cdots t_{n-1} \tag{6.4}$$

to the total weight.

(2) According to the above considerations, the contribution of the paths $P_0, P_1, \dots, P_{j-1}$ below equals the product of the their contributions between in the left substrip between the vertical lines $x = -1$ and $x = 0$ and their contributions in the right substrip between the vertical lines $x = 0$ and $x = +1$. Combinatorially, we find the same configurations in the left substrip and the right substrip. The only difference is that the "additional" horizontal steps have weight α in the left substrip, whereas they have weight β in the right substrip. Hence, if we denote the generating function of the configurations in the left substrip by $\mathrm{GF}_j(\alpha)$, then the total contribution coming from the strip between the vertical lines $x = -1$ and $x = +1$ equals

$$\mathrm{GF}_j(\alpha)\,\mathrm{GF}_j(\beta). \tag{6.5}$$

Let us concentrate on the contributions of $P_0, P_1, \dots, P_{j-1}$ in the left substrip, that is on $\mathrm{GF}_j(\alpha)$. In order to see that $\mathrm{GF}_j(\alpha) = f_j(\alpha)$, we set up a weight-preserving bijection between these contributions and the pillars that we introduced at the beginning of this section. In this bijection, we map an "ordinary" horizontal step on height h to a square of the first kind on height h, we map an "additional" horizontal step on height h to a square of the second kind on height h, and we map a "cross", that is a pair consisting of a down-step and an up-step that cross each other, with the down-step ending on height h, to a domino on height h. The left part of

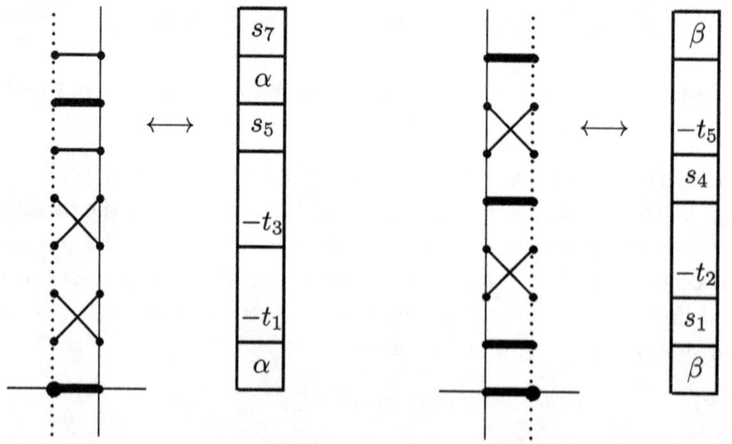

Fig. 3. Correspondence between local path configurations and pillars.

Fig. 3 shows this correspondence for the configuration in the left substrip of Fig. 2. (For the sake of completeness, the right part of Fig. 3 shows the correspondence for the configuration in the right substrip of Fig. 2. The reader is alerted that the weights of the squares of the second kind must be defined here to equal β in order to set up a weight-preserving correspondence.)

This is indeed a weight-preserving bijection. For the correspondence between horizontal steps and squares this is obvious, but it also works for the correspondence between crosses and dominoes: in a cross two paths change sides;[c] in the corresponding permutation $\sigma(\mathcal{P})$ this corresponds to a transposition, which has negative sign. Indeed, we have defined the weight of a domino on height h by $-t_h$, that is, with a negative sign in front. If we now remember (6.2), then we see that the proof of (2.5) is complete. □

Sketch of proof of Corollary 2. The two Hankel determinant evaluations in Corollary 2 can also be directly established by a non-intersecting lattice path model. Since the corresponding arguments are very similar to the previous proof, we content ourselves with giving just a rough sketch that essentially only consists in presenting the figures analogous to the ones in Figs. 2 and 3.

Here, the underlying graph is again the upper 2D integer lattice. However, in difference to the previous proof, there are no "ordinary" horizontal steps at all, while the "additional" horizontal steps are horizontal steps of length 2 from $(-2, y)$ to $(0, y)$ and from $(0, y)$ to $(2, y)$, for $y \geq 0$. The former horizontal steps have weight α, while the latter have weight β. As for the "ordinary" steps, up-steps have weight 1, and down-steps $(x, y+1) \to (x+1, y)$ have weight T_y.

[c]It is important to note that a cross is not a violation of the property of being non-intersecting: the "intersection" happens in a *non-lattice point*, that is, the corresponding paths do *not* have a vertex of the underlying graph in common.

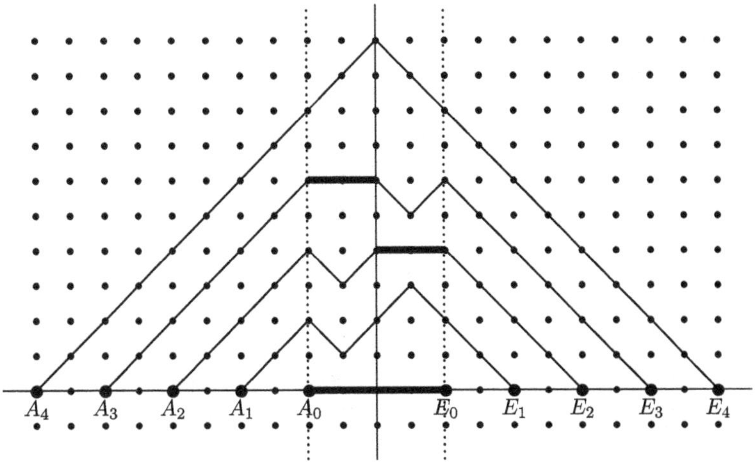

Fig. 4. Another family of non-intersecting lattice paths.

The non-intersecting lattice path model for the Hankel determinant in the numerator of (2.9) consists of families $\mathcal{P} = (P_0, P_1, \dots, P_{n-1})$ of non-intersecting lattice paths where P_i runs from $A_i = (-2i-2)$ to $E_{\sigma(i)}$, with $E_i = (2i+2, 0)$ and σ being some permutation of $\{0, 1, \dots, n-1\}$. See Fig. 4 for an example.

Here, the only freedom for the paths is in the strip between the vertical lines $x = -2$ and $x = +2$. Again, except for some top paths that have two up-steps followed by two down-steps in that strip, what happens in the substrips to the left and to the right of the y-axis is independent. We then map the configurations in these substrips to pillars, as is illustrated in Fig. 5 for the example in Fig. 4. The weight contribution of the uniquely determined parts to the left of $x = -2$ and to the right of $x = +2$ is exactly the right-hand side in (2.11) and thus equal to the denominator in (2.9).

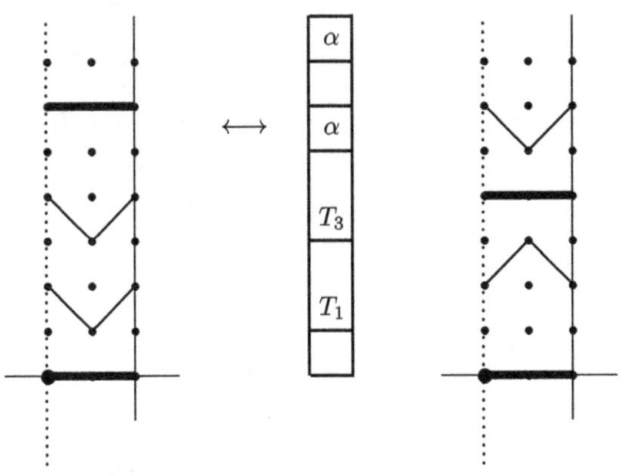

Fig. 5. Correspondence between local path configurations and pillars.

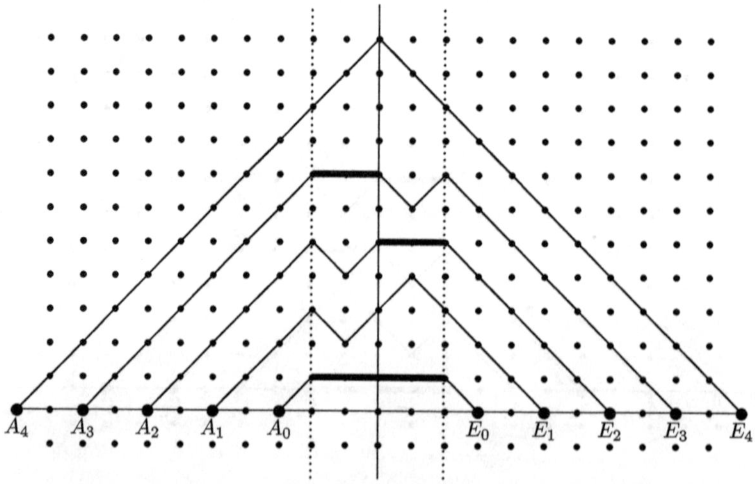

Fig. 6. Yet another family of non-intersecting lattice paths.

The combinatorial model for the Hankel determinant in the numerator of (2.10) is completely analogous to the previous one. Essentially, what one has to do is to raise all paths in the previous model by one unit, prepend an up-step and append a down-step to each path. We present a corresponding illustration in Fig. 6 and leave the details to the reader. □

7. Afterthoughts

A reader, after having seen [21] (presented here in (3.1) and (3.2)) and Theorem 1, will ask: what about the "next" Hankel determinant,

$$\det\left(\alpha\beta\gamma m_{i+j} + (\alpha\beta+\beta\gamma+\gamma\alpha)m_{i+j+1} + (\alpha+\beta+\gamma)m_{i+j+2} + m_{i+j+3}\right)_{i,j=0}^{n-1} \qquad ?$$

Nothing prevents us from attacking it in just the same way. As a matter of fact, the obvious extension of our combinatorial model from Sec. 6 in terms of non-intersecting lattice paths leads rather directly to the following result.

Theorem 5. *Let α, β, and γ be indeterminates. Then, for the setup of Sec. 2, for all positive integers n we have*

$$\frac{\det\left(\alpha\beta\gamma m_{i+j} + (\alpha\beta+\beta\gamma+\gamma\alpha)m_{i+j+1} + (\alpha+\beta+\gamma)m_{i+j+2} + m_{i+j+3}\right)_{i,j=0}^{n-1}}{\det(m_{i+j})_{i,j=0}^{n-1}}$$

$$= \sum_{0\le k\le j\le n} f_j(\alpha)f_k(\beta)f_k(\gamma)\left(f_{n-j}(\beta)\Big|_{\substack{s_l\to s_{l+j+1}\\ t_l\to t_{l+j+1}}}\right)\prod_{\ell=k}^{n-1} t_\ell$$

$$+ \sum_{0\le j<k\le n} f_j(\alpha)f_j(\beta)f_k(\gamma)\left(f_{n-k}(\beta)\Big|_{\substack{s_l\to s_{l+k+1}\\ t_l\to t_{l+k+1}}}\right)\prod_{\ell=j}^{n-1} t_\ell. \qquad (7.1)$$

Sketch of proof. We consider the same underlying graph as in the proof of Theorem 1 in Sec. 6, except that we do not only introduce *additional* horizontal steps between the vertical lines $x = -1$ and $x = 0$ (weighted by α) and between the vertical lines $x = 0$ and $x = +1$ (weighted by β) but also between the vertical lines $x = +1$ and $x = +2$, weighted by γ. The combinatorial model for the determinant in the numerator in the left-hand side of (7.1) consists of families $\mathcal{P} = (P_0, P_1, \ldots, P_{n-1})$ of non-intersecting lattice paths, where P_i runs from $A_i = (-i-1, 0)$ to $E_i = (\sigma(i) + 2, 0)$, $i = 0, 1, \ldots, n-1$, for some permutation σ that depends on $\mathcal{P}$. Figure 7 shows an example for $n = 9$. Again, we have indicated the additional horizontal edges by thick segments.

The combinatorial analysis of such families of non-intersecting lattice paths leads to the following conclusions:

(1) Being non-intersecting, the paths are uniquely determined to the left of the vertical line $x = -1$ and to the right of the vertical line $x = +2$. Only in the strip between these two vertical lines there is (some) freedom.
(2) In that strip, there may be some paths on top, $P_{n-1}, \ldots, P_m$, which have there an up-step followed by a horizontal step (ordinary or additional) followed by a down-step. (In Fig. 7, these are the two paths P_8 and P_7.)
(3) Below P_m, exactly one of the vertices of our graph (= lattice points with non-negative ordinate) on the vertical line $x = 0$ and exactly one of the vertices of our graph on the vertical line $x = +1$ is not taken by any of the paths. Let these vertices be $(0, j)$ and $(1, k)$, respectively. (In the example of Fig. 7, these are the points $(0, 6)$ and $(1, 7)$.) We must have $m = \max\{j, k\} + 1$.

Now let first $j < k$ (as in our example in Fig. 7). Then, necessarily, we must have a down-step from any lattice point $(0, y)$ with $j < y \leq k$ to $(1, y-1)$. In the substrip

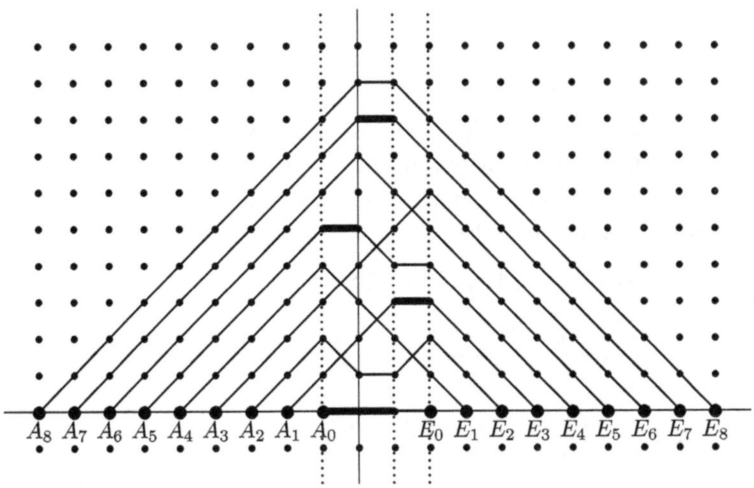

Fig. 7. Another family of non-intersecting Motzkin paths.

between the vertical lines $x = -1$ and $x = 0$ but *below* $y = j$, we find a configuration that is enumerated by $f_j(\alpha)$. Likewise in the substrip between $x = 0$ and $x = +1$ but *below* $y = j$, we find a configuration that is enumerated by $f_j(\beta)$, while in the substrip between $x = +1$ and $x = +2$ but *below* $y = k$, we find a configuration that is enumerated by $f_k(\gamma)$. Moreover, in the substrip between $x = 0$ and $x = +1$ but *above* $y = k$, we find a configuration that is enumerated by $f_{n-k}(\beta)$, where each s_l has to be replaced by s_{l+k+1} and each t_l by t_{l+k+1}. This explains the second sum on the right-hand side of (7.1).

The case where $j \geq k$ is disposed of similarly and it leads to the first sum on the right-hand side of (7.1). □

In the case where the s_i's and the t_i's are constant for $i \geq 1$, the right-hand side of (7.1) can again be evaluated in closed form in terms of Chebyshev polynomials. In order to have a succinct notation for the result, we introduce the umbral notation

$$U_\alpha^n \equiv U_n\left(\frac{\alpha+s}{2\sqrt{t}}\right), \qquad \text{for } n \geq 0.$$

Corollary 6. *Let α, β, and γ be indeterminates, and in the setup of Sec. 2 set $s_i \equiv s$ and $t_i \equiv t$ for $i \geq 1$. Then, for all positive integers n, we have*

$$\det\left(\alpha\beta\gamma m_{i+j} + (\alpha\beta+\beta\gamma+\gamma\alpha)m_{i+j+1} + (\alpha+\beta+\gamma)m_{i+j+2} + m_{i+j+3}\right)_{i,j=0}^{n-1}$$
$$= \frac{Num(\alpha,\beta,\gamma)}{(\alpha-\beta)(\beta-\gamma)(\gamma-\alpha)}, \tag{7.2}$$

where

$$\begin{aligned} Num(\alpha,\beta,\gamma) &= t^{(3n+3)/2}(U_\alpha - U_\beta)(U_\beta - U_\gamma)(U_\gamma - U_\alpha) \\ &\quad \times (1 - t^{-1/2}(s-s_0)U_\alpha^{-1} + t^{-1}(t-t_0)U_\alpha^{-2}) \\ &\quad \times (1 - t^{-1/2}(s-s_0)U_\beta^{-1} + t^{-1}(t-t_0)U_\beta^{-2}) \\ &\quad \times (1 - t^{-1/2}(s-s_0)U_\gamma^{-1} + t^{-1}(t-t_0)U_\gamma^{-2})U_\alpha^n U_\beta^n U_\gamma^n. \end{aligned}$$

Obviously, we could now again derive many Hankel determinant evaluations for the special cases that were discussed in Sec. 4. We refrain from doing so here in the interest of not going overboard at this point, and instead content ourselves with presenting a small list of examples:

$$\det(M_{i+j} + 3M_{i+j+1} + 3M_{i+j+2} + M_{i+j+3})_{i,j=0}^{n-1}$$
$$= \frac{(n+1)(n+2)^2(n+3)(2n+3)(2n+5)}{180}, \tag{7.3}$$

$$\det\left(-M_{i+j} + 3M_{i+j+1} - 3M_{i+j+2} + M_{i+j+3}\right)_{i,j=0}^{n-1}$$
$$= [n \text{ even}](-1)^{n/2}\left(\frac{n}{2}+1\right)^2, \tag{7.4}$$

$$\det(-M_{i+j}-M_{i+j+1}+M_{i+j+2}+M_{i+j+3})_{i,j=0}^{n-1}$$
$$=\begin{cases}(-1)^{n_0}(n_0+1)^2, & \text{if } n=2n_0,\\ (-1)^{n_0-1}\binom{2n_0+2}{3}, & \text{if } n=2n_0-1,\end{cases} \tag{7.5}$$

$$\det(-C_{i+j}+3C_{i+j+1}-3C_{i+j+2}+C_{i+j+3})_{i,j=0}^{n-1}$$
$$=\begin{cases}(-1)^{n_0}n_0(2n_0+1), & \text{if } n=3n_0-1,\\ (-1)^{n_0}(2n_0+1)^2, & \text{if } n=3n_0,\\ (-1)^{n_0}(n_0+1)(2n_0+1), & \text{if } n=3n_0+1,\end{cases} \tag{7.6}$$

$$\det(-C_{i+j+1}+3C_{i+j+2}-3C_{i+j+3}+C_{i+j+4})_{i,j=0}^{n-1}$$
$$=\begin{cases}(-1)^{n_0-1}n_0(2n_0+1), & \text{if } n=3n_0-1,\\ (-1)^{n_0}(n_0+1)(2n_0+1), & \text{if } n=3n_0,\\ (-1)^{n_0}(2n_0+2)^2, & \text{if } n=3n_0+1.\end{cases} \tag{7.7}$$

Interestingly, by chance (?) the evaluation in (7.3) turns out to be exactly the same as the evaluation of the Hankel determinant $\det(C_{i+j+4})_{i,j=0}^{n-1}$.

In the same way as we obtained Corollary 4 from Corollary 3, the above corollary entails that the expressions in (7.2) satisfy a linear recurrence of order 8. This generalizes [7, Theorem 4].

Corollary 7. *With the setup in Sec. 2, let $s_i \equiv s$ and $t_i \equiv t$ for $i \ge 1$. Then the scaled Hankel determinants*

$$t_0^{-(n-1)}t^{-\binom{n-1}{2}}\det(\alpha\beta\gamma m_{i+j}+(\alpha\beta+\beta\gamma+\gamma\alpha)m_{i+j+1}$$
$$+(\alpha+\beta+\gamma)m_{i+j+2}+m_{i+j+3})_{i,j=0}^{n-1} \tag{7.8}$$

satisfy a linear recurrence of order 8 with constant coefficients.

Clearly, one could continue in this manner. However, the combinatorial analysis would become more and more intricate. On the other hand, it should be noted that Corollaries 3 and 6 (as well as (3.2) with (4.2)) follow a pattern. On this basis, we make the following conjecture on the evaluation of the Hankel determinant of an arbitrary linear combination of path generating functions in the case where the s_i's and the t_i's are constant for $i \ge 1$. In order to state the result, we also employ umbral notation for the path generating functions, namely, $m^n \equiv m_n$ for all n.

Conjecture 8. *Let $\alpha_1, \alpha_2, \ldots, \alpha_d$ be indeterminates, and in the setup of Sec. 2 set $s_i \equiv s$ and $t_i \equiv t$ for $i \ge 1$. Then, for all positive integers d and n, we have*

$$\frac{\det\left(m^{i+j}\prod_{\ell=1}^{d}(\alpha_\ell+m)\right)_{i,j=0}^{n-1}}{\det(m_{i+j})_{i,j=0}^{n-1}}=\frac{Num(\alpha_1,\ldots,\alpha_d)}{\prod_{1\le i<j\le d}(\alpha_i-\alpha_j)}, \tag{7.9}$$

where

$$Num(\alpha_1, \dots, \alpha_d) = t^{\frac{1}{2}(dn+\binom{d}{2})} \prod_{1\le i<j\le d} (U_{\alpha_i} - U_{\alpha_j})$$
$$\times \prod_{i=1}^{d}(1 - t^{-1/2}(s - s_0)U_{\alpha_i}^{-1} + t^{-1}(t - t_0)U_{\alpha_i}^{-2})U_{\alpha_i}^{n}.$$

Obviously (compare with the proof of Corollary 4), if true, then this formula implies that the scaled determinants on the left-hand side of (7.9) satisfy a linear recurrence of order 2^d with constant coefficients. This would be a vast generalization of [7, Conjecture 5].

Note. After the first version of this paper was written, the second author discovered a general identity for Hankel determinants of linear combinations of moments of orthogonal polynomials of which Conjecture 8 is a special case, and that this identity somehow exists in the folklore of the theory of orthogonal polynomials but does not seem to be well-known. These findings will be reported in the forthcoming paper [19].

Acknowledgment

This paper was written while the second author was visiting the Institut Mittag–Leffler during the "Algebraic and Enumerative Combinatorics" programme (grant no. 2016-06596 of the Swedish Research Council) in Spring 2020, organized by Sara Billey, Petter Brändén, Sylvie Corteel and Svante Linusson. Research partially supported by the Austrian Science Foundation FWF (grant S50-N15) in the framework of the Special Research Program "Algorithmic and Enumerative Combinatorics.

References

[1] M. Aigner, Catalan and other numbers: A recurrent theme, in *Algebraic Combinatorics and Computer Science*, editors H. Crapo and D. Senato (Springer–Verlag, 2001) pp. 347–390.

[2] N. T. Cameron and A.C.M. Yip, Hankel determinants of sums of consecutive Motzkin numbers, *Linear Algebra Appl.* **434** (2011) 712–722.

[3] J. Cigler, Some relations between generalized Fibonacci and Catalan numbers, *Sitz. ber. d. Österr. Akad. Wiss. Math. Nat. wiss. Kl.* **211** (2002) 143–154.

[4] J. Cigler, A simple approach to some Hankel determinants, manuscript arXiv:0902.1650.

[5] J. Cigler, Hankel determinants of some polynomial sequences manuscript, arXiv:1211.0816.

[6] A. Cvetković, P. Rajković and M. Ivković , Catalan numbers, and Hankel transform, and Fibonacci numbers, *J. Integer Seq.* **5** (2002), Article ID: 02.1.3, 8 pp.

[7] M. Dougherty, C. French, B. Saderholm and W. Qian, Hankel transforms of linear combinations of Catalan numbers , *J. Integer Seq.* **14** (2011), Article ID: 11.5.1, 20 pp.

[8] M. Elouafi, A unified approach for the Hankel determinants of classical combinatorial numbers, *J. Math. Anal. Appl.* **431** (2015) 1253–1274.

[9] S.-P. Eu, T.-L. Wong and P.-L. Yen, Hankel determinants of sums of consecutive weighted Schröder numbers, *Linear Algebra Appl.* **437** (2012) 2285–2299.

[10] G. Gasper and M. Rahman, *Basic Hypergeometric Series*, 2nd edn., Encyclopedia of Mathematics and Its Applications, Vol. 96 (Cambridge University Press, 2004).
[11] I. M. Gessel and X. Viennot, Binomial determinants, paths, and hook length formulae, *Adv. Math.* **58** (1985) 300–321.
[12] I. M. Gessel and X. Viennot, Determinants, paths, and plane partitions, preprint (1989), Available at http://www.cs.brandeis.edu/~ira.
[13] F. Harary and R. C. Read, The enumeration of tree-like polyhexes, *Proc. Edinburgh Math. Soc.* (*2*) **17** (1970) 1–13.
[14] M. E. H. Ismail and D. Stanton, Classical orthogonal polynomials as moments, *Can. J. Math.* **49** (1997) 520–542.
[15] M. E. H. Ismail and D. Stanton, More orthogonal polynomials as moments, in *Mathematical Essays in Honor of Gian-Carlo Rota*, editors B. E. Sagan and R. P. Stanley, Progress in Mathematics, Vol. 161 (Birkhäuser, 1998), pp. 377–396.
[16] C. Krattenthaler, Advanced determinant calculus, *Sém. Lothar. Combin.* **42** ("The Andrews Festschrift") (1999), Article ID: B42q, 67 pp.
[17] C. Krattenthaler, Advanced determinant calculus: A complement, *Linear Algebra Appl.* **411** (2005) 68–166.
[18] C. Krattenthaler, Determinants of (generalized) Catalan numbers, *J. Statist. Plann. Inference* **140** (2010) 2260–2270.
[19] C. Krattenthaler, Hankel determinants of linear combinations of moments of orthogonal polynomials, II, preprint, arXiv:2101.04225.
[20] B. Lindström, On the vector representations of induced matroids, *Bull. London Math. Soc.* **5** (1973) 85–90.
[21] L. Mu, Y. Wang and Y.-N. Yeh, Hankel determinants of linear combinations of consecutive Catalan-like numbers, *Discrete Math.* **340** (2017) 3097–3103.
[22] N. J. A. Sloane, *The On-Line Encyclopedia of Integer Sequences*, http://oeis.org/.
[23] R. P. Stanley, *Enumerative Combinatorics*, Vol. 2 (Cambridge University Press, 1999).
[24] X. Viennot, *Une Théorie Combinatoire des Polynômes Orthogonaux Généraux* (UQAM, Montréal, Québec, 1983), Available at http://www.xavierviennot.org/xavier/polynomes_orthogonaux.html.

© 2024 World Scientific Publishing Company
https://doi.org/10.1142/9789811277375_0011

Polynomial analogues of restricted multicolor b-ary partition functions*

Karl Dilcher†

Department of Mathematics and Statistics, Dalhousie University
Halifax, Nova Scotia, B3H 4R2, Canada
dilcher@mathstat.dal.ca

Larry Ericksen

P.O. Box 172, Millville, NJ 08332-0172, USA
LE22@cornell.edu

Given an integer base $b \geq 2$, a number $\rho \geq 1$ of colors, and a finite sequence $\Lambda = (\lambda_1, \ldots, \lambda_\rho)$ of positive integers, we introduce the concept of a Λ-restricted ρ-colored b-ary partition of an integer $n \geq 1$. We also define a sequence of polynomials in $\lambda_1 + \cdots + \lambda_\rho$ variables, and prove that the nth polynomial characterizes all Λ-restricted ρ-colored b-ary partitions of n. In the process, we define a recurrence relation for the polynomials in question, obtain explicit formulas, and identify a factorization theorem.

Keywords: Binary partition; b-ary partition; restricted partition; polynomial analogue.

Mathematics Subject Classification 2020: 11P81, 11B37, 11B83

1. Introduction

The study of binary partitions and partition functions originated with Euler [10, p. 162ff.], but is still an active area of research. These concepts were later extended to b-ary and restricted b-ary partitions for arbitrary integer bases $b \geq 2$; some of the most important contributions were made, in chronological order, by Mahler [13], Churchhouse [3], Rødseth [15], Andrews [1], Reznick [14], and Dumont *et al.* [9]. Further details, putting these publication in perspective, can be found in [7].

More recently, Rødseth and Sellers [16] introduced and studied b-ary overpartitions, in analogy to ordinary overpartitions that had been introduced a little earlier by Corteel and Lovejoy [4]. Ordinary overpartitions were extended to restricted multicolor partitions by Keith [12], and colored b-ary partitions have recently been studied by Ulas and Żmija [18].

*This article originally appeared in *International Journal of Number Theory*, **17**(2), pp. 371–391 © 2021 World Scientific Publishing Company. DOI: 10.1142/S1793042120400096

†Corresponding author.

In almost all the papers mentioned above, the authors obtained interesting and important arithmetic properties of the partition functions in question, including Ramanujan-type congruences. The purpose of this paper is quite different. Building on earlier work, especially on our recent paper [7], we define a sequence of multivariable polynomials that characterize all individual partitions of an integer $n \geq 0$ (as opposed to just their numbers) for a large class of restricted colored b-ary partitions.

To make the results of this paper more transparent and easier to understand, we recall some earlier work, which may serve as an example. Given integers $b \geq 2$ and $\lambda \geq 1$, the representations

$$n = \sum_{j\geq 0} c_j b^j, \qquad c_j \in \{0, 1, \ldots, \lambda\}, \tag{1.1}$$

of an integer $n \geq 0$ are restricted b-ary partitions, and their numbers are denoted by $S_b^\lambda(n)$ in [7]. In particular, $S_b^{b-1}(n) = 1$ since it corresponds to the unique b-ary representation of n. In the base 2 case, $S_2^2(n)$ counts the number of hyperbinary representations of n, and $S_2^2(n) = s(n+1)$, where $\{s(n)\}$ is the well-known Stern sequence; see [14, p. 470]. Related to this, Bates and Mansour [2] and independently Stanley and Wilf [17] introduced polynomial analogues as refinements of $s(n+1)$; their coefficients count hyperbinary representations with certain properties.

Recently the current authors [5] refined this further by introducing the following sequence of bivariate polynomials.

Definition 1.1 ([5, Definition 4.1]). Let s and t be fixed positive integer parameters. We define the 2-parameter generalized Stern polynomials in the variables y and z by $\omega_{s,t}(0; y, z) = 0$, $\omega_{s,t}(1; y, z) = 1$, and for $n \geq 1$ by

$$\omega_{s,t}(2n; y, z) = y\,\omega_{s,t}(n; y^s, z^t), \tag{1.2}$$

$$\omega_{s,t}(2n+1; y, z) = z\,\omega_{s,t}(n; y^s, z^t) + \omega_{s,t}(n+1; y^s, z^t). \tag{1.3}$$

By comparing the recurrence relations (1.2), (1.3) with those of Stern's sequence, it is clear that $\omega_{s,t}(n; 1, 1) = s(n)$, independent of the parameters s, t. The usefulness of the polynomials $\omega_{s,t}(n; y, z)$, also relevant for this paper, lies in the following characterization of all hyperbinary expansions of a positive integer n, which we denote by $\mathbb{H}_n$.

Theorem 1.2 ([5, Theorem 4.2]). *For any integer $n \geq 1$, we have*

$$\omega_{s,t}(n+1; y, z) = \sum_{h\in\mathbb{H}_n} y^{p_h(s)} z^{q_h(t)}, \tag{1.4}$$

where for each hyperbinary expansion h of n, the exponents $p_h(s)$, $q_h(t)$ are polynomials in s and t, respectively, with only 0 and 1 as coefficients. Furthermore, if

$$p_h(s) = s^{\alpha_1} + \cdots + s^{\alpha_\mu}, \quad 0 \leq \alpha_1 < \cdots < \alpha_\mu, \quad \mu \geq 0, \tag{1.5}$$

$$q_h(t) = t^{\beta_1} + \cdots + t^{\beta_\nu}, \quad 0 \leq \beta_1 < \cdots < \beta_\nu, \quad \nu \geq 0, \tag{1.6}$$

then the hyperbinary expansion $h \in \mathbb{H}_n$ is given by

$$n = 2^{\alpha_1} + \cdots + 2^{\alpha_\mu} + (2^{\beta_1} + 2^{\beta_1}) + \cdots + (2^{\beta_\nu} + 2^{\beta_\nu}). \tag{1.7}$$

Example 1.3 ([5, Example 8]). With Definition 1.1, we compute

$$\omega_{s,t}(11;y,z)=z^{1+t^2}+y^s z^{t^2}+y^{s^2}z^{1+t}+y^{s^3}z+y^{s+s^3}.$$

If we choose the third term on the right, then according to Theorem 1.2 the corresponding hyperbinary expansion will be $n = 4+1+1+2+2$. The other four expansions can be "read off" analogously.

The sequence of polynomials $\omega_{s,t}(n;y,z)$ satisfies the following generating function.

Theorem 1.4 ([5, Theorem 4.4]). *For integers $s,t\geq 1$, we have*

$$\sum_{n=1}^{\infty}\omega_{s,t}(n;y,z)q^n=q\prod_{j=0}^{\infty}\left(1+y^{s^j}q^{2^j}+z^{t^j}q^{2^{j+1}}\right). \tag{1.8}$$

Finally, there is an explicit formula, for which we first need to recall some notation. We consider binomial coefficients modulo 2 as follows:

$$\binom{n}{k}^* \equiv \binom{n}{k} \pmod 2, \qquad \binom{n}{k}^* \in \{0,1\}.$$

Furthermore, let $n=\sum_{j\geq 0}c_j 2^j$, $c_j\in\{0,1\}$, be the binary expansion of an integer $n\geq 0$. Then for integers $t\geq 1$, we define

$$d_t(n):=\sum_{j\geq 0}c_j t^j. \tag{1.9}$$

We can now state a first version of the desired explicit formula.

Lemma 1.5 ([5, Theorem 4.3]). *For integers $s,t\geq 1$ and $n\geq 0$, we have*

$$\omega_{s,t}(n+1;y,z)=\sum_{k=0}^{\lfloor\frac{n}{2}\rfloor}\binom{n-k}{k}^* y^{d_s(n-2k)}z^{d_t(k)}. \tag{1.10}$$

As shown in [7], the identity (1.10) can be rewritten in a form that will extend more easily to a more general situation.

Theorem 1.6 ([7, Eq. (55)]). *For integers $s,t\geq 1$ and $n\geq 0$, we have*

$$\omega_{s,t}(n+1;y,z)=\sum_{k_1+2k_2=n}\binom{k_1+k_2}{k_2}^* y^{d_s(k_1)}z^{d_t(k_2)}. \tag{1.11}$$

Finally, the polynomials $\omega_{s,t}(n;y,z)$ also satisfy various identities, an example of which will follow. For greater ease of notation we drop the subscripts s,t and just write $\omega(n;y,z)$. We present a special case of [8, Theorem 4.2].

Theorem 1.7. *Let $\ell\geq 1$ and $k\geq 1$ be integers. Then for all integers j with $0\leq j\leq 2^\ell$, we have*

$$\begin{aligned}\omega(k\cdot 2^\ell+j;y,z) &= \omega(k+1;y^{s^\ell},z^{t^\ell})\cdot\omega(j;y,z) \\ &\quad +\omega(k;y^{s^\ell},z^{t^\ell})\cdot\left(\omega(2^\ell+j;y,z)-y^{s^\ell}\omega(j;y,z)\right).\end{aligned} \tag{1.12}$$

All this, from Definition 1.1 to Theorem 1.7, has been generalized, first to b-ary multivariable Stern polynomials in [8], and then further to polynomial analogues of restricted b-ary partition functions; see [7]. For general $b \geq 2$, some phenomena become visible that are trivial in the case $b = 2$. For instance, the general case of (1.12) leads to a factorization result that is meaningful only for $b \geq 3$; see [8, Corollary 4.3].

It is the purpose of this paper to obtain results for polynomial analogues of restricted colored b-ary partition functions, extending Theorems 1.2, 1.4, 1.6 and 1.7. We begin by defining, in Sec. 2, the concepts of restricted colored b-ary partitions, the corresponding partition functions, and their polynomial analogues. Section 3 contains the main result of this paper, namely a representation theorem that connects the restricted colored b-ary partitions of Definition 2.1 with the multivariate polynomials of Definition 2.4. In Sec. 4, we derive a recurrence relation for these polynomials, which we use in Sec. 5 to prove the representation theorem. In Secs. 6 and 7, we obtain an explicit formula and a factorization result, respectively, for the multivariate polynomials; we then end with some further remarks and examples in Sec. 8.

2. Colored b-ary Partition Functions

The concepts of a restricted or unrestricted colored b-ary partition can be seen as an extension of b-ary overpartitions that were introduced by Rødseth and Sellers [16]. According to their definition, a b-ary *overpartition* of an integer $n \geq 1$ is a sequence of nonnegative integer powers whose sum is n, and where the first occurrence of a power b may be overlined.

This means that each b-ary partition can have two types (or "colors") of parts, namely the "basic" type which may occur an arbitrary number of times, and an "overlined" type, which may occur at most once. This leads to the idea of assigning a fixed number $\rho \geq 1$ of colors in a more symmetric fashion, that is, such that all colors are restricted.

Definition 2.1. Let $b \geq 2$ and $\rho \geq 1$ be integers, and let $\Lambda = (\lambda_1, \dots, \lambda_\rho)$ be a finite integer sequence with $1 \leq \lambda_1 \leq \cdots \leq \lambda_\rho$. Then a Λ-restricted ρ-colored b-ary partition of an integer $n \geq 1$ is a set of powers b^j whose sum is n, and where for a given integer $j \geq 0$, up to λ_r parts b^j can be assigned the rth color, for $1 \leq r \leq \rho$.

In other words, each part b^j can be assigned the ℓth color at most λ_ℓ times, for $1 \leq \ell \leq \rho$. If $C_b^\Lambda(n)$ denotes the number of partitions of n as defined in Definition 2.1, then the generating function is

$$\sum_{n=0}^{\infty} C_b^\Lambda(n) q^n = \prod_{j=0}^{\infty} \left(1 + q^{b^j} + \cdots + q^{\lambda_1 \cdot b^j}\right) \cdots \left(1 + q^{b^j} + \cdots + q^{\lambda_\rho \cdot b^j}\right). \quad (2.1)$$

We now illustrate Definition 2.1 with a concrete example.

Example 2.2. Let $b = 2$ and $\Lambda = (2, 3)$, and thus $\rho = 2$. Then the Λ-restricted 2-colored binary partitions of $n = 4$ are

$4_2,\ 4_1,\ 2_2 + 2_2,\ 2_2 + 2_1,\ 2_1 + 2_1,\ 2_2 + 1_2 + 1_2,\ 2_2 + 1_2 + 1_1,\ 2_2 + 1_1 + 1_1,$

$2_1 + 1_2 + 1_2,\ 2_1 + 1_2 + 1_1,\ 2_1 + 1_1 + 1_1,\ 1_2 + 1_2 + 1_2 + 1_1,\ 1_2 + 1_2 + 1_1 + 1_1,$

where the subscripts denote the colors of the part, and by convention we write each partition in non-increasing order of its parts, and assign the colors also in non-increasing order. This is consistent with the notation of colored ordinary partitions used by Keith [12]. Counting the partitions in this example, we see that $C_2^\Lambda(4) = 13$. On the other hand, by (2.1) we have

$$\begin{aligned}\sum_{n=0}^{\infty} C_2^\Lambda(n) q^n &= \prod_{j=0}^{\infty}\left(1 + q^{2^j} + q^{2\cdot 2^j}\right)\left(1 + q^{2^j} + q^{2\cdot 2^j} + q^{3\cdot 2^j}\right)\\ &= 1 + 2q + 5q^2 + 7q^3 + 13q^4 + 17q^5 + 26q^6 + \cdots,\end{aligned}$$

which confirms that $C_2^\Lambda(4) = 13$.

When $\rho = 1$, Definition 2.1 reduces to that of usual restricted b-ary partitions as in (1.1). Their numbers were denoted by $S_b^\lambda(n)$ in [7], with generating function

$$\sum_{n=0}^{\infty} S_b^\lambda(n) q^n = \prod_{j=0}^{\infty}\left(1 + q^{b^j} + q^{2\cdot b^j} + \cdots + q^{\lambda\cdot b^j}\right). \tag{2.2}$$

When $\Lambda = (\lambda)$, then for all $n \geq 0$ we obviously have $C_b^\Lambda(n) = S_b^\lambda(n)$. The generating function (2.2) was then used in [7] as the basis for the following definition of a polynomial analogue.

Definition 2.3 ([7, Definition 1]). Let $b \geq 2$ and $\lambda \geq 1$ be integers, $T = (t_1, \ldots, t_\lambda)$ be a λ-tuple of positive integer parameters, and $Z = (z_1, \ldots, z_\lambda)$ a λ-tuple of variables. Then, we define the sequence of λ-variable polynomials $\omega_{b,T}^\lambda(n; Z)$ by the generating function

$$\sum_{n=0}^{\infty} \omega_{b,T}^\lambda(n+1; Z) q^n = \prod_{j=0}^{\infty}\left(1 + z_1^{t_1^j} q^{b^j} + z_2^{t_2^j} q^{2\cdot b^j} + \cdots + z_\lambda^{t_\lambda^j} q^{\lambda\cdot b^j}\right). \tag{2.3}$$

We immediately see that this extends both (1.8) and (2.2). Indeed, if $T = (s, t)$, $Z = (y, z)$, and $b = \lambda = 2$, then $\omega_{2,T}^2(n; Z) = \omega_{s,t}(n; y, z)$. Also, when $Z = (1, \ldots, 1)$ and T is arbitrary, then for any $b \geq 2$ and $\lambda \geq 1$, we have

$$\omega_{b,T}^\lambda(n + 1; (1, \ldots, 1)) = S_b^\lambda(n).$$

With (2.1) and (2.3) in mind, we are now ready to define the main object of study in this paper.

Definition 2.4. Let $b \geq 2$ and $\rho \geq 1$ be integers, and let $\Lambda = (\lambda_1, \ldots, \lambda_\rho)$ be a finite integer sequence with $1 \leq \lambda_1 \leq \cdots \leq \lambda_\rho$. Furthermore, let $\lambda := \lambda_1 + \cdots + \lambda_\rho$ and let $T = (t_{1,1}, \ldots, t_{1,\lambda_1}; \ldots; t_{\rho,1}, \ldots, t_{\rho,\lambda_\rho})$ be a λ-tuple of positive integer parameters, and $Z = (z_{1,1}, \ldots, z_{1,\lambda_1}; \ldots; z_{\rho,1}, \ldots, z_{\rho,\lambda_\rho})$ a λ-tuple of variables. Then, we define the sequence of λ-variable polynomials $\Omega_{b,T}^{\Lambda}(n; Z)$ by the generating function

$$\sum_{n=0}^{\infty} \Omega_{b,T}^{\Lambda}(n; Z) q^n = \prod_{j=0}^{\infty} \left(1 + z_{1,1}^{t_{1,1}^j} q^{b^j} + \cdots + z_{1,\lambda_1}^{t_{1,\lambda_1}^j} q^{\lambda_1 \cdot b^j}\right) \cdots \qquad (2.4)$$

$$\left(1 + z_{\rho,1}^{t_{\rho,1}^j} q^{b^j} + \cdots + z_{\rho,\lambda_\rho}^{t_{\rho,\lambda_\rho}^j} q^{\lambda_\rho \cdot b^j}\right).$$

Suppose that in Z we have $z_{i,j} = 1$ for all relevant indices i, j. Then for arbitrary T we have

$$\Omega_{b,T}^{\Lambda}(n; (1, \ldots, 1)) = C_b^{\Lambda}(n), \qquad n = 0, 1, \ldots. \qquad (2.5)$$

When $\rho = 1$, we set $\Lambda = (\lambda)$ and $t_{1,i} = t_i$, $z_{1,i} = z_i$ for $1 \le i \le \lambda$; then comparing (2.4) with (2.3) gives

$$\Omega_{b,T}^{\Lambda}(n; Z) = \omega_{b,T}^{\lambda}(n+1; Z), \qquad \rho = 1, \quad n = 0, 1, \ldots. \qquad (2.6)$$

The argument $n+1$ on the right of this last identity is due to the fact that the polynomials $\omega_{b,T}^{\lambda}(n+1; Z)$ came from the study of generalized Stern polynomials; see also Definition 1.1.

We conclude this section by illustrating Definition 2.4 with an example that can be seen as a continuation of Example 2.2.

Example 2.5. Let $b = 2$ and $\Lambda = (2, 3)$. To avoid double indices, we set $T = (s_1, s_2; t_1, t_2, t_3)$ and $Z = (y_1, y_2; z_1, z_2, z_3)$. Then the generating function (2.4) becomes

$$\sum_{n=0}^{\infty} \Omega_{b,T}^{\Lambda}(n; Z) q^n = \prod_{j=0}^{\infty} \left(1 + y_1^{s_1^j} q^{2^j} + y_2^{s_2^j} q^{2\cdot 2^j}\right) \left(1 + z_1^{t_1^j} q^{2^j} + z_2^{t_2^j} q^{2\cdot 2^j} + z_3^{t_3^j} q^{3\cdot 2^j}\right) \qquad (2.7)$$

The first few polynomials, obtained by expanding the right-hand side of (2.7), are shown in Table 1. Note that the number of terms of the polynomials in Table 1 are 1, 2, 5, 7, 13 and 17, respectively. This is consistent with (2.5) and Example 2.2.

Table 1. $\Omega_{b,T}^{\Lambda}(n; Z)$ as in Example 2.5, for $0 \le n \le 5$.

n	$\Omega_{b,T}^{\Lambda}(n; Z)$
0	1
1	$z_1 + y_1$
2	$z_1^{t_1} + z_2 + z_1 y_1 + y_1^{s_1} + y_2$
3	$z_1^{1+t_1} + z_3 + z_1^{t_1} y_1 + z_2 y_1 + z_1 y_1^{s_1} + y_1^{1+s_1} + z_1 y_2$
4	$z_1^{t_1^2} + z_1^{t_1} z_2 + z_2^{t_2} + z_1^{1+t_1} y_1 + z_3 y_1 + z_1^{t_1} y_1^{s_1}$
	$+z_2 y_1^{s_1} + y_1^{s_1^2} + z_1 y_1^{1+s_1} + z_1^{t_1} y_2 + z_2 y_2 + y_1^{s_1} y_2 + y_2^{s_2}$
5	$y_1^{1+s_1^2} + y_1 y_2^{s_2} + y_1^{s_1^2} z_1 + y_1^{s_1} y_2 z_1 + y_2^{s_2} z_1 + y_1^{1+s_1} z_1^{t_1} + y_1 z_1^{t_1^2} + y_1^{s_1} z_1^{1+t_1}$
	$+y_2 z_1^{1+t_1} + z_1^{1+t_1^2} + y_1^{1+s_1} z_2 + y_1 z_1^{t_1} z_2 + y_1 z_2^{t_2} + z_1 z_2^{t_2} + y_1^{s_1} z_3 + y_2 z_3 + z_1^{t_1} z_3$

We conclude this section with a result that is easily obtained by combining Definitions 2.3 and 2.4; it will be useful later, in Sec. 6.

Theorem 2.6. *Let $b \geq 2$ and $\rho \geq 1$ be integers, and let Λ, T, and Z be as in Definition 2.4. Furthermore, for each $\ell = 1, \dots, \rho$ let*

$$T_\ell = (t_{\ell,1}, \dots, t_{\ell,\lambda_\ell}) \quad \text{and} \quad Z_\ell = (z_{\ell,1}, \dots, z_{\ell,\lambda_\ell}).$$

Then for each integer $n \geq 0$, we have

$$\Omega_{b,T}^{\Lambda}(n; Z) = \sum_{n_1+\cdots+n_\rho=n} \omega_{b,T_1}^{\lambda_1}(n_1+1; Z_1) \cdots \omega_{b,T_\rho}^{\lambda_\rho}(n_\rho+1; Z_\rho). \tag{2.8}$$

Proof. We use the fact that the right-hand side of (2.4) is the product of $\rho \geq 1$ generating functions of the type that occurs on the right-hand side of the generating function (2.3). Equating coefficients of q^n, we see that the higher-order convolution on the right-hand side of (2.8) comes from the product of ρ appropriate power series from the left of (2.3), while the left-hand side of (2.8) comes from the left of the identity (2.4). □

We note that in the special case $\rho = 1$, the identity (2.8) reduces to (2.6).

3. A Representation Theorem

The main reason for introducing polynomial analogues for the Stern and b-ary Stern sequences [5, 6] is the fact that they can be used to characterize all hyperbinary and hyper b-ary representations. Theorem 1.2 and Example 1.3 may serve as an illustration of this. More generally, in [7] we used polynomial analogues for restricted b-ary partition functions to characterize all restricted b-ary partitions of a given integer.

In this section, we will extend these representation theorems to restricted *colored* b-ary partitions. We begin with an example.

Example 3.1. As in Example 2.2, we let $b = 2$ and $\Lambda = (2, 3)$. This time we consider the Λ-restricted 2-colored binary partitions of $n = 3$, namely

$$2_2 + 1_2, 2_2 + 1_1, 2_1 + 1_2, 2_1 + 1_1, 1_2 + 1_2 + 1_2, 1_2 + 1_2 + 1_1, 1_2 + 1_1 + 1_1. \tag{3.1}$$

On the other hand, from Table 1, we have

$$\Omega_{2,T}^{\Lambda}(3; Z) = z_1^{1+t_1} + z_3 + z_1^{t_1} y_1 + z_2 y_1 + z_1 y_1^{s_1} + y_1^{1+s_1} + z_1 y_2. \tag{3.2}$$

We rewrite this polynomial in the notation of Definition 2.4, namely with $T = (t_{1,1}, t_{1,2}; t_{2,1}, t_{2,2}, t_{2,3})$ and $Z = (z_{1,1}, z_{1,2}; z_{2,1}, z_{2,2}, z_{2,3})$. Then (3.2) becomes

$$\begin{aligned} \Omega_{2,T}^{\Lambda}(3; Z) = {} & z_{2,1}^{1+t_{2,1}} + z_{2,3} + z_{2,1}^{t_{2,1}} z_{1,1} + z_{2,2} z_{1,1} \\ & + z_{2,1} z_{1,1}^{t_{1,1}} + z_{1,1}^{1+t_{1,1}} + z_{2,1} z_{1,2}. \end{aligned} \tag{3.3}$$

Table 2. The monomials in (3.3).

Monomial	Partition	Rewritten
$z_{2,1}^{1+t_{2,1}}$	$1\cdot 2_2^0 + 1\cdot 2_2^1$	2_2+1_2
$z_{2,3}$	$3\cdot 2_2^0$	$1_2+1_2+1_2$
$z_{2,1}^{t_{2,1}} z_{1,1}$	$1\cdot 2_2^1 + 1\cdot 2_1^0$	2_2+1_1
$z_{2,2}z_{1,1}$	$2\cdot 2_2^0 + 1\cdot 2_1^0$	$1_2+1_2+1_1$
$z_{2,1}z_{1,1}^{t_{1,1}}$	$1\cdot 2_2^0 + 1\cdot 2_1^1$	2_1+1_2
$z_{1,1}^{1+t_{1,1}}$	$1\cdot 2_1^0 + 1\cdot 2_1^1$	2_1+1_1
$z_{2,1}z_{1,2}$	$1\cdot 2_2^0 + 2\cdot 2_1^0$	$1_2+1_1+1_1$

We assign each of the $C_2^{\Lambda}(3) = 7$ partitions in (3.1) to one of the seven monomials in (3.3) as shown in Table 2.

The one-to-one correspondence shown in Table 2 is a special case of the following representation theorem. Recall that $C_b^{\Lambda}(n)$ is the number of Λ-restricted colored b-ary partitions of n; see Definition 2.1.

Theorem 3.2. *Let $b \geq 2$ and $\rho \geq 1$ be integers, and let the finite sequences Λ, T, and Z be as in Definition* 2.4.

(a) *For any integer $n \geq 1$, we have*

$$\Omega_{b,T}^{\Lambda}(n; Z) = \sum_{h=1}^{C_b^{\Lambda}(n)} \prod_{\ell=1}^{\rho} z_{\ell,1}^{p_{h,\ell,1}(t_{\ell,1})} \cdots z_{\ell,\lambda_\ell}^{p_{h,\ell,\lambda_\ell}(t_{\ell,\lambda_\ell})}, \tag{3.4}$$

where for each $h = 1, \ldots, C_b^{\Lambda}(n)$, the exponents $p_{h,\ell,i}(t_{\ell,i})$, $1 \leq \ell \leq \rho$ and $1 \leq i \leq \lambda_\ell$, are polynomials with only 0 *and* 1 *as coefficients.*

(b) *We fix an $h = 1, \ldots, C_b^{\Lambda}(n)$, and for each pair (ℓ, i) with $1 \leq \ell \leq \rho$ and $1 \leq i \leq \lambda_\ell$ we write*

$$p_{h,\ell,i}(t_{\ell,i}) = t_{\ell,i}^{\tau_{\ell,i}(1)} + t_{\ell,i}^{\tau_{\ell,i}(2)} + \cdots + t_{\ell,i}^{\tau_{\ell,i}(\mu_{\ell,i})}, \quad \mu_{\ell,i} \geq 0. \tag{3.5}$$

Let the exponents $\tau_{\ell,i}(j)$, $j = 1, 2, \ldots$, form a strictly increasing finite sequence of nonnegative integers, and by convention we let $\mu_{\ell,i} = 0$ correspond to $p_{h,\ell,i}(t_{\ell,i}) = 0$. Then the Λ-restricted colored b-ary partition of n corresponding to the index-h monomial on the right-hand side of (3.4) *is*

$$\sum_{\ell=1}^{\rho} \sum_{i=1}^{\lambda_\ell} \left(i \cdot b_\ell^{\tau_{\ell,i}(1)} + \cdots + i \cdot b_\ell^{\tau_{\ell,i}(\mu_{\ell,i})} \right), \tag{3.6}$$

where the subscript ℓ in b_ℓ indicates the ℓth color of the part in question.

When $\rho = 1$, Theorem 3.2 becomes a special case of [7, Theorem 4]. By further restricting our attention to $\lambda_1 = \lambda = b$, we obtain [6, Theorem 11]; the case $b = 2$ of this is presented as Theorem 1.2.

For the proof of Theorem 3.2, we require recurrence relations, which will be derived in Sec. 4. We conclude the present section with an example.

Example 3.3. Once again we take $b = 2$ and $\Lambda = (2,3)$. From among the 17 terms of the entry for $\Omega_{2,T}^{\Lambda}(5;Z)$ in Table 1 we choose two monomials and rewrite them in terms of $T = (t_{1,1}, t_{1,2}; t_{2,1}, t_{2,2}, t_{2,3})$ and $Z = (z_{1,1}, z_{1,2}; z_{2,1}, z_{2,2}, z_{2,3})$, as in Example 3.1.

First, we consider

$$y_1^{1+s_1^2} = z_{1,1}^{1+t_{1,1}^2}.$$

Using (3.5) and (3.6), we see that this monomial corresponds to the colored binary partition $1 \cdot 2_1^0 + 1 \cdot 2_1^2$, which we rewrite as $4_1 + 1_1$.

Next, we consider the somewhat more interesting monomial

$$y_1 z_1^{t_1} z_2 = z_{1,1} \cdot z_{2,1}^{t_{2,1}} \cdot z_{2,2}.$$

Using (3.5) and (3.6) again, we get the partition $1 \cdot 2_1^0 + 1 \cdot 2_2^1 + 2 \cdot 2_2^0$, which we rewrite as $2_2 + 1_2 + 1_2 + 1_1$.

Finally, we note that the correspondence between (3.5) and (3.6) also leads to the entries of Table 2.

4. Recurrence Relations

In this section, we generalize the recurrence relations given by (1.2) and (1.3), and show that the polynomials $\Omega_{b,T}^{\Lambda}(n;Z)$ satisfy similar, though vastly extended recurrence relations. In the case of polynomial analogues of ordinary restricted b-ary partition functions, i.e. when $\rho = 1$, we obtained such recurrence relations in [7]. Rewriting the relevant theorem in terms of the polynomials $\Omega_{b,T}^{\Lambda}(n;Z)$ by using the identity (2.6), we have the following result.

Theorem 4.1 ([7, Theorem 5]). *Let $b \geq 2$ and $\lambda \geq 1$ be integers, and set $\Lambda = (\lambda)$, so that $\rho = 1$. Furthermore, let*

$$Z = (z_1, \dots, z_\lambda), \quad T = (t_1, \dots, t_\lambda), \quad Z^T = (z_1^{t_1}, \dots, z_\lambda^{t_\lambda}).$$

Then $\Omega_{b,T}^{\Lambda}(0;Z) = 1$, and for integers $n \geq 0$, we have

$$\Omega_{b,T}^{\Lambda}(bn+j;Z) = \sum_{k=0}^{\lfloor \lambda/b \rfloor} z_{kb+j} \Omega_{b,T}^{\Lambda}(n-k;Z^T) \quad (j = 0, \dots, b-1), \tag{4.1}$$

with the conventions that $z_0 = 1$, $z_\mu = 0$ for $\mu > \lambda$, and $\Omega_{b,T}^{\Lambda}(m;Z) = 0$ if $m < 0$.

Example 4.2. Let $\lambda = b = 2$. Then, writing out the cases $j = 0$ and $j = 1$ separately, the identity (4.1) reduces to

$$\Omega_{2,T}^{\Lambda}(2n;z_1,z_2) = z_0 \Omega_{2,T}^{\Lambda}(n;z_1^{t_1},z_2^{t_2}) + z_2 \Omega_{2,T}^{\Lambda}(n-1;z_1^{t_1},z_2^{t_2}), \tag{4.2}$$

$$\Omega_{2,T}^{\Lambda}(2n+1;z_1,z_2) = z_1 \Omega_{2,T}^{\Lambda}(n;z_1^{t_1},z_2^{t_2}) + z_3 \Omega_{2,T}^{\Lambda}(n-1;z_1^{t_1},z_2^{t_2}). \tag{4.3}$$

In this case, i.e. when $\Lambda = (2)$, we have $\Omega_{2,T}^{\Lambda}(n; z_1, z_2) = \omega_{t_1,t_2}(n+1; z_1, z_2)$, and also $z_0 = 1$ and $z_3 = 0$. Therefore (4.2) and (4.3) are identical with (1.3) and (1.2), respectively.

We now extend Theorem 4.1 to polynomial analogues of Λ-restricted *colored* b-ary partition functions.

Theorem 4.3. *Let $b \geq 2$ and $\rho \geq 1$ be integers, and let the finite sequences Λ, T, and Z be as in Definition 2.4. Then $\Omega_{b,T}^{\Lambda}(0; Z) = 1$, and for integers $n \geq 0$ we have*

$$\Omega_{b,T}^{\Lambda}(bn + j; Z) = \sum_{k=0}^{\lfloor \lambda/b \rfloor} Y_{bk+j} \Omega_{b,T}^{\Lambda}(n - k; Z^T) \qquad (j = 0, \dots b-1), \tag{4.4}$$

with the convention that $\Omega_{b,T}^{\Lambda}(m; Z) = 0$ if $m < 0$. The coefficients Y_ν, $0 \leq \nu \leq \lambda$, are given by

$$Y_\nu = \sum z_{1,i_1} z_{2,i_2} \cdots z_{\rho,i_\rho}, \qquad z_{1,0} = z_{2,0} = \cdots = z_{\rho,0} = 1, \tag{4.5}$$

with the sum taken over all $i_1, i_2, \dots, i_\rho$ with $i_1 + \cdots + i_\rho = \nu$ and $0 \leq i_\ell \leq \lambda_\ell$ for $\ell = 1, \dots, \rho$. In particular, $Y_0 = 1$, $Y_1 = z_{1,1} + z_{2,1} + \cdots + z_{\rho,1}$, $Y_\lambda = z_{1,\lambda_1} z_{2,\lambda_2} \cdots z_{\rho,\lambda_\rho}$, and $Y_\mu = 0$ when $\mu > \lambda$.

When $\rho = 1$, we see that $Y_\nu = z_{1,\nu} = z_\nu$ for $0 \leq \nu \leq \lambda$, and so Theorem 4.3 reduces to Theorem 4.1. Before proving Theorem 4.3, we present an example.

Example 4.4. Continuing with Examples 2.5 and 3.1, we let $b = 2$ and $\Lambda = (2, 3)$, with $Z = (y_1, y_2; z_1, z_2, z_3)$. For the sake of simplicity we delete the subscripts and superscripts of Ω; Theorem 4.3 then gives

$$\Omega(2n; Z) = Y_0 \Omega(n; Z^T) + Y_2 \Omega(n-1; Z^T) + Y_4 \Omega(n-2; Z^T), \tag{4.6}$$

$$\Omega(2n+1; Z) = Y_1 \Omega(n; Z^T) + Y_3 \Omega(n-1; Z^T) + Y_5 \Omega(n-2; Z^T), \tag{4.7}$$

where $Y_0 = 1$, $Y_1 = y_1 + z_1$, $Y_2 = z_2 + y_1 z_1 + y_2$, $Y_3 = z_3 + y_1 z_2 + y_2 z_1$, $Y_4 = y_1 z_3 + y_2 z_2$, and $Y_5 = y_2 z_3$. As a specific example we take (4.7) with $n = 2$. Then

$$\begin{aligned}
\Omega(5; Z) &= (z_1 + y_1) \cdot \Omega(2; Z^T) + (z_3 + y_1 z_2 + y_2 z_1) \cdot \Omega(1; Z^T) + y_2 z_3 \cdot \Omega(0; Z^T) \\
&= (z_1 + y_1) \cdot (z_1^{t_1^2} + z_2^{t_2} + z_1^{t_1} y_1^{s_1} + y_1^{s_1^2} + y_2^{s_2}) \\
&\quad + (z_3 + y_1 z_2 + y_2 z_1) \cdot (z_1^{t_1} + y_1^{s_1}) + y_2 z_3,
\end{aligned}$$

where we have used the first three entries in Table 1. Expanding the right-hand side of this last identity, we obtain the entry for $n = 5$ in Table 1.

Proof of Theorem 4.3. For ease of notation, we delete subscripts and superscripts from Ω. By manipulating the infinite product in (2.4), we get

$$\begin{aligned}
\sum_{m=0}^{\infty} \Omega(m;Z)q^m &= \prod_{j=0}^{\infty}\prod_{\ell=1}^{\rho}\left(1+z_{\ell,1}^{t_{\ell,1}^j}q^{b^j}+\cdots+z_{\ell,\lambda_\ell}^{t_{\ell,\lambda_\ell}^j}q^{\lambda_\ell\cdot b^j}\right)\\
&= \prod_{\ell=1}^{\rho}\left(1+z_{\ell,1}q+\cdots+z_{\ell,\lambda_\ell}q^{\lambda_\ell}\right)\\
&\quad\cdot\prod_{j=0}^{\infty}\prod_{\ell=1}^{\rho}\left(1+z_{\ell,1}^{t_{\ell,1}^{j+1}}q^{b^{j+1}}+\cdots+z_{\ell,\lambda_\ell}^{t_{\ell,\lambda_\ell}^{j+1}}q^{\lambda_\ell\cdot b^{j+1}}\right)\\
&= \left(1+Y_1q+\cdots+Y_\lambda q^\lambda\right)\\
&\quad\cdot\prod_{j=0}^{\infty}\prod_{\ell=1}^{\rho}\left(1+(z_{\ell,1}^{t_{\ell,1}})^{t_{\ell,1}^j}(q^b)^{b^j}+\cdots+(z_{\ell,\lambda_\ell}^{t_{\ell,\lambda}})^{t_{\ell,\lambda_\ell}^j}(q^b)^{\lambda_\ell\cdot b^j}\right).
\end{aligned}$$

We then obtain, again with (2.4),

$$\sum_{m=0}^{\infty}\Omega(m;Z)q^m = \left(1+Y_1q+\cdots+Y_\lambda q^\lambda\right)\sum_{\nu=0}^{\infty}\Omega(\nu;Z^T)q^{\nu b}. \tag{4.8}$$

With the conventions $Y_0=1$ and $Y_\mu=0$ for $\mu>\lambda$, we get

$$1+Y_1q+\cdots+Y_\lambda q^\lambda = \sum_{k=0}^{\lfloor\lambda/b\rfloor}\sum_{j=0}^{b-1}Y_{kb+j}q^{kb+j} = \sum_{k=0}^{\lfloor\lambda/b\rfloor}q^{kb}\sum_{j=0}^{b-1}Y_{kb+j}q^j. \tag{4.9}$$

Next, we write $m=bn+j$, $0\le j\le b-1$, in (4.8) and substitute (4.9) into (4.8). Then (4.8) becomes

$$\begin{aligned}
&\sum_{n=0}^{\infty}\sum_{j=0}^{b-1}\Omega(bn+j;Z)q^{bn+j} \qquad (4.10)\\
&\quad=\left(\sum_{k=0}^{\lfloor\lambda/b\rfloor}q^{kb}\sum_{j=0}^{b-1}Y_{bk+j}q^j\right)\cdot\sum_{\nu=0}^{\infty}\Omega(\nu;Z^T)q^{\nu b}\\
&\quad=\sum_{n=0}^{\infty}\sum_{j=0}^{b-1}\left(\sum_{k=0}^{\lfloor\lambda/b\rfloor}Y_{bk+j}\Omega(n-k;Z^T)\right)q^{nb+j},
\end{aligned}$$

where we have used a Cauchy product, with $k+\nu=n$. Finally, by equating coefficients of q^{nb+j} in the first and the third line of (4.10), we immediately get the desired identity (4.4). □

We finish this section by explicitly stating a functional equation that was derived as part of the proof of Theorem 4.3. We set

$$F(Z,q) := \sum_{m=0}^{\infty}\Omega_{b,T}^{\Lambda}(m;Z); \tag{4.11}$$

then the identity (4.8) can be rewritten as follows.

Corollary 4.5. *Let $b \geq 2$ and $\rho \geq 1$ be integers, let the finite sequences Λ, T, and Z be as in Definition 2.4, and the polynomials Y_ν, $1 \leq \nu \leq \lambda$, as in (4.5). Then the generating function $F(Z,q)$, defined by (4.11), satisfies the functional equation*

$$F(Z,q) = \left(1 + Y_1 q + \cdots + Y_\lambda q^\lambda\right) F(Z^T, q^b).$$

5. Proof of Theorem 3.2

In this section, we use the recurrence relation, Theorem 4.3, to prove our representation result, namely Theorem 3.2. In doing so, we follow the outline of the proof of [7, Theorem 4], although the details will be different. In order to simplify terminology, if there is no danger of ambiguity we will refer to a Λ-restricted ρ-colored b-ary partition, as defined in Definition 2.1, simply as "a partition of n".

We fix the integers $b \geq 2$ and $\rho \geq 1$, as well as the ρ-tuple Λ, integer λ, and the λ-tuples T and Z as in Definition 2.4. As we did in the proof of Theorem 4.3, we suppress the subscripts and superscripts of Ω. We proceed by induction on n.

(1) For the induction beginning we consider (4.4) with $n = 0$. Then

$$\Omega(j;Z) = Y_j, \qquad j = 0, 1, \ldots, b-1, \tag{5.1}$$

where by (4.5) we have

$$Y_j = \sum_{\substack{i_1+\cdots+i_\rho=j \\ 0\leq i_\ell \leq \lambda_\ell,\, \ell=1,\ldots,\rho}} z_{1,i_1} z_{2,i_2} \cdots z_{\rho,i_\rho}. \tag{5.2}$$

On the other hand, the only partitions of j, with $0 \leq j \leq b-1$, are

$$i_1 b_1^0 + i_2 b_2^0 + \cdots + i_\rho b_\rho^0, \qquad 0 \leq i_\ell \leq \lambda_\ell,\ \ell = 1, \ldots, \rho. \tag{5.3}$$

This is consistent with (3.4)–(3.6). Indeed, in view of (5.2), the identity (3.4) becomes

$$\Omega(j;Z) = \sum_{h=1}^{C_b^\Lambda(n)} \prod_{\ell=1}^{\rho} z_{\ell,i_\ell}^{p_{h,\ell,i_\ell}(t_{\ell,i_\ell})}.$$

This means that, in view of (3.5), for each $\ell = 1, \ldots, \rho$ we have

$$p_{h,\ell,i_\ell}(t_{\ell,i_\ell}) = 1, \quad \mu_{\ell,i_\ell} = 1, \quad \tau_{\ell,i_\ell} = 0,$$

and so by (3.6) the corresponding partition of j is

$$\sum_{\ell=1}^{\rho} i_\ell b_\ell^0,$$

which is consistent with (5.3). This completes the induction beginning.

(2) Next, we assume that Theorem 3.2 is true for all $m \leq nb-1$, for some $n \geq 1$. Our goal is to show that it is also true for $m = nb+j$, for all $j = 0, 1, \ldots, b-1$. To do so, we first fix an arbitrary j with $0 \leq j \leq b-1$. We then consider all partitions of $nb+j$, which can be obtained recursively as follows. We fix an integer $k \geq 0$ and

(a) Take all partitions of $n-k$ and multiply each power of b_ℓ (including the 0th power) by b_ℓ, $\ell = 1, 2, \ldots, \rho$, thus raising all powers by 1.
(b) To each of the "raised" partitions from (a), add all partitions of $kb + j$ of the form

$$kb + j = i_1 b_1^0 + i_2 b_2^0 + \cdots + i_\rho b_\rho^0, \qquad 0 \le i_\ell \le \lambda_\ell \quad \text{for} \quad \ell = 1, \ldots, \rho. \tag{5.4}$$

(c) Do (a) and (b) for all $k \ge 0$ that satisfy $kb + j \le \lambda$.

This procedure gives all partitions of $nb+j$ since $(n-k)\cdot b + (kb+j) = nb+j$. Also, the maximal k given by (c) is $\lfloor \lambda/b \rfloor$.

(3) Using the induction hypothesis and (3.4), we have

$$\Omega(n-k; Z) = \sum_{h=1}^{C_b^\Lambda(n-k)} \prod_{\ell=1}^{\rho} z_{\ell,1}^{p_{h,\ell,1}(t_{\ell,1})} \cdots z_{\ell,\lambda_\ell}^{p_{h,\ell,\lambda_\ell}(t_{\ell,\lambda_\ell})}, \tag{5.5}$$

with exponents $p_{h,\ell,1}(t_{\ell,1}), \ldots, p_{h,\ell,\lambda_\ell}(t_{\ell,\lambda_\ell})$ as in (3.5). In order to lift the partitions of $n-k$ to those of $(n-k)b$, which corresponds to step (a), all powers of $t_{\ell,i}$, $1 \le \ell \le \rho$ and $1 \le i \le \lambda_\ell$, are augmented by 1. This, in turn, means that (5.5) changes to $\Omega(n-k; Z^T)$.

Next, for each k with $0 \le k \le \lfloor \lambda/b \rfloor$, we multiply $\Omega(n-k; Z^T)$ by the monomial

$$z_{1,i_1} z_{2,i_2} \cdots z_{\rho,i_\rho}, \qquad 0 \le i_\ell \le \lambda_\ell \quad \text{for} \quad \ell = 1, \ldots, \rho,$$

which corresponds to an individual partition of $kb + j$ of the form (5.4) in step (b). This means that for each $\ell = 1, \ldots, \rho$, the exponents of z_{ℓ,i_ℓ} are increased by 1, which is again consistent with the duality between monomials and individual partitions. Now, taking into account all partitions of $kb+j$ of the form (5.4) means multiplying $\Omega(n-k; Z^T)$ by

$$\sum_{\substack{i_1+\cdots+i_\rho = kb+j \\ 0 \le i_\ell \le \lambda_\ell,\ \ell=1,\ldots,\rho}} z_{1,i_1} z_{2,i_2} \cdots z_{\rho,i_\rho}.$$

But, by (4.5), this sum is exactly Y_{kb+j}.

(4) Recall that steps 2 and 3 were done for an arbitrary fixed integer k with $0 \le k \le \lfloor \lambda/b \rfloor$. The polynomial corresponding to all partitions of $nb+j$ is therefore

$$\sum_{k=0}^{\lfloor \lambda/b \rfloor} Y_{bk+j} \Omega(n-k; Z^T),$$

which, by Theorem 4.3, is $\Omega(bn+j; Z)$. Since j was an arbitrarily chosen but fixed integer with $0 \le j \le b-1$, this concludes the proof of Theorem 3.2 by induction.

6. Explicit Formulas

In this section, we derive explicit formulas for our polynomials $\Omega_{b,T}^{\Lambda}(n;Z)$, thus generalizing Theorem 1.6. For greater clarity, we begin with an intermediate result for the binary case, i.e. the case $b=2$. The function $d_t(n)$ is defined as in (1.9), and the "starred" multinomial coefficient is defined to be the least nonnegative residue modulo 2 of the original multinomial coefficient.

Theorem 6.1 **([7, Corollary 17]).** *Let $\lambda\geq 1$ be an integer, and Z and T as in Definition 2.3. Then for $n\geq 0$, we have*

$$\omega_{2,T}^{\lambda}(n+1;Z)=\sum_{k_1+2k_2+\cdots+\lambda k_\lambda=n}\binom{k_1+\cdots+k_\lambda}{k_1,\ldots,k_\lambda}^{*}z_1^{d_{t_1}(k_1)}\cdots z_\lambda^{d_{t_\lambda}(k_\lambda)}. \tag{6.1}$$

We see that for $\lambda=2$ the identity (6.1) reduces to (1.11). On the other hand, Theorem 6.1 was extended in [7] to arbitrary bases $b\geq 2$. Since this more general case will be needed to obtain the main result of this section, we state it here, but without proof. We need some additional definitions with corresponding notations; see also [7].

First, we define the set

$$M_b:=\left\{\sum_{j\geq 0}c_jb^j\mid c_j\in\{0,1\}\right\}, \tag{6.2}$$

i.e. the set of all nonnegative integers whose b-ary digits are only 0 or 1. Clearly, we have $M_2=\mathbb{N}\cup\{0\}$.

Next, we extend (1.9) as follows. Let $k\in M_b$ with $k=\sum_{j\geq 0}c_jb^j$, $c_j\in\{0,1\}$. Then for an integer base $t\geq 1$ we define

$$d_t^b(k):=\sum_{j\geq 0}c_jt^j. \tag{6.3}$$

It is clear that $d_t^2(k)=d_t(k)$ and $d_b^b(k)=k$ for all integers $b\geq 2$ and $k\in M_b$.

Finally, we extend the multinomial coefficient modulo 2, as it is used in Theorem 6.1. If $k_1,\ldots,k_\lambda\in M_b$, we set

$$\binom{k_1+\cdots+k_\lambda}{k_1,\ldots,k_\lambda}_b^{*}:=\binom{d_2^b(k_1)+\cdots+d_2^b(k_\lambda)}{d_2^b(k_1),\ldots,d_2^b(k_\lambda)}^{*}, \tag{6.4}$$

with the right-hand side of (6.4) as defined earlier. The desired generalization of Theorem 6.1 is then as follows.

Theorem 6.2 **([7, Theorem 21]).** *Let $b\geq 2$ and $\lambda\geq 1$ be integers, and T and Z as in Definition 2.3. Then for $n\geq 0$ we have*

$$\omega_{b,T}^{\lambda}(n+1;Z)=\sum_{\substack{k_1+2k_2+\cdots+\lambda k_\lambda=n\\ k_1,\ldots,k_\lambda\in M_b}}\binom{k_1+\cdots+k_\lambda}{k_1,\ldots,k_\lambda}_b^{*}z_1^{d_{t_1}^b(k_1)}\cdots z_\lambda^{d_{t_\lambda}^b(k_\lambda)}. \tag{6.5}$$

We note that Theorem 6.2 immediately implies Theorem 6.1. Indeed, when $b = 2$, the multinomial coefficient defined in (6.4) reduces to the one in (6.1) since $d_2^2(k_i) = k_i$. Then in the $b = 2$ case we also have $d_{t_i}^b(k_i) = d_{t_i}(k_i)$, and the conditions concerning M_b become irrelevant.

We are now ready to state and prove the main result of this section.

Theorem 6.3. *Let $b \geq 2$ and $\rho \geq 1$ be integers, and let Λ, T, and Z be as in Definition 2.4. Furthermore, let*

$$D_{\ell,i} := d_{t_{\ell,i}}^b(k_{\ell,i}), \qquad 1 \leq i \leq \lambda_\ell \quad \textit{and} \quad 1 \leq \ell \leq \rho.$$

Then for each integer $n \geq 0$, we have

$$\Omega_{b,T}^{\Lambda}(n;Z) = \sum \prod_{\ell=1}^{\rho} \binom{k_{\ell,1} + \cdots + k_{\ell,\lambda_\ell}}{k_{\ell,1}, \ldots, k_{\ell,\lambda_\ell}}_b^* z_{\ell,1}^{D_{\ell,1}} \cdots z_{\ell,\lambda_\ell}^{D_{\ell,\lambda_\ell}}, \tag{6.6}$$

where the sum is taken over all integers $k_{\ell,i} \in M_b$ with $1 \leq i \leq \lambda_\ell$ and $1 \leq \ell \leq \rho$, such that

$$\sum_{\ell=1}^{\rho} \sum_{i=1}^{\lambda_\ell} i \cdot k_{\ell,i} = n. \tag{6.7}$$

Proof. We combine Theorem 2.6 with Theorem 6.2, where in (6.1) we replace n by n_ℓ, λ by λ_ℓ and add the additional subscript ℓ to k_i, t_i and z_i, where $\ell = 1, \ldots, \rho$. The right-hand side of (6.6) then follows immediately from the right-hand side of (2.8), and the condition (6.7) follows from the summation conditions in (2.8) and in (6.5). □

We conclude this section with two examples. The first one is our "running example", continuing Examples 2.5, 3.1 and 4.4.

Example 6.4. Let $b = 2$ and $\Lambda = (2, 3)$, and to avoid double indices, we set again $T = (s_1, s_2; t_1, t_2, t_3)$ and $Z = (y_1, y_2; z_1, z_2, z_3)$. We also set $h_i := k_{1,i}$, $i = 1, 2$, and $k_i := k_{2,i}$, $i = 1, 2, 3$. Then by Theorem 6.3, we have

$$\begin{aligned} \Omega_{2,T}^{\Lambda}(n;Z) \;=\; & \sum_{\substack{h_1+2h_2+k_1 \\ +2k_2+3k_3=n}} \binom{h_1+h_2}{h_1, h_2}^* \binom{k_1+k_2+k_3}{k_1, k_2, k_3}^* \\ & \times y_1^{d_{s_1}(h_1)} y_2^{d_{s_2}(h_2)} z_1^{d_{t_1}(k_1)} z_2^{d_{t_2}(k_2)} z_3^{d_{t_3}(k_3)}. \end{aligned} \tag{6.8}$$

Since $b = 2$, the condition in Theorem 6.3 that concerns M_b becomes irrelevant; see the remark following Theorem 6.2.

As a specific example we consider $n = 4$. By direct computation we find that the solutions $(h_1, h_2; k_1, k_2, k_3)$ of $h_1 + 2h_2 + k_1 + 2k_2 + 3k_3 = 4$ for which the binomial/trinomial coefficients in (6.8) are odd (i.e. the "starred" versions are 1) are as follows:

$$(0,0;4,0,0),(0,0;2,1,0),(0,0;0,2,0),(1,0;3,0,0),(1,0;0,0,1),$$
$$(2,0;2,0,0),(2,0;0,1,0),(4,0;0,0,0),(3,0;1,0,0),(0,1;2,0,0),$$
$$(0,1;0,1,0),(2,1;0,0,0),(0,2;0,0,0),$$

so that (6.8) becomes

$$\begin{aligned}\Omega_{2,T}^{\Lambda}(4;Z) &= z_1^{d_{t_1}(4)} + z_1^{d_{t_1}(2)}z_2^{d_{t_2}(1)} + z_2^{d_{t_2}(2)} + z_1^{d_{t_1}(3)}y_1^{d_{s_1}(1)} + z_3^{d_{t_3}(1)}y_1^{d_{s_1}(1)}\\ &\quad + z_1^{d_{t_1}(2)}y_1^{d_{s_1}(2)} + z_2^{d_{t_2}(1)}y_1^{d_{s_1}(2)} + y_1^{d_{s_1}(4)} + z_1^{d_{t_1}(1)}y_1^{d_{s_1}(3)}\\ &\quad + z_1^{d_{t_1}(2)}y_2^{d_{s_2}(1)} + z_2^{d_{t_2}(1)}y_2^{d_{s_2}(1)} + y_1^{d_{s_1}(2)}y_2^{d_{s_2}(1)} + y_2^{d_{s_2}(2)}\\ &= z_1^{t_1^2} + z_1^{t_1}z_2 + z_2^{t_2} + z_1^{1+t_1}y_1 + z_3y_1 + z_1^{t_1}y_1^{s_1} + z_2y_1^{s_1} + y_1^{s_1^2}\\ &\quad + z_1y_1^{1+s_1} + z_1^{t_1}y_2 + z_2y_2 + y_1^{s_1}y_2 + y_2^{s_2}.\end{aligned}$$

This is consistent with the entry for $\Omega_{2,T}^{\Lambda}(4;Z)$ in Table 1.

It is interesting to note that the smallest n for which all three entries in the trinomial coefficient in (6.8) are nonzero is $n=11$, with $k_1=4$, $k_2=2$, and $k_3=1$, giving

$$\binom{k_1+k_2+k_3}{k_1,k_2,k_3} = \binom{7}{4,2,1} = 105 \equiv 1 \pmod 2.$$

This, in turn, leads to the following monomial and corresponding colored binary partition:

$$z_1^{t_1^2}z_2^{t_2}z_3 \longleftrightarrow 11 = 2_2^2 + 2_2 + 2_2 + 2_2^0 + 2_2^0 + 2_2^0.$$

Example 6.5. We now choose $b=3$, but leave Λ, T, and Z as in Example 6.4. Then in analogy to (6.8), we get

$$\begin{aligned}\Omega_{3,T}^{\Lambda}(n;Z) &= \sum_{\substack{h_1+2h_2+k_1\\+2k_2+3k_3=n;\\h_1,\dots,k_3\in M_3}} \binom{h_1+h_2}{h_1,h_2}_3^* \binom{k_1+k_2+k_3}{k_1,k_2,k_3}_3^* \\ &\quad\times y_1^{d_{s_1}^3(h_1)}y_2^{d_{s_2}^3(h_2)}z_1^{d_{t_1}^3(k_1)}z_2^{d_{t_2}^3(k_2)}z_3^{d_{t_3}^3(k_3)}.\end{aligned}\tag{6.9}$$

This time we consider $n=6$ as a specific example. The conditions $h_1,\dots,k_3\in M_3$ mean, in particular, that 2 and 5 cannot occur among the solutions of $h_1+2h_2+k_1+2k_2+3k_3=6$. By direct computation, using the definition (6.4), we find that both of the modified binomial/trinomial coefficients in (6.9) are 1 when $(h_1,h_2;k_1,k_2,k_3)$ is of the form

$$(0,3;0,0,0),(3,1;1,0,0),(3,0;3,0,0),(0,1;4,0,0),$$
$$(4,0;0,1,0),(1,0;3,1,0),(0,0;0,3,0),(3,0;0,0,1),(0,0;3,0,1),$$

Table 3. The terms of $\Omega^{\Lambda}_{3,T}(6;Z)$ and corresponding partitions.

Monomial	Partition	Monomial	Partition
$y_2^{s_2}$	3_1+3_1	$y_1z_1^{t_1}z_2$	$1_1+3_2+1_2+1_2$
$y_1^{s_1}y_2z_1$	$3_1+1_1+1_1+1_2$	$z_2^{t_2}$	3_2+3_2
$y_1^{s_1}z_1^{t_1}$	3_1+3_2	$y_1^{s_1}z_3$	$3_1+1_2+1_2+1_2$
$y_2z_1^{1+t_1}$	$1_1+1_1+1_2+3_2$	$z_1^{t_1}z_3$	$3_2+1_2+1_2+1_2$
$y_1^{1+s_1}z_2$	$1_1+3_1+1_2+1_2$		

so that (6.9) becomes

$$\begin{aligned}\Omega^{\Lambda}_{3,T}(6;Z) &= y_2^{d^3_{s_2}(3)} + y_1^{d^3_{s_1}(3)}y_2^{d^3_{s_2}(1)}z_1^{d^3_{t_1}(1)} + y_1^{d^3_{s_1}(3)}z_1^{d^3_{t_1}(3)} + y_2^{d^3_{s_2}(1)}z_1^{d^3_{t_1}(4)}\\ &\quad + y_1^{d^3_{s_1}(4)}z_2^{d^3_{t_2}(1)} + y_1^{d^3_{s_1}(1)}z_1^{d^3_{t_1}(3)}z_2^{d^3_{t_2}(1)} + z_2^{d^3_{t_2}(3)} + y_1^{d^3_{s_1}(3)}z_3^{d^3_{t_3}(1)}\\ &\quad + z_1^{d^3_{t_1}(3)}z_3^{d^3_{t_3}(1)}\\ &= y_2^{s_2} + y_1^{s_1}y_2z_1 + y_1^{s_1}z_1^{t_1} + y_2z_1^{1+t_1} + y_1^{1+s_1}z_2 + y_1z_1^{t_1}z_2 + z_2^{t_2}\\ &\quad + y_1^{s_1}z_3 + z_1^{t_1}z_3.\end{aligned}$$

Finally, we illustrate in Table 3 how the polynomial $\Omega^{\Lambda}_{3,T}(6;Z)$ translates to the set of nine $(2,3)$-restricted 2-colored ternary partitions of $n=6$.

7. Some Product Identities

In dealing with polynomials in one or more variables with integer coefficients, questions of divisibility and irreducibility become relevant. For instance, in the case of multivariate b-ary Stern polynomials [8], we obtain an analogue of Theorem 1.7. We present a special case of this result, rewritten in the notation of the current paper. It corresponds to $\rho=1$ and $\Lambda=(\lambda)=(b)$.

Theorem 7.1 ([8, Corollary 4.3]). *Let $b\ge 2$, $\ell\ge 1$, and $n\ge 1$ be integers. Then*

$$\Omega^{(b)}_{b,T}(n\cdot b^{\ell}+j;Z)=\Omega^{(b)}_{b,T}(n;Z^{T^{\ell}})\cdot\Omega^{(b)}_{b,T}(j;Z),\qquad \frac{b^{\ell}-1}{b-1}\le j\le b^{\ell}-1. \tag{7.1}$$

In this section, we show that Theorem 7.1 can be extended to an arbitrary number of colors ρ and an arbitrary $\Lambda=(\lambda_1,\dots,\lambda_\rho)$. We use again the notations of Definition 2.4.

Theorem 7.2. *Let $b\ge 2$, $\ell\ge 1$, and $n\ge 1$ be integers. Then*

$$\Omega^{\Lambda}_{b,T}(n\cdot b^{\ell}+j;Z)=\Omega^{\Lambda}_{b,T}(n;Z^{T^{\ell}})\cdot\Omega^{\Lambda}_{b,T}(j;Z), \tag{7.2}$$

which in the case $b\le\lambda=\lambda_1+\cdots+\lambda_\rho$ holds for all integers j with

$$(\lambda-b+1)\frac{b^{\ell}-1}{b-1}\le j\le b^{\ell}-1. \tag{7.3}$$

When $b>\lambda$, then (7.2) holds for all j with $0\le j\le b^{\ell}-1$.

Proof. We proceed by induction on $\ell \geq 1$, and as usual we suppress the subscripts and superscripts of Ω. When $\ell = 1$ and $\lambda - b + 1 \leq j \leq b - 1$, then the recurrence relation (4.4) gives

$$\Omega(n \cdot b + j; Z) = Y_j \cdot \Omega(n; Z^T), \tag{7.4}$$

since $b + j \geq \lambda + 1$, so all other coefficients Y_{bk+j} in (4.4) vanish. For $n = 0$, the identity (7.4) gives $\Omega(j; Z) = Y_j$, and we get

$$\Omega(n \cdot b + j; Z) = \Omega(n; Z^T) \cdot \Omega(j; Z). \tag{7.5}$$

In the case $\lambda < b$, (7.5) holds for all $0 \leq j \leq b - 1$. This establishes the induction beginning.

Now, suppose that Theorem 7.2 holds for some $\ell \geq 1$. We wish to show that it also holds for $\ell + 1$. To do so, we begin by setting

$$j = r \cdot b + i, \quad \lambda - b + 1 \leq i \leq b - 1, \quad (\lambda - b + 1)\frac{b^\ell - 1}{b - 1} \leq r \leq b^\ell - 1. \tag{7.6}$$

This means that j satisfies the inequalities

$$\begin{aligned} j &\leq (b^\ell - 1)b + (b - 1) = b^{\ell+1} - 1, \\ j &\geq (\lambda - b + 1)\frac{b^\ell - 1}{b - 1}b + (\lambda - b + 1) = (\lambda - b + 1)\frac{b^{\ell+1} - 1}{b - 1}, \end{aligned}$$

as required. Now, we have

$$\begin{aligned} \Omega(n \cdot b^{\ell+1} + j; Z) &= \Omega((n \cdot b^\ell + r)b + i; Z) \\ &= Y_i \cdot \Omega(n \cdot b^\ell + r; Z^T) \qquad \text{by (7.4)} \\ &= Y_i \cdot \Omega(n; Z^{T^{\ell+1}}) \cdot \Omega(r; Z^T), \end{aligned}$$

where we have used the induction hypothesis (7.2). Reordering this last line, we have

$$\Omega(n \cdot b^{\ell+1} + j; Z) = \Omega(n; Z^{T^{\ell+1}}) \cdot Y_i \cdot \Omega(r; Z^T) = \Omega(n; Z^{T^{\ell+1}}) \cdot \Omega(j; Z),$$

where we have used (7.4) again. Note that in several steps the inequalities in (7.6) were essential. This completes the proof by induction. □

Example 7.3. We choose $b = 4$, $\Lambda = (2, 3)$, and leave T and Z as in Example 6.4. Using the recurrence relation in Theorem 4.3 we can compute $\Omega_{4,T}^{\Lambda}(n; Z)$ for $0 \leq n \leq 11$, and factor them when possible; see Table 4.

We note that the entries in Table 4 are consistent with Theorem 7.2. The entries for $n = 5, 8$, and 9 are irreducible and consist of 5, 7 and 8 monomials, respectively.

As far as Theorem 1.7 is concerned, we believe that it extends in this form to general $\rho \geq 1$ and $\Lambda = (\lambda_1, \ldots, \lambda_\rho)$ only when $1 \leq (\lambda + 1)/b < 2$, that is, when $b - 1 \leq \lambda_1 + \cdots + \lambda_\rho \leq 2b - 2$. This would be the case, for instance, in Example 7.3. Pursuing this question further would be beyond the scope of this paper.

Table 4. $\Omega^{\Lambda}_{4,T}(n;Z)$ with $\Lambda=(2,3)$, for $0\le n\le 11$.

n	$\Omega^{\Lambda}_{4,T}(n;Z)$	n	$\Omega^{\Lambda}_{4,T}(n;Z)$
0	1	6	$(y_1^{s_1}+z_1^{t_1})(y_2+y_1z_1+z_2)$
1	y_1+z_1	7	$(y_1^{s_1}+z_1^{t_1})(y_2z_1+y_1z_2+z_3)$
2	$y_2+y_1z_1+z_2$	8	$y_2^{s_2}+\cdots+y_1z_1^{t_1}z_3$
3	$y_2z_1+y_1z_2+z_3$	9	$y_1y_2^{s_2}+\cdots+y_2z_1^{t_1}z_3$
4	$y_1^{s_1}+z_1^{t_1}+y_2z_2+y_1z_3$	10	$(y_2^{s_2}+y_1^{s_1}z_1^{t_1}+z_2^{t_2})(y_2+y_1z_1+z_2)$
5	$y_1^{1+s_1}+\cdots+y_2z_3$	11	$(y_2^{s_2}+y_1^{s_1}z_1^{t_1}+z_2^{t_2})(y_2z_1+y_1z_2+z_3)$

8. Further Remarks

We recall that, by setting $Z=(1,\dots,1)$, we get the numerical sequences $C_b^{\Lambda}(n)=\Omega^{\Lambda}_{b,T}(n;(1,\dots,1))$; see (2.5). While it is not the purpose of this paper to study these sequences, we present one easy but rather surprising result.

Corollary 8.1. *Let $b\ge 2$ be an integer, and let $\Lambda=(b-1,\dots,b-1)$ with $\rho\ge 1$ elements. Then the number of Λ-restricted ρ-colored b-ary partitions is independent of b, namely*

$$C_b^{\Lambda}(n)=\binom{n+\rho-1}{\rho-1},\qquad n=0,1,\dots \tag{8.1}$$

Proof. We use the identity

$$\prod_{j=0}^{\infty}\left(1+q^{b^j}+q^{2\cdot b^j}+\cdots+q^{(b-1)\cdot b^j}\right)=\sum_{n=0}^{\infty}q^n=\frac{1}{1-q}, \tag{8.2}$$

which reflects the fact that each integer $n\ge 0$ has a unique b-ary expansion, or, in other words, $S_b^{b-1}(n)=1$. Now, we raise the left- and the right-most terms in (8.2) to the power ρ and use the well-known binomial expansion

$$(1-q)^{-\rho}=\sum_{n=0}^{\infty}\binom{n+\rho-1}{\rho-1}q^n$$

(see, e.g., [11, Eq. (1.3)]). Then the generating function (2.1) immediately gives the desired identity (8.1). □

Example 8.2. Let $\rho=3$ and $n=3$. When $b=2$, we have $\Lambda=(1,1,1)$, and the $\binom{5}{2}=10$ allowable partitions are

$2_3+1_3,\ 2_3+1_2,\ 2_3+1_1,\ 2_2+1_3,\ 2_2+1_2,\ 2_2+1_1,\ 2_1+1_3,\ 2_1+1_2,\ 2_1+1_1,$
$1_3+1_2+1_1.$

For $b=3$, and thus $\Lambda=(2,2,2)$, we have

$3_3,\ 3_2,\ 3_1,\ 1_3+1_3+1_2,\ 1_3+1_3+1_1,\ 1_3+1_2+1_2,\ 1_3+1_2+1_1,$
$1_3+1_1+1_1,\ 1_2+1_2+1_1,\ 1_2+1_1+1_1,$

and for any $b\ge 4$ we have the partitions

$1_3+1_3+1_3,\ 1_3+1_3+1_2,\ \dots,\ 1_2+1_1+1_1,\ 1_1+1_1+1_1.$

This last example is an instance of the more general situation where $n < b$, in which case all partitions have parts 1 with ρ different colors. These colors may be assigned $b - 1 \geq n$ times, which means that there are no restrictions. Therefore, we have a one-to-one correspondence between such partitions and all non-increasing sequences of length n consisting of integers $1, 2, \ldots, \rho$. The number of such sequences is a known combinatorial quantity, namely $\binom{n+\rho-1}{\rho-1}$, consistent with Corollary 8.1.

Finally, we return to the main topic of this paper, namely the multivariate polynomials that characterize all Λ-restricted ρ-colored b-ary partitions of an integer $n \geq 1$.

Example 8.3. As in Example 8.2, we let $\rho = 3$ and $n = 3$. Using any of the various methods described in this paper for computing the relevant polynomials, we obtain the following expressions:

(a) Let $b = 2$ and $\Lambda = (1, 1, 1)$. Then with $Z = (x_1; y_1; z_1)$ and $T = (r_1; s_1; t_1)$, we get

$$\begin{aligned} \Omega_{2,T}^{(1,1,1)}(3; Z) &= z_1^{1+t_1} + y_1 z_1^{t_1} + x_1 z_1^{t_1} + y_1^{s_1} z_1 + y_1^{1+s_1} \\ &\quad + x_1 y_1^{s_1} + x_1^{r_1} z_1 + x_1^{r_1} y_1 + x_1^{1+r_1} + x_1 y_1 z_1. \end{aligned} \tag{8.3}$$

(b) Let $b = 3$ and $\Lambda = (2, 2, 2)$. Then with $Z = (x_1, x_2; y_1, y_2; z_1, z_2)$ and $T = (r_1, r_2; s_1, s_2; t_1, t_2)$, we get

$$\begin{aligned} \Omega_{3,T}^{(2,2,2)}(3; Z) &= z_1^{t_1} + y_1^{s_1} + x_1^{r_1} + y_1 z_2 + x_1 z_2 \\ &\quad + y_2 z_1 + x_1 y_1 z_1 + x_2 z_1 + x_1 y_2 + x_2 y_1. \end{aligned} \tag{8.4}$$

(c) Let $b \geq 4$ and $\Lambda = (b-1, b-1, b-1)$. Then with $Z = (x_1, \ldots, x_{b-1}; y_1, \ldots, y_{b-1}; z_1, \ldots, z_{b-1})$, we get

$$\begin{aligned} \Omega_{b,T}^{\Lambda}(3; Z) &= z_3 + y_1 z_2 + x_1 z_2 + y_2 z_1 + x_1 y_1 z_1 \\ &\quad + x_2 z_1 + y_3 + x_1 y_2 + x_2 y_1 + x_3. \end{aligned} \tag{8.5}$$

Here, in (8.5), we did not specify T because no exponents other than 1 occur.

All these polynomials (8.3)–(8.5) consist of ten monomials, which is consistent with Corollary 8.1 and Example 8.2. Each of these polynomials characterizes the three sets of partitions listed in Example 8.2, in the order in which the respective monomials are given.

Acknowledgment

Research supported in part by the Natural Sciences and Engineering Research Council of Canada, Grant # 145628481.

References

[1] G. E. Andrews, Congruence properties of the m-ary partition function, *J. Number Theory* **3** (1971) 104–110.

[2] B. Bates and T. Mansour, The q-Calkin-Wilf tree, *J. Combin. Theory Ser. A* **118** (2011) 1143–1151.

[3] R. F. Churchhouse, Congruence properties of the binary partition function, *Proc. Cambridge Philos. Soc.* **66** (1969) 371–376.

[4] S. Corteel and J. Lovejoy, Overpartitions, *Trans. Amer. Math. Soc.* **356**(4) (2004) 1623–1635.

[5] K. Dilcher and L. Ericksen, Generalized Stern polynomials and hyperbinary representations, *Bull. Pol. Acad. Sci. Math.* **65** (2017) 11–28.

[6] K. Dilcher and L. Ericksen, Polynomials characterizing hyper b-ary representations, *J. Integer Seq.* **21** (2018), Article ID: 18.4.3, 11 pp.

[7] K. Dilcher and L. Ericksen, Polynomial analogues of restricted b-ary partition functions, *J. Integer Seq.* **22** (2019) Article ID: 19.3.2.

[8] K. Dilcher and L. Ericksen, Properties of multivariate b-ary Stern polynomials, *Ann. Comb.* **23**(3–4) (2019) 695–711.

[9] J. M. Dumont, N. Sidorov and A. Thomas, Number of representations related to a linear recurrent basis, *Acta Arith.* **88**(1) (1999) 371–396.

[10] L. Euler, De partitione numerorum, *Novi Comment. Acad. Scie. Petropo.* **3** (1753) 125–169, in *Opera Omnia*, Series 1, Vol. 2 (1753), pp. 254–294. The Euler Archive, `http://eulerarchive.maa.org`.

[11] H. W. Gould, *Combinatorial Identities*, revised edn. (Gould Publications, 1972).

[12] W. J. Keith, Restricted k-color partitions, *Ramanujan J.* **40** (2016) 71–92.

[13] K. Mahler, On a special functional equation, *J. London Math. Soc.* **15** (1940) 115–123.

[14] B. Reznick, Some binary partition functions, in *Analytic Number Theory: Proc. of a Conf. in Honor of Paul T. Bateman*, eds. B. C. Berndt *et al.* (Birkhäuser, 1990), pp. 451–477.

[15] Ö. Rödseth, Some arithmetical properties of m-ary partitions, *Proc. Cambridge Philos. Soc.* **68** (1970) 447–453.

[16] Ø. Rødseth and J. A. Sellers, On m-ary overpartitions, *Ann. Comb.* **9**(3) (2005) 345–353.

[17] R. P. Stanley and H. S. Wilf, Refining the Stern diatomic sequence, (2010), http://www-math.mit.edu/~rstan/papers/stern.pdf.

[18] M. Ulas and B. Żmija, On p-adic valuations of colored p-ary partitions, *Monatsh. Math.* **188**(2) (2019) 351–368.

© 2024 World Scientific Publishing Company
https://doi.org/10.1142/9789811277375_0012

On Hurwitz zeta function and Lommel functions*

Atul Dixit[†] and Rahul Kumar[‡]

Discipline of Mathematics
Indian Institute of Technology
Gandhinagar, Palaj, Gandhinagar 382355
Gujarat, India
[†]*adixit@iitgn.ac.in*
[‡]*rahul.kumr@iitgn.ac.in*

Dedicated to Professor Bruce C. Berndt on the occasion of his 80th birthday.

We obtain a new proof of Hurwitz's formula for the Hurwitz zeta function $\zeta(s,a)$ beginning with Hermite's formula. The aim is to reveal a nice connection between $\zeta(s,a)$ and a special case of the Lommel function $S_{\mu,\nu}(z)$. This connection is used to rephrase a modular-type transformation involving infinite series of Hurwitz zeta function in terms of those involving Lommel functions.

Keywords: Hurwitz zeta function; Lommel functions; Riemann zeta function; Hermite's formula; functional equation.

Mathematics Subject Classification 2020: 11M06, 11M35, 33C10, 33C47

1. Introduction

The Hurwitz zeta function $\zeta(s,a)$ is defined for $\mathrm{Re}(s) > 1$ and $a \in \mathbb{C}\backslash\{x \in \mathbb{R} : x \leq 0\}$ by [23, p. 36]

$$\zeta(s,a) := \sum_{n=0}^{\infty} \frac{1}{(n+a)^s}.$$

It is well known that for $\zeta(s,a)$, it can be analytically continued to the entire s-complex plane except for a simple pole at $s = 1$ with residue 1, and that $\zeta(s,1) = \zeta(s)$.

One of the fundamental results in the theory of $\zeta(s,a)$ is the following formula of Hurwitz [23, Eq. (2.17.3), p. 37].

*This article originally appeared in *International Journal of Number Theory*, **17**(2), pp. 393–404 © 2021 World Scientific Publishing Company. DOI: 10.1142/S1793042120400047

Theorem 1.1. *For* $0 < a \leq 1$ *and* $\text{Re}(s) < 0$,

$$\zeta(s,a) = \frac{2\Gamma(1-s)}{(2\pi)^{1-s}} \left\{ \sin\left(\frac{1}{2}\pi s\right) \sum_{n=1}^{\infty} \frac{\cos(2\pi na)}{n^{1-s}} + \cos\left(\frac{1}{2}\pi s\right) \sum_{n=1}^{\infty} \frac{\sin(2\pi na)}{n^{1-s}} \right\}. \tag{1.1}$$

The above result also holds[a] *for* $\text{Re}(s) < 1$ *if* $0 < a < 1$.

We note that when $a = 1$, the above formula reduces to the functional equation of $\zeta(s)$ [23, Eq. (2.1.1), p. 13] for $\text{Re}(s) < 0$, which can then be seen to be true for all $s \in \mathbb{C}$ by analytic continuation.

Several proofs of (1.1) are available in the literature. For example, Hurwitz himself obtained it by transforming the Mellin transform representation of $\zeta(s,a)$ as a loop integral and then evaluating the latter. This proof can be found, for example, in [23, p. 37]. Berndt [4, Sec. 5] found a short proof of (1.1) by using the boundedly convergent Fourier series of $\lfloor x \rfloor - x + \frac{1}{2}$. We refer the reader interested in knowing the various proofs of this formula to [11] and the references therein (see also [12]). In [11, Sec. 4], Kanemitsu, Tanigawa, Tsukada and Yoshimoto obtained a new proof of (1.1). Their proof commences with employing [11, Eq. (4.1)] (see also [10, Eq. (47)])

$$\begin{aligned} \zeta(s,a) &= \frac{1}{2}a^{-s} + \frac{a^{1-s}}{s-1} \\ &\quad + \sum_{n=1}^{\infty} \left\{ \frac{e^{-2\pi ina}}{(-2\pi ina)^{1-s}} \Gamma(1-s, -2\pi ina) + \frac{e^{2\pi ina}}{(2\pi ina)^{1-s}} \Gamma(1-s, 2\pi ina) \right\}, \end{aligned}$$

which is a special case of the Ueno–Nishizawa formula [24] and then invoking the Fourier series of the Dirac-delta function $\delta(s)$.

The aim of this paper is to give a yet another new proof of (1.1) beginning with Hermite's well-known formula for $\zeta(s,a)$ [17, Formula 25.11.29, p. 609], valid for $\text{Re}(a) > 0$ and $s \neq 1$:

$$\zeta(s,a) = \frac{1}{2}a^{-s} + \frac{a^{1-s}}{s-1} + 2\int_0^{\infty} \frac{\sin(s\tan^{-1}(x/a))\,dx}{(a^2+x^2)^{s/2}\,(e^{2\pi x}-1)}. \tag{1.2}$$

The novelty of this proof is that it reveals the connection between Hurwitz zeta function and the Lommel functions $s_{\mu,\nu}(z)$ and $S_{\mu,\nu}(z)$ which, to the best of our knowledge, seems to have been unnoticed before. The Lommel functions are defined

[a]See [2, Theorem 12.6, p. 257].

by [25, Eq. (10), p. 346]

$$s_{\mu,\nu}(z) = \frac{z^{\mu+1}}{(\mu-\nu+1)(\mu+\nu+1)}\,{}_1F_2\left(1; \frac{1}{2}\mu - \frac{1}{2}\nu + \frac{3}{2}, \frac{1}{2}\mu + \frac{1}{2}\nu + \frac{3}{2}; -\frac{1}{4}z^2\right). \tag{1.3}$$

and [25, Eq. (2), p. 347]

$$S_{\mu,\nu}(z) = s_{\mu,\nu}(z) + \frac{2^{\mu-1}\Gamma\left(\frac{\mu-\nu+1}{2}\right)\Gamma\left(\frac{\mu+\nu+1}{2}\right)}{\sin(\nu\pi)} \times \left\{\cos\left(\frac{1}{2}(\mu-\nu)\pi\right)J_{-\nu}(z) - \cos\left(\frac{1}{2}(\mu+\nu)\pi\right)J_\nu(z)\right\} \tag{1.4}$$

for $\nu \notin \mathbb{Z}$, and

$$S_{\mu,\nu}(z) = s_{\mu,\nu}(z) + 2^{\mu-1}\Gamma\left(\frac{\mu-\nu+1}{2}\right)\Gamma\left(\frac{\mu+\nu+1}{2}\right) \times \left\{\sin\left(\frac{1}{2}(\mu-\nu)\pi\right)J_\nu(z) - \cos\left(\frac{1}{2}(\mu-\nu)\pi\right)Y_\nu(z)\right\} \tag{1.5}$$

for $\nu \in \mathbb{Z}$, where $J_\nu(z)$ and $Y_\nu(z)$ are Bessel functions of the first and second kinds, respectively. The Lommel functions are the solutions of the Bessel differential equation in an inhomogeneous form [25, p. 345], namely,

$$z^2\frac{d^2y}{dz^2} + z\frac{dy}{dz} + (z^2 - \nu^2)y = z^{\mu+1}.$$

Lommel functions arise in mathematics, for example, in the theory of positive trigonometric sums [13]. Outside of mathematics, Lommel functions have been found to be very useful in physics, notably in mathematical physics. See, for example, [1, 7, 20, 22]. Lewis [14] studied a special case of $S_{\mu,\nu}(z)$, that is,

$$\mathcal{C}_s(z) = \sqrt{z}\Gamma(2s+1)S_{-2s-\frac{1}{2},\frac{1}{2}}(z), \tag{1.6}$$

and represented it in terms of the incomplete gamma function. Lewis and Zagier [15] represented the period functions for Maass wave forms with spectral parameter s in terms of an infinite series of $\mathcal{C}_s(z)$, and in the course of process they gave different representations for this special case of the Lommel function. See [15, Proposition 1, p. 214].

In this work, we require a new integral representation for this special case of the Lommel function $S_{\mu,\nu}(z)$ which, to the best of our knowledge, does not seem to have been explicitly stated anywhere including [15]. This is derived in Lemma 2.1.

Another ingredient needed in our proof of (1.1) is a recent result of Maširević [9, Theorem 2.1] (see also [3, Theorem 5.23, p. 176]) which states that for all $m \in \mathbb{N} \cup \{0\}$, $\nu \in \mathbb{R}$, $x \in (0, 2\pi)$ and $\mu > \max\{-\nu - 1, \nu - 2, -\frac{1}{2}\}$,

$$\sum_{k=1}^{\infty} \frac{s_{\mu,\nu}(kx)}{k^{2m+\mu+1}} = \frac{x^{\mu+1}}{4}\Gamma\left(\frac{1+\mu-\nu}{2}\right)\Gamma\left(\frac{1+\mu+\nu}{2}\right)$$
$$\times \left(\frac{(-1)^m \pi}{2\Gamma(m+1+(\mu-\nu)/2)\Gamma(m+1+(\mu+\nu)/2)} \left(\frac{x}{2}\right)^{2m-1} \right.$$
$$\left. + \sum_{n=0}^{m} \frac{(-1)^n \zeta(2m-2n)}{\Gamma(n+1+(1+\mu-\nu)/2)\Gamma(n+1+(1+\mu+\nu)/2)} \left(\frac{x}{2}\right)^{2n} \right). \tag{1.7}$$

2. A New Proof of Hurwitz's Formula Using Hermite's Formula (1.2)

Here we prove Theorem 1.1. To do that, however, we first need a lemma which evaluates an integral in terms of the Lommel function $S_{\mu,\nu}(z)$. This lemma seems to be new.

Lemma 2.1. *Let $s \in \mathbb{C}$. Let the Lommel function $S_{\mu,\nu}(z)$ be defined in* (1.4) *and* (1.5). *For $k \in \mathbb{N}$ and $a > 0$, we have*

$$\int_0^{\infty} \frac{e^{-2\pi k x} \sin(s \tan^{-1}(x/a))}{(a^2+x^2)^{s/2}}\, dx = s\sqrt{a}(2\pi k)^{s-\frac{1}{2}} S_{-s-\frac{1}{2},\frac{1}{2}}(2\pi a k). \tag{2.1}$$

Proof. We first prove the above result for $-1 < \mathrm{Re}(s) < 0$ and then extend it to all $s \in \mathbb{C}$ by analytic continuation.

Using the inverse Mellin transform representation of the exponential function, for $c_1 := \mathrm{Re}(\xi) > 0$ and $k > 0$,

$$\frac{1}{2\pi i} \int_{(c_1)} \frac{\Gamma(\xi)}{(2\pi k)^{\xi}} y^{-\xi}\, d\xi = e^{-2\pi k y}, \tag{2.2}$$

where here, and throughout the paper, $\int_{(d)}$ will always mean the integral $\int_{d-i\infty}^{d+i\infty}$.

Also, from [16, Formula 5.19, p. 193], for $-1 < c_2 := \mathrm{Re}(\xi) < \mathrm{Re}(s)$,

$$\frac{1}{2\pi i} \int_{(c_2)} \frac{\Gamma(s-\xi)\Gamma(\xi)}{\Gamma(s)} \sin\left(\frac{\pi\xi}{2}\right) a^{\xi-s} y^{-\xi} d\xi = \frac{\sin\left(s\tan^{-1}\left(\frac{y}{a}\right)\right)}{(y^2+a^2)^{\frac{s}{2}}}. \tag{2.3}$$

From (2.2), (2.3) and Parseval's identity [18, Eq. (3.1.11), p. 82]

$$\int_0^{\infty} g(x)h(x)\, dx = \frac{1}{2\pi i} \int_{(c)} \mathfrak{G}(1-s)\mathfrak{H}(s)\, ds,$$

where $\mathfrak{G}$ and $\mathfrak{H}$ are Mellin transforms of g and h respectively, for $c := \mathrm{Re}(\xi) < \mathrm{Re}(s)$ and $-1 < \mathrm{Re}(\xi) < 1$,

$$\begin{aligned}
2\int_0^\infty \frac{e^{-2\pi kx}\sin(s\tan^{-1}(x/a))}{(a^2+x^2)^{s/2}}\,dx &= \frac{2}{2\pi i}\int_{(c)} \frac{\Gamma(\xi)\Gamma(1-\xi)\Gamma(s-\xi)}{(2\pi k)^{1-\xi}\Gamma(s)} \\
&\quad \times \sin\left(\frac{\pi\xi}{2}\right) a^{\xi-s}\,d\xi \\
&= \frac{a^{-s}}{\Gamma(s)}\frac{1}{2\pi i}\int_{(c)} \frac{\pi\Gamma(s-\xi)}{\cos\left(\frac{\pi\xi}{2}\right)}(2\pi k)^{\xi-1}a^{\xi}\,d\xi \\
&= \frac{a^{-s}}{\Gamma(s)}\frac{1}{2\pi i}\int_{(c)} \Gamma\left(\frac{1+\xi}{2}\right)\Gamma\left(\frac{1-\xi}{2}\right) \\
&\quad \times \Gamma(s-\xi)(2\pi k)^{\xi-1}a^{\xi}\,d\xi,
\end{aligned}$$

where we used the reflection formula for the gamma function. Replacing ξ by $-\xi$, we see that for $c := \mathrm{Re}(\xi) > -\mathrm{Re}(s)$ and $-1 < \mathrm{Re}(\xi) < 1$,

$$\begin{aligned}
2\int_0^\infty \frac{e^{-2\pi kx}\sin(s\tan^{-1}(x/a))}{(a^2+x^2)^{s/2}}\,dx &= \frac{a^{-s}}{2\pi k\Gamma(s)} \\
\times \frac{1}{2\pi i}\int_{(c)} \Gamma\left(\frac{1+\xi}{2}\right)\Gamma\left(\frac{1-\xi}{2}\right)&\Gamma(s+\xi)(2\pi ak)^{-\xi}\,d\xi. \qquad (2.4)
\end{aligned}$$

To evaluate the integral on the right-hand side of the above equation, first consider

$$I_s(z) := \frac{1}{2\pi i}\int_{(c)} \Gamma\left(\frac{1+\xi}{2}\right)\Gamma\left(\frac{1-\xi}{2}\right)\Gamma(s+\xi)z^{-\xi}\,d\xi, \qquad (2.5)$$

where $c = \mathrm{Re}(\xi)$.

We evaluate the above integral first for $|z| < 1, z \notin (-1,0]$ and then extend it later to all $z \in \mathbb{C}\backslash(-\infty,0]$ by analytic continuation. Consider the contour formed by the line segments $[c-iT, c+iT]$, $[c+iT, -\lambda+iT]$, $[-\lambda+iT, -\lambda-iT]$ and $[-\lambda-iT, c-iT]$, where $\lambda \notin \mathbb{Z}, \lambda > 1$. Observe that the integrand on the right-hand side of (2.5) has simple poles at $\xi = -s-m, 0 \le m \le \lfloor \lambda - \mathrm{Re}(s)\rfloor$, and $\xi = -2n-1$, where $0 \le n \le \lfloor \frac{\lambda-1}{2}\rfloor$ due to $\Gamma(s+\xi)$ and $\Gamma(\frac{1+\xi}{2})$, respectively. (Note that since $-1 < \mathrm{Re}(s) < 0$, there will not be any pole of order 2.) The residues of the integrand at these poles can be easily calculated to be $\frac{(-1)^m z^{m+s}\pi}{m!\cos(\frac{\pi}{2}(m+s))}$ and $2(-1)^n z^{2n+1}\Gamma(-1-2n+s)$, respectively. By employing Stirling's formula in a vertical strip $\alpha \le c \le \beta$ [17, Formula **5.11.9**, p. 141], namely,

$$|\Gamma(c+iT)| = (2\pi)^{\frac{1}{2}}|T|^{c-\frac{1}{2}}e^{-\frac{1}{2}\pi|T|}\left(1+O\left(\frac{1}{|T|}\right)\right), \qquad (2.6)$$

as $|T| \to \infty$, we see that the integrals along horizontal segments go to 0 as $T \to \infty$.

Hence by Cauchy's residue theorem, we obtain

$$I_s(z) = \pi z^s \sum_{m=0}^{\lfloor \lambda - \mathrm{Re}(s) \rfloor} \frac{(-z)^m}{m! \cos\left(\frac{\pi}{2}(m+s)\right)} + 2z \sum_{n=0}^{\lfloor \frac{\lambda-1}{2} \rfloor} (-1)^n z^{2n} \Gamma(-1-2n+s)$$
$$+ \frac{1}{2\pi i} \int_{(-\lambda)} \Gamma\left(\frac{1+\xi}{2}\right) \Gamma\left(\frac{1-\xi}{2}\right) \Gamma(s+\xi) z^{-\xi} \, d\xi. \tag{2.7}$$

Next, we show that as $\lambda \to \infty$,

$$\frac{1}{2\pi i} \int_{(-\lambda)} \Gamma\left(\frac{1+\xi}{2}\right) \Gamma\left(\frac{1-\xi}{2}\right) \Gamma(s+\xi) z^{-\xi} \, d\xi \to 0. \tag{2.8}$$

By (2.6), we find that as $|t| \to \infty$,

$$\left| \Gamma\left(\frac{1 \pm \lambda - it}{2}\right) \right| = O_\lambda\left(|t|^{\pm \lambda/2} e^{-\frac{\pi}{4}|t|} \right), \tag{2.9}$$

and

$$|\Gamma(s - \lambda + it)| = O\left(|(t + \mathrm{Im}(s))|^{-\lambda + \mathrm{Re}(s) - 1/2} e^{-\frac{\pi}{2}|(t+\mathrm{Im}(s))|} \right). \tag{2.10}$$

Upon making change of variable $\xi = -\lambda + it$ and then using (2.9) and (2.10), we see that

$$\left| \int_{(-\lambda)} \Gamma\left(\frac{1+\xi}{2}\right) \Gamma\left(\frac{1-\xi}{2}\right) \Gamma(s+\xi) z^{-\xi} \, d\xi \right|$$
$$= \left| \frac{1}{i} \int_{-\infty}^{\infty} \Gamma\left(\frac{1-\lambda+it}{2}\right) \Gamma\left(\frac{1+\lambda-it}{2}\right) \Gamma(s-\lambda+it) z^{\lambda-it} \, dt \right|$$
$$= |z|^\lambda \int_{-M}^{M} O(1) \, dt + |z|^\lambda$$
$$\times \int_{|t| \geq M} O\left(|(t+\mathrm{Im}(s))|^{-\lambda+\mathrm{Re}(s)-1/2} e^{-\frac{\pi}{2}|t| - \frac{\pi}{2}|(t+\mathrm{Im}(s))|} \right) dt$$
$$= O(|z|^\lambda),$$

where M is a large enough positive real number. Since $|z| < 1$, as $\lambda \to \infty$, we arrive at (2.8). Therefore, by (2.7) and (2.8), for $|z| < 1, z \notin (-1, 0]$,

$$I_s(z) = \pi z^s \sum_{m=0}^{\infty} \frac{(-z)^m}{m! \cos\left(\frac{\pi}{2}(m+s)\right)} + 2z \sum_{n=0}^{\infty} (-1)^n z^{2n} \Gamma(-1-2n+s). \tag{2.11}$$

Now observe that both sides of above equation are analytic, as functions of z, in $\mathbb{C} \backslash (-\infty, 0]$. Therefore, (2.11) holds for all $z \in \mathbb{C} \backslash (-\infty, 0]$ by analytic continuation. Hence letting $z = 2\pi a k, a > 0, k \in \mathbb{N}$, in (2.11) and simplifying the resulting first sum by splitting it into two sums, one over even m and other over odd m and by rephrasing the resulting second sum into a ${}_1F_2$ using the reflection and duplication formulas of the gamma function, we obtain

$$\begin{aligned} I_s(2\pi ak) &= \frac{2\pi(2\pi ak)^s}{\sin(\pi s)}\sin\left(\frac{\pi}{2}(4ak+s)\right) \\ &\quad + 4\pi ak\Gamma(s-1){}_1F_2\left(1;1-\frac{s}{2},\frac{3-s}{2};-a^2\pi^2k^2\right). \end{aligned} \tag{2.12}$$

Equation (2.1) now follows from (1.3), (1.4), (2.4), (2.5) and (2.12). This completes the proof for $\mathrm{Re}(s)<0$. Now using an argument similar to that in [26, pp. 269–270], one can show that the left-hand side of (2.1) is an entire function of s. The right-hand side of (2.1) is also analytic in the whole s-complex plane except for possible poles at $s=1,2,3,\ldots$. However, as shown in [25, pp. 347–349], the Lommel function $S_{\mu,\nu}(z)$ has a limit when $\mu+\nu$ or $\mu-\nu$ are odd negative integers. Since the positive integer values of s render the $\mu+\nu$ and $\mu-\nu$ of our special case of the Lommel function, namely, $S_{-s-\frac{1}{2},\frac{1}{2}}(2\pi ak)$, to fall precisely in this category, we see that $s=1,2,3,\ldots$ are indeed removable singularities of the right-hand side. Therefore, both sides of (2.1) are entire functions of s, and hence the equality follows for all $s\in\mathbb{C}$ by analytic continuation. □

We are now ready to prove Theorem 1.1.

Proof of Theorem 1.1. For $a>0$ and $s\neq 1$, from (1.2),

$$\zeta(s,a)=\frac{1}{2}a^{-s}+\frac{a^{1-s}}{s-1}+2\sum_{k=1}^{\infty}\int_0^{\infty}\frac{e^{-2\pi kx}\sin(s\tan^{-1}(x/a))}{(a^2+x^2)^{s/2}}\,dx, \tag{2.13}$$

where the interchange of the order of summation and integration is justified by absolute and uniform convergence (see, for example, [21, Theorem 2.1, p. 30]). Invoking Lemma 2.1 in (2.13), we have, for $a>0$ and $s\neq 1$,

$$\begin{aligned} \zeta(s,a) &= \frac{1}{2}a^{-s}+\frac{a^{1-s}}{s-1}+2s\sqrt{a}(2\pi)^{s-\frac{1}{2}}\sum_{k=1}^{\infty}\frac{S_{-s-\frac{1}{2},\frac{1}{2}}(2\pi ak)}{k^{\frac{1}{2}-s}} \\ &= \frac{1}{2}a^{-s}+\frac{a^{1-s}}{s-1}+\frac{1}{\Gamma(s)}\sum_{k=1}^{\infty}\left\{2\sqrt{a}(2\pi)^{s-\frac{1}{2}}\Gamma(s+1)\frac{S_{-s-\frac{1}{2},\frac{1}{2}}(2\pi ak)}{k^{\frac{1}{2}-s}}\right. \\ &\quad \left.+\frac{(2\pi)^s}{\sin(\pi s)}\frac{\sin\left(\frac{\pi}{2}(4ak+s)\right)}{k^{1-s}}\right\}, \end{aligned} \tag{2.14}$$

where in the second step, we used (1.4). This is the first instance where the infinite series on the right-hand side of (1.1) makes its conspicuous presence.

We will also be needing the value $\zeta(0)=-1/2$ in our proof which can be readily obtained by letting $s=0$ and $a=1$ in (1.2) or in (2.13).

The proof is now divided into two cases.

Case 1: $0<a<1$.

We first prove the result for $s<0$ and then extend it by analytic continuation.

Let $m=0$, $\mu=-s-\frac{1}{2}$, $\nu=\frac{1}{2}$ and $x=2\pi a$ in (1.7) and then use $\zeta(0)=-1/2$ to deduce that for $s<0$ and $0<a<1$,

$$\sum_{k=1}^{\infty}\frac{S_{-s-\frac{1}{2},\frac{1}{2}}(2\pi ak)}{k^{\frac{1}{2}-s}}=\frac{(2\pi a)^{\frac{1}{2}-s}}{4}\Gamma\left(-\frac{s}{2}\right)\Gamma\left(\frac{1-s}{2}\right)$$
$$\times\left(\frac{\pi}{2\pi a\Gamma\left(\frac{1-s}{2}\right)\Gamma\left(1-\frac{s}{2}\right)}+\frac{-1/2}{\Gamma\left(1-\frac{s}{2}\right)\Gamma\left(\frac{3-s}{2}\right)}\right)$$
$$=\frac{(2\pi a)^{\frac{1}{2}-s}}{4}\frac{\sqrt{\pi}\Gamma(-s)}{2^{-s-1}}\left(\frac{2^{-s}}{2a\sqrt{\pi}\Gamma(1-s)}-\frac{2^{1-s}}{2\sqrt{\pi}\Gamma(2-s)}\right),$$

where in the last step, we used the duplication formula for the gamma function twice. Thus,

$$\sum_{k=1}^{\infty}\frac{S_{-s-\frac{1}{2},\frac{1}{2}}(2\pi ak)}{k^{\frac{1}{2}-s}}=\frac{-1}{2s\sqrt{a}(2\pi)^{s-\frac{1}{2}}}\left(\frac{1}{2}a^{-s}+\frac{a^{1-s}}{s-1}\right).\tag{2.15}$$

Hence from (2.14) and (2.15),

$$\zeta(s,a)=\frac{(2\pi)^s}{\Gamma(s)\sin(\pi s)}\sum_{k=1}^{\infty}\frac{\sin\left(\frac{\pi}{2}(4ak+s)\right)}{k^{1-s}},$$

which readily gives (1.1). This completes the proof of Theorem 1.1 for $s<0$. Since both sides of (1.1) are analytic for $\mathrm{Re}(s)<0$, we conclude that (1.1) is valid for $\mathrm{Re}(s)<0$.

Note that the series on the right-hand side of (1.1) are absolutely and uniformly convergent for $\mathrm{Re}(s)<0$. However, since $0<a<1$, they are uniformly convergent (but not absolutely) on $0<\mathrm{Re}(s)<1$ as can be seen using Dirichlet's test for uniform convergence. Then by Weierstrass' theorem on uniformly convergent sequences of analytic functions, we see that these series are analytic in $\mathrm{Re}(s)<1$ whence the right-hand side of (1.1) is analytic in $\mathrm{Re}(s)<1$. Since $\zeta(s,a)$ is also analytic in $\mathrm{Re}(s)<1$, we conclude that (1.1) actually holds for $\mathrm{Re}(s)<1$.

Case 2: $a=1$.

We first prove the result for $s<-1$ and then extend it to all complex s by analytic continuation. From (2.14) and the fact that $\zeta(s,1)=\zeta(s)$ for all $s\in\mathbb{C}$, we have

$$\zeta(s)=\frac{1}{2}+\frac{1}{s-1}+\frac{1}{\Gamma(s)}\sum_{k=1}^{\infty}\left\{2(2\pi)^{s-\frac{1}{2}}\Gamma(s+1)\frac{S_{-s-\frac{1}{2},\frac{1}{2}}(2\pi k)}{k^{\frac{1}{2}-s}}\right.$$
$$\left.+\frac{(2\pi)^s}{\sin(\pi s)}\frac{\sin\left(\frac{\pi}{2}(4k+s)\right)}{k^{1-s}}\right\}.\tag{2.16}$$

We now wish to evaluate in closed form the series $\sum_{k=1}^{\infty}\frac{S_{-s-\frac{1}{2},\frac{1}{2}}(2\pi k)}{k^{\frac{1}{2}-s}}$, which is indeed convergent for $s<-1$ since $\sum_{k=1}^{\infty}k^{s-1}\sin\left(\frac{\pi}{2}(4k+s)\right)$ converges for $s<-1$.

However, one should be careful as (1.7) cannot be applied here. This is because, it requires $x \in (0, 2\pi)$, whereas here we need x in (1.7) to be 2π. Thankfully, Maširević has also obtained the following result [9, Theorem 2.2] where $0 \leq x \leq 2\pi, m \in \mathbb{N}$ and $\mu > 0$:

$$\sum_{k=1}^{\infty} \frac{s_{\mu-\frac{3}{2},\frac{1}{2}}(kx)}{k^{2m+\mu-\frac{1}{2}}} = \frac{1}{2}(-1)^{m-1} x^{2m+\mu-\frac{1}{2}} \Gamma(\mu-1) \times \left(\frac{-\pi}{x\Gamma(2m+\mu)} + 2\sum_{n=0}^{m} \frac{(-1)^{n-1}\zeta(2n)}{\Gamma(2m+\mu+1-2n)x^{2n}} \right). \tag{2.17}$$

Even though x can equal 2π in the above result, note that m has to be a natural number, whereas, to evaluate the series $\sum_{k=1}^{\infty} \frac{s_{-s-\frac{1}{2},\frac{1}{2}}(2\pi k)}{k^{\frac{1}{2}-s}}$ using (2.17), we would require $m = 0$. To circumvent this problem, we first employ the well-known result [8, p. 946, Formula **8.575.1**],

$$s_{\mu+2,\nu}(z) = z^{\mu+1} - [(\mu+1)^2 - \nu^2] s_{\mu,\nu}(z). \tag{2.18}$$

Use (2.18) with $\mu = -s - 5/2$, $\nu = 1/2$ and $z = 2\pi k$ so that

$$\sum_{k=1}^{\infty} \frac{s_{-s-\frac{1}{2},\frac{1}{2}}(2\pi k)}{k^{\frac{1}{2}-s}} = \sum_{k=1}^{\infty} \frac{(2\pi k)^{-s-\frac{3}{2}}}{k^{\frac{1}{2}-s}} - \left\{ \left(s+\frac{3}{2}\right)^2 - \frac{1}{4} \right\} \sum_{k=1}^{\infty} \frac{s_{-s-\frac{5}{2},\frac{1}{2}}(2\pi k)}{k^{\frac{1}{2}-s}}. \tag{2.19}$$

We now transform the second series on the right-hand side of the above equation using (2.17). Let $m = 1$, $x = 2\pi$ and $\mu = -s - 1$ in (2.17) so that for $s < -1$,

$$\sum_{k=1}^{\infty} \frac{s_{-s-\frac{5}{2},\frac{1}{2}}(2\pi k)}{k^{\frac{1}{2}-s}} = \frac{1}{2}(2\pi)^{\frac{1}{2}-s}\Gamma(-s-2) \left(-\frac{1}{2\Gamma(1-s)} + \frac{1}{\Gamma(2-s)} + \frac{1}{12\Gamma(-s)} \right), \tag{2.20}$$

where we used $\zeta(0) = -1/2$ and $\zeta(2) = \sum_{k=1}^{\infty} \frac{1}{k^2} = \pi^2/6$.

Substitute (2.20) in (2.19) and simplify using the functional equation of the gamma function to arrive at

$$\sum_{k=1}^{\infty} \frac{s_{-s-\frac{1}{2},\frac{1}{2}}(2\pi k)}{k^{\frac{1}{2}-s}} = -\frac{1}{2s(2\pi)^{s-\frac{1}{2}}} \left(\frac{1}{2} + \frac{1}{s-1} \right). \tag{2.21}$$

Comparing the above equation with (2.15), we see that (2.15) holds for $a = 1$ too.

Using (2.21) in (2.16), we arrive at

$$\zeta(s) = \frac{2\Gamma(1-s)}{(2\pi)^{1-s}} \sin\left(\frac{\pi s}{2}\right) \zeta(1-s)$$

for $s < -1$. The result then follows for all complex s by analytic continuation. □

3. A Modular-Type Transformation Involving the Lommel Function $S_{-s-\frac{1}{2},\frac{1}{2}}(z)$

Modular-type transformations are the ones governed by the map $\alpha \to \beta$, where $\alpha\beta = 1$. An equivalent way to say this using the language of modular forms is that they consist of functions which transform nicely under $z \to -1/z$, $\mathrm{Im}(z) > 0$. But they may not transform nicely under $z \to z+1$, hence the nomenclature *modular-type* transformations. For a detailed survey on modular-type transformations, the reader is referred to [6].

The following modular-type transformation involving infinite series of Hurwitz zeta function was obtained by the first author in [5, Theorem 1.4] as a generalization of a transformation of Ramanujan [5, Theorem 1.1] on p. 220 of the Lost Notebook [19].

Theorem 3.1. *Let* $0 < \mathrm{Re}(s) < 2$. *Define* $\varphi(s,x)$ *by*

$$\varphi(s,x) := \zeta(s,x) - \frac{1}{2}x^{-s} + \frac{x^{1-s}}{1-s}.$$

If α *and* β *are any positive numbers such that* $\alpha\beta = 1$,

$$\begin{aligned}
&\alpha^{\frac{s}{2}}\left(\sum_{n=1}^{\infty}\varphi(s,n\alpha) - \frac{\zeta(s)}{2\alpha^s} - \frac{\zeta(s-1)}{(s-1)\alpha}\right)\\
&= \beta^{\frac{s}{2}}\left(\sum_{n=1}^{\infty}\varphi(s,n\beta) - \frac{\zeta(s)}{2\beta^s} - \frac{\zeta(s-1)}{(s-1)\beta}\right)\\
&= \frac{8(4\pi)^{\frac{s-4}{2}}}{\Gamma(s)}\int_0^{\infty}\Gamma\left(\frac{s-2+it}{4}\right)\Gamma\left(\frac{s-2-it}{4}\right)\Xi\left(\frac{t+i(s-1)}{2}\right)\\
&\quad\times\Xi\left(\frac{t-i(s-1)}{2}\right)\frac{\cos\left(\frac{1}{2}t\log\alpha\right)}{s^2+t^2}\,dt,
\end{aligned}$$

where $\Xi(t) := \xi(\frac{1}{2}+it)$ *with* $\xi(s) = \frac{1}{2}s(s-1)\pi^{-\frac{1}{2}s}\Gamma(\frac{1}{2}s)\zeta(s)$.

In view of the first equality in (2.14), the modular-type transformation in the above result can be rephrased in the following form.

Corollary 3.2. *Let* $0 < \mathrm{Re}(s) < 2$ *and let* $\sigma_s(n) = \sum_{d|n} d^s$. *Let* $S_{\mu,\nu}(z)$ *be defined in* (1.4). *If* α *and* β *are any positive numbers such that* $\alpha\beta = 1$,

$$\begin{aligned}
&\alpha^{\frac{s}{2}}\left(2s(2\pi)^{s-\frac{1}{2}}\sqrt{\alpha}\sum_{m=1}^{\infty}\sigma_{1-s}(m)m^{s-\frac{1}{2}}S_{-s-\frac{1}{2},\frac{1}{2}}(2\pi m\alpha) - \frac{\zeta(s)}{2\alpha^s} - \frac{\zeta(s-1)}{(s-1)\alpha}\right)\\
&= \beta^{\frac{s}{2}}\left(2s(2\pi)^{s-\frac{1}{2}}\sqrt{\beta}\sum_{m=1}^{\infty}\sigma_{1-s}(m)m^{s-\frac{1}{2}}S_{-s-\frac{1}{2},\frac{1}{2}}(2\pi m\beta) - \frac{\zeta(s)}{2\beta^s} - \frac{\zeta(s-1)}{(s-1)\beta}\right).
\end{aligned}$$

Proof. The result follows at once if we observe that from (2.14),

$$\begin{aligned}\sum_{n=1}^{\infty}\varphi(s,n\alpha) &= 2s(2\pi)^{s-\frac{1}{2}}\sqrt{\alpha}\sum_{n=1}^{\infty}\sqrt{n}\sum_{k=1}^{\infty}\frac{S_{-s-\frac{1}{2},\frac{1}{2}}(2\pi nk\alpha)}{k^{\frac{1}{2}-s}}\\ &= 2s(2\pi)^{s-\frac{1}{2}}\sqrt{\alpha}\sum_{n,k=1}^{\infty}n^{1-s}(nk)^{s-\frac{1}{2}}S_{-s-\frac{1}{2},\frac{1}{2}}(2\pi nk\alpha)\\ &= 2s(2\pi)^{s-\frac{1}{2}}\sqrt{\alpha}\sum_{m=1}^{\infty}\sigma_{1-s}(m)m^{s-\frac{1}{2}}S_{-s-\frac{1}{2},\frac{1}{2}}(2\pi m\alpha).\end{aligned}$$

□

Remark 1. The series in Corollary 3.2 should be compared with the series considered by Lewis and Zagier in [15, Eq. (2.11)], namely, $\sum_{n=1}^{\infty} n^{s-1/2}A_n\mathcal{C}_s(2\pi na)$, where $\mathcal{C}_s(z)$ is defined by (1.6).

Acknowledgments

We sincerely thank the referee for reading the paper in great detail and for giving nice suggestions. The first author's research is partially supported by the SERB MATRICS grant MTR/2018/000251. He sincerely thanks SERB for the support.

References

[1] N. T. Adelman, Y. Stavsky and E. Segal, Axisymmetric vibrations of radially polarized piezoelectric ceramic cylinders, *J. Sound Vib.* **38**(2) (1975) 245–254.

[2] T .M. Apostol, *Introduction to Analytic Number Theory* (Springer-Verlag, New York, 1998).

[3] Á. Baricz, D. Jankov Maširević and T. K. Pogány, *Series of Bessel and Kummer-Type Functions*, Lecture Notes in Mathematics, Vol. 2207 (Springer, Cham, 2017).

[4] B. C. Berndt, On the Hurwitz zeta-function, *Rocky Mountain J. Math.* **2**(1) (1972) 151–157.

[5] A. Dixit, Analogues of a transformation formula of Ramanujan, *Int. J. Number Theory* **7**(5) (2011) 1151–1172.

[6] A. Dixit, Modular-type transformations and integrals involving the Riemann Ξ-function, *Math. Student* **87**(3–4) (2018) 47–59.

[7] S. Goldstein, On the vortex theory of screw propellers, *Proc. R. Soc. Lond. A* **123** (1929) 440–465.

[8] I. S. Gradshteyn and I. M. Ryzhik (eds), *Table of Integrals, Series, and Products*, 7th edn., eds A. Jeffrey and D. Zwillinger (Academic Press, New York, 2007).

[9] D. Jankov Maširević, Summations of Schlomilch series containing some Lommel functions of the first kind terms, *Integral Transforms Spec. Funct.* **27**(2) (2015) 153–162.

[10] S. Kanemitsu, K. Chakraborty and H. Tsukada, Ewald expansions of a class of zeta-functions, Springer-Plus **5**(1) (2016), Article No. 99.

[11] S. Kanemitsu, Y. Tanigawa, H. Tsukada and M. Yoshimoto, Contributions to the theory of the Hurwitz zeta-function, *Hardy-Ramanujan J.* **30** (2007) 31–55.

[12] S. Kanemitsu and H. Tsukada, *Contributions to the Theory of Zeta-Functions. The Modular Relation Supremacy*, Series on Number Theory and its Applications, Vol. 10 (World Scientific, New Jersey, 2015).

[13] S. Koumandos and M. Lamprecht, The zeros of certain Lommel functions, *Proc. Amer. Math. Soc.* **140**(9) (2012) 3091–3100.

[14] J. Lewis, Spaces of holomorphic functions equivalent to the even Maass cusp form, *Invent. Math.* **127** (1997) 271–306.

[15] J. Lewis and D. Zagier, Period functions for Maass wave forms. I, *Ann. of Math.* **153**(1) (2001) 191–258.

[16] F. Oberhettinger, *Tables of Mellin Transforms* (Springer-Verlag, New York, 1974).

[17] F. W. J. Olver, D. W. Lozier, R. F. Boisvert and C. W. Clark (eds.), *NIST Handbook of Mathematical Functions* (Cambridge University Press, Cambridge, 2010).

[18] R. B. Paris and D. Kaminski, *Asymptotics and Mellin–Barnes Integrals*, Encyclopedia of Mathematics and its Applications, Vol. 85 (Cambridge University Press, Cambridge, 2001).

[19] S. Ramanujan, *The Lost Notebook and Other Unpublished Papers* (Narosa, New Delhi, 1988).

[20] M. R. Sitzer, Stress distribution in rotating aeolotropic laminated heterogeneous disc under action of a time-dependent loading, *Z. Angew. Math. Phys.* **36** (1985) 134–145.

[21] N. M. Temme, Special functions: An introduction to the classical functions of mathematical physics (Wiley-Interscience Publication, New York, 1996).

[22] B. K. Thomas and F. T. Chan, Glauber e^-+ He elastic scattering amplitude: A useful integral representation, *Phys. Rev. A* **8** (1973) 252–262.

[23] E. C. Titchmarsh, *The Theory of the Riemann Zeta Function* (Clarendon Press, Oxford, 1986).

[24] K. Ueno and M. Nishizawa, Quantum groups and zeta-functions, in *Quantum Groups: Formalism and Applications*, eds. Lubkierski *et al.* (Polish Scientific Publishers, PWN, Warsaw, 1995), pp. 115–126.

[25] G. N. Watson, *A Treatise on the Theory of Bessel Functions*, 2nd edn. (Cambridge University Press, London, 1994).

[26] E. T. Whittaker and G. N. Watson, *A Course of Modern Analysis*, 2nd edn. (Cambridge University Press, 1996).

© 2024 World Scientific Publishing Company
https://doi.org/10.1142/9789811277375_0013

An inequality between finite analogues of rank and crank moments*

Pramod Eyyunni[†,¶], Bibekananda Maji[†,‡,‖] and Garima Sood[†,§,**]

[†]*Discipline of Mathematics, Indian Institute of Technology Gandhinagar Palaj, Gandhinagar 382355, Gujarat, India*

[‡]*Discipline of Mathematics, Indian Institute of Technology Indore Simrol, Indore 453552, Madhya Pradesh, India*

[§]*Fakultät für Mathematik, Universität Wien Oskar-Morgenstern-Platz 1 A-1090 Wien, Austria*

[¶]*pramodeyy@gmail.com*
[‖]*bibek10iitb@gmail.com*
[**]*garimasood18@gmail.com*

Dedicated to Professor Bruce C. Berndt on the occasion of his 80th birthday

The inequality between rank and crank moments was conjectured and later proved by Garvan himself in 2011. Recently, Dixit and the authors introduced finite analogues of rank and crank moments for vector partitions while deriving a finite analogue of Andrews' famous identity for smallest parts function. In the same paper, they also conjectured an inequality between finite analogues of rank and crank moments, analogous to Garvan's conjecture. In this paper, we give a proof of this conjecture.

Keywords: Partitions; finite analogues; smallest parts function; moments inequality; symmetrized moments.

Mathematics Subject Classification 2020: 11P80, 11P81, 11P82, 05A17

1. Introduction

Let $p(n)$ denote the number of unrestricted partitions of a positive integer n. To give a combinatorial explanation of the famous congruences of Ramanujan for the partition function $p(n)$, namely, for $m \geq 0$,

$$p(5m+4) \equiv 0 \pmod 5, \quad p(7m+5) \equiv 0 \pmod 7,$$

*This article originally appeared in *International Journal of Number Theory*, **17**(2), pp. 405–423 © 2021 World Scientific Publishing Company. DOI: 10.1142/S1793042120400217
[‖]Corresponding author.

Dyson [16] defined the rank of a partition as the largest part minus the number of parts. He also conjectured that there must be another statistic, which he named 'crank', that would explain Ramanujan's third congruence, namely

$$p(11m+6) \equiv 0 \pmod{11}.$$

After a decade, Atkin and Swinnerton-Dyer [9] confirmed Dyson's observations for the first two congruences for $p(n)$. Also, in 1988, 'crank' was discovered by Andrews and Garvan [7]. An interesting thing to note is that by using the partition statistic 'crank', Andrews and Garvan were able to explain not only the third congruence but also the first two. Atkin and Garvan [8] found that the moments of ranks and cranks were important in the study of further partition congruences. In particular, they defined the kth moments of rank and crank, respectively, as follows:

$$N_k(n) = \sum_{m=-\infty}^{\infty} m^k N(m,n), \quad M_k(n) = \sum_{m=-\infty}^{\infty} m^k M(m,n),$$

where $N(m,n)$ and $M(m,n)$ denote the number of partitions of n with rank m and crank m, respectively. In 2008, Andrews [4] introduced the smallest parts function $\mathrm{spt}(n)$ as the total number of appearances of the smallest parts in all partitions of n and showed that

$$\mathrm{spt}(n) = np(n) - \frac{1}{2}N_2(n).$$

Using Dyson's identity [17, Theorem 5], i.e. $np(n) = \frac{1}{2}M_2(n)$, we can rewrite this as

$$\mathrm{spt}(n) = \frac{1}{2}M_2(n) - \frac{1}{2}N_2(n). \tag{1.1}$$

From this result, it is immediate that $M_2(n) > N_2(n)$. Garvan [20, Conjecture (1.1)] conjectured that

$$M_{2k}(n) > N_{2k}(n), \tag{1.2}$$

for all $k > 1$ and $n \geq 1$. Studying the asymptotic behavior of the difference $M_{2k}(n) - N_{2k}(n)$, Bringmann and Mahlburg [12] proved (1.2) for $k = 2, 4$, and subsequently, for each fixed k, the inequality was proved for sufficiently large n by Bringmann *et al.* [13]. Later, Garvan [20] himself proved his conjecture for all n and k with the help of a combinatorial interpretation for the difference between symmetrized crank and rank moments. Andrews [3] defined the kth symmetrized rank moment as

$$\eta_k(n) := \sum_{m=-n}^{n} \binom{m + \left\lfloor \frac{k-1}{2} \right\rfloor}{k} N(m,n).$$

Andrews [3, Theorem 2] showed that the odd moments are all identically zero and also obtained the generating function for even moments $\eta_{2k}(n)$, that is, for any

$k \geq 1$, we have

$$\sum_{n=1}^{\infty} \eta_{2k}(n) q^n = \frac{1}{(q)_\infty} \sum_{n=1}^{\infty} \frac{(-1)^{n-1}(1+q^n) q^{\frac{n(3n-1)}{2}+kn}}{(1-q^n)^{2k}} \tag{1.3}$$

$$= \frac{1}{(q)_\infty} \sum_{\substack{n=-\infty \\ n\neq 0}}^{\infty} \frac{(-1)^{n-1} q^{\frac{n(3n+1)}{2}+kn}}{(1-q^n)^{2k}}. \tag{1.4}$$

Analogous to the symmetrized rank moments $\eta_k(n)$, Garvan [21] introduced the kth symmetrized crank moment $\mu_k(n)$ in the study of the higher order spt-function $\mathrm{spt}_k(n)$. To be more specific,

$$\mu_k(n) := \sum_{m=-n}^{n} \binom{m + \left\lfloor \frac{k-1}{2} \right\rfloor}{k} M(m,n).$$

Analogous to (1.3) and (1.4), the generating function for the symmetrized crank moments was given by Garvan [21, Theorem (2.2)], that is, for any $k \geq 1$, we have

$$\sum_{n=1}^{\infty} \mu_{2k}(n) q^n = \frac{1}{(q)_\infty} \sum_{n=1}^{\infty} \frac{(-1)^{n-1}(1+q^n) q^{\frac{n(n-1)}{2}+kn}}{(1-q^n)^{2k}} \tag{1.5}$$

$$= \frac{1}{(q)_\infty} \sum_{\substack{n=-\infty \\ n\neq 0}}^{\infty} \frac{(-1)^{n-1} q^{\frac{n(n+1)}{2}+kn}}{(1-q^n)^{2k}}. \tag{1.6}$$

Garvan [21, Eq. (1.4)] also gave the following generating function for the symmetrized crank moments:

$$\sum_{n=1}^{\infty} \mu_{2k}(n) q^n = \frac{1}{(q)_\infty} \sum_{n_k \geq \cdots \geq n_2 \geq 1} \frac{q^{n_1+n_2+\cdots+n_k}}{(1-q^{n_1})^2(1-q^{n_2})^2 \cdots (1-q^{n_k})^2}. \tag{1.7}$$

One of the main results in [21], due to Garvan [21, Eq. (1.3)], which was instrumental in proving the inequality between rank and crank moments is as follows:

$$\sum_{n=1}^{\infty} (\mu_{2k}(n) - \eta_{2k}(n)) q^n$$
$$= \sum_{n_k \geq \cdots \geq n_2 \geq 1} \frac{q^{n_1+n_2+\cdots+n_k}}{(1-q^{n_1})^2(1-q^{n_2})^2 \cdots (1-q^{n_k})^2 (q^{n_1+1};q)_\infty}, \tag{1.8}$$

for any $k \geq 1$. One can easily check that for $k=1$, the above theorem reduces to (1.1). After this observation, Garvan defined higher order spt-function $\mathrm{spt}_k(n)$ as

$$\mathrm{spt}_k(n) := \mu_{2k}(n) - \eta_{2k}(n),$$

for all $k \geq 1$ and $n \geq 1$. He also gave a combinatorial interpretation of $\mathrm{spt}_k(n)$.

In the next subsection, we shall describe recent developments related to Andrews' identity (1.1) for the smallest parts function $\mathrm{spt}(n)$.

1.1. *Finite analogue of Andrews' spt-identity*

Ramanujan's identities are a constant source of inspiration for everyone and motivate us to do beautiful mathematics. Recently, Dixit and Maji [15] found a generalization of a q-series identity [22, p. 354], [10, p. 263, Entry 3] of Ramanujan and derived many partition theoretic implications from this generalization. They also established a new identity [15, Theorem 2.8] involving Fine's function $F(a, b; t)$ [18, p. 1] from which they were able to derive Andrews' identity (1.1) for $\mathrm{spt}(n)$. Very recently, together with Dixit, the authors found a finite analogue [14, Theorem 1.1] of the aforementioned generalization of Dixit and Maji [15, Theorem 2.1], whose special case gave a finite analogue of Andrews' spt-identity, namely.

Theorem 1.1 ([14, Theorem 2.4]). *For any natural numbers n, L, we have*

$$\mathrm{spt}(n, L) = \frac{1}{2}(M_{2,L}(n) - N_{2,L}(n)),$$

where $\mathrm{spt}(n, L)$ is the number of smallest parts in all partitions of n whose corresponding largest parts are less than or equal to L, and $M_{2,L}(n)$ and $N_{2,L}(n)$ are defined below.

In [14, p. 9, Eqs. (2.9) and (2.10)], for $k \geq 1$, we defined finite analogues of rank and crank moments for vector partitions as

$$N_{k,L}(n) := \sum_{m=-\infty}^{\infty} m^k N_{S_1}(m, n), \tag{1.9}$$

$$M_{k,L}(n) := \sum_{m=-\infty}^{\infty} m^k M_{S_2}(m, n), \tag{1.10}$$

where $N_{S_1}(m, n)$ and $M_{S_2}(m, n)$ are defined below in (2.1) and (2.5), respectively. From Theorem 1.1, it is immediate that $M_{2,L}(n) > N_{2,L}(n)$. Analogous to Garvan's conjecture (1.2), we gave the following conjecture on the inequality between the finite analogues of kth rank and crank moments, that is,

Conjecture 1.2 ([14, Conjecture 10.1]). *For any fixed natural number L and even $k \geq 2$,*

$$M_{k,L}(n) > N_{k,L}(n) \quad \text{for all } n \geq 1.$$

In this paper, our main goal is to prove Conjecture 1.2. For that reason, we define finite analogues of symmetrized rank and crank moments and their generating functions. The remainder of this paper is organized as follows. In Sec. 2, we present our main results. In Sec. 3, we collect all necessary results which will be useful throughout the paper. The generating functions of the finite analogues of the symmetrized rank and crank moments are proved in Sec. 4. In Sec. 5, we derive important results using Bailey's lemma and give a proof of Conjecture 1.2. We conclude the paper, by discussing further questions in Sec. 6.

2. Main Results: Finite Analogues of the kth Symmetrized Rank and Crank Moments

Before defining finite analogues of symmetrized rank and crank moments we need to recall certain definitions from [14, p. 7]. For the sake of completeness we reproduce them below.

Let $V_1 = \mathcal{D} \times \mathcal{P}$ denote a set of vector partitions, where $\mathcal{D}$ is the collection of all partitions with distinct parts and $\mathcal{P}$ is the collection of all unrestricted partitions. So an element $\vec{\pi}$ of V_1 is of the form (π_1, π_2), where the magnitude of $\vec{\pi}$ is given by $|\vec{\pi}| := |\pi_1| + |\pi_2|$. Let L be a positive integer. Then for any positive integer j with $1 \leq j \leq L$, set

$$\begin{aligned} S_1^{L,j} := \{\vec{\pi} \in V_1 : \pi_1 &\text{is either an empty partition or such that its parts lie} \\ &\text{in } [L-j+1, L] \text{ and } \pi_2 \text{ is an unrestricted partition} \\ &\text{with Durfee square of size } j\}. \end{aligned}$$

For simplicity, we refer to the set $S_1^{L,j}$ as S_1 in the remainder of the paper. For a vector partition $\vec{\pi} = (\pi_1, \pi_2)$ in V_1, let $w_r(\vec{\pi}) := (-1)^{\#(\pi_1)}$ be its weight and $\text{rank}(\vec{\pi}) := \text{rank}(\pi_2)$, its vector rank. Now define

$$N_{S_1}(m,n) := \sum_{j=1}^{L} N_{S_1}\left(m, n; \boxed{j}\right), \tag{2.1}$$

where

$$N_{S_1}\left(m, n; \boxed{j}\right) := \sum_{\substack{\vec{\pi} \in S_1, |\vec{\pi}|=n \\ \text{rank}(\vec{\pi})=m}} w_r(\vec{\pi}).$$

As observed in [14], as $L \to \infty$, $N_{S_1}(m,n)$ equals $N(m,n)$, the number of ordinary partitions of n with rank m. We are now ready to define the finite analogue of the kth symmetrized rank function. Let k, L be positive integers. Then for any $n \geq 1$,

$$\eta_{k,L}(n) := \sum_{m=-n}^{n} \binom{m + \left\lfloor \frac{k-1}{2} \right\rfloor}{k} N_{S_1}(m,n). \tag{2.2}$$

Proposition 2.1. *Let L be a positive integer and k be an odd positive integer. Then $\eta_{k,L}(n) = 0$ for all $n \geq 1$.*

This is straightforward from the fact that the finite analogues of all the odd rank moments $N_{k,L}(n)$ are zero. We now give an expression for the generating function of $\eta_{k,L}(n)$ for *even* k.

Theorem 2.2. *Let $L \in \mathbb{N}$. Then for any positive integer ν, we have*

$$\sum_{n=1}^{\infty} \eta_{2\nu,L}(n)q^n = (q)_L \sum_{n=1}^{L} \frac{(-1)^{n-1} q^{\frac{n(3n-1)}{2}+\nu n}(1+q^n)}{(q)_{L+n}(q)_{L-n}(1-q^n)^{2\nu}} \tag{2.3}$$

$$= (q)_L \sum_{\substack{n=-L \\ n\neq 0}}^{L} \frac{(-1)^{n-1} q^{\frac{n(3n+1)}{2}+\nu n}}{(q)_{L+n}(q)_{L-n}(1-q^n)^{2\nu}}. \tag{2.4}$$

Letting $L \to \infty$, we obtain the generating functions for the symmetrized rank moment, namely, (1.3) and (1.4). Next, we are going to define the finite analogue of the symmetrized crank moments. Again, for convenience, we recollect some definitions from [14, pp. 8–9].

Let V_2 denote the set of vector partitions $\mathcal{D} \times \mathcal{P} \times \mathcal{P}$. Denote an element $\vec{\pi}$ of V_2 by (π_1, π_2, π_3) so that the magnitude of $\vec{\pi}$ is $|\vec{\pi}| = |\pi_1| + |\pi_2| + |\pi_3|$.
For any positive integer L, we define the following set:

$$S_2 := \{\vec{\pi} \in V_2 : l(\pi_1), l(\pi_2), l(\pi_3) \leq L\}.$$

Define $w_c(\vec{\pi}) := (-1)^{\#(\pi_1)}$ to be the weight of the vector partition $\vec{\pi} = (\pi_1, \pi_2, \pi_3)$ and $\text{crank}(\vec{\pi}) := \#(\pi_2) - \#(\pi_3)$ be its vector crank. We define

$$M_{S_2}(m,n) := \sum_{\substack{\vec{\pi} \in S_2, |\vec{\pi}|=n \\ \text{crank}(\vec{\pi})=m}} w_c(\vec{\pi}). \tag{2.5}$$

Letting $L \to \infty$, we see that S_2 approaches the whole set V_2 and consequently $M_{S_2}(m,n)$ approaches $\sum_{\substack{\vec{\pi} \in V_2, |\vec{\pi}|=n \\ \text{crank}(\vec{\pi})=m}} w_c(\vec{\pi})$, which is the total number of weighted vector partitions of n with vector crank m, a quantity first studied by Garvan (see [19, p. 50]). By the work of Andrews and Garvan [7, Theorem 1], we know that this equals $M(m,n)$, the number of integer partitions of n with crank m.

We now define a finite analogue of the kth symmetrized crank moment. Let k, L be positive integers. Then for any $n \geq 1$,

$$\mu_{k,L}(n) := \sum_{m=-n}^{n} \binom{m + \left\lfloor \frac{k-1}{2} \right\rfloor}{k} M_{S_2}(m,n). \tag{2.6}$$

Proposition 2.3. *For any odd positive integer k, we have $\mu_{k,L}(n) = 0$ for all $n \geq 1$.*

This easily follows because all the odd crank moments $M_{k,L}(n)$ are zero. Analogous to Theorem 2.2, we derive the following result for the generating function of $\mu_{k,L}(n)$ for *even* k.

Theorem 2.4. *Let $L \in \mathbb{N}$. Then for any positive integer ν, one has*

$$\sum_{n=1}^{\infty} \mu_{2\nu,L}(n)q^n = (q)_L \sum_{n=1}^{L} \frac{(-1)^{n-1} q^{\frac{n(n-1)}{2}+\nu n}(1+q^n)}{(q)_{L+n}(q)_{L-n}(1-q^n)^{2\nu}} \tag{2.7}$$

$$= (q)_L \sum_{\substack{n=-L \\ n\neq 0}}^{L} \frac{(-1)^{n-1} q^{\frac{n(n+1)}{2}+\nu n}}{(q)_{L+n}(q)_{L-n}(1-q^n)^{2\nu}}. \tag{2.8}$$

One can easily observe that this result is a finite analogue of Eqs. (1.5) and (1.6) by letting $L \to \infty$. The next result provides us information about the generating function of the difference between finite analogues of symmetrized crank and rank moments.

Theorem 2.5. *Let $L \in \mathbb{N}$. Then for any positive integer k, we have*

$$\sum_{n=1}^{\infty} (\mu_{2k,L}(n) - \eta_{2k,L}(n))q^n$$

$$= \frac{1}{(q)_L} \sum_{L\geq n_k \geq \cdots \geq n_1 \geq 1} \frac{(q)_{n_1} q^{n_1+n_2+\cdots+n_k}}{(1-q^{n_1})^2(1-q^{n_2})^2 \ldots (1-q^{n_k})^2}. \tag{2.9}$$

This is a finite analogue of Garvan's result (1.8) for the generating function of the difference between symmetrized crank and rank moments.

Remark 2.6. If we substitute $k = 1$ in the above result, then we can obtain Theorem 1.1. Thus, we have $\mu_{2,L}(n) - \eta_{2,L}(n) = \mathrm{spt}(n, L)$. This suggests us to define a finite analogue of higher order spt-function as $\mathrm{spt}_k(n, L) := \mu_{2k,L}(n) - \eta_{2k,L}(n)$.

3. Preliminaries

In [14, Theorem 2.2], the authors together with Dixit, noted that the generating function of $N_{S_1}(m, n)$ is

$$R_{S_1}(z;q) := \sum_{n=1}^{\infty} \sum_{m=-\infty}^{\infty} N_{S_1}(m,n) z^m q^n = \sum_{j=1}^{L} \begin{bmatrix} L \\ j \end{bmatrix} \frac{q^{j^2}(q)_j}{(zq)_j(z^{-1}q)_j}. \tag{3.1}$$

We call (3.1) as the finite analogue of the rank generating function, for, letting $L \to \infty$ on both sides, gives the well-known result for the rank generating function (for more details, see [14, p. 8]),

$$\sum_{n=1}^{\infty} \sum_{m=-\infty}^{\infty} N(m,n) z^m q^n = \sum_{j=1}^{\infty} \frac{q^{j^2}}{(zq)_j(z^{-1}q)_j}.$$

Again, in [2, p. 252, Theorem 2.1], [6, Eq. (12.2.2), p. 263], Andrews showed that

$$\sum_{n=0}^{L} \begin{bmatrix} L \\ n \end{bmatrix} \frac{(q)_n q^{n^2}}{(zq)_n (z^{-1}q)_n} = \frac{1}{(q)_L} + (1-z) \sum_{n=1}^{L} \begin{bmatrix} L \\ n \end{bmatrix} \times \frac{(-1)^n (q)_n q^{n(3n+1)/2}}{(q)_{L+n}} \left(\frac{1}{1-zq^n} - \frac{1}{z-q^n} \right). \tag{3.2}$$

Now, we recall the crank generating function, that is,

$$\frac{(q)_\infty}{(zq)_\infty (z^{-1}q)_\infty} = \sum_{n=0}^{\infty} \sum_{m=-\infty}^{\infty} M(m,n) z^m q^n, \tag{3.3}$$

where $M(m,n)$ is the number of partitions of n with crank m. In [14, Theorem 2.3], it was proved that the generating function of $M_{S_2}(m,n)$ is

$$C_{S_2}(z;q) := \sum_{n=0}^{\infty} \sum_{m=-\infty}^{\infty} M_{S_2}(m,n) z^m q^n = \frac{(q)_L}{(zq)_L (z^{-1}q)_L}, \tag{3.4}$$

which is the finite analogue of (3.3). Andrews [2, p. 258, Theorem 4.1] showed that

$$\frac{(q)_L}{(zq)_L (z^{-1}q)_L} = \frac{1}{(q)_L} + (1-z) \sum_{n=1}^{L} \begin{bmatrix} L \\ n \end{bmatrix} \times \frac{(-1)^n (q)_n q^{n(n+1)/2}}{(q)_{n+L}} \left(\frac{1}{1-zq^n} - \frac{1}{z-q^n} \right). \tag{3.5}$$

Now, we collect some useful facts about Bailey pairs, see [5, p. 582]. A pair of sequences $(\alpha_n(a,q), \beta_n(a,q))$ is called a Bailey pair with parameters (a,q) if, for each non-negative integer n,

$$\beta_n(a,q) = \sum_{r=0}^{n} \frac{\alpha_r(a,q)}{(q;q)_{n-r} (aq;q)_{n+r}}. \tag{3.6}$$

Theorem 3.1 (Bailey's Lemma). *Suppose $(\alpha_n(a,q), \beta_n(a,q))$ is a Bailey pair with parameters (a,q). Then $(\alpha'_n(a,q), \beta'_n(a,q))$ is another Bailey pair with parameters (a,q), where*

$$\alpha'_n(a,q) = \frac{(\rho_1, \rho_2; q)_n}{(aq/\rho_1, aq/\rho_2; q)_n} \left(\frac{aq}{\rho_1 \rho_2} \right)^n \alpha_n(a,q)$$

and

$$\beta'_n(a,q) = \sum_{k=0}^{n} \frac{(\rho_1, \rho_2; q)_k (aq/\rho_1\rho_2; q)_{n-k}}{(aq/\rho_1, aq/\rho_2; q)_n (q;q)_{n-k}} \left(\frac{aq}{\rho_1 \rho_2} \right)^k \beta_k(a,q).$$

For more information on Bailey's Lemma, readers can see [1, Chap. 3]. We also require the following result:

$$\lim_{\rho_2 \to 1} \lim_{\rho_1 \to 1} \frac{1}{(1-\rho_1)(1-\rho_2)} \left(1 - \frac{(q)_k (q/\rho_1\rho_2)_k}{(q/\rho_1)_k (q/\rho_2)_k} \right) = \sum_{j=1}^{k} \frac{q^j}{(1-q^j)^2}. \tag{3.7}$$

4. Proofs of Theorems 2.2 and 2.4

Proof of Theorem 2.2. By definition (2.2) of $\eta_{k,L}(n)$, we know that

$$\eta_{2\nu,L}(n) := \sum_{m=-n}^{n} \binom{m+\nu-1}{2\nu} N_{S_1}(m,n).$$

From the definition (3.1) of $R_{S_1}(z;q)$, it follows at once that

$$\left(\frac{d^{2\nu}}{dz^{2\nu}} z^{\nu-1} R_{S_1}(z;q)\right)\Bigg|_{z=1}$$
$$= \sum_{n=1}^{\infty} \sum_{m=-\infty}^{\infty} (m+\nu-1)(m+\nu-2)\cdots(m-\nu+1)(m-\nu) N_{S_1}(m,n) q^n.$$

In other words,

$$\left(\frac{d^{2\nu}}{dz^{2\nu}} z^{\nu-1} R_{S_1}(z;q)\right)\Bigg|_{z=1} = (2\nu)! \sum_{n=1}^{\infty} \eta_{2\nu,L}(n) q^n.$$

Using Leibniz's chain rule, we get

$$\sum_{n=1}^{\infty} \eta_{2\nu,L}(n) q^n = \frac{1}{(2\nu)!} \sum_{j=0}^{\nu-1} \binom{2\nu}{j} (\nu-1)(\nu-2)\cdots(\nu-j) R_{S_1}^{(2\nu-j)}(1;q). \tag{4.1}$$

Thus, in order to obtain the generating function of $\eta_{2\nu,L}(n)$, it is sufficient to find the derivatives of $R_{S_1}(z;q)$ with respect to z at $z=1$. To this end, we wish to write $R_{S_1}(z;q)$ in a suitable form. Using (3.2) in the rightmost expression of (3.1), we deduce that

$$R_{S_1}(z;q) = \frac{1}{(q)_L} - 1 + (1-z)\sum_{n=1}^{L} \begin{bmatrix} L \\ n \end{bmatrix} \frac{(-1)^n (q)_n q^{n(3n+1)/2}}{(q)_{L+n}} \left(\frac{1}{1-zq^n} - \frac{1}{z-q^n}\right)$$
$$= -1 + \frac{1}{(q)_L}\left(1 + \sum_{n=1}^{L} \frac{(-1)^n (q)_L^2 q^{\frac{n(3n+1)}{2}}}{(q)_{L+n}(q)_{L-n}} \left(\frac{1-z}{1-zq^n} + \frac{1-z^{-1}}{1-z^{-1}q^n}\right)\right).$$

Splitting the summation in the right-hand side above, we get

$$1 + R_{S_1}(z;q) = \frac{1}{(q)_L}\left(1 + \sum_{n=1}^{L} \frac{(-1)^n (q)_L^2 q^{\frac{n(3n+1)}{2}}}{(q)_{L+n}(q)_{L-n}} \left(\frac{1-z}{1-zq^n}\right)\right.$$
$$\left. + \sum_{n=1}^{L} \frac{(-1)^n (q)_L^2 q^{\frac{n(3n+1)}{2}}}{(q)_{L+n}(q)_{L-n}} \left(\frac{1-z^{-1}}{1-z^{-1}q^n}\right)\right). \tag{4.2}$$

Making a change of variable from n to $-n$ in the rightmost summation in (4.2), we arrive at

$$1 + R_{S_1}(z;q) = \frac{1}{(q)_L} \sum_{n=-L}^{L} \frac{(-1)^n (q)_L^2 q^{\frac{n(3n+1)}{2}}}{(q)_{L+n}(q)_{L-n}} \left(\frac{1-z}{1-zq^n}\right).$$

We now take the derivatives of $1+R_{S_1}(z;q)$ with respect to z. Firstly, we obtain

$$R'_{S_1}(z;q)=\frac{-1}{(q)_L}\sum_{\substack{n=-L\\n\neq 0}}^{L}\frac{(-1)^n(q)_L^2q^{\frac{n(3n+1)}{2}}}{(q)_{L+n}(q)_{L-n}}\left(\frac{1-q^n}{(1-zq^n)^2}\right)$$

and so for $j\geq 1$,

$$R_{S_1}^{(j)}(z;q)=\frac{-j!}{(q)_L}\sum_{\substack{n=-L\\n\neq 0}}^{L}\frac{(-1)^n(q)_L^2q^{\frac{n(3n-1)}{2}+jn}}{(q)_{L+n}(q)_{L-n}}\left(\frac{1-q^n}{(1-zq^n)^{j+1}}\right). \tag{4.3}$$

Putting (4.3) in the right-hand side of (4.1), we have

$$\begin{aligned}\sum_{n=1}^{\infty}\eta_{2\nu,L}(n)q^n&=(q)_L\sum_{j=0}^{\nu-1}\binom{\nu-1}{j}\sum_{\substack{n=-L\\n\neq 0}}^{L}\frac{(-1)^{n-1}q^{\frac{n(3n-1)}{2}+(2\nu-j)n}}{(q)_{L+n}(q)_{L-n}}\left(\frac{1-q^n}{(1-q^n)^{2\nu-j+1}}\right)\\&=(q)_L\sum_{\substack{n=-L\\n\neq 0}}^{L}\frac{(-1)^{n-1}q^{\frac{n(3n-1)}{2}+2\nu n}}{(q)_{L+n}(q)_{L-n}(1-q^n)^{2\nu}}\sum_{j=0}^{\nu-1}\binom{\nu-1}{j}\frac{q^{-nj}}{(1-q^n)^{-j}}.\end{aligned}$$

Now, by an application of binomial theorem to the inner sum, we get

$$\sum_{n=1}^{\infty}\eta_{2\nu,L}(n)q^n=(q)_L\sum_{\substack{n=-L\\n\neq 0}}^{L}\frac{(-1)^{n-1}q^{\frac{n(3n+1)}{2}+\nu n}}{(q)_{L+n}(q)_{L-n}(1-q^n)^{2\nu}}.$$

We split the sum on the right-hand side into two parts, namely, from 1 to L and from $-L$ to -1.

$$\begin{aligned}\sum_{n=1}^{\infty}\eta_{2\nu,L}(n)q^n=(q)_L&\left(\sum_{n=1}^{L}\frac{(-1)^{n-1}q^{\frac{n(3n+1)}{2}+\nu n}}{(q)_{L+n}(q)_{L-n}(1-q^n)^{2\nu}}\right.\\&\left.+\sum_{n=-1}^{-L}\frac{(-1)^{n-1}q^{\frac{n(3n+1)}{2}+\nu n}}{(q)_{L+n}(q)_{L-n}(1-q^n)^{2\nu}}\right).\end{aligned}$$

Replace n by $-n$ in the rightmost sum to get

$$\sum_{n=1}^{\infty}\eta_{2\nu,L}(n)q^n=(q)_L\sum_{n=1}^{L}\frac{(-1)^{n-1}q^{\frac{n(3n-1)}{2}+\nu n}(1+q^n)}{(q)_{L+n}(q)_{L-n}(1-q^n)^{2\nu}},$$

which is nothing but (2.3). □

Proof of Theorem 2.4. We know from (2.6) that

$$\mu_{2\nu,L}(n):=\sum_{m=-n}^{n}\binom{m+\nu-1}{2\nu}M_{S_2}(m,n).$$

It follows, from the definition (3.4) of $C_{S_2}(z;q)$ and by an application of Leibniz's rule, that

$$\sum_{n=1}^{\infty} \mu_{2\nu,L}(n)q^n = \frac{1}{(2\nu)!}\sum_{j=0}^{\nu-1}\binom{2\nu}{j}(\nu-1)(\nu-2)\cdots(\nu-j)C_{S_2}^{(2\nu-j)}(1;q). \tag{4.4}$$

Using (3.5) in (3.4), we get

$$\begin{aligned} C_{S_2}(z;q) &= \frac{1}{(q)_L} + (1-z)\sum_{n=1}^{L}\begin{bmatrix} L \\ n \end{bmatrix}\frac{(-1)^n(q)_n q^{n(n+1)/2}}{(q)_{n+L}}\left(\frac{1}{1-zq^n} - \frac{1}{z-q^n}\right) \\ &= \frac{1}{(q)_L}\left(1 + \sum_{n=1}^{L}\frac{(-1)^n(q)_L^2 q^{\frac{n(n+1)}{2}}}{(q)_{L+n}(q)_{L-n}}\left(\frac{1-z}{1-zq^n} + \frac{1-z^{-1}}{1-z^{-1}q^n}\right)\right). \end{aligned}$$

Making a change of variable as in Theorem 2.2, we finally get

$$C_{S_2}(z;q) = \frac{1}{(q)_L}\sum_{n=-L}^{L}\frac{(-1)^n(q)_L^2 q^{\frac{n(n+1)}{2}}}{(q)_{L+n}(q)_{L-n}}\left(\frac{1-z}{1-zq^n}\right).$$

Hence, for $j \geq 1$, we have

$$C_{S_2}^{(j)}(z;q) = \frac{-j!}{(q)_L}\sum_{\substack{n=-L \\ n\neq 0}}^{L}\frac{(-1)^n(q)_L^2 q^{\frac{n(n-1)}{2}+jn}}{(q)_{L+n}(q)_{L-n}}\left(\frac{1-q^n}{(1-zq^n)^{j+1}}\right).$$

Substituting these derivative expressions in (4.4) and then by an application of binomial theorem, we obtain

$$\sum_{n=1}^{\infty} \mu_{2\nu,L}(n)q^n = (q)_L\sum_{\substack{n=-L \\ n\neq 0}}^{L}\frac{(-1)^{n-1}q^{\frac{n(n+1)}{2}+\nu n}}{(q)_{L+n}(q)_{L-n}(1-q^n)^{2\nu}}.$$

Splitting the sum into the ranges 1 to L and $-L$ to -1 and then making a variable change, we get

$$\sum_{n=1}^{\infty} \mu_{2\nu,L}(n)q^n = (q)_L\sum_{n=1}^{L}\frac{(-1)^{n-1}q^{\frac{n(n-1)}{2}+\nu n}(1+q^n)}{(q)_{L+n}(q)_{L-n}(1-q^n)^{2\nu}}.$$

□

5. Proof of Theorem 2.5 and Conjecture 1.2

Using Bailey's lemma, i.e. Theorem 3.1, we give a result which is essential for the proof of Conjecture 1.2.

Proposition 5.1. *Let $(\alpha_n(a,q),\beta_n(a,q))$ be a Bailey pair with $a=1$ and $\alpha_0=\beta_0=1$. We then have*

$$\sum_{L\geq n_k\geq\cdots\geq n_1\geq 1}\frac{(q)_{n_1}^2 q^{n_1+\cdots+n_k}\beta_{n_1}}{(1-q^{n_k})^2(1-q^{n_{k-1}})^2...(1-q^{n_1})^2}$$
$$=\sum_{L\geq n_k\geq\cdots\geq n_1\geq 1}\frac{q^{n_1+\cdots+n_k}}{(1-q^{n_k})^2(1-q^{n_{k-1}})^2...(1-q^{n_1})^2}$$
$$+\sum_{r=1}^{L}\frac{(q)_L^2}{(q)_{L-r}(q)_{L+r}}\frac{q^{kr}\alpha_r}{(1-q^r)^{2k}}.$$

Proof. Since $(\alpha_n(a,q),\beta_n(a,q))$ form a Bailey pair with $a=1$, we have the relation

$$\beta_n(1,q)=\sum_{r=0}^{n}\frac{\alpha_r(1,q)}{(q)_{n-r}(q)_{n+r}}.$$

By Bailey's Lemma, $(\alpha_n'(a,q),\beta_n'(a,q))$ is also a Bailey pair with parameters $(1,q)$. Hence, by (3.6),

$$\beta_n'(1,q)=\sum_{r=0}^{n}\frac{\alpha_r'(1,q)}{(q)_{n-r}(q)_{n+r}}.$$

Substituting the values of $\alpha_n'(a,q)$ and $\beta_n'(a,q)$ from Theorem 3.1, we get

$$\sum_{k=0}^{n}\frac{(\rho_1)_k(\rho_2)_k(q/\rho_1\rho_2)_{n-k}}{(q/\rho_1)_n(q/\rho_2)_n(q)_{n-k}}\left(\frac{q}{\rho_1\rho_2}\right)^k\beta_k(1,q)$$
$$=\sum_{k=0}^{n}\frac{(\rho_1)_k(\rho_2)_k}{(q/\rho_1)_k(q/\rho_2)_k(q)_{n-k}(q)_{n+k}}\left(\frac{q}{\rho_1\rho_2}\right)^k\alpha_k(1,q).$$

Separating the terms corresponding to $k=0$ in both the summations and multiplying throughout by $(q/\rho_1)_n(q/\rho_2)_n$,

$$\sum_{k=1}^{n}\frac{(\rho_1)_k(\rho_2)_k(q/\rho_1\rho_2)_{n-k}(q/\rho_1\rho_2)^k}{(q)_{n-k}}\beta_k(1,q)$$
$$=\frac{(q/\rho_1)_n(q/\rho_2)_n}{(q)_n^2}\left(1-\frac{(q)_n(q/\rho_1\rho_2)_n}{(q/\rho_1)_n(q/\rho_2)_n}\right)$$
$$+(q/\rho_1)_n(q/\rho_2)_n\sum_{k=1}^{n}\frac{(\rho_1)_k(\rho_2)_k(q/\rho_1\rho_2)^k}{(q/\rho_1)_k(q/\rho_2)_k(q)_{n-k}(q)_{n+k}}\alpha_k(1,q).$$

Dividing both sides by $(1-\rho_1)(1-\rho_2)$, then letting $\rho_1\to 1$, $\rho_2\to 1$ and using (3.7), we get

$$\sum_{k=1}^{n}(q)_{k-1}^2q^k\beta_k=\sum_{k=1}^{n}\frac{q^k}{(1-q^k)^2}+\sum_{k=1}^{n}\frac{(q)_n^2q^k\alpha_k}{(q)_{n-k}(q)_{n+k}(1-q^k)^2}.$$

This is the $k = 1$ case of the theorem. We are going to prove the theorem using induction. To this end, suppose that the theorem holds for $k = \ell - 1$. This means that

$$\sum_{L\geq n_\ell\geq\cdots\geq n_2\geq 1}\frac{(q)_{n_2}^2 q^{n_2+\cdots+n_\ell}\beta_{n_2}}{(1-q^{n_2})^2...(1-q^{n_\ell})^2}=\sum_{L\geq n_\ell\geq\cdots\geq n_2\geq 1}\frac{q^{n_2+\cdots+n_\ell}}{(1-q^{n_2})^2...(1-q^{n_\ell})^2}$$
$$+\sum_{r=1}^{L}\frac{(q)_L^2}{(q)_{L-r}(q)_{L+r}}\frac{q^{(\ell-1)r}\alpha_r}{(1-q^r)^{2(\ell-1)}}. \quad (5.1)$$

This equation is true for any Bailey pair $(\alpha_n(a,q),\beta_n(a,q))$ with $a = 1$ and $\alpha_0 = \beta_0 = 1$. Note that, since $\alpha'_0 = \alpha_0 = 1$ and $\beta'_0 = \beta_0 = 1$, (5.1) also holds for the Bailey pair $(\alpha'_n(a,q),\beta'_n(a,q))$. So, we replace (α_n,β_n) by (α'_n,β'_n) in (5.1) to get

$$\sum_{L\geq n_\ell\geq\cdots\geq n_2\geq 1}\frac{(q)_{n_2}^2 q^{n_2+\cdots+n_\ell}\beta'_{n_2}}{(1-q^{n_2})^2...(1-q^{n_\ell})^2}=\sum_{L\geq n_\ell\geq\cdots\geq n_2\geq 1}\frac{q^{n_2+\cdots+n_\ell}}{(1-q^{n_2})^2...(1-q^{n_\ell})^2}$$
$$+\sum_{r=1}^{L}\frac{(q)_L^2}{(q)_{L-r}(q)_{L+r}}\frac{q^{(\ell-1)r}\alpha'_r}{(1-q^r)^{2(\ell-1)}}.$$

We now substitute for α'_n and β'_n in terms of α_n and β_n using Bailey's Lemma

$$\sum_{\substack{L\geq n_\ell\geq\cdots\geq n_2\geq 1,\\ n_2\geq n_1\geq 0}}\frac{(q)_{n_2}^2 q^{n_2+\cdots+n_\ell}}{(1-q^{n_2})^2...(1-q^{n_\ell})^2}\frac{(\rho_1)_{n_1}(\rho_2)_{n_1}(q/\rho_1\rho_2)_{n_2-n_1}(q/\rho_1\rho_2)^{n_1}\beta_{n_1}}{(q/\rho_1)_{n_2}(q/\rho_2)_{n_2}(q)_{n_2-n_1}}$$
$$=\sum_{L\geq n_\ell\geq\cdots\geq n_2\geq 1}\frac{q^{n_2+\cdots+n_\ell}}{(1-q^{n_2})^2...(1-q^{n_\ell})^2}$$
$$+\sum_{r=1}^{L}\frac{(q)_L^2}{(q)_{L-r}(q)_{L+r}}\frac{q^{(\ell-1)r}(\rho_1)_r(\rho_2)_r(q/\rho_1\rho_2)^r\alpha_r}{(1-q^r)^{2(\ell-1)}(q/\rho_1)_r(q/\rho_2)_r}.$$

Again, separating the terms corresponding to $n_1 = 0$ from the sum on the left-hand side, then dividing both sides by $(1-\rho_1)(1-\rho_2)$, letting $\rho_1 \to 1$, $\rho_2 \to 1$ and using (3.7), we obtain

$$\sum_{L\geq n_\ell\geq\cdots\geq n_1\geq 1}\frac{(q)_{n_1}^2 q^{n_1+n_2+\cdots+n_\ell}\beta_{n_1}}{(1-q^{n_1})^2(1-q^{n_2})^2...(1-q^{n_\ell})^2}$$
$$=\sum_{L\geq n_\ell\geq\cdots\geq n_2\geq n_1\geq 1}\frac{q^{n_1+n_2+\cdots+n_\ell}}{(1-q^{n_1})^2(1-q^{n_2})^2...(1-q^{n_\ell})^2}$$
$$+\sum_{r=1}^{L}\frac{(q)_L^2 q^{\ell r}\alpha_r}{(q)_{L-r}(q)_{L+r}(1-q^r)^{2\ell}}.$$

This concludes the proof of the theorem by induction. □

Corollary 5.2.

$$(q)_L^2 \sum_{r=1}^{L} \frac{(-1)^{r-1} q^{\frac{r(r-1)}{2}+kr}(1+q^r)}{(q)_{L-r}(q)_{L+r}(1-q^r)^{2k}} = \sum_{L \ge n_k \ge \cdots \ge n_1 \ge 1} \frac{q^{n_1+n_2+\cdots+n_k}}{(1-q^{n_1})^2(1-q^{n_2})^2 \ldots (1-q^{n_k})^2}.$$

Proof. Consider the well-known Bailey pair below [1, pp. 27–28],

$$\alpha_n = \begin{cases} 1 & \text{if } n = 0, \\ (-1)^n q^{\frac{n(n-1)}{2}}(1+q^n) & \text{if } n \ge 1, \end{cases}$$

and

$$\beta_n = \begin{cases} 1 & \text{if } n = 0, \\ 0 & \text{if } n \ge 1. \end{cases}$$

Substituting the above Bailey pair in Proposition 5.1, we get

$$0 = \sum_{L \ge n_k \ge \cdots \ge n_1 \ge 1} \frac{q^{n_1+n_2+\cdots+n_k}}{(1-q^{n_1})^2(1-q^{n_2})^2 \ldots (1-q^{n_k})^2} + (q)_L^2 \sum_{r=1}^{L} \frac{(-1)^r q^{\frac{r(r-1)}{2}+kr}(1+q^r)}{(q)_{L-r}(q)_{L+r}(1-q^r)^{2k}}.$$

□

Corollary 5.3.

$$\sum_{n=1}^{\infty} \mu_{2k,L}(n) q^n = \frac{1}{(q)_L} \sum_{L \ge n_k \ge \cdots \ge n_1 \ge 1} \frac{q^{n_1+n_2+\cdots+n_k}}{(1-q^{n_1})^2(1-q^{n_2})^2 \ldots (1-q^{n_k})^2}. \tag{5.2}$$

Proof. Using Eq. (2.7) from Theorem 2.4 along with Corollary 5.2, we get this result. Note that this is a finite analogue of (1.7). □

Corollary 5.4.

$$\sum_{L \ge n_k \ge \cdots \ge n_1 \ge 1} \frac{(q)_{n_1} q^{n_1+n_2+\cdots+n_k}}{(1-q^{n_1})^2(1-q^{n_2})^2 \ldots (1-q^{n_k})^2} = \sum_{L \ge n_k \ge \cdots \ge n_1 \ge 1} \frac{q^{n_1+n_2+\cdots+n_k}}{(1-q^{n_1})^2(1-q^{n_2})^2 \ldots (1-q^{n_k})^2} + (q)_L^2 \sum_{r=1}^{L} \frac{(-1)^r q^{\frac{r(3r-1)}{2}+kr}(1+q^r)}{(q)_{L-r}(q)_{L+r}(1-q^r)^{2k}}.$$

Proof. Again we use a well-known Bailey pair [1, p. 28],

$$\alpha_n = \begin{cases} 1 & \text{if } n = 0, \\ (-1)^n q^{\frac{n(3n-1)}{2}}(1+q^n) & \text{if } n \ge 1, \end{cases} \quad \text{and} \quad \beta_n = \frac{1}{(q)_n}.$$

Putting the values of α_n and β_n in Theorem 5.1, we get the result. □

Now, we are ready to give a proof of Theorem 2.5.

Proof of Theorem 2.5. Divide both sides of (5.4) by $(q)_L$ to get

$$\frac{1}{(q)_L}\sum_{L\geq n_k\geq\cdots\geq n_1\geq 1}\frac{(q)_{n_1}q^{n_1+n_2+\cdots+n_k}}{(1-q^{n_1})^2(1-q^{n_2})^2...(1-q^{n_k})^2}$$

$$=\frac{1}{(q)_L}\sum_{L\geq n_k\geq\cdots\geq n_1\geq 1}\frac{q^{n_1+n_2+\cdots+n_k}}{(1-q^{n_1})^2(1-q^{n_2})^2...(1-q^{n_k})^2}$$

$$+(q)_L\sum_{r=1}^{L}\frac{(-1)^r q^{\frac{r(3r-1)}{2}+kr}(1+q^r)}{(q)_{L-r}(q)_{L+r}(1-q^r)^{2k}}.$$

Using (5.2) and (2.3), we get the desired result. □

Before going to the proof of Conjecture 1.2, we require one more concept, an analogue of Stirling numbers of the second kind, defined by Garvan [21]. He defined a sequence of polynomials

$$g_k(x)=\prod_{j=0}^{k-1}(x^2-j^2),\quad \text{for } k\geq 1$$

and a sequence of numbers $S^*(n,k)$ such that, for $n\geq 1$,

$$x^{2n}=\sum_{k=1}^{n}S^*(n,k)g_k(x). \tag{5.3}$$

Definition ([21, p. 249]). Define the sequence $S^*(n,k)$, for $1\leq k\leq n$, recursively by

(i) $S^*(1,1)=1$,
(ii) $S^*(n,k)=0$ if $k\leq 0$ or $k>n$,
(iii) $S^*(n+1,k)=S^*(n,k-1)+k^2S^*(n,k)$, for $1\leq k\leq n+1$.

From this definition, Garvan showed that the relation (5.3) indeed holds [21, Lemma 4.2]. Next, we link the finite analogues of rank and crank moments with their symmetrized counterparts via the numbers $S^*(n,k)$.

Proposition 5.5. *For any two positive integers k and L,*

$$\mu_{2k,L}(n)=\frac{1}{(2k)!}\sum_{m=-n}^{n}g_k(m)M_{S_2}(m,n), \tag{5.4}$$

$$\eta_{2k,L}(n)=\frac{1}{(2k)!}\sum_{m=-n}^{n}g_k(m)N_{S_1}(m,n), \tag{5.5}$$

$$M_{2k,L}(n) = \sum_{j=1}^{k}(2j)!S^*(k,j)\ \mu_{2j,L}(n), \tag{5.6}$$

$$N_{2k,L}(n) = \sum_{j=1}^{k}(2j)!S^*(k,j)\ \eta_{2j,L}(n). \tag{5.7}$$

Proof. By the definition of finite analogue of kth symmetrized crank moment, we know that

$$\begin{aligned}\mu_{2k,L}(n) &= \sum_{m=-n}^{n} \binom{m+k-1}{2k} M_{S_2}(m,n)\\ &= \frac{1}{(2k)!}\sum_{m=-n}^{n}(m^2-(k-1)^2)(m^2-(k-2)^2)\\ &\quad \times \cdots (m^2-1^2)m(m-k)M_{S_2}(m,n).\end{aligned}$$

By the definition of the polynomials g_k, this may be written as

$$\begin{aligned}\mu_{2k,L}(n) = \frac{1}{(2k)!}\sum_{m=-n}^{n} g_k(m)M_{S_2}(m,n)\\ - \frac{k}{(2k)!}\sum_{m=-n}^{n}(m^2-(k-1)^2)\cdots(m^2-1^2)mM_{S_2}(m,n).\end{aligned}$$

Since $M_{S_2}(m,n) = M_{S_2}(-m,n)$ [14, p. 9], the rightmost sum vanishes and we get (5.4). Similarly one can prove (5.5). Now, for the proof of (5.6), we start with the definition of $M_{2k,L}(n)$ (1.10), namely,

$$M_{2k,L}(n) = \sum_{m=-n}^{n} m^{2k}M_{S_2}(m,n).$$

We use (5.3) to substitute for m^{2k} and obtain

$$\begin{aligned}M_{2k,L}(n) &= \sum_{m=-n}^{n}\left(\sum_{j=1}^{k} S^*(k,j)g_j(m)\right)M_{S_2}(m,n)\\ &= \sum_{j=1}^{k} S^*(k,j)\sum_{m=-n}^{n} g_j(m)M_{S_2}(m,n)\\ &= \sum_{j=1}^{k}(2j)!S^*(k,j)\mu_{2j,L}(n),\end{aligned}$$

the last step following from (5.4). This completes the proof of (5.6) and on similar lines we can prove (5.7). □

We are now ready to prove the inequality for the finite analogues of rank and crank moments.

Proof of Conjecture 1.2. From (5.6) and (5.7), we get

$$M_{2k,L}(n) - N_{2k,L}(n) = \sum_{j=1}^{k}(2j)!S^*(k,j)\,(\mu_{2j,L}(n) - \eta_{2j,L}(n)). \tag{5.8}$$

From Theorem 2.5, we know

$$\sum_{n=1}^{\infty}(\mu_{2t,L}(n)q^n - \eta_{2t,L}(n))q^n$$

$$= \sum_{L\geq n_t\geq\cdots\geq n_1\geq 1} \frac{q^{n_1+n_2+\cdots+n_t}}{(1-q^{n_1})^2(1-q^{n_2})^2...(1-q^{n_t})^2(q^{n_1+1})_{L-n_1}}.$$

From the generating function, we infer that $\mu_{2t,L}(n) - \eta_{2t,L}(n) \geq 0$ for $n, t, L \geq 1$. Moreover, the numbers $S^*(k,j)$ are all positive, so from (5.8), we can write

$$M_{2k,L}(n) - N_{2k,L}(n) \geq 2(\mu_{2,L}(n) - \eta_{2,L}(n)) = 2\mathrm{spt}(n,L) > 0,$$

where the last equality follows from Remark 2.6. This finishes the proof of the conjecture. □

6. Concluding Remarks

In Remark 2.6, we defined a finite analogue of higher order spt-function as $\mathrm{spt}_k(n,L) := \mu_{2k,L}(n) - \eta_{2k,L}(n)$. A combinatorial interpretation of the higher order spt-function $\mathrm{spt}_k(n)$ was described by Garvan [21, p. 252]. Looking at the generating function (2.9) of the difference between finite analogues of the symmetrized moments and comparing it with (1.8), one can give a combinatorial interpretation of $\mathrm{spt}_k(n,L)$ on similar lines as that of Garvan's for $\mathrm{spt}_k(n)$, the only restriction being that the largest parts of the corresponding partitions are less than or equal to L.

Bringmann *et al.* [13] showed that, for any $k \geq 1$, as $n \to \infty$,

$$M_{2k}(n) \sim N_{2k}(n) \sim \alpha_{2k} n^k p(n),$$
$$M_{2k}(n) - N_{2k}(n) \sim \beta_{2k} n^{k-\frac{1}{2}} p(n),$$

where α_{2k}, β_{2k} are certain explicitly computable constants (see [13, p. 665, Corollary 1.4]). Since in this paper, we have proved the inequality for the finite analogues of rank and crank moments, it would be fascinating to find the asymptotic behavior of the finite analogues and their difference.

Given any prime $p > 3$ and for fixed positive integers k and j, Bringmann *et al.* [11, Corollary 1.3] established that there are infinitely many arithmetic progressions $An + B$ such that $\eta_{2k}(An + B) \equiv 0 \pmod{p^j}$. It would be worthwhile to see if such congruences exist for $\eta_{2k,L}(n)$.

A number of explicit congruences for higher order spt-functions were proved by Garvan [21, Theorems 6.1–6.3]. It would also be interesting to see if there exists a refinement of these congruences for $\mathrm{spt}_k(n, L)$.

Acknowledgments

The authors sincerely thank the referee for the careful reading of the paper and for giving important suggestions. We would also like to thank Prof. Atul Dixit for going through the manuscript and giving valuable suggestions. The first author wishes to thank Harish-Chandra Research Institute and IIT Gandhinagar for the conducive environment. The second author was a SERB National Post Doctoral Fellow (NPDF) supported by the fellowship PDF/2017/000370. The third author was supported partially by the IIT Gandhinagar and by SERB ECR Grant ECR/2015/000070 of Prof. Atul Dixit.

References

[1] G. E. Andrews, *q-Series: Their Development and Application in Analysis, Number Theory, Combinatorics, Physics, and Computer Algebra*, CBMS Regional Conference Series in Mathematics, Vol. 66 (American Mathematical Society, Providence, RI, 1986).

[2] G. E. Andrews, *Ramanujan and Partial Fractions*, Contributions to the History of Indian Mathematics (Hindustan Book Agency, New Delhi, 2005).

[3] G. E. Andrews, Partitions, Durfee symbols, and the Atkin–Garvan moments of ranks, *Invent. Math.* **169** (2007) 37–73.

[4] G. E. Andrews, The number of smallest parts in the partitions on n, *J. Reine Angew. Math.* **624** (2008) 133–142.

[5] G. E. Andrews, R. Askey and R. Roy, *Special Functions*, Encyclopedia of Mathematics and its Applications, Vol. 71 (Cambridge University Press, Cambridge, 1999).

[6] G. E. Andrews and B. C. Berndt, *Ramanujan's Lost Notebook, Part I* (Springer, New York, 2005).

[7] G. E. Andrews and F. G. Garvan, Dyson's crank of a partition, *Bull. Amer. Math. Soc.* (*N.S.*) **18** (1988) 167–171.

[8] A. O. L. Atkin and F. Garvan, Relations between the ranks and cranks of partitions, *Ramanujan J.* **7** (2003) 343–366.

[9] A. O. L. Atkin and H. P. F. Swinnerton-Dyer, Some properties of partitions, *Proc. London Math. Soc. Ser. 3* **4** (1954) 84–106.

[10] B. C. Berndt, *Ramanujan's Notebooks, Part IV* (Springer-Verlag, New York, 1991).

[11] K. Bringmann, F. Garvan and K. Mahlburg, Partition statistics and quasiharmonic Maass forms, *Int. Math. Res. Not.* **2009**(1) (2009) 63–97.

[12] K. Bringmann, K. Mahlburg, Inequalities between ranks and cranks, *Proc. Amer. Math. Soc.* **137** (2009) 2567–2574.

[13] K. Bringmann, K. Mahlburg and R. C. Rhoades, Asymptotics for rank and crank moments, *Bull. London Math. Soc.* **43** (2011) 661–672.

[14] A. Dixit, P. Eyyunni, B. Maji and G. Sood, Untrodden pathways in the theory of the restricted partition function $p(n,N)$, preprint (2018), arXiv:1812.01424.

[15] A. Dixit and B. Maji, Partition implications of a three-parameter q-series identity, *Ramanujan J.* **52** (2020) 323–358.

[16] F. J. Dyson, Some guesses in the theory of partitions, *Eureka* (*Cambridge*) **8** (1944) 10–15.
[17] F. J. Dyson, Mappings and symmetries of partitions, *J. Combin. Theory Ser. A* **51** (1989) 169–180.
[18] N. J. Fine, *Basic Hypergeometric Series and Applications*, Mathematical Surveys and Monographs (American Mathematical Society, 1989).
[19] F. G. Garvan, New combinatorial interpretations of Ramanujan's partition congruences mod 5, 7 and 11, *Trans. Amer. Math. Soc.* **305** (1988) 47–77.
[20] F. G. Garvan, Congruences for Andrews' smallest parts partition function and new congruence for Dyson's rank, *Int. J. Number Theory* **6** (2010) 281–309.
[21] F. G. Garvan, Higher order spt-function, Adv. Math. **228** (2011) 241–265.
[22] S. Ramanujan, *Notebooks of Srinivasa Ramanujan, Vol.* II (Tata Institute of Fundamental Research, Mumbai, 2012).

© 2024 World Scientific Publishing Company
https://doi.org/10.1142/9789811277375_0014

Asymptotic expansions, partial theta functions, and radial limit differences of mock modular and modular forms*

Amanda Folsom

Department of Mathematics and Statistics
Amherst College, Amherst, MA 01002, USA
afolsom@amherst.edu

Dedicated to Bruce Berndt, in honor of his 80*th birthday.*

In 1920, Ramanujan studied the asymptotic differences between his mock theta functions and modular theta functions, as q tends towards roots of unity singularities radially from within the unit disk. In 2013, the bounded asymptotic differences predicted by Ramanujan with respect to his mock theta function $f(q)$ were established by Ono, Rhoades, and the author, as a special case of a more general result, in which they were realized as special values of a quantum modular form. Our results here are threefold: we realize these radial limit differences as special values of a partial theta function, provide full asymptotic expansions for the partial theta function as q tends towards roots of unity radially, and explicitly evaluate the partial theta function at roots of unity as simple finite sums of roots of unity.

Keywords: Mock theta functions; mock modular forms; Ramanujan; partial theta functions; asymptotic expansions; radial limits.

Mathematics Subject Classification 2020: 11F37, 11F99, 33D15, 33D70

1. Introduction and Statement of Results

Ramanujan's last letter to Hardy, written in 1920 [2, 22], explores the asymptotic properties of his mock theta functions. Specifically, in the letter, Ramanujan states without proof that as q approaches an even order 2κ root of unity radially from within the unit disk, that his mock theta function

$$f(q) := \sum_{n=0}^{\infty} \frac{q^{n^2}}{(-q;q)_n^2}$$

(defined using the q-Pochhammer symbol $(a;q)_n = \prod_{j=0}^{n-1}(1-aq^j)$ $(n \in \mathbb{N} \cup \{\infty\})$) satisfies

$$f(q) - (-1)^{\kappa}(1-q)(1-q^3)(1-q^5)\cdots(1-2q+2q^4-\cdots) = O(1). \tag{1.1}$$

*This article originally appeared in *International Journal of Number Theory*, **17**(2), pp. 425–434 © 2021 World Scientific Publishing Company. DOI: 10.1142/S1793042120400126

It is not difficult to see that the even order 2κ roots of unity are singularities of the mock theta function $f(q)$, and that the function $b(q)$ appearing in (1.1), defined by $b(q) := (1-q)(1-q^3)(1-q^5)\cdots(1-2q+2q^4-\cdots)$, is a modular form (up to multiplication by a suitable power of q and under the usual modular change of variable $q = e^{2\pi i\tau}, \tau \in \mathbb{H} := \{z \in \mathbb{C} \,|\, \mathrm{Im}(z) > 0\}$). Thus, Ramanujan's claim in (1.1) may be interpreted as stating that the modular forms $\pm b(q)$ carve out the exponential singularities of the mock theta function $f(q)$.

Decades later, the truth of (1.1) was proved, first in [15] as a special case of a more general result, in which the implied $O(1)$ constants are realized as special values of quantum modular forms, and the proof of which uses the more recent theory of mock modular forms. (See [23; 4, Chap. 21] for more on quantum modular forms, and [4, Part 2] for more on mock modular forms.) This more general result, namely, [15, Theorem 1.2], which proves (1.1) as a special case, replaces $f(q)$ (respectively, $b(q)$) by the more general mock modular (respectively, modular) partition rank (respectively, crank) generating function $R(w;q)$ (respectively, $C(w;q)$). Note that $f(q) = R(-1;q)$ (respectively, $b(q) = C(-1;q)$). The functions R and C are given by

$$R(w;q) = \sum_{n=0}^{\infty} \frac{q^{n^2}}{(wq;q)_n(w^{-1}q;q)_n}, \quad C(w;q) = \frac{(q;q)_\infty}{(wq;q)_\infty(w^{-1}q;q)_\infty}.$$

With the w-variable fixed to be suitable roots of unity ζ_b^a (with $\zeta_N := e^{2\pi i/N}$), the $O(1)$ constants implied in the generalization of (1.1) in [15, Theorem 1.2] as q radially tends towards roots of unity ζ_k^h are realized (up to an explicit constant multiple) as specializations (at $(w;q) = (\zeta_b^a;\zeta_k^h)$) of the quantum modular strongly unimodal sequence rank generating function $U(w;q)$,[a] given by

$$U(w;q) = \sum_{n=0}^{\infty} (wq;q)_n(w^{-1}q;q)_n q^{n+1}.$$

Precisely, we have the following theorem.

Theorem 1.1 ([15, Theorem 1.2]). *Let $1 \le a < b$ and $1 \le h < k$ be integers with $\gcd(a,b) = \gcd(h,k) = 1$ and $b \,|\, k$. If $h' \in \mathbb{Z}$ satisfies $hh' \equiv -1 \pmod{k}$ then we have that*

$$\lim_{t\to 0^+} \left(R(\zeta_b^a;\zeta_k^h e^{-2\pi t}) - \zeta_{b^2}^{-a^2h'k} C(\zeta_b^a;\zeta_k^h e^{-2\pi t})\right) = -(1-\zeta_b^a)(1-\zeta_b^{-a})U(\zeta_b^a;\zeta_k^h).$$

Ramanujan's claim (1.1) (with an added explicit implied $O(1)$ constant) is deduced from [15, Theorem 1.2] (stated above) by setting $(a,b) = (1,2)$. See also [8, 10, 12–14, 17, 20, 24] for more recent related work.

[a]We caution the reader that the function U is defined with slightly different normalizations (using the same notation for the function) in different sources. Here, we have used the definition from [15].

Remark 1.2. The above theorem is stated slightly differently than in [15] but is equivalent: namely, we have replaced the variable q by $\zeta_k^h e^{-2\pi t}$, and have replaced the limit as $q \to \zeta_k^h$ radially from within the unit disk by the limit as $t \to 0^+$.

As alluded to above, the functions R, C, and U are now well known to possess different types of modular properties, namely, they are mock modular, modular, and quantum modular, respectively, when appropriately normalized and specialized in q and w. When expanded as two-variable series in q and w, these functions are also well-known to be combinatorial generating functions for partition ranks, partition cranks, and ranks of strongly unimodal sequences, respectively. (See, for example, [4, 15] for more.)

At the June 2019 conference "Analytic and Combinatorial Number Theory: The Legacy of Ramanujan" in honor of Bruce Berndt at the University of Illinois Urbana-Champaign, Peter Paule (RISC, Johannes Kepler University, Linz, Austria) asked the author after her lecture on this subject about the existence of full asymptotic expansions for implied constant values in the $O(1)$ expressions in (1.1) as q tends to roots of unity radially within the unit disk [19]. One of the main results of this paper is to establish these asymptotic expansions; in fact, we do so as a special case of a more general result given in Theorem 1.3, which builds from and extends results in [15].

Our Theorem 1.3 is stated in terms of the two-variable partial theta function ϑ, (with $q = e^{2\pi i\tau}$ and $w = e^{2\pi iz}$), defined by

$$\widetilde{\vartheta}(z;\tau) := \sum_{n=0}^{\infty} \chi_{12}(n) q^{\frac{n^2}{24}} w^{\frac{n}{2}},$$

where $\chi_{12}(n) := \left(\frac{12}{n}\right)$ is defined by the Kronecker symbol. Although the case $z=0$ (which corollaryresponds to $w = 1$) is irrelevant in Theorem 1.3, we remark for context that $\widetilde{\vartheta}(0;\tau) = \eta(24\tau)$, the weight 1/2 modular η-function. For more general z, the function $\widetilde{\vartheta}$ may be described as a (two-variable) partial, or false, theta function, and is related to certain holomorphic Eichler integrals, the latter of which have been of historical importance in the theory of modular forms [11, 21]. More recently, such functions have been shown to play important roles in the theory of quantum modular forms, mathematical physics, representation theory, and the intersections of these areas (for example, see [3–7, 9, 15, 23]).

In the three parts of Theorem 1.3 and Corollary 1.4, we realize the radial limit differences studied by Ramanujan and later, more generally, in [15], as special values of the partial theta function $\widetilde{\vartheta}$ defined above; we provide full asymptotic expansions for the partial theta function $\widetilde{\vartheta}$ as q tends towards roots of unity radially; and we explicitly evaluate the partial theta function $\widetilde{\vartheta}$ at roots of unity as simple finite sums of roots of unity, thereby producing new simple expressions for the radial limit differences in question.

Throughout, $B_k(x)$ denotes the kth Bernoulli polynomial.

Theorem 1.3. *Let $1 \le a < b$ and $1 \le h < k$ be integers with $\gcd(a,b) = \gcd(h,k) = 1$ and $b \,|\, k$, so that $bb' = k$ for some integer b'. Let $h' \in \mathbb{Z}$ satisfy $hh' \equiv -1 \pmod{k}$. The following are true:*

(i) *We have that*

$$\lim_{t\to 0^+} \left(R(\zeta_b^a; \zeta_k^h e^{-2\pi t}) - \zeta_{b^2}^{-a^2h'k} C(\zeta_b^a; \zeta_k^h e^{-2\pi t})\right)$$
$$= -2i\zeta_{24k}^h \sin\left(\frac{\pi a}{b}\right) \widetilde{\vartheta}\left(\frac{a}{b}; -\frac{h}{k}\right). \tag{1.2}$$

(ii) *The partial theta function $\widetilde{\vartheta}$ appearing on the right-hand side of (1.2) has the asymptotic expansion as $t \to 0^+$*

$$\widetilde{\vartheta}\left(\frac{a}{b}; -\frac{h}{k} + \frac{12it}{\pi}\right) \sim \sum_{r=0}^{\infty} L(-2r, d)\frac{(-t)^r}{r!},$$

where the L-values are given by

$$L(-r,d) = -\frac{(12k)^r}{r+1}\sum_{n=1}^{12k} d(n) B_{r+1}\left(\frac{n}{12k}\right), \quad (r = 0, 1, 2, \ldots)$$

with

$$d(n) = d_{a,b,h,k}(n) := \chi_{12}(n)\zeta_{24k}^{-hn^2}\zeta_{2b}^{an}.$$

(iii) *In particular, the value $\widetilde{\vartheta}(\frac{a}{b}; -\frac{h}{k})$ appearing on the right-hand side of (1.2) may be explicitly computed as*

$$\widetilde{\vartheta}\left(\frac{a}{b}; -\frac{h}{k}\right) = -\frac{1}{12k}\sum_{n=1}^{12k} nd(n),$$

with $d(n)$ as in part (ii). *Combining this with part* (i), *we have that*

$$\lim_{t\to 0^+} \left(R(\zeta_b^a; \zeta_k^h e^{-2\pi t}) - \zeta_{b^2}^{-a^2h'k} C(\zeta_b^a; \zeta_k^h e^{-2\pi t})\right)$$
$$= (6k)^{-1} i\zeta_{24k}^h \sin\left(\frac{\pi a}{b}\right)\sum_{n=1}^{12k} nd(n).$$

From Theorem 1.3 with $(a, b) = (1, 2)$ and $k = 2\kappa$, we obtain results directly pertaining to Ramanujan's original radial limit (1.1) and his mock theta function $f(q)$.

Corollary 1.4. *Let $\kappa \in \mathbb{N}$. The following are true:*

(i) *We have that*

$$\lim_{t\to 0^+} \left(f(\zeta_{2\kappa}^h e^{-2\pi t}) - (-1)^\kappa b(\zeta_{2\kappa}^h e^{-2\pi t})\right) = -2i\zeta_{48\kappa}^h \widetilde{\vartheta}\left(\frac{1}{2}; -\frac{h}{2\kappa}\right).$$

(ii) *The partial theta function* $\widetilde{\vartheta}$ *appearing on the right-hand side of* (1.2) *has the asymptotic expansion as* $t \to 0^+$

$$\widetilde{\vartheta}\left(\frac{1}{2}; -\frac{h}{2\kappa} + \frac{12it}{\pi}\right) \sim \sum_{r=0}^{\infty} L(-2r, d)\frac{(-t)^r}{r!},$$

where the L*-values are given by*

$$L(-r, d) = -\frac{(24\kappa)^r}{r+1}\sum_{n=1}^{24\kappa} d(n) B_{r+1}\left(\frac{n}{24\kappa}\right), \quad (r = 0, 1, 2, \dots)$$

with

$$d(n) = d_{1,2,h,2\kappa} = d_{h,\kappa}(n) := \chi_{12}(n)\zeta_{48\kappa}^{-hn^2} i^n.$$

(iii) *In particular, the value* $\widetilde{\vartheta}(\frac{1}{2}; -\frac{h}{2\kappa})$ *appearing on the right-hand side of* (1.2) *may be explicitly computed as*

$$\widetilde{\vartheta}\left(\frac{1}{2}; -\frac{h}{2\kappa}\right) = -\frac{1}{24\kappa}\sum_{n=1}^{24\kappa} n d(n),$$

with $d(n)$ *as in part* (ii). *Combining this with part* (i), *we have that*

$$\lim_{t\to 0^+} (f(\zeta_{2\kappa}^h e^{-2\pi t}) - (-1)^\kappa b(\zeta_{2\kappa}^h e^{-2\pi t})) = (12\kappa)^{-1} i\zeta_{48\kappa}^h \sum_{n=1}^{24\kappa} n d(n).$$

Remark 1.5. It is interesting to compare the explicit expressions given on the right-hand sides of the radial limits in Theorem 1.3 (and Corollary 1.4) with those given in [15, Theorem 1.2 (and Theorem 1.1)]. From Theorem 1.3, we have that

$$\begin{aligned} &\lim_{t\to 0^+} (R(\zeta_b^a; \zeta_k^h e^{-2\pi t}) - \zeta_{b^2}^{-a^2h'k} C(\zeta_b^a; \zeta_k^h e^{-2\pi t})) \\ &\quad = (6k)^{-1} i\zeta_{24k}^h \sin\left(\frac{\pi a}{b}\right) \sum_{n=1}^{12k} n\chi_{12}(n)\zeta_{24k}^{-hn^2}\zeta_{2b}^{an}. \end{aligned} \tag{1.3}$$

On the other hand, from [15, Theorem 1.2], we have that

$$\begin{aligned} &\lim_{t\to 0^+} (R(\zeta_b^a; \zeta_k^h e^{-2\pi t}) - \zeta_{b^2}^{-a^2h'k} C(\zeta_b^a; \zeta_k^h e^{-2\pi t})) \\ &\quad = -(1-\zeta_b^a)(1-\zeta_b^{-a}) \sum_{n=0}^{c(a,b,h,k)} (\zeta_b^a\zeta_k^h; \zeta_k^h)_n (\zeta_b^{-a}\zeta_k^h; \zeta_k^h)_n \zeta_k^{h(n+1)}, \end{aligned} \tag{1.4}$$

where $c(a, b, h, k)$ is a non-negative integer (which can be explicitly computed) depending on a, b, h and k.

Without Theorem 1.3 and [15, Theorem 1.2], it is not obvious that the finite sum of roots of unity on the right-hand sides of (1.3) and the finite q-hypergeometric sum at roots of unity on the right-hand side of (1.4) are equal. We state this as an open problem.

Open Problem. Give a direct proof of the fact that the finite sum on the right-hand side of (1.3) equals the finite sum on the right-hand side of (1.4), under the hypotheses given, avoiding theorems such as Theorem 1.3 and [15, Theorem 1.2] (and their proofs and corollaries), and instead using, for example, elements from the theory of q-hypergeometric series, partial theta functions, or other elementary or direct methods.

Example. Let $(a,b) = (1,2)$ and $(h,k) = (5,6)$. Then by part (ii) of Theorem 1.3 (or part (ii) of Corollary 1.4), as $t \to 0^+$,

$$\widetilde{\vartheta}\left(\frac{1}{2}; -\frac{5}{6} + \frac{12it}{\pi}\right) \sim -\sum_{r=0}^{\infty} \frac{(72)^{2r}}{2r+1} \sum_{n=1}^{72} \chi_{12}(n)\zeta_{144}^{-5n^2} i^n B_{2r+1}\left(\frac{n}{72}\right) \frac{(-t)^r}{r!}.$$

By part (iii) of Theorem 1.3 (or part (iii) of Corollary 1.4),

$$\vartheta\left(\frac{1}{2}, -\frac{5}{6}\right) = -\frac{1}{72} \sum_{n=1}^{72} n\chi_{12}(n)\zeta_{144}^{-5n^2} i^n.$$

Combining this with part (i) of Theorem 1.3 (or part (i) of Corollary 1.4), we have that Ramanujan's radial limit satisfies

$$\begin{aligned}\lim_{t\to 0^+} (f(\zeta_6^5 e^{-2\pi t}) - (-1)^\kappa b(\zeta_6^5 e^{-2\pi t})) &= (36)^{-1} i \zeta_{144}^5 \sum_{n=1}^{72} n\chi_{12}(n)\zeta_{144}^{-5n^2} i^n \\ &\approx -2 + 3.4641i.\end{aligned}$$

On the other hand, using [15, Theorem 1.2] (see (1.4)), we have that this same radial limit satisfies

$$\begin{aligned}\lim_{t\to 0^+} (f(\zeta_6^5 e^{-2\pi t}) - (-1)^\kappa b(\zeta_6^5 e^{-2\pi t})) &= -4 \sum_{n=0}^{2} (-\zeta_6^5; \zeta_6^5)_n^2 \zeta_6^{5(n+1)} \\ &\approx -2 + 3.4641i.\end{aligned}$$

2. Proofs

Proof of Theorem 1.3. We begin by establishing (i). From [14, Corollary 1.4], an earlier result of Ki *et al.*, and the author, we see that the right-hand side of claimed radial limit may be expressed as the value $-(1-\zeta_b^a)(1-\zeta_b^{-a})F(\zeta_b^a; \zeta_k^{-h})$, where

$$F(w;q) := \sum_{n=0}^{\infty} w^{n+1}(wq;q)_n.$$

We next invoke a result of Hikami on certain difference equations, namely [16, Theorem 8], which shows after some simplification that $(1-w)w^{-\frac{1}{2}}q^{\frac{1}{24}}F(w;q) = \widetilde{\vartheta}(z;\tau)$, with $w = e^{2\pi i z}$ and $q = e^{2\pi i\tau}$. Replacing $w = \zeta_b^a$ (hence $z = a/b$) and $q = \zeta_k^{-h}$ (hence $\tau = -h/k$), we have that the right-hand side of the claimed radial limit is

$$-\zeta_{24k}^h\zeta_{2b}^a(1-\zeta_b^{-a})\widetilde{\vartheta}\left(\frac{a}{b};-\frac{h}{k}\right) = -2i\zeta_{24k}^h\sin\left(\frac{\pi a}{b}\right)\widetilde{\vartheta}\left(\frac{a}{b};-\frac{h}{k}\right)$$

as claimed.

Now, we establish (ii) Using the definition of $\widetilde{\vartheta}$, for any $t>0$, with hypotheses as given on a,b,h,k, we have that

$$\widetilde{\vartheta}\left(\frac{a}{b};-\frac{h}{k}+\frac{12it}{\pi}\right)=\sum_{n=0}^{\infty}d(n)e^{-n^2t},$$

where the coefficients $d(n)$ are as defined in Theorem 1.3. Towards the proof of (ii), we will show (below) that the coefficients $d(n)$ are periodic with period $12k$ and have mean value zero, in order to apply the following result of Lawrence and Zagier.

Proposition 2.1 ([18, p. 98]). *Let $C:\mathbb{Z}\to\mathbb{C}$ be a periodic function with mean value 0. Then the associated L-series $L(s,C)=\sum_{n=1}^{\infty}C(n)n^{-s}$ $(\mathrm{Re}(s)>1)$ extends holomorphically to all of $\mathbb{C}$ and the function $\sum_{n=1}^{\infty}C(n)e^{-n^2t}$ $(t>0)$ has the asymptotic expansion*

$$\sum_{n=1}^{\infty}C(n)e^{-n^2t}\sim\sum_{r=0}^{\infty}L(-2r,C)\cdot\frac{(-t)^r}{r!}$$

as $t\to 0^+$. The L-values $L(-r,C)$ are given explicitly by

$$L(-r,C)=-\frac{M^r}{r+1}\sum_{n=1}^{M}C(n)B_{r+1}\left(\frac{n}{M}\right)\qquad(r=0,1,\ldots)$$

where M is any period of the function $C(n)$.

First, it is clear by definition, and using the fact that $b\,|\,k$, that the $d(n)$ are periodic mod $12k$. To prove that the $d(n)$ have mean value 0, we begin by using the definition of χ_{12} and re-write

$$\begin{aligned}\sum_{n=0}^{12k-1}d(n)&=\sum_{\pm}\left(\sum_{\substack{n=0\\ n\equiv\pm1\,(\mathrm{mod}\,12)}}^{12k-1}\zeta_{24k}^{-hn^2}\zeta_{2b}^{an}-\sum_{\substack{n=0\\ n\equiv\pm5\,(\mathrm{mod}\,12)}}^{12k-1}\zeta_{24k}^{-hn^2}\zeta_{2b}^{an}\right)\\
&=\sum_{\pm}\left(\sum_{n=\frac{1\mp1}{2}}^{k-\frac{1\pm1}{2}}\zeta_{24k}^{-h(12n\pm1)^2}\zeta_{2b}^{a(12n\pm1)}-\sum_{n=\frac{1\mp1}{2}}^{k-\frac{1\pm1}{2}}\zeta_{24k}^{-h(12n\pm5)^2}\zeta_{2b}^{a(12n\pm5)}\right)\\
&=\sum_{\pm}\left(\sum_{n\,(\mathrm{mod}\,k)}\zeta_{24k}^{-h(12n\pm1)^2}\zeta_{2b}^{a(12n\pm1)}\right.\\
&\qquad\left.-\sum_{n\,(\mathrm{mod}\,k)}\zeta_{24k}^{-h(12n\pm5)^2}\zeta_{2b}^{a(12n\pm5)}\right)\end{aligned}\tag{2.1}$$

where we have also used that the summands $\zeta_{24k}^{h(12n\pm\nu)^2}\zeta_{2b}^{a(12n\pm\nu)}$, where $\nu\in\{1,5\}$, are periodic mod k (again using that $b\,|\,k$).

Case 1. $\gcd(k,6)>1$. In this case, we will show that each of the four sums on n in (2.1) are identically zero. Up to a constant multiple, each of the sums on n, where $\nu\in\{1,5\}$, and b' is such that $bb'=k$, may be re-written as $G(-6h,6ab'\mp h\nu,k)$, where $G(A,B,C)$ denotes the Gauss sum

$$G(A,B,C):=\sum_{n \bmod C}\zeta_C^{An^2+Bn}.$$

It is well known that if $\gcd(A,C)=g>1$ and $g\nmid B$, that $G(A,B,C)=0$ [1]. We will use this fact here, to show that each of the four sums on n are identically zero. Here, we have that $\gcd(A,C)=\gcd(6h,k)=\gcd(6,k)=g>0$ by hypothesis. Suppose $g\,|\,B=6ab'\mp h\nu$. Then since $g\,|\,6$, and $\gcd(g,\nu)=1$, we have that $g\,|\,h$. But $g\,|\,k$ as well, so this is a contradiction, because $\gcd(h,k)=1$ and $g>1$. Hence, $g\nmid 6ab'\mp h\nu$ and $G(6h,6ab'\mp h,k)=0$.

Case 2. $\gcd(k,6)=1$. In this case, we write, using that the sum (at the start of (2.2)) may be taken over any set of representatives mod k,

$$\sum_{n\,(\mathrm{mod}\,k)}\zeta_{24k}^{-h(12n\pm5)^2}\zeta_{2b}^{a(12n\pm5)}=\sum_{n\,(\mathrm{mod}\,k)}\zeta_{24k}^{-h(12(n\mp\alpha)\pm5)^2}\zeta_{2b}^{a(12(n\mp\alpha)\pm5)},\tag{2.2}$$

where α is any integer satisfying $3\alpha\equiv1\pmod{k}$ – a number we know exists in this case since $\gcd(3,k)=1$. After some expanding and simplifying in the exponents of the roots of unity in the summands, and writing $b=k/b'$, we rewrite (2.2) as

$$\zeta_k^{\pm2ab'(1-3\alpha)}\zeta_{3k}^{-2h(1-3\alpha)^2}\sum_{n\,(\mathrm{mod}\,k)}\zeta_{24k}^{-h(12n\pm1)^2}\zeta_{2k}^{ab'(12n\pm1)}\zeta_{3k}^{\mp h(12n\pm1)(1-3\alpha)}$$

$$=\zeta_k^{\pm2ab'(1-3\alpha)}\zeta_k^{-h(3\alpha-1)(2\alpha-1)}\sum_{n\,(\mathrm{mod}\,k)}\zeta_{24k}^{-h(12n\pm1)^2}\zeta_{2k}^{ab'(12n\pm1)}\zeta_k^{\mp4hn(1-3\alpha)}$$

$$=\sum_{n\,(\mathrm{mod}\,k)}\zeta_{24k}^{-h(12n\pm1)^2}\zeta_{2k}^{ab'(12n\pm1)},\tag{2.3}$$

where we have used that $3\alpha\equiv1\pmod{k}$ to obtain the last line in (2.3). Substituting (2.2) for (2.3) (and replacing $b'=k/b$), (2.1) becomes $\sum_{\pm}0=0$.

Having established the periodic and mean value 0 nature of the $d(n)$, we invoke the proposition [18, Proposition, p. 98] stated above, which yields, as $t\to0^+$,

$$\widetilde{\vartheta}\left(\frac{a}{b};-\frac{h}{k}+\frac{12it}{\pi}\right)\sim\sum_{r=0}^{\infty}L(-2r,d)\frac{(-t)^r}{r!},$$

with L-values $L(-r,d)$ as defined in Theorem 1.3, as claimed.

Part (iii) of Theorem 1.3 follows from the asymptotic expansion established in part (ii), by letting $t\to0^+$. To see this, using the definition of $B_1(x)$ and the previously established facts that the $d(n)$ have period $12k$ and mean value 0, we have that that

$$\widetilde{\vartheta}\left(\frac{a}{b};-\frac{h}{k}\right) = -\sum_{n=1}^{12k} d(n)B_1\left(\frac{n}{12k}\right) = -\sum_{n=1}^{12k} d(n)\left(\frac{n}{12k}-\frac{1}{2}\right) = -\frac{1}{12k}\sum_{n=1}^{12k} nd(n)$$

as claimed. □

Acknowledgments

The author is grateful for the support of NSF Grant DMS-1901791. The author thanks Professor Peter Paule for suggesting some of the topics of study in this paper, as well as Professor Bruce Berndt and the University of Illinois for hosting the stimulating conference at which this paper originated. The author also thanks the anonymous referee for helpful suggestions and comments on the initial version of this manuscript.

References

[1] B. C. Berndt and R. J. Evans, *Gauss and Jacobi Sums*, Canadian Mathematical Society Series of Monographs and Advanced Texts (A Wiley-Interscience Publication, John Wiley & Sons, Inc., New York, 1998), pp. xii+583.

[2] B. C. Berndt and R. A. Rankin, *Ramanujan*, History of Mathematics, Vol. 9 (American Mathematical Society, Providence, RI; London Mathematical Society, London, 1995), Letters and commentary.

[3] K. Bringmann, A. Folsom and A. Milas, Asymptotic behavior of partial and false theta functions arising from Jacobi forms and regularized characters, *J. Math. Phys.* **58** (2017) 011702.

[4] K. Bringmann, A. Folsom, K. Ono and L. Rolen, *Harmonic Maass Forms and Mock Modular Forms: Theory and Applications*, Vol. 64 (American Mathematical Society Colloquium Publications, Providence, RI, 2017), p. 391.

[5] K. Bringmann, A. Folsom and R. C. Rhoades, Partial theta functions and mock modular forms as q-hypergeometric series, *Ramanujan J.* **29** (2012) 295–310.

[6] K. Bringmann, J. Kaszian and A. Milas, Higher depth quantum modular forms, multiple Eichler integrals, and $\mathfrak{sl}_3$ false theta functions, *Res. Math. Sci.* **6** (2019) 20.

[7] K. Bringmann and A. Milas, W-algebras, higher rank false theta functions and quantum dimensions, *Selecta Math.* **23** (2017) 1249–1278.

[8] K. Bringmann and L. Rolen, Radial limits of mock theta functions, *Res. Math. Sci.* **2** (2015) 17.

[9] K. Bringmann and L. Rolen, Half-integral weight Eichler integrals and quantum modular forms, *J. Number Theory* **161** (2016) 240–254.

[10] D. Choi, S. Lim and R. C. Rhoades, Mock modular forms and quantum modular forms, *Proc. Amer. Math. Soc.* **144**(6) (2016) 2337–2349.

[11] M. Eichler, Eine Verallgemeinerung der Abelschen Integrale, *Math. Z.* **67** (1957) 267–298.

[12] A. Folsom, Quantum Jacobi forms in number theory, topology, and mathematical physics, *Res. Math. Sci.* **6**(3) (2019) 34.

[13] A. Folsom, Asymptotics and Ramanujans mock theta functions: Then and now, *Philos. Trans. Roy. Soc. A* **378**(2163) (2020) 13.

[14] A. Folsom, C. Ki, Y. N. Truong Vu and B. Yang, Strange combinatorial quantum modular forms, *J. Number Theory* **170** (2017) 315–346.

[15] A. Folsom, K. Ono and R. C. Rhoades, Mock theta functions and quantum modular forms, *Forum Math. Pi* **1** (2013) e2–27.

[16] K. Hikami, Difference equation of the colored Jones polynomial for torus knot, *Internat. J. Math.* **15**(9) (2004) 959–965.

[17] M.-J. Jang and S. Löbrich, Radial limits of the universal mock theta function g_3, *Proc. Amer. Math. Soc.* **145** (2017) 925–935.

[18] R. Lawrence and D. Zagier, Modular forms and quantum invariants of 3-manifolds, *Asian J. Math.* **3** (1999) 93–107.

[19] P. Paule, *Private Communication, Analytic and Combinatorial Number Theory: The Legacy of Ramanujan* (University of Illinois, Urbana-Champaign, 2019).

[20] R. C. Rhoades, On Ramanujan's definition of mock theta functions, *Proc. Natl. Acad. Sci. USA* **110** (2013) 759–7594.

[21] G. Shimura, Sur les integrales attachées aux formes automorphe, *J. Math. Soc. Japan* **11** (1959) 291–311.

[22] G. N. Watson, The final problem: An account of the mock theta functions, *J. London Math. Soc.* **11**(1) (1936) 55–80.

[23] D. Zagier, *Quantum Modular Forms*, *Quanta of Maths*, Clay Mathematics Proceedings, Vol. 11 (American Mathematical Society, Providence, RI, 2010), pp. 659–675.

[24] W. Zudilin, On three theorems of Folsom, Ono and Rhoades, *Proc. Amer. Math. Soc.* **143**(4) (2015) 1471–1476.

© 2024 World Scientific Publishing Company
https://doi.org/10.1142/9789811277375_0015

On conjectures of Koike and Somos for modular identities for the Rogers–Ramanujan functions*

Chadwick Gugg

Department of Mathematics
Georgia Southwestern State University
800 GSW State University Drive, Americus, GA 31709, USA
Chadwick.Gugg@gsw.edu

In this paper, we prove modular identities conjectured by Koike and Somos for the Rogers–Ramanujan functions. Our methods focus on approaches that Ramanujan could have employed, including the theory of modular equations, dissections of identities, and methods derived from the approaches of Watson and Rogers.

Keywords: Rogers–Ramanujan functions; Ramanujan's notebooks; modular identities; theta functions; Ramanujan's 40 identities.

Mathematics Subject Classification 2020: 11P84, 33D90

1. Introduction

The Rogers–Ramanujan functions are defined for $|q| < 1$ by

$$G(q) := \sum_{n=0}^{\infty} \frac{q^{n^2}}{(q;q)_n} \quad \text{and} \quad H(q) := \sum_{n=0}^{\infty} \frac{q^{n(n+1)}}{(q;q)_n},$$

where $(a;q)_0 := 1$ and, for $n \geq 1$,

$$(a;q)_n := \prod_{j=0}^{n-1}(1 - aq^j).$$

These functions satisfy the famous Rogers–Ramanujan identities [22, 26, 28, pp. 214–216]

$$G(q) = \frac{1}{(q;q^5)_\infty (q^4;q^5)_\infty} \quad \text{and} \quad H(q) = \frac{1}{(q^2;q^5)_\infty (q^3;q^5)_\infty}, \tag{1.1}$$

*This article originally appeared in *International Journal of Number Theory*, **17**(2), pp. 435–472
© 2021 World Scientific Publishing Company. DOI: 10.1142/S1793042120400138

where

$$(a;q)_\infty := \lim_{n\to\infty} (a;q)_n, \quad |q| < 1.$$

At the end of his brief communication [23, 26, p. 213] announcing his proofs of the Rogers–Ramanujan identities (1.1), Ramanujan remarked, "I have now found an algebraic relation between $G(q)$ and $H(q)$, viz.:

$$H(q)\{G(q)\}^{11} - q^2G(q)\{H(q)\}^{11} = 1 + 11q\{G(q)H(q)\}^6.$$

Another noteworthy formula is

$$H(q)G(q^{11}) - q^2G(q)H(q^{11}) = 1.$$

Each of these formulae is the simplest of a large class". Ramanujan did not reveal how he had discovered or proved the identities. They are two identities from a set of 40 identities for $G(q)$ and $H(q)$ that he never published.

In the years since then, the 40 identities have been established in a series of papers by Rogers [29] in 1921, Watson [39] in 1933, Bressoud [9, 10] in 1977, Biagioli [6] in 1989, and, most recently, Yesilyurt [41, 42] in 2009 and 2010. It should be remarked that the complete list was not brought before the mathematical community until 1975, when Birch [7] found Watson's handwritten copy of them in the Oxford University Library. Although all the identities can be proved by the theory of modular forms, the method employed by Biagioli, it is more instructive to find proofs that Ramanujan might have found. Outside of the theory of modular forms, Rogers's method, which has been extended by both Bressoud and Yesilyurt, is the only general method that has been devised to prove the identities. Further proofs of many of the identities, as well as a history of the 40 identities up to 1997, can be found in the Memoir [4] of Berndt *et al.*, while a more recent account can be found in the work of Andrews and Berndt [2].

Motivated by connections to Thompson series, Koike [21] used a computer to discover empirically several new relations for $G(q)$ and $H(q)$ in 2003; his searches also yielded many identities already on Ramanujan's list. Some of Koike's identities also expressed relations for the Rogers–Ramanujan functions directly in terms of Thompson series. In 2007 and 2008, Bringmann and Swisher [11, 12] put Koike's identities on a rigorous foundation using the theory of modular forms. In a similar spirit to Koike's approach, Somos [38] has written PARI-GP code in order to discover further relations for the Rogers–Ramanujan functions.

The purpose of this paper is to provide new proofs for various identities of Koike and Somos. Our approach utilizes a variety of methods in the spirit of Ramanujan, including even–odd dissections of the Rogers–Ramanujan functions originally due to Watson, the theory of modular equations, and methods from the theory of q-series and theta functions. As a consequence of our approach, further new relations for the Rogers–Ramanujan and related functions are found.

2. Definitions and Preliminaries

Ramanujan's general theta function is defined by

$$f(a,b) := \sum_{n=-\infty}^{\infty} a^{n(n+1)/2} b^{n(n-1)/2}, \qquad |ab| < 1.$$

Basic properties satisfied by $f(a,b)$ include [3, Entry 18, p. 34]

$$f(a,b) = f(b,a), \tag{2.1}$$

$$f(1,a) = 2f(a,a^3), \tag{2.2}$$

$$f(-1,a) = 0, \tag{2.3}$$

and, if n is an integer,

$$f(a,b) = a^{n(n+1)/2} b^{n(n-1)/2} f(a(ab)^n, b(ab)^{-n}). \tag{2.4}$$

Property (2.1) is used without mention in the sequel. The function $f(a,b)$ satisfies the well-known Jacobi Triple Product Identity [3, Entry 19, p. 35],

$$f(a,b) = (-a;ab)_\infty(-b;ab)_\infty(ab;ab)_\infty. \tag{2.5}$$

The three most important special cases of (2.5) are [3, Entry 22, p. 36]

$$\phi(q) := f(q,q) = \sum_{n=-\infty}^{\infty} q^{n^2} = (-q;q^2)_\infty^2(q^2;q^2)_\infty, \tag{2.6}$$

$$\psi(q) := f(q,q^3) = \sum_{n=0}^{\infty} q^{n(n+1)/2} = \frac{(q^2;q^2)_\infty}{(q;q^2)_\infty}, \tag{2.7}$$

and

$$f(-q) := f(-q,-q^2) = \sum_{n=-\infty}^{\infty} (-1)^n q^{n(3n-1)/2} = (q;q)_\infty. \tag{2.8}$$

The latter equality in (2.8) is Euler's pentagonal number theorem. Note that $q^{1/24}f(-q) = \eta(\tau)$, where $q = e^{2\pi i\tau}$, $\operatorname{Im}\tau > 0$, and $\eta(\tau)$ is the Dedekind eta function. Also, after Ramanujan, define

$$\chi(q) := (-q;q^2)_\infty. \tag{2.9}$$

Lemma 2.1 ([3, Entry 29, p. 45]). *If $ab = cd$ with $|ab| < 1$, then*

$$f(a,b)f(c,d) + f(-a,-b)f(-c,-d) = 2f(ac,bd)f(ad,bc) \tag{2.10}$$

and

$$f(a,b)f(-c,-d) - f(-a,-b)f(-c,-d) = 2af\left(\frac{b}{c},\frac{c}{b}abcd\right)f\left(\frac{b}{d},\frac{b}{d}abcd\right). \tag{2.11}$$

Next, we state the Quintuple Product Identity, a version of which can be found on page 207 of Ramanujan's Lost Notebook. For a history of this important identity, see the work of Cooper [15].

Lemma 2.2 ([3, Eq. (38.2), p. 80; 1, Entry 1.2.1, p. 14]).

$$f(B^3q, q^5/B^3) - B^2 f(q/B^3, B^3q^5) = f(-q^2)\frac{f(-B^2, -q^2/B^2)}{f(Bq, q/B)}. \tag{2.12}$$

By (1.1), (2.5), and (2.8), we easily deduce the following lemma.

Lemma 2.3. *We have*

$$G(q) = \frac{f(-q^2, -q^3)}{f(-q)}, \quad H(q) = \frac{f(-q, -q^4)}{f(-q)}, \tag{2.13}$$

and

$$G(q)H(q) = \frac{f(-q^5)}{f(-q)}. \tag{2.14}$$

Schröter [33; 3, pp. 65–72] developed useful representations for a product of two theta functions as a sum of m products of pairs of theta functions. An elegant generalization of Schröter's work has been discovered by Blecksmith *et al.* [8, Theorem 2]. Recently, Cao [13, 14] has proved a generalization of Schröter's work that includes as a special case the theorem of Blecksmith *et al.* Below, we translate [8, Theorem 2] into Ramanujan's notation. Define, for $\epsilon \in \{0, 1\}$ and $|ab| < 1$,

$$f_\epsilon(a, b) = \sum_{n=-\infty}^{\infty} (-1)^{\epsilon n}(ab)^{n^2/2}(a/b)^{n/2}.$$

Theorem 2.4 ([8, Theorem 2]). *Let a, b, c, and d denote positive numbers with $|ab|$, $|cd| < 1$. Suppose that there exist positive integers α, β, and m such that*

$$(ab)^\beta = (cd)^{\alpha(m-\alpha\beta)}.$$

Let $\epsilon_1, \epsilon_2 \in \{0, 1\}$, and define $\delta_1, \delta_2 \in \{0, 1\}$ by

$$\delta_1 \equiv \epsilon_1 - \alpha\epsilon_2 \pmod 2 \quad \text{and} \quad \delta_2 \equiv \beta\epsilon_1 + p\epsilon_2 \pmod 2,$$

respectively, where $p = m - \alpha\beta$. Then, if R denotes any complete residue system modulo m,

$$\begin{aligned} f_{\epsilon_1}(a, b)f_{\epsilon_2}(c, d) &= \sum_{r\in R}(-1)^{\epsilon_2 r}c^{r(r+1)/2}d^{r(r-1)/2} \\ &\quad \times f_{\delta_1}\left(\frac{a(cd)^{\alpha(\alpha+1-2r)/2}}{c^\alpha}, \frac{b(cd)^{\alpha(\alpha+1+2r)/2}}{d^\alpha}\right) \\ &\quad \times f_{\delta_2}\left(\frac{(b/a)^{\beta/2}(cd)^{p(m+1-2r)/2}}{c^p}, \frac{(a/b)^{\beta/2}(cd)^{p(m+1+2r)/2}}{d^p}\right). \end{aligned}$$

Lastly, we introduce some definitions from the theory of modular equations. There are many definitions of a modular equation in the literature. For example, see the books by Rankin [27, p. 76] and Schoeneberg [32, pp. 144–142]. We give the definition of modular equation as understood by Ramanujan. To that end, we first define the complete elliptic integral of the first kind, $K(k)$, by [40, pp. 499–500]

$$K := K(k) := \int_0^{\pi/2} \frac{d\phi}{\sqrt{1-k^2\sin^2\phi}}$$

$$= \frac{\pi}{2}\sum_{n=0}^{\infty} \frac{\left(\frac{1}{2}\right)_n^2}{(n!)^2} k^{2n} = \frac{\pi}{2}\,{}_2F_1\left(\frac{1}{2},\frac{1}{2};1;k^2\right) \qquad (0<k<1),$$

where ${}_2F_1$ denotes the ordinary or Gaussian hypergeometric function,

$$(\alpha)_0 := 1,$$

and

$$(\alpha)_n := \alpha(\alpha+1)\cdots(\alpha+n-1)$$

for each positive integer n. The number k is called the modulus of K, and $k' := \sqrt{1-k^2}$ is called the complementary modulus.

Let K, K', L, and L' denote complete elliptic integrals of the first kind associated with the moduli k, k', l, and l', respectively, where $0 < k, l < 1$. Suppose that

$$\frac{L'}{L} = n\frac{K'}{K} \tag{2.15}$$

holds for some positive integer n. A relation between k and l induced by (2.15) is called a modular equation of degree n.

Following Ramanujan, set

$$\alpha = k^2 \quad \text{and} \quad \beta = l^2.$$

If α and β satisfy a modular equation of degree n, we say that β has degree n over α.

If $q = \exp(-\pi K'/K)$, one of the most fundamental relations in the theory of elliptic functions is given by [3, p. 102]

$$\phi^2(q) = {}_2F_1\left(\frac{1}{2},\frac{1}{2};1;k^2\right) = \frac{2}{\pi}\int_0^{\pi/2} \frac{d\theta}{\sqrt{1-k^2\sin^2\theta}} = \frac{2}{\pi}K(k).$$

Ramanujan derived an extensive catalog of formulas [3, pp. 122–124] giving "evaluations" of $f(\pm q)$, $\phi(\pm q)$, $\psi(\pm q)$, and $\chi(\pm q)$ at various powers of the arguments in terms of

$$z := z_1 = \frac{2}{\pi}K(k) = \phi^2(q), \quad \alpha, \quad \text{and} \quad q.$$

If q is replaced by q^n, then the evaluations are given in terms of

$$z_n := \frac{2}{\pi}K(l) = \phi^2(q^n), \quad \beta, \quad \text{and} \quad q^n,$$

where β has degree n over α. We require in the sequel the following evaluations.

Lemma 2.5 (3, Entries 11(i), (iv); 12(ii)–(iv), (vi), pp. 123–124). *The following identities hold:*

$$\psi(q) = \sqrt{\frac{1}{2}z_1}\left(\alpha\frac{1}{q}\right)^{1/8}, \quad \psi(q^4) = \frac{1}{2}\sqrt{\frac{1}{2}z_1}\left((1-\sqrt{1-\alpha})\frac{1}{q}\right)^{1/2},$$

$$f(-q) = \sqrt{z_1}2^{-1/6}(1-\alpha)^{1/6}\left(\alpha\frac{1}{q}\right)^{1/24},$$

$$f(-q^2) = \sqrt{z_1}2^{-1/3}\left(\alpha(1-\alpha)\frac{1}{q}\right)^{1/12},$$

$$f(-q^4) = \sqrt{z_1}4^{-1/3}(1-\alpha)^{1/24}\left(\alpha\frac{1}{q}\right)^{1/6},$$

and

$$\chi(-q) = 2^{1/6}(1-\alpha)^{1/12}\left(\alpha\frac{1}{q}\right)^{-1/24}.$$

3. Further Required Identities

Lemma 3.1 is due originally to Watson [39]; the lemma provides even–odd dissections of the Rogers–Ramanujan functions. Watson's proof is based on identities for the expressions $G(q)G(q^4) \pm H(q)H(q^4)$, for which the author provided a new proof in [18, Theorem 3.1(i), (ii)]. For a different proof of Watson's Lemma, see [4, pp. 19–20].

Lemma 3.1. *We have*

$$G(q) = \frac{f(-q^8)}{f(-q^2)}(G(q^{16}) + qH(-q^4)), \tag{3.1}$$

$$H(q) = \frac{f(-q^8)}{f(-q^2)}(q^3H(q^{16}) + G(-q^4)). \tag{3.2}$$

In the work that follows, we make use of several known identities for the Rogers–Ramanujan functions. For convenience, we list some of the required results here.

Theorem 3.2 ([4, Entries 3.2, 3.3, 3.6–3.8, 3.11, 3.12, 3.15, 3.16, 3.20, 3.21, pp. 7–9]). *The following identities hold:*

$$G(q)G(q^4) + qH(q)H(q^4) = \frac{\phi(q)}{f(-q^2)}, \tag{3.3}$$

$$G(q)G(q^4) - qH(q)H(q^4) = \frac{\phi(q^5)}{f(-q^2)}, \tag{3.4}$$

$$G(q)G(q^9) + q^2H(q)H(q^9) = \frac{f^2(-q^3)}{f(-q)f(-q^9)}, \tag{3.5}$$

$$G(q^2)G(q^3) + qH(q^2)H(q^3) = \frac{\chi(-q^3)}{\chi(-q)}, \tag{3.6}$$

$$G(q^6)H(q) - qG(q)H(q^6) = \frac{\chi(-q)}{\chi(-q^3)}, \tag{3.7}$$

$$G(q^8)H(q^3) - qG(q^3)H(q^8) = \frac{\chi(-q)\chi(-q^4)}{\chi(-q^3)\chi(-q^{12})}, \tag{3.8}$$

$$G(q)G(q^{24}) + q^5H(q)H(q^{24}) = \frac{\chi(-q^3)\chi(-q^{12})}{\chi(-q)\chi(-q^4)}, \tag{3.9}$$

$$G(q^3)G(q^7) + q^2H(q^3)H(q^7) = G(q^{21})H(q) - q^4G(q)H(q^{21}) \tag{3.10}$$

$$= \frac{1}{2\sqrt{q}}\chi(q^{1/2})\chi(-q^{3/2})\chi(q^{7/2})\chi(-q^{21/2}) - \frac{1}{2\sqrt{q}}\chi(-q^{1/2})\chi(q^{3/2})\chi(-q^{7/2})\chi(q^{21/2}), \tag{3.11}$$

$$G(q^2)G(q^{13}) + q^3H(q^2)H(q^{13}) = G(q^{26})H(q) - q^5G(q)H(q^{26}) \tag{3.12}$$

$$= \sqrt{\frac{\chi(-q^{13})}{\chi(-q)} - q\frac{\chi(-q)}{\chi(-q^{13})}}, \tag{3.13}$$

$$G(q)H(-q) + G(-q)H(q) = \frac{2\psi(q^2)}{f(-q^2)}, \tag{3.14}$$

$$G(q)H(-q) - G(-q)H(q) = \frac{2q\psi(q^{10})}{f(-q^2)}. \tag{3.15}$$

Theorem 3.3 ([5, Eqs. 5.4, 5.6; 4.31, 4.36, 4.28, 4.39]). *The following identities hold:*

$$G(q^{14})G(q^{16}) + q^6H(q^{14})H(q^{16}) = \frac{1}{2}\chi(-q^4)\{\chi(q)\chi(-q^7) + \chi(-q)\chi(q^7)\}, \tag{3.16}$$

$$G(q^{112})H(q^2) - q^{22}H(q^{112})G(q^2) = \frac{1}{2}\chi(-q^{28})\{\chi(q)\chi(-q^7) + \chi(-q)\chi(q^7)\}, \tag{3.17}$$

$$G(q^{42})H(-q^2) + q^8H(q^{42})G(-q^2) = \frac{1}{2q}\frac{\chi(-q^{12})}{\chi(q^2)}\{\chi(q)\chi(-q^{21}) - \chi(-q)\chi(q^{21})\}, \tag{3.18}$$

$$G(q^6)G(-q^{14}) - q^4H(q^6)H(-q^{14}) = \frac{1}{2q^3}\frac{\chi(-q^{84})}{\chi(q^{14})}\{\chi(q^3)\chi(-q^7) - \chi(-q^3)\chi(q^7)\} \tag{3.19}$$

$$G(q^8)G(q^{42}) + q^{10}H(q^8)H(q^{42}) = \frac{1}{2}\frac{\chi(-q^4)}{\chi(q^6)}\{\chi(q)\chi(-q^{21}) + \chi(-q)\chi(q^{21})\}, \tag{3.20}$$

$$G(q^{56})H(q^6) - q^{10}H(q^{56})G(q^6) = \frac{1}{2}\frac{\chi(-q^{28})}{\chi(q^{42})}\{\chi(q^3)\chi(-q^7) + \chi(-q^3)\chi(q^7)\}. \tag{3.21}$$

Theorem 3.4 ([5, Theorem 1.1]). *Let*

$$\begin{aligned}
B(q) &:= G(q^{12})H(-q^7) + qG(-q^7)H(q^{12}),\\
C(q) &:= G(q)G(q^{84}) + q^{17}H(q)H(q^{84}),\\
V(q) &:= H(-q)G(q^{21}) + q^4G(-q)H(q^{21}),\\
W(q) &:= G(q^4)G(q^{21}) + q^5H(q^4)H(q^{21}),\\
Z(q) &:= H(q^3)G(q^{28}) - q^5G(q^3)H(q^{28}),\\
Y(q) &:= G(q^3)G(-q^7) - q^2H(q^3)H(-q^7).
\end{aligned}$$

Then,

$$\frac{C(q^2)}{Y(-q^2)} = \frac{V(-q^2)}{B(-q^2)} = \frac{C(q)}{B(q)} = \frac{f(-q^{12})f(-q^{14})}{f(-q^2)f(-q^{84})} \tag{3.22}$$

and

$$\frac{Z(-q)}{W(q)} = \frac{Z(q)}{W(-q)} = \frac{Y(q^2)}{W(q^2)} = \frac{Z(q^2)}{V(q^2)} = \frac{f(-q^4)f(-q^{42})}{f(-q^6)f(-q^{28})}. \tag{3.23}$$

Huang derived an identity that belongs to the same class of identities as those in Theorem 3.4, but is different from any identity in (3.22) or (3.23). In particular, Huang proved the following theorem.

Theorem 3.5 ([19, Eq. 2.28]). *With $W(q)$ and $C(q)$ defined as in the preceding theorem, we have*

$$\begin{aligned}
W(q)C(q) = \frac{1}{2q}\Bigg\{&\frac{f^3(-q^2)f^2(-q^3)f(-q^{12})f(-q^{14})}{f^2(-q)f^2(-q^4)f^2(-q^6)f(-q^{84})}\\
&- \frac{f(-q^6)f^2(-q^7)f(-q^{28})f^3(-q^{42})}{f(-q^4)f^2(-q^{14})f^2(-q^{21})f^2(-q^{84})}\Bigg\}.
\end{aligned} \tag{3.24}$$

Last, we make a definition. The identities for the Rogers–Ramanujan functions that we consider usually involve $G(q)$ and $H(q)$ in the combinations

$$G(q^a)G(q^b) + q^{(a+b)/5}H(q^a)H(q^b)$$

or

$$G(q^a)H(q^b) - q^{(a-b)/5}G(q^b)H(q^a).$$

Set

$$N := ab.$$

Following Koike [21], we call N the level of the identity.

4. Identities Connecting the Rogers–Ramanujan Functions at the Arguments q and $-q$

Identity (4.1) below was discovered empirically by Somos [38], and communicated to the author. We offer, to our knowledge, the first proof of this result. The second identity (4.2) is a natural companion found by the author.

Theorem 4.1. *We have*

$$\begin{aligned} G(-q^2)H(-q^2)&[G(q)H(q) - G(-q)H(-q)] \\ &= G(q^2)H(q^2)[G(q)H(-q) - G(-q)H(q)] \end{aligned} \tag{4.1}$$

and

$$\begin{aligned} G(q^4)H(q^4)&[G(q)H(q) + G(-q)H(-q)] \\ &= G(q^2)G(-q^2)H(q^2)H(-q^2)[G(q)H(-q) + G(-q)H(q)]. \end{aligned} \tag{4.2}$$

Proof of (4.1). From (3.3) and (3.4), we easily deduce that

$$G(q) = \frac{\phi(q) + \phi(q^5)}{2G(q^4)f(-q^2)} \quad \text{and} \quad H(q) = \frac{\phi(q) - \phi(q^5)}{2qH(q^4)f(-q^2)}. \tag{4.3}$$

By [24; 3, Entry 10(iv), p. 262], we know that

$$\phi^2(q) - \phi^2(q^5) = 4qf(q, q^9)f(q^3, q^7). \tag{4.4}$$

By (2.11) with $a = q$, $b = q^9$, $c = q^3$, and $d = q^7$, we find that

$$f(q, q^9)f(q^3, q^7) - f(-q, -q^9)f(-q^3, -q^7) = 2qf(q^6, q^{14})f(q^2, q^{18}). \tag{4.5}$$

Applying (4.3)–(4.5), we deduce that

$$\begin{aligned} &G(q)H(q) - G(-q)H(-q) \\ &\quad = \frac{(\phi(q) + \phi(q^5))(\phi(q) - \phi(q^5))}{4qG(q^4)H(q^4)f^2(-q^2)} + \frac{(\phi(-q) + \phi(-q^5))(\phi(-q) - \phi(-q^5))}{4qG(q^4)H(q^4)f^2(-q^2)} \\ &\quad = \frac{1}{4qG(q^4)H(q^4)f^2(-q^2)}[(\phi^2(q) - \phi^2(q^5)) + (\phi^2(-q) - \phi^2(-q^5))] \\ &\quad = \frac{4q}{4qG(q^4)H(q^4)f^2(-q^2)}[f(q, q^9)f(q^3, q^7) - f(-q, -q^9)f(-q^3, -q^7)] \\ &\quad = \frac{2q}{G(q^4)H(q^4)f^2(-q^2)}f(q^6, q^{14})f(q^2, q^{18}). \end{aligned} \tag{4.6}$$

Similarly, by (4.3), [3, Entry 25(iii), p. 40], namely,

$$\phi(q)\phi(-q) = \phi^2(-q^2), \tag{4.7}$$

and by (4.4), we find that

$$\begin{aligned}
&G(q)H(-q) - G(-q)H(q) \\
&\quad = \frac{(\phi(q)+\phi(q^5))(\phi(-q)-\phi(-q^5))}{-4qG(q^4)H(q^4)f^2(-q^2)} - \frac{(\phi(-q)+\phi(-q^5))(\phi(q)-\phi(q^5))}{4qG(q^4)H(q^4)f^2(-q^2)} \qquad (4.8)\\
&\quad = \frac{-2}{4qG(q^4)H(q^4)f^2(-q^2)}[\phi(q)\phi(-q) - \phi(q^5)\phi(-q^5)] \\
&\quad = \frac{-2}{4qG(q^4)H(q^4)f^2(-q^2)}[\phi^2(-q^2) - \phi^2(-q^{10})] \\
&\quad = \frac{2q}{G(q^4)H(q^4)f^2(-q^2)} f(-q^2,-q^{18})f(-q^6,-q^{14}). \qquad (4.9)
\end{aligned}$$

Employing (4.6) and (4.9) in (4.1), replacing q^2 by q and simplifying, we deduce that (4.1) is equivalent to

$$\frac{G(-q)H(-q)}{G(q)H(q)} = \frac{f(-q,-q^9)f(-q^3,-q^7)}{f(q,q^9)f(q^3,q^7)}. \tag{4.10}$$

We prove (4.10). By (2.14),

$$\frac{G(-q)H(-q)}{G(q)H(q)} = \frac{f(-q)f(q^5)}{f(-q^5)f(q)}. \tag{4.11}$$

By the Jacobi Triple Product Identity (2.5) and by (2.8),

$$\begin{aligned}
f(-q,-q^9)f(-q^3,-q^7) &= (q;q^{10})_\infty(q^3;q^{10})_\infty(q^7;q^{10})_\infty(q^9;q^{10})_\infty(q^{10};q^{10})_\infty^2 \\
&= \frac{(q;q^2)_\infty(q^{10};q^{10})_\infty^2}{(q^5;q^{10})_\infty} = \frac{(q;q)_\infty(q^{10};q^{10})_\infty^3}{(q^2;q^2)_\infty(q^5;q^5)_\infty} \\
&= \frac{f(-q)f^3(-q^{10})}{f(-q^2)f(-q^5)}. \qquad (4.12)
\end{aligned}$$

Replacing q by $-q$ in (4.12), we see that

$$f(q,q^9)f(q^3,q^7) = \frac{f(q)f^3(-q^{10})}{f(-q^2)f(q^5)}. \tag{4.13}$$

Combining (4.12) and (4.13), we arrive at

$$\frac{f(-q,-q^9)f(-q^3,-q^7)}{f(q,q^9)f(q^3,q^7)} = \frac{f(-q)f(q^5)}{f(-q^5)f(q)}. \tag{4.14}$$

Comparing (4.11) with (4.14), we easily deduce the truth of (4.10). This completes the proof of (4.1). □

Proof of (4.2). The proof is similar to the proof of (4.1). By (2.10) with $a = q$, $b = q^9$, $c = q^3$, and $d = q^7$,

$$f(q, q^9)f(q^3, q^7) + f(-q, -q^9)f(-q^3, -q^7) = 2f(q^4, q^{16})f(q^8, q^{12}). \qquad (4.15)$$

Apply (4.3), (4.4), and (4.15), and proceed as in the proof of (4.1) to deduce that

$$G(q)H(q) + G(-q)H(-q) = \frac{2}{G(q^4)H(q^4)f^2(-q^2)}f(q^4, q^{16})f(q^8, q^{12}). \qquad (4.16)$$

Similarly, apply (4.3) and (4.4) to discover that

$$\begin{aligned} &G(q)H(-q) + G(-q)H(q) \\ &\quad = \frac{1}{2qG(q^4)H(q^4)f^2(-q^2)}[\phi(q)\phi(-q^5) - \phi(-q)\phi(q^5)]. \end{aligned} \qquad (4.17)$$

By (2.6),

$$\phi(q)\phi(-q^5) - \phi(-q)\phi(q^5) = f(q, q)f(-q^5, -q^5) - f(-q, -q)f(q^5, q^5). \qquad (4.18)$$

To simplify the expression in (4.18), we first apply Theorem 2.4 with $a = b = q$, $c = d = q^5$, $\alpha = 1$, $\beta = 5$, $\epsilon_1 = 0$, $\epsilon_2 = 1$, and $m = 6$. Using also (2.3) and (2.4), we find that

$$\begin{aligned} &f(q, q)f(-q^5, -q^5) \\ &\quad = f(-q^6, -q^6)f(-q^{30}, -q^{30}) - q^5 f(-q^{-4}, -q^{16})f(-q^{20}, -q^{40}) \\ &\qquad + q^{20}f(-q^{-14}, -q^{26})f(-q^{10}, -q^{50}) - q^{45}f(-q^{-24}, -q^{36})f(-1, -q^{60}) \\ &\qquad + q^{80}f(-q^{-34}, -q^{46})f(-q^{-10}, -q^{70}) - q^{125}f(-q^{-44}, -q^{56})f(-q^{-20}, -q^{80}) \\ &\quad = f(-q^6, -q^6)f(-q^{30}, -q^{30}) + 2qf(-q^4, -q^8)f(-q^{20}, -q^{40}) \\ &\qquad + 2q^4 f(-q^2, -q^{10})f(-q^{10}, -q^{50}). \end{aligned} \qquad (4.19)$$

Replace q by $-q$ in (4.19) and subtract the result from (4.19) to arrive at

$$f(q, q)f(-q^5, -q^5) - f(-q, -q)f(q^5, q^5) = 4qf(-q^4, -q^8)f(-q^{20}, -q^{40}). \qquad (4.20)$$

Applying the Jacobi Triple Product Identity (2.5) and writing the resulting products in the common base q^{60}, it is easy to show that

$$f(-q^4, -q^8)f(-q^{20}, -q^{40}) = f(-q^4, -q^{16})f(-q^8, -q^{12}). \qquad (4.21)$$

Hence,

$$f(q, q)f(-q^5, -q^5) - f(-q, -q)f(q^5, q^5) = 4qf(-q^4, -q^{16})f(-q^8, -q^{12}). \qquad (4.22)$$

Employ (4.18) and (4.22) to rewrite (4.17) in the form

$$G(q)H(-q) + G(q)H(-q) = \frac{2f(-q^4, -q^{16})f(-q^8, -q^{12})}{G(q^4)H(q^4)f^2(-q^2)}. \qquad (4.23)$$

By (4.16) and (4.23), we see that (4.2) is equivalent to

$$G(q^4)H(q^4)\left[\frac{2f(q^4,q^{16})f(q^8,q^{12})}{G(q^4)H(q^4)f^2(-q^2)}\right]$$
$$= G(q^2)G(-q^2)H(q^2)H(-q^2)\left[\frac{2f(-q^4,-q^{16})f(-q^8,-q^{12})}{G(q^4)H(q^4)f^2(-q^2)}\right], \tag{4.24}$$

or, after replacing q^2 by q and simplifying,

$$\frac{f(q^2,q^8)f(q^4,q^6)}{f(-q^2,-q^8)f(-q^4,-q^6)} = \frac{G(q)G(-q)H(q)H(-q)}{G(q^2)H(q^2)}. \tag{4.25}$$

By the Jacobi Triple Product Identity,

$$\begin{aligned} f(q^2,q^8)f(q^4,q^6) &= (-q^2;q^{10})_\infty(-q^4;q^{10})_\infty(-q^6;q^{10})_\infty(-q^8;q^{20})_\infty(q^{10};q^{10})^2_\infty \\ &= \frac{(-q^2;q^2)_\infty}{(-q^{10};q^{10})_\infty}(q^{10};q^{10})^2_\infty = \frac{(q^4;q^4)_\infty(q^{10};q^{10})^3_\infty}{(q^2;q^2)_\infty(q^{20};q^{20})_\infty} \\ &= \frac{f(-q^4)f^3(-q^{10})}{f(-q^2)f(-q^{20})}. \end{aligned} \tag{4.26}$$

By (4.21),

$$f(-q^2,-q^8)f(-q^4,-q^6) = f(-q^2,-q^4)f(-q^{10},-q^{20}) = f(-q^2)f(-q^{10}). \tag{4.27}$$

Combine (4.26) and (4.27) to see that

$$\frac{f(q^2,q^8)f(q^4,q^6)}{f(-q^2,-q^8)f(-q^4,-q^6)} = \frac{f(-q^4)f^2(-q^{10})}{f^2(-q^2)f(-q^{20})}. \tag{4.28}$$

Apply (2.14) followed by (2.5) and (2.8) to find that

$$\frac{G(q)G(-q)H(q)H(-q)}{G(q^2)H(q^2)} = \frac{f(-q^5)f(q^5)f(-q^2)}{f(-q)f(q)f(-q^{10})} = \frac{f(-q^4)f^2(-q^{10})}{f^2(-q^2)f(-q^{20})}. \tag{4.29}$$

Using (4.28) and (4.29), we readily deduce the truth of (4.25), and thus complete the proof of (4.2). □

Next, we record several natural corollaries of our work.

Corollary 4.2. *We have*

$$\frac{G(q)H(-q)-G(-q)H(q)}{G(q)H(q)-G(-q)H(-q)} = \frac{\phi(-q^2)\phi(-q^{20})}{\phi(-q^4)\phi(-q^{10})} \tag{4.30}$$

and

$$\frac{G(q)H(-q)+G(-q)H(q)}{G(q)H(q)+G(-q)H(-q)} = \frac{\phi(-q^4)\phi(-q^{10})}{\chi(-q^{10})\chi(-q^{20})}. \tag{4.31}$$

Proof. To prove (4.30), rearrange (4.1) to see that

$$\frac{G(q)H(-q)-G(-q)H(q)}{G(q)H(q)-G(-q)H(-q)} = \frac{G(-q^2)H(-q^2)}{G(q^2)H(q^2)} = \frac{f(q^{10})f(-q^2)}{f(q^2)f(-q^{10})}, \tag{4.32}$$

where we have employed (2.14) to obtain the right-most expression in (4.32). Utilizing (2.5), (2.6), (2.8), and (2.9), we readily complete the proof. Similarly (4.31) follows from (4.2); we omit the details. □

Corollary 4.3. *We have*

$$G(q)H(q) - G(-q)H(-q) = 2q\frac{\psi(q^{10})}{f(-q^2)}\frac{f(q^2)f(-q^{10})}{f(-q^2)f(q^{10})}, \tag{4.33}$$

$$G(q)H(q) + G(-q)H(-q) = 2\frac{\psi(q^2)}{f(-q^2)}\frac{\chi(-q^{10})f(q^{10})}{\chi(-q^2)f(q^2)}. \tag{4.34}$$

Proof. To prove (4.34), rewrite (4.2) and apply (4.29) with q replaced by q^2, (3.14), (2.5), (2.8), and (2.9) in order to deduce that

$$\begin{aligned} G(q)H(q) + G(-q)H(-q) &= \frac{G(q^2)H(q^2)G(-q^2)H(-q^2)}{G(q^4)H(q^4)} \\ &\quad \times [G(q)H(-q) + G(-q)H(q)] \\ &= \frac{f(-q^8)f^2(-q^{20})}{f^2(-q^4)f(-q^{40})}\left[2\frac{\psi(q^2)}{f(-q^2)}\right] \\ &= 2\frac{\psi(q^2)}{f(-q^2)}\frac{\chi(-q^{10})f(q^{10})}{\chi(-q^2)f(q^2)}. \end{aligned} \tag{4.35}$$

Similarly, (4.33) follows from (4.1), (4.11), (3.15), (2.5), and (2.8). □

We also have the following elegant theta function identities.

Corollary 4.4. *We have*

$$f(-q^5)f(q) - f(q^5)f(-q) = 2q\phi(-q^4)\psi(-q^{10}), \tag{4.36}$$

$$f(-q^5)f(q) + f(q^5)f(-q) = 2\psi(-q^2)\phi(-q^{20}). \tag{4.37}$$

Proof. To prove (4.36), apply (2.14) to (4.33). Thus,

$$\frac{f(-q^5)}{f(-q)} - \frac{f(q^5)}{f(q)} = 2q\frac{\psi(q^{10})}{f(-q^2)}\frac{f(q^2)f(-q^{10})}{f(-q^2)f(q^{10})}, \tag{4.38}$$

whence we deduce that

$$f(-q^5)f(q) - f(q^5)f(-q) = 2qf(q)f(-q)\frac{\psi(q^{10})}{f(-q^2)}\frac{f(q^2)f(-q^{10})}{f(-q^2)f(q^{10})}. \tag{4.39}$$

With the help of (2.5)–(2.8), we see that the right-hand sides of (4.36) and (4.39) are equal, and thus we deduce (4.36). Similarly, (4.37) follows from (4.34) and (2.5)–(2.9). This completes the proof. □

5. An Identity for the Level 9

The following theorem was conjectured by Koike [21] and first proved by Bringmann and Swisher [11] using the theory of modular forms.

Theorem 5.1.

$$(G(q^9)G(q)+q^2H(q^9)H(q))^6=\frac{f^3(-q)}{f^3(-q^9)}+27q^2\frac{f^3(-q^9)}{f^3(-q)}+9q. \tag{5.1}$$

Proof. Recall (3.5), namely,

$$G(q^9)G(q)+q^2H(q^9)H(q)=\frac{f^2(-q^3)}{f(-q)f(-q^9)}. \tag{5.2}$$

Raising both sides of (5.2) to the sixth power and comparing the result with (5.1), we see that it suffices to show that

$$\frac{f^3(-q)}{f^3(-q^9)}+27q^2\frac{f^3(-q^9)}{f^3(-q)}+9q=\frac{f^{12}(-q^3)}{f^6(-q)f^6(-q^9)}. \tag{5.3}$$

To prove (5.3), we employ a second result of Ramanujan, namely [3, Entry 1(iv), p. 345],

$$3+\frac{f^3(-q^{1/3})}{q^{1/3}f^3(-q^3)}=\left(27+\frac{f^{12}(-q)}{qf^{12}(-q^3)}\right)^{1/3}. \tag{5.4}$$

Replace $q^{1/3}$ by q in (5.4) and cube both sides. Multiply by $q^3f^6(-q^9)/f^6(-q)$ and simplify to deduce (5.3). This completes the proof. □

6. An Identity for the Level 56

The next result was conjectured by Koike [21] and first proved in [11].

Theorem 6.1.

$$\begin{aligned}&\{G(q^{56})H(q)-q^{11}H(q^{56})G(q)\}\{G(q^7)G(q^8)+q^3H(q^7)H(q^8)\}\\&=\frac{\chi(-q^4)\chi(-q^{28})}{\chi(-q)\chi(-q^7)}-q.\end{aligned} \tag{6.1}$$

Proof. Multiplying (3.16) with (3.17) and replacing q^2 by q, we deduce that

$$\begin{aligned}&\{G(q^{56})H(q)-q^{11}H(q^{56})G(q)\}\{G(q^7)G(q^8)+q^3H(q^7)H(q^8)\}\\&=\frac{1}{4}\chi(-q^2)\chi(-q^{14})\{\chi(q^{1/2})\chi(-q^{7/2})+\chi(-q^{1/2})\chi(q^{7/2})\}^2.\end{aligned} \tag{6.2}$$

Comparing (6.2) with (6.1), we see that it is sufficient to show that

$$\frac{1}{4}\chi(-q^2)\chi(-q^{14})\{\chi(q^{1/2})\chi(-q^{7/2})+\chi(-q^{1/2})\chi(q^{7/2})\}^2=\frac{\chi(-q^4)\chi(-q^{28})}{\chi(-q)\chi(-q^7)}-q. \tag{6.3}$$

By (2.5)–(2.9) (see also [3, Entry 24(iii), p. 39] for the second representation of $\chi(q)$), we readily see that

$$\chi(q)=\sqrt{\frac{\phi(q)}{f(-q^2)}}=\frac{f(-q^2)}{\psi(-q)}. \tag{6.4}$$

For the expression inside the brackets in (6.3), we apply the first equality in (6.4) with q replaced by $\pm q^{1/2}$ and by $\pm q^{7/2}$. We also apply the second representation for $\chi(q)$ in (6.4) to the two χ functions outside the brackets on the left-hand side of (6.3). We thus write (6.3) in the equivalent form

$$\begin{aligned}\frac{1}{4}\frac{f(-q^4)f(-q^{28})}{\psi(q^2)\psi(q^{14})}&\left\{\sqrt{\frac{\phi(q^{1/2})\phi(-q^{7/2})}{f(-q)f(-q^7)}}+\sqrt{\frac{\phi(-q^{1/2})\phi(q^{7/2})}{f(-q)f(-q^7)}}\right\}^2\\&=\frac{\chi(-q^4)\chi(-q^{28})}{\chi(-q)\chi(-q^7)}-q.\end{aligned} \tag{6.5}$$

By (2.5)–(2.6) (see also [3, p. 40, Entry 25(iii)]),

$$\phi(q)\phi(-q)=\phi^2(-q^2). \tag{6.6}$$

Multiplying (6.5) by $\psi(q^2)\psi(q^{14})f(-q)f(-q^7)/\left(f(-q^4)f(-q^{28})\right)$, expanding out the square on the left-hand side of (6.5), and simplifying with the use of (6.6), we find that

$$\begin{aligned}&\frac{1}{4}\{\phi(q^{1/2})\phi(-q^{7/2})+\phi(-q^{1/2})\phi(q^{7/2})\}+\frac{1}{2}\phi(-q)\phi(-q^7)\\&\quad=\frac{\psi(q^2)\psi(q^{14})f(-q)f(-q^7)}{f(-q^4)f(-q^{28})}\left\{\frac{\chi(-q^4)\chi(-q^{28})}{\chi(-q)\chi(-q^7)}-q\right\}.\end{aligned} \tag{6.7}$$

With the help of (2.5)–(2.9), we deduce that

$$\frac{\psi(q^2)\psi(q^{14})f(-q)f(-q^7)}{f(-q^4)f(-q^{28})}\cdot\frac{\chi(-q^4)\chi(-q^{28})}{\chi(-q)\chi(-q^7)}=\phi(-q^4)\phi(-q^{28}) \tag{6.8}$$

and

$$\frac{\psi(q^2)\psi(q^{14})f(-q)f(-q^7)}{f(-q^4)f(-q^{28})}=\psi(-q)\psi(-q^7). \tag{6.9}$$

Employing (6.8) and (6.9) in (6.7) and then replacing q^2 by q, we deduce that (6.3) is equivalent to the theta function identity

$$\begin{aligned}&\frac{1}{4}\{\phi(q^{1/4})\phi(-q^{7/4})+\phi(-q^{1/4})\phi(q^{7/4})\}+\frac{1}{2}\phi(-q^{1/2})\phi(-q^{7/2})\\&\quad=\phi(-q^2)\phi(-q^{14})-q^{1/2}\psi(-q^{1/2})\psi(-q^{7/2}).\end{aligned} \tag{6.10}$$

We prove (6.10). To that end, we first convert (6.10) into a modular equation. Let β be of the seventh degree in α. Employing [3, Entries 10(iii), (vii)–(ix); Entry

11(vii), pp. 122–123], we find after performing some minor simplifications that (6.10) is equivalent to the modular equation

$$\frac{1}{4}\{(1+\alpha^{1/4})(1-\beta^{1/4})+(1-\alpha^{1/4})(1+\beta^{1/4})\}+\frac{1}{2}(1-\sqrt{\alpha})^{1/2}(1-\sqrt{\beta})^{1/2}$$
$$=(1-\alpha)^{1/8}(1-\beta)^{1/8}-\frac{1}{\sqrt{2}}(1-\sqrt{\alpha})^{1/4}(1-\sqrt{\beta})^{1/4}(\alpha\beta)^{1/16}. \tag{6.11}$$

To prove (6.11), we require two modular equations of degree seven stated by Ramanujan, namely [3, Entry 19(i), p. 314]

$$(\alpha\beta)^{1/8}+\{(1-\alpha)(1-\beta)\}^{1/8}=1, \tag{6.12}$$

$$\left(\frac{1}{2}(1+(\alpha\beta)^{1/2}+\{(1-\alpha)(1-\beta)\}^{1/2})\right)^{1/2}=1-\{\alpha\beta(1-\alpha)(1-\beta)\}^{1/8}. \tag{6.13}$$

Ramanujan recorded a large number of elegant modular equations of degree seven in his notebooks [3, pp. 314–315; 1, pp. 387–388, 391–392]. The seventh-order modular equation (6.12) is due originally to Guetzlaff [17] in 1834. Schröter in 1854 [33–35] and Fiedler [16] in 1885 also proved this modular equation. More complicated modular equations of degree seven have been discovered by Schläfli [31], Klein [20], Sohncke [36, 37] and Russell [30].

Beginning with (6.12), multiply through by $\{(1-\alpha)(1-\beta)\}^{1/8}$, apply (6.12) to the right-hand side of the resulting equation, and then apply (6.13) and rearrange the results. We thus deduce the sequence of equalities

$$\{(1-\alpha)(1-\beta)\}^{1/4}+\{\alpha\beta(1-\alpha)(1-\beta)\}^{1/8}=\{(1-\alpha)(1-\beta)\}^{1/8},$$
$$\{(1-\alpha)(1-\beta)\}^{1/4}+\{\alpha\beta(1-\alpha)(1-\beta)\}^{1/8}=1-(\alpha\beta)^{1/8},$$

and

$$\{(1-\alpha)(1-\beta)\}^{1/4}+(\alpha\beta)^{1/8}=\sqrt{\frac{1}{2}(1+\sqrt{\alpha\beta}+\sqrt{(1-\alpha)(1-\beta)})}. \tag{6.14}$$

Now,

$$\frac{1}{2}[(1+\sqrt{\alpha})^{1/2}(1+\sqrt{\beta})^{1/2}+(1-\sqrt{\alpha})^{1/2}(1-\sqrt{\beta})^{1/2}]$$
$$=\sqrt{\frac{1}{2}(1+\sqrt{\alpha\beta}+\sqrt{(1-\alpha)(1-\beta)})}, \tag{6.15}$$

as can readily be seen by squaring both sides. Inserting the left-hand side of (6.15) into (6.14), we deduce that

$$\{(1-\alpha)(1-\beta)\}^{1/4}+(\alpha\beta)^{1/8}$$
$$=\frac{1}{2}[(1+\sqrt{\alpha})^{1/2}(1+\sqrt{\beta})^{1/2}+(1-\sqrt{\alpha})^{1/2}(1-\sqrt{\beta})^{1/2}]. \tag{6.16}$$

Observe that by squaring and rearranging (6.17), we readily deduce that (6.16) is equivalent to the beautiful identity

$$\sqrt{2}(\alpha\beta)^{1/16}=[(1+\sqrt{\alpha})(1+\sqrt{\beta})]^{1/4}-[(1-\sqrt{\alpha})(1-\sqrt{\beta})]^{1/4}. \tag{6.17}$$

Multiply (6.17) by $-\frac{1}{2}\cdot(1-\sqrt{\alpha})^{1/4}(1-\sqrt{\beta})^{1/4}$, add $(1-\alpha)^{1/4}(1-\beta)^{1/4}$ to both sides of the resulting equation, and simplify. We thus deduce that

$$\begin{aligned}&\{(1-\alpha)(1-\beta)\}^{1/8}\cdot\{(1-\alpha)(1-\beta)\}^{1/8}\\&\quad-\frac{1}{\sqrt{2}}(1-\sqrt{\alpha})^{1/4}(1-\sqrt{\beta})^{1/4}(\alpha\beta)^{1/16}\\&=\frac{1}{2}(1-\alpha)^{1/4}(1-\beta)^{1/4}+\frac{1}{2}(1-\sqrt{\alpha})^{1/2}(1-\sqrt{\beta})^{1/2}.\end{aligned}\tag{6.18}$$

By (6.12), the first term on the left-hand side of (6.18) is equal to

$$\{(1-\alpha)(1-\beta)\}^{1/8}\cdot[1-(\alpha\beta)^{1/8}].\tag{6.19}$$

Inserting (6.19) into (6.18) and rearranging, we find that

$$\begin{aligned}&\frac{1}{2}\{(1-\alpha)(1-\beta)\}^{1/4}+\{\alpha\beta(1-\alpha)(1-\beta)\}^{1/8}\\&\quad+\frac{1}{2}(1-\sqrt{\alpha})^{1/2}(1-\sqrt{\beta})^{1/2}\\&=\{(1-\alpha)(1-\beta)\}^{1/8}-\frac{1}{\sqrt{2}}[(1-\sqrt{\alpha})(1-\sqrt{\beta})]^{1/4}(\alpha\beta)^{1/16}.\end{aligned}\tag{6.20}$$

Squaring (6.12), dividing all terms by 1/2 and rearranging, we deduce that

$$\frac{1}{2}\{(1-\alpha)(1-\beta)\}^{1/4}+\{\alpha\beta(1-\alpha)(1-\beta)\}^{1/8}=\frac{1}{2}-\frac{1}{2}(\alpha\beta)^{1/4}.\tag{6.21}$$

Employing (6.21), we find that the left-hand side of (6.20) is equal to

$$\frac{1}{2}-\frac{1}{2}(\alpha\beta)^{1/4}+\frac{1}{2}(1-\sqrt{\alpha})^{1/2}(1-\sqrt{\beta})^{1/2},$$

which, after a modicum of algebra, is readily seen to be equal to

$$\begin{aligned}&\frac{1}{4}[(1+\alpha^{1/4})(1-\beta^{1/4})+(1-\alpha^{1/4})(1+\beta^{1/4})]\\&\quad+\frac{1}{2}(1-\sqrt{\alpha})^{1/2}(1-\sqrt{\beta})^{1/2}.\end{aligned}\tag{6.22}$$

Replacing the left-hand side of (6.20) with (6.22), we deduce (6.11). This completes the proof. □

7. Identities for the Level 84

The next theorem was discovered empirically on a computer by Koike [21] and independently by Somos [38].

Theorem 7.1.

$$\frac{G(q^{84})G(q)+q^{17}H(q^{84})H(q)}{G(q^{21})G(q^4)+q^5H(q^{21})H(q^4)}=\frac{G(q^{28})H(q^3)-q^5H(q^{28})G(q^3)}{G(q^{12})H(q^7)-qH(q^{12})G(q^7)}\tag{7.1}$$

and

$$G(q^{84})G(q)+q^{17}H(q^{84})H(q)=\frac{G(q^{42})H(q^2)-q^8H(q^{42})G(q^2)}{G(q^{12})H(q^7)-qH(q^{12})G(q^7)}. \tag{7.2}$$

Proof. In the notation of Theorem 3.4, (7.1) reads as

$$\frac{C(q)}{W(q)}=\frac{Z(q)}{B(-q)}. \tag{7.3}$$

Applying the first equality in (3.22), the third equality of (3.23) with q^2 replaced by $-q^2$, and the first equality of (3.23) with q replaced by q^2, we readily deduce that

$$C(q^2)B(-q^2)=Y(-q^2)V(-q^2)=W(-q^2)Z(-q^2)=W(q^2)Z(q^2). \tag{7.4}$$

Replacing q^2 by q in (7.4) and rearranging, we arrive at (7.3). This completes the proof of (7.1).

To prove (7.2), we first define

$$T(q):=G(q^{21})H(q)-q^4H(q^{21})G(q). \tag{7.5}$$

Then, using the notation of Theorem 3.4 and (7.5), we cast (7.2) in the equivalent form

$$B(-q^2)C(q^2)=T(q^4). \tag{7.6}$$

By repeated applications of Theorem 3.4, we conclude that

$$\begin{aligned}B(-q^2)&=\frac{f(-q^2)f(-q^{84})}{f(-q^{12})f(-q^{14})}V(-q^2)\\&=\frac{f(-q^2)f(-q^{84})}{f(-q^{12})f(-q^{14})}\cdot\frac{f(q^6)f(-q^{28})}{f(-q^4)f(q^{42})}Z(-q^2)\\&=\frac{f(-q^2)f(-q^{84})}{f(-q^{12})f(-q^{14})}\cdot\frac{f(q^6)f(-q^{28})}{f(-q^4)f(q^{42})}\cdot\frac{f(-q^8)f(-q^{84})}{f(-q^{12})f(-q^{56})}W(q^2).\end{aligned} \tag{7.7}$$

From (2.5) and (2.8) we may show that

$$f(q)=\frac{f^3(-q^2)}{f(-q)f(-q^4)}. \tag{7.8}$$

Hence, employing (7.8) in (7.7), we find that

$$\begin{aligned}B(-q^2)C(q^2)&=W(q^2)C(q^2)\\&\quad\times\left[\frac{f(-q^2)f(-q^8)f(-q^{12})f(-q^{28})f(-q^{42})f(-q^{168})}{f(-q^4)f(-q^6)f(-q^{14})f(-q^{24})f(-q^{56})f(-q^{84})}\right].\end{aligned} \tag{7.9}$$

By Theorem 3.5 with q replaced by q^2,

$$\begin{aligned}W(q^2)C(q^2)=\frac{1}{2q^2}&\left\{\frac{f^3(-q^4)f^2(-q^6)f(-q^{24})f(-q^{28})}{f^2(-q^2)f^2(-q^8)f^2(-q^{12})f(-q^{168})}\right.\\&\left.-\frac{f(-q^{12})f^2(-q^{14})f(-q^{56})f^3(-q^{84})}{f(-q^8)f^2(-q^{28})f^2(-q^{42})f^2(-q^{168})}\right\}.\end{aligned} \tag{7.10}$$

Combining (7.9) with (7.10), simplifying with the help of (2.5), (2.8), and (2.9), and finally applying (3.11) and the definition of $T(q)$, we deduce that

$$\begin{aligned}B(-q^2)C(q^2) &= \frac{1}{2q^2}\left\{\frac{f^2(-q^4)}{f(-q^2)f(-q^8)}\cdot\frac{f(-q^6)}{f(-q^{12})}\cdot\frac{f^2(-q^{28})}{f(-q^{14})f(-q^{56})}\cdot\frac{f(-q^{42})}{f(-q^{84})}\right.\\ &\quad\left.-\frac{f(-q^2)}{f(-q^4)}\cdot\frac{f^2(-q^{12})}{f(-q^6)f(-q^{24})}\cdot\frac{f(-q^{14})}{f(-q^{28})}\cdot\frac{f^2(-q^{84})}{f(-q^{42})f(-q^{168})}\right\}\\ &= \frac{1}{2q^2}\{\chi(q^2)\chi(-q^6)\chi(q^{14})\chi(-q^{42})-\chi(-q^2)\chi(q^6)\chi(-q^{14})\chi(q^{42})\}\\ &= T(q^4). \qquad (7.11)\end{aligned}$$

This completes the proof. □

We record some interesting factorization theorems as corollaries of Theorem 7.1 and the work in [5].

Corollary 7.2. *The following factorizations hold:*

$$\begin{aligned}&\{G(q^3)G(-q^7)-q^2H(q^3)H(-q^7)\}\{G(q^{21})H(-q)+q^4H(q^{21})G(-q)\}\\ &\quad=\{G(-q^3)G(q^7)-q^2H(-q^3)H(q^7)\}\{G(-q^{21})H(q)+q^4H(-q^{21})G(q)\}\\ &\quad=G(q^{42})H(q^2)-q^8H(q^{42})G(q^2)\\ &\quad=G(q^6)G(q^{14})+q^4H(q^6)H(q^{14}). \qquad (7.12)\end{aligned}$$

Proof. Recall the notation of Theorem 3.4 as well as the definition of $T(q)$, (7.5). By Theorem 3.4,

$$B(-q^2)C(q^2)=V(-q^2)Y(-q^2)=W(-q^2)Z(-q^2)=W(q^2)Z(q^2)=V(q^2)Y(q^2). \qquad (7.13)$$

Combining (7.13) with (7.6), we find that

$$V(q^2)Y(q^2)=V(-q^2)Y(-q^2)=T(q^4). \qquad (7.14)$$

Replacing q^2 by q in (7.14), we deduce the first two equalities in (7.12). The third equality in (7.12) is simply (3.10) with q replaced by q^2. □

Corollary 7.3. *The following factorizations hold:*

$$\begin{aligned}&\frac{1}{2q^2}\frac{\chi(-q^{12})\chi(-q^{84})}{\chi(q^2)\chi(q^{14})}\{\chi(q)\chi(-q^{21})-\chi(-q)\chi(q^{21})\}\\ &\qquad\times\{\chi(q^3)\chi(-q^7)-\chi(-q^3)\chi(q^7)\}\\ &\quad=\frac{q^2\chi(-q^4)\chi(-q^{28})}{2\chi(q^6)\chi(q^{42})}\{\chi(q)\chi(-q^{21})+\chi(-q)\chi(q^{21})\}\\ &\qquad\times\{\chi(q^3)\chi(-q^7)+\chi(-q^3)\chi(q^7)\}\\ &\quad=\{\chi(q^2)\chi(-q^6)\chi(q^{14})\chi(-q^{42})-\chi(-q^2)\chi(q^6)\chi(-q^{14})\chi(q^{42})\}. \qquad (7.15)\end{aligned}$$

Proof. Employing (3.18), (3.19), (7.14), and (3.11), we readily deduce that the first and third expressions in Corollary 7.3 are equal. Similarly, from (3.20), (3.21), (7.13), (7.14), and (3.11), we verify that the second and third expressions in Corollary 7.3 are equal. This completes the proof. □

8. An Identity for the Level 96

Koike [21] conjectured Theorem 8.1, and Bringmann and Swisher [11] gave the first proof.

Theorem 8.1. *We have*

$$\begin{aligned}&\{G(q^{48})G(q^2)+q^{10}H(q^{48})H(q^2)\}\{G(q^{32})G(q^3)+q^7H(q^{32})H(q^3)\}\\&\quad=\{G(q^{96})H(q)-q^{19}H(q^{96})G(q)\}.\end{aligned}$$

Proof. Let us define $K(q)$, $L(q)$, and $M(q)$ by

$$\begin{aligned}K(q)&:=G(q^{24})G(q)+q^5H(q^{24})H(q),\\L(q)&:=G(q^{96})H(q)-q^{19}H(q^{96})G(q),\\M(q)&:=G(q^{32})G(q^3)+q^7H(q^{32})H(q^3).\end{aligned}$$

Then a slight rearrangement of Theorem 8.1 reads

$$K(q^2)=\frac{L(q)}{M(q)}. \tag{8.1}$$

Employing (3.1) and (3.2), we find that

$$\begin{aligned}L(q)&=\frac{f(-q^8)}{f(-q^2)}\{G(q^{96})(q^3H(q^{16})+G(-q^4))\\&\quad-q^{19}H(q^{96})(G(q^{16})+qH(-q^4))\}\\&=\frac{f(-q^8)}{f(-q^2)}\{q^3(G(q^{96})H(q^{16})-q^{16}H(q^{96})G(q^{16}))\\&\quad+(G(q^{96})G(-q^4)-q^{20}H(q^{96})H(-q^4))\}.\end{aligned} \tag{8.2}$$

Applying (3.7) with q replaced by q^{16} and (3.9) with q replaced by $-q^4$, we conclude from (8.2) that

$$L(q)=\frac{f(-q^8)}{f(-q^2)}\left\{q^3\frac{\chi(-q^{16})}{\chi(-q^{48})}+\frac{\chi(q^{12})\chi(-q^{48})}{\chi(q^4)\chi(-q^{16})}\right\}. \tag{8.3}$$

In the definition of $M(q)$ above, employ (3.1) and (3.2), each with q replaced by q^3. Then apply (3.6) with q replaced by q^{16} and (3.8) with q replaced by $-q^4$, arguing as in (8.3), in order to conclude that

$$M(q)=\frac{f(-q^{24})}{f(-q^6)}\left\{q^3\frac{\chi(q^4)\chi(-q^{16})}{\chi(q^{12})\chi(-q^{48})}+\frac{\chi(-q^{48})}{\chi(-q^{16})}\right\}. \tag{8.4}$$

A slight simplification of (8.4) yields

$$M(q)=\frac{f(-q^{24})\chi(q^4)}{f(-q^6)\chi(q^{12})}\left\{q^3\frac{\chi(-q^{16})}{\chi(-q^{48})}+\frac{\chi(q^{12})\chi(-q^{48})}{\chi(q^4)\chi(-q^{16})}\right\}. \tag{8.5}$$

Combining (8.3) and (8.5), we arrive at

$$\frac{L(q)}{M(q)}=\frac{f(-q^6)f(-q^8)\chi(q^{12})}{f(-q^2)f(-q^{24})\chi(q^4)}. \tag{8.6}$$

By (3.9) with q replaced by q^2, we see that

$$K(q^2)=\frac{\chi(-q^6)\chi(-q^{24})}{\chi(-q^2)\chi(-q^8)}. \tag{8.7}$$

With the use of (2.5), (2.8), and (2.9), we readily verify that

$$\frac{f(-q^6)f(-q^8)\chi(q^{12})}{f(-q^2)f(-q^{24})\chi(q^4)}=\frac{\chi(-q^6)\chi(-q^{24})}{\chi(-q^2)\chi(-q^8)}. \tag{8.8}$$

Combining (8.6)–(8.8), we deduce (8.1). This completes the proof. □

9. Identities for the Level 104

Next, we prove relations for the level 104. Our primary result is Theorem 9.1, which was originally conjectured by Koike [21] and proved first in [11]. The proof of this identity is harder than the proofs considered thus far. At the end, we record some interesting corollaries yielded by our methods.

Theorem 9.1. *We have*

$$\begin{aligned}&\{G(q^{13})H(q^8)-qH(q^{13})G(q^8)\}\{G(q)G(q^{104})+q^{21}H(q)H(q^{104})\}\\&\quad=G(q^{52})H(q^2)-q^{10}H(q^{52})H(q^2).\end{aligned} \tag{9.1}$$

Before we begin, we establish the following lemmas.

Lemma 9.2. *We have*

$$f(-q,-q^9)=G(-q)H(q^4)f(-q^2), \tag{9.2}$$

$$f(-q^3,-q^7)=H(-q)G(q^4)f(-q^2), \tag{9.3}$$

$$\phi(-q^5)=[G(-q)G(q^4)+qH(-q)H(q^4)]f(-q^2), \tag{9.4}$$

$$\frac{G(q)}{H(q)}f(-q^5)=f(-q^7,-q^8)+qf(-q^2,-q^{13}), \tag{9.5}$$

$$\frac{H(q)}{G(q)}f(-q^5)=f(-q^4,-q^{11})-qf(-q,-q^{14}), \tag{9.6}$$

Proof. Relations (9.2)–(9.3) follow from (2.5), (2.8), and (2.13). Replace q by $-q$ in (3.4) to obtain (9.4).

By [3, Entry 10(i), (ii), p. 379] (also by the Quintuple Product Identity (2.12)), we know that

$$f(-q^7,-q^8)+qf(-q^2,-q^{13})=\frac{f(-q^2,-q^3)}{f(-q,-q^4)}f(-q^5) \tag{9.7}$$

and

$$f(-q^4,-q^{11})-qf(-q,-q^{14})=\frac{f(-q,-q^4)}{f(-q^2,-q^3)}f(-q^5). \tag{9.8}$$

Applying (2.13) to each of (9.7) and (9.8), we deduce (9.5) and (9.6), respectively. This completes the proof. □

Lemma 9.3. *The following relations hold:*

$$\begin{aligned}&\{G(q^{26})H(-q)+q^5H(q^{26})G(-q)\}\\&\quad\times\{G(q^4)G(q^{26})+q^6H(q^4)H(q^{26})\}f(-q^2)f(-q^{26})\\&=f(-q^3,-q^7)\{f(-q^{182},-q^{208})+q^{26}f(-q^{52},-q^{338})\}+q^{11}f(-q,-q^9)\\&\quad\times\{f(-q^{104},-q^{286})-q^{26}f(-q^{26},-q^{364})\}+q^5\phi(-q^5)f(-q^{130})\end{aligned} \tag{9.9}$$

and

$$\begin{aligned}&-\{G(q^{52})H(q^2)-q^{10}G(q^2)H(q^{52})\}\\&\quad\times\{G(q^2)G(-q^{13})-q^3H(q^2)H(-q^{13})\}f(-q^2)f(-q^{26})\\&=q^3f(-q^{39},-q^{91})\{f(-q^8,-q^{22})-q^2f(-q^2,-q^{28})\}+q^{10}f(-q^{13},-q^{117})\\&\quad\times\{f(-q^{14},-q^{16})+q^2f(-q^4,-q^{26})\}-\phi(-q^{65})f(-q^{10}).\end{aligned} \tag{9.10}$$

Proof. Applying (9.2)–(9.6) to the right-hand side of (9.9) and simplifying, we conclude that

$$\begin{aligned}&f(-q^3,-q^7)\{f(-q^{182},-q^{208})+q^{26}f(-q^{52},-q^{338})\}+q^{11}f(-q,-q^9)\\&\quad\times\{f(-q^{104},-q^{286})-q^{26}f(-q^{26},-q^{364})\}+q^5\phi(-q^5)f(-q^{130})\\&=G(q^4)H(-q)f(-q^2)\left\{\frac{G(q^{26})}{H(q^{26})}f(-q^{130})\right\}\\&\quad+q^{11}G(-q)H(q^4)f(-q^2)\left\{\frac{H(q^{26})}{G(q^{26})}f(-q^{130})\right\}\\&\quad+q^5[G(-q)G(q^4)+qH(-q)H(q^4)]f(-q^2)f(-q^{130})\end{aligned}$$

$$
\begin{aligned}
&= f(-q^2)f(-q^{130})\left\{\frac{H(-q)G(q^4)G(q^{26})}{H(q^{26})}+q^{11}\frac{G(-q)H(q^4)H(q^{26})}{G(q^{26})}\right\}\\
&\quad + q^5 f(-q^2)f(-q^{130})\frac{G(q^{26})H(q^{26})}{G(q^{26})H(q^{26})}[G(-q)G(q^4)+qH(-q)H(q^4)]\\
&= \frac{f(-q^2)f(-q^{130})}{G(q^{26})H(q^{26})}[\{G^2(q^{26})G(q^4)H(-q)+q^{11}H^2(q^{26})H(q^4)G(-q)\}\\
&\quad + q^5[G(-q)G(q^4)G(q^{26})H(q^{26})+qH(-q)H(q^4)G(q^{26})H(q^{26})]]\\
&= \frac{f(-q^2)f(-q^{130})}{G(q^{26})H(q^{26})}\\
&\quad \times [(G(q^{26})H(-q)+q^5H(q^{26})G(-q))(G(q^4)G(q^{26})+q^6H(q^4)H(q^{26}))]\\
&= f(-q^2)f(-q^{26})\times[(G(q^{26})H(-q)+q^5H(q^{26})G(-q))(G(q^4)G(q^{26})\\
&\quad + q^6H(q^4)H(q^{26}))],
\end{aligned}
\tag{9.11}
$$

where we used (2.14) with q replaced by q^{26} in the final line. This establishes (9.9). The proof of (9.10) is analogous; we omit the details. □

Let us define, by (2.13),

$$
g(q) := f(-q)G(q) = f(-q^2,-q^3) \quad \text{and} \quad h(q) := f(-q)H(q) = f(-q,-q^4). \tag{9.12}
$$

Lemma 9.4. *The following relations hold:*

$$
\frac{g(q)}{\chi(-q)} = f(-q^{13},-q^{17})+qf(-q^7,-q^{23}), \tag{9.13}
$$

$$
\frac{h(q)}{\chi(-q)} = f(-q^{11},-q^{19})+q^3f(-q,-q^{29}). \tag{9.14}
$$

Proof. In (2.12), replace q by q^5 and set $B=-q^4$. After one application of (2.4), we arrive at

$$
\frac{f(-q^2,-q^8)}{f(-q,-q^9)}f(-q^{10}) = f(-q^{13},-q^{17})+qf(-q^7,-q^{23}). \tag{9.15}
$$

With the help of (2.5), (2.9), and (9.12), the left-hand side of (9.15) is readily seen to be equal to $g(q)/\chi(-q)$. Hence, (9.13) is proved. The proof of (9.14) is analogous; we omit the details. □

Now, we prove Theorem 9.1.

Proof of Theorem 9.1. First, we derive identities for the two bracketed expressions in (9.1). Employing (3.1) and (3.2) in (3.13), we deduce that

$$\begin{aligned}
&\sqrt{\frac{\chi(-q^{13})}{\chi(-q)} - q\frac{\chi(-q)}{\chi(-q^{13})}} \\
&\quad = G(q^{26})H(q) - q^5H(q^{26})G(q) \\
&\quad = \frac{f(-q^8)}{f(-q^2)}\{G(q^{26})(q^3H(q^{16}) + G(-q^4)) - q^5H(q^{26})(G(q^{16}) + qH(-q^4))\} \\
&\quad = \frac{f(-q^8)}{f(-q^2)}\{q^3(G(q^{26})H(q^{16}) - q^2H(q^{26})G(q^{16})) \\
&\qquad + (G(-q^4)G(q^{26}) - q^6H(-q^4)H(q^{26}))\}. \qquad (9.16)
\end{aligned}$$

Observe that the final expression in (9.16) gives us an odd–even dissection of the first expression in (9.16). Therefore, considering odd parts in (9.16), we find that

$$\begin{aligned}
&G(q^{26})H(q^{16}) - q^2H(q^{26})G(q^{16}) \\
&\quad = \frac{f(-q^2)}{2q^3f(-q^8)}\left(\sqrt{\frac{\chi(-q^{13})}{\chi(-q)} - q\frac{\chi(-q)}{\chi(-q^{13})}} - \sqrt{\frac{\chi(q^{13})}{\chi(q)} + q\frac{\chi(q)}{\chi(q^{13})}}\right). \qquad (9.17)
\end{aligned}$$

Similarly, starting from (3.13) and employing (3.1) and (3.2), both with q replaced by q^{13}, we see that

$$\begin{aligned}
\sqrt{\frac{\chi(-q^{13})}{\chi(-q)} - q\frac{\chi(-q)}{\chi(-q^{13})}} &= G(q^2)G(q^{13}) + q^3H(q^2)H(q^{13}) \\
&= \frac{f(-q^{104})}{f(-q^{26})}\{G(q^2)(G(q^{208}) + q^{13}H(-q^{52})) \\
&\quad + q^3H(q^2)\,(q^{39}H(q^{208}) + G(-q^{52}))\} \\
&= \frac{f(-q^{104})}{f(-q^{26})}\{(G(q^2)G(q^{208}) + q^{42}H(q^2)H(q^{208})) \\
&\quad + q^3(G(-q^{52})H(q^2) + q^{10}H(-q^{52})G(q^2))\}. \qquad (9.18)
\end{aligned}$$

We observe that the last expression in (9.18) yields another odd–even dissection of $\sqrt{\frac{\chi(-q^{13})}{\chi(-q)} - q\frac{\chi(-q)}{\chi(-q^{13})}}$. Considering even parts in (9.18), we deduce that

$$\begin{aligned}
&G(q^2)G(q^{208}) + q^{42}H(q^2)H(q^{208}) \\
&\quad = \frac{f(-q^{26})}{2f(-q^{104})}\left(\sqrt{\frac{\chi(-q^{13})}{\chi(-q)} - q\frac{\chi(-q)}{\chi(-q^{13})}} + \sqrt{\frac{\chi(q^{13})}{\chi(q)} + q\frac{\chi(q)}{\chi(q^{13})}}\right). \qquad (9.19)
\end{aligned}$$

Multiplying together (9.17) and (9.19), we obtain

$$\{G(q^{26})H(q^{16}) - q^2H(q^{26})G(q^{16})\}\{G(q^2)G(q^{208}) + q^{42}H(q^2)H(q^{208})\}$$
$$= \frac{f(-q^2)f(-q^{26})}{4q^3f(-q^8)f(-q^{104})}\left[\left(\frac{\chi(-q^{13})}{\chi(-q)} - q\frac{\chi(-q)}{\chi(-q^{13})}\right) - \left(\frac{\chi(q^{13})}{\chi(q)} + q\frac{\chi(q)}{\chi(q^{13})}\right)\right]. \tag{9.20}$$

Replacing q by q^2 in (9.1) and comparing the resulting equation with (9.20), we see that, in order to prove (9.1), it suffices to prove that

$$G(q^{104})H(q^4) - q^{20}H(q^{104})G(q^4)$$
$$= \frac{f(-q^2)f(-q^{26})}{4q^3f(-q^8)f(-q^{104})}\left[\left(\frac{\chi(-q^{13})}{\chi(-q)} - q\frac{\chi(-q)}{\chi(-q^{13})}\right) - \left(\frac{\chi(q^{13})}{\chi(q)} + q\frac{\chi(q)}{\chi(q^{13})}\right)\right]. \tag{9.21}$$

By (9.12), $f(-q^4)f(-q^{104})$ times the left-hand side of (9.21) is equal to

$$g(q^{104})h(q^4) - q^{20}h(q^{104})g(q^4). \tag{9.22}$$

Utilizing (2.5) and (2.7)–(2.9), we deduce that $f(-q^4)f(-q^{104})$ times the right-hand side of (9.21) is equal to

$$f(-q^4)f(-q^{104}) \cdot \frac{f(-q^2)f(-q^{26})}{4q^3f(-q^8)f(-q^{104})}$$
$$\times\left[\left(\frac{f(-q^{13})}{f(-q^{26})}\cdot\frac{f(-q^2)}{f(-q)} - q\frac{f(-q)}{f(-q^2)}\cdot\frac{f(-q^{26})}{f(-q^{13})}\right)\right.$$
$$\left. - \left(\frac{f^2(-q^{26})}{f(-q^{13})f(-q^{52})}\cdot\frac{f(-q)f(-q^4)}{f^2(-q^2)} + q\frac{f^2(-q^2)}{f(-q)f(-q^4)}\cdot\frac{f(-q^{13})f(-q^{52})}{f^2(-q^{26})}\right)\right]$$
$$= \frac{1}{4q^3}\frac{f(-q^4)}{f(-q^8)}\left[\frac{f^2(-q^2)}{f(-q)}\cdot f(-q^{13}) - qf(-q)\cdot\frac{f^2(-q^{26})}{f(-q^{13})}\right.$$
$$\left. - \frac{f(-q)f(-q^4)}{f(-q^2)}\cdot\frac{f^3(-q^{26})}{f(-q^{13})f(-q^{52})} - q\frac{f^3(-q^2)}{f(-q)f(-q^4)}\cdot\frac{f(-q^{13})f(-q^{52})}{f(-q^{26})}\right]$$
$$= \frac{1}{4q^3}\chi(-q^4)[\psi(q)f(-q^{13}) - qf(-q)\psi(q^{13}) - \psi(-q)f(q^{13}) - qf(q)\psi(-q^{13})]. \tag{9.23}$$

Therefore, combining (9.22) and (9.23), we see that, in order to prove (9.1), it is sufficient to prove the beautiful identity

$$\frac{1}{\chi(-q^4)}(g(q^{104})h(q^4) - q^{20}h(q^{104})g(q^4))$$
$$= \frac{1}{4q^3}[\psi(q)f(-q^{13}) - \psi(-q)f(q^{13}) - qf(-q)\psi(q^{13}) - qf(q)\psi(-q^{13})]. \tag{9.24}$$

To prove (9.24), we first provide alternative representations for the expressions $\psi(q)f(-q^{13}) - \psi(-q)f(q^{13})$ and $-qf(-q)\psi(q^{13}) - qf(q)\psi(-q^{13})$. To that end, we apply Theorem 2.4 with the parameters $\epsilon_1 = 0$, $\epsilon_2 = 1$, $a = 1$, $b = q$, $c = q^{13}$, $d = q^{26}$, $\alpha = 1$, $\beta = 39$ and $m = 40$. Accordingly, we deduce that

$$\begin{aligned} &f(1,q)f(-q^{13},-q^{26}) \\ &\quad = \sum_{r=-20}^{19} (-1)^r q^{13r(3r-1)/2} f(-q^{26-39r},-q^{14+39r}) f(-q^{806-39r},-q^{754+39r}). \end{aligned} \tag{9.25}$$

Applying Theorem 2.4 with the same set of parameters as for (9.25), except now with $\epsilon_1 = 1$ and $\epsilon_2 = 0$, we deduce that

$$\begin{aligned} &f(-1,-q)f(q^{13},q^{26}) \\ &\quad = \sum_{r=-20}^{19} q^{13r(3r-1)/2} f(-q^{26-39r},-q^{14+39r}) f(-q^{806-39r},-q^{754+39r}). \end{aligned} \tag{9.26}$$

Combining (9.25)–(9.26), along with the identities obtained from them by replacing q by $-q$, we see that

$$\begin{aligned} &f(1,q)f(-q^{13},-q^{26}) - f(1,-q)f(q^{13},-q^{26}) \\ &\quad + f(-1,-q)f(q^{13},q^{26}) - f(-1,q)f(-q^{13},q^{26}) = \sum_{i=0}^{3} S_i, \end{aligned} \tag{9.27}$$

where

$$\begin{aligned} S_i = \sum_{\substack{-20\le r\le 19 \\ r\equiv i \pmod 4}} &[\{((-1)^r+1)q^{13r(3r-1)/2} \\ &\times f(-q^{26-39r},-q^{14+39r}) f(-q^{806-39r},-q^{754+39r})\} \\ &- \{((-1)^r+1)(-q)^{13r(3r-1)/2} \\ &\times f(-(-q)^{26-39r},-(-q)^{14+39r}) f(-(-q)^{806-39r},-(-q)^{754+39r})\}]. \end{aligned} \tag{9.28}$$

It is easy to show directly that

$$S_0 = S_1 = S_3 = 0. \tag{9.29}$$

Indeed, for $i = 0$, replacing r by $4s$, we see that

$$\begin{aligned} S_0 = \sum_{\substack{-20\le r\le 19 \\ r\equiv 0 \pmod 4}} &[\{((-1)^r+1)q^{13r(3r-1)/2} \\ &\times f(-q^{26-39r},-q^{14+39r}) f(-q^{806-39r},-q^{754+39r})\} \end{aligned}$$

$$
\begin{aligned}
&\quad - \{((-1)^r + 1)(-q)^{13r(3r-1)/2} \\
&\quad \times f(-(-q)^{26-39r}, -(-q)^{14+39r}) f(-(-q)^{806-39r}, -(-q)^{754+39r})\}] \\
&= \sum_{s=-5}^{4} [\{(1+1)q^{26s(12s-1)} \\
&\quad \times f(-q^{-156s+26}, -q^{156s+14}) f(-q^{-156s+806}, -q^{156s+754})\} \\
&\quad - \{(1+1)q^{26s(12s-1)} \\
&\quad \times f(-q^{-156s+26}, -q^{156s+14}) f(-q^{-156s+806}, -q^{156s+754})\}] \\
&= \sum_{s=-5}^{4} 0 = 0.
\end{aligned} \tag{9.30}
$$

The proofs that $S_i = 0$ for $i = 1$ and $i = 3$ are analogous.

Now we address S_2. Replace r by $4s + 2$ in (9.28) in order to deduce that

$$
\begin{aligned}
S_2 &= \sum_{\substack{-20 \le r \le 19 \\ r \equiv 2 \pmod 4}} [\{((-1)^r + 1)q^{13r(3r-1)/2} \\
&\quad \times f(-q^{26-39r}, -q^{14+39r}) f(-q^{806-39r}, -q^{754+39r})\} \\
&\quad - \{((-1)^r + 1)(-q)^{13r(3r-1)/2} \\
&\quad \times f(-(-q)^{26-39r}, -(-q)^{14+39r}) f(-(-q)^{806-39r}, -(-q)^{754+39r})\}] \\
&= \sum_{s=-5}^{4} [2q^{13(2s+1)(12s+5)} f(-q^{-156s-52}, -q^{156s+92}) f(-q^{-156s+728}, -q^{156s+832}) \\
&\quad - 2(-1)q^{13(2s+1)(12s+5)} f(-q^{-156s-52}, -q^{156s+92}) f(-q^{-156s+728}, -q^{156s+832})] \\
&= 4 \sum_{s=-5}^{4} q^{13(2s+1)(12s+5)} f(-q^{-156s-52}, -q^{156s+92}) f(-q^{-156s+728}, -q^{156s+832}).
\end{aligned} \tag{9.31}
$$

Simplifying (9.31) with the help of (2.4) and (2.3), we obtain

$$
\begin{aligned}
S_2 = 4(&q^{6435} f(-q^{728}, -q^{-688}) f(-q^{1508}, -q^{52}) \\
&+ q^{3913} f(-q^{572}, -q^{-532}) f(-q^{1352}, -q^{208}) \\
&+ q^{2015} f(-q^{416}, -q^{-376}) f(-q^{1196}, -q^{364}) \\
&+ q^{741} f(-q^{260}, -q^{-220}) f(-q^{1040}, -q^{520})
\end{aligned}
$$

$$
\begin{aligned}
&+q^{91}f(-q^{104},-q^{-64})f(-q^{884},-q^{676})\\
&+q^{65}f(-q^{-52},-q^{92})f(-q^{728},-q^{832})+q^{663}f(-q^{-208},-q^{248})f(-q^{572},-q^{988})\\
&+q^{1885}f(-q^{-364},-q^{404})f(-q^{416},-q^{1144})\\
&+q^{3731}f(-q^{-520},-q^{560})f(-q^{260},-q^{1300})\\
&+q^{6201}f(-q^{-676},-q^{716})f(-q^{104},-q^{1456}))\\
&=4q^{171}f(-q^{8},-q^{32})f(-q^{52},-q^{1508})+4q^{105}f(-q^{12},-q^{28})f(-q^{208},-q^{1352})\\
&+4q^{55}f(-q^{16},-q^{24})f(-q^{364},-q^{1196})+4q^{21}f(-q^{20},-q^{20})f(-q^{520},-q^{1040})\\
&+4q^{3}f(-q^{16},-q^{24})f(-q^{676},-q^{884})+4qf(-q^{12},-q^{28})f(-q^{728},-q^{832})\\
&+4q^{15}f(-q^{8},-q^{32})f(-q^{572},-q^{988})+4q^{45}f(-q^{4},-q^{36})f(-q^{416},-q^{1144})\\
&+4q^{91}f(-q^{40},-1)f(-q^{260},-q^{1300})-4q^{149}f(-q^{4},-q^{36})f(-q^{104},-q^{1456})\\
&=4q^{3}f(-q^{16},-q^{24})\{f(-q^{676},-q^{884})+q^{52}f(-q^{364},-q^{1196})\}\\
&+4q^{15}f(-q^{8},-q^{32})\{f(-q^{572},-q^{988})+q^{156}f(-q^{52},-q^{1508})\}\\
&+4qf(-q^{12},-q^{28})\{f(-q^{728},-q^{832})+q^{104}f(-q^{208},-q^{1352})\}\\
&+4q^{45}f(-q^{4},-q^{36})\{f(-q^{416},-q^{1144})-q^{104}f(-q^{104},-q^{1456})\}\\
&+4q^{21}f(-q^{20},-q^{20})f(-q^{520},-q^{1040}). \qquad (9.32)
\end{aligned}
$$

By (2.3),

$$f(-1,-q)f(q^{13},q^{26})=f(-1,q)f(-q^{13},q^{26})=0. \qquad (9.33)$$

Thus, utilizing (2.7), (2.2), (2.8), and (9.33), and then applying (9.27), (9.29), and (9.32), we conclude that

$$
\begin{aligned}
&\psi(q)f(-q^{13})-\psi(-q)f(q^{13})\\
&=\frac{1}{2}(f(1,q)f(-q^{13},-q^{26})-f(1,-q)f(q^{13},-q^{26})\\
&\quad+f(-1,-q)f(q^{13},q^{26})-f(-1,q)f(-q^{13},q^{26}))\\
&=\frac{1}{2}\sum_{i=0}^{3}S_i=\frac{1}{2}S_2\\
&=2q^{3}f(-q^{16},-q^{24})\{f(-q^{676},-q^{884})+q^{52}f(-q^{364},-q^{1196})\}\\
&\quad+2q^{15}f(-q^{8},-q^{32})\{(-q^{572},-q^{988})+q^{156}f(-q^{52},-q^{1508})\}\\
&\quad+2qf(-q^{12},-q^{28})\{f(-q^{728},-q^{832})+q^{104}f(-q^{208},-q^{1352})\}
\end{aligned}
$$

$$+2q^{45}f(-q^4,-q^{36})\{f(-q^{416},-q^{1144})-q^{104}f(-q^{104},-q^{1456})\}$$
$$+2q^{21}f(-q^{20},-q^{20})f(-q^{520},-q^{1040}). \tag{9.34}$$

Now, by (9.12) with q replaced by q^8, and by (9.13) and (9.14), each with q replaced by q^{52},

$$2q^3f(-q^{16},-q^{24})\{f(-q^{676},-q^{884})+q^{52}f(-q^{364},-q^{1196})\}$$
$$+2q^{15}f(-q^8,-q^{32})\{(-q^{572},-q^{988})+q^{156}f(-q^{52},-q^{1508})\}$$
$$=\frac{2q^3}{\chi(-q^{52})}(g(q^8)g(q^{52})+q^{12}h(q^8)h(q^{52})). \tag{9.35}$$

By (2.6) and (2.8), we know that $\phi(-q^{20}) = f(-q^{20},-q^{20})$ and $f(-q^{520}) = f(-q^{520},-q^{1040})$. Therefore, applying (9.9) with q replaced by q^4, we deduce that

$$2qf(-q^{12},-q^{28})\{f(-q^{728},-q^{832})+q^{104}f(-q^{208},-q^{1352})\}$$
$$+2q^{45}f(-q^4,-q^{36})\{f(-q^{416},-q^{1144})-q^{104}f(-q^{104},-q^{1456})\}$$
$$+2q^{21}f(-q^{20},-q^{20})f(-q^{520},-q^{1040})$$
$$=2qf(-q^8)f(-q^{104})[G(q^{104})H(-q^4)+q^{20}H(q^{104})G(-q^4)]$$
$$\times[G(q^{16})G(q^{104})+q^{24}H(q^{16})H(q^{104})]. \tag{9.36}$$

Consequently, employing (9.35)–(9.36), we deduce from (9.34) that

$$\psi(q)f(-q^{13})-\psi(-q)f(q^{13})$$
$$=\frac{2q^3}{\chi(-q^{52})}(g(q^8)g(q^{52})+q^{12}h(q^8)h(q^{52}))$$
$$+2qf(-q^8)f(-q^{104})[G(q^{104})H(-q^4)+q^{20}H(q^{104})G(-q^4)]$$
$$\times[G(q^{16})G(q^{104})+q^{24}H(q^{16})H(q^{104})]. \tag{9.37}$$

Next, we derive an analogous representation for

$$-qf(-q)\psi(q^{13})-qf(q)\psi(-q^{13}).$$

Applying Theorem 2.4 with the parameters $\epsilon_1=0$, $\epsilon_2=1$, $a=1$, $b=q^{13}$, $c=q$, $d=q^2$, $\alpha=13$, $\beta=3$ and $m=40$, we deduce that

$$f(1,q^{13})f(-q,-q^2)$$
$$=\sum_{r=-20}^{19}(-1)^rq^{r(3r-1)/2}f(-q^{260-39r},-q^{260+39r})f(-q^{80-3r},-q^{40+3r}). \tag{9.38}$$

Applying Theorem 2.4 with the same set of parameters as for (9.38), except now with $\epsilon_1 = 1$ and $\epsilon_2 = 0$, we find that

$$f(-1,-q^{13})f(q,q^2) = \sum_{r=-20}^{19} q^{r(3r-1)/2} f(-q^{260-39r},-q^{260+39r})f(-q^{80-3r},-q^{40+3r}). \tag{9.39}$$

Combining (9.38)–(9.39), along with the identities obtained by replacing q by $-q$ in each of them, we find that

$$f(1,q^{13})f(-q,-q^2)+f(1,-q^{13})f(q,-q^2) + f(-1,-q^{13})f(q,q^2)+f(-1,q^{13})f(-q,q^2) = \sum_{i=0}^{3} S_i, \tag{9.40}$$

where

$$\begin{aligned} S_i = \sum_{\substack{-20\le r\le 19\\ r\equiv i \pmod 4}} &[\{((-1)^r+1)q^{r(3r-1)/2} \\ &\times f(-q^{260-39r},-q^{260+39r})f(-q^{80-3r},-q^{40+3r})\} \\ &+\{((-1)^r+1)(-q)^{r(3r-1)/2} \\ &\times f(-(-q)^{260-39r},-(-q)^{260+39r})f(-(-q)^{80-3r},-(-q)^{40+3r})\}]. \end{aligned}$$

It is easy to show directly that

$$S_1 = S_2 = S_3 = 0. \tag{9.41}$$

In S_0, replace r by $4s$, and simplify with the help of (2.4) and (2.3). Accordingly, we deduce that

$$\begin{aligned} S_0 = \sum_{\substack{-20\le r\le 19\\ r\equiv 0 \pmod 4}} &[\{((-1)^r+1)q^{r(3r-1)/2} \\ &\times f(-q^{260-39r},-q^{260+39r})f(-q^{80-3r},-q^{40+3r})\} \\ &+\{((-1)^r+1)(-q)^{r(3r-1)/2} \\ &\times f(-(-q)^{260-39r},-(-q)^{260+39r})f(-(-q)^{80-3r},-(-q)^{40+3r})\}] \\ = 4\sum_{s=-5}^{4} &q^{2s(12s-1)} f(-q^{-156s+260},-q^{156s+260})f(-q^{-12s+80},-q^{12s+40}) \\ = 4(&q^{610}f(-q^{1040},-q^{-520})f(-q^{140},-q^{-20}) \\ &+q^{392}f(-q^{884},-q^{-364})f(-q^{128},-q^{-8}) \\ &+q^{222}f(-q^{728},-q^{-208})f(-q^{116},-q^{4}) \end{aligned}$$

$$
\begin{aligned}
&+q^{100}f(-q^{572},-q^{-52})f(-q^{104},-q^{16})+q^{26}f(-q^{416},-q^{104})f(-q^{92},-q^{28})\\
&+f(-q^{260},-q^{260})f(-q^{80},-q^{40})+q^{22}f(-q^{104},-q^{416})f(-q^{68},-q^{52})\\
&+q^{92}f(-q^{-52},-q^{572})f(-q^{56},-q^{64})+q^{210}f(-q^{-208},-q^{728})f(-q^{44},-q^{76})\\
&+q^{376}f(-q^{-364},-q^{884})f(-q^{32},-q^{88}))\\
={}&4q^{70}f(-q^{520},-1)f(-q^{20},-q^{100})+4q^{20}f(-q^{364},-q^{156})f(-q^{8},-q^{112})\\
&-4q^{14}f(-q^{208},-q^{312})f(-q^{4},-q^{116})-4q^{48}f(-q^{52},-q^{468})f(-q^{16},-q^{104})\\
&+4q^{26}f(-q^{416},-q^{104})f(-q^{92},-q^{28})+4f(-q^{260},-q^{260})f(-q^{80},-q^{40})\\
&+4q^{22}f(-q^{104},-q^{416})f(-q^{68},-q^{52})-4q^{40}f(-q^{52},-q^{468})f(-q^{56},-q^{64})\\
&-4q^{2}f(-q^{208},-q^{312})f(-q^{44},-q^{76})-4q^{12}f(-q^{156},-q^{364})f(-q^{32},-q^{88})\\
={}&4q^{22}f(-q^{104},-q^{416})\{f(-q^{68},-q^{52})+q^{4}f(-q^{28},-q^{92})\}\\
&-4q^{2}f(-q^{208},-q^{312})\{f(-q^{44},-q^{76})+q^{12}f(-q^{4},-q^{116})\}\\
&-4q^{12}f(-q^{156},-q^{364})\{f(-q^{32},-q^{88})-q^{8}f(-q^{8},-q^{112})\}\\
&-4q^{40}f(-q^{52},-q^{468})\{f(-q^{56},-q^{64})+q^{8}f(-q^{16},-q^{104})\}\\
&+4f(-q^{260},-q^{260})f(-q^{40},-q^{80}).
\end{aligned}
\tag{9.42}
$$

Observe that, by (2.3),

$$f(-1,-q^{13})f(q,q^2)=f(-1,q^{13})f(-q,q^2)=0. \tag{9.43}$$

Thus, utilizing (2.7), (2.2), (2.8) and (9.43), and then applying (9.40), (9.41), and (9.42), we conclude that

$$
\begin{aligned}
&-qf(-q)\psi(q^{13})-qf(q)\psi(-q^{13})\\
&=-\frac{q}{2}(f(-q,-q^2)f(1,q^{13})+f(q,-q^2)f(1,-q^{13})\\
&\quad+f(-1,-q^{13})f(q,q^2)+f(-1,q^{13})f(-q,q^2))\\
&=-\frac{q}{2}\sum_{i=0}^{3}S_i=-\frac{q}{2}S_0\\
&=-2q^{23}f(-q^{104},-q^{416})\{f(-q^{68},-q^{52})+q^{4}f(-q^{28},-q^{92})\}\\
&\quad+2q^{3}f(-q^{208},-q^{312})\{f(-q^{44},-q^{76})+q^{12}f(-q^{4},-q^{116})\}\\
&\quad+2q^{13}f(-q^{156},-q^{364})\{f(-q^{32},-q^{88})-q^{8}f(-q^{8},-q^{112})\}\\
&\quad+2q^{41}f(-q^{52},-q^{468})\{f(-q^{56},-q^{64})+q^{8}f(-q^{16},-q^{104})\}\\
&\quad-2qf(-q^{260},-q^{260})f(-q^{40},-q^{80}).
\end{aligned}
\tag{9.44}
$$

Now, by (9.12) with q replaced by q^{104}, and by (9.13) and (9.14), each with q replaced by q^4,

$$\begin{aligned}
&-2q^{23}f(-q^{104},-q^{416})\{f(-q^{68},-q^{52})+q^4f(-q^{28},-q^{92})\}\\
&\quad+2q^3f(-q^{208},-q^{312})\{f(-q^{44},-q^{76})+q^{12}f(-q^4,-q^{116})\}\\
&\quad=\frac{2q^3}{\chi(-q^4)}(g(q^{104})h(q^4)-q^{20}h(q^{104})g(q^4)). \qquad (9.45)
\end{aligned}$$

By (2.6) and (2.8), we know that $\phi(-q^{260}) = f(-q^{260},-q^{260})$ and $f(-q^{40}) = f(-q^{40},-q^{80})$. Hence, applying (9.10) with q replaced by q^4, we deduce that

$$\begin{aligned}
&2q^{13}f(-q^{156},-q^{364})\{f(-q^{32},-q^{88})-q^8f(-q^8,-q^{112})\}\\
&\quad+2q^{41}f(-q^{52},-q^{468})\{f(-q^{56},-q^{64})+q^8f(-q^{16},-q^{104})\}\\
&\quad-2qf(-q^{260},-q^{260})f(-q^{40},-q^{80})\\
&\quad=-2q\{G(q^{208})H(q^8)-q^{40}G(q^8)H(q^{208})\}\\
&\quad\times\{G(q^8)G(-q^{52})-q^{12}H(q^8)H(-q^{52})\}f(-q^8)f(-q^{104}). \qquad (9.46)
\end{aligned}$$

Consequently, employing (9.45)–(9.46), we deduce from (9.44) that

$$\begin{aligned}
&-qf(-q)\psi(q^{13})-qf(q)\psi(-q^{13})\\
&\quad=\frac{2q^3}{\chi(-q^4)}(g(q^{104})h(q^4)-q^{20}h(q^{104})g(q^4))\\
&\quad-2qf(-q^8)f(-q^{104})[G(q^{208})H(q^8)-q^{40}G(q^8)H(q^{208})]\\
&\quad\times[G(q^8)G(-q^{52})-q^{12}H(q^8)H(-q^{52})]. \qquad (9.47)
\end{aligned}$$

Hence, combining our representations for $\psi(q)f(-q^{13})-\psi(-q)f(q^{13})$ and $-qf(-q)\psi(q^{13})-qf(q)\psi(-q^{13})$, namely (9.37) and (9.47), we conclude that

$$\begin{aligned}
&\frac{1}{4q^3}[\psi(q)f(-q^{13})-\psi(-q)f(q^{13})-qf(-q)\psi(q^{13})-qf(q)\psi(-q^{13})]\\
&\quad=\frac{1}{2\chi(-q^{52})}(g(q^8)g(q^{52})+q^{12}h(q^8)h(q^{52}))\\
&\quad+\frac{1}{2q^2}f(-q^8)f(-q^{104})[G(q^{104})H(-q^4)+q^{20}H(q^{104})G(-q^4)]\\
&\quad\times[G(q^{16})G(q^{104})+q^{24}H(q^{16})H(q^{104})]\\
&\quad+\frac{1}{2\chi(-q^4)}(g(q^{104})h(q^4)-q^{20}h(q^{104})g(q^4))\\
&\quad-\frac{1}{2q^2}f(-q^8)f(-q^{104})[G(q^{208})H(q^8)-q^{40}G(q^8)H(q^{208})]\\
&\quad\times[G(q^8)G(-q^{52})-q^{12}H(q^8)H(-q^{52})]. \qquad (9.48)
\end{aligned}$$

By (3.12), first with q replaced by $-q^4$, and second with q replaced by q^8, we deduce, respectively, that

$$G(q^{104})H(-q^4)+q^{20}H(q^{104})G(-q^4)=G(q^8)G(-q^{52})-q^{12}H(q^8)H(-q^{52}) \tag{9.49}$$

and

$$G(q^{16})G(q^{104})+q^{24}H(q^{16})H(q^{104})=G(q^{208})H(q^8)-q^{40}G(q^8)H(q^{208}). \tag{9.50}$$

Hence, the second and fourth terms on the right-hand side of (9.48) cancel. Thus, we may rewrite (9.48) in the simpler form

$$\begin{aligned}\frac{1}{4q^3}&[\psi(q)f(-q^{13})-\psi(-q)f(q^{13})-qf(-q)\psi(q^{13})-qf(q)\psi(-q^{13})]\\&=\frac{1}{2\chi(-q^{52})}(g(q^8)g(q^{52})+q^{12}h(q^8)h(q^{52}))\\&\quad+\frac{1}{2\chi(-q^4)}(g(q^{104})h(q^4)-q^{20}h(q^{104})g(q^4)).\end{aligned} \tag{9.51}$$

Employing (9.12) in (3.12), we find that

$$\frac{g(q^2)g(q^{13})}{f(-q^2)f(-q^{13})}+q^3\frac{h(q^2)h(q^{13})}{f(-q^2)f(-q^{13})}=\frac{g(q^{26})h(q)}{f(-q)f(-q^{26})}-q^5\frac{h(q^{26})g(q)}{f(-q)f(-q^{26})}. \tag{9.52}$$

Therefore, multiplying (9.52) by $f(-q^2)f(-q^{26})$ and applying (2.5), (2.8), and (2.9), we conclude that

$$\frac{1}{\chi(-q^{13})}(g(q^2)g(q^{13})+q^3h(q^2)h(q^{13}))=\frac{1}{\chi(-q)}(g(q^{26})h(q)-q^5h(q^{26})g(q)),$$

whence we deduce that

$$\frac{1}{\chi(-q^{52})}(g(q^8)g(q^{52})+q^{12}h(q^8)h(q^{52}))=\frac{1}{\chi(-q^4)}(g(q^{104})h(q^4)-q^{20}h(q^{104})g(q^4)). \tag{9.53}$$

Employing (9.53) in (9.51), we complete the derivation of (9.24). This then completes the proof of a fascinating identity! □

Corollary 9.5.

$$
\begin{aligned}
&G(q^{104})H(q^4) - q^{20}H(q^{104})G(q^4)\\
&\quad = G(q^8)G(q^{52}) + q^{12}H(q^8)H(q^{52})\\
&\quad = \sqrt{\frac{\chi(-q^{52})}{\chi(-q^4)} - q^4\frac{\chi(-q^4)}{\chi(-q^{52})}}\\
&\quad = \frac{1}{4q^3f(-q^8)f(-q^{104})} \times [\psi(q)f(-q^{13}) - \psi(-q)f(q^{13})\\
&\qquad - q(f(-q)\psi(q^{13}) + f(q)\psi(-q^{13}))]. \qquad (9.54)
\end{aligned}
$$

Proof. The first two equalities are immediate from (3.12) and (3.13). Next, divide (9.24) by $f(-q^4)f(-q^{104})$ to reveal that

$$
\begin{aligned}
&\frac{1}{\chi(-q^4)}\left(\frac{g(q^{104})h(q^4)}{f(-q^4)f(-q^{104})} + q^{21}\frac{h(q^{104})h(q^4)}{f(-q^4)f(-q^{104})}\right)\\
&\quad = \frac{1}{4q^3f(-q^4)f(-q^{104})}[\psi(q)f(-q^{13}) - \psi(-q)f(q^{13})\\
&\qquad - q(f(-q)\psi(q^{13}) + f(q)\psi(-q^{13}))]. \qquad (9.55)
\end{aligned}
$$

By (2.5), (2.8), and (2.9),

$$
\chi(-q^4) = \frac{f(-q^4)}{f(-q^8)}. \qquad (9.56)
$$

By (9.12),

$$
\frac{g(q^{104})h(q^4)}{f(-q^4)f(-q^{104})} + q^{20}\frac{h(q^{104})h(q^4)}{f(-q^4)f(-q^{104})} = G(q^{104})H(q^4) + q^{21}H(q^{104})G(q^4). \qquad (9.57)
$$

Employing (9.56)–(9.57) in (9.55), we deduce that the first and final expressions in (9.54) are equal. This then completes the proof. □

We close this section with a curious corollary that provides alternative formulations for the expressions appearing in Lemma 9.3. Let S denote a subset of the rational numbers and $\sum_{n\in S} a_n q^n$ a generic q-series. Let t be a fixed rational number. We define an operator $\mathcal{L}_t$ acting on $\sum_{n\in S} a_n q^n$ by $\mathcal{L}_t(\sum_{n\in S} a_n q^n) = \sum_{n\in S'} a_n q^n$, where $S' \subseteq S$ is the set of all rational numbers in S that are congruent modulo 1 to t.

Corollary 9.6. *The following relations hold:*

$$\begin{aligned}
&\frac{1}{q^{8/5}}\mathcal{L}_{3/5}([\phi(q^{-52/5})+\phi(-q^{260})]f(-q^{8/5}))\\
&\quad=-2\{G(q^{208})H(q^{8})-q^{40}G(q^{8})H(q^{208})\}\\
&\qquad\times\{G(q^{8})G(-q^{52})-q^{12}H(q^{8})H(-q^{52})\}f(-q^{8})f(-q^{104})\\
&\quad=-2\{G(q^{104})H(-q^{4})+q^{20}H(q^{104})G(-q^{4})\}\\
&\qquad\times\{G(q^{16})G(q^{104})+q^{24}H(q^{16})H(q^{104})\}f(-q^{8})f(-q^{104})\\
&\quad=\frac{1}{q^{4/5}}\mathcal{L}_{4/5}([\phi(-q^{4/5})+\phi(-q^{20})]f(-q^{104/5})).
\end{aligned}\tag{9.58}$$

Proof. In [3, Entry 10(ii), p. 262], add $2\phi(q^5)$ to both sides of the identity and replace q by $-q$. We consequently find that

$$\phi(-q^{1/5})+\phi(-q^{5})=-2q^{1/5}f(-q^{3},-q^{7})+2q^{4/5}f(-q,-q^{9})+2\phi(-q^{5}).\tag{9.59}$$

In [3, Entry 31, p. 48], set $a=-q^{1/5}$, $b=-q^{2/5}$, and $n=5$. After two applications of (2.4), we deduce that

$$\begin{aligned}
f(-q^{1/5})&=f(-q^{7},-q^{8})-q^{1/5}f(-q^{5},-q^{10})+qf(-q^{2},-q^{13})\\
&\quad-q^{2/5}f(-q^{4},-q^{11})+q^{7/5}f(-q,-q^{14}).
\end{aligned}\tag{9.60}$$

Replacing q by q^{52} in (9.59), we find that

$$\begin{aligned}
&\phi(-q^{52/5})+\phi(-q^{260})\\
&\quad=-2q^{52/5}f(-q^{156},-q^{364})+2q^{208/5}f(-q^{52},-q^{468})+2\phi(-q^{260}).
\end{aligned}\tag{9.61}$$

Replacing q by q^8 in (9.60) and also employing (2.8), we see that

$$\begin{aligned}
f(-q^{8/5})&=f(-q^{56},-q^{64})-q^{8/5}f(-q^{40})+q^{8}f(-q^{16},-q^{104})\\
&\quad-q^{16/5}f(-q^{32},-q^{88})+q^{56/5}f(-q^{8},-q^{112}).
\end{aligned}\tag{9.62}$$

Hence,

$$\begin{aligned}
&[\phi(q^{-52/5})+\phi(-q^{260})]f(-q^{8/5})\\
&\quad=-2q^{52/5}f(-q^{56},-q^{64})f(-q^{156},-q^{364})\\
&\qquad+2q^{208/5}f(-q^{56},-q^{64})f(-q^{52},-q^{468})+2f(-q^{56},-q^{64})\phi(-q^{260})\\
&\qquad+2q^{12}f(-q^{40})f(-q^{156},-q^{364})-2q^{216/5}f(-q^{40})f(-q^{52},-q^{468})
\end{aligned}$$

$$
\begin{aligned}
&- 2q^{8/5}f(-q^{40})\phi(-q^{260}) - 2q^{92/5}f(-q^{16},-q^{104})f(-q^{156},-q^{364})\\
&+ 2q^{248/5}f(-q^{16},-q^{104})f(-q^{52},-q^{468}) + 2q^{8}f(-q^{16},-q^{104})\phi(-q^{260})\\
&+ 2q^{68/5}f(-q^{32},-q^{88})f(-q^{156},-q^{364})\\
&- 2q^{224/5}f(-q^{32},-q^{88})f(-q^{52},-q^{468}) - 2q^{16/5}f(-q^{32},-q^{88})\phi(-q^{260})\\
&- 2q^{108/5}f(-q^{8},-q^{112})f(-q^{156},-q^{364})\\
&+ 2q^{264/5}f(-q^{8},-q^{112})f(-q^{52},-q^{468})\\
&+ 2q^{56/5}f(-q^{8},-q^{112})\phi(-q^{260}).
\end{aligned}
\tag{9.63}
$$

From (9.63), we deduce that

$$
\begin{aligned}
&\frac{1}{2q^{8/5}}\mathcal{L}_{3/5}([\phi(q^{-52/5})+\phi(-q^{260})]f(-q^{8/5}))\\
&\quad = q^{40}f(-q^{56},-q^{64})f(-q^{52},-q^{468}) - f(-q^{40})\phi(-q^{260})\\
&\qquad + q^{48}f(-q^{16},-q^{104})f(-q^{52},-q^{468}) + q^{12}f(-q^{32},-q^{88})f(-q^{156},-q^{364})\\
&\qquad - q^{20}f(-q^{8},-q^{112})f(-q^{156},-q^{364}).
\end{aligned}
\tag{9.64}
$$

Now, replace q by q^4 in (9.10), and observe that the right-hand side of the resulting equation matches the right-hand side of (9.64). Hence,

$$
\begin{aligned}
&\frac{1}{2q^{8/5}}\mathcal{L}_{3/5}([\phi(q^{-52/5})+\phi(-q^{260})]f(-q^{8/5}))\\
&\quad = -\{G(q^{208})H(q^{8}) - q^{40}G(q^{8})H(q^{208})\}\\
&\qquad \times \{G(q^{8})G(-q^{52}) - q^{12}H(q^{8})H(-q^{52})\}f(-q^{8})f(-q^{104}).
\end{aligned}
\tag{9.65}
$$

This implies the first equality in (9.58).

Starting from (9.59) and (9.60) and proceeding analogously, but using (9.9) in place of (9.10), we derive that

$$
\begin{aligned}
&\frac{1}{2q^{4/5}}\mathcal{L}_{4/5}([\phi(-q^{4/5})+\phi(-q^{20})]f(-q^{104/5}))\\
&\quad = -\{G(q^{104})H(-q^{4}) + q^{20}H(q^{104})G(-q^{4})\}\\
&\qquad \times \{G(q^{16})G(q^{104}) + q^{24}H(q^{16})H(q^{104})\}f(-q^{8})f(-q^{104}),
\end{aligned}
\tag{9.66}
$$

from which we deduce the third equality in (9.58).

Employing (9.49) and (9.50), we readily deduce that the expressions on the right-hand sides of (9.65) and (9.66) are equal. Therefore, the second equality in (9.58) holds. This completes the proof. □

Acknowledgments

The author thanks Bruce Berndt for introducing him to the beauty of the Rogers-Ramanujan functions, and for welcoming him into the q-series family of mathematicians. Berndt's kindness and support over the years are warmly appreciated.

References

[1] G. E. Andrews and B. C. Berndt, *Ramanujan's Lost Notebook, Part I* (Springer, New York, 2005).

[2] G. E. Andrews and B. C. Berndt, *Ramanujan's Lost Notebook, Part III* (Springer, New York, 2010).

[3] B. C. Berndt, *Ramanujan's Notebooks III* (Springer-Verlag, New York, 1991).

[4] B. C. Berndt, G. Choi, Y.-S. Choi, H. Hahn, B. P. Yeap, A. J. Yee, H. Yesilyurt and J. Yi, Ramanujan's forty identities for the Rogers–Ramanujan functions, *Mem. Amer. Math. Soc.* **188**(880) (2007) 1–96.

[5] B. C. Berndt and H. Yesilyurt, New identities for the Rogers–Ramanujan functions, *Acta Arith.* **120** (2005) 395–413.

[6] A. J. F. Biagioli, A proof of some identities of Ramanujan using modular forms, *Glasg. Math. J.* **31** (1989) 271–295.

[7] B. J. Birch, A look back at Ramanujan's notebooks, *Math. Proc. Cambridge Philos. Soc.* **78** (1975) 73–79.

[8] R. Blecksmith, J. Brillhart and I. Gerst, A fundamental modular identity and some applications, *Math. Comp.* **61** (1993) 83–95.

[9] D. Bressoud, *Proof and generalization of certain identities conjectured by Ramanujan*, Ph.D. thesis, Temple University (1977).

[10] D. Bressoud, Some identities involving Rogers–Ramanujan-type functions, *J. London Math. Soc.* **16**(2) (1977) 9–18.

[11] K. Bringmann and H. Swisher, On a conjecture of Koike on identities between Thompson series and Rogers–Ramanujan functions, *Proc. Amer. Math. Soc.* **135**(8) (2007) 2317–2326.

[12] K. Bringmann and H. Swisher, Erratum to "on a conjecture of Koike on identities between Thompson series and Rogers–Ramanujan functions", *Proc. Amer. Math. Soc.* **136**(4) (2008) 1501.

[13] Z. Cao, *Product identities for theta functions*, Ph.D. thesis, University of Illinois at Urbana-Champaign (2008).

[14] Z. Cao, Integer matrix exact covering systems and product identities for theta functions, *Int. Math. Res. Not.* **2011**(19) (2010) 4471–4514,

[15] S. Cooper, The quintuple product identity, *Int. J. Number Theory* **2**(1) (2006) 115–161.

[16] E. Fiedler, Ueber eine besondere Classe irrationaler Modulargleichungen der elliptischen Functionen, *Vierteljahrsschr. Naturforsch. Gesell. Zürich* **12** (1885) 129–229.

[17] C. Guetzlaff, Aequatio modularis pro transformatione functionum ellipticarum septimi ordinis, *J. Reine Angew. Math.* **12** (1834) 173–177.

[18] C. Gugg, Two modular equations for squares of the Rogers–Ramanujan functions with applications, *Ramanujan J.* **18**(2) (2009) 183–207.

[19] S.-S. Huang, On modular relations for the Göllnitz–Gordon functions with applications to partitions, *J. Number Theory* **68** (1998) 178–216.

[20] F. Klein, Ueber die Transformation der elliptischen Functionen und die Auflösung der Gleichungen fünften Grades, *Math. Ann.* **14** (1879) 111–172.

[21] M. Koike, Thompson series and Ramanujan's identities, in *Galois Theory and Modular Forms*, eds. K. Hashimoto, K. Miyake and H. Nakamura (Kluwer, Dordrecht, 2003), pp. 367–374.

[22] S. Ramanujan, Proof of certain identities in combinatory analysis, *Proc. Cambridge Philos. Soc.* **19** (1919) 214–216.

[23] S. Ramanujan, Algebraic relations between certain infinite products, *Proc. London Math. Soc.* **2** (1920) xviii.

[24] S. Ramanujan, *Notebooks (2 Volumes)* (Tata Instititute of Fundamental Research, Bombay, 1957).

[25] S. Ramanujan, *The Lost Notebook and Other Unpublished Papers* (Narosa, New Delhi, 1988).

[26] S. Ramanujan, *Collected Papers* (Cambridge University Press, Cambridge, 1927); reprinted by Chelsea, New York, 1962; reprinted by the American Mathematical Society, Providence, RI, 2000.

[27] R. A. Rankin, *Modular Forms and Functions* (Cambridge University Press, Cambridge, 1977).

[28] L. J. Rogers, Second memoir on the expansion of certain infinite products, *Proc. London Math. Soc.* **25** (1894) 318–343.

[29] L. J. Rogers, On a type of modular relation, *Proc. London Math. Soc.* **19** (1921) 387–397.

[30] R. Russell, On $\kappa\lambda - \kappa'\lambda'$ modular equations, *Proc. London Math. Soc.* **19**(2) (1920) 387–397.

[31] L. Schläfli, Bewies der Hermiteschen Verwandlungsstafeln für die elliptischen Modularfunctionen, *J. Reine Angew. Math.* **72** (1870) 360–369.

[32] B. Schoeneberg, *Elliptic Modular Functions* (Springer-Verlag, New York, 1974).

[33] H. Schröter, *De aequationibus modularibus*, Dissertatio Inauguralis, Albertina Litterarum Universitate, Regiomonti (1854).

[34] H. Schröter, Extrait d'une lettre adressee A. M. Liouville, *J. Math.* **3**(2) (1858) 258–264.

[35] H. Schröter, Beiträge zur Theorie der Elliptischen Funktionen, *Acta Math.* **5** (1894) 205–208.

[36] L. A. Sohncke, Aequationes modulares pro transformatione functionum ellipticarum et undecimi et decimi tertii et decimi septimi ordinis, *J. Reine Angew. Math.* **12** (1834) 178.

[37] L. A. Sohncke, Aequationes modulares pro transformatione functionum ellipticarum, *J. Reine Angew. Math.* **16** (1837) 97–130.

[38] M. Somos, private communications (2008).

[39] G. N. Watson, Proof of certain identities in combinatory analysis, *J. Indian Math. Soc.* **20** (1933) 57–69.

[40] E. T. Whittaker and G. N. Watson, *A Course of Modern Analysis*, 4th edn. (University Press, Cambridge, 1966).

[41] H. Yesilyurt, A generalization of a modular identity of Rogers, *J. Number Theory* **129**(6) (2009) 1256–1271.

[42] H. Yesilyurt, Elementary proofs of some identities of Ramanujan for the Rogers–Ramanujan functions, *J. Math. Anal. Appl.* **388**(1) (2012) 420–434.

© 2024 World Scientific Publishing Company
https://doi.org/10.1142/9789811277375_0016

Proof of a rational Ramanujan-type series for $1/\pi$. The fastest one in level 3*

Jesús Guillera
Department of Mathematics, University of Zaragoza
Pedro Cerbuna 12, Zaragoza 50009, Spain
jguillera@gmail.com

Using a modular equation of level 3 and degree 23 due to Chan and Liaw, we prove the fastest known (conjectured to be the fastest one) convergent rational Ramanujan-type series for $1/\pi$ of level 3.

Keywords: Hypergeometric series; Ramanujan-type series for $1/\pi$; elliptic modular functions; Russell-type modular equations.

Mathematics Subject Classification 2020: 33E05, 33C05, 33C20, 11F03

1. The Method

A prerequisite to understand this paper well is familiarity with the notation and the method developed in [9]. This method is based on an original idea of Wan [10], whose paper was in turn influenced by some ideas of [8]. Using the notation

$$F_s(\alpha) = {}_2F_1\left(\begin{matrix} \frac{1}{s}, 1-\frac{1}{s} \\ 1 \end{matrix} \,\middle|\, \alpha \right), \quad s \in \{6,4,3,2\}, \quad \ell = 4\sin^2\frac{\pi}{s},$$

we proved in [9] the following result.

Theorem 1.1. *Let*

$$F_s(\alpha) = m(\alpha,\beta)F_s(\beta), \quad A(\alpha,\beta) = 0,$$

be a transformation of modular origin and degree $1/d$, and let $\beta_0 = 1-\alpha_0$ be a solution of $A(\alpha_0,\beta_0)=0$, and $m_0 = m(\alpha_0,\beta_0)$. Then, if

$$m_0 = \frac{1}{\sqrt{d}}, \quad \text{or} \quad m_0 = \frac{\sqrt{4d-\ell}}{2d} + \frac{\sqrt{\ell}}{2d}i,$$

*This article originally appeared in *International Journal of Number Theory*, **17**(2), pp. 473–477 © 2021 World Scientific Publishing Company. DOI: 10.1142/S1793042120400242

we have

$$\sum_{n=0}^{\infty} \frac{\left(\frac{1}{2}\right)_n \left(\frac{1}{s}\right)_n \left(1-\frac{1}{s}\right)_n}{(1)_n^3} (a+bn)\, z^n = \frac{1}{\pi},$$

where

$$z = 4\alpha_0\beta_0, \quad b = 2(1-2\alpha_0)\sqrt{\frac{d}{\ell}}, \quad a = -2\alpha_0\beta_0 \frac{m_0'}{\alpha_0'} \frac{d}{\sqrt{\ell}}. \tag{1.1}$$

or

$$z = 4\alpha_0\beta_0, \quad b = 2\,(1-2\alpha_0)\sqrt{\frac{d}{\ell}-\frac{1}{4}}, \quad a = -2\alpha_0\beta_0 \frac{m_0'}{\alpha_0'} \frac{d}{\sqrt{\ell}}, \tag{1.2}$$

respectively. The ′ means differentiation with respect to the variable that we choose as independent.

2. The Formula $(A, 3, 23)$

In this paper, we prove the Ramanujan-type formula

$$\sum_{n=0}^{\infty} \frac{\left(\frac{1}{2}\right)_n \left(\frac{1}{3}\right)_n \left(\frac{2}{3}\right)_n}{(1)_n^3} (14151n+827) \frac{(-1)^n}{500^{2n}} = \frac{1500\sqrt{3}}{\pi}. \tag{2.1}$$

As far as I know no other complete proof of this series has been provided in the literature. It is the formula $(A, 3, 23)$ in the notation of [9], we will also refer to it as $3A23$, and is the fastest known (conjectured to be the fastest one) convergent rational Ramanujan-type series for $1/\pi$ of level $\ell = 3$. It was discovered by Chan *et al.* [7, Eq. 1.19]. As $z_0 = 4\alpha_0\beta_0$, where $\beta_0 = 1-\alpha_0$, we get

$$\alpha_0 = \frac{1}{2} - \frac{53\sqrt{89}}{1000}, \quad \beta_0 = 1-\alpha_0 = \frac{1}{2} + \frac{53\sqrt{89}}{1000}.$$

Then, with a numerical approximation of 20 digits, we have

$$m_0 = \frac{F_3(\alpha_0)}{F_3(\beta_0)} \approx 0.20508654634905660459 + 0.037653278425410375946\, i,$$

which we identify (for example, using the `minpoly` Maple command) as

$$m_0 = \frac{\sqrt{89}}{46} + \frac{\sqrt{3}}{46} i.$$

A goal of the paper is to produce the value of m_0 rigorously (instead of relying on a numerical guess). Observe that m_0 is of the form

$$m_0 = \frac{\sqrt{4d-\ell}}{2d} + \frac{\sqrt{\ell}}{2d} i, \quad |m_0| = \frac{1}{\sqrt{d}},$$

with $d = 23$. Hence, for proving (2.1) with our method, we need a transformation of degree $1/d$ with $d = 23$ for the level $\ell = 3$, and with such a transformation we can prove it rigorously. This is done in Sec. 3 using a modular equation in which the algebraic relation $A(\alpha,\beta)=0$ is written using two auxiliary variables u and v, in the following way:

$$u^k=\alpha\beta, \quad v^k=(1-\alpha)(1-\beta), \quad P(u,v)=0, \tag{2.2}$$

where $P(u,v)$ is a polynomial in u and v, and k is a positive integer.

3. The Proof of $(A,3,23)$

We use a modular equation of level 3 and degree 23 due to Chan and Liaw [6, Corollary 3.7], in which we have replaced u and v with u^2 and v^2, respectively. It is

$$u^{12}=\alpha\beta, \quad v^{12}=(1-\alpha)(1-\beta), \quad P(u,v)=0, \tag{3.1}$$

where

$$\begin{aligned}P(u,v)&=(u^8+v^8)-12\sqrt{3}(u^7v+uv^7)-87(u^6v^2+u^2v^6)-84\sqrt{3}(u^5v^3+u^3v^5)\\&\quad-160(u^4v^4)-2(u^4+v^4)-15\sqrt{3}(u^3v+uv^3)-48(u^2v^2)+1.\end{aligned} \tag{3.2}$$

If we let $\beta=1-\alpha$, then we see that $u^{12}=v^{12}$. If we choose

$$v=\left(\frac{\sqrt{3}}{2}+\frac{1}{2}i\right)u, \tag{3.3}$$

and replace it in $P(u,v)=0$, we have the equation

$$250(1-\sqrt{3}i)u^8-\frac{95}{2}(1+\sqrt{3}i)u^4+1=0, \tag{3.4}$$

which factors as

$$\frac{1-\sqrt{3}\,i}{16}(10u^4+1-\sqrt{3}\,i)(20u^2-\sqrt{3}+i)(20u^2+\sqrt{3}-i)=0.$$

One solution is

$$u_0=\frac{\sqrt{15}+\sqrt{5}}{20}-\frac{\sqrt{15}-\sqrt{5}}{20}i, \quad v_0=\frac{\sqrt{15}+\sqrt{5}}{20}+\frac{\sqrt{15}-\sqrt{5}}{20}i.$$

Then, from

$$\alpha_0(1-\alpha_0)=u_0^{12}=-10^{-6}, \quad \beta_0=1-\alpha_0,$$

we get

$$\alpha_0=\frac{1}{2}-\frac{53}{1000}\sqrt{89}, \quad \beta_0=\frac{1}{2}+\frac{53}{1000}\sqrt{89}.$$

Differentiating $P(u,v)=0$ with respect to u at $u=u_0$, we find

$$v_0'=-\frac{294573\sqrt{3}+82573\,i}{516854}.$$

Then, differentiating $P(u,v)=0$ twice with respect to u at $u=u_0$, we get

$$v_0''=\frac{617592000\sqrt{5}(-1+i)}{(447\sqrt{3}+659i)^3}.$$

From (3.1), we see that

$$u^{12} = \alpha\beta, \quad u^{12} - v^{12} + 1 = \alpha + \beta. \tag{3.5}$$

Differentiating (3.5) with respect to u at $u = u_0$, we can solve for α_0' and β_0'. Then, differentiating (3.5) twice with respect to u at $u = u_0$, we can solve for α_0'' and β_0''. As the multiplier is given by

$$m^2 = \frac{1}{d}\frac{\beta(1-\beta)}{\alpha(1-\alpha)}\frac{\alpha'}{\beta'}, \tag{3.6}$$

see [9], replacing the already known values at $u = u_0$, we get

$$m_0 = \sqrt{\frac{1}{23}\frac{\alpha_0'}{\beta_0'}} = \frac{\sqrt{89}}{46} + \frac{\sqrt{3}}{46}i = \frac{\sqrt{4d-\ell}}{2d} + \frac{\sqrt{\ell}}{2d}i. \tag{3.7}$$

Taking logarithms in (3.6), differentiating with respect to u, and dividing by α', we get

$$\frac{m'}{\alpha'} = \frac{m}{2\alpha'}\left(\frac{\beta'}{\beta} - \frac{\beta'}{1-\beta} - \frac{\alpha'}{\alpha} + \frac{\alpha'}{1-\alpha} + \frac{\alpha''}{\alpha'} - \frac{\beta''}{\beta'}\right),$$

and from it, we obtain

$$\frac{m_0'}{\alpha_0'} = \frac{827000}{69}.$$

Finally, using the formulas

$$z = 4\alpha_0\beta_0, \quad b = 2(1-2\alpha_0)\sqrt{\frac{d}{\ell} - \frac{1}{4}}, \quad a = -2\alpha_0\beta_0\frac{m_0'}{\alpha_0'}\frac{d}{\sqrt{\ell}}, \tag{3.8}$$

we obtain

$$z = \frac{-1}{500^2}, \qquad b = \frac{4717}{1500}\sqrt{3}, \qquad a = \frac{827}{4500}\sqrt{3},$$

and we are done.

4. Ramanujan-Type Series for $1/\pi$ and Modular Equations

In a completely similar way, we can prove other Ramanujan-type series for $1/\pi$ using modular equations [1–3, 6], written in the form (2.2). For example, for proving $(A, 2, 7)$, we can use the modular equation

$$u^8 = \alpha\beta, \quad v^8 = (1-\alpha)(1-\beta), \quad P(u,v) = 0,$$

where

$$P(u,v) = (u^4 + v^4) + 8\sqrt{2}(u^3v + uv^3) + 20u^2v^2 - 1,$$

see [2; 3, Theorem 10.3]. For proving $(A, 3, 11)$, we can use

$$u^{12} = \alpha\beta, \quad v^{12} = (1-\alpha)(1-\beta), \quad P(u,v) = 0,$$

where

$$P(u,v) = (u^4 + v^4) + 3\sqrt{3}(u^3v + uv^3) + 6u^2v^2 - 1,$$

see [2; 3, Theorem 7.8], and we can get the proofs of the formulas $(A, 3, 5)$ and $(P, 3, 5)$, with the modular equation

$$u^6 = \alpha\beta, \quad v^6 = (1-\alpha)(1-\beta), \quad u^2 + v^2 + 3uv - 1 = 0,$$

see [2; 3, Theorem 7.6]. Note that all the above examples correspond to rational series. Of course, one can use the same method to prove irrational instances.

5. Conclusion

As we pointed out in [9], the main aspect of our method is that the modular equations used to prove the Ramanujan-type alternating series for $1/\pi$ have a much lower degree than those in other methods. It would be interesting to analyze in depth this phenomenon. The reader is invited to compare the values of d in the tables in [9] with the respective values of N given in the tables in [5]. We hope that the method developed here will attract others to find, for example, new modular equations of levels $\ell \geq 5$ and apply them to prove non-hypergeometric series of Ramanujan–Sato type for $1/\pi$ [4, 5]. Of course, a generalization of the method would be needed for that.

References

[1] B. C. Berndt, *Ramanujan's Notebooks, Part III* (Springer-Verlag, New York, 1991).

[2] B. C. Berndt, *Ramanujan's Notebooks, Part V* (Springer-Verlag, New York, 1998).

[3] B. C. Berndt, S. Bhargava and F. G. Garvan, Ramanujan's theories elliptic functions to alternative bases, *Trans. Amer. Math. Soc.* **347** (1995) 4163–4244.

[4] H. H. Chan, S. H. Chan and Z. Liu, Domb's numbers and Ramanujan–Sato type series for $1/\pi$, *Trans. Amer. Math. Soc.* **186** (2004) 396–410.

[5] H. H. Chan and S. Cooper, Rational analogues of Ramanujan's series for $1/\pi$, *Math. Proc. Camb. Phil. Soc.* **153** (2012) 361–383.

[6] H. H. Chan and W.-C. Liaw, On Russell-type modular equations, *Canad. J. Math.* **52** (2000) 31–36.

[7] H. H. Chan, W.-C. Liaw and V. Tan, Ramanujan's class invariant λ_n and a new class of series for $1/\pi$, *J. London Math. Soc.* (*2*) **64** (2001) 93–106.

[8] H. H. Chan, J. Wan and W. Zudilin, Legendre polynomials and Ramanujan-type series for $1/\pi$, *Israel J. Math.* **194** (2013) 183–207.

[9] J. Guillera, A method for proving Ramanujan's series for $1/\pi$, *Ramanujan J.* **52** (2020) 421–431.

[10] J. G. Wan, Series for $1/\pi$ using Legendre's relation, *Integral Transforms Spec. Funct.* **25** (2014) 1–14.

© 2024 World Scientific Publishing Company
https://doi.org/10.1142/9789811277375_0017

Poles of triple product L-functions involving monomial representations*

Heekyoung Hahn
Department of Mathematics
Duke University
Durham, North Carolina 27708, USA
hahn@math.duke.edu

In this paper, we study the order of the pole of the triple tensor product L-functions $L(s, \pi_1 \times \pi_2 \times \pi_3, \otimes^3)$ for cuspidal automorphic representations π_i of $\mathrm{GL}_{n_i}(\mathbb{A}_F)$ in the setting where one of the π_i is a monomial representation. In the view of Brauer theory, this is a natural setting to consider. The results provided in this paper give crucial examples that can be used as a point of reference for Langlands' beyond endoscopy proposal.

Keywords: Langlands L-functions; automorphic representations; automorphic induction; Rankin–Selberg L-functions.

Mathematics Subject Classification 2020: 11F70, 11F66

1. Introduction

Let G be a reductive group over a number field F and let $\mathbb{A}_F$ be the adeles of F. For a given representation

$$^{L}G \to \mathrm{GL}_n(\mathbb{C}), \tag{1.1}$$

the Langlands functoriality conjectures [18] predict there should be a corresponding transfer of automorphic representations of $G(\mathbb{A}_F)$ to automorphic representations of $\mathrm{GL}_n(\mathbb{A}_F)$. The image of functorial transfers of automorphic representations can conjectually be characterized in terms of L-functions. This is the crux of the Langlands' beyond endoscopy proposal [20]. Roughly speaking, it states that if π is a functorial transfer from G, then $L(s, \pi, r \otimes \eta)$ has a pole at $s = 1$ for some character $\eta : F^\times \backslash \mathbb{A}_F^\times \to \mathbb{C}^\times$ whenever a representation r detects the Zariski closure of the image of the L-map (1.1) in the sense of [10] and [11] (see also [12] for related work).

A great deal of work on the properties of the L-functions has been done when $r = \mathrm{Sym}^2$ (see [1, 4, 9] for example). There are similar results for $r = \Lambda^2$ (see [7, 8]

*This article originally appeared in *International Journal of Number Theory*, **17**(2), pp. 479–486 © 2021 World Scientific Publishing Company. DOI: 10.1142/S1793042120400291

for instance). Therefore, the tensor product L-function associated to $r = \otimes^2 = \mathrm{Sym}^2 \oplus \Lambda^2$ is relatively well-understood. In fact, via Rankin–Selberg theory, one knows the analytic properties of the tensor product Langlands L-functions

$$L(s, \pi_1 \times \pi_2, \otimes^2) = L(s, \pi_1 \times \pi_2),$$

where π_1 (respectively, π_2) is a cuspidal automorphic representation of $\mathrm{GL}_m(\mathbb{A}_F)$ (respectively, $\mathrm{GL}_n(\mathbb{A}_F)$), the L-function on the right is the Rankin–Selberg L-function and the L-function on the left is the Langlands L-function associated to

$$\otimes^2 : {}^L(\mathrm{GL}_m \times \mathrm{GL}_n) \to {}^L\mathrm{GL}_{mn},$$

the representation induced by the usual tensor product (see also [6] for related work on Rankin–Selberg transfers). Apart from this case, however, at this point very little is known about the general picture, even in the case of triple tensor products to be defined below. Thus, the community is in dire need of concrete examples.

We will concentrate on the next simplest case, that of triple tensor product L-functions

$$L(s, \pi_1 \times \pi_2 \times \pi_3) := L(s, \pi_1 \times \pi_2 \times \pi_3, \otimes^3) \tag{1.2}$$

associated to the triple tensor product

$$\otimes^3 : {}^L(\mathrm{GL}_{n_1} \times \mathrm{GL}_{n_2} \times \mathrm{GL}_{n_3}) \to {}^L\mathrm{GL}_{n_1 n_2 n_3}.$$

There is no known integral representation of these L-functions, so when we write $L(s, \pi_1 \times \pi_2 \times \pi_3, \otimes^3)$ we mean the Langlands L-function. It is well-defined since the local Langlands correspondence is known for GL_n [13, 15, 19, 21].

We will further impose the restriction that π_3 is induced from a Hecke character. Under this assumption, it is not hard to deduce from the literature that $L(s, \pi_1 \times \pi_2 \times \pi_3)$ has a meromorphic continuation to the plane as we explain in Theorem 2.1 below. Our goal is to investigate the order of the pole of $L(s, \pi_1 \times \pi_2 \times \pi_3)$ in this setting.

An important remark should be made here: the restriction that π_3 is a monomial representation is motivated by Brauer theory. Brauer theory states that any representation of a finite group is a $\mathbb{Z}$-linear combination of monomial representations, that is, representations induced from characters of subgroups. In other words, this theory indicates that monomial representations are "enough" to study representations with finite image.

To state our results, for $i = 1, 2$, we let π_i be cuspidal automorphic representations of $A_{\mathrm{GL}_{n_i}} \backslash \mathrm{GL}_{n_i}(\mathbb{A}_F)$, where $A_{\mathrm{GL}_{n_i}}$ is the neutral component of the real points of the greatest $\mathbb{Q}$-split torus in the center of $\mathrm{Res}_{F/\mathbb{Q}}\mathrm{GL}_{n_i}$. Let χ be a Hecke character $\chi : A_{\mathbb{G}_m} K^\times \backslash \mathbb{A}_K^\times \to \mathbb{C}^\times$, where K/F is any cyclic extensions of number fields of prime degree p. Denote by π_{iK} the base changes to K. It is well-known from [2, Chap. 3] that π_{iK} is an automorphic representation of $A_{\mathrm{GL}_{n_i}} \backslash \mathrm{GL}_{n_i}(\mathbb{A}_K)$. We study the order of the pole at $s = 1$ of the triple tensor product L-function

$$L(s, \pi_1 \times \pi_2 \times \mathrm{AI}(\chi)), \tag{1.3}$$

where $\mathrm{AI}(\chi)$ is the automorphic induction of χ. Here for any cuspidal automorphic representation π' of $A_{\mathrm{GL}_n}\backslash\mathrm{GL}_n(\mathbb{A}_K)$ and cyclic extension K/F of degree p, we denote by $\mathrm{AI}(\pi')$ the automorphic induction of π' (see [2, Definition 6.1]). It is an isobaric automorphic representation of $A_{\mathrm{GL}_{np}}\backslash\mathrm{GL}_{np}(\mathbb{A}_F)$. Its existence is proven in [2]. We say that a cuspidal automorphic representation π of $A_{\mathrm{GL}_{np}}\backslash\mathrm{GL}_{np}(\mathbb{A}_F)$ is induced from K if $\pi \cong \mathrm{AI}(\pi')$ for some cuspidal automorphic representation π' of $A_{\mathrm{GL}_n}\backslash\mathrm{GL}_n(\mathbb{A}_K)$.

In our first result, we discuss a setting when the triple tensor product L-function has at most a simple pole.

Theorem 1.1. *Let K/F be a cyclic extension of number fields of prime degree, let $\chi : A_{\mathbb{G}_m}K^\times\backslash\mathbb{A}_K^\times \to \mathbb{C}^\times$ be a Hecke character, and for $i = 1, 2$, let π_i be cuspidal automorphic representations of $A_{\mathrm{GL}_{n_i}}\backslash\mathrm{GL}_{n_i}(\mathbb{A}_F)$. Suppose that one of the π_{iK} is a cuspidal automorphic representation of $A_{\mathrm{GL}_{n_i}}\backslash\mathrm{GL}_{n_i}(\mathbb{A}_K)$. Then one has that*

$$\mathrm{ord}_{s=1}L(s, \pi_1 \times \pi_2 \times \mathrm{AI}(\chi)) = \begin{cases} 1 & \text{if } \pi_{1K} \cong \pi_{2K}^\vee \otimes \chi^{-1} \\ 0 & \text{otherwise.} \end{cases}$$

Next result gives a bound on the order of the pole.

Theorem 1.2. *Let K/F be a cyclic extension of number fields of prime degree p, let $\chi : A_{\mathbb{G}_m}K^\times\backslash\mathbb{A}_K^\times \to \mathbb{C}^\times$ be a Hecke charcater, and for $i = 1, 2$, let π_i be cuspidal automorphic representations of $A_{\mathrm{GL}_{n_i}}\backslash\mathrm{GL}_{n_i}(\mathbb{A}_F)$. Suppose that π_1 and π_2 both are induced from K. Then the order of the pole of $L(s, \pi_1 \times \pi_2 \times \mathrm{AI}(\chi))$ at $s = 1$ is at most p.*

As consequences of Theorems 1.1 and 1.2, one can deduce when these L-functions do not have a pole at $s = 1$ and what the exact order of its pole is. This is in Proposition 2.2 and in Corollary 2.4, respectively.

We close by the introduction with a few remarks. The interplay between finite group theory and Artin L-functions enriched both representation theory and number theory and led to important advances, including Artin–Brauer theory. Similarly, Langlands functoriality motivates beautiful and concrete problems in the representation theory of algebraic groups and algebraic combinatorics. Many of these problems have not received the attention they deserve, but represent fertile ground for future work.

Lastly, this paper is dedicated to my Ph.D. advisor, B. C. Berndt, who has guided me throughout my career mathematically and personally. By contributing this work to his birthday conference proceedings, I give him my sincere thanks for his endless support.

2. Proof of the Results

We begin this section by stating Theorem 2.1 below which will be used throughout the paper. To state it, let π_i be cuspidal automorphic representations of

$A_{\mathrm{GL}_{n_i}}\backslash\mathrm{GL}_{n_i}(\mathbb{A}_F)$ as above, let χ be a Hecke character $\chi : A_{\mathbb{G}_m}K^\times\backslash\mathbb{A}_K^\times \to \mathbb{C}^\times$, where K/F is any cyclic extensions of number fields of prime degree. Denote by $\mathrm{AI}(\chi)$ the automorphic induction of χ and π_{iK} the base changes to K. Then one has the following:

Theorem 2.1. *The Langlands L-function $L(s, \pi_1 \times \pi_2 \times \mathrm{AI}(\chi))$ defined as in* (1.2) *satisfies*

$$L(s, \pi_1 \times \pi_2 \times \mathrm{AI}(\chi)) = L(s, \pi_{1K} \times \pi_{2K} \times \chi). \tag{2.1}$$

In particular, the Langlands L-function on the left is in fact a Rankin–Selberg L-function and hence admits a meromorphic continuation to the plane.

Proof. Assume that K/F is inert or ramified at a place v of F and let w be the prime of K above v. To ease notation, we let

$$\mathrm{Ind}_{K_w}^{F_v} := \mathrm{Ind}_{W'_{K_w}}^{W'_{F_v}},$$

where W'_{F_v} and W'_{K_w} are the Weil–Deligne groups of F_v and K_w, respectively.

By the definition of L-functions of triple tensor product in terms of local Langlands correspondence for GL_n and the compatibility of the local Langlands correspondence with induction [14, 16], one has that

$$L(s, \pi_{1v} \times \pi_{2v} \times \mathrm{AI}(\chi)_v) = L(s, \sigma_1 \otimes \sigma_2 \otimes \mathrm{Ind}_{K_w}^{F_v}(\chi_w)),$$

where

$$\sigma_i : W'_{F_v} \to {}^L\mathrm{GL}_{n_i}$$

are the Langlands' parameters for π_i.

Now, recall the basic fact

$$\mathrm{Ind}(V|_H \otimes W) \cong V \otimes \mathrm{Ind}(W), \tag{2.2}$$

where H is a finite index subgroup of G, V is a G-module, and W is a H-module. Then, we have that

$$\begin{aligned} L(s, \sigma_1 \otimes \sigma_2 \otimes \mathrm{Ind}_{K_w}^{F_v}(\chi_w)) &= L(s, \mathrm{Ind}_{K_w}^{F_v}(\sigma_1|_{\mathrm{Gal}_{K_w}} \otimes \sigma_2|_{\mathrm{Gal}_{K_w}} \otimes \chi_w)) & (2.3)\\ &= L(s, \sigma_1|_{\mathrm{Gal}_{K_w}} \otimes \sigma_2|_{\mathrm{Gal}_{K_w}} \otimes \chi_w) & (2.4)\\ &= L(s, \pi_{1K_w} \times \pi_{2K_w} \times \chi_w). & (2.5) \end{aligned}$$

Here for the first equality (2.3), we use (2.2), for the second equality (2.4), we use the fact that L-functions are invariant under induction, in the last equality (2.5), we use that the local Langlands correspondence for GL_n is compatible with base change [14].

For the case where K/F is split at a place v of F, a similar but simper proof holds. □

Remark 2.1. For the purposes of this paper, we only are interested in poles of L-functions. Hence we could just fix a finite set of places S of F including the infinite places and all ramified places and work with partial L-functions. The analogue of the identity of Theorem 2.1 is more or less obvious for partial L-functions. However, it is more elegant to work with entire L-functions and make use of the fundamental work of Harris–Taylor and Henniart on the local Langlands correspondence.

If one of π_{1K} and π_{2K} is a cuspidal automorphic representation of $\mathrm{GL}_{n_i}(\mathbb{A}_K)$, then the corresponding L-function has at most simple pole.

Theorem 2.2. *Let K/F be a cyclic extension of number fields of prime degree, let $\chi : A_{\mathbb{G}_m}K^\times\backslash\mathbb{A}_K^\times \to \mathbb{C}^\times$ be a Hecke character, and for $i = 1, 2$, let π_i be cuspidal automorphic representations of $A_{\mathrm{GL}_{n_i}}\backslash\mathrm{GL}_{n_i}(\mathbb{A}_F)$. Suppose that one of the π_{iK} is a cuspidal automorphic representation of $A_{\mathrm{GL}_{n_i}}\backslash\mathrm{GL}_{n_i}(\mathbb{A}_K)$. Then one has that*

$$\mathrm{ord}_{s=1}L(s, \pi_1 \times \pi_2 \times \mathrm{AI}(\chi)) = \begin{cases} 1 & \text{if } \pi_{1K} \cong \pi_{2K}^\vee \otimes \chi^{-1} \\ 0 & \text{otherwise.} \end{cases}$$

Proof. By Theorem 2.1, one has that

$$L(s, \pi_1 \times \pi_2 \times \mathrm{AI}(\chi)) = L(s, \pi_{1K} \times (\pi_{2K} \otimes \chi)).$$

Assume that π_{1K} is a cuspidal automorphic representation of $A_{\mathrm{GL}_{n_1}}\backslash\mathrm{GL}_{n_1}(\mathbb{A}_K)$. Then by the Rankin–Selberg theory (see [3] for instance), one knows that $L(s, \pi_{1K} \times (\pi_{2K} \otimes \chi))$ has a simple pole when

$$\pi_{1K} \cong \pi_{2K}^\vee \otimes \chi^{-1},$$

otherwise holomorphic. This completes the proof. □

Under what conditions on π_1 and π_2, do the triple tensor product L-functions have higher order poles? If so, how big it can be? From Theorem 2.2, both π_{1K} and π_{2K} have to be noncuspidal. This implies that π_1 and π_2 are both induced from K [2, Chap. 3]. In the following result, we give an upper bound on the order of the pole of triple tensor L-functions involving monomials:

Theorem 2.3. *Let K/F be a cyclic extension of number fields of prime degree p, let $\chi : A_{\mathbb{G}_m}K^\times\backslash\mathbb{A}_K^\times \to \mathbb{C}^\times$ be a Hecke character, and for $i = 1, 2$, let π_i be cuspidal automorphic representations of $A_{\mathrm{GL}_{n_i}}\backslash\mathrm{GL}_{n_i}(\mathbb{A}_F)$. Suppose that π_1 and π_2 both are induced from K. Then the order of the pole of $L(s, \pi_1 \times \pi_2 \times \mathrm{AI}(\chi))$ at $s = 1$ is at most p.*

Proof. Suppose that π_1 and π_2 are induced from K (and hence p divides both n_1 and n_2) and denote by $\boxplus$ the isobaric sum (see [5, Chap. 10] for example). Let $\langle\sigma\rangle = \mathrm{Gal}(K/F)$. Then it is known from [2, Theorems 4.2 and Theorem 5.1] that

$$\pi_{1K} = \boxplus_\sigma \sigma^j(\theta_1) \quad \text{and} \quad \pi_{2K} = \boxplus_\sigma \sigma^j(\theta_2),$$

where θ_i are cuspidal automorphic representation of $A_{\mathrm{GL}_{n_i/p}}\backslash\mathrm{GL}_{n_i/p}(\mathbb{A}_K)$. Therefore, one has that

$$\begin{aligned} L(s, \pi_{1K} \times (\pi_{2K} \otimes \chi)) &= L(s, \boxplus_\sigma \sigma^j(\theta_1) \times (\boxplus_\sigma \sigma^j(\theta_2) \otimes \chi)) \\ &= L(s, \boxplus_\sigma \sigma^j(\theta_1) \times \boxplus_\sigma (\sigma^j(\theta_2) \otimes \chi)) \\ &= \prod_{j,k} L(s, \sigma^j(\theta_1) \times (\sigma^k(\theta_2) \otimes \chi)) \end{aligned}$$

for $0 \leq j, k \leq (p-1)$. Hence

$$\mathrm{ord}_{s=1} L(s, \pi_{1K} \times (\pi_{2K} \otimes \chi)) = \sum_{j,k} \mathrm{ord}_{s=1} L(s, \sigma^j(\theta_1) \times (\sigma^k(\theta_2) \otimes \chi)).$$

Therefore, one has to count all the possible pairs (j, k), $0 \leq j, k \leq (p-1)$ satisfying

$$(\sigma^j(\theta_2))^\vee \otimes \chi^{-1} \cong \sigma^k(\theta_1). \tag{2.6}$$

The key fact we use is neither θ_1 nor θ_2 can be invariant under $\mathrm{Gal}(K/F)$. Otherwise, π_1 and π_2 would be noncuspidal, contradicting our assumptions (see [2, Theorem 6.2, Chap. 3]).

Consider the $p \times p$ matrix whose entries are all the possible pairs of (j, k), namely,

$$A = \begin{bmatrix} (0,0) & (0,1) & (0,2) & \cdots & (0,p-1) \\ (1,0) & (1,1) & (1,2) & \cdots & (1,p-1) \\ (2,0) & (2,1) & (2,2) & \cdots & (2,p-1) \\ & & \ddots & & \\ (p-1,0) & (p-1,1) & (p-1,2) & \cdots & (p-1,p-1) \end{bmatrix}.$$

We claim that no two pairs in a column or in a row of the matrix A can satisfy (2.6) simultaneously: Suppose that two pairs (j, k) and (j, m) in the jth row of A with $k > m$ satisfy

$$(\sigma^j(\theta_2))^\vee \otimes \chi^{-1} \cong \sigma^k(\theta_1) \quad \text{and} \quad (\sigma^j(\theta_2))^\vee \otimes \chi^{-1} \cong \sigma^m(\theta_1).$$

This would imply that $\sigma^{k-m}(\theta_1) \cong \theta_1$, which contradicts to the fact that θ_1 can not be invariant under $\mathrm{Gal}(K/F)$. Similarly, no two pairs in the kth column of A satisfy (2.6) simultaneously. Therefore, the number of pairs (j, k) satisfying (2.6) is at most p. □

The following result is to state when one trivially knows that the triple tensor L-functions do not have a pole at $s = 1$:

Proposition 2.2. *Let K/F be a cyclic extension of number fields of prime degree p, let $\chi : A_{\mathbb{G}_m} K^\times \backslash \mathbb{A}_K^\times \to \mathbb{C}^\times$ be a Hecke character, and for $i = 1, 2$, let π_i be cuspidal automorphic representations of $A_{\mathrm{GL}_{n_i}}\backslash\mathrm{GL}_{n_i}(\mathbb{A}_F)$. Suppose that π_1 and π_2 both are induced from K with $n_1 \neq n_2$. Then $L(s, \pi_1 \times \pi_2 \times \mathrm{AI}(\chi))$ does not have a pole at $s = 1$.*

Proof. Suppose that both π_1 and π_2 are induced from K and $n_1 \neq n_2$. In particular, p divides both n_1 and n_2. As in the proof of Theorem 2.3, by [2, Theorem 4.2, Theorem 5.1], we have that

$$\pi_{1K} = \boxplus_\sigma \sigma(\theta_1), \quad \text{and} \quad \pi_{2K} = \boxplus_{\sigma'} \sigma'(\theta_2),$$

where $\sigma, \sigma' \in \mathrm{Gal}(K/F)$ and θ_i are cuspidal automorphic representation of $A_{\mathrm{GL}_{n_i/p}} \backslash \mathrm{GL}_{n_i/p}(\mathbb{A}_K)$. Note that $\frac{n_1}{p} \neq \frac{n_2}{p}$. Therefore

$$\sigma(\theta_2)^\vee \otimes \chi^{-1} \not\cong \sigma'(\theta_1)$$

for any $\sigma, \sigma' \in \mathrm{Gal}(K/F)$, and hence $L(s, \sigma(\theta_1) \times (\sigma'(\theta_2) \otimes \chi))$ does not have a pole at $s = 1$ again by the theory of Rankin–Selberg [3]. This completes the claim. □

The proof of Theorem 2.3 above provides in fact an explicit formula of the exact order of the pole of the triple tensor product L-functions in a special setting.

Corollary 2.4. *Let K/F be a cyclic extension of number fields of prime degree p with $\mathrm{Gal}(K/F) = \langle \sigma \rangle$ and let $\chi : A_{\mathbb{G}_m} K^\times \backslash \mathbb{A}_K^\times \to \mathbb{C}^\times$ be a Hecke character. For $i = 1, 2$, let π_i be cuspidal automorphic representations of $A_{\mathrm{GL}_{n_i}} \backslash \mathrm{GL}_{n_i}(\mathbb{A}_F)$ such that $\pi_{iK} = \boxplus_\sigma \sigma^j(\theta_i)$ for cuspidal automorphic representations θ_i of $A_{\mathrm{GL}_{n_i/p}} \backslash \mathrm{GL}_{n_i/p}(\mathbb{A}_K)$. If ℓ is the number of pairs $(j,k), 0 \leq j,k \leq (p-1)$ satisfying*

$$(\sigma^j(\theta_2))^\vee \otimes \chi^{-1} \cong \sigma^k(\theta_1), \tag{2.7}$$

then ℓ is the order of the pole of $L(s, \pi_1 \times \pi_2 \times \mathrm{AI}(\chi))$ at $s = 1$.

We close this paper with a final remark. One might ask if the triple tensor product L-functions can ever achieve the maximum order of the pole. When $p = 2$, it is known from Ikeda [17] that the triple tensor product L-functions have at most simple pole. Therefore, in some cases, it will not happen.

Acknowledgments

The author is grateful to J. R. Getz for his constant support throughout this project and helping with the edit of this paper. The author is also thankful for an anonymous referee's careful reading of the first draft of this paper.

References

[1] J. Arthur, *The Endoscopic Classification of Representations: Orthogonal and Symplectic Groups*, Vol. 61 (American Mathematical Society Colloquium Publications, 2013).

[2] J. Arthur and L. Clozel, *Simple Algebras, Base Change, and the Advanced Theory of the Trace Formula*, Annals of Mathematics Studies, Vol. 120 (Princeton University Press, 1989).

[3] J. Cogdell, L-functions and converse theorems for GL_n, in *Automorphic Forms and Applications*, IAS/Park City Mathematics Series, Vol. 12 (American Mathematical Society, 2007), pp. 97–177.

[4] J. Cogdell, H. Kim, I. Piatetski-Shapiro and F. Shahidi, Functoriality for classical groups, *Publ. Math. Inst. Hautes Études Sci.* **99** (2004) 163–233.

[5] J. R. Getz and H. Hahn, *An Introduction to Automorphic Representations with a View Towards Trace Formulae*, Springer GTM series, https://sites.duke.edu/heekyounghahn/, to appear.

[6] J. R. Getz and J. Klassen, Isolating Rankin–Selberg lifts, *Proc. Amer. Math. Soc.* **143**(8) (2015) 3319–3329.

[7] D. Ginzburg, D. Jiang and S. Rallis, On nonvanishing of the central value of the Rankin–Selberg L-functions, *J. Amer. Math. Soc.* **17**(3) (2004) 679–722.

[8] D. Ginzburg, D. Jiang and S. Rallis, On nonvanishing of the central value of the Rankin–Selberg L-functions II, in *Automorphic Represenations, L-Functions and Applications: Progress and Prospect*, Vol. 11 (Ohio State University, Mathematics Research Institute Publications, de Gruyter, Berlin, 2005), pp. 157–191.

[9] D. Ginzburg, S. Rallis and D. Soundry, Generic automorphic forms on $\mathrm{SO}(2n+1)$: Functorial lift to $\mathrm{GL}(2n)$, endoscopy, and base change, *Internat. Math. Res. Not. (14)* **2001** (2001) 729–764.

[10] H. Hahn, On tensor third L-functions of automorphic representations of $\mathrm{GL}_n(\mathbb{A}_F)$, *Proc. Amer. Math. Soc.* **144** (12) (2016) 5061–5069.

[11] H. Hahn, On classical groups detected by the triple tensor products and the Littlewood–Richardson semigroup, *Res. Number Theory* **2**(1) (2016) 1–12.

[12] H. Hahn, J. Huh, E. Lim and J. Sohn, From partition identities to a combinatorial approach to explicit Satake inversion, *Ann. Comb.* **22**(3) (2018) 543–562.

[13] M. Harris and R. Taylor, *The Geometry and Cohomology of Some Simple Shimura Varieties*, Annals of Mathematics Studies, Vol. 151 (Princeton University Press, 2001). With an appendix by Vladimir G. Berkovich.

[14] G. Henniart, Le point sur la conjecture de Langlands pour $\mathrm{GL}(N)$ sur un corps local, in *Séminaire De Théorie Des Nombres, Paris 1983–84*, Progress in Mathematics Series, Vol. 59 (Birkhäuser, 1985), pp. 115–131.

[15] G. Henniart, Une preuve simple des conjectures de Langlands pour $\mathrm{GL}(n)$ sur un corps p-adique, *Invent. Math.* **139**(2) (2000) 439–455.

[16] G. Henniart, Induction automorphe pour $\mathrm{GL}(n, \mathbb{C})$, *J. Funct. Anal.* **258** (2010) 3082–3096.

[17] T. Ikeda, On the location of poles of the triple L-functions, *Compos. Math.* **83**(2) (1992) 187–237.

[18] R. P. Langlands, *Letter to André Weil*, http://publications.ias.edu/rpl/section/21.

[19] R. P. Langlands, On the classification of irreducible representations of real algebraic groups, in *Representation Theory and Harmonic Analysis on Semisimple Lie Groups*, Mathematical Surveys and Monographs, Vol. 31 (American Mathematical Society, 1989), pp. 101–170.

[20] R. P. Langlands, Beyond endoscopy, in *Contributions to Automorphic Forms, Geometry, and Number Theory: A Volume in Honor of Joseph Shalika* (Johns Hopkins University Press, 2004).

[21] P. Scholze, The local Langlands correspondence for GL_n over p-adic fields, *Invent. Math.* **192**(3) (2013) 663–715.

© 2024 World Scientific Publishing Company
https://doi.org/10.1142/9789811277375_0018

Polynomials and reciprocals of Eisenstein series*

Bernhard Heim†

German University of Technology in Oman (GUtech)
PO Box 1816, Athaibah PC 130, Sultanate of Oman

Faculty of Mathematics, Computer Science, and Natural Sciences
11 RWTH Aachen University, 52056 Aachen, Germany
bernhard.heim@rwth-aachen.de

Markus Neuhauser

German University of Technology in Oman (GUtech)
PO Box 1816, Athaibah PC 130, Sultanate of Oman

University (KIU), Youth Avenue, Turn 5/7 Kutaisi, 4600 Georgia
markus.neuhauser@kiu.edu.ge

Hardy and Ramanujan introduced the Circle Method to study the Fourier expansion of certain meromorphic modular forms on the upper complex half-plane. These led to asymptotic results for the partition numbers and proven and unproven formulas for the coefficients of the reciprocals of Eisenstein series E_k, especially of weight 4. Berndt *et al.* finally proved them all. Recently, Bringmann and Kane generalized Petersson's approach via Poincaré series, to handle the general case. We introduce a third approach. We attach recursively defined polynomials to reciprocals of Eisenstein series. This provides easy access to the signs of the Fourier coefficients of reciprocals of Eisenstein series, sheds some light on reciprocals of E_k of general weight, and provides some upper and lower bounds for their growth.

Keywords: Eisenstein series; Fourier coefficients; meromorphic modular forms; polynomials; recurrence relations; Ramanujan.

Mathematics Subject Classification 2020: 11F30, 11M36, 26C10, 05A17, 11B37

1. Introduction

Since the times of Hardy and Ramanujan, the study of the q-expansion of reciprocals of weakly holomorphic modular forms has played an important role in the development of mathematics.

*This article originally appeared in *International Journal of Number Theory*, **17**(2), pp. 487–497
© 2021 World Scientific Publishing Company. DOI: 10.1142/S1793042120400199
†Corresponding author.

Hardy and Ramanujan's fundamental work [7] on partition numbers $p(n)$, considered as the coefficients of the reciprocal of the Dedekind eta function

$$\eta(\tau) = q^{\frac{1}{24}} \prod_{n=1}^{\infty} (1 - q^n),$$

where $q = e^{2\pi i \tau}$ throughout and τ in the upper complex half-plane, gave birth to the Circle Method. A partition of a positive integer n is any non-increasing sequence of positive integers whose sum is n. The number of possible partitions on n is denoted $p(n)$ [13].

Hardy and Ramanujan derived the famous asymptotic formula as $n \to \infty$:

$$p(n) \sim \frac{1}{4n\sqrt{3}} e^{\pi\sqrt{\frac{2n}{3}}}. \tag{1.1}$$

In [8] they applied their method to other forms, especially to the reciprocal of Eisenstein series E_k of weight $k = 6$ (we refer to Sec. 2 for a definition). Ramanujan [15] stated additional formulas for other meromorphic functions, including $1/E_4$, and other quotients of Eisenstein series, e.g. E_4/E_6. The claims for meromorphic forms with simple poles were proven by Bialek in his thesis written under the supervision of Berndt (see, for example [2]), and also Berndt *et al.* [3] for meromorphic forms with second-order poles. Recently, also exact formulas for meromorphic modular forms of higher-order have been studied by Bringmann and Kane [4], extending the work of Petersson [14]. Finally, the general case was finished [5].

We recall the asymptotic formula in the case of $1/E_k$ (for $k = 4$), where we denote the coefficients by $\beta_{k,n}$. Let $\rho = e^{i\pi/3}$. Then

$$\beta_{4,n} \sim (-1)^n \frac{3}{E_6(\rho)} e^{\pi n\sqrt{3}}. \tag{1.2}$$

The Circle Method and Petersson's approach via Poincaré series depend on the modular property of the involved meromorphic functions. Although both approaches are able to handle higher-order poles, they depend heavily on the pole-order and distribution. For an excellent introduction to this research and recent results we refer to [2–4]. In this paper, we offer a new approach of a more analytic and combinatorial flavor in the spirit of Rota [12] and the work of Erdős [6], Asai, Kaneko, and Ninomiya [1], and recently [9].

Properties of the Fourier coefficients of $1/E_4$ have been a central theme of many research papers. Besides the asymptotic growth (1.2), the signs of the coefficients have been studied without providing an explicit formula (Circle Method). In [1] (see Sec. 5.1) a proof of $(-1)^n\beta_{4,n} > 0$, involving Eichler integrals and generalized Faber polynomials, has been indicated.

In this paper, we attach to each E_k, families of polynomials $Q_n^{\sigma_{k-1}}(x)$ of degree n and non-negative coefficients. Let $\sigma_\ell(n) = \sum_{d|n} d^\ell$. Moreover, let B_k denote the kth Bernoulli number. We study the root distribution of these polynomials. We show, for example, that $|\frac{8}{B_4}| > |\alpha|$ for any root α of $Q_n^{\sigma_3}(x)$ for any n. The polynomials have the property that $\beta_{k,n} = Q_n^{\sigma_{k-1}}(2k/B_k)$. Hence, $\beta_{4,n}$ has the same sign as

$(-1)^n$. The method detects that the sign distribution of the coefficients of $1/E_k$ is different, for $k \geq 12$ and $k \equiv 0 \pmod 4$ and E_4, E_6, E_8. In [7], a list of Fourier coefficients of $1/E_6$ is provided. We have extended this list to coefficients of $1/E_4$, $1/E_6$, $1/E_4^2$, and $1/E_4^3$. Note that $E_4^2 = E_8$, but $E_4^3 \neq E_{12}$. We further provide a procedure to obtain lower and upper bounds for $\beta_{k,n}$. We illustrate our results for $\beta_{4,n}$.

Main Example: Let $\beta_{4,n}$ be the Fourier coefficients of $1/E_4$. Then

$$2\, e^{n \ln(120)} < (-1)^n \beta_{4,n} < \frac{2}{3}\, e^{n \ln(360)}. \tag{1.3}$$

Note that $\ln(120) < \pi\sqrt{3} < \ln(360)$.

Outline. Section 2 gives a short summary of properties of Eisenstein series. In Sec. 3, the growth conditions and non-vanishing properties of the q-expansion of certain meromorphic functions attached to Eisenstein series E_k of weight k are studied. This includes Theorem 1, improvements of some upper bounds, and a direct application towards powers of $1/E_4$. In Sec. 4, some special results dependent on the parity of $k/2$ are obtained. Finally, tables of the coefficients of $1/E_4$, $1/E_6$, $1/E_4^2$, and E_4^3 are provided which extend the tables of Hardy and Ramanujan [8].

2. Eisenstein Series

In this section, we recall some basic definitions and properties of Eisenstein series.

Let $k > 2$ be an even integer. Then, the standard Eisenstein series of weight k with respect to $\mathrm{SL}_2(\mathbb{Z})$ is provided by the q-expansion

$$E_k(\tau) = 1 - \frac{2k}{B_k} \sum_{n=1}^{\infty} \sigma_{k-1}(n)\, q^n, \quad (q = e^{2\pi i \tau}), \tag{2.1}$$

where τ is in the upper complex half-plane $\mathbb{H}$. Eisenstein series E_k are modular forms of weight k. They satisfy the functional equation

$$E_k\left(\frac{a\tau + b}{c\tau + d}\right) = (c\tau + d)^k\, E_k(\tau)$$

for all $\left(\begin{smallmatrix} a & b \\ c & d \end{smallmatrix}\right) \in \mathrm{SL}_2(\mathbb{Z})$. It is well-known that the zeros of the Eisenstein series in the fundamental domain lie on the unit circle [16]. In other words, let j be Klein's invariant and τ_0 a zero of an Eisenstein series. Then $j(\tau_0) \in [j(\rho), j(i)] = [0, 1728]$. Note that (2.1) defines also the Eisenstein series E_2, which is a quasimodular form. In Ramanujan's notation $P(q) = E_2(\tau)$, $Q(q) = E_4(\tau)$, and $R(q) = E_6(\tau)$ (see [2], introduction):

$$E_2(\tau) = 1 - 24 \sum_{n=1}^{\infty} \sigma_1(n)\, q^n, \tag{2.2}$$

$$E_4(\tau) = 1 + 240 \sum_{n=1}^{\infty} \sigma_3(n)\, q^n, \tag{2.3}$$

$$E_6(\tau) = 1 - 504 \sum_{n=1}^{\infty} \sigma_5(n)\, q^n. \tag{2.4}$$

It is convenient to also consider meromorphic and invariant functions $f(\tau+1) = f(\tau)$ with respect to q. Thus, we can consider $1/E_k$ as a function in q around $q = 0$.

3. Reciprocals of Eisenstein Series and Polynomials

We are interested in the q-expansion of certain meromorphic functions with a complex parameter x attached to normalized arithmetic functions g, regular at $q = 0$, which are meromorphic modular forms for specifically chosen x.

3.1. *On the q-expansion of a certain type of meromorphic functions*

We start with the definition of a new type of meromorphic functions.

Definition 3.1. Let $g : \mathbb{N} \to \mathbb{N}$ with $g(1) = 1$. We assume that $\sum_{n=1}^{\infty} g(n)q^n$ has a positive radius of convergence. Let $x, q \in \mathbb{C}$ and $|q| < 1$, such that the following meromorphic function is regular around $q = 0$:

$$f(g, x, q) = \frac{1}{1 - x \sum_{n=1}^{\infty} g(n)q^n}. \tag{3.1}$$

We denote the nth coefficient of the q-expansion by $Q_n^g(x)$.

Remark 3.2. (1) The $Q_n^{\sigma_{k-1}}(\frac{2k}{B_k})$ denote the Fourier coefficients of the reciprocal of E_k for k even and positive.

(2) The $Q_n^g(x)$ are polynomials in x of degree n.

(3) The polynomials $Q_n^g(x)$ satisfy the recurrence relation

$$Q_n^g(x) = x \sum_{k=1}^{n} g(k) Q_{n-k}^g(x), \quad Q_0^g(x) = 1.$$

This follows from (3.1) by multiplying the right-hand side denominator with the q-expansion of the left-hand side.

(4) The coefficients of $Q_n^g(x)$ are non-negative numbers. This implies that the real roots of $Q_n^g(x)$ are non-positive.

(5) The polynomials $Q_n^g(x)$ are integer-valued.

Theorem 3.3. *Let k be a natural number and $k \geq 2$. Let $g(n) = \sigma_k(n)$. Then, there exists a positive constant κ_k such that for all complex numbers $|z| > \kappa_k$ and all $n \in \mathbb{N}_0$ we have the non-vanishing property $Q_n^g(z) \neq 0$ and*

$$\frac{|z|}{2} |Q_{n-1}^g(z)| < |Q_n^g(z)| < \frac{3|z|}{2} |Q_{n-1}^g(z)|. \tag{3.2}$$

Moreover, we have for κ_k the upper bound:

$$\kappa_k \leq 2\left(1 - \frac{1}{\sqrt[k+1]{1+(4\cdot k!)^{-1}}}\right)^{-1}. \tag{3.3}$$

Proof. We recall the following result from [9].

Let g be a normalized arithmetic function. Let $G(T) = \sum_{n=2}^{\infty} |g(n)|\, T^{n-1}$ be a positive radius of convergence. Let $\kappa_g > 0$ with $G(2/\kappa_g) < 1/2$ and $|x| > \kappa_g$. Then for all $n \in \mathbb{N}$ we have

$$\frac{|x|}{2} |Q_{n-1}^g(x)| < |Q_n^g(x)|. \tag{3.4}$$

This implies $Q_n^g(x) \neq 0$ for $|x| > \kappa_g$, since $Q_1^g(x) = x$. From the proof of [9, Theorem 2] we can also deduce that

$$|Q_n(x) - xQ_{n-1}(x)| < \frac{1}{2}|xQ_{n-1}(x)|.$$

The simple observation

$$|Q_n(x)| \leq |xQ_{n-1}(x)| + |Q_n(x) - xQ_{n-1}(x)| \tag{3.5}$$

finally leads to $|Q_n^g(z)| < \frac{3|z|}{2}|Q_{n-1}^g(z)|$. To complete the proof, it remains to determine a suitable positive q, such that

$$G_k(q) = \sum_{n=1}^{\infty} \sigma_k(n+1) q^n \leq \frac{1}{2}.$$

Note that $\sigma_k(n+1) \leq 2(n+1)^k \leq 2 \cdot k!\binom{n+k}{k}$ implies for positive q the following inequality:

$$\begin{aligned} G_k(q) &\leq 2 \cdot k! \sum_{n=1}^{\infty} \binom{n+k}{k} q^n = 2\left(-k! + \left(\frac{\mathrm{d}}{\mathrm{d}q}\right)^k \frac{1}{1-q}\right) \\ &= 2 \cdot k!((1-q)^{-k-1} - 1) = \frac{1}{2} \end{aligned}$$

for $q = 1 - \frac{1}{\sqrt[k+1]{1+(4\cdot k!)^{-1}}}$. □

For x real and sufficiently small, the values of $Q_n^{\sigma_k}(x)$ have the same signs as x^n.

Corollary 3.4. *Let $x \in \mathbb{R}$ and $x < -\kappa_k < 0$. Then,*

$$(-1)^n\, Q_n^{\sigma_k}(x) > 0. \tag{3.6}$$

In the proof, we used the estimation $n^k \leq \sigma_k(n) \leq 2n^k$, $(k \geq 2)$. This follows from

$$\sigma_k(n) \leq n^k \left(1 + \int_1^n t^{-k}\, \mathrm{d}t\right) = n^k \left(2 - \frac{n^{1-k}}{k-1}\right) \leq 2n^k. \tag{3.7}$$

Theorem 3.3 leads to first upper bounds for κ_k. We record the first values.

Table 1. Upper bounds for κ_k.

k	2	3	4
κ_k	51.95	196.97	965.99
k	5	6	7
κ_k	5767.00	40328.00	322569.00
k	8	9	10
κ_k	2903050.00	29030411.00	319334412.00
k	11	12	13
κ_k	3832012813.00	49816166414.00	697426329615.00
k	14	15	16
κ_k	697426329615.00	10461394944016.00	167382319104016.78

3.2. *Optimization of the upper bound*

Theorem 3.3 can be applied directly to the reciprocal of E_4, which leads to growth and non-vanishing results of $Q_n^{\sigma_3}(-240)$. Note that the upper bound $\kappa_5 \leq 5767$ is too weak to apply Theorem 3.3, since the relevant x value is equal to 504. Nevertheless, the following lemma shows that κ_5 can be chosen such that $\kappa_5 < 504$. Thus, Theorem 3.3 can be applied.

Lemma 3.5. *We have* $\kappa_5 \leq 280$.

Proof. Using (3.7), we have the upper bound

$$\begin{aligned} G_5(q) &\leq 33q + 240 \sum_{n=2}^{\infty} \binom{n+5}{5} q^n \\ &= 33q + 2(-120 - 720q + 120(1-q)^{-6}) \\ &= -240 - 1407q + 240(1-q)^{-6} < \frac{1}{2} \end{aligned}$$

for $q = \frac{1}{140}$. □

3.3. *Additional information*

A further refinement of the techniques related to the proof of Theorem 3.3 and the following result leads to an improvement for the upper bounds of κ_k (including $k = -1, 0, 1$). For the convenience of the reader, we record them in the next table. We can deduce that $Q_n^{\sigma_1}(24) > 12\, Q_{n-1}^{\sigma_1}(24)$. It is obvious that $Q_n^{\sigma_1}(24)$ is always positive.

Theorem 3.6 ([10]). *Define*

$$G(q) = G_k(q) = \sum_{m=2}^{\infty} |\sigma_{k-1}(m+1) - \sigma_{k-1}(2)\sigma_{k-1}(m)| q^m$$

with positive radius of convergence R. Suppose that there is $0 < q_k < 1$ *such that* $G_k(q_k) \leq \frac{1}{250}$ *and* $\kappa_{k-1} \leq \frac{250}{249}\left(\frac{1}{q_k} + \sigma_{k-1}(2)\right) < \frac{2k}{|B_k|}$, *then the absolute value of the*

Table 2. New bounds.

k	-1	0	1	2	3	4	5	6	7
κ_k	5.06	7.03	10.9	18.5	33.8	64.8	128	255	509
k	8	9	10	11	12	13	14	15	16
κ_k	1020	2050	4110	8230	16500	33000	66100	131000	262000

Table 3. Coefficients with Bernoulli numbers.

k	2	4	6	8	10	12	14	16	18
$\frac{2k}{B_k}$	24	-240	504	-480	264	$-\frac{65520}{691}$	24	$-\frac{16320}{3617}$	$\frac{28728}{43867}$
$\approx$	24	-240	504	-480	264	-94.819	24	-4.512	0.655

nth coefficient of $1/E_k$ *can be estimated by*

$$\left|\frac{2k}{B_k}\right| \left(\left(1 \pm \frac{1}{250} \left| \frac{2k/B_k}{2k/B_k + \sigma_{k-1}(2)} \right| \right) \left| \frac{2k}{B_k} + \sigma_{k-1}(2) \right| \right)^{n-1}. \tag{3.8}$$

Remark 3.7. Let $g_k(n) = \sigma_k(n)$. Let $\alpha_{k,N}$ be the *minimal* root of all polynomials $Q_n^{g_k}(x)$ for $n \leq N$. Here, we consider all roots of all polynomials $Q_n^{g_k}(x)$ for $n \leq N$ and determine the smallest real part. It can be observed that in all considered cases, the minimal root is already real. Let $N = 1000$. Then, we obtain by numerical calculations:

$$\begin{aligned} \alpha_{1,N} &= -9.722, \\ \alpha_{3,N} &= -29.282, \\ \alpha_{5,N} &= -115.826, \\ \alpha_{7,N} &= -479.995. \end{aligned}$$

3.4. *Application*

Let $g(n) = \sigma_{k-1}(n)$ for $k \geq 0$ even. Evaluating the polynomials $Q_n^g(x)$ at $x = \frac{2k}{B_k}$ provides the nth coefficient of the reciprocal of $1/E_k$. Furthermore, assuming that x is positive, we obtain that $Q_n^g(x) > 0$. This is the case for $k \equiv 2 \pmod 4$. Let $x < -\kappa_{k-1}$. We then have $(-1)^n Q_n^g(x) > 0$. This is the case for $1/E_4$. This implies that also all coefficients of $1/E_4^m$, $(m \in \mathbb{N})$, are non-vanishing and have strict sign changes.

4. Sign Changes for General $1/E_k$

Theorem 4.1. *Let* $k \geq 2$ *be an even integer. We denote by* $\beta_{k,n}$ *the nth Fourier coefficient of the reciprocal Eisenstein series*:

$$\frac{1}{E_k(\tau)} = \sum_{n=0}^{\infty} \beta_{k,n}\, q^n \tag{4.1}$$

around $q = 0$. Then, we have:

(1) *Let $k \equiv 2 \pmod 4$. Then $\beta_{k,n} > 0$.*
(2) *Let $k \in \{4, 8\}$. Then $(-1)^n \beta_{k,n} > 0$.*
(3) *Let $k \equiv 0 \pmod 4$ and $k \geq 12$. Then $\beta_{k,1}\beta_{k,2} > 0$.*

Proof.

(1) This is clear, since $E_k(\tau) = 1 - \sum_{n=1}^{\infty} A_n q^n$, where $A_n > 0$ for all n.
(2) We have shown that $Q_n^{\sigma_3}(z) \neq 0$ for all $z \in \mathbb{C}$ with $|z| > \kappa_3 = 196.97$, see Table 1. Hence, $(-1)^n Q_n^{\sigma_3}(x_0) > 0$ for $x_0 = \frac{8}{B_4} = -240$. Since $E_8(\tau) = (E_4(\tau))^2$, the property is proven.
(3) Let $k \equiv 0 \pmod 4$. Then $x_0 = \frac{2k}{B_k} < 0$. We have $Q_1^{\sigma_{k-1}}(x_0) = x_0 < 0$ and

$$Q_2^{\sigma_{k-1}}(x_0) = x_0(x_0 + \sigma_{k-1}(2)) = x_0 \left(\frac{2k}{B_k} + 1 + 2^{k-1} \right).$$

Note that $|B_k/(2k)| > 1$ for $k \geq 20$. We refer to [11] (Sec. 15) and some explicit calculations for small k (see also Table 3). This implies that $\frac{2k}{B_k} + 1 + 2^{k-1}$ is positive and thus, $Q_2^{\sigma_{k-1}}(x_0) < 0$. □

In particular, (3) implies that $\kappa_{k-1} \geq \frac{-2k}{B_k}$ for $k \equiv 0 \pmod 4$ and $k \geq 12$. Not all real roots can be located in the interval $(\frac{2k}{B_k}, 0]$. Otherwise, we would have strict sign changes of the coefficients $\beta_{k,n}$. We finally note the following result.

Corollary 4.2. *Let k be an even positive integer. Then $Q_n^{\sigma_{k-1}}\left(\frac{2k}{B_k}\right) \neq 0$.*

Proof. We only have to consider $k \equiv 0 \pmod 4$ and $k \geq 12$. Note that $\frac{2k}{B_k} \notin \mathbb{Z}$ because $|B_k/(2k)| > 1$ for $k \geq 20$. The cases $k = 12$ and $k = 16$ follow from the fact, that $|B_k/(2k)|$ has an irregular prime in the numerator. We have already observed that $Q_n^{\sigma_{k-1}}(x)$ is a normalized polynomial of degree n and integer coefficients. Thus, the roots have to be algebraic integers by the Theorem of Gauss. This implies the non-vanishing. □

Remark 4.3. We have checked the signs of $\beta_{12,n}$ and obtained for $n \leq 163$, the following pattern: $\beta_{12,n} < 0$ for $n \equiv 1, 2 \pmod 4$ and $\beta_{12,n} > 0$ in the remaining cases. Nevertheless, the pattern stops at $n = 164$, which demonstrates the impact of the roots of $Q_n^{\sigma_{11}}(x)$ on the signs.

5. Coefficients of $1/E_4$, $1/E_6$, $1/E_4^2$, and $1/E_4^3$

Hardy and Ramanujan [8] recorded the first 13 coefficients of $1/E_6$ to illustrate their integral property and growth. For the convenience of the reader, we also record the first coefficients of $1/E_4$ and $1/E_4^2$. In contrast to the case $k = 6$, we have strict sign changes.

Table 4. nth coefficent of $1/E_4$.

n	
0	1
1	−240
2	55440
3	−12793920
4	2952385680
5	−681306078240
6	157221316739520
7	−36281112432850560
8	8372395974330234000
9	−1932052510261208053680
10	445849302141400152457440
11	−102886230661038692118348480
12	23742498662203277988768469440

Table 5. nth coefficent of $1/E_6$.

n	
0	1
1	504
2	270648
3	144912096
4	77599626552
5	41553943041744
6	22251789971649504
7	11915647845248387520
8	6380729991419236488504
9	3416827666558895485479576
10	1829682703808504464920468048
11	979779820147442370107345764512
12	524663917940510191509934144603104

Table 6. nth coefficient of $1/E_4^2$.

n	
0	1
1	−480
2	168480
3	−52199040
4	15119446560
5	−4198347132480
6	1132514464199040
7	−299116847254053120
8	77742157641008378400
9	−19951615350261029163360
10	5068304275307482667436480
11	−1276700988345016720650917760
12	319335920420038500407104106880

Table 7. nth coefficient of $1/E_4^3$.

n	
0	1
1	−720
2	339120
3	−132039360
4	46081214640
5	−14974899930720
6	4627836408778560
7	−1377759164154871680
8	398508058352289409200
9	−112648427646194257313040
10	31252327416307233967209120
11	−8536592939398421710286859840
12	2301363255613811638678456000320

5.1. *Further work on* $1/E_4$ *and other quotients*

Asai *et al.* [1], Theorem 1, proved that the zeros of the Faber polynomials are simple and lie in the interval (0,1728). As an application, they showed that the coefficients of the reciprocal of Klein's invariant j as a function in q are strictly alternating and non-zero (see also [9] for alternative proof).

In the last section of their paper, they indicated how to generalize the result to Green's function in the sense of Eichler. They prove, for example, that the coefficients of E_6/E_4 have strictly alternating signs. They indicated that their Theorem 1 also holds for more general Faber polynomials. "*. . . though each evaluation becomes rather complicated. We can therefore obtain similar results for the signs of the Fourier expansions of various meromorphic modular forms, such as* $1/E_4, \Delta/E_4$*, . . .*". Here, Δ denotes the discriminant function, the unique cusp form of weight 12.

Acknowledgments

The authors would like to thank Robert Tröger for useful discussions and for providing numerical data. We would like to thank Kathrin Bringmann and the anonymous referee for many useful comments. The research was partly funded by the German University of Technology in Oman Grant: SG-18/19-009-RF-BH-EH-MN Modeling of recursively defined functions and scientific computations (PI B. Heim).

References

[1] T. Asai, M. Kaneko and H. Ninomiya, Zeros of certain modular functions and an application, *Comment. Math. Univ. St. Paul.* **46** (1997) 93–101.

[2] B. Berndt and P. Bialek, On the power series coefficients of certain quotients of Eisenstein series, *Trans. Amer. Math. Soc.* **357** (2005) 4379–4412.

[3] B. Berndt, P. Bialek and A. Yee, Formulas of Ramanujan for the power series coefficients of certain quotients of Eisenstein series, *Int. Math. Res. Not.* **21** (2002) 1077–1109.

[4] K. Bringmann and B. Kane, Ramanujan and coefficients of meromorphic modular forms, *J. Math. Pures Appl.* **107** (2017) 100–122.

[5] K. Bringmann and B. Kane, Fourier coefficients of meromorphic modular forms and a question of Petersson, preprint.

[6] P. Erdős, On an elementary proof of some asymptotic formulas in the theory of partitions, *Ann. Math.* **43** (1942) 437–450.

[7] G. Hardy and S. Ramanujan, Asymptotic formulae in combinatory analysis, *Proc. London Math. Soc.* **17** (1918) 75–118.

[8] G. Hardy and S. Ramanujan, On the coefficients in the expansion of certain modular functions, *Proc. Roy. Soc. A* **95** (1918) 144–155.

[9] B. Heim and M. Neuhauser, On the reciprocal of Klein's absolute j-invariant and sign changes, *Res. Number Theory* **6** (2020) 1–8.

[10] B. Heim and M. Neuhauser, On the growth and zeros of polynomials attached to arithmetic functions, submitted.

[11] K. Ireland and M. Rosen, *A Classical Introduction to Modern Number Theory,* Graduate Texts in Mathematics, Vol. 84 (Springer, 1993).

[12] J. Kung, G.-C. Rota and C. Yan, *Combinatorics: The Rota Way* (Cambridge University Press, 2009).

[13] K. Ono, *The Web of Modularity: Arithmetic of the Coefficients of Modular Forms and q-Series,* Vol. 102 (Conference Board of Mathematical Sciences, 2003).

[14] H. Petersson, Konstruktion der Modulformen und der zu gewissen Grenzkreisgruppen gehörigen automorphen Formen von positiver reeller Dimension und die vollständige Bestimmung ihrer Fourierkoeffizienten, *S.-B. Heidelberger Akad. Wiss. Math.-Nat. Kl.* **1950** (1950) 415–474.

[15] S. Ramanujan, *The Lost Notebook and Other Unpublished Papers* (Narosa, New Delhi, 1988).

[16] F. K. C. Rankin and H. P. F. Swinnerton-Dyer, On the zeros of Eisenstein Series, *Bull. London Math. Soc.* **2** (1970) 169–170.

© 2024 World Scientific Publishing Company
https://doi.org/10.1142/9789811277375_0019

Arithmetic properties of septic partition functions*

Timothy Huber[†], Mayra Huerta[‡] and Nathaniel Mayes[§]
School of Mathematical and Statistical Sciences
University of Texas Rio Grande Valley
Edinburg, Texas 78539, USA
[†]*timothy.huber@utrgv.edu*
[‡]*mayra.huerta@utrgv.edu*
[§]*nathaniel.mayes01@utrgv.edu*

Congruences and related identities are derived for a set of colored and weighted partition functions whose generating functions generate the graded algebra of integer weight modular forms of level seven. The work determines a general strategy for identifying and proving identities and associated congruences for modular forms on the principal congruence subgroup of level 7. Ramanujan's partition congruence modulo 7 serves as a prototype for the process used to prove new congruences for modular forms of level 7.

Keywords: Partitions; congruences; modular forms; theta functions.

Mathematics Subject Classification 2020: 11P82, 11P83, 11F33

1. Introduction

The integer partition function $p(n)$, defined for $|q| < 1$, by

$$\sum_{n=0}^{\infty} p(n)q^n = \frac{1}{(q;q)_\infty}, \quad (a;q)_n = \prod_{k=0}^{n-1}(1 - aq^n), \quad (a;q)_\infty = \lim_{n\to\infty}(a;q)_n, \tag{1.1}$$

satisfies congruences discovered by Ramanujan [14]

$$p(5n+4) \equiv 0 \pmod 5, \quad p(7n+5) \equiv 0 \pmod 7, \quad p(11n+6) \equiv 0 \pmod{11}.$$

In this paper, arithmetic properties are developed for a set of infinite products spanning the vector space of modular forms of weight one on subgroups of $SL_2(\mathbb{Z})$ of level seven. Congruences are derived for the functions that count colored partitions according to weights determined by the parity of the number of parts in congruence classes modulo 7. The arithmetic functions are termed "septic partition functions" because of their association with modular forms of level 7. The functions are described in terms of three marked parameters.

*This article originally appeared in *International Journal of Number Theory*, **17**(2), pp. 499–512 © 2021 World Scientific Publishing Company. DOI: 10.1142/S179304212040031X
[†]Corresponding author.

Definition 1.1. Let $P_{a_1^*,a_2^*,a_3^*}(n)$ enumerate the number of colored partitions λ of n with the following conditions:

- Parts congruent to 0 modulo 7 appear with 2 colors, are distinct, and weighted by $(-1)^{\ell_0(\lambda)}$, where $\ell_0(\lambda)$ counts parts congruent to 0 (mod 7).
- Unmarked subscripts a_i indicate that parts congruent to $i, 7-i \pmod 7$ appear with a_i colors.
- Marked a_i^* indicate parts congruent to $i, 7-i \pmod 7$ appear with a_i-colors, are distinct, and contribute $(-1)^{\ell_i(\lambda)}$, where $\ell_i(\lambda)$ is the number of parts congruent to $i, 7-i \pmod 7$.

The definition is a natural arithmetic interpretation for Fourier coefficients of the corresponding infinite products. For example

$$\sum_{n=0}^{\infty} P_{3^*,1,3}(n)q^n = \frac{(q^7;q^7)_\infty^2 (q;q^7)_\infty^3 (q^6;q^7)_\infty^3}{(q^2;q^7)_\infty (q^5;q^7)_\infty (q^3;q^7)_\infty^3 (q^4;q^7)_\infty^3}.$$

Radu [13] proved $p(An+B) \not\equiv 0 \pmod{2,3}$ for all positive integers A, B. In contrast, the septic partition functions satisfy many congruences modulo 2 and 3.

Theorem 1.2. *The following congruences are satisfied*

$$P_{0,1^*,2}(7n+1) \equiv P_{2,0,1^*}(7n+1) \equiv P_{1^*,2,0}(7n+2) \equiv 0 \pmod 2, \tag{1.2}$$

$$P_{1^*,2,0}(7n+3) \equiv P_{0,1^*,2}(7n+4) \equiv P_{2,0,1^*}(7n+6) \equiv 0 \pmod 2, \tag{1.3}$$

$$P_{3,3,^*,1}(7n+3) \equiv P_{3^*,1,3}(7n+4) \equiv P_{1,3,3^*}(7n+5) \equiv 0 \pmod 2, \tag{1.4}$$

$$P_{0,1^*,2}(7n+5) \equiv P_{1^*,2,0}(7n+4) \equiv P_{2,0,1^*}(7n+3) \equiv 0 \pmod 3, \tag{1.5}$$

$$P_{3^*,1,3}(7n+1) \equiv P_{3,3,^*,1}(7n+5) \equiv P_{1,3,3^*}(7n+6) \equiv 0 \pmod 3, \tag{1.6}$$

$$P_{1,3,3^*}(7n+2) \equiv P_{3,3,^*,1}(7n+4) \equiv P_{3^*,1,3}(7n+6) \equiv 0 \pmod 4. \tag{1.7}$$

Ramanujan's analytic proof of the congruence for the classical partition function modulo 7 involves a dissections of the generating function for 7-core partitions [15]

$$\frac{(q^7;q^7)_\infty^7}{(q;q)_\infty} := \sum_{n=0}^{\infty} t(n)q^n, \quad t(7n+5) \equiv 0 \pmod 7. \tag{1.8}$$

When an additional factor of q^2 is included, the generating function for $t(n)$ is a modular form of weight 3 and level 7 with Jacobi symbol nebentypus [11]. Theorem 2.1 shows that the septic partition functions appearing in Theorem 1.2 correspond to modular forms of weight one and level 7 that provide a factorization of (1.8)

$$\frac{(q^7;q^7)_\infty^7}{(q;q)_\infty} = \sum_{n=0}^{\infty} P_{a_1^*,a_2,a_3}(n)q^n \times \sum_{n=0}^{\infty} P_{a_3,a_1^*,a_2}(n)q^n \times \sum_{n=0}^{\infty} P_{a_2,a_3,a_1^*}(n)q^n. \tag{1.9}$$

This paper formulates decomposition formulas and congruences for a class of functions, including those listed in (1.8). The results link congruences modulo 2 and 3 for weight-one dissection components to congruences modulo 7 for counting functions associated to modular forms of higher weight such as $t(n)$ and $p(n)$. The dissections in Theorem 2.2 generate a multitude of identities featured in Theorem 3.1 between septic partition functions. Theorem 1.2 catalogues the congruences resulting from these identities. The decomposition formulas yield expansions for dissections of $p(n)$ and $t(n)$ in Corollary 3.2. The dissections also yield congruences for extensions of the septic partition functions. Since the associated generating functions generate the graded algebra of modular forms of level 7, the procedure outlined here determines a general strategy for identifying and proving congruences and related identities for modular forms of level 7. The paper concludes with identities in Corollary 3.6 for products formulated in terms of the septic partition functions.

The congruences in Theorem 1.2 result from the identities between spanning sets for the vector space of weight-one modular forms of level seven. Relations with alternative spanning sets produce representations in terms of divisor sums $\sigma_\chi(n) = \sum_{d|n} \chi(d)$.

Theorem 1.3. *If χ is the Dirichlet character modulo 7 satisfying $\chi(3) = e^{\pi i/3}$,*

$$\begin{aligned} P_{1,0,0}(n) &= -P_{0,1^*,2}(7n+2) = \frac{1}{2}P_{1^*,2,0}(7n+2) = \frac{1}{3}P_{2,0,1^*}(7n+3) \\ &= \frac{3-i\sqrt{3}}{6}\sigma_\chi(7n+3), \\ P_{0,1,0}(n) &= P_{2,0,1^*}(7n+5) = \frac{1}{2}P_{0,1^*,2}(7n+4) = \frac{1}{3}P_{1^*,2,0}(7n+4) \\ &= \frac{3+i\sqrt{3}}{6}\sigma_\chi(7n+5), \\ P_{0,0,1}(n) &= -P_{1^*,2,0}(7n+5) = \frac{1}{2}P_{2,0,1^*}(7n+6) = -\frac{1}{3}P_{0,1^*,2}(7n+5) \\ &= -\frac{\sqrt{3}}{3}\sigma_\chi(7n+6). \end{aligned}$$

Theorem 1.4. *Let $\left(\frac{n}{7}\right)$ denote the Jacobi symbol modulo 7. Then*

$$P_{2,1^*,0}(n) \equiv \sum_{d|7n+1} \left(\frac{d}{7}\right) \pmod{2}, \tag{1.10}$$

$$P_{0,2,1^*}(n) \equiv \sum_{d|7n+4} \left(\frac{d}{7}\right) \pmod{2}, \tag{1.11}$$

$$P_{1^*,0,2}(n) \equiv \sum_{d|7n+9} \left(\frac{d}{7}\right) \pmod{2}. \tag{1.12}$$

Throughout the paper, we assume basic knowledge of the theory of modular forms available from [1, 2, 12, 16] and employ standard notation $\Gamma_0(7), \Gamma_1(7)$, and

$\Gamma(7)$ to denote relevant congruence subgroups of $SL_2(\mathbb{Z})$ consisting of the matrices $\left(\begin{smallmatrix} * & * \\ 0 & * \end{smallmatrix}\right)$, $\left(\begin{smallmatrix} 1 & * \\ 0 & 1 \end{smallmatrix}\right)$, and $\left(\begin{smallmatrix} 1 & 0 \\ 0 & 1 \end{smallmatrix}\right)$, respectively.

2. A Characterization of Modular Forms of Level 7

A function $f(\tau)$ of the complex upper half plane $\mathbb{H}$ is a modular form of weight k for a finite index subgroup Γ of $SL_2(\mathbb{Z})$ if $f(\tau)$ is holomorphic in $\mathbb{H}$ and at the cusps of Γ and

$$f\left(\frac{a\tau+b}{c\tau+d}\right) = (c\tau+d)^k f(\tau) \quad \text{for all } M := \begin{pmatrix} a & b \\ c & d \end{pmatrix} \in \Gamma.$$

To verify the holomorphicity condition at the cusps, we need to show that $f(M\tau)/(c\tau+d)^k$ is holomorphic at $i\infty$ for each $M \in SL_2(\mathbb{Z})$. The vector space $\mathcal{M}_k(\Gamma)$ of weight k modular forms for a subgroup Γ of the full modular group is known to be finite dimensional. In what follows, we construct a basis of weight one forms in terms of the Dedekind eta function

$$\eta(\tau) = q^{1/24}(q;q)_\infty, \quad q = e^{2\pi i\tau} \tag{2.1}$$

and the theta constants given in product form by

$$\theta\begin{bmatrix} r/N \\ 1 \end{bmatrix}(N\tau) = \exp\left(\frac{\pi i r}{2N}\right) q^{r^2/(8N)} (q^{(N-r)/2}; q^N)_\infty (q^{(N+r)/2}; q^N)_\infty (q^N; q^N)_\infty. \tag{2.2}$$

Let $N = 7$ be fixed. Then for each odd positive integer $1 \le r \le 5$, let $T_r(\tau)$ denote

$$T_r(\tau) = \exp\left(\frac{\pi i r}{14}\right) \eta^3(7\tau) \cdot \theta^{-1}\begin{bmatrix} r/7 \\ 1 \end{bmatrix}(7\tau). \tag{2.3}$$

Theorem 2.1. *Define the following subscripted vectors:*

$$\vec{v}_0 = \left\langle \frac{T_1^2}{T_3}, \frac{T_3^2}{T_5}, \frac{T_5^2}{T_1} \right\rangle_0, \quad \vec{v}_1 = \left\langle \frac{T_5^2}{T_3}, \frac{T_3 T_1}{T_5} \right\rangle_1, \quad \vec{v}_2 = \left\langle \frac{T_1^2}{T_5}, \frac{T_3 T_5}{T_1} \right\rangle_2, \tag{2.4}$$

$$\vec{v}_3 = \langle T_5 \rangle_3, \quad \vec{v}_4 = \left\langle \frac{T_3^2}{T_1}, \frac{T_5 T_1}{T_3} \right\rangle_4, \quad \vec{v}_5 = \langle T_3 \rangle_5, \quad \vec{v}_6 = \langle T_1 \rangle_6. \tag{2.5}$$

(1) *Components of* (2.4)–(2.5) *form a basis for* $\mathcal{M}_1(\Gamma(7))$.
(2) *Components of* $\vec{v}_0$ *form a basis for* $\mathcal{M}_1(\Gamma_1(7))$.
(3) *Components of* $\vec{v}_j$ *have the form*

$$q^{j/7} \sum_{n=0}^{\infty} a_n q^n, \quad a_n \in \mathbb{Z}.$$

(4) *Components of* $\vec{v}_0$ *generate the graded ring of modular forms for* $\Gamma_1(7)$.
(5) *Components of* $\vec{v}_j$, $0 \le j \le 6$, *generate the ring of modular forms for* $\Gamma(7)$.

Proof. Part 3 may be verified from the definition of T_i in (2.3). The relevant transformation formulas for parts (1) and (2) may be deduced from the transformation formulas for $\eta(\tau)$ and those for the theta functions given in [3, Lemma A.1], [9] under a suitable set of generators for $\Gamma(7)$ and $\Gamma_1(7)$. In particular, the transformation formula for $\eta(\tau)$ on $SL_2(\mathbb{Z})$ can be applied to derive transformations for $\eta(7\tau)$ under action by $\Gamma_0(7)$ since

$$7\frac{a\tau+b}{c\tau+d}=\frac{a(7\tau)+7b}{(c/7)(7\tau)+d}\quad\text{and}\quad\det\begin{pmatrix}a&b\\c&d\end{pmatrix}=\det\begin{pmatrix}a&7b\\c/7&d\end{pmatrix}.$$

From the product expansions of T_i given by (2.1)–(2.2), any quotient of powers of the T_i is holomorphic on $\mathbb{H}$. However, we must check the holomorphicity of each vector component at the cusps. The action of $SL_2(\mathbb{Z})$ may be characterized from its Bruhat decomposition as a disjoint union [10, p. 179]

$$SL_2(\mathbb{Z})=\Gamma_0(7)\cup\Gamma_0(7)S\Gamma_0(7),\quad S=\begin{pmatrix}0&-1\\1&0\end{pmatrix}. \tag{2.6}$$

To trace the action on $T_i(\tau)$, we employ functions considered by Ibukiyama [5]. Define

$$\chi_i(\tau)=\exp\left(-\frac{\pi i r}{14}\right)\theta\begin{bmatrix}r/7\\1\end{bmatrix}(7\tau)\eta^{-3/7}(\tau). \tag{2.7}$$

The product representations for the T_i in terms of theta quotients may be used to show that with an appropriately chosen branch of the square root

$$T_1=\sqrt{\chi_1\chi_3^3\chi_5^3},\quad T_3=\sqrt{\chi_1^3\chi_3\chi_5^3},\quad T_5=\sqrt{\chi_1^3\chi_3^3\chi_5}. \tag{2.8}$$

For $M=\left(\begin{smallmatrix}a&b\\c&d\end{smallmatrix}\right)\in SL_2(\mathbb{Z})$, define the action on $m\in\mathbb{Q}^2$ by $M\circ m=\left(\begin{smallmatrix}d&-c\\-b&a\end{smallmatrix}\right)m+\left(\begin{smallmatrix}c&d\\a&b\end{smallmatrix}\right)$. If $M=\left(\begin{smallmatrix}a&b\\7c&d\end{smallmatrix}\right)\in\Gamma_0(N)$, the transformation formula for the theta constants may be written

$$\theta\begin{bmatrix}r/7\\1\end{bmatrix}\left(\frac{a(7\tau)+7b}{c(7\tau)+d}\right)=\kappa\cdot(c(7\tau)+d)^{1/2}\theta\left[\begin{pmatrix}a&7b\\c&d\end{pmatrix}^{-1}\circ\begin{pmatrix}r/7\\1\end{pmatrix}\right](7\tau), \tag{2.9}$$

for some κ depending only on M and r. Note that

$$\begin{pmatrix}a&7b\\c&d\end{pmatrix}^{-1}\circ\begin{pmatrix}r/7\\1\end{pmatrix}=\begin{pmatrix}(ar+7c-7ca)/7\\rb+d-7bd\end{pmatrix}. \tag{2.10}$$

If ac is even, then $ar+7c-7ca$ is odd, since either a or c is odd. If ac is odd, then ar, $7c$, and $7ca$ are odd, so $ar+7c-7ca$ is odd. Similarly, we can see that $rb+d-7db$ is odd. Therefore, for some odd s, the vector in (2.10) is congruent to

$$\begin{pmatrix}s/7\\1\end{pmatrix}\pmod 1. \tag{2.11}$$

This means that the action of $\Gamma_0(7)$ permutes the theta constants up to a constant, possibly different in each case. The action of $\Gamma_0(7)$ fixes $\eta(7\tau)$ up to a constant.

Hence, the action of $\Gamma_0(7)$ permutes χ_i, $i = 1,2,3$ up to a constant. Therefore by (2.8), action by $\Gamma_0(7)$ permutes $T_i(\tau)$ up to a constant. Ibukiyama [5, Theorem 1.2] shows that the image of $\chi_i(\tau)$ under the action of S is a linear combination of χ_t, t odd $1 \le t \le 3$ with non-vanishing coefficients. Since the q-expansion of χ_1 starts from a constant, and χ_3, χ_5 start from some positive power of q, it follows that the image of χ_t under the action of S does not vanish at $i\infty$. Hence, the components of (2.4)–(2.5) are holomorphic at the cusps of $\Gamma_0(7)$. To complete the proof of (1) and (2), note the dimension formulas [2] $\dim M_1(\Gamma_1(7)) = 3$ and $\dim M_1(\Gamma(7)) = 12$ and use the q-expansions to prove linear independence. The claim for $\Gamma_1(7)$ is proven in [4]. The claim for $\Gamma(7)$ in part (4) follows from part (1) and [6]. □

Theorem 2.2. *Let* $(b_1,\ldots,b_k)_j = \vec{v}_j(\tau).(b_1,\ldots,b_k)$. *Then*

$$\begin{aligned}\frac{T_1^2}{T_3}\left(\frac{\tau}{7}\right) &= (1,0,0)_0 + (1,1)_1 + (-2,0)_2 + (-1)_3 + (-1,3)_4\\ &\quad + (2)_5 + (-3)_6, \end{aligned}\tag{2.12}$$

$$\begin{aligned}\frac{T_3^2}{T_5}\left(\frac{\tau}{7}\right) &= (0,1,0)_0 + (1,-3)_1 + (1,-1)_2 + (2)_3 + (-2,0)_4\\ &\quad + (3)_5 + (-1)_6, \end{aligned}\tag{2.13}$$

$$\frac{T_5^2}{T_1}\left(\frac{\tau}{7}\right) = (0,0,1)_0 + (2,0)_1 + (1,3)_2 + (3)_3 + (1,1)_4 + (1)_5 + (2)_6, \tag{2.14}$$

$$\begin{aligned}\frac{T_3T_1^3}{T_5^3}\left(\frac{\tau}{7}\right) &= (1,-1,0)_0 + (0,4)_1 + (-3,1)_2 + (-3)_3 + (1,3)_4\\ &\quad + (-1)_5 + (-2)_6, \end{aligned}\tag{2.15}$$

$$\begin{aligned}\frac{T_1T_5^3}{T_3^3}\left(\frac{\tau}{7}\right) &= (1,0,1)_0 + (3,1)_1 + (-1,3)_2 + (2)_3 + (0,4)_4\\ &\quad + (3)_5 + (-1)_6, \end{aligned}\tag{2.16}$$

$$\begin{aligned}\frac{T_5T_3^3}{T_1^3}\left(\frac{\tau}{7}\right) &= (0,-1,1)_0 + (1,3)_1 + (0,4)_2 + (1)_3 + (3,1)_4\\ &\quad + (-2)_5 + (3)_6. \end{aligned}\tag{2.17}$$

Proof. The left side of (2.12)–(2.14) are the vector components of $\vec{v}_0$ in Theorem 2.1 evaluated at $\tau/7$. It is straightforward to check that if $g(\tau)$ is a modular form for $\Gamma_0(7)$, then $g(\tau/7)$ is a modular form of the same weight for $\Gamma(7)$. Therefore, by Theorem 2.1, the left sides of (2.12)–(2.14) are elements of $\mathcal{M}_1(\Gamma(7))$. Hence, each expression on the left of (2.12)–(2.14) may be expressed as a linear combination of the basis elements from (2.4)–(2.5). The decomposition formulas listed in (2.15)–(2.17) follow similarly from the fact that the quotients $T_5T_3^3/T_1^3$, $T_3T_1^3/T_5^3$, $T_1T_5^3/T_3^3$ constitute an alternative basis for $M_1(\Gamma_1(7))$. This can be verified using a procedure similar to that in the proof of Theorem 2.1. □

3. Arithmetic Consequences

The congruences in Theorem 1.2 result from a set of identities relating the septic partition functions. The identities follow by equating terms with indices in specific congruence classes modulo 7 in the $q^{1/7}$ expansions on both sides of (2.12)–(2.17). The congruences of Theorem 1.2 are consequences.

Corollary 3.1. *For each nonnegative integer n*

$$P_{0,1^*,2}(7n+8) = -2P_{1^*,0,2}(n) \tag{3.1}$$

$$P_{0,1^*,2}(7n+2) = -P_{1,0,0}(n), \tag{3.2}$$

$$P_{0,1^*,2}(7n+4) = 2P_{0,1,0}(n), \tag{3.3}$$

$$P_{0,1^*,2}(7n+5) = -3P_{0,0,1}(n), \tag{3.4}$$

$$P_{0,1^*,2}(7n+6) = P_{0,1^*,2}(n), \tag{3.5}$$

$$P_{1^*,2,0}(7n+2) = 2P_{1,0,0}(n), \tag{3.6}$$

$$P_{1^*,2,0}(7n+3) = -2P_{0,2,1^*}(n), \tag{3.7}$$

$$P_{1^*,2,0}(7n+4) = 3P_{0,1,0}(n), \tag{3.8}$$

$$P_{1^*,2,0}(7n+5) = -P_{0,0,1}(n), \tag{3.9}$$

$$P_{1^*,2,0}(7n+6) = P_{1^*,2,0}(n), \tag{3.10}$$

$$P_{2,0,1^*}(7n) = P_{2,0,1^*}(n), \tag{3.11}$$

$$P_{2,0,1^*}(7n+1) = 2P_{2,1^*,0}(n), \tag{3.12}$$

$$P_{2,0,1^*}(7n+3) = 3P_{1,0,0}(n), \tag{3.13}$$

$$P_{2,0,1^*}(7n+5) = P_{0,1,0}(n), \tag{3.14}$$

$$P_{2,0,1^*}(7n+6) = 2P_{0,0,1}(n), \tag{3.15}$$

$$P_{3^*,1,3}(7n+13) = 4P_{1^*,1,1}(n), \tag{3.16}$$

$$P_{3^*,1,3}(7n+1) = -3P_{1,0,0}(n), \tag{3.17}$$

$$P_{3^*,1,3}(7n+3) = -P_{0,1,0}(n), \tag{3.18}$$

$$P_{3^*,1,3}(7n+4) = -2P_{0,0,1}(n), \tag{3.19}$$

$$P_{3,3^*,1}(7n+3) = 2P_{1,0,0}(n), \tag{3.20}$$

$$P_{3,3^*,1}(7n+4) = 4P_{1,1^*,1}(n), \tag{3.21}$$

$$P_{3,3^*,1}(7n+5) = 3P_{0,1,0}(n), \tag{3.22}$$

$$P_{3,3^*,1}(7n+6) = -P_{0,0,1}(n), \tag{3.23}$$

$$P_{1,3,3^*}(7n+2) = 4P_{1,1,1^*}(n), \tag{3.24}$$

$$P_{1,3,3^*}(7n+3) = P_{1,0,0}(n), \tag{3.25}$$

$$P_{1,3,3^*}(7n+5) = -2P_{0,1,0}(n), \tag{3.26}$$

$$P_{1,3,3^*}(7n+6) = 3P_{0,0,1}(n). \tag{3.27}$$

Proof. For $a, b, c \in \mathbb{Z}$, Eq. (2.3) implies

$$q^{-(6a+5b+3c)/7} T_1^a T_3^b T_5^c = (q, q^6; q^7)_\infty^{-c} (q^2, q^5; q^7)_\infty^{-b} (q^3, q^4; q^7)_\infty^{-a} (q^7; q^7)^{2(a+b+c)}. \tag{3.28}$$

If $a + b + c = 1$, then

$$T_1^a T_3^b T_5^c = q^{(6a+5b+3c)/7} \sum_{n=0}^{\infty} P_{\hat{c},\hat{b},\hat{a}}(n) q^n, \quad \hat{k} = \begin{cases} k^* & \text{if } k < 0, \\ k, & \text{otherwise} \end{cases} \tag{3.29}$$

In particular,

$$\frac{T_5^2}{T_1}(\tau) = \sum_{n=0}^{\infty} P_{2,0,1^*}(n) q^n, \quad \frac{T_5^2}{T_3}(\tau) = q^{1/7} \sum_{n=0}^{\infty} P_{2,1^*,0}(n) q^n.$$

As a power series identity in $q^{1/7}$, Eq. (2.14) takes the form

$$\sum_{n=0}^{\infty} P_{2,0,1^*}(n) q^{n/7} = \sum_{n=0}^{\infty} P_{2,0,1^*}(n) q^n + 2q^{1/7} \sum_{n=0}^{\infty} P_{2,1^*,0}(n) q^n + O(q^{2/7}). \tag{3.30}$$

Therefore, upon equating terms of index congruent to 1 modulo 7 in the $q^{1/7}$ expansion of both sides of (2.14), we obtain

$$\sum_{n=0}^{\infty} P_{2,0,1^*}(7n+1) q^{(7n+1)/7} = 2 \sum_{n=0}^{\infty} P_{2,1^*,0}(n) q^{(7n+1)/7}. \tag{3.31}$$

Hence, (3.31) implies

$$P_{2,0,1^*}(7n+1) = 2P_{2,1^*,0}(n).$$

The remaining identities are derived similarly by equating coefficients of terms corresponding to vectors on the right sides of (2.12)–(2.17) containing a single nonzero element. □

Corollary 3.2. *Let $p(n)$ and $t(n)$ be as defined in the introduction. Then*

$$p(7n+5) \equiv t(7n+5) \equiv 0 \pmod{7}. \tag{3.32}$$

Proof. Equation (3.28) implies that

$$q^2 \frac{(q^7; q^7)_\infty^7}{(q; q)_\infty} = (T_1 T_2 T_3)(\tau). \tag{3.33}$$

Replace q by $q^{1/7}$ in (3.33) to derive

$$q^{2/7}\frac{(q;q)_\infty^7}{(q^{1/7};q^{1/7})_\infty} = q^{2/7}(q;q)_\infty^7 \sum_{n=0}^{\infty} p(n)q^{n/7} = (T_1T_3T_5)(\tau/7). \tag{3.34}$$

Expand the right side of (3.34) using (2.12)–(2.14) and simplify using Klein's relation [7, 8]

$$\frac{T_1^2T_3}{T_5} + \frac{T_3^2T_5^2}{T_1} - \frac{T_1T_5^2}{T_3} = 0. \tag{3.35}$$

Isolate terms of index congruent to 0 modulo seven in the $q^{1/7}$-expansion on both sides of (3.34) to get

$$\begin{aligned} q^{2/7}&\sum_{n=0}^{\infty} t(7n+5)q^{(7n+5)/7} \\ &= q^{2/7}(q;q)_\infty^7 \sum_{n=0}^{\infty} p(7n+5)q^{(7n+5)/7} \\ &= 7\left(\frac{2T_3^4}{T_1} + \frac{T_1^2T_3^3}{T_5^2} - \frac{2T_1^4}{T_5} - 4T_1T_3T_5 + \frac{T_1^3T_5^2}{T_3^2} - \frac{T_3^2T_5^3}{T_1^2} + \frac{2T_5^4}{T_3}\right). \end{aligned} \tag{3.36}$$

The claimed congruences follow by noting that the expression on the right side of (3.36) is a power series in q. □

Congruences modulo 3 result from relations between linear combinations of the partition functions.

Corollary 3.3. *For each nonnegative integer n*

$$P_{0,1^*,2}(7n+2) \equiv -P_{0,2,1^*}(n) \pmod 3, \tag{3.37}$$

$$P_{0,1^*,2}(7n+2) \equiv P_{2,1^*,0}(n+1) \pmod 3, \tag{3.38}$$

$$P_{2,0,1^*}(7n+9) \equiv P_{1^*,0,2}(n) \pmod 3, \tag{3.39}$$

$$P_{3^*,1,3}(7n) \equiv P_{1,1,1^*}(n) \pmod 3, \tag{3.40}$$

$$P_{3^*,1,3}(7n+2) \equiv P_{0,2,1^*}(n) \pmod 3, \tag{3.41}$$

$$P_{1,3,3^*}(7n+1) \equiv P_{2,1^*,0}(n) \pmod 3, \tag{3.42}$$

$$P_{1,3,3^*}(7n+4) \equiv P_{1,1^*,1}(n) \pmod 3, \tag{3.43}$$

$$P_{3,3^*,1}(7n+8) \equiv P_{1^*,1,1}(n) \pmod 3, \tag{3.44}$$

$$P_{3,3^*,1}(7n+9) \equiv -P_{1^*,0,2}(n) \pmod 3. \tag{3.45}$$

Proof. Isolating powers of $q^{1/7}$ on both sides of (2.14) congruent to 2 modulo 7 yields

$$\sum_{n=0}^{\infty} P_{2,0,1^*}(n)q^{(7n+2n)/7} = \sum_{n=1}^{\infty} P_{1^*,0,2}(n-1)q^{(7n+2)/7} + 3\sum_{n=0}^{\infty} P_{1,1,1^*}(n)q^{(7n+2)/7}.$$

Equating coefficients implies

$$P_{0,1^*,2}(7n+2) = 3P_{1,1^*,1}(n) - P_{0,2,1^*}(n), \quad n \geq 1.$$

Congruence (3.39) follows from this identity. The remaining congruences result from

$$P_{1^*,2,0}(7n+7) = P_{2,1^*,0}(n+1) - 3P_{1^*,1,1}(n), \tag{3.46}$$

$$P_{2,0,1^*}(7n+9) = P_{1^*,0,2}(n) + 3P_{1,1,1^*}(n+1), \tag{3.47}$$

$$P_{3^*,1,3}(7n+7) = P_{1,1,1^*}(n+1) - 3P_{1^*,0,2}(n), \tag{3.48}$$

$$P_{3^*,1,3}(7n+2) = P_{0,2,1^*}(n) + 3P_{1,1^*,1}(n), \tag{3.49}$$

$$P_{1,3,3^*}(7n+18) = P_{2,1^*,0}(n+1) + 3P_{1^*,1,1}(n), \tag{3.50}$$

$$P_{1,3,3^*}(7n+4) = 3P_{0,2,1^*}(n) + P_{1,1^*,1}(n), \tag{3.51}$$

$$P_{3,3^*,1}(7n+8) = 3P_{2.1^*,0}(n+1) + P_{1^*,1,1}(n), \tag{3.52}$$

$$P_{3,3^*,1}(7n+9) = 3P_{1,1,1^*}(n+1) - P_{1^*,0,2}(n). \tag{3.53}$$

Each relation is derived similarly by equating relevant terms in the $q^{1/7}$ expansions on both sides of the dissection formulas from Theorem 2.2. □

Recall that the vector components of $\vec{v}_0$ in Theorem 2.1 form a basis for $\mathcal{M}_1(\Gamma_1(7))$. An alternative basis is given in terms of Eisenstein series defined for the 3 Dirichlet characters χ modulo 7

$$E_{\chi,k}(\tau) = 1 + \frac{2}{L(1-k,\chi)} \sum_{n=1}^{\infty} \chi(n) \frac{n^{k-1}q^n}{1-q^n}, \tag{3.54}$$

where $L(1-k,\chi)$ is the analytic continuation of the associated Dirichlet L-series. The congruences in Theorem 1.3 follow from a dissection of $E_{\chi,(\frac{\cdot}{7})}(\tau)$ in terms of the basis for $\mathcal{M}_1(\Gamma_1(7))$ from Theorem 2.1. These dissections, in turn, yield the identities of Corollary 3.4.

Corollary 3.4.

$$\sum_{d|7n+1} \left(\frac{d}{7}\right) = P_{2,1^*,0}(n) + 2P_{1^*,1,1}(n-1), \tag{3.55}$$

$$\sum_{d|7n+2} \left(\frac{d}{7}\right) = 2P_{1,1,1^*}(n) - P_{1^*,0,2}(n-1), \tag{3.56}$$

$$\sum_{d|7n+4} \left(\frac{d}{7}\right) = P_{0,2,1^*}(n) + 2P_{1,1^*,1}(n), \tag{3.57}$$

$$\sum_{d|7n+j} \left(\frac{d}{7}\right) = 0, \quad j = 3, 5, 6. \tag{3.58}$$

Proof. The identities follow from representing $E_{\chi,(\frac{\cdot}{7})}(\tau/7) \in \mathcal{M}_1(\Gamma(7))$ in terms of the basis given in Theorem 2.1 and equating dissection components. □

Several additional congruences result from relations with Eisenstein series.

Corollary 3.5. *For each nonnegative integer n,*

$$P_{0,1^*,2}(n-1) - P_{1^*,2,0}(n-1) + P_{2,0,1^*}(n) \equiv 0 \pmod{2}, \tag{3.59}$$

$$P_{3^*,1,3}(n-2) + P_{3,3^*,1}(n) + P_{1,3,3^*}(n) \equiv 0 \pmod{4}. \tag{3.60}$$

Proof. Equation (3.59) follows from the relation

$$1 + 2\sum_{n=1}^{\infty} \left(\frac{n}{7}\right) \frac{q^n}{1-q^n} = \frac{T_1^2}{T_3} - \frac{T_3^2}{T_5} + \frac{T_5^2}{T_3}. \tag{3.61}$$

This implies

$$2\sum_{d|n} \left(\frac{d}{7}\right) = P_{0,1^*,2}(n-1) - P_{1^*,2,0}(n-1) + P_{2,0,1^*}(n). \tag{3.62}$$

Equation (3.60) is similarly derived from

$$4\sum_{d|n} \left(\frac{d}{7}\right) = P_{3^*,1,3}(n-2) + P_{3,3^*,1}(n) + P_{1,3,3^*}(n). \tag{3.63}$$

Equation (3.63) results from the expansion of $E_{\chi,(\frac{\cdot}{7})}(\tau)$ in terms of the alternative basis for $\mathcal{M}_1(\Gamma_1(7))$ given by $T_5T_3^3/T_1^3$, $T_3T_1^3/T_5^3$, $T_1T_5^3/T_3^3$. □

We conclude the paper with congruences resulting from dissections of products and powers of the generating functions from Theorem 2.1. In order to simplify subsequent references, we extend the notation given in Definition 1.1 for the septic theta functions to 4 marked parameters,

$$P_{a_0,a_1,a_2,a_3}(n), \tag{3.64}$$

where the additional initial subscript a_0 refers to parts in the partition λ congruent to 0 modulo 7; unmarked subscripts a_i, for $0 \leq i \leq 3$, indicate that parts congruent to $i, 7-i \pmod{7}$ appear with a_i colors; and marked a_i^* indicate parts congruent to $i, 7-i \pmod{7}$ are a_i-colored, distinct, and contribute $(-1)^{\ell_i(\lambda)}$ to the enumeration, where $\ell_i(\lambda)$ is the number of parts congruent to $i, 7-i \pmod 7$. In particular, using the notation of (3.29)

$$(q^7;q^7)_\infty^{-a_0}(q,q^6;q^7)_\infty^{-a_1}(q^2,q^5;q^7)_\infty^{-a_2}(q^3,q^4;q^7)_\infty^{-a_3} = \sum_{n=0}^{\infty} P_{\hat{a_0},\hat{a_1},\hat{a_2},\hat{a_3}}(n)q^n. \tag{3.65}$$

Corollary 3.6. *For each nonnegative integer n,*

$$P_{4^*,0,2^*,4}(7n+1) \equiv 0 \pmod{14}, \tag{3.66}$$

$$P_{4^*,2^*,4,0}(7n+4) \equiv 0 \pmod{14}, \tag{3.67}$$

$$P_{4^*,4,0,2^*}(7n+5) \equiv 0 \pmod{14}, \tag{3.68}$$

$$P_{6^*,0,3^*,6}(7n+5) \equiv 0 \pmod{7}, \tag{3.69}$$

$$P_{6^*,0,3^*,6}(7n+1) \equiv 0 \pmod{21}, \tag{3.70}$$

$$P_{6^*,3^*,6,0}(7n+6) \equiv 0 \pmod{7}, \tag{3.71}$$

$$P_{6^*,3^*,6,0}(7n+5) \equiv 0 \pmod{21}, \tag{3.72}$$

$$P_{6^*,6,0,3^*}(7n+4) \equiv 0 \pmod{7}, \tag{3.73}$$

$$P_{6^*,6,0,3^*}(7n+2) \equiv 0 \pmod{21}, \tag{3.74}$$

$$P_{6^*,1^*,0,4}(7n+2) \equiv 0 \pmod{7}, \tag{3.75}$$

$$P_{6^*,1^*,0,4}(7n+4) \equiv 0 \pmod{7^2}, \tag{3.76}$$

$$P_{6^*,2^*,3,2}(7n+4) \equiv 0 \pmod{7}, \tag{3.77}$$

$$P_{6^*,2^*,3,2}(7n+6) \equiv 0 \pmod{7}, \tag{3.78}$$

$$P_{6^*,2,2^*,3}(7n+5) \equiv 0 \pmod{7}, \tag{3.79}$$

$$P_{6^*,2,2^*,3}(7+6n) \equiv 0 \pmod{7}, \tag{3.80}$$

$$P_{6^*,4,1^*,0}(7n+5) \equiv 0 \pmod{7}, \tag{3.81}$$

$$P_{6^*,4,1^*,0}(7n+6) \equiv 0 \pmod{7^2}, \tag{3.82}$$

$$P_{6^*,0,4,1^*}(7n+1) \equiv 0 \pmod{7}, \tag{3.83}$$

$$P_{6^*,0,4,1^*}(7n+5) \equiv 0 \pmod{7^2}, \tag{3.84}$$

$$P_{6^*,3,0,0}(7n+3) \equiv 0 \pmod{7}, \tag{3.85}$$

$$P_{6^*,3,0,0}(7n+6) \equiv 0 \pmod{7}, \tag{3.86}$$

and

$$P_{4^*,2,1^*,1}(n) = P_{4^*,1^*,1,2}(n) + P_{4^*,1,2,1^*}(n). \tag{3.87}$$

Proof. Apply the dissection formulas in Theorem 2.2 to the square of the first component of $\vec{v}_0$ in Theorem 2.1, namely

$$q^2 \sum_{n=0}^{\infty} P_{4^*,0,2^*,4}(n) q^n = \left(q \sum_{n=1}^{\infty} P_{0,1^*,2}(n) q^n \right)^2 . \tag{3.88}$$

In particular, replace q by $q^{1/7}$ in (3.88) and apply (2.12) and Klein's relation (3.35) to derive

$$q^{2/7} \sum_{n=0}^{\infty} P_{4^*,0,2^*,4}(7n+1) q^{(7n+1)/7} = 14 \left(T_3^2 - 2 \frac{T_1^2 T_5}{T_3} \right). \tag{3.89}$$

The congruence claimed by (3.66) follows by equating coefficients in the power series expansion in $q^{1/7}$ on both sides of (3.89). The remaining congruences are derived similarly. Equation (3.87) is the combinatorial interpretation of Klein's relation (3.35). □

Acknowledgments

The authors wish to thank Tomoyoshi Ibukiyama for his generous help and valuable suggestions.

References

[1] J. H. Bruinier, G. van der Geer, G. Harder and D. Zagier, *The 1-2-3 of Modular Forms*, Universitext Lectures from the Summer School on Modular Forms and their Applications held in Nordfjordeid, ed. K. Ranestad (Springer-Verlag, Berlin, 2008).

[2] F. Diamond and J. Shurman, *A First Course in Modular Forms*, Graduate Texts in Mathematics, Vol. 228 (Springer-Verlag, New York, 2005).

[3] W. Duke, Continued fractions and modular functions, *Bull. Amer. Math. Soc. (N.S.)* **42**(2) (2005) 137–162 (electronic).

[4] T. Huber, D. Lara and E. Melendez, Balanced modular parametrizations in *Frontiers in Orthogonal Polynomials and q-series*, Contemporary Mathematics and its Applications: Monographs Expositions and Lecture Notes, Vol. 1 (World Scientific Publishing, Hackensack, NJ, 2018), pp. 319–342.

[5] T. Ibukiyama, Modular forms of rational weights and modular varieties, *Abh. Math. Sem. Univ. Hamburg* **70** (2000) 315–339.

[6] K. Khuri-Makdisi, Moduli interpretation of Eisenstein series, *Int. J. Number Theory* **8**(3) (2012) 715–748.

[7] F. Klein, Ueber die Transformation siebenter Ordnung der elliptischen Functionen, *Math. Ann.* **14**(3) (1878) 428–471.

[8] F. Klein, On the order-seven transformation of elliptic functions, in *The Eightfold Way*, Mathematical Sciences Research Institute Publications, Vol. 35 (Cambridge University Press, Cambridge, 1999), pp. 287–331. Translated from the German and with an introduction by S. Levy.

[9] M. I. Knopp, *Modular Functions in Analytic Number Theory* (Markham Publishing Chicago, Ill., 1970).

[10] S. Lang, *Introduction to Modular Forms*, Grundlehren der Mathematischen Wissenschaften Fundamental Principles of Mathematical Sciences, Vol. 222 (Springer-Verlag, Berlin, 1995). With appendixes by D. Zagier and W. Feit, Corrected reprint of the 1976 original.

[11] K. Ono, On the positivity of the number of t-core partitions, *Acta Arith.* **66**(3) (1994) 221–228.

[12] K. Ono, *The Web of Modularity: Arithmetic of the Coefficients of Modular Forms and q-series*, CBMS Regional Conference Series in Mathematics, Published for the Conference Board of the Mathematical Sciences, Washington, DC, Vol. 102 (American Mathematical Society, Providence, RI, 2004).

[13] C.-S. Radu, Proof of a conjecture by Ahlgren and Ono on the non-existence of certain partition congruences, *Trans. Amer. Math. Soc.* **365**(9) (2013) 4881–4894.

[14] S. Ramanujan, Congruence properties of partitions, *Math Z.* **9** (1921) 147–153.

[15] S. Ramanujan, Some properties of $p(n)$, the number of partitions of n, *Proc. Cambridge Philos. Soc.* **19** (1919) 207–210]; *Collected Papers of Srinivasa Ramanujan* (AMS Chelsea Publications, Providence, RI, 2000), pp. 210–213.

[16] W. Stein, *Modular Forms: A Computational Approach*, Graduate Studies in Mathematics, Vol. 79 (American Mathematical Society, Providence, RI, 2007). With an appendix by P. E. Gunnells.

© 2024 World Scientific Publishing Company
https://doi.org/10.1142/9789811277375_0020

A twisted generalization of the classical Dedekind sum*

Brad Isaacson
New York City College of Technology, Mathematics Department
300 Jay St., Brooklyn, NY 11201, USA
bisaacson@citytech.cuny.edu

In this paper, we express three different, yet related, character sums in terms of generalized Bernoulli numbers. Two of these sums are generalizations of sums introduced and studied by Berndt and Arakawa–Ibukiyama–Kaneko in the context of the theory of modular forms. A third sum generalizes a sum already studied by Ramanujan in the context of theta function identities. Our methods are elementary, relying only on basic facts from algebra and number theory.

Keywords: Character sums; cotangent sums; exponential sums; generalized Bernoulli numbers; trigonometric sums.

Mathematics Subject Classification 2020: 11L03

1. Introduction

Let χ be a primitive Dirichlet character with conductor $f > 1$. Let h be a positive integer prime to f. Let m, n be integers such that $m, n \geq 0$. We put $\zeta = \exp(2\pi i/f)$. In this paper, we obtain formulas expressing the closely related character sums

$$M_{m,n}(h, \chi) = \sum_{a=1}^{f-1} \frac{\chi(a)}{(\zeta^{ha} - 1)^m (\zeta^a - 1)^n},$$

$$S_{m,n}(h, \chi; e_1, \dots, e_{m+n}) = \sum_{a_1, \dots, a_{m+n}=1}^{f-1} \chi\left(h \sum_{i=1}^{m} a_i + \sum_{j=1}^{n} a_{m+j}\right) a_1^{e_1} \dots a_{m+n}^{e_{m+n}},$$

$$c_{m,n}(h, \chi) = \sum_{a=1}^{f-1} \cot^m\left(\frac{ha}{f}\right) \cot^n\left(\frac{a}{f}\right) \chi(a),$$

by generalized Bernoulli numbers using only elementary methods from algebra and number theory. Observe that $M_{1,1}(h, \chi)$ without the character is the classical

*This article originally appeared in *International Journal of Number Theory*, **17**(2), pp. 513–528 © 2021 World Scientific Publishing Company. DOI: 10.1142/S1793042120400059

Dedekind sum $s(h,f)$ up to trivial constants. For this reason, we refer to the sum $M_{m,n}(h,\chi)$ as a twisted generalization of the classical Dedekind sum.

The sums $M_{m,n}(h,\chi)$ and $S_{m,n}(h,\chi;e_1,\ldots,e_{m+n})$ are natural generalizations of sums introduced and studied by Berndt [6] and Arakawa–Ibukiyama–Kaneko[3], where formulas for $M_{m,n}(h,\chi)$ and $S_{m,n}(h,\chi;e_1,\ldots,e_{m+n})$ in the case of $h=1$ were obtained. In [15], the present author obtained formulas for $M_{m,1}(h,\chi)$ and $S_{m,1}(h,\chi;e_1,\ldots,e_{m+n})$, and in this paper, we treat the general case. These types of sums are of interest due to their connection with the action of the symplectic group $Sp(4,\mathbb{F}_p)$ on the space of Siegel cusp forms of degree 2 belonging to the principal congruence subgroup $\Gamma(p)$. While studying this action by the holomorphic Lefshetz fixed-point theorem of Atiyah–Singer, traces of this action give rise to certain exponential sums which are variations of the sums $M_{m,n}(h,\chi)$ and $S_{m,n}(h,\chi;e_1,\ldots,e_{m+n})$. For examples, we refer the interested reader to [2, 3, 9–16, 19, 20].

The sum $c_{m,n}(h,\chi)$ is a variation of trigonometric sums first investigated by Ramanujan in connection to certain theta function identities. Similar sums were later studied by Berndt *et al.* [4, 8] using complex contour integration and Beck–Halloran [5] using discrete Fourier analysis.

Before stating our results, we need more notation. Let $\tau(\chi)$ denote the Gaussian sum $\tau(\chi)=\sum_{a=0}^{f-1}\chi(a)\zeta^a$. For $x\in\mathbb{Q}$ with denominator prime to f, we extend the definition of χ by multiplicativity. Following Snyder [18], we define the generalized Bernoulli functions $B_{k,\chi}(x)$ by the generating function

$$\sum_{a=0}^{f-1}\frac{\chi(a+x)te^{(a+x)t}}{e^{ft}-1}=\sum_{k=0}^{\infty}B_{k,\chi}(x)\frac{t^k}{k!},$$

where $B_{k,\chi}=B_{k,\chi}(0)$ are the generalized Bernoulli numbers. Let $\overline{B}_{k,\chi}(x)$ be the periodic generalized Bernoulli functions given by $\overline{B}_{k,\chi}(x)=B_{k,\chi}(x-[x])$, where $[x]$ denotes the greatest integer not exceeding x.

We first express the sum $M_{m,n}(h,\chi)$ by periodic generalized Bernoulli functions $\overline{B}_{j,\overline{\chi}}(x)$.

Theorem 3.3. *Let χ,f,h,m,n be as above. We have*

$$M_{m,n}(h,\chi)=\frac{(-1)^{m+n-1}\overline{\chi}(h)\tau(\chi)}{(m+n-1)!}\sum_{j=1}^{m+n}\sum_{c=0}^{n(h-1)}C_h(n,c)$$

$$\times\left(\sum_{r=0}^{m+n-j}\frac{(-1)^r}{r+j}\begin{bmatrix}m+n\\r+j\end{bmatrix}\binom{r+j}{j}\left(\frac{c}{h}\right)^r\right)\overline{B}_{j,\overline{\chi}}\left(\frac{c}{h}\right),$$

where $\left[{n \atop m}\right]$ *are Stirling numbers of the first kind (Stirling's cycle numbers), and*

$$C_h(n,c)=\begin{cases}1 & \text{if } n=0,\\ \displaystyle\sum_{j=0}^{[c/h]}(-1)^j\binom{n}{j}\binom{n-1+c-hj}{n-1} & \text{if } n\geq 1.\end{cases}$$

For any positive integer u with $u\mid h$, denote by h_u the u-primary part of h, that is, the maximum integer which divides h and is prime to u. We denote by $Y(u)$ the set of primitive Dirichlet characters with conductor u. Let ϕ denote the Euler phi function. The next theorem expresses periodic generalized Bernoulli functions $\overline{B}_{k,\chi}(c/h)$ by generalized Bernoulli numbers. This was proved in [15] and is a generalization of [10, Theorem 2].

Theorem 1.1 ([15]). *Let χ, f, h be as above. For any integer c prime to h, we have*

$$\overline{B}_{k,\chi}\left(\frac{c}{h}\right)=\frac{\overline{\chi}(h)}{\phi(h)h^{k-1}}\sum_{u\mid h}\sum_{\delta\in Y(u)}\left(\overline{\delta}(c)B_{k,\delta\chi}\prod_{\substack{q\mid h_u\\ q\ prime}}(1-q^{k-1}\chi(q)\delta(q))\right).$$

The relationship between the sums $S_{m,n}(h,\chi;e_1,\ldots,e_{m+n})$ and $M_{m,n}(h,\chi)$ can be seen by the following theorem.

Theorem 4.1. *Let χ, f, h, m, n be as above. We have*

$$S_{m,n}(h,\chi;e_1,\ldots,e_{m+n})=\frac{(-1)^{e_1+\cdots+e_{m+n}}}{\tau(\overline{\chi})}$$
$$\times\sum_{\substack{1\leq m_j\leq e_j\\ 1\leq l_j\leq e_j-m_j+1\\ 1\leq j\leq m+n}}\left(\prod_{i=1}^{m+n}(-f)^{l_i}\binom{e_i}{l_i}\left\{{e_i-l_i+1 \atop m_i}\right\}(m_i-1)\right)$$
$$\times M_{m_1+\cdots+m_m,m_{m+1}+\cdots+m_{m+n}}(h,\overline{\chi}),$$

where $\left\{{n \atop m}\right\}$ *are Stirling numbers of the second kind (Stirling's subset numbers).*

The relationship between the sums $c_{m,n}(h,\chi)$ and $M_{m,n}(h,\chi)$ can be seen by the following theorem.

Theorem 5.1. *Let χ, f, h, m, n be as above. We have*

$$c_{m,n}(h,\chi)=i^{m+n}\sum_{j=0}^{m}\sum_{k=0}^{n}2^{j+k}\binom{m}{j}\binom{n}{k}M_{j,k}(h,\chi).$$

By virtue of the above theorems, we can express the sums $M_{m,n}(h,\chi)$, $S_{m,n}(h,\chi;e_1,\ldots,e_{m+n})$ and $c_{m,n}(h,\chi)$ by generalized Bernoulli numbers. As a corollary, we obtain a formula for cotangent power sums considered by Apostol [1].

Corollary 5.2. *Let χ, f, m be as above. We have*

$$\sum_{a=1}^{f-1} \cot^m\left(\frac{\pi a}{f}\right)\chi(a) = -i^m\tau(\chi)\sum_{k=1}^{m}\frac{1}{k}\left(\sum_{j=k}^{m}\frac{(-2)^j\binom{m}{j}\left[{j \atop k}\right]}{(j-1)!}\right)B_{k,\overline{\chi}}.$$

Corollary 5.2 resolves the computational problem of having to evaluate cotangent power sums recursively, as discussed in [17].

2. Preliminaries

In this section, we discuss all relevant background information needed for the proofs of the theorems given in Sec 1. We omit several proofs which can be found in the literature.

The Bernoulli polynomials $B_k(x)$ are defined by using the generating function

$$\frac{te^{tx}}{e^t-1} = \sum_{k=0}^{\infty} B_k(x)\frac{t^k}{k!}.$$

Corresponding to the Bernoulli polynomials $B_k(x)$, we have two types of periodic Bernoulli functions $\overline{B}_k(x), P_k(x)$ given by

$$\overline{B}_k(x) = B_k(\{x\}),$$

$$P_k(x) = \begin{cases} 0 & \text{if } k=1,\ x\in\mathbb{Z}, \\ \overline{B}_k(x) & \text{otherwise,} \end{cases}$$

where $\{x\}$ denotes the fractional part of x. We note that $\overline{B}_k(x) = P_k(x)$ unless $k=1$ and $x \in \mathbb{Z}$, in which case $\overline{B}_k(x) = -1/2$ and $P_k(x) = 0$.

Let χ be a primitive Dirichlet character with conductor f. For integers a, b with b prime to f, we extend the definition of χ by multiplicativity by defining $\chi(a/b) = \chi(ab^{-1})$. For $x \in \mathbb{Q}$ with denominator prime to f, we follow Snyder [18] and define the generalized Bernoulli functions $B_{k,\chi}(x)$ by the generating function

$$\sum_{a=0}^{f-1}\frac{\chi(a+x)te^{(a+x)t}}{e^{ft}-1} = \sum_{k=0}^{\infty} B_{k,\chi}(x)\frac{t^k}{k!}.$$

Note that if χ is trivial, then $B_{k,\chi}(x)$ reduces to the ordinary Bernoulli polynomials $B_k(x)$. This definition leads to the formula

$$B_{k,\chi}(x) = f^{k-1}\sum_{a=0}^{f-1} B_k\left(\frac{a+x}{f}\right)\chi(a+x),$$

or equivalently, for any positive integer h prime to f and any integer c, we have

$$B_{k,\chi}\left(\frac{c}{h}\right) = \overline{\chi}(h)f^{k-1}\sum_{a=0}^{f-1} B_k\left(\frac{ha+c}{hf}\right)\chi(ha+c).$$

Corresponding to the generalized Bernoulli functions $B_{k,\chi}(x)$, we have the periodic generalized Bernoulli functions $\overline{B}_{k,\chi}(x)$ given by $\overline{B}_{k,\chi}(x) = B_{k,\chi}(\{x\})$.

Remark. The more common formulation of generalized Bernoulli functions is the one given by Berndt [7] which does not contain shifted character values, namely,

$$B_{k,\chi}(x) = f^{k-1} \sum_{a=0}^{f-1} B_k\left(\frac{a+x}{f}\right) \overline{\chi}(a).$$

For the purposes of this paper, we will use Snyder's definition.

We define the generalized Bernoulli numbers by $B_{k,\chi} = B_{k,\chi}(0)$, or equivalently,

$$B_{k,\chi} = f^{k-1} \sum_{a=0}^{f-1} B_k\left(\frac{a}{f}\right) \chi(a).$$

We state the following without proof which appears as Lemma 8.11 in [3].

Proposition 2.1 ([3]). Let e be a natural number and b be an integer not divisible by f. Then, we have

$$\sum_{a=1}^{f-1} a^e \zeta^{ab} = (-1)^e \sum_{m=1}^{e} \frac{(m-1)!}{(\zeta^b - 1)^m} \sum_{l=1}^{e-m+1} (-1)^l f^l \binom{e}{l} \left\{ {e-l+1 \atop m} \right\}.$$

We prepare two propositions which are used in Lemma 3.2.

Proposition 2.2. *Let x be a rational number with denominator relatively prime to f. Then, we have*

$$B_{j,\chi}(x) = \begin{cases} \overline{B}_{j,\chi}(x) + j \displaystyle\sum_{k=0}^{[x]-1} (k+\{x\})^{j-1} \chi(k+\{x\}) & \text{if } x \geq 0, \\ \overline{B}_{j,\chi}(x) - j \displaystyle\sum_{k=1}^{-[x]} (-k+\{x\})^{j-1} \chi(-k+\{x\}) & \text{if } x < 0. \end{cases}$$

Proof. We only prove the case where $x \geq 0$ as the case where $x < 0$ can be proven similarly. Suppose $x \geq 0$. Writing $B_{j,\chi}(x)$ as a telescoping sum, we have

$$B_{j,\chi}(x) = f^{j-1} \sum_{a=0}^{f-1} B_j\left(\frac{a+\{x\}}{f}\right) \chi(a+\{x\}, \chi)$$

$$+ f^{j-1} \sum_{k=0}^{[x]-1} \left(B_j\left(\frac{f+k+\{x\}}{f}\right) - B_j\left(\frac{k+\{x\}}{f}\right)\right) \chi(k+\{x\}).$$

Since $B_j(x+1) - B_j(x) = jx^{j-1}$, we get

$$B_{j,\chi}(x) = \overline{B}_{j,\chi}(x) + j \sum_{k=0}^{[x]-1} (k+\{x\})^{j-1} \chi(k+\{x\}).$$

□

Proposition 2.3. *Let h, k be positive integers with $k \geq 2$. Let c_0 be any integer with $0 \leq c_0 \leq h-1$. Let s be an integer with $1 \leq s \leq [k(h-1)/h]$ and l be an integer such that $0 \leq l \leq s-1$. Then, we have*

$$\sum_{j=1}^{k}\sum_{r=0}^{k-j}(-1)^r \begin{bmatrix} k \\ r+j \end{bmatrix} \binom{r+j-1}{j-1} \left(\frac{c_0}{h}+s\right)^r \left(\frac{c_0}{h}+l\right)^{j-1} = 0.$$

Proof. Let S denote the left-hand side of the assertion of the lemma. Interchanging the order of summation, taking $u = j+r$, and interchanging the order of summation again, we get

$$S = \sum_{u=1}^{k} \begin{bmatrix} k \\ u \end{bmatrix} (l-s)^{u-1} = \frac{(l-s)^{(k)}}{l-s},$$

where $(x)^{(k)}$ is the rising factorial given by $(x)^{(k)} = x(x+1)\cdots(x+k-1)$. Clearly $(l-s)^{(k)} = 0$, and the lemma is proved. □

3. The Evaluation of $M_{m,n}(h,\chi)$

In this section, we obtain formulas expressing $M_{m,n}(h,\chi)$ by generalized Bernoulli numbers. All of the results in this section will be obtained using only elementary methods from algebra and number theory.

Let χ be a primitive Dirichlet character with conductor $f > 1$. Let h be a positive integer prime to f. Let m, n be integers such that $m, n \geq 0$. We put $\zeta = \exp(2\pi i/f)$. Let $\tau(\chi)$ denote the Gaussian sum $\tau(\chi) = \sum_{a=0}^{f-1} \chi(a)\zeta^a$. We recall the definition of $M_{m,n}(h,\chi)$:

$$M_{m,n}(h,\chi) = \sum_{a=1}^{f-1} \frac{\chi(a)}{(\zeta^{ha}-1)^m(\zeta^a-1)^n}.$$

We begin by expressing the sum $M_{m,n}(h,\chi)$ by generalized Bernoulli functions $\overline{B}_{k,\chi}(x)$. We first prepare a lemma.

For $k, s \in \mathbb{Z}$ with $k \geq 1$, we define the character sum function $P_{k,s,\chi}(t)$ by

$$P_{k,s,\chi}(t) = \sum_{a=0}^{f-1} \frac{\chi(a)\zeta^{as}}{(\zeta^a e^t - 1)^k}.$$

In [15], we proved the following.

Proposition 3.1 ([15]). *Let χ, h, k be as above. For any integer c, we have*

$$P_{k,ch^{-1},\chi}(0) = \frac{(-1)^{k-1}\tau(\chi)}{(k-1)!} \sum_{j=1}^{k} \left(\sum_{r=0}^{k-j} \frac{(-1)^r}{r+j} \begin{bmatrix} k \\ r+j \end{bmatrix} \binom{r+j}{j} \left(\frac{c}{h}\right)^r \right) B_{j,\overline{\chi}}\left(\frac{c}{h}\right),$$

where h^{-1} is regarded as an element of $(\mathbb{Z}/f\mathbb{Z})^*$.

In Proposition 3.1, it turns out that we can replace $B_{j,\chi}(c/h)$ by $\overline{B}_{j,\chi}(c/h)$ for any integer c such that $0 \leq c \leq k(h-1)$.

Lemma 3.2. *Let χ, h, k be as above. For any integer c such that $0 \leq c \leq k(h-1)$, we have*

$$P_{k,ch^{-1},\chi}(0) = \frac{(-1)^{k-1}\tau(\chi)}{(k-1)!}\sum_{j=1}^{k}\left(\sum_{r=0}^{k-j}\frac{(-1)^r}{r+j}\begin{bmatrix}k\\ r+j\end{bmatrix}\binom{r+j}{j}\left(\frac{c}{h}\right)^r\right)\overline{B}_{j,\overline{\chi}}\left(\frac{c}{h}\right).$$

Proof. Since $B_{j,\overline{\chi}}(c/h) = \overline{B}_{j,\overline{\chi}}(c/h)$ for $0 \leq c \leq h-1$, the assertion follows immediately from Proposition 3.1 in this case. Thus we assume $h \leq c \leq k(h-1)$ with $k \geq 2$. We write $c = c_0 + sh$ for integers c_0, s with $0 \leq c_0 \leq h-1$ and $1 \leq s \leq [k(h-1)/h]$. By Propositions 3.1 and 2.2, it remains to show

$$\frac{(-1)^{k-1}\tau(\chi)}{(k-1)!}\sum_{j=1}^{k}\left(\sum_{r=0}^{k-j}\frac{(-1)^r}{r+j}\begin{bmatrix}k\\ r+j\end{bmatrix}\binom{r+j}{j}\left(\frac{c_0}{h}+s\right)^r\right)$$

$$\times j\sum_{l=0}^{s-1}\left(l+\frac{c_0}{h}\right)^{j-1}\overline{\chi}\left(l+\frac{c_0}{h}\right) = 0,$$

but this follows immediately from Proposition 2.3. □

Theorem 3.3. *Let χ, f, h, m, n be as above. We have*

$$M_{m,n}(h,\chi) = \frac{(-1)^{m+n-1}\overline{\chi}(h)\tau(\chi)}{(m+n-1)!}\sum_{j=1}^{m+n}\sum_{c=0}^{n(h-1)}C_h(n,c)$$

$$\times\left(\sum_{r=0}^{m+n-j}\frac{(-1)^r}{r+j}\begin{bmatrix}m+n\\ r+j\end{bmatrix}\binom{r+j}{j}\left(\frac{c}{h}\right)^r\right)\overline{B}_{j,\overline{\chi}}\left(\frac{c}{h}\right),$$

where

$$C_h(n,c) = \begin{cases} 1 & \text{if } n = 0,\\ \displaystyle\sum_{j=0}^{[c/h]}(-1)^j\binom{n}{j}\binom{n-1+c-hj}{n-1} & \text{if } n \geq 1.\end{cases}$$

Proof. The theorem is clear in the case of $m, n = 0$ as both sides are identically zero. Thus we assume that $m+n > 0$. Letting $C_h(n,c)$ be defined as in the theorem, we have

$$(1+\zeta^a+\cdots+\zeta^{(h-1)a})^n = \sum_{c=0}^{n(h-1)} C_h(n,c)\zeta^{ca}.$$

For $f \nmid a$, we have

$$\frac{\zeta^{ha}-1}{\zeta^a-1} = \sum_{c=0}^{h-1} \zeta^{ca}.$$

Hence,

$$\begin{aligned} M_{m,n}(h,\chi) &= \sum_{a=1}^{f-1} \frac{\chi(a)}{(\zeta^{ha}-1)^m(\zeta^a-1)^n} \\ &= \sum_{a=1}^{f-1} \frac{\chi(a)\left(\sum_{c=0}^{h-1}\zeta^{ca}\right)^n}{(\zeta^{ha}-1)^{m+n}} \\ &= \sum_{c=0}^{n(h-1)} C_h(n,c) \sum_{a=1}^{f-1} \frac{\chi(a)\zeta^{ca}}{(\zeta^{ha}-1)^{m+n}} \\ &= \overline{\chi}(h) \sum_{c=0}^{n(h-1)} C_h(n,c) P_{m+n,ch^{-1},\chi}(0). \end{aligned}$$

The assertion of the theorem then follows by Lemma 3.2. □

We now give examples expressing $M_{m,n}(h,\chi)$ by generalized Bernoulli numbers.

Proposition 3.4. *Let $f > 1$ be a natural number prime to 4 and let χ be a primitive Dirichlet character with conductor f. We denote by δ the unique primitive Dirichlet character modulo 4. We also put $\zeta = exp(2\pi i/f)$. Then, we get*

(i)

$$M_{0,1}(4,\chi) = \chi(4)M_{1,0}(4,\chi) = \sum_{a=1}^{f-1} \frac{\chi(a)}{(\zeta^a-1)} = \tau(\chi)B_{1,\overline{\chi}},$$

(ii)

$$M_{0,2}(4,\chi) = \chi(4)M_{2,0}(4,\chi) = \sum_{a=1}^{f-1} \frac{\chi(a)}{(\zeta^a-1)^2} = -\tau(\chi)\left\{B_{1,\overline{\chi}} + \frac{1}{2}B_{2,\overline{\chi}}\right\},$$

(iii)

$$\begin{aligned} M_{1,1}(4,\chi) &= \sum_{a=1}^{f-1} \frac{\chi(a)}{(\zeta^{4a}-1)(\zeta^a-1)} \\ &= -\tau(\chi)\left\{\frac{1+\overline{\chi}(4)}{2}B_{1,\overline{\chi}} + \frac{1}{4}B_{1,\delta\overline{\chi}} + \frac{1}{8}B_{2,\overline{\chi}}\right\}, \end{aligned}$$

(iv)

$$M_{2,1}(4,\chi)=\sum_{a=1}^{f-1}\frac{\chi(a)}{(\zeta^{4a}-1)^2(\zeta^a-1)}$$
$$=\tau(\chi)\left\{\frac{13-\overline{\chi}(2)+20\overline{\chi}(4)}{32}B_{1,\overline{\chi}}+\frac{1}{4}B_{1,\delta\overline{\chi}}+\frac{1+2\overline{\chi}(4)}{8}B_{2,\overline{\chi}}\right.$$
$$\left.+\frac{1}{32}B_{2,\delta\overline{\chi}}+\frac{1}{96}B_{3,\overline{\chi}}\right\},$$

(v)

$$M_{1,2}(4,\chi)=\sum_{a=1}^{f-1}\frac{\chi(a)}{(\zeta^{4a}-1)(\zeta^a-1)^2}$$
$$=\tau(\chi)\left\{\frac{7+\overline{\chi}(2)}{8}B_{1,\overline{\chi}}+\frac{1}{4}B_{1,\delta\overline{\chi}}+\frac{3}{8}B_{2,\overline{\chi}}+\frac{1}{24}B_{3,\overline{\chi}}\right\}.$$

Proof. Clearly $B_{j,\overline{\chi}}(0/4)=B_{j,\overline{\chi}}$, and by Theorem 1.1, we have

$$B_{j,\overline{\chi}}(1/4)=\frac{\chi(4)}{2\cdot 4^{j-1}}((1-2^{j-1}\overline{\chi}(2))B_{j,\overline{\chi}}+B_{j,\delta\overline{\chi}}),$$
$$B_{j,\overline{\chi}}(2/4)=\overline{B}_{j,\overline{\chi}}(1/2)=\frac{\chi(2)}{2^{j-1}}(1-2^{j-1}\overline{\chi}(2))B_{j,\overline{\chi}},$$
$$B_{j,\overline{\chi}}(3/4)=\frac{\chi(4)}{2\cdot 4^{j-1}}((1-2^{j-1}\overline{\chi}(2))B_{j,\overline{\chi}}-B_{j,\delta\overline{\chi}})\quad (j\in\mathbb{Z},\ j\geq 0).$$

Thus the assertions (i)–(v) follow from Theorem 3.3. □

4. The Evaluation of $S_{m,n}(h,\chi;e_1,\ldots,e_{m+n})$

In this section, we obtain formulas expressing $S_{m,n}(h,\chi;e_1,\ldots,e_{m+n})$ by generalized Bernoulli numbers. We keep the notation used previously. We begin by establishing the relationship between the sums $S_{m,n}(h,\chi;e_1,\ldots,e_{m+n})$ and $M_{m,n}(h,\chi)$.

Theorem 4.1. *Let χ,f,h,m,n be as above. We have*

$$S_{m,n}(h,\chi;e_1,\ldots,e_{m+n})=\frac{(-1)^{e_1+\cdots+e_{m+n}}}{\tau(\overline{\chi})}$$
$$\times\sum_{\substack{1\leq m_j\leq e_j\\ 1\leq l_j\leq e_j-m_j+1\\ 1\leq j\leq m+n}}\left(\prod_{i=1}^{m+n}(-f)^{l_i}\binom{e_i}{l_i}\left\{{e_i-l_i+1 \atop m_i}\right\}(m_i-1)!\right)$$
$$\times M_{m_1+\cdots+m_m,m_{m+1}+\cdots+m_{m+n}}(h,\overline{\chi}),$$

where $\left\{{n \atop m}\right\}$ are Stirling numbers of the second kind (Stirling's subset numbers).

Proof. Since

$$\chi\left(h\sum_{i=1}^{m}a_i+\sum_{j=1}^{n}a_{m+j}\right)=\frac{1}{\tau(\overline{\chi})}\sum_{b=1}^{f-1}\overline{\chi}(b)\zeta^{(h\sum_{i=1}^{m}a_i+\sum_{j=1}^{n}a_{m+j})b},$$

we have

$$S_{m,n}(h,\chi;e_1,\ldots,e_{m+n})$$

$$=\frac{1}{\tau(\overline{\chi})}\sum_{b=1}^{m-1}\overline{\chi}(b)\sum_{a_1,\ldots,a_{m+n}=1}^{m-1}a_1^{e_1}\cdots a_{m+n}^{e_{m+n}}\zeta^{b(h\sum_{i=1}^{m}a_i+\sum_{j=1}^{n}a_{m+j})}.$$

The assertion of the theorem then follows by applying Proposition 2.1 to each of the sums over a_j $(j=1,\ldots,m+n)$. □

Remark. In the case where $e_j=1$ for $j=1,\ldots,m+n$, we get

$$S_{m,n}(h,\chi;1,\ldots,1)=\frac{f^{m+n}}{\tau(\overline{\chi})}M_{m,n}(h,\overline{\chi}).$$

We now give examples expressing $S_{m,n}(h,\chi;e_1,\ldots,e_{m+n})$ by generalized Bernoulli numbers.

Proposition 4.2. *Let $f>1$ be a natural number prime to 4 and let χ be a primitive Dirichlet character with conductor f. We denote by δ the unique primitive Dirichlet character modulo 4. Then, we get*

(i)

$$S_{1,1}(4,\chi;1,1)=\sum_{a,b=1}^{f-1}\chi(4a+b)ab=-f^2\left\{\frac{1+\chi(4)}{2}B_{1,\chi}+\frac{1}{4}B_{1,\delta\chi}+\frac{1}{8}B_{2,\chi}\right\},$$

(ii)

$$S_{1,1}(4,\chi;2,1)=\sum_{a,b=1}^{f-1}\chi(4a+b)a^2b$$

$$=-f^2\left\{\frac{8f-3-\chi(2)+(8f+4)\chi(4)}{16}B_{1,\chi}+\frac{f}{4}B_{1,\delta\chi}\right.$$

$$\left.+\frac{f+4\chi(4)}{8}B_{2,\chi}+\frac{1}{16}B_{2,\delta\chi}+\frac{1}{48}B_{3,\chi}\right\},$$

(iii)

$$S_{1,1}(4,\chi;1,2) = \sum_{a,b=1}^{f-1} \chi(4a+b)ab^2$$

$$= -f^2\left\{\frac{2f+3+\chi(2)+(2f-4)\chi(4)}{4}B_{1,\chi} + \frac{f}{4}B_{1,\delta\chi}\right.$$

$$\left. + \frac{f+4}{8}B_{2,\chi} + \frac{1}{12}B_{3,\chi}\right\}.$$

Proof. By Theorem 4.1, we have

$$S_{1,1}(4,\chi;1,1) = \frac{f^2}{\tau(\overline{\chi})}M_{1,1}(4,\overline{\chi}),$$

$$S_{1,1}(4,\chi;2,1) = \frac{f^2}{\tau(\overline{\chi})}((f-2)M_{1,1}(4,\overline{\chi}) - 2M_{2,1}(4,\overline{\chi})),$$

$$S_{1,1}(4,\chi;1,2) = \frac{f^2}{\tau(\overline{\chi})}((f-2)M_{1,1}(4,\overline{\chi}) - 2M_{1,2}(4,\overline{\chi})).$$

Thus the assertions (i)–(iii) follow from Proposition 3.4. □

5. The Evaluation of $c_{m,n}(h,\chi)$

In this section, we obtain formulas expressing $c_{m,n}(h,\chi)$ by generalized Bernoulli numbers. We keep the notation used previously. We begin by establishing the relationship between the sums $c_{m,n}(h,\chi)$ and $M_{m,n}(h,\chi)$.

Theorem 5.1. *Let χ, f, h, m, n be as above. We have*

$$c_{m,n}(h,\chi) = i^{m+n}\sum_{j=0}^{m}\sum_{k=0}^{n} 2^{j+k}\binom{m}{j}\binom{n}{k}M_{j,k}(h,\chi).$$

Proof. For $m,n=0$, the assertion is clear since both sides are identically zero. Thus we assume $m+n>0$. Without loss of generality, we further assume $n>0$. We recall the discrete Fourier transform of $P_1(a/f)\,(a\in\mathbb{Z})$,

$$P_1\left(\frac{a}{f}\right) = -\frac{1}{2if}\sum_{b=1}^{f-1}\cot\left(\frac{\pi b}{f}\right)\zeta^{ba}. \tag{5.1}$$

We put

$$T_{m,n}(h,\chi) = \sum_{a_1,\ldots,a_{m+n}(f)} P_1\left(\frac{a_1}{f}\right)\cdots P_1\left(\frac{a_{m+n}}{f}\right)\chi\left(h\sum_{i=1}^{m}a_i + \sum_{j=1}^{n}a_{m+j}\right),$$

where $a_1, a_2, \ldots, a_{m+n}$ each run over a complete residue system modulo f.

We begin by expressing $M_{m,n}(h,\chi)$ by $T_{j,k}(h,\chi)$. Noting that $a/f = P_1(a/f)+1/2$ for $1 \le a \le f-1$, we have

$$\begin{aligned}
M_{m,n}(h,\chi) &= \frac{\tau(\chi)}{f^{m+n}} S_{m,n}(h,\overline{\chi};1,\ldots,1)\\
&= \tau(\chi) \sum_{a_1,\ldots,a_{m+n}=1}^{f-1} \left(P_1\left(\frac{a_1}{f}\right)+\frac{1}{2}\right)\cdots\left(P_1\left(\frac{a_{m+n}}{f}\right)+\frac{1}{2}\right)\\
&\quad \times \overline{\chi}\left(h\sum_{i=1}^{m} a_i + \sum_{j=1}^{n} a_{m+j}\right)\\
&= \tau(\chi)\sum_{k=0}^{m+n}\left(\frac{-1}{2}\right)^k \sum_{j=0}^{k}\binom{m}{j}\binom{n}{k-j} T_{m-j,n-k+j}(h,\overline{\chi}).
\end{aligned}$$

Interchanging the order of summation and taking $r=m-j$, $s=n-k+j$, we get

$$M_{m,n}(h,\chi) = \tau(\chi)\left(\frac{-1}{2}\right)^{m+n} \sum_{r=0}^{m}\sum_{s=0}^{n}(-2)^{r+s}\binom{m}{r}\binom{n}{s} T_{r,s}(h,\overline{\chi}). \tag{5.2}$$

We next express the sum $T_{m,n}(h,\overline{\chi})$ by $c_{m,n}(h,\chi)$. Taking the discrete Fourier transform (5.1) of each of the P_1's in $T_{m,n}(h,\overline{\chi})$, we have

$$\begin{aligned}
T_{m,n}(h,\chi) &= \left(\frac{-1}{2if}\right)^{m+n} \sum_{b_1,\ldots,b_{m+n}=1}^{f-1} \cot\left(\frac{\pi b_1}{f}\right)\cdots\cot\left(\frac{\pi b_{m+n}}{f}\right)\\
&\quad \times \sum_{a_1,\ldots,a_{m+n}(f)} \zeta^{b_1a_1+\cdots+b_{m+n}a_{m+n}}\,\overline{\chi}\left(h\sum_{i=1}^{m} a_i + \sum_{j=1}^{n} a_{m+j}\right).
\end{aligned}$$

Replacing a_{m+n} by $a_{m+n} - \left(h\sum_{i=1}^{m} a_i + \sum_{j=1}^{n-1} a_{m+j}\right)$ and evaluating the Gaussian sum over a_{m+n}, we get

$$\begin{aligned}
T_{m,n}(h,\chi) &= \left(\frac{-1}{2if}\right)^{m+n} \tau(\overline{\chi}) \sum_{b_1,\ldots,b_{m+n}=1}^{f-1} \cot\left(\frac{\pi b_1}{f}\right)\cdots\cot\left(\frac{\pi b_{m+n}}{f}\right)\chi(b_{m+n})\\
&\quad \times \sum_{a_1,\ldots,a_{m+n-1}(f)} \zeta^{\sum_{i=1}^{m} a_i(b_i - hb_{m+n}) + \sum_{j=1}^{n-1} a_{m+j}(b_{m+j}-b_{m+n})}.
\end{aligned}$$

Since

$$\sum_{a(f)} \zeta^{an} = \begin{cases} f & \text{if } n \equiv 0(f),\\ 0 & \text{if } n \not\equiv 0(f),\end{cases}$$

we obtain

$$T_{m,n}(h,\overline{\chi}) = \left(\frac{-1}{2if}\right)^{m+n} f^{m+n-1}\tau(\overline{\chi})c_{m,n}(h,\chi). \tag{5.3}$$

From (5.2) and (5.3), noting that $\tau(\chi)\tau(\overline{\chi}) = \chi(-1)f$, we get

$$M_{m,n}(h,\chi) = \left(\frac{-1}{2}\right)^{m+n} \chi(-1)\sum_{j=0}^{m}\sum_{k=0}^{n}(-i)^{j+k}\binom{m}{j}\binom{n}{k}c_{j,k}(h,\chi).$$

Since $c_{m,n}(h,\chi) = (-1)^{m+n}\chi(-1)c_{m,n}(h,\chi)$ for parity reasons, this simplifies to

$$M_{m,n}(h,\chi) = \left(\frac{-1}{2}\right)^{m+n} \sum_{j=0}^{m}\sum_{k=0}^{n} i^{j+k}\binom{m}{j}\binom{n}{k}c_{j,k}(h,\chi).$$

Let S denote the right-hand side of the assertion of the theorem. Plugging this expression of $M_{m,n}(h,\chi)$ into S, we get

$$S = i^{m+n}\sum_{j=0}^{m}\sum_{k=0}^{n}(-1)^{j+k}\binom{m}{j}\binom{n}{k}\sum_{r=0}^{j}\sum_{s=0}^{k} i^{r+s}\binom{j}{r}\binom{k}{s}c_{r,s}(h,\chi).$$

Interchanging the order of summations and taking $t = j - r$, $u = k - s$, we have

$$S = i^{m+n}\sum_{r=0}^{m}\sum_{s=0}^{n}(-i)^{r+s}\binom{m}{r}\binom{n}{s}c_{r,s}(h,\chi)\sum_{t=0}^{m-r}\sum_{u=0}^{n-k}(-1)^{t+u}\binom{m-r}{t}\binom{n-s}{u}.$$

The inner two sums vanish unless $r = m$, $k = n$, in which case we have $S = c_{m,n}(h,\chi)$, as desired. □

As a corollary to Theorem 5.1, we obtain a formula for cotangent power sums considered by Apostol [1].

Corollary 5.2. *Let χ, f, m be as above. We have*

$$\sum_{a=1}^{f-1}\cot^m\left(\frac{\pi a}{f}\right)\chi(a) = -i^m\tau(\chi)\sum_{k=1}^{m}\frac{1}{k}\left(\sum_{j=k}^{m}\frac{(-2)^j\binom{m}{j}\left[{j \atop k}\right]}{(j-1)!}\right)B_{k,\overline{\chi}}.$$

Proof. By Theorem 5.1, we have

$$c_{m,0}(1,\chi) = i^m\sum_{j=0}^{m}2^j\binom{m}{j}M_{j,0}(1,\chi), \tag{5.4}$$

where from Theorem 3.3, we have

$$M_{j,0}(1,\chi) = \frac{(-1)^{j-1}\tau(\chi)}{(j-1)!} \sum_{k=1}^{j} \frac{1}{k} \begin{bmatrix} j \\ k \end{bmatrix} B_{k,\overline{\chi}}.$$

Plugging these values of $M_{j,0}(1,\chi)$ into (5.4) and interchanging the order of summation, we get the assertion of the corollary. □

Corollary 5.2 resolves the computational problem of having to evaluate cotangent power sums recursively, as discussed in [17].

Remark. A consequence of Corollary 5.2 is that

$$\sum_{j=k}^{m} \frac{(-2)^j \binom{m}{j} \left[\begin{smallmatrix} j \\ k \end{smallmatrix}\right]}{(j-1)!} = 0$$

whenever k and m are of opposite parity.

We now give examples expressing $c_{m,n}(h,\chi)$ by generalized Bernoulli numbers.

Proposition 5.3. *Let $f > 1$ be a natural number prime to 4 and let χ be a primitive Dirichlet character with conductor f. We denote by δ the unique primitive Dirichlet character modulo 4. Then, we get*

(i)

$$c_{1,1}(4,\chi) = \sum_{j=1}^{f-1} \cot\left(\frac{4\pi j}{f}\right) \cot\left(\frac{\pi j}{f}\right) \chi(j) = \tau(\chi)\left\{ B_{1,\delta\overline{\chi}} + \frac{1}{2} B_{2,\overline{\chi}} \right\},$$

(ii)

$$c_{2,1}(4,\chi) = \sum_{j=1}^{f-1} \cot^2\left(\frac{4\pi j}{f}\right) \cot\left(\frac{\pi j}{f}\right) \chi(j)$$

$$= -i\tau(\chi)\left\{ \frac{5-\overline{\chi}(2)+4\overline{\chi}(4)}{4} B_{1,\overline{\chi}} + \frac{1}{4} B_{2,\delta\overline{\chi}} + \frac{1}{12} B_{3,\overline{\chi}} \right\},$$

(iii)

$$c_{1,2}(4,\chi) = \sum_{j=1}^{f-1} \cot\left(\frac{4\pi j}{f}\right) \cot^2\left(\frac{\pi j}{f}\right) \chi(j)$$

$$= -i\tau(\chi)\left\{ (5+\overline{\chi}(2)-2\overline{\chi}(4)) B_{1,\overline{\chi}} + \frac{1}{3} B_{3,\overline{\chi}} \right\}.$$

Proof. From Theorem 5.1, we have

$$c_{1,1}(4,\chi) = -2\{M_{1,0}(4,\chi) + M_{0,1}(4,\chi) + 2M_{1,1}(4,\chi)\},$$

$$c_{2,1}(4,\chi) = -2i\{M_{0,1}(4,\chi) + 2M_{1,0}(4,\chi) + 2M_{0,2}(4,\chi) + 4M_{1,1}(4,\chi)$$
$$+ 2M_{2,0}(4,\chi) + 4M_{2,1}(4,\chi)\},$$

$$c_{1,2}(4,\chi) = -2i\{2M_{0,1}(4,\chi) + M_{1,0}(4,\chi) + 2M_{0,2}(4,\chi) + 4M_{1,1}(4,\chi)$$
$$+ 2M_{2,0}(4,\chi) + 4M_{1,2}(4,\chi)\}.$$

Thus the assertions (i)–(iii) follow from Proposition 3.4. □

Acknowledgments

I wish to thank Robert Sczech for his insightful comments and helpful suggestions. Support for this project was provided by a PSC-CUNY Award, jointly funded by The Professional Staff Congress and The City University of New York.

This paper is dedicated to Bruce C. Berndt on the occasion of his 80th birthday.

References

[1] T. Apostol, Dirichlet L-functions and character power sums. *J. Number Theory* **2** (1970) 223–234.

[2] T. Arakawa, Special values of L-functions associated with the space of quadratic forms and the representation of $Sp(2,\mathbb{F}_p)$ in the space of Siegel cusp forms, in *Automorphic Forms and Arithmetic Varieties*, Advanced Studies in Pure Mathematics, Vol. 15 (Kinokuniya and Academic Press, 1989), pp. 99–169.

[3] T. Arakawa, T. Ibukiyama and M. Kaneko, *Bernoulli Numbers and Zeta Functions* (Springer, 2014).

[4] M. Beck, B. C. Berndt, O. Chan and A. Zaharescu, Determinations of analogues of Gauss sums and other trigonometric sums, *Int. J. Number Theory* **1**(3) (2005) 333–356.

[5] M. Beck and M. Halloran, Finite trigonometric character sums via discrete fourier analysis, *Int. J. Number Theory* **6**(1) (2010) 51–67.

[6] B. C. Berndt, An elementary proof of some character sum identities of Apostol, *Glasgow Math. J.* **14**(1) (1973) 50–53.

[7] B. C. Berndt, Character analogues of the Poisson and Euler–MacLaurin summation formulas with applications, *J. Number Theory* **7** (1975) 413–445.

[8] B. C. Berndt and A. Zaharescu, Finite trigonometric sums and class numbers, *Math. Ann.* **330** (2003) 551–575.

[9] K. Hashimoto, Representations of the finite symplectic group $Sp(4,\mathbb{F}_p)$ in the spaces of Siegel modular forms, *Contemp. Math.* **53** (1986) 253–276.

[10] T. Ibukiyama, On some elementary character sums, *Comment. Math. Univ. St. Pauli* **47**(1) (1998) 7–13.

[11] T. Ibukiyama and H. Saito, On L-functions of ternary zero forms and exponential sums of Lee and Weintraub, *J. Number Theory* **48**(2) (1994) 252–257.

[12] B. Isaacson, The Tsukano conjectures on exponential sums, *Osaka J. Math.* **57**(3) (2020) 543–561.

[13] B. Isaacson, Character sums of Lee and Weintraub, *J. Number Theory* **191** (2018) 316–344.
[14] B. Isaacson, Special values of Ibukiyama-Saito *L*-functions, *Kyushu J. Math.* **72**(2) (2018) 343–373.
[15] B. Isaacson, On a generalization of a theorem of Ibukiyama, *Comment. Math. Univ. St. Pauli* **67**(1) (2019) 1–16.
[16] R. Lee and S. H. Weintraub, A generalization of a theorem of Hecke to the Siegel space of degree two, *Contemp. Math.* **96** (1989) 243–259.
[17] X. Lin and W. Zhang, Identities involving trigonometric functions and Bernoulli numbers, *Appl. Math. Comput.* **331** (2018) 228–294.
[18] C. Snyder, p-adic interpolation of Dedekind sums, *Bull. Austral. Math. Soc.* **38**(2) (1988) 293–301.
[19] R. Tsushima, The spaces of Siegel cusp forms of degree two and the representation of Sp(2, Fp), *Proc. Japan Acad.* **60**(6) (1984) 209–221.
[20] R. Tsushima, Dimension formula for the spaces of Siegel cusp forms and a certain exponential sum, *Mem. Inst. Sci. Tech. Meiji Univ.* **36** (1997) 1–56.

© 2024 World Scientific Publishing Company
https://doi.org/10.1142/9789811277375_0021

Stability of Asai local factors for GL(2)*

Yeongseong Jo[†] and M. Krishnamurthy[‡]

Department of Mathematics, The University of Iowa
14 MacLean Hall, Iowa City, IA 52242-1419, USA
[†]*yeongseong-jo@uiowa.edu*
[‡]*muthu-krishnamurthy@uiowa.edu*

On the occasion of Bruce C. Berndt's 80th birthday

Let F be a non-archimedean local field of characteristic not equal to 2 and let E/F be a quadratic algebra. We prove the stability of local factors attached to irreducible admissible (complex) representations of $\mathrm{GL}(2, E)$ via the Rankin–Selberg method under highly ramified twists. This includes both the Asai as well as the Rankin–Selberg local factors attached to pairs. Our method relies on expressing the gamma factor as a Mellin transform using Bessel functions.

Keywords: Asai local factors; Bessel functions; Howe vectors.

Mathematics Subject Classification 2020: 11F70, 11F85, 22E50

1. Introduction and Statement of the Main Result

Let $\mathbf{G} = \mathrm{GL}_2$ and let F be a non-archimedean local field whose $\mathrm{char}(F) \neq 2$. Let q denote the cardinality of the residue field. Let E be a quadratic algebra over F. We consider the local γ-factor attached to an irreducible admissible (complex) representation Π of $\mathbf{G}(E \otimes_F F)$ by the Rankin–Selberg method [9, 11, 13]. Let $\gamma(s, \Pi, \chi, \psi)$ denote this γ-factor, where χ is a character (smooth) of $F^\times$ and $\psi = \psi_F$ is a fixed nontrivial character of F. Then

$$\gamma(s, \Pi, \chi, \psi) = \begin{cases} \gamma(s, (\pi \otimes \chi) \times \sigma, \psi) & \text{Case (1) } E = F \times F, \\ \gamma_{\mathrm{As}}(s, \Pi, \chi, \psi) & \text{Case (2) } E \text{ is a field,} \end{cases}$$

where $\gamma(s, (\pi \otimes \chi) \times \sigma, \psi)$ is the Rankin–Selberg γ-factor attached to the representation $\Pi = (\pi \otimes \chi) \times \sigma$ of $\mathrm{GL}_2(F) \times \mathrm{GL}_2(F)$ [11], and $\gamma_{\mathrm{As}}(s, \Pi, \chi, \psi)$ is the *Asai*

*This article originally appeared in *International Journal of Number Theory*, **17**(3), pp. 529–546 © 2021 World Scientific Publishing Company. DOI: 10.1142/S1793042120400035
[†]Corresponding author.

γ-factor studied by Flicker [9]. In this latter situation, if χ_E denotes a character of $E^\times$ whose restriction to $F^\times$ is χ, then $\gamma_{\mathrm{As}}(s, \Pi, \chi, \psi)$ is the Asai γ-factor attached to the representation $\Pi \otimes \chi_E$ of $\mathbf{G}(E)$. By definition, it is independent of the choice of χ_E. In Case (1), we consider Π_1 and Π_2 of the form $\Pi_1 = \pi_1 \otimes \sigma$ and $\Pi_2 = \pi_2 \otimes \sigma$, respectively, where $\pi_i, i = 1, 2$, and σ are representations of $\mathrm{GL}_2(F)$ such that π_1 and π_2 share the same central character. In Case (2), we consider two representations Π_1 and Π_2 of $\mathrm{GL}_2(E)$ whose central characters have the same restriction to $F^\times$. In this paper, we prove the following.

Theorem 1.1. *Suppose Π_1 and Π_2 are irreducible admissible representations of $\mathrm{GL}_2(E \otimes F)$ as above sharing the same central character ω when restricted to $F^\times$. Then there exists an integer N so that*

$$\gamma(s, \Pi_1, \chi, \psi) = \gamma(s, \Pi_2, \chi, \psi)$$

for all characters χ whose conductor $n(\chi) > N$. In other words, the local γ-factor $\gamma(s, \Pi, \chi, \psi)$ is stable for χ sufficiently highly ramified. In this situation,

$$L(s, \Pi_1, \chi) = L(s, \Pi_2, \chi) \equiv 1,$$

and consequently the epsilon factor stabilize as well:

$$\varepsilon(s, \Pi_1, \chi, \psi) = \varepsilon(s, \Pi_2, \chi, \psi).$$

In the split case, the above theorem is not new, it is in fact proved by Jacquet and Shalika [12] for $\mathrm{GL}_n \times \mathrm{GL}_n$, $n \geq 2$. They first prove the unequal rank case, i.e. $\mathrm{GL}_n \times \mathrm{GL}_m, n \neq m$, and then deduce the equal rank case by embedding $\mathrm{GL}_n \times \mathrm{GL}_n \subset \mathrm{GL}_n \times \mathrm{GL}_{n+1}$, in order to avoid using the more complicated Rankin–Selberg integral in the equal rank case. On the other hand, our proof (for $\mathrm{GL}_2 \times \mathrm{GL}_2$) follows from a direct manipulation of the underlying local integral. Also, when $\mathrm{char}(F) = 0$ and E is a field, the above Theorem follows from [6, Theorem A] since the stability property for Artin γ-factors is known [8]. The proof there is global-to-local whereas the proof here, using the theory of Howe vectors [10], is purely local. To the best of our knowledge, Theorem 1.1 is new in the positive characteristic case. More importantly, we believe the approach here will be applicable to other situations where stability remains an open problem (see Remark). We would also like to point out the recent works [1, 4, 16] on comparing the different notions of Asai local factors using global-to-local methods.

We now outline our proof of Theorem 1.1 which is inspired by the works of Baruch [3] and Zhang [19]. For an irreducible admissible representation Π of $\mathrm{GL}_2(E \otimes F)$, recall the local zeta integral

$$Z(s, W, \Phi, \chi) = \int_{\mathrm{U}(F)\backslash \mathbf{G}(F)} W(g)\Phi(e_2 g)\chi(\det(g))|\det(g)|_F^s dg,$$

where W lies in the Whittaker model of Π, and Φ is a Schwartz–Bruhat function on F^2. Also, we have $\widetilde{W}$, the Whittaker function dual to W, which belongs to the Whittaker model of the contragredient representation $\widetilde{\Pi}$. If $\widehat{\Phi}$ denotes the Fourier transform of Φ relative to ψ, then there is the associated dual zeta integral $Z(1-s, \widetilde{W}, \widehat{\Phi}, \chi^{-1})$. The gamma factor $\gamma(s, \Pi, \chi, \psi)$ is a rational function of q^{-s} satisfying the relation

$$Z(1-s, \widetilde{W}, \widehat{\Phi}, \chi^{-1}) = \gamma(s, \Pi, \chi, \psi) Z(s, W, \Phi, \chi)$$

for all pairs (W, Φ).

For $m \gg 0$ suitably large and an appropriate Φ, we show that $Z(s, W_m, \Phi, \chi)$ is a nonzero constant independent of s and that it only depends on π through its central character. This is done in Proposition 3.2. Here, W_m is the Howe Whittaker function as in (2.2) on the group $\mathrm{GL}_2(E \otimes F)$. Following [20], we use the notation $B_m(g, W)$ to denote $W_m(g)$, since it is also a partial Bessel function in the sense of [7]. Let Π_1, Π_2 and ω be as in Theorem 1.1. Taking Whittaker functions W_{Π_1}, W_{Π_2} in their Whittaker models, respectively, let $(W_{\Pi_1})_m = W_{\Pi_1,m}$ and $(W_{\Pi_2})_m = W_{\Pi_2,m}$ denote the corresponding Howe Whittaker functions. First, we show that the difference $Z(1-s, \widetilde{W}_{\Pi_1,m}, \widehat{\Phi}, \chi^{-1}) - Z(1-s, \widetilde{W}_{\Pi_2,m}, \widehat{\Phi}, \chi^{-1})$ (see (3.3)) is

$$\approx \int_{F^\times} [B_m(t(a,1)w, W_{\Pi_1}) - B_m(t(a,1)w, W_{\Pi_2})](\omega\chi)^{-1}(a)|a|_F^{-s} d^\times a,$$

where the implied constant depends only on ω, χ, ψ and m. We exploit properties of Howe Whittaker functions to arrive at this equality, particularly Proposition 2.5, which controls the support of the function $B_m(\cdot, W_{\Pi_1}) - B_m(\cdot, W_{\Pi_2})$ away from the identity.

We note that m in the above expression depends on the conductor $n(\chi)$ of χ. At this point, we split the proof into two cases according to whether E is a field or not. Using properties of the functions W_m and the full Bessel function, we show that the function $a \mapsto B_m(t(a,1)w, W_{\Pi_1}) - B_m(t(a,1)w, W_{\Pi_2})$ is in fact independent of m for sufficiently large m. This in turn implies that the above integral is zero for a suitably highly ramified χ, thus concluding the proof of our Theorem. (The claim about the L-function may be proved along the lines of [12, Proposition 5.1].) These are all explained in Sec. 3. We also summarize the required properties of W_m in Sec. 2.3.

Remark 1.2. We hope to generalize our method to prove the stability of the Rankin–Selberg Asai, Sym^2 and $\wedge^2$ gamma factors for GL_n. This is nontrivial — the main obstacle being the analogue of Proposition 2.5 for GL_n (see [2, Lemma 6.2.2]) which introduces other relevant Weyl elements, besides the identity and the long element, while analyzing the partial Bessel function $B_m(\cdot, W_1 - W_2)$. We speculate the techniques of [7] to be useful in this regard.

2. Preliminaries

2.1. *Asai local factors*

Let $\mathbf{G} = \mathrm{GL}_2$. For any local field F, we write $\mathfrak{o}_F$ to denote its ring of integers and $\mathfrak{p}_F$ for the maximal ideal in $\mathfrak{o}_F$. Let $\mathrm{val}_F : F^\times \to \mathbb{Z}$ be the associated discrete valuation map. We fix a uniformizer ϖ_F, so that $\mathfrak{p}_F = (\varpi_F)$ and $\mathrm{val}_F(\varpi_F) = 1$. The corresponding absolute value $|x|_F = q^{-\mathrm{val}_F(x)}$ is the normalized absolute of F with $|\varpi_F|_F = q^{-1}$, where $q = |\mathfrak{o}_F/\mathfrak{p}_F|$ is the cardinality of the residue field. For the algebraic group $\mathbf{G} = \mathrm{GL}_2$, let $\mathrm{B} = \mathrm{TU}$ denote the Borel subgroup of upper triangular matrices, where

$$\mathrm{T} = \left\{ t(a,b) := \begin{pmatrix} a & \\ & b \end{pmatrix} \middle| a, b \in \mathbb{G}_m \right\}$$

is the maximal torus consisting of diagonal matrices and

$$\mathrm{U} = \left\{ u(x) := \begin{pmatrix} 1 & x \\ & 1 \end{pmatrix} \middle| x \in \mathbb{G}_a \right\}$$

is the unipotent radical of B. Let Z denote the center of GL_2, and let A denote the subtorus

$$\mathrm{A} = \{t(a,1) \,|\, a \in \mathbb{G}_m\}.$$

Let

$$\overline{\mathrm{U}} = \left\{ \overline{u}(x) := \begin{pmatrix} 1 & \\ x & 1 \end{pmatrix} \middle| x \in \mathbb{G}_a \right\}$$

be the unipotent subgroup opposed to U. We write

$$w = \begin{pmatrix} & 1 \\ 1 & \end{pmatrix}$$

to denote the long Weyl element in GL_2. Recall the Bruhat decomposition

$$\mathbf{G}(F) = \mathrm{B}(F) \cup \mathrm{B}(F)w\mathrm{U}(F),$$

with uniqueness of expression, i.e. every $g \notin \mathrm{B}(F)$ has a unique expression of the form $g = bwu, b \in \mathrm{B}(F), u \in \mathrm{U}(F)$. Put $\mathrm{K} = \mathrm{K}_F = \mathrm{GL}_2(\mathfrak{o}_F)$, the standard maximal compact subgroup of $\mathbf{G}(F)$.

Let F be a non-archimedean local field whose $\mathrm{char}(F) \neq 2$. Let E be a quadratic algebra over F with the associated trace map $\mathrm{tr}_{E/F}$. There are two possibilities for E : (1) $E = F \times F$ and (2) E is a field. We fix a nontrivial additive character $\psi = \psi_F$ of F. In Case (1), it is convenient to set $\xi = (1, -1)$, and in Case (2), we fix an element $\xi \in E - F$ such that $\mathrm{tr}_{E/F}(\xi) = 0$. We define the nontrivial additive character ψ_E of E by $\psi_E(x) = \psi(\mathrm{tr}_{E/F}(\xi x))$, then ψ_E is trivial on F. (Note that in the split case $\psi_E = (\psi, \psi^{-1})$.) The character ψ_E defines a non-degenerate character of $\mathrm{U}(E)$ via $u(x) \mapsto \psi_E(x)$. Similarly, ψ defines one for $\mathrm{U}(F)$.

Suppose π is an irreducible admissible representation of $\mathrm{GL}_2(F)$. If π is *generic*, we write $\mathcal{W}(\pi, \psi)$ to denote the associated Whittaker model relative to ψ. If π is not generic, then π is of the form $\pi = \eta_F \circ \det$ for some character η_F of $F^\times$ and may be realized as the *Langlands quotient* of the (normalized) induced representation

$(\rho, I(\eta_F|\cdot|_F^{1/2}, \eta_F|\cdot|_F^{-1/2}))$. We fix a nonzero Whittaker functional λ on this space and form

$$\mathcal{W}(\rho,\psi) = \{W_v(g) = \lambda(\rho(g)v) \,|\, v \in I(\eta_F|\cdot|_F^{1/2}, \eta_F|\cdot|_F^{-1/2}), g \in \mathrm{GL}_2(F)\}$$

with the natural action of $\mathrm{GL}_2(F)$ by right translation. This will then play the role of the Whittaker model for π and by abuse of notation we set $\mathcal{W}(\pi,\psi) := \mathcal{W}(\rho,\psi)$. Let π^ι denote the representation of $\mathrm{GL}_2(F)$ with $\pi^\iota(g) = \pi({}^tg^{-1})$ on the same underlying space V. It is well known that if π is irreducible, π^ι is isomorphic to its contragredient representation $\widetilde{\pi}$. If $\widetilde{W}(g) := W(w\ {}^tg^{-1})$, then $\widetilde{W}$ lies in $\mathcal{W}(\pi^\iota, \psi^{-1})$.

Suppose Π is an irreducible admissible representation of $\mathrm{GL}_2(E \otimes F)$ with the associated Whittaker model $\mathcal{W}(\Pi, \psi_E)$, that is,

- Case (1) $\Pi = \pi \otimes \sigma$, where π and σ are irreducible admissible representations of $\mathrm{GL}_2(F)$, and $\mathcal{W}(\Pi,\psi_E) = \mathcal{W}(\pi,\psi) \otimes \mathcal{W}(\sigma,\psi^{-1})$.
- Case (2) Π is an irreducible admissible representation of $\mathrm{GL}_2(E)$ with the associated Whittaker model $\mathcal{W}(\Pi,\psi_E)$.

In Case (1), we note that $\mathcal{W}(\Pi,\psi_E)$ is generated as a linear space by functions W of the form

$$W(g_1,g_2) = W_1(g_1)W_2(g_2), \quad W_1 \in \mathcal{W}(\pi,\psi), \quad W_2 \in \mathcal{W}(\sigma,\psi^{-1}),\ g_1,g_2 \in \mathrm{GL}_2(F).$$

In this situation, we write $W(g) = W(g,g)$ to denote its value on the diagonal subgroup $\mathrm{GL}_2(F) \hookrightarrow \mathrm{GL}_2(E\otimes F)$.

Let $C_c^\infty(F^2)$ be the space of locally constant, compactly supported functions $\Phi: F^2 \to \mathbb{C}$. We write $e_1 = (1,0)$ and $e_2 = (0,1)$ to denote the standard basis for F^2. The Fourier transform of $\Phi \in C_c^\infty(F^2)$ with respect to ψ is given by

$$\widehat{\Phi}(y) = \widehat{\Phi}_\psi(y) = \int_{F^2} \Phi(x)\psi(x\cdot {}^ty)dx,$$

where dx is the self-dual measure for which the Fourier inversion formula takes the form $\widehat{\widehat{\Phi}}(x) = \Phi(-x)$.

For $W \in \mathcal{W}(\Pi,\psi_E)$, $\Phi \in C_c^\infty(F^2)$, a smooth character χ of $F^\times$, and $s \in \mathbb{C}$, consider the zeta integral [9, 13]

$$Z(s,W,\Phi,\chi) = \int_{\mathrm{U}(F)\backslash \mathbf{G}(F)} W(g)\Phi(e_2g)\chi(\det(g))|\det(g)|_F^s dg,$$

where $\mathbf{G}(F) \hookrightarrow \mathbf{G}(E)$ embedded diagonally. In Case (2), if χ_E is a character of $E^\times$ whose restriction to $F^\times$ is χ, then this is the *Flicker* integral [9] attached to the representation $\pi \otimes \chi_E$ and it clearly only depends on the restriction of χ.

For each (s,W,Φ) as above, the integral $Z(s,W,\Phi,\chi)$ converges absolutely when $\Re(s)$ is sufficiently large and defines a rational function of q^{-s}. The collection of all such integrals span a $\mathbb{C}[q^s,q^{-s}]$-fractional ideal of $\mathbb{C}(q^{-s})$. Further, there is a function $\gamma(s,\Pi,\chi,\psi,\xi)\in\mathbb{C}(q^{-s})$ satisfying the local functional equation

$$Z(1-s,\widetilde{W},\widehat{\Phi},\chi^{-1})=\gamma(s,\Pi,\chi,\psi,\xi)Z(s,W,\Phi,\chi). \tag{2.1}$$

We refer the reader to [11, Theorem 2.7; 13, Theorem 2] for proofs of these properties. We write

$$\gamma(s,\Pi,\chi,\psi,\xi)=\begin{cases}\gamma(s,(\pi\otimes\chi)\times\sigma,\psi) & \text{in Case (1),}\\ \gamma_{\mathrm{As}}(s,\Pi,\chi,\psi,\xi) & \text{in Case (2),}\end{cases}$$

to denote the Rankin–Selberg γ-factor and the *Asai* γ-factor, respectively. From loc.cit., we also have the corresponding Asai L-factor $L_{\mathrm{As}}(s,\Pi,\chi)$ and the Asai ε-factor $\varepsilon_{\mathrm{As}}(s,\Pi,\chi,\psi,\xi)$ satisfying

$$\varepsilon_{\mathrm{As}}(s,\Pi,\chi,\psi,\xi)=\gamma_{\mathrm{As}}(s,\Pi,\chi,\psi,\xi)\frac{L_{\mathrm{As}}(s,\Pi,\chi)}{L_{\mathrm{As}}(1-s,\Pi^{\iota},\chi^{-1})},$$

and the Rankin–Selberg L-factor $L(s,(\pi\otimes\chi)\times\sigma)$ and ε-factor $\varepsilon(s,(\pi\otimes\chi)\times\sigma,\psi)$ satisfying

$$\varepsilon(s,(\pi\otimes\chi)\times\sigma,\psi)=\gamma(s,(\pi\otimes\chi)\times\sigma,\psi)\frac{L(s,(\pi\otimes\chi)\times\sigma)}{L(1-s,(\pi^{\iota}\otimes\chi^{-1})\times\sigma^{\iota})}.$$

2.2. *Dependence on the pair (ψ,ξ)*

We end this Section by examining the dependence of the local γ-factor on the pair (ψ,ξ). A different choice of (ψ',ξ') results in elements $a,b\in F^{\times}$, satisfying $\psi'(x)=\psi(ax), x\in F$, and $\xi'=b\xi$. Let ω_{Π} denote the central character of Π and put $\omega=\omega_{\Pi}|_{F^{\times}}$. In Case (1), $\omega=\omega_{\pi}\omega_{\sigma}$; and in Case (2), $\omega=\omega_{\Pi}|_{F^{\times}}$. We have the following well-known fact whose proof we include here for the sake of completeness.

Lemma 2.1. *For $a,b\in F^{\times}$ as above, we have*

$$\gamma(s,(\pi\otimes\chi)\times\sigma,\chi,\psi')=\omega^{2}(a)\chi^{4}(a)|a|^{4s-2}\gamma(s,(\pi\otimes\chi)\times\sigma,\chi,\psi),$$
$$\gamma_{\mathrm{As}}(s,\Pi,\chi,\psi',\xi)=\omega^{2}(a)\chi^{4}(a)|a|^{4s-2}\gamma_{\mathrm{As}}(s,\Pi,\chi,\psi,\xi),$$

and

$$\gamma_{\mathrm{As}}(s,\Pi,\chi,\psi,\xi')=\omega(b)\chi^{2}(b)|b|^{2s-1}\gamma_{\mathrm{As}}(s,\Pi,\chi,\psi,\xi).$$

Proof. We give the proof in the non-split case. A similar proof works in the split case as well. Let $W'(g)=W(t(a,1)g)$, then one checks that $W'\in\mathcal{W}(\Pi,\psi'_{E})$ and the map $W\mapsto W'$ is a bijection from $\mathcal{W}(\Pi,\psi_{E})$ to $\mathcal{W}(\Pi,\psi'_{E})$. It is easy to see that

$$\begin{aligned}Z(s,W',\Phi,\chi)&=\int_{\mathrm{U}(F)\backslash\mathbf{G}(F)}W'(g)\Phi(e_{2}g)\chi(\det(g))|\det(g)|_{F}^{s}dg\\&=\chi^{-1}(a)|a|_{F}^{1-s}Z(s,W,\Phi,\chi).\end{aligned}$$

For $\Phi \in C_c^\infty(F^2)$, if $\widehat{\Phi}_{\psi'}$ denotes the Fourier transform with respect to ψ', then

$$\widehat{\Phi}_{\psi'}(x,y) = |a|_F \widehat{\Phi}_{\psi}(ax, ay).$$

Thus (with $\widehat{\Phi} = \widehat{\Phi}_{\psi}$)

$$\begin{aligned} &Z(1-s, \widetilde{W}', \widehat{\Phi}_{\psi'}) \\ &\quad = \omega(a)|a|_F \int_{\mathrm{U}(F)\backslash \mathbf{G}(F)} \widetilde{W}(t(a,1)g)\widehat{\Phi}(e_2 t(a,a)g)\chi^{-1}(\det(g))|\det(g)|_F^{1-s} dg \\ &\quad = \omega^2(a)\chi^2(a)|a|_F^{2s-1} \int_{\mathrm{U}(F)\backslash \mathbf{G}(F)} \widetilde{W}(t(a,1)g)\widehat{\Phi}(e_2 g)\chi^{-1}(\det(g))|\det(g)|_F^{1-s} dg \\ &\quad = \omega^2(a)\chi^3(a)|a|_F^{3s-1} Z(1-s, \widetilde{W}, \widehat{\Phi}). \end{aligned}$$

The first assertion now follows from the local functional equation (2.1). A similar calculation works for the second assertion. One only needs to observe that, changing $\xi \mapsto \xi'$ but keeping ψ unchanged, alters the additive character with respect to which the Whittaker model is considered but has no effect on the Fourier transform. □

Henceforth, we fix a $\psi = \psi_F$ so that it is unramified, and also fix a choice of ξ in Case (2) so that the corresponding ψ_E is also unramified and we suppress ξ in the subsequent notation. Thus we write $\gamma(s, \Pi, \chi, \psi)$ to denote

$$\gamma(s, \Pi, \chi, \psi) = \begin{cases} \gamma(s, (\pi \otimes \chi) \times \sigma, \psi) & \text{in Case (1)}, \\ \gamma_{\mathrm{As}}(s, \Pi, \chi, \psi) & \text{in Case (2)}. \end{cases}$$

2.3. *Howe vectors and partial Bessel functions*

We review the theory of Howe vectors [10] for GL_2 over a local field F which was subsequently studied by Baruch [2] in a more general context. Let ψ be an additive character of F of conductor 0, i.e. ψ is trivial on $\mathfrak{o}_F$ while $\psi|_{\mathfrak{p}_F^{-1}} \neq 1$. Suppose (π, V) is an irreducible admissible representation of $\mathrm{GL}_2(F)$ with the associated Whittaker model $\mathcal{W} = \mathcal{W}(\pi, \psi)$ relative to ψ.

For $m \geq 1$, let K_m be the mth congruence subgroup, i.e. $\mathrm{K}_m = \mathrm{I} + \mathrm{M}_2(\mathfrak{p}_F^m)$. On K_m, define the function τ_m by the formula

$$\tau_m(k) = \psi(\varpi_F^{-2m} k_{1,2}),$$

where $(k_{i,j})$ is the matrix of k. One checks that τ_m is a linear (unitary) character of K_m, trivial on K_{2m}. Put

$$d_m = \begin{pmatrix} \varpi_F^{-2m} & \\ & 1 \end{pmatrix},$$

and let $\mathrm{J}_m = d_m \mathrm{K}_m d_m^{-1}$. Then J_m is given by

$$\mathrm{J}_m = \begin{pmatrix} 1 + \mathfrak{p}_F^m & \mathfrak{p}_F^{-m} \\ \mathfrak{p}_F^{3m} & 1 + \mathfrak{p}_F^m \end{pmatrix}.$$

Define a character φ_m of J_m by $\varphi_m(j) = \tau_m(d_m^{-1} j d_m)$, $j \in \mathrm{J}_m$.

It is well known (cf. [5]) that K_m is *decomposed* with respect to B, i.e. the product map

$$\mathrm{K}_m \cap \mathrm{U}(F) \times \mathrm{K}_m \cap \mathrm{T}(F) \times \mathrm{K}_m \cap \overline{\mathrm{U}}(F) \longrightarrow \mathrm{K}_m$$

is a bijection, in fact, a homeomorphism of topological spaces. Since d_m is diagonal, it follows that the group J_m is also decomposed with respect to B. Further,

$$d_m(\mathrm{K}_m \cap \mathrm{U}(F))d_m^{-1} \supset \mathrm{K}_m \cap \mathrm{U}(F) \quad \text{and} \quad d_m(\mathrm{K}_m \cap \overline{\mathrm{U}}(F))d_m^{-1} \subset \mathrm{K}_m \cap \overline{\mathrm{U}}(F),$$

i.e. conjugation by d_m enlarges the *upper* part $\mathrm{K}_m \cap \mathrm{U}(F)$ of K_m while shrinking its *lower* part $\mathrm{K}_m \cap \overline{\mathrm{U}}(F)$. Let us put $\mathrm{J}_m^u = \mathrm{J}_m \cap \mathrm{U}(F)$ and $\mathrm{J}_m^\ell = \mathrm{J}_m \cap \overline{\mathrm{U}}(F)$. $\{\mathrm{J}_m^u\}_{m \geq 0}$ becomes an exhaustive filtration by compact open subgroup of $\mathrm{U}(F)$. One checks that φ_m and ψ both agree on J_m^u.

For $m \geq 1$, consider the function f_m in the Hecke algebra of $\mathbf{G}(F)$ given by

$$f_m = \begin{cases} \dfrac{1}{\mathrm{vol}(\mathrm{J}_m)} \varphi_m^{-1} & \text{on } \mathrm{J}_m \\ 0 & \text{outside of } \mathrm{J}_m. \end{cases}$$

For v in V, put $v_m = \lambda(f_m)v$. Explicitly, $v_m = \frac{1}{\mathrm{vol}(\mathrm{J}_m)} \int_{\mathrm{J}_m} \varphi_m^{-1}(j)(\lambda(j)v)dj$.

Definition 2.2. Fix a v satisfying $\lambda(v) = 1$ and let l be an integer so that K_l fixes v. Then the vector v_m is called a Howe vector of π if $m \geq l$.

An important property of Howe vector v_m is that it is also given by the formula

$$v_m = \frac{1}{\mathrm{vol}(\mathrm{J}_m^u)} \int_{\mathrm{J}_m^u} (\lambda(u)v)\psi^{-1}(u)du.$$

To see this, one has to utilize the decomposition $\mathrm{J}_m = \mathrm{J}_m^u \cdot (\mathrm{J}_m \cap \overline{\mathrm{B}}(F))$ and observe that $\mathrm{J}_m \cap \overline{\mathrm{B}}(F) \subset \mathrm{K}_m$ fixes v for $m \geq l$. For $W \in \mathcal{W}$, we define $W_m \in \mathcal{W}$ by

$$W_m(g) = \frac{1}{\mathrm{vol}(\mathrm{J}_m^u)} \int_{\mathrm{J}_m^u} W(gu)\psi^{-1}(u)du. \tag{2.2}$$

Note, if $W = W_v$ and v_m is a Howe vector, then $W_m = W_{v_m}$ is the corresponding Howe Whittaker function. We collect the important properties of the functions W_m in the following Lemma. (cf. [3, Lemma 5.2]).

Lemma 2.3. *Choose $W \in \mathcal{W}$ so that $W(\mathrm{I}) = 1$. Let l be such that $\rho(\mathrm{K}_l)W = W$, where ρ denotes right translation. Then we have*

(1) $W_m(\mathrm{I}) = 1$;
(2) *If $m \geq l$, then $W_m(gj) = \varphi_m(j)W_m(g)$ for all $j \in \mathrm{J}_m$;*
(3) *If $m \geq k$, then*

$$W_m(g) = \frac{1}{\mathrm{vol}(\mathrm{J}_m^u)} \int_{\mathrm{J}_m^u} W_k(gu)\psi^{-1}(u)du.$$

We also have the following Lemma (cf. [19, Lemma 5.2]) concerning the support of W_m on the diagonal torus.

Lemma 2.4. *Let $m \geq l$ and W_m be as in Lemma 2.3. Then we have*

(1) $W_m(t(a,1)) \neq 0 \iff a \in 1 + \mathfrak{p}_F^m$.
(2) *If $W_m(t(a,1)w) \neq 0$, then $a \in \mathfrak{p}_F^{-3m}$.*

Proof. For $x \in \mathfrak{p}_F^{-m}$ we have $u(x) \in \mathrm{J}_m^u$. From the relation

$$\begin{pmatrix} a & \\ & 1 \end{pmatrix}\begin{pmatrix} 1 & x \\ & 1 \end{pmatrix} = \begin{pmatrix} 1 & ax \\ & 1 \end{pmatrix}\begin{pmatrix} a & \\ & 1 \end{pmatrix},$$

we obtain $\psi(x)W_m(t(a,1)) = \psi(ax)W_m(t(a,1))$. If $a \notin 1 + \mathfrak{p}_F^m$, we have $(1-a)x \notin \mathfrak{o}_F \subset \ker(\psi)$ for some $x \in \mathfrak{p}_F^{-m}$. Thus we get that $W_m(t(a,1)) = 0$. If $a \in 1 + \mathfrak{p}_F^m$, then $t(a,1) \in \mathrm{J}_m^u$. Our result follows from Lemma 2.3 that $W_m(t(a,1)) = \varphi_m(t(a,1))W_m(\mathrm{I}) = \varphi_m(0) = 1$.

To see property (2), for $x \in \mathfrak{p}_F^{3m}$, it is easy to see that $\overline{u}(x) \in \mathrm{J}_m^\ell$. Using

$$\begin{pmatrix} a & \\ & 1 \end{pmatrix} w \begin{pmatrix} 1 & \\ x & 1 \end{pmatrix} = \begin{pmatrix} 1 & ax \\ & 1 \end{pmatrix}\begin{pmatrix} a & \\ & 1 \end{pmatrix} w,$$

we observe that $W_m(t(a,1)w) = \psi(ax)W_m(t(a,1)w)$. The conclusion follows from the argument similar to that in property (1). □

By virtue of Lemma 2.3, W_m is what one calls a *partial Bessel function*, in the sense that,

$$W_m(u_1 g u_2) = \psi(u_1)\psi(u_2)W_m(g), \quad u_1 \in \mathrm{U}(F), \quad u_2 \in \mathrm{J}_m^u, \quad g \in \mathrm{GL}_2(F).$$

We use the notation $B_m(g, W) = W_m(g)$, $g \in \mathbf{G}(F)$. The main result that we need regarding these partial Bessel functions is [19, Lemma 5.3] (slightly paraphrased here).

Proposition 2.5. *Let π_1 and π_2 be irreducible admissible representation of $\mathrm{GL}_2(F)$ with the same central character. Let l be an integer so that $W_1 \in \mathcal{W}(\pi_1, \psi)$ and $W_2 \in \mathcal{W}(\pi_2, \psi)$ are Whittaker functions fixed by K_l. Assume $W_i(\mathrm{I}) = 1, i = 1, 2$. Then, for $m \geq 3l$, the function given by the difference*

$$B_m(g, W_1) - B_m(g, W_2)$$

is supported on $g \in \mathrm{B}(F)w\mathrm{J}_m^u$.

We also need the full Bessel function for the representation π. If $W_v \in \mathcal{W}(\pi, \psi)$ then the integral

$$\ell(v, x) = \int_{\mathrm{J}_m^u} W_v\left(\begin{pmatrix} x & \\ & 1 \end{pmatrix} w \begin{pmatrix} 1 & u \\ & 1 \end{pmatrix}\right)\psi^{-1}(u)du$$

converges in the sense that it stabilizes for large m depending on x (see [17, Lemma 4.1]). It defines a Whittaker functional on V for fixed $x \in F$, as does the functional $v \mapsto W_v(\mathrm{I})$. It follows from the uniqueness of Whittaker functionals that there is a constant proportionality $j_\pi(x)$ satisfying the relation

$$j_\pi(x)W_v(\mathrm{I}) = \int_{\mathrm{J}^u_m} W_v\left(\begin{pmatrix} x & \\ & 1 \end{pmatrix} w \begin{pmatrix} 1 & u \\ & 1 \end{pmatrix}\right) \psi^{-1}(u)du.$$

The function $x \mapsto j_\pi(x)$ is called *the Bessel function* (attached to w) and its basic properties were studied by Soudry [17, Lemma 4.2] which we recall in the following. The partial Bessel function introduced above (which is not quasi-invariant on the right under the full unipotent subgroup) is related to the full Bessel function.

Proposition 2.6. *Let l and $W \in \mathcal{W}$ be as in Lemma* 2.3.

(1) *For $n \geq \frac{3}{2}l$, we have*

$$\mathrm{vol}(\mathrm{J}^u_{3n})W_{3n}(t(x,1)w) = j_\pi(x) \quad \textit{for all } |x| \leq q^{6n}.$$

(2) *For any non-negative integer n, $x \in \mathfrak{p}_F^{-6n}$ and $a \in 1 + \mathfrak{p}_F^{3n}$, we have*

$$j_\pi(xa) = j_\pi(x).$$

3. Proof of Stability of Asai Local Gamma Factor for GL_2

Before computing integrals, we fix our choice of Haar measures. If H is any of the unimodular groups $\mathbf{G}(F), \mathrm{U}(F), \overline{\mathrm{U}}(F)$ or K, we normalize the Haar measure on H so that $\mathrm{vol}(\mathrm{H}\cap\mathrm{K}) = 1$. Since $\mathrm{U}(F), \overline{\mathrm{U}}(F) \simeq F$ and $\mathrm{A}(F) \simeq F^\times$, we may identify the measure on $\mathrm{U}(F)$ and $\overline{\mathrm{U}}(F)$ with the additive measure dx on F, and the measure on $\mathrm{A}(F)$ with the multiplicative measure $d^\times a$ on $F^\times$ (likewise, for $\mathrm{Z}(F) \simeq F^\times$). Our normalization is such that $\mathrm{vol}(\mathfrak{o}_F, dx) = 1$ and $\mathrm{vol}(\mathfrak{o}_F^\times, d^\times a) = 1 - 1/q$. In particular, $d^\times a$ assigns $1 + \mathfrak{p}_F^m$ to q^{-m}. If we use the coordinates $g = u(x)zt(a,1)k$, the Haar measure dg on $\mathbf{G}(F)$ can then be decomposed as

$$dg = \delta_{\mathrm{B}}^{-1}(a)dxd^\times zd^\times adk,$$

where δ_{B} is the modulus character. If $C_c^\infty(\mathbf{G}(F))$ denotes the space of smooth functions of compact support on $\mathbf{G}(F)$, then for $f \in C_c^\infty(\mathbf{G}(F))$, there is a constant $c = c(\mathrm{B})$ that only depends on B, so that we have the following integration formula [18] (also see [14, Proposition 8.45]):

$$\int_{\mathbf{G}(F)} f(g)dg = c\int_{\mathrm{U}(F)\times\mathrm{T}(F)\times\overline{\mathrm{U}}(F)} f(ut\bar{u})\delta_{\mathrm{B}}(t)^{-1}d\bar{u}dtdu. \tag{3.1}$$

We will use this formula in our calculations (of course, one can also normalize the Haar measures so that $c = 1$ in the above formula).

A (smooth) character η of $F^\times$ is said to be unramified if it is trivial on $\mathfrak{o}_F^\times$. If η is unramified, we set its conductor to be 0; otherwise, the conductor of η is the smallest positive integer d for which η is trivial upon restriction to $1+\mathfrak{p}_F^d$. We write $n(\eta)$ to denote the conductor of η. For $a \in F$, recall the Gauss sum

$$\tau(\eta, \psi, a) = \int_{\mathfrak{o}_F^\times} \eta^{-1}(u)\psi(au)d^\times u$$

of η relative to ψ. We have the following (cf. [5, Sec. 23.6], for example).

Lemma 3.1. *For η ramified,*

$$\int_{\mathfrak{o}_F^\times} \eta^{-1}(u)\psi(\varpi_F^k u)d^\times u \begin{cases} = 0 & \text{if } k \neq -n(\eta), \\ \neq 0 & \text{if } k = -n(\eta). \end{cases}$$

For a set $A \subset F$, we denote by $\mathbb{1}_A$ the characteristic function of A. For integers $i \geq 0$, $j > 0$, put

$$\Phi_{i,j}(x, y) = \mathbb{1}_{\mathfrak{p}_F^i}(x)\mathbb{1}_{1+\mathfrak{p}_F^j}(y).$$

We retain the notation of Sec. 2.1, recall that Π is an irreducible representation of $\mathrm{GL}_2(E \otimes F)$ and we write $\Pi = \pi \otimes \sigma$ in Case (1). For $m \geq 1$, let $\mathrm{K}_m \leq \mathrm{GL}_2(E \otimes F)$ be the mth congruence subgroup defined by

$$\mathrm{K}_m = \begin{cases} \begin{pmatrix} 1+\mathfrak{p}_F^m & \mathfrak{p}_F^m \\ \mathfrak{p}_F^m & 1+\mathfrak{p}_F^m \end{pmatrix} \times \begin{pmatrix} 1+\mathfrak{p}_F^m & \mathfrak{p}_F^m \\ \mathfrak{p}_F^m & 1+\mathfrak{p}_F^m \end{pmatrix} & \text{Case (1)}, \\ \begin{pmatrix} 1+\mathfrak{p}_E^m & \mathfrak{p}_E^m \\ \mathfrak{p}_E^m & 1+\mathfrak{p}_E^m \end{pmatrix} & \text{Case (2)}. \end{cases}$$

We then have the corresponding subgroups J_m^u and J_m^ℓ of $\mathrm{GL}_2(E \otimes F)$. Choose a Whittaker function $W_\Pi \in \mathcal{W}(\Pi, \psi_E)$ satisfying $W_\Pi(\mathrm{I}) = 1$ and let l be a positive integer such that $\rho(\mathrm{K}_l)W_\Pi = W_\Pi$. For $m \geq l$, we form the corresponding Howe Whittaker function $(W_\Pi)_m = W_{\Pi,m} \in \mathcal{W}(\Pi, \psi_E)$. In Case (1), if $W_\Pi = WW', W \in \mathcal{W}(\pi, \psi), W' \in \mathcal{W}(\sigma, \psi^{-1})$, then

$$W_{\Pi,m}(g_1, g_2) = W_m(g_1)W'_m(g_2).$$

Proposition 3.2. *Let $n = n(\omega\chi^2)$ and choose $m \geq \max\{l, 2n(\chi)\}$. Then for $i \geq 3m$,*

$$Z(s, W_{\Pi,m}, \Phi_{i,n}, \chi) = \begin{cases} cq^{-i-n-m} & \text{if } E = F \times F, \\ cq^{-i-n}\mathrm{vol}((1+\mathfrak{p}_E^m) \cap F^\times) & \text{if } E \text{ is a field.} \end{cases}$$

Proof. Put $\Phi = \Phi_{i,n}$. According to (3.1), we have

$$Z(s, W_{\Pi,m}, \Phi, \chi)$$
$$= c\int_{F^\times}\int_F W_{\Pi,m}(t(a,1)\overline{u}(x))\chi(a)|a|_F^{s-1}$$

$$\times\left(\int_{F^\times}\Phi(e_2t(a,1)\overline{u}(x)z)\omega(z)\chi^2(z)|z|_F^{2s}d^\times z\right)dxd^\times a$$

$$=c\int_{F^\times}\int_F W_{\Pi,m}(t(a,1)\overline{u}(x))\chi(a)|a|_F^{s-1}$$

$$\times\left(\int_{F^\times}\Phi(e_2\overline{u}(x)z)\omega(z)\chi^2(z)|z|_F^{2s}d^\times z\right)dxd^\times a.$$

The inner integral reduces to

$$\int_{F^\times}\Phi(e_2\overline{u}(x)z)\omega(z)\chi^2(z)|z|_F^{2s}d^\times z=\int_{F^\times}\mathbb{1}_{\mathfrak{p}_F^i}(xz)\mathbb{1}_{1+\mathfrak{p}_F^n}(z)d^\times z$$

$$=\begin{cases}q^{-n} & \text{if } x\in\mathfrak{p}_F^i,\\ 0 & \text{otherwise.}\end{cases}$$

Thus

$$Z(s,W_{\Pi,m},\Phi_{i,n},\chi)=cq^{-n}\int_{F^\times}\int_{\mathfrak{p}_F^i}W_{\Pi,m}(t(a,1)\overline{u}(x))\chi(a)|a|_F^{s-1}dxd^\times a.$$

Since $i\geq 3m$, $x\in\mathfrak{p}_F^i\subset\mathfrak{p}_F^{3m}\Rightarrow\overline{u}(x)\in\mathrm{J}_m$. But, according to Lemma 2.3, the function $W_{\Pi,m}(\cdot)$ is right invariant under $\overline{u}(x)$ for $x\in\mathfrak{p}_F^i$. Therefore,

$$Z(s,W_{\Pi,m},\Phi,\chi)=cq^{-i-n}\int_{F^\times}W_{\Pi,m}(t(a,1))\chi(a)|a|_F^{s-1}d^\times a$$

$$=\begin{cases}cq^{-i-n}\displaystyle\int_{(1+\mathfrak{p}_F^m)}\chi(a)|a|_F^{s-1}d^\times a & \text{Case (1)},\\ cq^{-i-n}\displaystyle\int_{(1+\mathfrak{p}_E^m)\cap F^\times}\chi(a)|a|_F^{s-1}d^\times a & \text{Case (2)}.\end{cases}$$

Here, the second equality follows from Lemma 2.4. Now, since $m\geq 2n(\chi)$ and $((1+\mathfrak{p}_E^m)\cap F^\times)\subset(1+\mathfrak{p}_F^{[\frac{m}{2}]})$, we obtain the formula. □

By our choice of the measure on F, for the characteristic function $\mathbb{1}_{\mathfrak{p}_F^i}$, we have

$$\widehat{\mathbb{1}}_{\mathfrak{p}_F^i}(y)=q^{-i}\mathbb{1}_{\mathfrak{p}_F^{-i}}(y),$$

and

$$\widehat{\mathbb{1}}_{1+\mathfrak{p}_F^j}(y)=\int_F\mathbb{1}_{1+\mathfrak{p}_F^j}(x)\psi(xy)dx$$

$$=\psi(y)\int_F\mathbb{1}_{\mathfrak{p}_F^j}(x)\psi(xy)dx=q^{-j}\psi(y)\mathbb{1}_{\mathfrak{p}_F^{-j}}(y).$$

Hence,

$$\widehat{\Phi}_{i,j}(x,y)=q^{-i-j}\psi(y)\mathbb{1}_{\mathfrak{p}_F^{-j}}(y)\mathbb{1}_{\mathfrak{p}_F^{-i}}(x). \tag{3.2}$$

Next, we consider the dual side. Suppose Π_1 and Π_2 are irreducible admissible representations of $\mathrm{GL}_2(E \otimes F)$ having the same central character when restricted to $F^\times$. For $j = 1, 2$, fix $W_{\Pi_j} \in \mathcal{W}(\Pi_j, \psi_E)$ as above, and form the corresponding Howe Whittaker functions $W_{\Pi_1,m}$ and $W_{\Pi_2,m}$, respectively.

Lemma 3.3. *Choose $m \geq \max\{l, n, 2n(\chi)\}$ with $n = n(\omega\chi^2) > 0$. Keeping notations as in Proposition 3.2, for $i \geq 3m$, we have*

$$\begin{aligned} &Z(1-s, \widetilde{W}_{\Pi_1,m}, \widehat{\Phi}_{i,n}, \chi^{-1}) - Z(1-s, \widetilde{W}_{\Pi_2,m}, \widehat{\Phi}_{i,n}, \chi^{-1}) \\ &\quad = c_{\omega,\chi,\psi} \int_{F^\times} [B_m(t(a,1)w, W_{\Pi_1}) - B_m(t(a,1)w, W_{\Pi_2})](\omega\chi)^{-1}(a)|a|_F^{-s} d^\times a, \end{aligned} \tag{3.3}$$

where

$$c_{\omega,\chi,\psi} = \begin{cases} cq^{-i-n+m+2n(1-s)}(\omega\chi^2)(\varpi_F^n)\tau(\omega\chi^2, \psi, \varpi_F^{-n}) & \textit{Case } (1), \\ cq^{-i-n+2n(1-s)}(\omega\chi^2)(\varpi_F^n)\tau(\omega\chi^2, \psi, \varpi_F^{-n})\mathrm{vol}(\mathfrak{p}_E^{-m} \cap F, dx) & \textit{Case } (2). \end{cases}$$

Proof. For $j = 1, 2$, using (3.1) again, we see that $Z(1-s, \widetilde{W}_{\Pi_j,m}, \widehat{\Phi}_{i,n}, \chi^{-1})$ equals

$$\begin{aligned} &c \int_{F^\times} \int_F \widetilde{W}_{\Pi_j,m}(t(a,1)\overline{u}(x))\chi^{-1}(a)|a|_F^{-s} \\ &\quad \times \left(\int_{F^\times} \widehat{\Phi}_{i,n}(e_2 t(a,1)\overline{u}(x)z)\omega^{-1}(z)\chi^{-2}(z)|z|_F^{2-2s} d^\times z \right) dx d^\times a. \end{aligned} \tag{3.4}$$

Substituting for the Fourier transform (3.2), the inner integral becomes

$$G_{i,n}(x) = \begin{cases} q^{-i-n} \displaystyle\int_{|z|_F \leq q^n} \psi(z)\omega^{-1}(z)\chi^{-2}(z)|z|_F^{2(1-s)} d^\times z & \text{if } |x|_F \leq q^{i-n}, \\ q^{-i-n} \displaystyle\int_{|z|_F \leq \frac{q^i}{|x|_F}} \psi(z)\omega^{-1}(z)\chi^{-2}(z)|z|_F^{2(1-s)} d^\times z & \text{if } |x|_F > q^{i-n}. \end{cases}$$

We may re-write the first integral in the above expression as a sum over *shells* to obtain

$$\begin{aligned} &q^{-i-n} \int_{|z|_F \leq q^n} \psi(z)(\omega\chi^2)(z^{-1})|z|_F^{2(1-s)} d^\times z \\ &\quad = q^{-i-n} \sum_{k=-n}^{\infty} \int_{\mathfrak{p}_F^k - \mathfrak{p}_F^{k+1}} \psi(z)(\omega\chi^2)(z^{-1})(q^{-k})^{2(1-s)} d^\times z \\ &\quad = q^{-i-n} \sum_{k=-n}^{\infty} q^{-2k(1-s)}(\omega\chi^2)(\varpi_F^{-k})\tau(\omega\chi^2, \psi, \varpi_F^k). \end{aligned}$$

Since $\omega\chi^2$ is a ramified character of $F^\times$ with conductor n, by Lemma 3.1, we obtain

$$G_{i,n}(x) = q^{-i-n+2n(1-s)}(\omega\chi^2)(\varpi_F^n)\tau(\omega\chi^2,\psi,\varpi_F^{-n}) \quad \text{if } |x|_F \le q^{i-n}. \tag{3.5}$$

Insert $\widetilde{W}_{\Pi_j,m}(t(a,1)\overline{u}(x)) = W_{\Pi_j,m}(t(1,a^{-1})wu(-x))$ in (3.4). It follows from Proposition 2.5 that

$$B_m(t(1,a^{-1})wu(-x),W_{\Pi_1}) - B_m(t(1,a^{-1})wu(-x),W_{\Pi_2}) = 0$$

for $x \notin \mathfrak{p}_E^{-m} \cap F$. Hence

$$\begin{aligned} &Z(1-s,\widetilde{W}_{\Pi_1,m},\widehat{\Phi}_{i,n},\chi^{-1}) - Z(1-s,\widetilde{W}_{\Pi_2,m},\widehat{\Phi}_{i,n},\chi^{-1}) \\ &\quad = c\int_{F^\times}\int [B_m(t(a,1)wu(-x),W_{\Pi_1}) - B_m(t(a,1)wu(-x),W_{\Pi_2})](\omega\chi)^{-1}(a) \\ &\qquad \times |a|_F^{-s} G_{i,n}(x)dxd^\times a, \end{aligned} \tag{3.6}$$

where the x-integral is over $\mathfrak{p}_F^{-m}$ in Case (1), and over $\mathfrak{p}_E^{-m} \cap F$ in Case (2). On the other hand, since $i \ge 3m$, we may choose $m \ge n$ so that $x \in \mathfrak{p}_E^{-m} \cap F \Rightarrow x \in \mathfrak{p}_F^{-(i-n)}$. By Lemma 2.3, part (2), both $W_{\Pi_1,m}$ and $W_{\Pi_2,m}$ are invariant under right translation by elements in J_m^u. Thus collecting (3.5) and (3.6), we get the desired conclusion. □

Next, we seek to remove the dependence of the integrand in (3.3) on m. A smooth function φ on a subtorus $\mathrm{A} \subset \mathrm{T}$ is said to be *uniformly smooth* if there exists a fixed compact open subgroup $\mathrm{A}_0 \subset \mathrm{A}$ such that $\varphi(aa_0) = \varphi(a)$ for all $a_0 \in \mathrm{A}_0$ and $a \in \mathrm{A}$. We do this on a case-by-case basis.

3.1. *Split case*

$E = F \times F$.

Here, $\Pi_1 = \pi_1 \otimes \sigma$ and $\Pi_2 = \pi_2 \otimes \sigma$. Hence $\omega = \omega_{\pi_1}\omega_\sigma = \omega_{\pi_2}\omega_\sigma$. Suppose $W_{\Pi_1,m} = W_{1,m}W'_m$ and $W_{\Pi_2,m} = W_{2,m}W'_m$, where $W_{j,m} \in \mathcal{W}(\pi_j,\psi)$, $j = 1,2$, and $W'_m \in \mathcal{W}(\sigma,\psi^{-1})$.

Proposition 3.4. *Let ω,χ,n,l and rest of the notation be as in Lemma 3.3.*

(1) *There exists a suitably large integer $m \gg 0$ such that, for $i \ge 3m$, we have*

$$\begin{aligned} &Z(1-s,\widetilde{W}_{1,m},\widetilde{W}'_m,\widehat{\Phi}_{i,n},\chi^{-1}) - Z(1-s,\widetilde{W}_{2,m},\widetilde{W}'_m,\widehat{\Phi}_{i,n},\chi^{-1}) \\ &\quad = cq^{-i-n+m+2n(1-s)}(\omega\chi^2)(\varpi_F^n)\tau(\omega\chi^2,\psi,\varpi_F^{-n})\frac{\mathrm{vol}(\mathrm{J}_{3l}^u)}{[\mathrm{vol}(\mathrm{J}_m^u)]^2} \\ &\qquad \times \int_{\mathfrak{p}_F^{-9l}} [B_{3l}(t(a,1)w,W_1) - B_{3l}(t(a,1)w,W_2)] j_\sigma(a)(\omega\chi)^{-1}(a)|a|_F^{-s}d^\times a. \end{aligned}$$

The integral can run over $F^\times$.

(2) (*Uniform smoothness*) *The function*

$$a \mapsto [B_{3l}(t(a,1)w, W_1) - B_{3l}(t(a,1)w, W_2)]j_\sigma(a)\omega^{-1}(a)|a|_F^{-s}$$

is uniformly smooth for $a \in \mathfrak{p}_F^{-9l}$.

Proof. We use part (3) of Lemma 2.3 with $m \geq 3l$, to obtain

$$\begin{aligned}&\int_{F^\times} [B_m(t(a,1)w, W_{\Pi_1}) - B_m(t(a,1)w, W_{\Pi_2})] \\ &\quad \times (\omega\chi)^{-1}(a)|a|_F^{-s} d^\times a \\ &= \int_{F^\times} [(W_1)_m(t(a,1)w) - (W_2)_m(t(a,1)w)] \\ &\quad \times W'_m(t(a,1)w)(\omega\chi)^{-1}(a)|a|_F^{-s} d^\times a \\ &= \frac{\mathrm{vol}(\mathrm{J}^u_{3l})}{\mathrm{vol}(\mathrm{J}^u_m)} \int_{F^\times} [(W_1)_{3l}(t(a,1)w) - (W_2)_{3l}(t(a,1)w)] \\ &\quad \times W'_m(t(a,1)w)(\omega\chi)^{-1}(a)|a|_F^{-s} d^\times a. \end{aligned} \tag{3.7}$$

Here, we have used Proposition 2.5 in deducing the second equality. By Lemma 2.4(2), we get

$$\begin{aligned}&\int_{F^\times} [B_m(t(a,1)w, W_{\Pi_1}) - B_m(t(a,1)w, W_{\Pi_2})](\omega\chi)^{-1}(a)|a|_F^{-s} d^\times a \\ &= \frac{\mathrm{vol}(\mathrm{J}^u_{3l})}{\mathrm{vol}(\mathrm{J}^u_m)} \int_{\mathfrak{p}_F^{-9l}} [(W_1)_{3l}(t(a,1)w) - (W_2)_{3l}(t(a,1)w)] \\ &\quad \times W'_m(t(a,1)w)(\omega\chi)^{-1}(a)|a|_F^{-s} d^\times a.\end{aligned}$$

Choosing $m = 3n$ with $n \geq 2l$, we apply Proposition 2.6(1) to obtain

$$\mathrm{vol}(\mathrm{J}^u_m)W'_m(t(a,1)w) = j_\sigma(a) \quad \text{for } |a|_F \leq q^{2m}.$$

Then inserting this into the above expression we get

$$\begin{aligned}&\int_{F^\times} [B_m(t(a,1)w, W_{\Pi_1}) - B_m(t(a,1)w, W_{\Pi_2})](\omega\chi)^{-1}(a)|a|_F^{-s} d^\times a \\ &= \frac{\mathrm{vol}(\mathrm{J}^u_{3l})}{[\mathrm{vol}(\mathrm{J}^u_m)]^2} \int_{\mathfrak{p}_F^{-9l}} [B_{3l}(t(a,1)w, W_1) - B_{3l}(t(a,1)w, W_2)] \\ &\quad \times j_\sigma(a)(\omega\chi)^{-1}(a)|a|_F^{-s} d^\times a. \end{aligned} \tag{3.8}$$

Plugging (3.8) into (3.3), we obtain the desired formula.

For the second assertion, it follows from Proposition 2.6(2) that for $a \in \mathfrak{p}_F^{-9l}$, the function is right invariant under $1 + \mathfrak{p}_F^{6l}$. □

Putting Propositions 3.2, 3.4 and (2.1) together, we get

$$\gamma(s,(\pi_1\otimes\chi)\times\sigma,\psi)-\gamma(s,(\pi_2\otimes\chi)\times\sigma,\psi)$$
$$=\beta_l d_{\omega,\chi,\psi}\int_{\mathfrak{p}_F^{-9l}}[B_{3l}(t(a,1)w,W_1)-B_{3l}(t(a,1)w,W_2)]j_\sigma(a)(\omega\chi)^{-1}(a)|a|_F^{-s}d^\times a,$$

where $\beta_l=\mathrm{vol}(\mathrm{J}_{3l}^u)$ and $d_{\omega,\chi,\psi}=q^{2n(1-s)}(\omega\chi^2)(\varpi_F^n)\tau(\omega\chi^2,\psi,\varpi_F^{-n})\neq 0$. By uniform smoothness, the integral on the right-hand side is 0 for χ satisfying $n(\chi)>6l$. Since $L(s,(\pi\otimes\chi)\times\sigma)=1$ for a highly ramified χ according to [12, Proposition 5.1], this concludes the proof of Theorem 1.1 in the split case.

3.2. *Non-split case*

E is a field.

Proposition 3.5. *In the set-up of Lemma 3.3, put $W_i=W_{\Pi_i}$ and $W_{i,m}=(W_i)_m$, $i=1,2$.*

(1) *There exists a suitably large integer $m\gg 0$ such that, for $i\geq 3m$, we have*

$$Z(1-s,\widetilde{W}_{1,m},\widehat{\Phi}_{i,n},\chi^{-1})-Z(1-s,\widetilde{W}_{2,m},\widehat{\Phi}_{i,n},\chi^{-1})$$
$$=cq^{-i-n+2n(1-s)}(\omega\chi^2)(\varpi_F^n)\tau(\omega\chi^2,\psi,\varpi_F^{-n})\mathrm{vol}(\mathfrak{p}_E^{-m}\cap F,\,dx)\frac{\mathrm{vol}(\mathrm{J}_{3l}^u)}{\mathrm{vol}(\mathrm{J}_m^u)}$$
$$\times\int_{F^\times}[B_{3l}(t(a,1)w,W_1)-B_{3l}(t(a,1)w,W_2)](\omega\chi)^{-1}(a)|a|_F^{-s}d^\times a.$$

(2) *(Uniform smoothness) The function*

$$a\mapsto[B_{3l}(t(a,1)w,W_1)-B_{3l}(t(a,1)w,W_2)]\omega^{-1}(a)|a|_F^{-s}$$

is uniformly smooth.

Proof. As in (3.7), we obtain

$$B_m(t(a,1)w,W_1)-B_m(t(a,1)w,W_2)$$
$$=\frac{1}{\mathrm{vol}(\mathrm{J}_m^u)}\int_{\mathrm{J}_m^u}[B_{3l}(t(a,1)wu,W_1)-B_{3l}(t(a,1)wu,W_2)]\psi_E^{-1}(u)du$$
$$=\frac{1}{\mathrm{vol}(\mathrm{J}_m^u)}\int_{\mathrm{J}_{3l}^u}[B_{3l}(t(a,1)wu,W_1)-B_{3l}(t(a,1)wu,W_2)]\psi_E^{-1}(u)du$$
$$=\frac{\mathrm{vol}(\mathrm{J}_{3l}^u)}{\mathrm{vol}(\mathrm{J}_m^u)}[B_{3l}(t(a,1)w,W_1)-B_{3l}(t(a,1)w,W_2)] \tag{3.9}$$

for $m\geq 3l$. Plugging (3.9) into (3.3), we arrive at the Proposition.

For the second assertion, we note that the function

$$a\mapsto B_{3l}(t(a,1)w,W_1)-B_{3l}(t(a,1)w,W_2)$$

is right invariant under $1+\mathfrak{p}_F^{3l}$ and is consequently uniformly smooth relative to m. □

Collecting Propositions 3.2, 3.5 and (2.1), we obtain

$$\gamma_{\mathrm{As}}(s,\Pi_1,\chi,\psi)-\gamma_{\mathrm{As}}(s,\Pi_2,\chi,\psi)$$
$$=\beta_l d_{\omega,\chi,\psi}\int_{F^\times}[B_{3l}(t(a,1)w,W_1)-B_{3l}(t(a,1)w,W_2)](\omega\chi)^{-1}(a)|a|_F^{-s}d^\times a, \tag{3.10}$$

where β_l is the volume factor $\beta_l=\frac{\mathrm{vol}(\mathfrak{p}_E^{-m}\cap F)\mathrm{vol}(\mathrm{J}_{3l}^{u})}{\mathrm{vol}(1+\mathfrak{p}_E^{m}\cap F^\times)\mathrm{vol}(\mathrm{J}_{m}^{u})}=\mathrm{vol}(\mathrm{J}_{3l}^{u})$ and

$$d_{\omega,\chi,\psi}=q^{2n(1-s)}(\omega\chi^2)(\varpi_F^n)\tau(\omega\chi^2,\psi,\varpi_F^{-n})$$

is a nonzero constant. Here, we used the fact that $\varpi_F\mathfrak{o}_E=\varpi_E\mathfrak{o}_E$ which gives $q_E=q^2$ if E/F is unramified. If E/F is ramified, $\varpi_F\mathfrak{o}_E=\varpi_E^2\mathfrak{o}_E$ so we have $q_E=q$. But the integral on the right-hand side of (3.10) is 0 for a suitable highly ramified χ. For instance, if χ is such that $n(\chi)>3l$, then the integral vanishes thus giving us stability of Asai γ-factor under highly ramified twists.

To prove the corresponding statement for L- and ε-factors, we only need to prove stability for the L-function. With [15, Corollary 4.3] in hand concerning the poles of the Asai L-function, this can be proved exactly as in Jacquet and Shalika [12, Proposition 5.1].

Acknowledgments

We thank Roger Howe for sending us a copy of [10]. We also thank the anonymous referee for useful comments.

References

[1] U. K. Anandavardhanan, R. Kurinczuk, N. Matringe, V. Sécherre and S. Stevens, Galois self-dual cuspidal types and Asai local factors, to appear in *J. Eur. Math. Soc.* (2019); https://arXiv.org/abs/1807.07755.

[2] E. M. Baruch, Local factors attached to representations of p-adic groups and strong multiplicity one, Ph.D. Thesis, Yale University (1995), 83 pp.

[3] E. M. Baruch, On the gamma factors attached to representations of U(2, 1) over a p-adic field, *Israel J. Math.* **102** (1997) 317–345.

[4] R. Beuzart-Plessis, Archimedean theory and ε-factors for the Asai Rankin–Selberg integrals, preprint (2018); https://arXiv.org/abs/1812.00053.

[5] C. J. Bushnell and G. Henniart, *The Local Langlands Conjecture for* GL(2), Grundlehren der Mathematischen Wissenschaften [Fundamental Principles of Mathematical Sciences], Vol. 335 (Springer-Verlag, Berlin, 2006), xii+347 pp.

[6] S.-Y. Chen, Y. Cheng and I. Ishikawa, Gamma factors for Asai representations of GL_2, *J. Number Theory* **209** (2020) 83–146.

[7] J. W. Cogdell, F. Shahidi and T.-L. Tsai, Local Langlands correspondence for GL_n and the exterior and symmetric square ε-factors, *Duke Math. J.* **166**(11) (2017) 2053–2132.

[8] P. Deligne and G. Henniart, Sur la variation, par torsion, des constantes locales d'équations fonctionnelles de fonctions L, *Invent. Math.* **64**(1) (1981) 89–118.

[9] Y. Z. Flicker, Twisted tensors and Euler products, *Bull. Soc. Math. France* **116**(3) (1988) 295–313.

[10] R. Howe, Classification of irreducible representations of $\mathrm{GL}_2(F)$, preprint, I.H.E.S., Bures-sur-Yvette, France (1978).

[11] H. Jacquet, I. I. Piatetskii-Shapiro and J. A. Shalika, Rankin–Selberg convolutions, *Amer. J. Math.* **105**(2) (1983) 367–464.

[12] H. Jacquet and J. Shalika, A lemma on highly ramified ϵ-factors, *Math. Ann.* **271**(3) (1985) 319–332.

[13] A. C. Kable, Asai L-functions and Jacquet's conjecture, *Amer. J. Math.* **126**(4) (2004) 789–820.

[14] A. W. Knapp, *Lie Groups Beyond an Introduction*, Progress in Mathematics, Vol. 140, 2nd edn. (Birkhäuser Boston, Inc., Boston, MA, 2002).

[15] N. Matringe, Conjectures about distinction and local Asai L-functions, *Int. Math. Res. Not.* **2009**(9) (2009) 1699–1741.

[16] D. Shankman, Local Langlands correspondence for Asai L-functions and epsilon factors, preprint (2018); https://arXiv.org/abs/1810.11852.

[17] D. Soudry, The L and γ factors for generic representations of $\mathrm{GSp}(4,k) \times \mathrm{GL}(2,k)$ over a local non-Archimedean field k, *Duke Math. J.* **51**(2) (1984) 355–394.

[18] J.-L. Waldspurger, La formule de Plancherel pour les groupes p-adiques (d'après Harish-Chandra), *J. Inst. Math. Jussieu* **2**(2) (2003) 235–333.

[19] Q. Zhang, A local converse theorem for $\mathrm{U}(1,1)$, *Int. J. Number Theory* **13**(8) (2017) 1931–1981.

[20] Q. Zhang, A local converse theorem for Sp_{2r}, *Math. Ann.* **372**(1–2) (2018) 451–488.

© 2024 World Scientific Publishing Company
https://doi.org/10.1142/9789811277375_0022

Local densities of diagonal integral ternary quadratic forms at odd primes*

Edna Jones

Mathematics Department,
Rutgers, The State University of New Jersey
110 Frelinghuysen Road, Piscataway
New Jersey 08854-8019, USA
elj44@math.rutgers.edu

Dedicated to Bruce Berndt on the occasion of his 80th birthday

We give formulas for local densities of diagonal integral ternary quadratic forms at odd primes. Exponential sums and quadratic Gauss sums are used to obtain these formulas. These formulas (along with 2-adic densities and Siegel's mass formula) can be used to compute the representation numbers of certain ternary quadratic forms.

Keywords: Ternary quadratic forms; local densities; quadratic Gauss sums.

Mathematics Subject Classification 2020: 11E08, 11E20, 11D85, 11L05, 11T23

1. Introduction

Mathematicians have been interested in sums of squares and quadratic forms for several centuries. For example, Hilbert stated in his 11th problem, "[S]olve a given quadratic equation with algebraic numerical coefficients in any number of variables by integral or fractional numbers belonging to the algebraic realm of rationality determined by the coefficients" [6, p. 458].

In this paper, we add to our understanding of quadratic forms by giving formulas for local densities of diagonal integral ternary quadratic forms at odd primes. By a diagonal integral ternary quadratic form, we mean a quadratic form of the form $ax^2 + by^2 + cz^2$, where a, b, c are integers. Let Q be the diagonal integral ternary quadratic form $Q(\mathbf{v}) = ax^2 + by^2 + cz^2$, where a, b, and c are integers and $\mathbf{v} = (x, y, z)^\top$.

Let m be an integer. A necessary condition for $Q(\mathbf{v}) = m$ to have a solution in the integers is that the congruence

$$Q(\mathbf{v}) = ax^2 + by^2 + cz^2 \equiv m \pmod{n} \tag{1.1}$$

*This article originally appeared in *International Journal of Number Theory*, **17**(3), pp. 547–575 © 2021 World Scientific Publishing Company. DOI: 10.1142/S1793042120400357

has a solution for every positive integer n. This leads us to count the number of solutions to (1.1). For a positive integer n, we define the *local representation number* $r_n(m,Q)$ as

$$r_n(m,Q) = \#\{\mathbf{v} \in (\mathbb{Z}/n\mathbb{Z})^3 : Q(\mathbf{v}) \equiv m \pmod{n}\}.$$

The Chinese Remainder Theorem tells us that $r_{n_1 n_2}(m,Q) = r_{n_1}(m,Q) \cdot r_{n_2}(m,Q)$ whenever n_1 and n_2 are relatively prime, so it suffices to study the local representation numbers in which n is a prime power. For a prime p, we use a p-adic density to encode information about $r_{p^k}(m,Q)$ for all positive integers k. Given a prime p, let $\mathbb{Z}_p$ denote the set of p-adic integers with the usual Haar measure, and define the *local (representation) density* $\alpha_p(m,Q)$ at the prime p by

$$\alpha_p(m,Q) = \lim_{U \to \{m\}} \frac{\mathrm{Vol}_{\mathbb{Z}_p^3}(Q^{-1}(U))}{\mathrm{Vol}_{\mathbb{Z}_p}(U)},$$

where U runs over a sequence of open sets in $\mathbb{Z}_p$ containing m, $\mathrm{Vol}_{\mathbb{Z}_p^3}(Q^{-1}(U))$ is the volume of $Q^{-1}(U)$ in $\mathbb{Z}_p^3$, and $\mathrm{Vol}_{\mathbb{Z}_p}(U)$ is the volume of U in $\mathbb{Z}_p$. The local density $\alpha_p(m,Q)$ is also called a *p-adic density.* It can be shown that

$$\alpha_p(m,Q) = \lim_{k \to \infty} \frac{r_{p^k}(m,Q)}{p^{2k}}. \tag{1.2}$$

(See [4, p. 368; 5, Lemma 2.7.2, p. 37].)

As can be seen by (1.2), a p-adic density encodes some information about $r_{p^k}(m,Q)$ for all positive integers k. In fact, in some cases, local densities are almost enough to compute the number of integral solutions to $Q(\mathbf{v}) = m$. Define the *representation number* $r(m,Q)$ by

$$r(m,Q) = \#\{\mathbf{v} \in \mathbb{Z}^3 : Q(\mathbf{v}) = m\}.$$

In 1935, Siegel [9] proved a formula (now often called Siegel's mass formula) that can be used to compute the weighted average of the representation numbers of certain positive definite quadratic forms. We state a specialized version of Siegel's mass formula here.

Theorem 1.1 (Specialized Version of Siegel's Mass Formula). *Let m be an integer and Q be a positive definite quadratic form of rank 3. Let $\{Q_j\}$ be a complete set representatives for classes in the same genus as Q. Then*

$$\frac{\displaystyle\sum_j \frac{r(m,Q_j)}{\#O(Q_j)}}{\displaystyle\sum_j \frac{1}{\#O(Q_j)}} = \alpha_{\mathbb{R}}(m,Q) \prod_{p \text{ prime}} \alpha_p(m,Q), \tag{1.3}$$

where $O(Q_j)$ is the orthogonal group of Q_j over $\mathbb{Z}$ and $\alpha_{\mathbb{R}}(m,Q)$ is the local density at the archimedean place. More explicitly, $\alpha_{\mathbb{R}}(m,Q) = \lim_{U\to\{m\}} \frac{\mathrm{Vol}_{\mathbb{R}^3}(Q^{-1}(U))}{\mathrm{Vol}_{\mathbb{R}}(U)}$, where U runs over a sequence of open sets in $\mathbb{R}$ containing m, $\mathrm{Vol}_{\mathbb{R}^3}(Q^{-1}(U))$ is the volume of $Q^{-1}(U)$ in $\mathbb{R}^3$, and $\mathrm{Vol}_{\mathbb{R}}(U)$ is the volume of U in $\mathbb{R}$.

For more information about Siegel's mass formula in English, see [10].

For an appropriate quadratic form Q, Theorem 1.1 implies that the representation number of $r(m, Q)$ is zero (that is, there is no solution to $Q(\mathbf{v}) = m$) if there is a local density that is zero. Therefore, showing a local density is zero is a method of proving that $Q(\mathbf{v}) = m$ does not have a solution.

A number of formulas for local densities have been stated over the years. In 1998, Yang [11] computed some local densities, but some work is needed to determine whether one of his formulas is equal to zero. In 2004, Hanke [4] did some computations of local representation numbers, but it also can be difficult to tell when some of his formulas are equal to zero due to their recursive nature. The formulas for the local densities that we will obtain can be easily computed with only knowing the prime factorizations of a, b, c, and m and the values of certain Legendre symbols. Therefore, it is relatively easy for one to state when one of these formulas is zero.

Some other formulas for local densities are explicit but are not very general when it comes to diagonal ternary quadratic forms. In 1935, Siegel [9, 3] computed some local densities. Two of Siegel's lemmas [9, Hilfssatz 12 and Hilfssatz 13, pp. 539–542] imply that, if p is a prime that does not divide $2abcm$, then $\alpha_p(m, Q) = 1 + \frac{1}{p}\left(\frac{-abcm}{p}\right)$, where $\left(\frac{\cdot}{p}\right)$ is the Legendre symbol. We generalize this result so that the odd prime p can divide $2abcm$. Our results also generalize to any diagonal integral ternary quadratic form the following theorem proved by Berkovich and Jagy.

Theorem 1.2 ([1, Theorem 3.1, p. 262]). *Let p be an odd prime and u be any integer with $\left(\frac{-u}{p}\right) = -1$. Let $Q(\mathbf{v}) = ux^2 + py^2 + upz^2$. Suppose m is a nonzero integer and $m = m_0 p^{m_1}$, where $\gcd(m_0, p) = 1$. Then*

$$\alpha_p(m, Q) = \begin{cases} p^{-m_1/2}\left(1 - \left(\dfrac{-m_0}{p}\right)\right) & \text{if } m_1 \text{ is even,} \\[2ex] p^{(-m_1+1)/2}\left(1 + \dfrac{1}{p}\right) & \text{if } m_1 \text{ is odd.} \end{cases}$$

In this paper, only local densities at odd primes are computed since computing 2-adic densities of quadratic forms tends to be more complicated. Our main result is the following theorem.

Theorem 1.3. *Let Q be the integral quadratic form $ax^2 + by^2 + cz^2$, where a, b, and c are integers. Let p be an odd prime. Suppose $p \nmid a$, $b = b_0 p^{b_1}$, and $c = c_0 p^{c_1}$, where $b_1 \leq c_1$, $\gcd(b_0, p) = 1$, and $\gcd(c_0, p) = 1$.*

Suppose m is a nonzero integer and $m = m_0 p^{m_1}$, where $\gcd(m_0, p) = 1$.

If $m_1 < b_1$, then

$$\alpha_p(m, Q) = \begin{cases} p^{m_1/2}\left(1 + \left(\dfrac{am_0}{p}\right)\right) & \text{if } m_1 \text{ is even,} \\[2ex] 0 & \text{if } m_1 \text{ is odd.} \end{cases} \tag{1.4}$$

If $b_1 \le m_1 < c_1$, then

$$\alpha_p(m,Q) = \begin{cases} p^{b_1/2}\left(1 - \frac{1}{p}\left(\frac{-ab_0}{p}\right)^{m_1+1} + \left(1-\frac{1}{p}\right)\right. \\ \quad \cdot \left(\frac{m_1-b_1}{2} + \frac{(-1)^{m_1}-1}{4} + \left(\frac{-ab_0}{p}\right)\right. \\ \quad \left.\left.\cdot \left(\frac{m_1-b_1}{2} + \frac{1-(-1)^{m_1}}{4}\right)\right)\right) & \text{if } b_1 \text{ is even,} \\ p^{(b_1-1)/2}\left(1 + \left(\frac{a}{p}\right)^{m_1+1}\left(\frac{b_0}{p}\right)^{m_1}\left(\frac{m_0}{p}\right)\right) & \text{if } b_1 \text{ is odd.} \end{cases} \tag{1.5}$$

If $m_1 \ge c_1$ and b_1 is even, then

$$\alpha_p(m,Q) = \begin{cases} p^{b_1/2}\left(1 + \frac{1}{p} + p^{-m_1/2+c_1/2-1}\right. \\ \quad \cdot \left(\left(\frac{-ab_0c_0m_0}{p}\right) - 1\right) \\ \quad + \left(1-\frac{1}{p}\right)\left(\frac{c_1-b_1}{2}\right. & \text{if } c_1 \text{ and } m_1 \\ \quad \left.\left. + \left(\frac{-ab_0}{p}\right)\frac{c_1-b_1}{2}\right)\right) & \text{are even,} \\ p^{b_1/2}\left(\left(1+\frac{1}{p}\right)\left(1 - p^{-(m_1+1)/2+c_1/2}\right)\right. \\ \quad + \left(1-\frac{1}{p}\right)\left(\frac{c_1-b_1}{2} + \left(\frac{-ab_0}{p}\right)\right. & \text{if } c_1 \text{ is even and} \\ \quad \left.\left. \cdot \frac{c_1-b_1}{2}\right)\right) & m_1 \text{ is odd,} \\ p^{b_1/2}\left(1 - p^{-m_1/2+(c_1-1)/2}\left(\frac{-ab_0}{p}\right)\right. \\ \quad \cdot \left(1+\frac{1}{p}\right) + \frac{1}{p}\left(\frac{-ab_0}{p}\right) \\ \quad + \left(1-\frac{1}{p}\right)\left(\frac{c_1-b_1-1}{2}\right. & \text{if } c_1 \text{ is odd and } m_1 \\ \quad \left.\left. + \left(\frac{-ab_0}{p}\right)\frac{c_1-b_1+1}{2}\right)\right) & \text{is even,} \\ p^{b_1/2}\left(1 + p^{-(m_1+1)/2+(c_1-1)/2}\right. \\ \quad \cdot \left(\left(\frac{c_0m_0}{p}\right) - \left(\frac{-ab_0}{p}\right)\right) \\ \quad + \frac{1}{p}\left(\frac{-ab_0}{p}\right) + \left(1-\frac{1}{p}\right) & \text{if } c_1 \text{ and } m_1 \\ \quad \left. \cdot \left(\frac{c_1-b_1-1}{2} + \left(\frac{-ab_0}{p}\right)\frac{c_1-b_1+1}{2}\right)\right) & \text{are odd.} \end{cases} \tag{1.6}$$

If $m_1 \geq c_1$ *and* b_1 *is odd, then*

$$\alpha_p(m,Q) = \begin{cases} p^{(b_1-1)/2}\left(1+\left(\frac{-ac_0}{p}\right)\right. \\ \qquad -p^{-m_1/2+c_1/2}\left(1+\frac{1}{p}\right) \\ \qquad \left.\cdot\left(\frac{-ac_0}{p}\right)\right) & \text{if } c_1 \text{ and } m_1 \text{ are even,} \\ p^{(b_1-1)/2}\left(1+\left(\frac{-ac_0}{p}\right)\right. \\ \qquad +p^{-(m_1+1)/2+c_1/2}\left(\left(\frac{b_0 m_0}{p}\right)\right. \\ \qquad \left.\left.-\left(\frac{-ac_0}{p}\right)\right)\right) & \text{if } c_1 \text{ is even and } m_1 \text{ is odd,} \\ p^{(b_1-1)/2}\left(1+\left(\frac{-b_0c_0}{p}\right)\right. \\ \qquad +p^{-m_1/2+(c_1-1)/2}\left(\left(\frac{am_0}{p}\right)\right. \\ \qquad \left.\left.-\left(\frac{-b_0c_0}{p}\right)\right)\right) & \text{if } c_1 \text{ is odd and } m_1 \text{ is even,} \\ p^{(b_1-1)/2}\left(1+\left(\frac{-b_0c_0}{p}\right)\right. \\ \qquad -p^{(-m_1+c_1)/2}\left(1+\frac{1}{p}\right) \\ \qquad \left.\cdot\left(\frac{-b_0c_0}{p}\right)\right) & \text{if } c_1 \text{ and } m_1 \text{ are odd.} \end{cases} \tag{1.7}$$

Furthermore,

$$\alpha_p(0,Q) = \begin{cases} p^{b_1/2}\left(1+\frac{1}{p}+\left(1-\frac{1}{p}\right)\left(\frac{c_1-b_1}{2}\right.\right. \\ \qquad \left.\left.+\left(\frac{-ab_0}{p}\right)\frac{c_1-b_1}{2}\right)\right) & \text{if } b_1 \text{ and } c_1 \text{ are even,} \\ p^{b_1/2}\left(1+\frac{1}{p}\left(\frac{-ab_0}{p}\right)\right. \\ \qquad +\left(1-\frac{1}{p}\right)\left(\frac{c_1-b_1-1}{2}\right. \\ \qquad \left.\left.+\left(\frac{-ab_0}{p}\right)\frac{c_1-b_1+1}{2}\right)\right) & \text{if } b_1 \text{ is even and } c_1 \text{ is odd,} \\ p^{(b_1-1)/2}\left(1+\left(\frac{-ac_0}{p}\right)\right) & \text{if } b_1 \text{ is odd and } c_1 \text{ is even,} \\ p^{(b_1-1)/2}\left(1+\left(\frac{-b_0c_0}{p}\right)\right) & \text{if } b_1 \text{ and } c_1 \text{ are odd.} \end{cases} \tag{1.8}$$

Remark 1.4. Theorem 1.3, namely (1.4) and (1.5), can be used to compute p-adic densities (p odd) of binary quadratic forms of the form $ax^2 + by^2$ by taking c_1 to be infinity for all odd primes p.

If Q is positive definite and in a genus containing exactly one class, then the left-hand side of Siegel's mass formula (1.3) simplifies to the representation number $r(m, Q)$. In 1939, Jones and Pall [7, p. 167] proved that there are 82 primitive quadratic forms of the form $ax^2 + by^2 + cz^2$ with $0 < a \leq b \leq c$ such that each is in a genus containing exactly one class. In 1971, Lomadze [8] computed the representation numbers for these 82 quadratic forms using the singular series. In the future, the author would like to compute 2-adic densities and use these in conjunction with Theorem 1.3 to evaluate the right-hand side of Siegel's mass formula (1.3) explicitly. In addition to giving another proof of Lomadze's evaluation of the representation numbers of the 82 primitive quadratic forms mentioned by Jones and Pall, this would give an explicit formula for the weighted average appearing in Theorem 1.1.

Traditionally, Hensel's lemma has been used to compute local densities. (For example, see [4].) In fact, Hensel's lemma can be used to show that the sequence

$$\left\{ \frac{r_{p^k}(m, Q)}{p^{2k}} \right\}_{k \geq 1} \tag{1.9}$$

is eventually constant if $m \neq 0$; that is, if $m \neq 0$, then there exists a K such that if $k \geq K$, then

$$\frac{r_{p^k}(m, Q)}{p^{2k}} = \frac{r_{p^K}(m, Q)}{p^{2K}}.$$

Without using Hensel's lemma, we show in this paper that the sequence in (1.9) is eventually constant if p is odd and $m \neq 0$. If $m \neq 0$ and p is odd, our proofs show that the subsequence

$$\left\{ \frac{r_{p^k}(m, Q)}{p^{2k}} \right\}_{k \geq m_1 + 1}$$

is a constant sequence, where m_1 is the same as in Theorem 1.3.

The remainder of this paper sets up and gives a proof of Theorem 1.3. Section 2 relates local representation numbers to exponential sums and quadratic Gauss sums. Section 3 states some lemmas that are useful in the proof of Theorem 1.3. A proof of Theorem 1.3 is given in Sec. 4.

2. Local Representation Numbers, Exponential Sums, and Quadratic Gauss Sums

We compute local densities by computing certain local representation numbers. In this section, we give some formulas for local representation numbers using Gauss sums and exponential sums. Throughout this paper, we abbreviate $e^{2\pi i w}$ as $\mathrm{e}(w)$.

For an integer a and a positive integer q, the *quadratic Gauss sum* $g(a;q)$ *over* $\mathbb{Z}/q\mathbb{Z}$ is defined by

$$g(a;q) := \sum_{j \pmod q} \mathrm{e}\left(\frac{aj^2}{q}\right) = \sum_{j\in\mathbb{Z}/q\mathbb{Z}} \mathrm{e}\left(\frac{aj^2}{q}\right) = \sum_{j=0}^{q-1} \mathrm{e}\left(\frac{aj^2}{q}\right).$$

Unless otherwise specified, the term "Gauss sum" is taken to refer to a quadratic Gauss sum.

We begin this section by stating a formula for an exponential sum that comes up frequently in number theory. This formula follows from the fact that the sum is also a geometric sum.

Lemma 2.1. *Let $a, q \in \mathbb{Z}$ and $q > 0$. Then*

$$\sum_{t=0}^{q-1} \mathrm{e}\left(\frac{at}{q}\right) = \begin{cases} q & \text{if } a \equiv 0 \pmod q, \\ 0 & \text{otherwise.} \end{cases}$$

Using the previous lemma, we prove the following theorem, which shows why Gauss sums are useful in computing local representation numbers.

Theorem 2.2. *Let a, b, and c be integers, and let n be a positive integer. Let $Q(\mathbf{v}) = ax^2 + by^2 + cz^2$. Then*

$$r_n(m,Q) = \frac{1}{n}\sum_{t=0}^{n-1} \mathrm{e}\left(\frac{-mt}{n}\right) g(at;n)\, g(bt;n)\, g(ct;n).$$

Proof. By Lemma 2.1,

$$\frac{1}{n}\sum_{t=0}^{n-1} \mathrm{e}\left(\frac{(Q(\mathbf{v})-m)t}{n}\right) = \begin{cases} 1 & \text{if } Q(\mathbf{v}) \equiv m \pmod n, \\ 0 & \text{otherwise.} \end{cases}$$

Therefore,

$$\begin{aligned} r_n(m,Q) &= \sum_{\mathbf{v}\in(\mathbb{Z}/n\mathbb{Z})^3} \frac{1}{n}\sum_{t=0}^{n-1} \mathrm{e}\left(\frac{(Q(\mathbf{v})-m)t}{n}\right) \\ &= \sum_{x=0}^{n-1}\sum_{y=0}^{n-1}\sum_{z=0}^{n-1} \frac{1}{n}\sum_{t=0}^{n-1} \mathrm{e}\left(\frac{(ax^2+by^2+cz^2-m)t}{n}\right) \\ &= \frac{1}{n}\sum_{t=0}^{n-1} \mathrm{e}\left(\frac{-mt}{n}\right)\sum_{x=0}^{n-1} \mathrm{e}\left(\frac{atx^2}{n}\right)\sum_{y=0}^{n-1} \mathrm{e}\left(\frac{bty^2}{n}\right)\sum_{z=0}^{n-1} \mathrm{e}\left(\frac{ctz^2}{n}\right) \\ &= \frac{1}{n}\sum_{t=0}^{n-1} \mathrm{e}\left(\frac{-mt}{n}\right) g(at;n)\, g(bt;n)\, g(ct;n). \end{aligned}$$

□

The next result is a special case of Theorem 2.2, describing what happens when n is a prime power.

Corollary 2.3. *Let a, b, and c be integers, let k be a positive integer, and let p be a prime. Let $Q(\mathbf{v}) = ax^2 + by^2 + cz^2$. Then*

$$r_{p^k}(m,Q) = \frac{1}{p^k}\sum_{t=0}^{p^k-1} \mathrm{e}\left(\frac{-mt}{p^k}\right) g\big(at;p^k\big)\, g\big(bt;p^k\big)\, g\big(ct;p^k\big)\,. \tag{2.1}$$

This may also be written as

$$r_{p^k}(m,Q) = p^{2k} + \frac{1}{p^k}\sum_{t=1}^{p^k-1} \mathrm{e}\left(\frac{-mt}{p^k}\right) g\big(at;p^k\big)\, g\big(bt;p^k\big)\, g\big(ct;p^k\big)\,. \tag{2.2}$$

Knowing this, it would be useful to evaluate Gauss sums. Most formulas for a Gauss sum $g\big(a;p^k\big)$ assume that a is coprime to the prime p. In order to remove this assumption, we use the following lemma to transform sums of periodic functions into sums with fewer summands. The lemma follows from the periodicity of the function.

Lemma 2.4. *Let $f : \mathbb{Z} \to \mathbb{C}$ be periodic modulo n. If m is a positive integer, then*

$$\sum_{t\in\mathbb{Z}/nm\mathbb{Z}} f(t) = \sum_{t=0}^{nm-1} f(t) = m \sum_{t\in\mathbb{Z}/n\mathbb{Z}} f(t) = m\sum_{t=0}^{n-1} f(t).$$

Using the previous lemma, we relate $g\big(a;p^k\big)$ to another Gauss sum $g\big(a_0;p^k\big)$, where a_0 is coprime to p.

Lemma 2.5. *Suppose k is a positive integer, p is a prime, and $a \neq 0$. Let $a = a_0p^\ell$ so that $\gcd(a_0,p) = 1$. If $\ell \le k$, then*

$$g\big(a;p^k\big) = p^\ell g\big(a_0;p^{k-\ell}\big)\,.$$

Proof. Using the definition of a quadratic Gauss sum and Lemma 2.4,

$$\begin{aligned} g\big(a;p^k\big) &= \sum_{j=0}^{p^k-1} \mathrm{e}\left(\frac{aj^2}{p^k}\right) = \sum_{j=0}^{p^k-1} \mathrm{e}\left(\frac{a_0p^\ell j^2}{p^k}\right) = \sum_{j=0}^{p^k-1} \mathrm{e}\left(\frac{a_0j^2}{p^{k-\ell}}\right) \\ &= p^\ell \sum_{j=0}^{p^{k-\ell}-1} \mathrm{e}\left(\frac{a_0j^2}{p^{k-\ell}}\right) = p^\ell g\big(a_0;p^{k-\ell}\big)\,. \end{aligned}$$

□

The following lemma states a formula for a Gauss sum $g\big(a;p^k\big)$ when a is coprime to the prime p. The lemma is a special case of Theorem 1.5.2 in [2, p. 26].

Lemma 2.6. *Suppose k is a positive integer and p is an odd prime. Suppose $\gcd(a,p) = 1$. Then*

$$g\big(a;p^k\big) = p^{k/2}\left(\frac{a}{p^k}\right)\varepsilon_{p^k},$$

where $\left(\frac{\cdot}{p^k}\right)$ *is the Jacobi symbol and*

$$\varepsilon_{p^k} = \begin{cases} 1 & \text{if } p^k \equiv 1 \pmod 4, \\ i & \text{if } p^k \equiv 3 \pmod 4. \end{cases}$$

Extending the previous lemma, the following lemma helps us evaluate a Gauss sum $g(a;q)$ when q is an odd prime power.

Lemma 2.7. *Suppose k is a positive integer, p is an odd prime, and $a \neq 0$. Let $a = a_0 p^\ell$ so that* $\gcd(a_0, p) = 1$. *Then*

$$g(a;p^k) = \begin{cases} p^k & \text{if } k \leq \ell, \\ p^{(k+\ell)/2}\left(\dfrac{a_0}{p^{k-\ell}}\right)\varepsilon_{p^{k-\ell}} & \text{if } k > \ell, \end{cases}$$

where ε_{p^k} is as defined in Lemma 2.6.

Proof. Suppose $k \leq \ell$. By the definition of a quadratic Gauss sum,

$$g(a;p^k) = \sum_{j=0}^{p^k-1} \mathrm{e}\left(\frac{aj^2}{p^k}\right) = \sum_{j=0}^{p^k-1} \mathrm{e}\left(a_0 p^{\ell-k} j^2\right) = \sum_{j=0}^{p^k-1} 1 = p^k.$$

Suppose $k > \ell$. By Lemma 2.5, $g(a;p^k) = p^\ell g(a_0;p^{k-\ell})$. We apply Lemma 2.6 to see that

$$g(a;p^k) = p^\ell p^{(k-\ell)/2}\left(\frac{a_0}{p^{k-\ell}}\right)\varepsilon_{p^{k-\ell}} = p^{(k+\ell)/2}\left(\frac{a_0}{p^{k-\ell}}\right)\varepsilon_{p^{k-\ell}}. \qquad \square$$

Remark 2.8. A simple calculation shows that $\varepsilon_{p^k}^2 = \left(\frac{-1}{p^k}\right) = \left(\frac{-1}{p}\right)^k$. Also, $\varepsilon_{p^k}^4 = 1$.

We now provide a link between Gauss sums and exponential sums.

Lemma 2.9. *Suppose p is an odd prime and $a \in \mathbb{Z}$. Then*

$$g(a;p) = \sum_{t=0}^{p-1}\left(1 + \left(\frac{t}{p}\right)\right)\mathrm{e}\left(\frac{at}{p}\right). \tag{2.3}$$

If $a \not\equiv 0 \pmod p$, then

$$g(a;p) = \sum_{t=0}^{p-1}\left(\frac{t}{p}\right)\mathrm{e}\left(\frac{at}{p}\right) = \sum_{t=1}^{p-1}\left(\frac{t}{p}\right)\mathrm{e}\left(\frac{at}{p}\right).$$

Proof. Let t be an integer. As noted by Cohen [3, p. 27], the number of solutions modulo p of the congruence

$$j^2 \equiv t \pmod p$$

is $1 + \left(\frac{t}{p}\right)$. Therefore,

$$g(a;p) = \sum_{j=0}^{p-1} \mathrm{e}\left(\frac{aj^2}{p}\right) = \sum_{t=0}^{p-1} \left(1 + \left(\frac{t}{p}\right)\right) \mathrm{e}\left(\frac{at}{p}\right).$$

When $a \not\equiv 0 \pmod{p}$, we see that

$$g(a;p) = \sum_{t=0}^{p-1} \left(\frac{t}{p}\right) \mathrm{e}\left(\frac{at}{p}\right)$$

follows from (2.3) and Lemma 2.1.

The last equality in Lemma 2.9 follows from the fact that $\left(\frac{0}{p}\right) = 0$. □

3. Some Supporting Lemmas

This section contains many lemmas used in our computations for local densities at odd primes. This first lemma gives the value of a certain exponential sum and follows immediately from Lemma 2.1.

Lemma 3.1. *Let $a, q \in \mathbb{Z}$ and $q > 0$. If $a \not\equiv 0 \pmod{q}$, then*

$$\sum_{t=1}^{q-1} \mathrm{e}\left(\frac{at}{q}\right) = -1.$$

This next lemma says that the value of a Gauss sum can be determined by certain residue classes and valuations. For a prime p, we say $p^\ell \parallel a$ if $p^\ell \mid a$ but $p^{\ell+1} \nmid a$.

Lemma 3.2. *Suppose k is a positive integer, p is an odd prime, $a \neq 0$, and $b \neq 0$. Suppose $p^\ell \parallel a$ and $p^\ell \parallel b$ for some $\ell \geq 0$ so that there exist $a_0, b_0 \in \mathbb{Z}$ such that $\gcd(a_0, p) = 1$, $\gcd(b_0, p) = 1$, $a = a_0 p^\ell$, and $b = b_0 p^\ell$. If $a_0 \equiv b_0 \pmod{p}$, then*

$$g(a; p^k) = g(b; p^k).$$

Proof. Since $a_0 \equiv b_0 \pmod{p}$, we have

$$\left(\frac{a_0}{p^{k-\ell}}\right) = \left(\frac{a_0}{p}\right)^{k-\ell} = \left(\frac{b_0}{p}\right)^{k-\ell} = \left(\frac{b_0}{p^{k-\ell}}\right).$$

By Lemma 2.7, we see that $g(a; p^k) = g(b; p^k)$. □

Now the sums of powers of Legendre symbols will come up repeatedly in our computations. This next lemma helps us compute them and follows from the fact that the Legendre symbol is a nontrivial real Dirichlet character modulo p.

Lemma 3.3. *Let p be an odd prime, $k \geq 1$, and m be an integer. Then*

$$\sum_{t \in (\mathbb{Z}/p^k\mathbb{Z})^*} \left(\frac{t}{p}\right)^m = \begin{cases} p^k \left(1 - \dfrac{1}{p}\right) & \text{if } m \text{ is even,} \\ 0 & \text{if } m \text{ is odd.} \end{cases}$$

Sums with Legendre symbols and exponentials occur frequently in calculations of local representation numbers. The following lemma helps us compute them.

Lemma 3.4. *Let p be an odd prime, n be an integer, $k \geq 1$, and m be an integer relatively prime to p, i.e.,* $\gcd(m, p) = 1$. *Then*

$$\sum_{t \in (\mathbb{Z}/p^k\mathbb{Z})^*} \mathrm{e}\left(\frac{mt}{p}\right)\left(\frac{t}{p}\right)^n = \sum_{t \in \mathbb{Z}/p^k\mathbb{Z}} \mathrm{e}\left(\frac{mt}{p}\right)\left(\frac{t}{p}\right)^n$$

$$= \begin{cases} -p^{k-1} & \text{if } n \text{ is even,} \\ p^{k-1/2}\left(\dfrac{m}{p}\right)\varepsilon_p & \text{if } n \text{ is odd.} \end{cases}$$

Proof. Because $\left(\dfrac{t}{p}\right) = 0$ if $\gcd(t, p) > 1$, we have

$$\sum_{t \in (\mathbb{Z}/p^k\mathbb{Z})^*} \mathrm{e}\left(\frac{mt}{p}\right)\left(\frac{t}{p}\right)^n = \sum_{t \in \mathbb{Z}/p^k\mathbb{Z}} \mathrm{e}\left(\frac{mt}{p}\right)\left(\frac{t}{p}\right)^n.$$

Since $f(t) = \mathrm{e}\left(\frac{mt}{p}\right)\left(\frac{t}{p}\right)^n$ is periodic modulo p, by Lemma 2.4,

$$\sum_{t \in \mathbb{Z}/p^k\mathbb{Z}} \mathrm{e}\left(\frac{mt}{p}\right)\left(\frac{t}{p}\right)^n = p^{k-1}\sum_{t=0}^{p-1} \mathrm{e}\left(\frac{mt}{p}\right)\left(\frac{t}{p}\right)^n.$$

If n is even, then

$$p^{k-1}\sum_{t=0}^{p-1} \mathrm{e}\left(\frac{mt}{p}\right)\left(\frac{t}{p}\right)^n = p^{k-1}\sum_{t=1}^{p-1} \mathrm{e}\left(\frac{mt}{p}\right) = -p^{k-1}$$

by Lemma 3.1.

If n is odd, then

$$p^{k-1}\sum_{t=0}^{p-1} \mathrm{e}\left(\frac{mt}{p}\right)\left(\frac{t}{p}\right)^n = p^{k-1}\sum_{t=0}^{p-1} \left(\frac{t}{p}\right)\mathrm{e}\left(\frac{mt}{p}\right) = p^{k-1}g(m; p)$$

by Lemma 2.9. By Lemma 2.6,

$$p^{k-1}\sum_{t=0}^{p-1} \mathrm{e}\left(\frac{mt}{p}\right)\left(\frac{t}{p}\right)^n = p^{k-1}p^{1/2}\left(\frac{m}{p}\right)\varepsilon_p = p^{k-1/2}\left(\frac{m}{p}\right)\varepsilon_p$$

if n is odd. □

The following lemma gives the sum of certain powers of Legendre symbols that are useful in computing local densities.

Lemma 3.5. *Let n_1, n_2, and r be integers such that $0 \leq n_1 \leq n_2 \leq k$ and $\gcd(r,p) = 1$. Then*

$$\sum_{\tau=k-n_2}^{k-n_1} \left(\frac{r}{p}\right)^{k-\tau} = \frac{n_2-n_1+1}{2} + \frac{(-1)^{n_1}+(-1)^{n_2}}{4} + \left(\frac{r}{p}\right)\left(\frac{n_2-n_1+1}{2} - \frac{(-1)^{n_1}+(-1)^{n_2}}{4}\right).$$

Proof. Let $\tau_1 = k - \tau$. Then

$$\sum_{\tau=k-n_2}^{k-n_1} \left(\frac{r}{p}\right)^{k-\tau} = \sum_{\tau_1=n_1}^{n_2} \left(\frac{r}{p}\right)^{\tau_1} = \sum_{\substack{\tau_1=n_1 \\ \tau_1 \text{ is even}}}^{n_2} 1 + \sum_{\substack{\tau_1=n_1 \\ \tau_1 \text{ is odd}}}^{n_2} \left(\frac{r}{p}\right).$$

By a simple counting argument, one can see that the number of even integers in the set $\{n \in \mathbb{Z} : n_1 \leq n \leq n_2\}$ is $\frac{n_2-n_1+1}{2} + \frac{(-1)^{n_1}+(-1)^{n_2}}{4}$. In a similar manner, we can see that the number of odd integers in the set $\{n \in \mathbb{Z} : n_1 \leq n \leq n_2\}$ is $\frac{n_2-n_1+1}{2} - \frac{(-1)^{n_1}+(-1)^{n_2}}{4}$. From this, the lemma follows. □

4. Proof of Theorem 1.3

Since we have stated some supporting lemmas in the previous section, we can now give a proof of Theorem 1.3 in this section. This section is split up into a proof of the formulas of $\alpha_p(m,Q)$ when $m \neq 0$ and a proof of the formulas for $\alpha_p(0,Q)$. In this section, let a, b, b_0, b_1, c, c_0, c_1, and p be the same as in the statement of Theorem 1.3.

4.1. *Proof of formulas for $\alpha_p(m,Q)$ for nonzero m*

This subsection proves the formulas of $\alpha_p(m,Q)$ when $m \neq 0$. In this subsection, let m, m_0, and m_1 be the same as in the statement of Theorem 1.3. Suppose $k > m_1$ for this entire subsection and its lemmas. Towards computing $\alpha_p(m,Q)$, we compute $r_{p^k}(m,Q)$ and then take the appropriate limit.

By (2.2),

$$r_{p^k}(m,Q) = p^{2k} + \frac{1}{p^k}\sum_{t=1}^{p^k-1} \mathrm{e}\left(\frac{-m_0 p^{m_1} t}{p^k}\right) g(at;p^k)\, g(b_0p^{b_1}t;p^k)\, g(c_0p^{c_1}t;p^k).$$

Let $t = t_0 p^{\tau}$, where $0 \leq \tau \leq k-1$ and $t_0 \in (\mathbb{Z}/p^{k-\tau}\mathbb{Z})^*$. Then

$$r_{p^k}(m,Q) = p^{2k} + \frac{1}{p^k}\sum_{\tau=0}^{k-1} s_{k,\tau}, \tag{4.1}$$

where

$$s_{k,\tau} = \sum_{t_0 \in (\mathbb{Z}/p^{k-\tau}\mathbb{Z})^*} \mathrm{e}\left(\frac{-m_0 t_0 p^{m_1+\tau}}{p^k}\right) g\left(a t_0 p^{\tau}; p^k\right) g\left(b_0 t_0 p^{b_1+\tau}; p^k\right) g\left(c_0 t_0 p^{c_1+\tau}; p^k\right).$$

Now the problem of computing $r_{p^k}(m, Q)$ becomes the problem of computing $s_{k,\tau}$ for $0 \le \tau \le k-1$. Our first lemma tells us that we only need to be concerned with $m_1 + 1$ values of τ.

Lemma 4.1. *If $0 \le \tau \le k - m_1 - 2$, then $s_{k,\tau} = 0$.*

Proof. Suppose that $0 \le \tau \le k-m_1-2$. Then let $t_0 = t_1 + t_2 p$, where $1 \le t_1 \le p-1$ and $0 \le t_2 \le p^{k-\tau-1} - 1$, so by Lemma 3.2,

$$\begin{aligned}
s_{k,\tau} &= \sum_{t_1=1}^{p-1} \sum_{t_2=0}^{p^{k-\tau-1}-1} \mathrm{e}\left(\frac{-m_0(t_1+t_2 p)p^{m_1+\tau}}{p^k}\right) g\left(a(t_1+t_2 p)p^{\tau}; p^k\right) \\
&\quad \cdot g\left(b_0(t_1+t_2 p)p^{b_1+\tau}; p^k\right) g\left(c_0(t_1+t_2 p)p^{c_1+\tau}; p^k\right) \\
&= \sum_{t_1=1}^{p-1} \sum_{t_2=0}^{p^{k-\tau-1}-1} \mathrm{e}\left(\frac{-m_0 t_1}{p^{k-m_1-\tau}}\right) \mathrm{e}\left(\frac{-m_0 t_2}{p^{k-m_1-1-\tau}}\right) g\left(a t_1 p^{\tau}; p^k\right) \\
&\quad \cdot g\left(b_0 t_1 p^{b_1+\tau}; p^k\right) g\left(c_0 t_1 p^{c_1+\tau}; p^k\right) \\
&= \sum_{t_1=1}^{p-1} \mathrm{e}\left(\frac{-m_0 t_1}{p^{k-m_1-\tau}}\right) g\left(a t_1 p^{\tau}; p^k\right) g\left(b_0 t_1 p^{b_1+\tau}; p^k\right) \\
&\quad \cdot g\left(c_0 t_1 p^{c_1+\tau}; p^k\right) \sum_{t_2=0}^{p^{k-\tau-1}-1} \mathrm{e}\left(\frac{-m_0 t_2}{p^{k-m_1-1-\tau}}\right).
\end{aligned} \tag{4.2}$$

By Lemmas 2.4 and 2.1 and the condition $0 \le \tau \le k - m_1 - 2$,

$$\sum_{t_2=0}^{p^{k-\tau-1}-1} \mathrm{e}\left(\frac{-m_0 t_2}{p^{k-m_1-1-\tau}}\right) = p^{m_1} \sum_{t_2=0}^{p^{k-m_1-\tau-1}-1} \mathrm{e}\left(\frac{-m_0 t_2}{p^{k-m_1-1-\tau}}\right) = p^{m_1} \cdot 0 = 0.$$

Thus, by (4.2), we obtain $s_{k,\tau} = 0$ for $0 \le \tau \le k - m_1 - 2$. □

In our next lemma, we evaluate $s_{k,\tau}$ when τ is greater than or equal to $k - \min(m_1, b_1)$.

Lemma 4.2. *If $k - \min(m_1, b_1) \le \tau \le k-1$, then*

$$s_{k,\tau} = \begin{cases} p^{3k+(k-\tau)/2}\left(1 - \dfrac{1}{p}\right) & \text{if } k-\tau \text{ is even,} \\[2ex] 0 & \text{if } k-\tau \text{ is odd.} \end{cases}$$

Proof. Suppose that $k - \min(m_1, b_1) \leq \tau \leq k-1$. Then $p^k \mid -m_0 t_0 p^{m_1+\tau}$ for all $t_0 \in (\mathbb{Z}/p^{k-\tau}\mathbb{Z})^*$ and $k \leq b_1 + \tau \leq c_1 + \tau$. Therefore, for all $t_0 \in (\mathbb{Z}/p^{k-\tau}\mathbb{Z})^*$, we have $\mathrm{e}\Big(\frac{-m_0 t_0 p^{m_1+\tau}}{p^k}\Big) = 1$. By Lemma 2.7,

$$s_{k,\tau} = \sum_{t_0 \in (\mathbb{Z}/p^{k-\tau}\mathbb{Z})^*} p^{(k+\tau)/2} \left(\frac{at_0}{p^{k-\tau}}\right) \varepsilon_{p^{k-\tau}} p^{2k}$$

$$= \varepsilon_{p^{k-\tau}} p^{5k/2+\tau/2} \left(\frac{a}{p}\right)^{k-\tau} \sum_{t_0 \in (\mathbb{Z}/p^{k-\tau}\mathbb{Z})^*} \left(\frac{t_0}{p}\right)^{k-\tau}.$$

By Lemma 3.3,

$$s_{k,\tau} = \begin{cases} \varepsilon_{p^{k-\tau}} p^{5k/2+\tau/2} \left(\frac{a}{p}\right)^{k-\tau} p^{k-\tau}\left(1 - \frac{1}{p}\right) & \text{if } k-\tau \text{ is even,} \\ 0 & \text{if } k-\tau \text{ is odd.} \end{cases}$$

Because $\varepsilon_{p^{k-\tau}} = \left(\frac{a}{p}\right)^{k-\tau} = 1$ if $k - \tau$ is even, we have

$$s_{k,\tau} = \begin{cases} p^{3k+(k-\tau)/2}\left(1 - \frac{1}{p}\right) & \text{if } k-\tau \text{ is even,} \\ 0 & \text{if } k-\tau \text{ is odd.} \end{cases}$$

□

The following lemma gives the value of the sum $\sum_{\tau=k-\min(m_1,b_1)}^{k-1} s_{k,\tau}$.

Lemma 4.3. *Let $n_1 = \min(m_1, b_1)$. Then*

$$\sum_{\tau=k-n_1}^{k-1} s_{k,\tau} = \sum_{\substack{\tau=k-n_1 \\ k-\tau \text{ is even}}}^{k-1} p^{3k+(k-\tau)/2}\left(1 - \frac{1}{p}\right) = p^{3k}(p^{\lfloor n_1/2 \rfloor} - 1), \tag{4.3}$$

where $\lfloor x \rfloor$ is the greatest integer less than or equal to x.

Proof. The first equality in (4.3) follows from Lemma 4.2.

It remains to prove the second equality in (4.3). Let $\tau_1 = \frac{k-\tau}{2}$. Then

$$\sum_{\substack{\tau=k-n_1 \\ k-\tau \text{ is even}}}^{k-1} p^{3k+(k-\tau)/2}\left(1 - \frac{1}{p}\right) = p^{3k}\left(1 - \frac{1}{p}\right) \sum_{\tau_1=1}^{\lfloor n_1/2 \rfloor} p^{\tau_1}.$$

The lemma follows by noting that the latter sum is a geometric sum that equals

$$\frac{p(p^{\lfloor n_1/2 \rfloor} - 1)}{p-1}.$$

□

4.1.1. $\alpha_p(m, Q)$ *when* $m_1 < b_1$

We start with computing formulas for $\alpha_p(m, Q)$ when $m_1 < b_1$ by computing $r_{p^k}(m, Q)$ when $m_1 < b_1$. The next lemma helps us compute $r_{p^k}(m, Q)$ when $m_1 < b_1$ by determining the value of $s_{k,k-m_1-1}$ when $m_1 < b_1$.

Lemma 4.4. *If* $m_1 < b_1$, *then*

$$s_{k,k-m_1-1} = \begin{cases} p^{3k+m_1/2}\left(\dfrac{am_0}{p}\right) & \text{if } m_1 \text{ is even,} \\ -p^{3k+m_1/2-1/2} & \text{if } m_1 \text{ is odd.} \end{cases}$$

Proof. Since $m_1 < b_1$, we have $k \le b_1 + k - m_1 - 1 \le c_1 + k - m_1 - 1$. Thus, by Lemma 2.7,

$$\begin{aligned} s_{k,k-m_1-1} &= \sum_{t_0 \in (\mathbb{Z}/p^{m_1+1}\mathbb{Z})^*} \mathrm{e}\left(\frac{-m_0 t_0}{p}\right) p^{(2k-m_1-1)/2}\left(\frac{at_0}{p^{m_1+1}}\right)\varepsilon_{p^{m_1+1}}p^{2k} \\ &= p^{3k-m_1/2-1/2}\varepsilon_{p^{m_1+1}}\left(\frac{a}{p}\right)^{m_1+1}\sum_{t_0 \in (\mathbb{Z}/p^{m_1+1}\mathbb{Z})^*} \mathrm{e}\left(\frac{-m_0 t_0}{p}\right)\left(\frac{t_0}{p}\right)^{m_1+1}. \end{aligned}$$

By Lemma 3.4,

$$\begin{aligned} s_{k,k-m_1-1} &= \begin{cases} p^{3k-m_1/2-1/2}\varepsilon_p\left(\dfrac{a}{p}\right)p^{m_1+1/2}\left(\dfrac{-m_0}{p}\right)\varepsilon_p & \text{if } m_1 \text{ is even,} \\ -p^{3k-m_1/2-1/2}p^{m_1} & \text{if } m_1 \text{ is odd,} \end{cases} \\ &= \begin{cases} p^{3k+m_1/2}\left(\dfrac{am_0}{p}\right) & \text{if } m_1 \text{ is even,} \\ -p^{3k+m_1/2-1/2} & \text{if } m_1 \text{ is odd,} \end{cases} \end{aligned}$$

since $\varepsilon_p^2 = \left(\frac{-1}{p}\right)$. □

Using Lemmas 4.1, 4.3, and 4.4, we can compute $r_{p^k}(m, Q)$ when $m_1 < b_1$. By using the lemmas and (4.1), we see that if $m_1 < b_1$, then

$$\begin{aligned} r_{p^k}(m, Q) &= p^{2k} + \frac{1}{p^k}s_{k,k-m_1-1} + \frac{1}{p^k}\sum_{\tau=k-m_1}^{k-1} s_{k,\tau} \\ &= \begin{cases} p^{2k} + p^{2k+m_1/2}\left(\dfrac{am_0}{p}\right) + p^{2k}(p^{\lfloor m_1/2 \rfloor} - 1) & \text{if } m_1 \text{ is even,} \\ p^{2k} - p^{2k+m_1/2-1/2} + p^{2k}(p^{\lfloor m_1/2 \rfloor} - 1) & \text{if } m_1 \text{ is odd,} \end{cases} \end{aligned}$$

$$= \begin{cases} p^{2k}\left(1 + p^{m_1/2}\left(\frac{am_0}{p}\right) + (p^{m_1/2} - 1)\right) & \text{if } m_1 \text{ is even,} \\ p^{2k}(1 - p^{(m_1-1)/2} + (p^{(m_1-1)/2} - 1)) & \text{if } m_1 \text{ is odd.} \end{cases}$$

$$= \begin{cases} p^{2k+m_1/2}\left(1 + \left(\frac{am_0}{p}\right)\right) & \text{if } m_1 \text{ is even,} \\ 0 & \text{if } m_1 \text{ is odd.} \end{cases}$$

By dividing by p^{2k} and computing the limit as $k \to \infty$, we obtain (1.4) if $m_1 < b_1$.

4.1.2. $\alpha_p(m, Q)$ *when* $b_1 \leq m_1 < c_1$

We start with computing formulas for $\alpha_p(m, Q)$ when $b_1 \leq m_1 < c_1$ by computing $r_{p^k}(m, Q)$ when $b_1 \leq m_1 < c_1$. The next lemmas help us compute $r_{p^k}(m, Q)$ when $b_1 \leq m_1 < c_1$. We first determine the value of $s_{k,k-m_1-1}$ when $b_1 \leq m_1 < c_1$.

Lemma 4.5. *If* $b_1 \leq m_1 < c_1$, *then*

$$s_{k,k-m_1-1} = \begin{cases} -p^{3k+b_1/2-1}\left(\frac{-ab_0}{p}\right)^{m_1+1} & \text{if } b_1 \text{ is even,} \\ p^{3k+(b_1-1)/2}\left(\frac{a}{p}\right)^{m_1+1}\left(\frac{b_0}{p}\right)^{m_1}\left(\frac{m_0}{p}\right) & \text{if } b_1 \text{ is odd.} \end{cases}$$

Proof. Since $b_1 \leq m_1 < c_1$, we have $b_1 + k - m_1 - 1 < k \leq c_1 + k - m_1 - 1$. Thus, by Lemma 2.7,

$$\begin{aligned} s_{k,k-m_1-1} = &\sum_{t_0 \in (\mathbb{Z}/p^{m_1+1}\mathbb{Z})^*} \mathrm{e}\left(\frac{-m_0 t_0}{p}\right) p^{(2k-m_1-1)/2}\left(\frac{at_0}{p^{m_1+1}}\right)\varepsilon_{p^{m_1+1}} \\ &\cdot p^{(2k+b_1-m_1-1)/2}\left(\frac{b_0 t_0}{p^{m_1-b_1+1}}\right)\varepsilon_{p^{m_1-b_1+1}} p^k \\ = &\, p^{3k+b_1/2-m_1-1}\left(\frac{a}{p}\right)^{m_1+1}\left(\frac{b_0}{p}\right)^{m_1-b_1+1} \\ &\cdot \varepsilon_{p^{m_1+1}}\varepsilon_{p^{m_1-b_1+1}} \sum_{t_0 \in (\mathbb{Z}/p^{m_1+1}\mathbb{Z})^*} \mathrm{e}\left(\frac{-m_0 t_0}{p}\right)\left(\frac{t_0}{p}\right)^{-b_1}. \end{aligned}$$

Suppose b_1 is even. Then $\left(\frac{b_0}{p}\right)^{-b_1} = \left(\frac{t_0}{p}\right)^{-b_1} = 1$ and $\varepsilon_{p^{m_1-b_1+1}} = \varepsilon_{p^{m_1+1}}$. Thus,

$$\begin{aligned} s_{k,k-m_1-1} &= p^{3k+b_1/2-m_1-1}\left(\frac{a}{p}\right)^{m_1+1}\left(\frac{b_0}{p}\right)^{m_1+1}\varepsilon^2_{p^{m_1+1}} \sum_{t_0 \in (\mathbb{Z}/p^{m_1+1}\mathbb{Z})^*} \mathrm{e}\left(\frac{-m_0 t_0}{p}\right) \\ &= -p^{3k+b_1/2-m_1-1}\left(\frac{-ab_0}{p}\right)^{m_1+1} p^{m_1} = -p^{3k+b_1/2-1}\left(\frac{-ab_0}{p}\right)^{m_1+1} \end{aligned}$$

by Lemma 3.4.

Now suppose that b_1 is odd. Then $\varepsilon_{p^{m_1+1}}\varepsilon_{p^{m_1-b_1+1}} = \varepsilon_p$, and

$$s_{k,k-m_1-1} = p^{3k+b_1/2-m_1-1}\left(\frac{a}{p}\right)^{m_1+1}\left(\frac{b_0}{p}\right)^{m_1}\varepsilon_p \sum_{t_0\in(\mathbb{Z}/p^{m_1+1}\mathbb{Z})^*} \mathrm{e}\left(\frac{-m_0t_0}{p}\right)\left(\frac{t_0}{p}\right)$$
$$= p^{3k+b_1/2-m_1-1}\left(\frac{a}{p}\right)^{m_1+1}\left(\frac{b_0}{p}\right)^{m_1}\varepsilon_p^2 p^{m_1+1/2}\left(\frac{-m_0}{p}\right)$$

by Lemma 3.4. Thus, if b_1 is odd, then

$$s_{k,k-m_1-1} = p^{3k+(b_1-1)/2}\left(\frac{a}{p}\right)^{m_1+1}\left(\frac{b_0}{p}\right)^{m_1}\left(\frac{m_0}{p}\right). \qquad \square$$

The next lemma gives a formula for $s_{k,\tau}$ when $b_1 \le m_1 < c_1$ and $k - m_1 \le \tau < k - b_1$.

Lemma 4.6. *If $b_1 \le m_1 < c_1$ and $k - m_1 \le \tau < k - b_1$, then*

$$s_{k,\tau} = \begin{cases} p^{3k+b_1/2}\left(\frac{-ab_0}{p}\right)^{k-\tau}\left(1-\frac{1}{p}\right) & \text{if } b_1 \text{ is even,} \\ 0 & \text{if } b_1 \text{ is odd.} \end{cases}$$

Proof. Since $b_1 \le m_1 < c_1$ and $k - m_1 \le \tau < k - b_1$, we have $\tau + b_1 < k \le \tau + m_1 < \tau + c_1$. Therefore, by Lemma 2.7,

$$s_{k,\tau} = \sum_{t_0\in(\mathbb{Z}/p^{k-\tau}\mathbb{Z})^*} p^{(k+\tau)/2}\left(\frac{at_0}{p^{k-\tau}}\right)\varepsilon_{p^{k-\tau}} p^{(k+b_1+\tau)/2}\left(\frac{b_0t_0}{p^{k-b_1-\tau}}\right)\varepsilon_{p^{k-b_1-\tau}} p^k$$
$$= p^{2k+\tau+b_1/2}\varepsilon_{p^{k-\tau}}\varepsilon_{p^{k-b_1-\tau}}\left(\frac{a}{p}\right)^{k-\tau}\left(\frac{b_0}{p}\right)^{k-b_1-\tau}\sum_{t_0\in(\mathbb{Z}/p^{k-\tau}\mathbb{Z})^*}\left(\frac{t_0}{p}\right)^{-b_1}.$$

By Lemma 3.3,

$$s_{k,\tau} = \begin{cases} p^{2k+\tau+b_1/2}\varepsilon_{p^{k-\tau}}^2\left(\frac{ab_0}{p}\right)^{k-\tau} p^{k-\tau}\left(1-\frac{1}{p}\right) & \text{if } b_1 \text{ is even,} \\ 0 & \text{if } b_1 \text{ is odd,} \end{cases}$$
$$= \begin{cases} p^{3k+b_1/2}\left(\frac{-ab_0}{p}\right)^{k-\tau}\left(1-\frac{1}{p}\right) & \text{if } b_1 \text{ is even,} \\ 0 & \text{if } b_1 \text{ is odd.} \end{cases} \qquad \square$$

Using Lemmas 4.1, 4.3, 4.5, 4.6, and 3.5, we can compute $r_{p^k}(m,Q)$ when $b_1 \leq m_1 < c_1$. By using the lemmas and (4.1), we see that if $b_1 \leq m_1 < c_1$, then

$$\begin{aligned}
r_{p^k}(m,Q) &= p^{2k} + \frac{1}{p^k} s_{k,k-m_1-1} + \frac{1}{p^k} \sum_{\tau=k-m_1}^{k-b_1-1} s_{k,\tau} + \frac{1}{p^k} \sum_{\tau=k-b_1}^{k-1} s_{k,\tau} \\
&= p^{2k} + \frac{1}{p^k} s_{k,k-m_1-1} + \frac{1}{p^k} \sum_{\tau=k-m_1}^{k-b_1-1} s_{k,\tau} + p^{2k}(p^{\lfloor b_1/2 \rfloor} - 1) \\
&= \begin{cases} p^{2k} - p^{2k+b_1/2-1} \left(\frac{-ab_0}{p}\right)^{m_1+1} \\ \quad + \sum\limits_{\tau=k-m_1}^{k-b_1-1} p^{2k+b_1/2} \left(\frac{-ab_0}{p}\right)^{k-\tau} \\ \quad \cdot \left(1 - \frac{1}{p}\right) + p^{2k}(p^{b_1/2} - 1) & \text{if } b_1 \text{ is even,} \\ p^{2k} + p^{2k+(b_1-1)/2} \left(\frac{a}{p}\right)^{m_1+1} \left(\frac{b_0}{p}\right)^{m_1} \\ \quad \cdot \left(\frac{m_0}{p}\right) + p^{2k}(p^{(b_1-1)/2} - 1) & \text{if } b_1 \text{ is odd.} \end{cases} \\
&= \begin{cases} p^{2k+b_1/2} \left(1 - \frac{1}{p}\left(\frac{-ab_0}{p}\right)^{m_1+1} + \left(1 - \frac{1}{p}\right) \right. \\ \quad \cdot \left(\frac{m_1-b_1}{2} + \frac{(-1)^{m_1}-1}{4} + \left(\frac{-ab_0}{p}\right)\right. \\ \quad \left.\left. \cdot \left(\frac{m_1-b_1}{2} + \frac{1-(-1)^{m_1}}{4}\right)\right)\right) & \text{if } b_1 \text{ is even,} \\ p^{2k+(b_1-1)/2} \left(1 + \left(\frac{a}{p}\right)^{m_1+1} \left(\frac{b_0}{p}\right)^{m_1} \left(\frac{m_0}{p}\right)\right) & \text{if } b_1 \text{ is odd.} \end{cases}
\end{aligned}$$

By dividing by p^{2k} and computing the limit as $k \to \infty$, we obtain (1.5) if $b_1 \leq m_1 < c_1$.

4.1.3. $\alpha_p(m,Q)$ *when* $m_1 \geq c_1$

We start with computing formulas for $\alpha_p(m,Q)$ when $m_1 \geq c_1$ by computing $r_{p^k}(m,Q)$ when $m_1 \geq c_1$. The next lemmas help us compute $r_{p^k}(m,Q)$ when $m_1 \geq c_1$. We begin with computing $s_{k,k-m_1-1}$ when $m_1 \geq c_1$.

Lemma 4.7. *Suppose $m_1 \geq c_1$.*
If $b_1 \equiv c_1 \pmod 2$, then

$$s_{k,k-m_1-1} = \begin{cases} p^{3k-m_1/2+(b_1+c_1)/2-1} \left(\frac{am_0}{p}\right)\left(\frac{-b_0c_0}{p}\right)^{b_1+1} & \text{if } m_1 \text{ is even,} \\ -p^{3k-(m_1+1)/2+(b_1+c_1)/2-1} \left(\frac{-b_0c_0}{p}\right)^{b_1} & \text{if } m_1 \text{ is odd.} \end{cases}$$

If $b_1 \not\equiv c_1 \pmod 2$, then

$$s_{k,k-m_1-1} = \begin{cases} -p^{3k-m_1/2+(b_1+c_1-1)/2-1} \left(\frac{-a}{p}\right)\left(\frac{b_0}{p}\right)^{b_1+1} \\ \quad \cdot \left(\frac{c_0}{p}\right)^{b_1} & \text{if } m_1 \text{ is even,} \\ p^{3k-(m_1+1)/2+(b_1+c_1-1)/2} \left(\frac{b_0}{p}\right)^{b_1}\left(\frac{c_0}{p}\right)^{b_1+1} \\ \quad \cdot \left(\frac{m_0}{p}\right) & \text{if } m_1 \text{ is odd.} \end{cases}$$

Proof. Since $m_1 \geq c_1$, we have $b_1 + k - m_1 - 1 \leq c_1 + k - m_1 - 1 \leq k - 1 < k$. Thus, by Lemma 2.7,

$$\begin{aligned} s_{k,k-m_1-1} = & \sum_{t_0 \in (\mathbb{Z}/p^{m_1+1}\mathbb{Z})^*} \mathrm{e}\left(\frac{-m_0t_0}{p}\right) p^{(2k-m_1-1)/2}\left(\frac{at_0}{p^{m_1+1}}\right) \\ & \cdot \varepsilon_{p^{m_1+1}} p^{(2k+b_1-m_1-1)/2} \cdot \left(\frac{b_0t_0}{p^{m_1-b_1+1}}\right) \varepsilon_{p^{m_1-b_1+1}} p^{(2k+c_1-m_1-1)/2} \\ & \cdot \left(\frac{c_0t_0}{p^{m_1-c_1+1}}\right) \varepsilon_{p^{m_1-c_1+1}} \\ = {} & p^{3k-3m_1/2-3/2+(b_1+c_1)/2} \varepsilon_{p^{m_1+1}} \varepsilon_{p^{m_1-b_1+1}} \varepsilon_{p^{m_1-c_1+1}} \left(\frac{a}{p}\right)^{m_1+1} \\ & \cdot \left(\frac{b_0}{p}\right)^{m_1-b_1+1} \cdot \left(\frac{c_0}{p}\right)^{m_1-c_1+1} \sum_{t_0 \in (\mathbb{Z}/p^{m_1+1}\mathbb{Z})^*} \mathrm{e}\left(\frac{-m_0t_0}{p}\right) \\ & \cdot \left(\frac{t_0}{p}\right)^{m_1-b_1-c_1+1}. \end{aligned}$$

Suppose that $b_1 \equiv c_1 \pmod 2$. Then

$$s_{k,k-m_1-1} = p^{3k-3m_1/2-3/2+(b_1+c_1)/2} \varepsilon_{p^{m_1+1}} \varepsilon^2_{p^{m_1-b_1+1}} \left(\frac{a}{p}\right)^{m_1+1} \cdot \left(\frac{b_0}{p}\right)^{m_1-b_1+1} \cdot \left(\frac{c_0}{p}\right)^{m_1-b_1+1} \sum_{t_0 \in (\mathbb{Z}/p^{m_1+1}\mathbb{Z})^*} \mathrm{e}\left(\frac{-m_0 t_0}{p}\right) \cdot \left(\frac{t_0}{p}\right)^{m_1+1}$$

$$= p^{3k-3m_1/2-3/2+(b_1+c_1)/2} \varepsilon_{p^{m_1+1}} \left(\frac{a}{p}\right)^{m_1+1} \left(\frac{-b_0 c_0}{p}\right)^{m_1-b_1+1} \cdot \sum_{t_0 \in (\mathbb{Z}/p^{m_1+1}\mathbb{Z})^*} \mathrm{e}\left(\frac{-m_0 t_0}{p}\right) \left(\frac{t_0}{p}\right)^{m_1+1}.$$

By Lemma 3.4,

$$s_{k,k-m_1-1} = \begin{cases} p^{3k-3m_1/2-3/2+(b_1+c_1)/2} \varepsilon_p \left(\frac{a}{p}\right) \left(\frac{-b_0 c_0}{p}\right)^{b_1+1} \\ \quad \cdot p^{m_1+1/2} \left(\frac{-m_0}{p}\right) \varepsilon_p & \text{if } m_1 \text{ is even,} \\ -p^{3k-3m_1/2-3/2+(b_1+c_1)/2} \left(\frac{-b_0 c_0}{p}\right)^{b_1} p^{m_1} & \text{if } m_1 \text{ is odd,} \end{cases}$$

which simplifies to what is stated in Lemma 4.7 for $b_1 \equiv c_1 \pmod 2$.

Now suppose $b_1 \not\equiv c_1 \pmod 2$. Then $\varepsilon_{p^{m_1+1}} \varepsilon_{p^{m_1-b_1+1}} \varepsilon_{p^{m_1-c_1+1}} = \varepsilon^2_{p^{m_1+1}} \varepsilon_{p^{m_1}} = \left(\frac{-1}{p}\right)^{m_1+1} \varepsilon_{p^{m_1}}$ and

$$s_{k,k-m_1-1} = p^{3k-3m_1/2-3/2+(b_1+c_1)/2} \left(\frac{-1}{p}\right)^{m_1+1} \varepsilon_{p^{m_1}} \left(\frac{a}{p}\right)^{m_1+1} \left(\frac{b_0}{p}\right)^{m_1-b_1+1} \cdot \left(\frac{c_0}{p}\right)^{m_1-b_1} \sum_{t_0 \in (\mathbb{Z}/p^{m_1+1}\mathbb{Z})^*} \mathrm{e}\left(\frac{-m_0 t_0}{p}\right) \left(\frac{t_0}{p}\right)^{m_1}$$

$$= p^{3k-3m_1/2+(b_1+c_1-1)/2-1} \varepsilon_{p^{m_1}} \left(\frac{-a}{p}\right)^{m_1+1} \left(\frac{b_0}{p}\right)^{m_1+b_1+1} \cdot \left(\frac{c_0}{p}\right)^{m_1+b_1} \sum_{t_0 \in (\mathbb{Z}/p^{m_1+1}\mathbb{Z})^*} \mathrm{e}\left(\frac{-m_0 t_0}{p}\right) \left(\frac{t_0}{p}\right)^{m_1}.$$

By Lemma 3.4,

$$s_{k,k-m_1-1} = \begin{cases} -p^{3k-3m_1/2+(b_1+c_1-1)/2-1}\left(\frac{-a}{p}\right) & \\ \quad \cdot \left(\frac{b_0}{p}\right)^{b_1+1}\left(\frac{c_0}{p}\right)^{b_1} p^{m_1} & \text{if } m_1 \text{ is even,} \\ p^{3k-3m_1/2+(b_1+c_1-1)/2-1}\varepsilon_p\left(\frac{b_0}{p}\right)^{b_1} & \\ \quad \cdot \left(\frac{c_0}{p}\right)^{b_1+1} p^{m_1+1/2}\left(\frac{-m_0}{p}\right)\varepsilon_p & \text{if } m_1 \text{ is odd,} \end{cases}$$

which simplifies to what is stated in Lemma 4.7 for $b_1 \not\equiv c_1 \pmod 2$. □

The next lemma gives the value of $s_{k,\tau}$ when $m_1 \geq c_1$ and $k - m_1 \leq \tau < k - c_1$.

Lemma 4.8. *If $m_1 \geq c_1$ and $k - m_1 \leq \tau < k - c_1$, then*

$$s_{k,\tau} = \begin{cases} p^{3k-(k-\tau)/2+(b_1+c_1)/2}\left(\frac{-b_0c_0}{p}\right)^{b_1} & \textit{if } b_1 \equiv c_1 \pmod 2 \textit{ and } k-\tau \\ \quad \cdot \left(1-\frac{1}{p}\right) & \textit{is even,} \\ p^{3k+(b_1+c_1-(k-\tau))/2}\left(\frac{-a}{p}\right)\left(\frac{b_0}{p}\right)^{b_1+1} & \textit{if } b_1 \not\equiv c_1 \pmod 2 \textit{ and } k-\tau \\ \quad \cdot \left(\frac{c_0}{p}\right)^{b_1}\left(1-\frac{1}{p}\right) & \textit{is odd,} \\ 0 & \textit{otherwise.} \end{cases}$$

Proof. Since $m_1 \geq c_1$ and $k-m_1 \leq \tau < k-c_1$, we have $\tau+b_1 \leq \tau+c_1 < k \leq m_1+\tau$. Therefore, by Lemma 2.7,

$$\begin{aligned} s_{k,\tau} &= \sum_{t_0 \in (\mathbb{Z}/p^{k-\tau}\mathbb{Z})^*} p^{(k+\tau)/2}\left(\frac{at_0}{p^{k-\tau}}\right)\varepsilon_{p^{k-\tau}} p^{(k+b_1+\tau)/2}\left(\frac{b_0t_0}{p^{k-b_1-\tau}}\right)\varepsilon_{p^{k-b_1-\tau}} \\ &\quad \cdot p^{(k+c_1+\tau)/2}\left(\frac{c_0t_0}{p^{k-c_1-\tau}}\right)\varepsilon_{p^{k-c_1-\tau}} \\ &= p^{(3k+3\tau+b_1+c_1)/2}\varepsilon_{p^{k-\tau}}\varepsilon_{p^{k-b_1-\tau}}\varepsilon_{p^{k-c_1-\tau}}\left(\frac{a}{p}\right)^{k-\tau}\left(\frac{b_0}{p}\right)^{k-b_1-\tau} \\ &\quad \cdot \left(\frac{c_0}{p}\right)^{k-c_1-\tau} \sum_{t_0 \in (\mathbb{Z}/p^{k-\tau}\mathbb{Z})^*}\left(\frac{t_0}{p}\right)^{k-\tau-b_1-c_1}. \end{aligned}$$

If $b_1 \equiv c_1 \pmod 2$, then, by Lemma 3.3,

$$s_{k,\tau} = \begin{cases} p^{(3k+3\tau+b_1+c_1)/2}\varepsilon_{p^{b_1}}^2 \left(\dfrac{b_0}{p}\right)^{b_1} \\ \quad \cdot \left(\dfrac{c_0}{p}\right)^{b_1} p^{k-\tau}\left(1-\dfrac{1}{p}\right) & \text{if } k-\tau \text{ is even,} \\ 0 & \text{if } k-\tau \text{ is odd,} \end{cases}$$

$$= \begin{cases} p^{3k-(k-\tau)/2+(b_1+c_1)/2}\left(\dfrac{-b_0c_0}{p}\right)^{b_1} \\ \quad \cdot \left(1-\dfrac{1}{p}\right) & \text{if } k-\tau \text{ is even,} \\ 0 & \text{if } k-\tau \text{ is odd.} \end{cases}$$

If $b_1 \not\equiv c_1 \pmod 2$, then, by Lemma 3.3,

$$s_{k,\tau} = \begin{cases} 0 & \text{if } k-\tau \text{ is even,} \\ p^{(3k+3\tau+b_1+c_1)/2}\varepsilon_p^2 \left(\dfrac{a}{p}\right)\left(\dfrac{b_0}{p}\right)^{b_1+1} \\ \quad \cdot \left(\dfrac{c_0}{p}\right)^{b_1} p^{k-\tau}\left(1-\dfrac{1}{p}\right) & \text{if } k-\tau \text{ is odd,} \end{cases}$$

$$= \begin{cases} 0 & \text{if } k-\tau \text{ is even,} \\ p^{3k+(b_1+c_1-(k-\tau))/2}\left(\dfrac{-a}{p}\right)\left(\dfrac{b_0}{p}\right)^{b_1+1} \\ \quad \cdot \left(\dfrac{c_0}{p}\right)^{b_1}\left(1-\dfrac{1}{p}\right) & \text{if } k-\tau \text{ is odd.} \end{cases}$$

□

We would like to compute the sum $\sum_{\tau=k-m_1}^{k-c_1-1} s_{k,\tau}$ when $m_1 \geq c_1$. In order to do that, we make use of some geometric-like sums given in the next lemma. We use the notation $\lceil x \rceil$ for the least integer greater than or equal to x. The details of the proof are omitted.

Lemma 4.9. *Let n_1 and n_2 be integers such that $0 \leq n_1 \leq n_2 \leq k$. Then*

$$\sum_{\substack{\tau=k-n_2 \\ k-\tau \text{ is even}}}^{k-n_1} p^{-(k-\tau)/2}\left(1-\frac{1}{p}\right) = p^{-\lceil n_1/2\rceil}\left(1-p^{-\lfloor n_2/2\rfloor+\lceil n_1/2\rceil-1}\right)$$

and

$$\sum_{\substack{\tau=k-n_2 \\ k-\tau \text{ is odd}}}^{k-n_1} p^{-(k-\tau)/2}\left(1-\frac{1}{p}\right) = p^{1/2-\lceil (n_1+1)/2\rceil}\left(1-p^{-\lfloor (n_2+1)/2\rfloor+\lceil (n_1+1)/2\rceil-1}\right).$$

Using previous lemmas, we now compute the sum $\sum_{\tau=k-m_1}^{k-c_1-1} s_{k,\tau}$ when $m_1 \geq c_1$.

Lemma 4.10. *If* $m_1 \geq c_1$, *then*

$$\sum_{\tau=k-m_1}^{k-c_1-1} s_{k,\tau} = \begin{cases} p^{3k+(b_1+c_1)/2-\lceil (c_1+1)/2\rceil}\left(\dfrac{-b_0c_0}{p}\right)^{b_1} \\ \qquad \cdot\left(1-p^{-\lfloor m_1/2\rfloor+\lceil (c_1+1)/2\rceil-1}\right) & \text{if } b_1 \equiv c_1 \pmod 2, \\ p^{3k+(b_1+c_1-1)/2-\lceil c_1/2\rceil}\left(\dfrac{-a}{p}\right)\left(\dfrac{b_0}{p}\right)^{b_1+1} \\ \qquad \cdot\left(\dfrac{c_0}{p}\right)^{b_1}\cdot\left(1-p^{-\lfloor (m_1+1)/2\rfloor+\lceil c_1/2\rceil}\right) & \text{if } b_1 \not\equiv c_1 \pmod 2. \end{cases}$$

Proof. By Lemmas 4.8 and 4.9, if $b_1 \equiv c_1 \pmod 2$, then

$$\begin{aligned} \sum_{\tau=k-m_1}^{k-c_1-1} s_{k,\tau} &= p^{3k+(b_1+c_1)/2}\left(\frac{-b_0c_0}{p}\right)^{b_1}\sum_{\substack{\tau=k-m_1 \\ k-\tau \text{ is even}}}^{k-c_1-1} p^{-(k-\tau)/2}\left(1-\frac{1}{p}\right) \\ &= p^{3k+(b_1+c_1)/2}\left(\frac{-b_0c_0}{p}\right)^{b_1} p^{-\lceil (c_1+1)/2\rceil} \\ &\quad \cdot\left(1-p^{-\lfloor m_1/2\rfloor+\lceil (c_1+1)/2\rceil-1}\right), \end{aligned}$$

which simplifies to what is stated in Lemma 4.10 for $b_1 \equiv c_1 \pmod 2$.

By Lemmas 4.8 and 4.9, if $b_1 \not\equiv c_1 \pmod 2$, then

$$\begin{aligned} \sum_{\tau=k-m_1}^{k-c_1-1} s_{k,\tau} &= p^{3k+(b_1+c_1)/2}\left(\frac{-a}{p}\right)\left(\frac{b_0}{p}\right)^{b_1+1}\left(\frac{c_0}{p}\right)^{b_1}\sum_{\substack{\tau=k-m_1 \\ k-\tau \text{ is odd}}}^{k-c_1-1} p^{-(k-\tau)/2}\left(1-\frac{1}{p}\right) \\ &= p^{3k+(b_1+c_1+1)/2}\left(\frac{-a}{p}\right)\left(\frac{b_0}{p}\right)^{b_1+1}\left(\frac{c_0}{p}\right)^{b_1} p^{-\lceil c_1/2\rceil-1} \\ &\quad \cdot\left(1-p^{-\lfloor (m_1+1)/2\rfloor+\lceil c_1/2\rceil}\right), \end{aligned}$$

which simplifies to what is stated in Lemma 4.10 for $b_1 \not\equiv c_1 \pmod 2$. □

We now compute $s_{k,\tau}$ when $m_1 \geq c_1$ and $k - c_1 \leq \tau < k - b_1$.

Lemma 4.11. *If $m_1 \geq c_1$ and $k - c_1 \leq \tau < k - b_1$, then*

$$s_{k,\tau} = \begin{cases} p^{3k+b_1/2} \left(\frac{-ab_0}{p}\right)^{k-\tau} \left(1 - \frac{1}{p}\right) & \text{if } b_1 \text{ is even}, \\ 0 & \text{if } b_1 \text{ is odd}. \end{cases}$$

Proof. Since $m_1 \geq c_1$ and $k-c_1 \leq \tau < k-b_1$, we have $b_1+\tau < k \leq c_1+\tau \leq m_1+\tau$. Therefore, by Lemma 2.7,

$$s_{k,\tau} = \sum_{t_0 \in (\mathbb{Z}/p^{k-\tau}\mathbb{Z})^*} p^{(k+\tau)/2} \left(\frac{at_0}{p^{k-\tau}}\right) \varepsilon_{p^{k-\tau}} p^{(k+b_1+\tau)/2} \left(\frac{b_0 t_0}{p^{k-b_1-\tau}}\right) \varepsilon_{p^{k-b_1-\tau}} p^k.$$

By the same methods used in the proof of Lemma 4.6, we have the result of this lemma. □

The above lemma together with Lemma 3.5 now immediately gives the following.

Lemma 4.12. *If $m_1 \geq c_1$, then*

$$\sum_{\tau=k-c_1}^{k-b_1-1} s_{k,\tau} = \begin{cases} p^{3k+b_1/2} \left(1 - \frac{1}{p}\right) \left(\frac{c_1 - b_1}{2} + \frac{(-1)^{c_1} - 1}{4} \right. \\ \qquad \left. + \left(\frac{-ab_0}{p}\right) \left(\frac{c_1 - b_1}{2} + \frac{1 - (-1)^{c_1}}{4}\right)\right) & \text{if } b_1 \text{ is even}, \\ 0 & \text{if } b_1 \text{ is odd}. \end{cases}$$

Using Lemmas 4.1, 4.3, 4.7, 4.10, and 4.12, we now compute $r_{p^k}(m, Q)$ when $m_1 \geq c_1$. By using Lemmas 4.1 and 4.3 and (4.1), we see that

$$\begin{aligned} r_{p^k}(m, Q) &= p^{2k} + \frac{1}{p^k} s_{k,k-m_1-1} + \frac{1}{p^k} \sum_{\tau=k-m_1}^{k-c_1-1} s_{k,\tau} + \frac{1}{p^k} \sum_{\tau=k-c_1}^{k-b_1-1} s_{k,\tau} + \frac{1}{p^k} \sum_{\tau=k-b_1}^{k-1} s_{k,\tau} \\ &= p^{2k} + \frac{1}{p^k} s_{k,k-m_1-1} + \frac{1}{p^k} \sum_{\tau=k-m_1}^{k-c_1-1} s_{k,\tau} + \frac{1}{p^k} \sum_{\tau=k-c_1}^{k-b_1-1} s_{k,\tau} \\ &\quad + p^{2k}(p^{\lfloor b_1/2 \rfloor} - 1) \\ &= \frac{1}{p^k} s_{k,k-m_1-1} + \frac{1}{p^k} \sum_{\tau=k-m_1}^{k-c_1-1} s_{k,\tau} + \frac{1}{p^k} \sum_{\tau=k-c_1}^{k-b_1-1} s_{k,\tau} + p^{2k+\lfloor b_1/2 \rfloor}. \end{aligned}$$

Substituting the identities from Lemmas 4.7, 4.10, and 4.12 and simplifying, we deduce that

$$r_{p^k}(m,Q)=\begin{cases} p^{2k+b_1/2}\Big(1+\dfrac{1}{p}+p^{-m_1/2+c_1/2-1} \\ \quad\cdot\Big(\Big(\dfrac{-ab_0c_0m_0}{p}\Big)-1\Big) \\ \quad+\Big(1-\dfrac{1}{p}\Big)\Big(\dfrac{c_1-b_1}{2} \\ \quad+\Big(\dfrac{-ab_0}{p}\Big)\dfrac{c_1-b_1}{2}\Big)\Big) & \text{if } c_1 \text{ and } m_1 \text{ are even,} \\ p^{2k+b_1/2}\Big(\Big(1+\dfrac{1}{p}\Big)\big(1-p^{-(m_1+1)/2+c_1/2}\big) \\ \quad+\Big(1-\dfrac{1}{p}\Big)\Big(\dfrac{c_1-b_1}{2} & \text{if } c_1 \text{ is even and } m_1 \\ \quad+\Big(\dfrac{-ab_0}{p}\Big)\dfrac{c_1-b_1}{2}\Big)\Big) & \text{is odd,} \\ p^{2k+b_1/2}\Big(1-p^{-m_1/2+(c_1-1)/2}\Big(\dfrac{-ab_0}{p}\Big) \\ \quad\cdot\Big(1+\dfrac{1}{p}\Big)+\dfrac{1}{p}\Big(\dfrac{-ab_0}{p}\Big) \\ \quad+\Big(1-\dfrac{1}{p}\Big)\Big(\dfrac{c_1-b_1-1}{2} & \text{if } c_1 \text{ is odd and } m_1 \\ \quad+\Big(\dfrac{-ab_0}{p}\Big)\dfrac{c_1-b_1+1}{2}\Big)\Big) & \text{is even,} \\ p^{2k+b_1/2}\big(1+p^{-(m_1+1)/2+(c_1-1)/2} \\ \quad\cdot\Big(\Big(\dfrac{c_0m_0}{p}\Big)-\Big(\dfrac{-ab_0}{p}\Big)\Big) \\ \quad+\dfrac{1}{p}\Big(\dfrac{-ab_0}{p}\Big)+\Big(1-\dfrac{1}{p}\Big) \\ \quad\cdot\Big(\dfrac{c_1-b_1-1}{2}+\Big(\dfrac{-ab_0}{p}\Big)\dfrac{c_1-b_1+1}{2}\Big)\Big) & \text{if } c_1 \text{ and } m_1 \text{ are odd,} \end{cases}$$

if b_1 is even and

$$r_{p^k}(m,Q) = \begin{cases} p^{2k+(b_1-1)/2}\left(1+\left(\frac{-ac_0}{p}\right)\right. \\ \qquad \left.-\,p^{-m_1/2+c_1/2}\left(1+\frac{1}{p}\right)\left(\frac{-ac_0}{p}\right)\right) & \text{if } c_1 \text{ and } m_1 \text{ are even,} \\ p^{2k+(b_1-1)/2}\left(1+\left(\frac{-ac_0}{p}\right)\right. \\ \qquad +p^{-(m_1+1)/2+c_1/2}\left(\left(\frac{b_0m_0}{p}\right)\right. \\ \qquad \left.\left.-\left(\frac{-ac_0}{p}\right)\right)\right) & \text{if } c_1 \text{ is even and } m_1 \text{ is odd,} \\ p^{2k+(b_1-1)/2}\left(1+\left(\frac{-b_0c_0}{p}\right)\right. \\ \qquad +p^{-m_1/2+(c_1-1)/2}\left(\left(\frac{am_0}{p}\right)\right. \\ \qquad \left.\left.-\left(\frac{-b_0c_0}{p}\right)\right)\right) & \text{if } c_1 \text{ is odd and } m_1 \text{ is even,} \\ p^{2k+(b_1-1)/2}\left(1+\left(\frac{-b_0c_0}{p}\right)\right. \\ \qquad -p^{(-m_1+c_1)/2}\left(1+\frac{1}{p}\right) \\ \qquad \left.\cdot\left(\frac{-b_0c_0}{p}\right)\right) & \text{if } c_1 \text{ and } m_1 \text{ are odd,} \end{cases}$$

if b_1 is odd. By dividing by p^{2k} and computing the limit as $k \to \infty$, we obtain (1.6) and (1.7) if $m_1 \geq c_1$.

4.2. *Proof of formulas for* $\alpha_p(0, Q)$

We now compute $\alpha_p(0,Q)$. Let k be a positive integer. Towards computing $\alpha_p(0,Q)$, we compute $r_{p^k}(0,Q)$ and then take the appropriate limit. By (2.2),

$$r_{p^k}(0,Q) = p^{2k} + \frac{1}{p^k}\sum_{t=1}^{p^k-1} g(at;p^k)\, g(b_0p^{b_1}t;p^k)\, g(c_0p^{c_1}t;p^k)\,.$$

Let $t = t_0 p^\tau$, where $0 \le \tau \le k-1$ and $t_0 \in (\mathbb{Z}/p^{k-\tau}\mathbb{Z})^*$. Then

$$\begin{aligned} r_{p^k}(0,Q) &= p^{2k} + \frac{1}{p^k}\sum_{\tau=0}^{k-1} s_{k,\tau} \\ &= p^{2k} + \frac{1}{p^k}\sum_{\tau=0}^{k-c_1-1} s_{k,\tau} + \frac{1}{p^k}\sum_{\tau=k-c_1}^{k-b_1-1} s_{k,\tau} + \frac{1}{p^k}\sum_{\tau=k-b_1}^{k-1} s_{k,\tau}, \end{aligned} \tag{4.4}$$

where

$$s_{k,\tau} = \sum_{t_0\in(\mathbb{Z}/p^{k-\tau}\mathbb{Z})^*} g\left(at_0p^\tau;p^k\right) g\left(b_0t_0p^{b_1+\tau};p^k\right) g\left(c_0t_0p^{c_1+\tau};p^k\right).$$

Doing calculations similar to those in Lemmas 4.2 and 4.3, we notice that the result of Lemma 4.3 is applicable to the sum $\sum_{\tau=k-b_1}^{k-1} s_{k,\tau}$ in (4.4) if we take $b_1 = \min(m_1, b_1)$. Therefore,

$$\begin{aligned} r_{p^k}(0,Q) &= p^{2k} + \frac{1}{p^k}\sum_{\tau=0}^{k-c_1-1} s_{k,\tau} + \frac{1}{p^k}\sum_{\tau=k-c_1}^{k-b_1-1} s_{k,\tau} + p^{2k}\left(p^{\lfloor b_1/2\rfloor} - 1\right) \\ &= \frac{1}{p^k}\sum_{\tau=0}^{k-c_1-1} s_{k,\tau} + \frac{1}{p^k}\sum_{\tau=k-c_1}^{k-b_1-1} s_{k,\tau} + p^{2k+\lfloor b_1/2\rfloor}. \end{aligned}$$

A quick check shows that the result of Lemma 4.12 is applicable to the sum $\sum_{\tau=k-c_1}^{k-b_1-1} s_{k,\tau}$ in (4.4), so

$$r_{p^k}(0,Q) = \begin{cases} \dfrac{1}{p^k}\displaystyle\sum_{\tau=0}^{k-c_1-1} s_{k,\tau} + p^{2k+b_1/2} \\ \quad + p^{2k+b_1/2}\left(1 - \dfrac{1}{p}\right) \\ \quad \cdot \left(\dfrac{c_1-b_1}{2} + \dfrac{(-1)^{c_1}-1}{4} + \left(\dfrac{-ab_0}{p}\right)\right. \\ \quad \left. \cdot \left(\dfrac{c_1-b_1}{2} + \dfrac{1-(-1)^{c_1}}{4}\right)\right) & \text{if } b_1 \text{ is even,} \\ \dfrac{1}{p^k}\displaystyle\sum_{\tau=0}^{k-c_1-1} s_{k,\tau} + p^{2k+(b_1-1)/2} & \text{if } b_1 \text{ is odd.} \end{cases}$$

Doing calculations similar to those in Lemmas 4.8 and 4.10, we see that Lemma 4.10 is applicable to the sum $\sum_{\tau=0}^{k-c_1-1} s_{k,\tau}$ in (4.4) with $m_1 = k$. Thus,

$$r_{p^k}(0,Q) = \begin{cases} p^{2k+b_1/2-1}(1-p^{-\lfloor k/2\rfloor + c_1/2}) \\ \quad + p^{2k+b_1/2} + p^{2k+b_1/2}\left(1-\frac{1}{p}\right) \\ \quad \cdot \left(\frac{c_1-b_1}{2} + \left(\frac{-ab_0}{p}\right)\frac{c_1-b_1}{2}\right) & \text{if } b_1 \text{ and } c_1 \text{ are even,} \\ p^{2k+b_1/2-1}\left(\frac{-ab_0}{p}\right)(1-p^{-\lfloor (k+1)/2\rfloor + (c_1+1)/2}) \\ \quad + p^{2k+b_1/2} + p^{2k+b_1/2}\left(1-\frac{1}{p}\right) \\ \quad \cdot \left(\frac{c_1-b_1-1}{2}\right. & \text{if } b_1 \text{ is even and } c_1 \\ \quad \left. + \left(\frac{-ab_0}{p}\right)\frac{c_1-b_1+1}{2}\right) & \text{is odd,} \\ p^{2k+(b_1-1)/2}\left(\frac{-ac_0}{p}\right) \\ \quad \cdot (1-p^{-\lfloor (k+1)/2\rfloor + c_1/2}) & \text{if } b_1 \text{ is odd and } c_1 \\ \quad + p^{2k+(b_1-1)/2} & \text{is even,} \\ p^{2k+(b_1-1)/2}\left(\frac{-b_0c_0}{p}\right) \\ \quad \cdot (1-p^{-\lfloor k/2\rfloor + (c_1+1)/2-1}) \\ \quad + p^{2k+(b_1-1)/2} & \text{if } b_1 \text{ and } c_1 \text{ are odd.} \end{cases}$$

By dividing by p^{2k} and taking the limit as $k \to \infty$, we obtain (1.8).

Acknowledgments

I thank Matthew Young and the National Science Foundation Research Experience for Undergraduates at Texas A&M (funded under NSF DMS-1156589) for introducing me to local densities of quadratic forms. I thank Kenneth Williams for asking me to look at local densities of quadratic forms again. I also thank John Rickert, William Duke, Ranier Schulze-Pillot, Alex Kontorovich, Jonathan Hanke, and Gene Kopp for their helpful discussions. I thank Meinhard Peters for reading my paper and bringing Lomadze's work to my attention. I thank Kenneth Williams, Gene Kopp, Matthew Young, and Bruce Berndt for reading my paper and making great suggestions that improved the article. I especially thank the anonymous referee for thoroughly reading my manuscript and making excellent suggestions that improved the article.

Much of this material is based upon work supported by the National Science Foundation Graduate Research Fellowship under NSF DGE-1842213.

References

[1] A. Berkovich and W. C. Jagy, On representation of an integer as the sum of three squares and ternary quadratic forms with the discriminants p^2, $16p^2$, *J. Number Theory* **132**(1) (2012) 258–274.

[2] B. C. Berndt, R. J. Evans and K. S. Williams, *Gauss and Jacobi Sums* (John Wiley & Sons, 1998).

[3] H. Cohen, *A Course in Computational Algebraic Number Theory*, Graduate Texts in Mathematics, Vol. 138 (Springer-Verlag, 1993).

[4] J. Hanke, Local densities and explicit bounds for representability by a quadratic form, *Duke Math. J.* **124**(2) (2004) 351–388.

[5] J. Hanke, Quadratic forms and automorphic forms, preprint (2011), arXiv:1105.5759v3.

[6] D. Hilbert, Mathematical problems, *Bull. Amer. Math. Soc.* **8**(10) (1902) 437–479.

[7] B. W. Jones and G. Pall, Regular and semi-regular positive ternary quadratic forms, *Acta Math.* **70**(1) (1939) 165–191.

[8] G. A. Lomadze, Formulae for the number of representations of numbers by all primitive positive ternary diagonal quadratic forms that belong to one-class genera, *Sakharth. SSR Mecn. Akad. Math. Inst. Šrom.* **40**(1) (1971) 140–179.

[9] C. L. Siegel, Über die analytische theorie der quadratischem formen, *Ann. of Math. (2)* **36**(3) (1935) 527–606.

[10] C. L. Siegel, *Lectures on the Analytical Theory of Quadratic Forms*, notes by M. Ward, rev. edn. (Institute for Advanced Study, 1949).

[11] T. Yang, An explicit formula for local densities of quadratic forms, *J. Number Theory*, **72**(2) (1998) 309–356.

© 2024 World Scientific Publishing Company
https://doi.org/10.1142/9789811277375_0023

Divisibility properties of the Fourier coefficients of (mock) modular functions and Ramanujan*

Soon-Yi Kang
Department of Mathematics
Kangwon National University
Chuncheon 24341, Korea
sy2kang@kangwon.ac.kr

We survey divisibility properties of the Fourier coefficients of modular functions inspired by Ramanujan. Then using recent results of the generalized Hecke operator on harmonic Maass functions and known divisibility of Fourier coefficients of modular functions, we establish congruence relations of the Fourier coefficients of certain modular functions and mock modular functions of various levels.

Keywords: Congruence; Fourier coefficients; Hecke operator; modular function; mock modular form; Ramanujan.

Mathematics Subject Classification 2020: 11F03, 11F12, 11F25, 11F30, 11F33

1. Introduction

Fourier coefficients of modular forms have rich arithmetic properties and give a wide range of applications. It is in the work of S. Ramanujan where we find important such examples and the rudiment of the arithmetic theory of modular forms.

In his paper [24–26], Ramanujan stated and proved his three famous congruences for the partition function $p(n)$, namely,

$$p(5n+4) \equiv 0 \pmod{5}, \tag{1.1}$$

$$p(7n+5) \equiv 0 \pmod{7}, \tag{1.2}$$

$$p(11n+6) \equiv 0 \pmod{11}, \tag{1.3}$$

where n is any non-negative integer. The partition function is closely related with the τ-function that was introduced by Ramanujan in his important paper [23] via

$$\sum_{n=1}^{\infty} \tau(n) q^n = q \prod_{n=1}^{\infty} (1-q^n)^{24} = \left(\prod_{n=1}^{\infty} (1-q^n)^{25} \right) \sum_{n=0}^{\infty} p(n) q^{n+1}.$$

*This article originally appeared in *International Journal of Number Theory*, **17**(3), pp. 577–589 © 2021 World Scientific Publishing Company. DOI: 10.1142/S1793042120400333

He proved a congruence for $\tau(n)$ as well:

$$\tau(n) \equiv \sigma_{11}(n) \pmod{691}, \tag{1.4}$$

where $\sigma_k(n)$ is the sum of kth powers of divisors of n. Ramanujan's observations on $\tau(n)$ such as

$$\tau(p^{\alpha+2}) = \tau(p)\tau(p^{\alpha+1}) - p^{11}\tau(p^{\alpha})$$

for prime p and $\alpha \geq 0$ led to the development of Hecke theory. The proofs of (1.1)–(1.3) in [26] employ Eisenstein series. In [6], Berndt also uses Eisenstein series and gives simple and unified proofs for (1.1)–(1.3). In fact, the proofs in [26] were extracted from an unpublished paper of Ramanujan on $p(n)$ and $\tau(n)$ by Hardy after Ramanujan died in 1920. Berndt and Ono [7] edited this paper with an extensive commentary and references to the literature.

Ramanujan [24] conjectured that there are further congruences for $p(n)$ modulo powers of the primes 5, 7 and 11 and left a sketch of proofs of the conjecture for all powers of 5 and 7 in his paper [7]. Watson [27] completed the proofs. The case for all powers of 11 was proved by Atkin [3], who followed Lehner's approach developed in [15, 18]. For more history and discoveries on congruences for $p(n)$ including two groundbreaking work by Atkin [4] and Ahlgren and Ono [1], see [5, 20–22].

It is too extensive to discuss all divisibility properties of Fourier coefficients of modular forms inspired by Ramanujan. We rather focus on the divisibility of those relating with modular functions. It was Lehner who started investigating the divisibility properties of the Fourier coefficients of a modular function, in particular, the modular j-invariant. For τ with $\operatorname{Im}\tau > 0$ and $q = \exp(2\pi i\tau)$, the classical modular j-invariant $j(\tau) = q^{-1} + 744 + \sum_{n=1}^{\infty} b(n)q^n$ is a modular function on the full modular group, that is, a weight 0 weakly holomorphic modular form of level 1. The Fourier coefficients of this fundamental function in modular form theory and number theory are now known as the graded dimensions of the Monster module. Expanding the methods used in his proofs of Ramanujan's partition congruences, Lehner showed in [16] that

$$b(5^{\alpha+1}n) \equiv 0 \pmod{5^{\alpha}}, \tag{1.5}$$

$$b(7^{\alpha}n) \equiv 0 \pmod{7^{\alpha}}, \tag{1.6}$$

$$b(11^{2}n) \equiv 0 \pmod{11^{2}}, \tag{1.7}$$

for any positive integers n and α. In [3], Atkin generalized (1.7) to all positive powers α:

$$b(11^{\alpha}n) \equiv 0 \pmod{11^{\alpha}}. \tag{1.8}$$

As he claimed in [3, p. 22], congruence in (1.8) is best possible in the sense that $b(11^{\alpha}) \not\equiv 0 \pmod{11^{\alpha+1}}$. Lehner [17] further showed that the coefficients $b(n)$ are highly divisible by small primes dividing n by proving that

$$b(2^{\alpha}3^{\beta}5^{\gamma}7^{\delta}n) \equiv 0 \pmod{2^{3\alpha+8}3^{2\beta+3}5^{\gamma+1}7^{\delta}} \tag{1.9}$$

for any positive integers α, β, γ and δ. He also claimed that the Fourier coefficients of certain modular functions of higher level also satisfy the same divisibility properties. Recently, these results have been generalized to a modular function of level of genus zero. Griffin [9] showed that similar congruences with (1.9) hold for all elements of the canonical basis $j_m(\tau) := q^{-m} + O(q)$ for the space of weakly holomorphic modular functions on the full modular group. Note that $j_0(\tau) = 1$ and $j_1(\tau) = j(\tau) - 744$. Applying (1.8) into the method used in [9], one can easily obtain a prime 11 analog to the results for smaller primes in [9].

Theorem 1.1. *Let $j_m(\tau) = q^{-m} + \sum_{n=1}^{\infty} b(m,n)q^n$ be an element of the canonical basis for the space of weakly holomorphic modular functions on the full modular group. If $m = 11^\alpha m'$ for a non-negative integer α and $(11, m') = 1$, then for any positive integers n and $\beta > \alpha$,*

$$b(11^\alpha m', 11^\beta n) \equiv 0 \pmod{11^{(\beta-\alpha)}}. \tag{1.10}$$

Anderson and Jenkins [2] generalized Lehner's claim on modular functions of higher level by proving the following theorem.

Theorem 1.2 ([2, Theorem 2]). *Let $p \in \{2,3,5,7\}$, and let*

$$f_{p,m}(\tau) = q^{-m} + \sum_{n=0}^{\infty} b_p(m,n)q^n$$

be an element of the canonical basis for the space of level p modular functions which are holomorphic at 0. If $m = p^\alpha m'$ for a non-negative integer α and $(m',p) = 1$, then for any positive integer n and $\beta > \alpha$,

$$\begin{aligned} b_2(2^\alpha m', 2^\beta n) &\equiv 0 \pmod{2^{3(\beta-\alpha)+8}}, \\ b_3(3^\alpha m', 3^\beta n) &\equiv 0 \pmod{3^{2(\beta-\alpha)+3}}, \\ b_5(5^\alpha m', 5^\beta n) &\equiv 0 \pmod{5^{(\beta-\alpha)+1}}, \\ b_7(7^\alpha m', 7^\beta n) &\equiv 0 \pmod{7^{(\beta-\alpha)}}. \end{aligned} \tag{1.11}$$

Divisibility properties of Fourier coefficients of modular functions of level of genus zero are further investigated in a series of papers by Jenkins and his collaborators in [10–12]. They showed that many of the Fourier coefficients of the basis elements for the space of modular functions of level N of genus zero are divisible by high powers of the prime dividing the level N.

Unlike modular curves of level 2, 3, 5 and 7, the curve of level 11 is of genus 1 not of genus 0. Hence, there is no Hauptmodul on this curve. In order to prove congruence relations relating with 11, Atkin and Lehner, each found a pair of two generators to construct a basis for the space of modular functions of level 11. In [14], the author with Jeon and Kim constructed reduced form bases for the spaces of the modular functions that are holomorphic away from the cusp at $i\infty$ for all levels. In particular, each basis element function with level N of genus 1 is given by

$$\begin{cases} f_{N,0}(\tau) = 1, \\ f_{N,m}(\tau) = q^{-m} + a_N(m,-1)q^{-1} + \sum_{n=1}^{\infty} a_N(m,n)q^n \quad (m \geq 2) \end{cases} \tag{1.12}$$

with integer coefficients. Using Atkin's basis and its properties discovered in [3], we establish a divisibility property of the Fourier coefficients of $f_{11,m}(\tau)$ modulo powers of 11.

Theorem 1.3. *Let $f_{11,m}(\tau)$ be an element of the basis for the space of level 11 modular functions which are holomorphic at 0 with the Fourier expansion in (1.12). If $m = 11^{\alpha}m' \geq 2$ for a non-negative integer α and $(11, m') = 1$, then for any positive integers n and $\beta > \alpha$,*

$$a_{11}(11^{\alpha}m', 11^{\beta}n) \equiv 0 \pmod{11^{(\beta-\alpha)}}. \tag{1.13}$$

According to numerical experiments, such beautiful and simple congruences appearing in (1.9) through (1.13) do not seem to hold for levels 17 or 19. However, a very general congruence relation of Fourier coefficients of modular functions of level of positive genus is found in [14, Theorem 1.11]. We state only its genus 1 version here.

Theorem 1.4. *Let $N \in \{11, 14, 15, 17, 19, 20, 21, 24, 27, 32, 36, 49\}$. Then for any prime p and positive integers α, m and n with m, $n \not\equiv 0 \pmod{p}$, we have*

$$a_N(p^{\alpha}m, n) + a_N(m,-1)a_N(p^{\alpha}, n) \equiv 0 \pmod{p^{\alpha}}. \tag{1.14}$$

Ramanujan's mock theta functions are holomorphic parts of harmonic Maass forms. In this spirit and following Zagier, we call the holomorphic part of a harmonic Maass function (i.e. a weight zero harmonic Maass form) by a mock modular function. For each level N, there are certain analytic continuations of Niebur–Poincaré series which form a basis for the space of harmonic Maass functions of level N. We denote these basis elements by $j_{N,m}$. These are considered to be generalizations of the j_m functions, because $j_{1,m}(\tau) = j_m(\tau) + 24\sigma_1(m)$ for any non-negative integer m [13, p. 99]. The harmonic Maass function $j_{N,m}$ is decomposed into [19, Theorem 1; 8, Proposition 3.1]

$$j_{N,m}(\tau) = j^h_{N,m}(\tau) + j^{nh}_{N,m}(\tau), \tag{1.15}$$

where the Fourier expansion of the mock modular function $j^h_{N,m}$ is given by

$$j^h_{N,m}(\tau) = q^{-m} + c_N(m,0) + \sum_{n=1}^{\infty} c_N(m,n)q^n \tag{1.16}$$

and its corresponding non-holomorphic part $j^{nh}_{N,m}$ can be written as

$$j^{nh}_{N,m}(\tau) = -\overline{q}^m + \sum_{n=1}^{\infty} c_N(m,-n)\overline{q}^n.$$

The functions j_m form a Hecke system, that is, for the normalized Hecke operator $T(m)$, they satisfy that

$$j_m(\tau) = j_1|T(m)(\tau). \tag{1.17}$$

In [14], a Hecke system for $j_{N,m}$ is constructed with a suitable generalization of the Hecke operator in (1.17). By abuse of notation, we may write for any positive integers N and m [14, Corollary 1.4]

$$j_{N,m}(\tau) = j_{N,1}|T(m)(\tau). \tag{1.18}$$

Using this property and its applications, one can obtain several congruences for the Fourier coefficients of mock modular function $j^h_{N,m}$.

Theorem 1.5. *Let p be a prime and $c_p(m,n)$ be the Fourier coefficients of $j^h_{p,m}(\tau)$, the mock modular function of level p given in* (1.16). *Then*

(1) $pc_p(m,pn) - c_p(pm,n) \in \mathbb{Z}$;
(2) *if $p \nmid n$ and $c_p(m,pn) \in \mathbb{Z}$, then $c_p(pm,n) \equiv 0 \pmod{p}$.*

Since $j_{N,m}$ is weakly holomorphic when the level N is of genus 0, for each $p \in \{2,3,5,7\}$, $j_{p,m} = j^h_{p,m}$, and also it equals $f_{p,m}$ in Theorem 1.2 up to a constant. Utilizing congruences in [9] into the identity induced from (1.18) gives congruences of $c_p(m,n)$ for the primes p, but they can be obtained from Theorem 1.2 as well. The following congruence for the Fourier coefficients of a mock modular form of level 11 that we derive from Theorem 1.5 using Theorem 1.1 is new.

Corollary 1.6. *If $m,n \not\equiv 0 \pmod{11}$ and non-negative integers α and β satisfy $\beta > \alpha+1$, then*

$$11c_{11}(11^\alpha m, 11^{\beta+1}n) - c_{11}(11^{\alpha+1}m, 11^\beta n) \equiv 0 \pmod{11^{\beta-\alpha-1}}. \tag{1.19}$$

Furthermore, by the relation between $f_{N,m}(\tau)$ and $j_{N,m}(\tau)$ provided in (3.2) below, one can find more congruences for $c_p(m,n)$ from Theorems 1.3 and 1.4.

The rest of the paper is organized as follows. In Sec. 2, we introduce the *p-plication formula* for the harmonic Maass function $j_{N,m}$ proved in [14] and congruence relations found in [9]. Then we prove Theorem 1.5 and Corollary 1.6. We also prove Theorem 1.1 following the method in [9]. In Sec. 3, we present the basis element $f_{N,m}$ and discuss shortly its congruence property in Theorem 1.4. In Sec. 4, utilizing the U-operator and applying Atkin's results in [3], we prove Theorem 1.3.

2. Proofs of Theorems 1.1 and 1.5 and Corollary 1.6

For any positive integer N, let $M^!_k(N)$ (respectively, $H_k(N)$) denote the space of weight k weakly holomorphic modular forms (respectively, harmonic Maass forms) on $\Gamma_0(N)$. Also, let $M^{!,\infty}_k(N)$ (respectively, $M^{!,0}_k(N)$) be its subspace of modular forms whose poles are supported only at infinity (respectively, 0) and holomorphic at other cusps. For a prime p, the U_p operator that acts on a complex-valued function on $\mathbb{H}$ is defined by

$$U_p f(\tau) = \frac{1}{p}\sum_{i=0}^{p-1} f\left(\frac{\tau+i}{p}\right) \tag{2.1}$$

and $U_p^{n+1} f(\tau) = U_p(U_p^n f(\tau))$ for $n \geq 1$. Also, for a non-negative integer i, the *p-plication of* $j_{N,m}$ is given by

$$j_{N,m}^{(p^i)} := j_{\frac{N}{(N,p^i)},m}, \tag{2.2}$$

where $j_{N,\frac{m}{p}} = 0$ if $p \nmid m$. In [14], the following *p-plication formula* is established, which was essential in constructing the generalized Hecke operator for $j_{N,m}$ satisfying (1.18).

Proposition 2.1 ([14, Theorem 1.1]). *For any prime* p, *we have that*

$$p(U_p j_{N,m})(\tau) + j_{N,m}^{(p)}(p\tau) = j_{N,pm}(\tau) + p j_{N,\frac{m}{p}}^{(p)}(\tau). \tag{2.3}$$

Proof of Theorem 1.5. Let $N = p$ in (2.3) so that after rearrangement of terms, we have by (2.2) that

$$p(U_p j_{p,m})(\tau) - j_{p,pm}(\tau) = p j_{1,\frac{m}{p}}(\tau) - j_{1,m}(p\tau). \tag{2.4}$$

Note that the function on the right-hand side of (2.4) is weakly holomorphic, because $j_{1,m} = j_m + 24\sigma_1(m)$. Hence, we can decompose (2.4) into two identities:

$$p(U_p j_{p,m}^h)(\tau) - j_{p,pm}^h(\tau) = p j_{1,\frac{m}{p}}(\tau) - j_{1,m}(p\tau) \tag{2.5}$$

and

$$p(U_p j_{p,m}^{nh})(\tau) = j_{p,pm}^{nh}(\tau). \tag{2.6}$$

By comparing the coefficients of q^n in both sides of (2.5), we find that

$$p c_p(m, pn) - c_p(pm, n) = p c_1\left(\frac{m}{p}, n\right) - c_1\left(m, \frac{n}{p}\right). \tag{2.7}$$

As the coefficients of $j_{1,m}(\tau)$ are integers, so are the Fourier coefficients of the mock modular function $p(U_p j_{p,m}^h)(\tau) - j_{p,pm}^h(\tau)$ in the left-hand side of (2.7). Theorem 1.5 immediately follows from (2.7). □

Applying known congruences of $c_1(m,n) = b(m,n)$ $(n > 0)$ into (2.7), one can obtain many congruences for $p c_p(m, pn) - c_p(pm, n)$. In particular, we recall Griffin's congruences on the Fourier coefficients of $j_m(\tau)$, the most general congruences for $j_m(\tau)$ known. He used Zagier duality between the Fourier coefficients of weights k and $2-k$ weakly holomorphic modular forms and properties of Hecke operator.

Theorem 2.2 ([9, Theorem 2.1]). *For each* $p \in \{2,3,5,7\}$, *let* $\alpha, \beta \geq 0$, $\gamma = |\alpha - \beta|$ *and* $m, n \not\equiv 0 \pmod{p}$. *Then for* $p = 2$:

$$\begin{aligned} b(2^{\alpha}m, 2^{\beta}n) & \\ &\equiv -2^{3\gamma+8}3^{\gamma-1}m\sigma_7(m)\sigma_7(n) \pmod{2^{3\gamma+13}} \quad \text{if } \beta > \alpha, \\ &\equiv -2^{4\gamma+8}3^{\gamma-1}m\sigma_7(m)\sigma_7(n) \pmod{2^{4\gamma+13}} \quad \text{if } \alpha > \beta, \\ &\equiv 20m\sigma_7(m)\sigma_7(n) \pmod{2^7} \quad \text{if } \gamma = 0 \text{ and } mn \equiv 1 \pmod{8}, \\ &\equiv \frac{1}{2}m\sigma_1(m)\sigma_1(n) \pmod{2^3} \quad \text{if } \gamma = 0 \text{ and } mn \equiv 3 \pmod{8}, \\ &\equiv -12m\sigma_7(m)\sigma_7(n) \pmod{2^8} \quad \text{if } \gamma = 0 \text{ and } mn \equiv 5 \pmod{8}. \end{aligned}$$

For $p = 3$:

$$\begin{aligned} b(3^{\alpha}m, 3^{\beta}n) & \\ &\equiv \mp 3^{2\gamma+3}10^{\gamma-1}\frac{\sigma_1(m)\sigma_1(n)}{n} \pmod{3^{2\gamma+6}} \quad \text{if } \beta > \alpha \text{ and } mn \equiv \pm 1 \pmod{3}, \\ &\equiv \mp 3^{3\gamma+3}10^{\gamma-1}\frac{\sigma_1(m)\sigma_1(n)}{n} \pmod{3^{3\gamma+6}} \quad \text{if } \alpha > \beta \text{ and } mn \equiv \pm 1 \pmod{3}, \\ &\equiv 2 \cdot 3^3\frac{\sigma_1(m)\sigma_1(n)}{n} \pmod{3^7} \quad \text{if } \gamma = 0 \text{ and } mn \equiv \pm 1 \pmod{3}. \end{aligned}$$

For $p = 5$:

$$\begin{aligned} b(5^{\alpha}m, 5^{\beta}n) & \\ &\equiv -5^{\gamma+1}3^{\gamma-1}m^2n\sigma_1(m)\sigma_1(n) \pmod{5^{\gamma+2}} \quad \text{if } \beta > \alpha, \\ &\equiv -5^{2\gamma+1}3^{\gamma-1}m^2n\sigma_1(m)\sigma_1(n) \pmod{5^{2\gamma+2}} \quad \text{if } \alpha > \beta, \\ &\equiv 10m^2n\sigma_1(m)\sigma_1(n) \pmod{5^2} \quad \text{if } \gamma = 0 \text{ and } \left(\frac{mn}{5}\right) = -1. \end{aligned}$$

For $p = 7$:

$$\begin{aligned} b(7^{\alpha}m, 7^{\beta}n) &\equiv 7^{\gamma}5^{\gamma-1}m^2n\sigma_3(m)\sigma_3(n) \pmod{7^{\gamma+1}} \quad \text{if } \beta > \alpha, \\ &\equiv 7^{2\gamma}5^{\gamma-1}m^2n\sigma_3(m)\sigma_3(n) \pmod{7^{2\gamma+1}} \quad \text{if } \alpha > \beta, \\ &\equiv 2m^2n\sigma_3(m)\sigma_3(n) \pmod{7} \quad \text{if } \gamma = 0 \text{ and } \left(\frac{mn}{7}\right) = 1. \end{aligned}$$

Utilizing congruences in Theorem 2.2 into (2.7) for $p \in \{2,3,5,7\}$, one can find congruences of $c_p(m,n)$ for corresponding p's, which are immediate consequences of those in Theorem 1.2.

Proof of Corollary 1.6. Similarly, Corollary 1.6 follows from Theorem 1.1 and (2.7). □

Proof of Theorem 1.1. Recall from standard formulas for the action of the Hecke operator that if $(k,m) = 1$, then

$$j_m|T(k)(\tau) = \sum_n \left(\sum_{d|(k,n)} \frac{k}{d} b\left(m, \frac{kn}{d^2}\right) \right) q^n.$$

Hence, we have

$$b(mk,n)=\sum_{d|(k,n)}\frac{k}{d}b\left(m,\frac{kn}{d^2}\right).$$

Taking $m=1$ and $k=m'$ not divisible by 11, we find by using (1.8) that

$$\begin{aligned} b(m',11^{\beta}n) &= \sum_{d|(m',n)}\frac{m'}{d}b\left(1,11^{\beta}\frac{m'n}{d^2}\right) \\ &= \sum_{d|(m',n)}\frac{m'}{d}b\left(11^{\beta}\frac{m'n}{d^2}\right) \equiv 0 \pmod{11^{\beta}}. \end{aligned} \tag{2.8}$$

On the other hand, if we apply $T(11^{\alpha})$ to $j_{m'}(\tau)$, then we obtain

$$b(11^{\alpha}m',11^{\beta}n)=\sum_{d|11^{\alpha}}\frac{11^{\alpha}}{d}b\left(m',n\frac{11^{\alpha+\beta}}{d^2}\right). \tag{2.9}$$

The theorem now follows from (2.8) and (2.9). □

3. Basis Elements $f_{N,m}$ and Theorem 1.4

Let $N=11,14,15,17,19,20,21,24,27,32,36,49$. These are the values of N precisely when the modular curves $X_0(N)$ are of genus 1. For each of these N, there are two canonical generators of $\mathbb{C}$-algebra $M_0^{!,\infty}(N)$ of the form $q^{-2}+O(q^{-1})$ and $q^{-3}+O(q^{-1})$, because the infinity point is not a Weierstrass point of the Riemann surface $X_0(N)$. Using these, one can construct a basis for $M_0^{!,\infty}(N)$ which consists of unique modular functions with integral Fourier coefficients in the form of $f_{N,m}=q^{-m}+a_N(m,-1)q^{-1}+\sum_{n=1}^{\infty}a_N(m,n)q^n\in M_0^{!,\infty}(N)$ with $f_{N,0}(\tau)=1$, $f_{N,1}(\tau)=0$ and $a_N(1,-1)=-1$. For example, when $N=11$, the smallest N for which $X_0(N)$ is of genus 1, two generators of $\mathbb{C}(X_0(11))$ are found by Yang [28]. They are

$$X=q^{-2}+2q^{-1}+4+5q+8q^2+q^3+7q^4-11q^5+\cdots$$

and

$$Y=q^{-3}+3q^{-2}+7q^{-1}+12+17q+26q^2+19q^3+37q^4-15q^5+\cdots.$$

We let $f_{11,2}:=X-4$ and $f_{11,3}:=Y-3X$. One can continue to construct modular functions of the form $f_{11,m}=q^{-m}+O(q^{-1})=Q_m(X,Y)$, where $Q_m(X,Y)$ is a polynomial in X and Y that eliminates the terms between q^{-m} and q^{-1} and also the constant term if there is any. The first few of $f_{11,m}(\tau)$ are given by

$$\begin{aligned} f_{11,0} &= 1, \\ f_{11,1} &= q^{-1} - q^{-1} + 0q + 0q^2 + 0q^3 + 0q^4 + 0q^5 + \cdots = 0, \\ f_{11,2} &= q^{-2} + 2q^{-1} + 5q + 8q^2 + q^3 + 7q^4 - 11q^5 + \cdots = X - 4, \\ f_{11,3} &= q^{-3} + q^{-1} + 2q + 2q^2 + 16q^3 + 16q^4 + 18q^5 + \cdots = Y - 3X, \\ f_{11,4} &= q^{-4} - 2q^{-1} + 6q + 3q^2 + 18q^3 - 42q^4 + \cdots = X^2 - 4Y - 4X - 36, \\ f_{11,5} &= q^{-5} - q^{-1} - 14q - 16q^2 + 34q^3 + \cdots = XY - 2X^2 + X + 7Y + 60, \\ &\vdots \end{aligned} \tag{3.1}$$

Using properties of the generalized Hecke operator appearing in (1.18), several congruences of the Fourier coefficients of $f_{N,m}(\tau)$ are proved in [14]. Theorem 1.4 is one of them, which gives a strong divisibility property of Fourier coefficients of modular functions that holds for arbitrary prime powers.

Let $J_{N,m}(\tau) := j_{N,m}(\tau) - c_N(m,0)$ and define $J_{N,0}(\tau) := 1$. As $H_0(N)$ is generated by $j_{N,m}$, each of $f_{N,m}$ can be written as a linear combination of $j_{N,m}$'s. Specifically,

$$f_{N,m} = J_{N,m} + a_N(m,-1)J_{N,1}. \tag{3.2}$$

We also define $f_{N,\frac{m}{p}} = 0$ unless p divides m. For more congruences of the Fourier coefficients of modular functions of arbitrary level, see [14].

4. Proof of Theorem 1.3

The method of proof given in this section is very similar to those in [2, 3, 16, 17]. In [3], Atkin constructed bases G_m's for $M_0^{!,\infty}(11)$ and g_m's for $M_0^{!,0}(11)$.

Lemma 4.1 ([3, Lemma 3]). *For all integers $m \geq 2$, $G_m(\tau)$ and $g_m(\tau)$ satisfy the following properties:*

(1) $G_m(\tau) \in M_0^{!,\infty}(11)$ *and* $g_m(\tau) \in M_0^{!,0}(11)$;
(2) $G_m(-1/11\tau) = 11^{\theta(m)} g_m(\tau)$, *where* $\theta(m) = 6k+2,3,4,6,6$ *according as* $m = 5k+2,3,4,5,6$ $(k \geq 0)$;
(3) *the Fourier coefficients of* $G_m(\tau)$ *have integral coefficients with leading term* q^{-m};
(4) *the Fourier coefficients of* $g_m(\tau)$ *have integral coefficients with leading term* $q^{\psi(m)}$, *where* $\psi(m) = 5k+1,2,3,5,4$ *according as* $m = 5k+2,3,4,5,6$ $(k \geq 0)$.

Further, the function $B(\tau) = G_2(\tau)g_2(\tau) - 12 \in M_0^!(11)$ *has simple poles at* $\tau = 0$ *and* $\tau = i\infty$ *such that* $B(-1/11\tau) = B(\tau)$ *and the Fourier coefficients of* $B(\tau)$ *have integral coefficients with leading term* q^{-1}.

According to [3, Lemma 4], $G_m(\tau)$, $11^{\theta(m)}g_m(\tau)$ and $B(\tau)$ form a basis for $M_0^!(11)$. Thus, if $F(\tau) \in M_0^!(11)$ has a pole of order M at $\tau = 0$ and a pole of order K at $\tau = i\infty$, then

$$F(\tau) = \sum_{r=2}^{K} \lambda_{-r}G_r(\tau) + \lambda_{-1}B(\tau) + \lambda_0 + \sum_{r=2}^{M} \lambda_r 11^{\theta(r)}g_r(\tau) \tag{4.1}$$

for some constants λ_r $(-K \le r \le M)$.

The Fourier expansions of $G_m(\tau)$, $g_m(\tau)$ and $B(\tau)$ are found in [3, Table 1]. We note that $G_2(\tau) = f_{11,2}(\tau) - 12 = X - 16$ and $G_3(\tau) = f_{11,3}(\tau) - 3f_{11,2}(\tau) + 24 = Y - 6X + 36$. Also, note that $g_5(\tau) = \frac{\eta^{12}(11\tau)}{\eta^{12}(\tau)}$ and $G_5(\tau) = \frac{\eta^{12}(\tau)}{\eta^{12}(11\tau)}$, where $\eta(\tau)$ is the Dedekind eta-function $\eta(\tau) = q^{1/24}\prod_{n=1}^{\infty}(1-q^n)$. In general, if we write $G_m(\tau) = q^{-m} + \sum_{\ell=-m+1}^{\infty} a_m(\ell)q^{\ell}$, then

$$G_m(\tau) = f_{11,m}(\tau) + a_m(0) + \sum_{\ell=2}^{m-1} a_m(-\ell)f_{11,\ell}(\tau) \quad (m \ge 2). \tag{4.2}$$

Conversely, we can also express the element $f_{11,m}(\tau)$ of the reduced form basis of $M_0^{!,\infty}(11)$ as a linear combination of $G_\ell(\tau)$'s with $2 \le \ell \le m$. Thus, the congruence (1.13) is equivalent to the same congruence of coefficients $a_m(\ell)$ of $G_m(\tau)$. More precisely, $a_{11^\alpha m'}(11^\beta \ell) \equiv 0 \pmod{11^{\beta-\alpha}}$.

In [3], Atkin showed the divisibility property of the Fourier coefficients of g_m's. Using this and the fact $U_{11}j(\tau) \in M_0^{!,0}$ with a pole of order 121 at $\tau = 0$, he proved (1.8). From now till the end of the paper, we use the notation $U := U_{11}$.

Lemma 4.2 **([15, Theorem 8 and (8.81)]).** *Let* $f(\tau) \in M_0^!(11)$. *Then*

(1) $Uf(\tau) \in M_0^!(11)$;
(2) $11(Uf)(-\frac{1}{11\tau}) = 11(Uf)(11\tau) + f(-\frac{1}{121\tau}) - f(\tau)$.

Lemma 4.3 **([3, Corollary, p. 21]).** *Define* $\xi(2) = 0, \xi(3) = 1$ *and* $\xi(n) = 5k+1, 3, 3, 4, 5$ *according as* $n = 5k+4, 5, 6, 7, 8$ $(k \ge 0)$, *We denote by* $\mathcal{S}$ *the class of functions* $F(\tau)$ *with* $F(\tau) = \sum_{n=2}^{M} \lambda_n 11^{\xi(n)}g_n(\tau)$. *Then whenever* $F(\tau) \in \mathcal{S}$, *we have* $11^{-1}UF(\tau) \in \mathcal{S}$.

We now let $m = 11^\alpha m'$ when $(11, m') = 1$ and want to prove the congruence $a_{11}(m, 11^{\alpha+\gamma}n) \equiv 0 \pmod{11^\gamma}$ for any positive integer γ by using induction on α. For brevity, we let $f_m := f_{11,m}$. Consider

$$H_m(\tau) := 11(Uf_m)(\tau) - 11f_{m/11}(\tau). \tag{4.3}$$

Here, $f_{m/11}(\tau) = 0$ when $11 \nmid m$. By Lemma 4.2(2), we have

$$\begin{aligned} H_m\left(-\frac{1}{11\tau}\right) &= 11(Uf_m)\left(-\frac{1}{11\tau}\right) - 11f_{m/11}\left(-\frac{1}{11\tau}\right) \\ &= 11(Uf_m)(11\tau) + f_m\left(-\frac{1}{121\tau}\right) \\ &\quad - f_m(\tau) - 11f_{m/11}\left(-\frac{1}{11\tau}\right) \\ &= 11q^{-m} + O(1) + O(1) - q^{-m} - O(q^{-1}) - O(1) \\ &= 10q^{-m} + O(q^{-1}). \end{aligned} \tag{4.4}$$

Apparently, when $\alpha = 0$, $H_m(\tau)$ is holomorphic at $\tau = i\infty$ and has a pole of order m at $\tau = 0$. Hence, it follows from (4.1) that

$$Uf_m(\tau) = \lambda_0 + \sum_{r=2}^{m} \lambda_r 11^{\theta(r)-1} g_r(\tau)$$

for some constants λ_0 and λ_r $(2 \leq r \leq m)$. Thus, every coefficient of q^{11n} in $f_m(\tau)$ is a multiple of 11. Moreover, by Lemma 4.3, after applying U-operator γ times, we find that

$$a_{11}(m, 11^\gamma n) \equiv 0 \pmod{11^\gamma}$$

for each positive integer γ.

Nextly, if $\alpha = 1$ and $m' = 1$, i.e. $m = 11$, $H_{11}(\tau)$ has a simple pole at $\tau = i\infty$ and a pole of order 11 at $\tau = 0$ so that we obtain from (4.1) that

$$Uf_{11}(\tau) = \lambda_{-1}B(\tau) + \lambda_0 + \sum_{r=2}^{11} \lambda_r 11^{\theta(r)-1} g_r(\tau) \tag{4.5}$$

for some constants λ_r $(-1 \leq r \leq 11)$. As $UB(\tau) + 5 \in \mathcal{S}$ by [3, p. 22], every coefficient of q^{11n} in $B(\tau)$ is a multiple of 11. We hence deduce from (4.5) that every coefficient of q^{121n} in $f_{11}(\tau)$ is a multiple of 11. Again using Lemma 4.3 repeatedly, we find that

$$a_{11}(11, 11^{\gamma+1} n) \equiv 0 \pmod{11^\gamma}.$$

Now, assume the congruence holds for all m of the form $m = 11^a m'$ with $a < \alpha$. Then it remains to prove that it holds for $m = 11^\alpha m'$ as well. In this general case, $H_m(\tau)$ has a pole of order $m/11$ at $\tau = i\infty$ and a pole of order m at $\tau = 0$. Therefore, by (4.1)

$$Uf_m(\tau) = \sum_{r=2}^{m/11} \lambda_{-r} G_r(\tau) + \lambda_{-1}B(\tau) + \lambda_0 + f_{m/11}(\tau) + \sum_{r=2}^{m} \lambda_r 11^{\theta(r)-1} g_r(\tau)$$

for some constants λ_r $(-m/11 \leq r \leq m)$. By induction hypothesis and the arguments so far, we see that every coefficient of $q^{11^{\alpha+\gamma-1}n}$ in $Uf_m(\tau)$ is a multiple of 11^γ, which implies the desired result

$$a_{11}(m, 11^{\alpha+\gamma}n) \equiv 0 \pmod{11^\gamma}.$$

Replacing γ by $\beta - \alpha$, we have (1.13).

Acknowledgments

The author is grateful to the referee for valuable comments that helped to improve the paper. The author was supported by the National Research Foundation of Korea (NRF) Grant (NRF-2016R1D1A1B01012258).

References

[1] S. Ahlgren and K. Ono, Congruence properties for the partition function, *Proc. Natl. Acad. Sci. USA* **98**(23) (2001) 12882–12884.

[2] N. Anderson and P. Jenkins, Divisibility properties of coefficients of level p modular functions for genus zero primes, *Proc. Amer. Math. Soc.* **141**(1) (2012) 41–53.

[3] A. O. L. Atkin, Proof of a conjecture of Ramanujan, *Glasg. Math. J.* **8** (1967) 14–32.

[4] A. O. L. Atkin, Multiplicative congruence properties and density problems for $p(n)$, *Proc. London Math. Soc. (3)* **18** (1968) 563–576.

[5] B. C. Berndt, Commentary on Ramanujan's collected papers, in *Collected Papers of Srinivasa Ramanujan*, eds. G. H. Hardy, P. V. Seshu Aiyar and B. M. Wilson (AMS Chelsea Publishing, 2000), pp. 357–426.

[6] B. C. Berndt, Ramanujan's congruences for the partition function modulo 5, 7, and 11, *Int. J. Number Theory* **3** (2007) 349–354.

[7] B. C. Berndt and K. Ono, Ramanujan's unpublished manuscript on the partition and tau functions with proofs and commentary, in *The Andrews Festschrift* (*Maratea*, 1998), eds. D. Foata and G.-N. Han (Springer, 2001), pp. 39–110.

[8] K. Bringmann, B. Kane, S. Löbrich, K. Ono and L. Rolen, On divisors of modular forms, *Adv. Math.* **329** (2018) 541–554.

[9] M. Griffin, Divisibility properties of coefficients of weight 0 weakly holomorphic modular forms, *Int. J. Number Theory* **7**(4) (2011) 933–941.

[10] V. Iba, P. Jenkins and M. Warnick, Divisibility properties of coefficients of modular functions in genus zero levels, *Integers* **19** (2019) Paper No. A7, 17 pp.

[11] P. Jenkins and D. J. Thornton, Congruences for coefficients of modular functions, *Ramanujan J.* **38**(3) (2015) 619–628.

[12] P. Jenkins and D. J. Thornton, Weakly holomorphic modular forms in prime power levels of genus zero, *Integers* **18** (2018) Paper No. A18, 12 pp.

[13] D. Jeon, S.-Y. Kang and C. H. Kim, Cycle integrals of a sesqui-harmonic Maass form of weight zero, *J. Number Theory* **141** (2014) 92–108.

[14] D. Jeon, S.-Y. Kang and C. H. Kim, Hecke system of harmonic Maass functions and its applications to modular curves of higher genera (submitted for publication).

[15] J. Lehner, Ramanujan identities involving the partition function for the moduli 11^α, *Amer. J. Math.* **65**(3) (1943) 492–520.

[16] J. Lehner, Divisibility properties of the Fourier coefficients of the modular invariant $j(\tau)$, *Amer. J. Math.* **71**(1) (1949) 136–148.

[17] J. Lehner, Further congruence properties of the Fourier coefficients of the modular invariant $j(\tau)$, *Amer. J. Math.* **71**(2) (1949) 373–386.

[18] J. Lehner, Proof of Ramanujan's partition congruence for the modulus 11^3, *Proc. Amer. Math. Soc.* **1** (1950) 172–181.

[19] D. Niebur, A class of nonanalytic automorphic functions, *Nagoya Math. J.* **52** (1973) 133–145.

[20] K. Ono, *The Web of Modularity: Arithmetic of the Coefficients of Modular Forms and q-Series*, CBMS Regional Conference Series in Mathematics, Vol. 102 (American Mathematical Society, Providence, RI, 2004).

[21] K. Ono, Unearthing the visions of a master: Harmonic Maass forms and number theory, in *Proc. 2008 Harvard-MIT Current Developments in Mathematics* (International Press, 2009), pp. 347–454.

[22] P. Paul and C.-S. Radu, A unified algorithm framework for Ramanujan's congruences modulo powers of 5, 7 and 11 (preprint), Available at http://www3.risc.jku.at/publications/download/risc_5760/PP_submission.pdf.

[23] S. Ramanujan, On certain arithmetical functions, *Trans. Cambridge Philos. Soc.* **22** (1916) 159–184.

[24] S. Ramanujan, Some properties of $p(n)$, the number of partitions of n, *Proc. Cambridge Philos. Soc.* **19** (1919) 207–210; *Collected Papers of Srinivasa Ramanujan*, No. 210–213 (AMS Chelsea Publishing, Providence, RI, 2000).

[25] S. Ramanujan, Congruence properties of partitions, *Proc. London Math. Soc.* **18** (1920) xix; *Collected Papers of Srinivasa Ramanujan*, No. 230 (AMS Chelsea Publishing, Providence, RI, 2000).

[26] S. Ramanujan, Congruence properties of partitions, *Math. Z.* **9**(1–2) (1921) 147–153; *Collected Papers of Srinivasa Ramanujan*, No. 232–238 (AMS Chelsea Publishing, Providence, RI, 2000).

[27] G. N. Watson, Ramanujan's Vermutung über Zerfällungsanzahlen, *J. Reine Angew. Math.* **179** (1938) 97–128.

[28] Y. Yang, Defining equations of modular curves, *Adv. Math.* **204** (2006) 481–508.

© 2024 World Scientific Publishing Company
https://doi.org/10.1142/9789811277375_0024

Restricted k-color partitions, II*

William J. Keith

Department of Mathematical Sciences
Michigan Technological University
1400 Townsend Drive Houghton
MI 49931, USA
wjkeith@mtu.edu

We consider (k, j)-colored partitions, partitions in which k colors exist but at most j colors may be chosen per size of part. In particular these generalize overpartitions. Advancing previous work, we find new congruences, including in previously unexplored cases where k and j are not coprime, as well as some noncongruences. As a useful aside, we give the apparently new generating function for the number of partitions in the $N \times M$ box with a given number of part sizes, and extend to multiple colors a conjecture of Dousse and Kim on unimodality in overpartitions.

Keywords: Partitions; colored partitions; overpartitions; congruences.

Mathematics Subject Classification 2020: 05A17, 11P83

1. Introduction

Two classes of partition-theoretic objects that have excited much research are overpartitions, defined in 2004 in [4] and since studied in [5, 6, 11] among many other papers; and k-colored partitions or multipartitions, which have a longer history, being of interest to see Ramanujan [3] and many authors since: see [2] for a good survey. Precise definitions for these are given below.

In this paper, second in a series beginning in [7], we further study a unification and generalization of these two classes, (k, j)-colored partitions. Think of overpartitions as 2-colored partitions in which only one color may be used for each part size: the (k, j)-colored partitions are the natural generalization in which, of k colors, only j may be used for a given part size. The usual k-colored partitions are (k, k)-colored partitions, and overpartitions are $(2, 1)$-colored partitions: consider the two colors to be "overlined" and "not overlined", and regard each color as applying to all of a part size when the last element is overlined or not.

Overpartitions exhibit many congruences and symmetries. Many instances of the (k, j)-colored partitions with particular values of k and j were shown in [7] to

*This article originally appeared in *International Journal of Number Theory*, **17**(3), pp. 591–601 © 2021 World Scientific Publishing Company. DOI: 10.1142/S1793042120400151

possess such features. Most of the pairs (k, j) studied in that paper were coprime, since those were immediately accessible to more elementary machinery. Emphasis was given to $j = 1$ and $j = k - 1$. Here, we consider some non-coprime cases, which, though more intricate, exhibit new internal symmetries and connections to other functions such as divisor sums. As with the previous paper, the congruences proved herein are certainly samples of a wide and rich field of similar results.

In addition to the above results we consider $(k, 1)$-colored partitions in the $M \times N$ box, and generalize a conjecture of Dousse and Kim on unimodality of overpartitions. In order to do so we require a theorem, apparently new, giving the generating function for the number of partitions in the $M \times N$ box in which exactly r different part sizes appear. This completes a trio of theorems on partitions with specified numbers of part sizes, with the unrestricted case by MacMahon [8] and elaborated by Andrews [1], and the singly restricted case in which sizes must be less than N given just in 2016 by Merca in [9].

The main theorems of this paper are as follows.

Denote by $c_{k,j}(n)$ the number of (k, j)-colored partitions of n.

Theorem 1.1. *For* $j \in \{2, 5, 8, 9\}$, *we have* $c_{9,j}(3n+2) \equiv 0 \pmod{27}$ *for all* $n \geq 0$. *For* $j \in \{3, 6\}$, *we have* $c_{9,j}(9n + 2) \equiv 0 \pmod{27}$ *for all* $n \geq 0$. *For* $j \in \{1, 4, 7\}$, *we have that if* $c_{9,j}(3^{\ell}n + 2) \equiv 0 \pmod{27}$ *for all* $n \geq 0$, *then* ℓ *is large, and does not exist for* $j = 1$.

Denote by $\overline{p}(n)$ the overpartition function. Then we have the following theorem.

Theorem 1.2.

$$c_{2p,p}(n) \equiv \overline{p}\left(\frac{n}{p}\right) + p \sum_{\lambda \in S} 2^{r(\lambda)}(t(\lambda) + 1)y(\lambda)^{-1} \pmod{p^2},$$

where S *is a certain subset of partitions of* n, $t(\lambda)$, $y(\lambda)$ *are statistics on these partitions, and inversion is taken mod* p.

Given $\vec{t} = (t_1, \ldots, t_k)$, let $p_{M\times N}(\vec{t}; n)$ denote the number of $(k, 1)$-colored partitions of n, i.e. k-colored partitions where only one color is chosen per size of part, with at most N parts, all of size at most M, in which t_i part sizes are of color i for all i. Then we have the following theorem.

Theorem 1.3. *Fix* M *and* N *dimensions of a box and fix a number of colors* k. *We have the generating function*

$$\sum_{n,\vec{t}} p_{M\times N}(\vec{t}; n)x_1^{t_1} \ldots x_k^{t_k} q^n = \sum_{r=0}^{\min(M,N)} (x_1 + \cdots + x_k)^r f_{(M\times N;r)}(q),$$

where $\vec{t}$ *runs over nonnegative integer vectors in* ${\mathbb{Z}_{\geq 0}}^k$ *and* $f_{(M\times N;r)}(q)$ *is the generating function for partitions in the* $M \times N$ *box with exactly* r *different part sizes.*

We record this latter generating function as being of significant independent interest.

Theorem 1.4.

$$f_{(M\times N);r}(q) = \sum_{r_0=r}^{\min(M,N)} \binom{r_0}{r}(-1)^{r_0-r}q^{\binom{r_0+1}{2}}\begin{bmatrix} M \\ r_0 \end{bmatrix}_q \begin{bmatrix} M+N-r_0 \\ N-r_0 \end{bmatrix}_q.$$

Finally, we propose a generalization of a unimodality conjecture of Dousse and Kim on overpartitions in the box.

Conjecture 1.5. *Fix n. Set any one $x_i = 1$ and consider the remaining $(n-1)$-dimensional array of coefficients in the x_j on q^n in $\sum_{n,\vec{t}} p_{M\times N}(t_1,\dots,t_k;n)x_1^{t_1}\dots x_k^{t_k}q^n$. Any linear cut through this array is a unimodal sequence.*

This generalizes the conjecture of Dousse and Kim on overpartitions, which is the $k=2$ case in which x_2 is set to 1 and the remaining one-dimensional sequence of coefficients is conjectured to be unimodal.

2. Background and Definitions

The most common notations we will use for partitions are either the sum notation, $\lambda = \lambda_1 + \lambda_2 + \cdots \vdash N$, with the λ_i the parts of the partition; or the frequency notation, $\lambda = \lambda_1^{e_1}\lambda_2^{e_2}\dots$, meaning that part size λ_1 appears e_1 times, etc. Context should always make it clear which notation is being used, as we never talk about exponents for parts.

Overpartitions are partitions in which the last instance of a given part may be overlined, or not. The overpartitions of 3 are

$$3, \quad \overline{3}, \quad 2+1, \quad \overline{2}+1, \quad 2+\overline{1}, \quad \overline{2}+\overline{1}, \quad 1+1+1, \quad 1+1+\overline{1}.$$

The generating function for $\overline{p}(n)$, the number of overpartitions of n, is

$$\overline{P}(q) := \sum_{n=0}^{\infty}\overline{p}(n)q^n = \prod_{k=1}^{\infty}\frac{1+q^k}{1-q^k} = \prod_{k=1}^{\infty}\frac{1-q^{2k}}{(1-q^k)^2}.$$

The k-colored partitions, also known as multipartitions, are partitions in which parts may appear in k different types known as colors, with the order of colors not mattering (i.e. by convention we list colors within a part in weakly descending order). For instance, denoting colors by subscripts, the 2-colored partitions of 3 are

$$3_2, \quad 3_1, \quad 2_2+1_2, \quad 2_2+1_1, \quad 2_1+1_2, \quad 2_1+1_1, \quad 1_2+1_2+1_2, \quad 1_2+1_2+1_1,$$
$$1_2+1_1+1_1, \quad 1_1+1_1+1_1.$$

The generating function of $c_k(n)$, the number of k-colored partitions of n, is

$$C_k(q) := \sum_{n=0}^{\infty} c_k(n)q^n = \prod_{n=1}^{\infty}\frac{1}{(1-q^n)^k}. \tag{2.1}$$

In the language of colored partitions, overpartitions are 2-colored partitions with the restriction that only one color may be used per size of part. This leads to the natural generalization in which, of k colors, only j may be used per size of part, which we call (k,j)-colored partitions.

For our final example, the set of $(3,1)$-colored partitions of 3 is

$$3_3, 3_2, 3_1, 2_3+1_3, 2_3+1_2, 2_3+1_1, 2_2+1_3, 2_2+1_2, 2_2+1_1,$$

$$2_1+1_3, 2_1+1_2, 2_1+1_1, 1_3+1_3+1_3, 1_2+1_2+1_2, 1_1+1_1+1_1.$$

So for instance, among 3-colored partitions of 3, the only partitions not in the above set are those such as $1_3+1_2+1_2$ in which several parts of size 1 have different colors. The $(3,2)$-colored partitions of 3 include all 3-colored partitions of 3 except $1_3+1_2+1_1$.

Call $c_{k,j}(n)$ the number of (k,j)-colored partitions of n. Their generating function, as derived in [7], is

$$C_{k,j}(q) := \sum_{n=0}^{\infty} c_{k,j}(n)q^n = \frac{1}{(q)_\infty^j} \prod_{n=1}^{\infty} \left(\sum_{i=0}^{j} \binom{k-j+i-1}{i} q^{in} \right). \tag{2.2}$$

Here, we have used the standard notation

$$(a;q)_n = (1-a)(1-aq)\cdots(1-aq^{n-1}), \quad (a;q)_\infty = \lim_{n\to\infty}(a;q)_n,$$

$$(a)_n = (a;q)_n, \quad (q)_\infty := (q;q)_\infty.$$

Partitions in which there are at most N parts, each of size at most M, are said to fit in the $M \times N$ box, and have the q-binomial coefficients as their generating function:

$$\begin{bmatrix} M+N \\ N \end{bmatrix}_q = \frac{(q)_{M+N}}{(q)_M (q)_N}.$$

MacMahon [8], and later Andrews [1], studied the generating function for otherwise unrestricted partitions with exactly r part sizes:

$$N_r(q) = \frac{1}{(q;q)_\infty} \sum_{n=0}^{\infty} (-1)^{n-r} \frac{\binom{n}{r} q^{\binom{n+1}{2}}}{(q;q)_n}.$$

Just in 2016, Merca [9] produced the generating function for partitions with r part sizes restricted by largest part size:

$$N_{r,\le M}(q) = \frac{1}{(q;q)_M} \sum_{j=r}^{M} (-1)^{j-r} q^{\binom{j+1}{2}} \binom{j}{r} \begin{bmatrix} M \\ j \end{bmatrix}_q.$$

Our Theorem 1.4 doubly restricts the partitions and is thus the completing third member in this line. It is easily seen that Theorem 1.4 becomes Merca's identity in the limit $N \to \infty$.

Finally, when for two power series $f(q) = \sum_{n=0}^{\infty} a(n)q^n$ and $g(q) = \sum_{n=0}^{\infty} b(n)q^n$ we write $f(q) \equiv_m g(q)$, we mean $a(n) \equiv b(n) \pmod{m}$ for all n.

3. Proofs of the Main Results

3.1. *Proof of Theorem* 1.1

Begin by noticing that, given a partition λ, there are a large number of (k, j)-colored partitions which assign different colors to part sizes in λ. Many of these can be grouped and ignored when proving congruences such as this theorem. For instance, among all assignments of colors to a partition in which the part size 100 appears five times, those in which this part size is assigned two colors, the first coloring three parts and the next coloring 2, can be grouped in the $\binom{9}{2}$ ways to choose these colors.

Recall that for $i \not\equiv 0 \pmod 3$, $\binom{9}{i} \equiv 0 \pmod 9$, while $\binom{9}{3} \equiv \binom{9}{6} \equiv 3 \pmod 9$. Hence, for $j < 9$, $(9, j)$-colored partitions with three or more distinct sizes of part contribute groups of size divisible by 27 to the total, and so we need only consider partitions in which one or two part sizes appear.

For $j = 2$, we are considering 9-colored partitions in which only one or two colors may be used per size of part.

If a part size has one color, there are nine options; if a part size has two colors, there are $\binom{9}{2} = 36 \equiv 0 \pmod 9$ options that assign the same number of parts to the first and second. Ergo, if a partition λ has two or more part sizes, then the associated $(9, 2)$-colored partitions number 0 mod 81.

This reduces the problem to observing $\lambda = \lambda_1^{e_1} \vdash N$, partitions into just one part size, which contribute

$$\sum_{e_1 | N} \binom{9}{1}\binom{e_1 - 1}{0} + \binom{9}{2}\binom{e_1 - 1}{1} \equiv 9\sigma_1(N) \pmod{27}.$$

Here, $\sigma_1(n) = \sum_{d|n} d$, the usual sum-of-divisors function.

But for $N \equiv 2 \pmod 3$, N has at least one prime factor $p_i \equiv 2 \pmod 3$ appearing to an odd power, so $\sigma_1(N) \equiv 0 \pmod 3$, and so $9\sigma_1(N) \equiv 0 \pmod{27}$.

For $j = 5$, partitions with two distinct part sizes contribute a nonzero amount mod 27 only if both part sizes include exactly 3 of the 9 possible colors. Let $\lambda = \lambda_1^{e_1}\lambda_2^{e_2}$. The number of $(9, 5)$-colored partitions of this type contributed is

$$\binom{9}{3}\binom{9}{3}\binom{e_1 - 1}{2}\binom{e_2 - 1}{2}.$$

But $\binom{k}{2} \equiv 0 \pmod 3$ unless $k \equiv 2 \pmod 3$. If $e_1 - 1 \equiv e_2 - 1 \equiv 2 \pmod 3$, then $e_1 \equiv e_2 \equiv 0 \pmod 3$, which is impossible since $N \not\equiv 0 \pmod 3$. Hence, partitions of this type contribute a number divisible by 27 to the total.

It remains to consider partitions into exactly one part size, which contribute a total of

$$\sum_{e_1 | N} \sum_{i=1}^{5} \binom{9}{i}\binom{e_1 - 1}{i - 1} = \sum_{e_1 | N} \frac{3}{4}\left(76 - 162e_1 + 133{e_1}^2 - 42{e_1}^3 + 7{e_1}^4\right).$$

All divisors e_1 of $N \equiv 2 \pmod 3$ are either 1 or 2 mod 3, and furthermore these come in pairs. That is, if $e_1 \equiv 1 \pmod 3$, the complementary divisor $N/e_1 \equiv 2$

(mod 3). It is quickly seen that if $e_1 \equiv 1 \pmod 3$, then the sum above is congruent to 9 mod 27, while if $e_1 \equiv 2 \pmod 3$, the sum is 18 mod 27. Hence, the contributions of complementary divisors sum to 0 mod 27.

These exhaust all cases for $j = 5$ and so the theorem is proved for this case.

The $j = 8$ case is established with a similar argument; $\binom{9}{6} = \binom{9}{3}$, and the binomials $\binom{e_i - 1}{5}$ behave the same way as $\binom{e_i - 1}{2}$ mod 3. The polynomial $\sum_{i=1}^{8} \binom{9}{i}\binom{e_1 - 1}{i-1}$ has the same property for complementary divisors as discussed earlier.

Interestingly, the case $j = 9$ can be proved from the case $j = 8$: it was shown in [7] that the generating functions $C_{k,k-1}$ are η-quotients, and in fact we have

$$C_{9,8}(q) = \frac{(q^9; q^9)_\infty}{(q;q)_\infty^9}, \quad C_{9,9}(q) = \frac{1}{(q;q)_\infty^9}.$$

This means that the coefficients $c_{9,8}(n)$ and $c_{9,9}(n)$ can be related by the (9-magnified) pentagonal number recurrence:

$$c_{9,8}(n) = c_{9,9}(n) - c_{9,9}(n-9) - c_{9,9}(n-18) + c_{9,9}(n-45) + c_{9,9}(n-63) - \cdots.$$

If $n \equiv 2 \pmod 3$, then $c_{9,8}(n) \equiv 0 \pmod{27}$. Now solve for $c_{9,9}(n)$ and induct on n; since $n - 9k \equiv 2 \pmod 3$ as well, all terms in the sum are 0 mod 27 and the theorem follows. This is perhaps a relatively rare case of a combinatorial proof of a congruence for a multipartition function above the number of colors itself.

For $j = 3$ and $j = 6$, we refer to the generating function.

$$C_{9,3}(q) \equiv_{27} \prod_{n=1}^{\infty} \frac{1 + 6q^n - 6q^{2n} + 2q^{3n}}{(1-q^n)^3} = \prod_{n=1}^{\infty} \left(-2 + \frac{3}{(1-q^n)^3}\right)$$

$$= \prod_{n=1}^{\infty} \left(1 + \sum_{k=1}^{\infty} 3\binom{k+2}{2} q^{kn}\right).$$

Here, we use the classical fact that $\frac{1}{(1-x)^j} = \sum_{k=0}^{\infty} \binom{k+j-1}{j-1} x^k$.

Portions of the argument are similar. Again, we only need to look at parts of one or two part sizes.

Modulo 27, the contribution of $3\binom{k+2}{2}$ to the coefficient on q^n is the number of partitions of n satisfying the following conditions, multiplied by associated factors:

- exactly two part sizes both repeated a multiple of three times, multiplied by 9;
- exactly one part size repeated 1 or 5 mod nine times, multiplied by 9;
- exactly one part size repeated 2 or 4 mod nine times, multiplied by 18; and
- exactly one part size repeated a multiple of three times, multiplied by 3.

But for $n \equiv 2 \pmod 9$, the first and fourth are empty sets, and the other two divisors pair up as in the previous argument, and thus the congruence holds.

For $j = 6$, we have

$$C_{9,6}(q) = \prod_{n=1}^{\infty} \frac{1 + 3q^n + 6q^{2n} + 10q^{3n} + 15q^{4n} + 21q^{5n} + 28q^{6n}}{(1-q^n)^6}$$

$$\equiv_{27} \prod_{n=1}^{\infty} \left(1 + \frac{9q^n - 9q^{2n} + 30q^{3n}}{(1-q^n)^6}\right) \equiv_{27} \prod_{n=1}^{\infty} \left(1 + \frac{9q^n}{(1-q^n)^5} + \frac{3q^{3n}}{(1-q^n)^6}\right)$$

$$\equiv_{27} \prod \left(1 + 9\sum_{k=1}^{\infty} \binom{k+3}{4} q^{kn} + 3\sum_{k=1}^{\infty} \binom{k+2}{5} q^{kn}\right).$$

We have used the same identity as before, with the index suitably shifted by the numerators of each term.

Any partition counted with the $9\binom{k+3}{4}$ term contributing can only have one part size. The binomial $\binom{k+3}{4}$ is 1 mod 3 if k is 2 or 8 mod 9, and 2 mod 3 if k is 1 or 7 mod 9; these divisors pair up as before. For those with $3\binom{k+2}{5}$ contributing, see the previous discussion once again.

Finally, for $j \in \{1, 4, 7\}$, calculation of early cases suffices to eliminate many small ℓ; for the case $j = 1$ specifically, it suffices to note that $c_{9,1}(2) = 18 \not\equiv 0 \pmod{27}$, and hence no arithmetic progression starting with 2 will be 0 mod 27. This concludes the proof of the theorem.

Remark. Smoot, in a preprint at [10], has implemented an algorithm of Cristian-Silviu Radu for the verification of certain η-quotient identities, which effectively can take an arithmetic progression such as $3n + 2$, a proposed modulus, and a generating function such as those for $j = 8$ and $j = 9$, and return a confirmation of whether the proposed congruence exists. Upon presentation of this theorem at the Berndt Conference, he was able to verify the $j = 8$ congruence with a few minutes of calculation. Although, it is not applicable to the other congruences listed here, the algorithm certainly bodes likely to be useful in the field more generally.

3.2. *Proof of Theorem* 1.2

Among the families of $C_{k,j}$ that generalize overpartitions, two that were studied in [7] were $C_{k,1}$ and $C_{k,k-1}$, both of which turn out to be η-quotients. Another family that we might consider is $C_{2j,j}$. For ease of calculation, we restrict ourselves to the case where j is a prime p.

Expand $C_{2p,p}$ in the basic generating function, Eq. (2.2). It is immediate that, when p is prime, we get

$$C_{2p,p}(q) \equiv_p \prod_{n=1}^{\infty} \frac{1 + q^{pn}}{1 - q^{pn}},$$

which is just the overpartition function, magnified by p. This easily gives us the relations

$$p \mid n \Rightarrow c_{2p,p}(n) \equiv \overline{p}(n/p) \pmod{p} \quad \text{otherwise } c_{2p,p}(n) \equiv 0 \pmod{p}.$$

But we can say a bit more mod p^2. Consider $C_{2p,p}$ and expand each binomial coefficient as integers mod p^2. For instance, in the denominator we expand $(1-q^n)^p$, from which we get terms such as $\binom{p}{2} = \frac{p(p-1)}{2} \equiv -p \cdot 2^{-1} \pmod{p^2}$, and $-\binom{p}{3} = -\frac{p(p-1)(p-2)}{3!} \equiv -p \cdot 3^{-1} \pmod{p^2}$. From the numerator, each $\binom{p+i-1}{i} \equiv p \cdot i^{-1} \pmod{p^2}$. Thus, the generating function reduces to

$$C_{2p,p} \equiv_{p^2} \prod_{n=1}^{\infty} \frac{1 + pq^n + 2^{-1}pq^{2t} + \cdots + (p-1)^{-1}pq^{(p-1)n} + q^{pn}}{1 - pq^n - 2^{-1}pq^{2t} - \cdots - (p-1)^{-1}pq^{(p-1)n} - q^{pn}},$$

where inverses are taken mod p^2.

Expand each factor. For illustration, let

$$X = pq^n + 2^{-1}pq^{2t} + \cdots + (p-1)^{-1}pq^{(p-1)n} + q^{pn}.$$

Then we are expanding $\frac{1+X}{1-X} = 1 + 2X + 2X^2 + \cdots$, which is precisely the form of each factor in the overpartition generating function, Eq. (2.1). However, in X^k, almost all terms are 0 mod p^2 unless they involve only the terms q^{pn}, or exactly one factor chosen from among the other terms.

Thus mod p^2, each factor expands as

$$1 + \sum_{t=1}^{\infty} 2q^{ptn} + \sum_{t=0}^{\infty} (2t+2)p(1^{-1}q^{(tp+1)n} + \cdots + (p-1)^{-1}q^{(tp+(p-1))n}).$$

Here, each exponent $tp+y$ is to be combinatorially interpreted as the number of times n appears in the partition. This means that every part size contributes a factor of 2 to a weighting, and $2^{\#\{\text{part sizes of } \lambda\}}$ is exactly the number of overpartitions of shape λ. Thus, the first term tells us that partitions in which all part sizes are repeated a number of times divisible by p, which (ignoring colors) are in bijection with the partitions of n/p by dividing the multiplicities, contribute a total of $\overline{p}(n/p)$ to the coefficient $c_{2p,p}(n)$ mod p^2.

If a part is repeated $tp+y$ times, $y \not\equiv 0 \pmod{p}$, such parts give an additional factor of $p(t+1)y^{-1}$ to the weight of the partition in the count. Hence, only partitions which have at most one such part can contribute a nonzero amount mod p^2.

Let S be the set of λ in which at most one part size is repeated a number of times not divisible by p, and let $t(\lambda)$ and $y(\lambda)$ be the associated values for such partitions. Say $r(\lambda)$ is the number of part sizes of λ. Taking the convention that the number of partitions, or overpartitions, of a noninteger amount is zero, we can refine the earlier result thus

$$c_{2p,p}(n) \equiv_{p^2} \overline{p}\left(\frac{n}{p}\right) + p \sum_{\lambda \in S} 2^{r(\lambda)}(t(\lambda)+1)y(\lambda)^{-1}.$$

The theorem is proved.

Remark. By isolating the part size repeated a nonzero number of times modulo p, one could probably further refine this theorem with some additional information to get a more elegant formula.

3.3. *Proof of Theorems* 1.3 *and* 1.4

Each r term on the right-hand side of Theorem 1.3 is counting the partitions in which exactly r part sizes appear; the multinomial term $(x_1 + \cdots + x_k)^r$ distributes the colors among the part sizes in the usual way. The meat of the theorem, then, is in Theorem 1.4.

To prove the latter, what we need is the enumeration of overpartitions in the $M \times N$ box given by Dousse and Kim in [5]: letting $s(\overline{\lambda})$ be the number of overlined parts of an overpartition $\overline{\lambda}$, they give

$$\sum_{\overline{\lambda} \in M \times N} t^{s(\overline{\lambda})} q^{|\overline{\lambda}|} = \sum_{k=0}^{\min(M,N)} t^k q^{\binom{k+1}{2}} \begin{bmatrix} M \\ k \end{bmatrix}_q \begin{bmatrix} M+N-k \\ N-k \end{bmatrix}_q .$$

From our viewpoint, this is just setting $x_1 = t$, $x_2 = 1$ in Theorem 1.3, where this identity becomes

$$\sum_{r=0}^{\min(M,N)} (t+1)^r f_{(M \times N;r)}(q) = \sum_{k=0}^{\min(M,N)} t^k q^{\binom{k+1}{2}} \begin{bmatrix} M \\ k \end{bmatrix}_q \begin{bmatrix} M+N-k \\ N-k \end{bmatrix}_q .$$

Making the substitution $t \to t-1$, we have

$$\sum_{r=0}^{\min(M,N)} t^r f_{(M \times N;r)}(q) = \sum_{k=0}^{\min(M,N)} (t-1)^k q^{\binom{k+1}{2}} \begin{bmatrix} M \\ k \end{bmatrix}_q \begin{bmatrix} M+N-k \\ N-k \end{bmatrix}_q .$$

Extract coefficients on t^r and we have the claim of the theorem. Apply this formula to Theorem 1.3 to get the explicit generating function.

4. Future Directions

The congruences described in Theorem 1.1 are certainly just one example of a large family. The study of congruences in η-quotients, as suggested by the remark concerning the work of Radu, has reached a high peak concerning the verification of any particular desired congruence, and much is known about congruences for overpartitions specifically, including larger infinite families. Since (k, j)-colored partitions were originally proposed as a generalization of overpartitions, further developing the connections between these and the better-known overpartitions could be a useful line of investigation.

Still, most of the functions $C_{k,j}$ are not η-quotients, and so different tools will be necessary to obtain information about their congruences. Where η-quotients have the power of well-developed modular form machinery that can be brought to bear, it seems that more combinatorial thinking may be fruitful in producing more information about (k, j)-colored partitions. This could be a productive field of advancement.

An interesting conjecture arises when considering the coefficients in Theorem 1.3. Dousse and Kim conjecture that the coefficient of q^n in

$$\sum_{k=0}^{\min(M,N)} t^k q^{\binom{k+1}{2}} \begin{bmatrix} M \\ k \end{bmatrix}_q \begin{bmatrix} M+N-k \\ N-k \end{bmatrix}_q$$

is unimodal in t.

This appears to generalize to more colors: if x_k is set to 1, the array of the remaining coefficients appears to be unimodal in any linear cut.

For example, consider the 11×7 box with three colors, and look at the polynomial in x_1, x_2, x_3 which is the coefficient of q^n, $n = 14$. Set $x_3 = 1$. The coefficients of $x_1^r x_2^c$, indexing row and column in order, are

$$\begin{matrix} 101 & 291 & 300 & 129 & 19 \\ 291 & 600 & 387 & 76 & 0 \\ 300 & 387 & 114 & 0 & 0 \\ 129 & 76 & 0 & 0 & 0 \\ 19 & 0 & 0 & 0 & 0 \end{matrix}\,.$$

The antidiagonals are of course Pascal's triangle multiplied by the entries of the first column, as is immediate from Theorem 1.3. The rows (and columns, by symmetry) and diagonals also appear to be unimodal. Even slant cuts through the array — here, the sequence $(300, 387, 19)$ is the $(2, 1)$ cut starting from position $(3, 1)$ — appear to have this property. Perhaps Dousse and Kim's conjecture could be fruitfully set in this larger context.

References

[1] G. Andrews, Stacked lattice boxes, *Ann. Comb.* **3** (1999) 115–130.

[2] G. Andrews, A survey of multipartitions: Congruences and identities, in *Surveys in Number Theory*, Developments in Mathematics (Diophantine Approximation: Festschrift for Wolfgang Schmidt), Vol. 17 (Springer, New York, NY, 2008), pp. 1–19, doi:10.1007/978-0-387-78510-3_1.

[3] B. C. Berndt, C. Gugg and S. Kim, Ramanujan's elementary method in partition congruences, in *Partitions, q-Series, and Modular Forms*, eds. K. Alladi and F. Garvan, Developments in Mathematics, Vol. 23 (Springer, New York, NY, 2012), pp. 13–21, doi: 10.1007/978-1-4614-0028-8_2.

[4] S. Corteel and J. Lovejoy, Overpartitions, *Trans. Amer. Math. Soc.* **356** (2004) 1623–1635.

[5] J. Dousse and B. Kim, An overpartition analogue of q-binomial coefficients, II: Combinatorial proofs and (q, t)-log concavity, *J. Combin. Theory Ser. A* **158** (2018) 228–253. doi:10.1016/j.jcta.2018.03.011.

[6] M. D. Hirschhorn and J. A. Sellers, An infinite family of overpartition congruences modulo 12, *Integers* **5** (2005) A20.

[7] W. J. Keith, Restricted k-colored partitions. *Ramanujan J.* **40**(1) (2016) 71–92, doi:10.1007/s11139-015-9704-x.

[8] P. A. MacMahon, Divisors of numbers and their continuations in the theory of partitions, *Proc. London Math. Soc.* **1** (1919) 75–113.

[9] M. Merca, A note on the partitions involving parts of k different magnitudes, *J. Number Theory* **162** (2016) 23–34, https://www.sciencedirect.com/science/article/pii/S0022314X15003637.

[10] N. Smoot, An implementation of Radu's Ramanujan–Kolberg algorithm, to appear *Journal of Symbolic Computation.* arXiv preprint: arxiv.org/abs/2004.03949.

[11] E. X. W. Xia and O. X. M. Yao, New Ramanujan-like congruences modulo powers of 2 and 3 for overpartitions, *J. Number Theory* **133** (2013) 1932–1949.

© 2024 World Scientific Publishing Company
https://doi.org/10.1142/9789811277375_0025

On weighted overpartitions related to some q-series in Ramanujan's lost notebook*

Byungchan Kim
School of Liberal Arts
Seoul National University of Science and Technology
232 Gongneung-ro, Nowon-gu, Seoul 01811, Republic of Korea
bkim4@seoultech.ac.kr

Eunmi Kim†
Institute of Mathematical Sciences
Ewha Womans University, 52 Ewhayeodae-gil
Seodaemun-gu, Seoul 03760, Republic of Korea
ekim67@ewha.ac.kr

Jeremy Lovejoy
Centre National de la Recherche Scientifique
Université Denis Diderot — Paris 7
Case 7014, 75205 Paris Cedex 13, France
lovejoy@math.cnrs.fr

Dedicated to Bruce C. Berndt on his 80th birthday

Motivated by certain q-series of Ramanujan, we examine two overpartition difference functions. We give both combinatorial and asymptotic formulas for the differences and show that they are always positive. We also briefly discuss similar differences for some other types of partitions. Our main tools are elementary q-series transformations and Ingham's Tauberian theorem.

Keywords: Partitions; overpartitions; q-series; inequalities; asymptotics.

Mathematics Subject Classification 2020: 11P81, 33D15

*This article originally appeared in *International Journal of Number Theory*, **17**(3), pp. 603–619
© 2021 World Scientific Publishing Company. DOI: 10.1142/S1793042120400345
†Corresponding author.

1. Introduction

Recall the usual q-series notation

$$(a)_n = (a;q)_n = \prod_{k=1}^{n}(1 - aq^{k-1}),$$

valid for $n \in \mathbb{N}_0 \cup \{\infty\}$. Let $g(q)$ be the q-series defined by

$$g(q) = \sum_{n\geq 0} \frac{q^{n(n+1)/2}}{(q)_n^2} = 1 + q + q^2 + 2q^3 + 2q^4 + 3q^5 + 4q^6 + \cdots. \tag{1.1}$$

This series appears in Ramanujan's last letter to Hardy as an example of a function which does *not* have the mock theta property (see [24, Eq. (C), p. 58]). It also appears in Ramanujan's lost notebook, in the identity [4, Entry 1.4.9]

$$\sum_{n\geq 0} \frac{q^{n(n+1)/2}}{(q)_n^2} = \frac{(-q)_\infty}{(q)_\infty} \sum_{n\geq 0} \frac{(-1)^n q^{n(n+1)/2}}{(q^2;q^2)_n}. \tag{1.2}$$

As was typically the case, Ramanujan did not discuss the combinatorial significance of his identity or of the coefficients of $g(q)$. Many years later, Andrews interpreted $g(q)$ as the generating function for the number of gradual stacks with summit of size n [1, Eq. (3.5)]. Bringmann and Mahlburg [12] also studied these stacks, proving an asymptotic formula for the coefficients of $g(q)$. Taking a partition-theoretic approach, Berndt *et al.* examined the combinatorics of the identity (1.2) and gave a bijective proof using partition pairs [8, Theorem 5.2].

Ramanujan also recorded a "signed" version of $g(q)$,

$$\begin{aligned} g^s(q) &= \sum_{n\geq 0} \frac{(-1)^n q^{n(n+1)/2}}{(-q;q)_n^2} \\ &= 1 - q + 2q^2 - 2q^3 + 2q^4 - 4q^5 + 5q^6 - 3q^7 + \cdots. \end{aligned} \tag{1.3}$$

This series appears on the left-hand side of a notoriously difficult identity from the lost notebook [4, Entry (7.4.1)],

$$\sum_{n\geq 0} \frac{(-1)^n q^{n(n+1)/2}}{(-q;q)_n^2} = \sum_{n\geq 0} \frac{(-q)^{n(n+1)/2}}{(-q^2;q^2)_n} - 2\sum_{n\geq 1} \frac{(-1)^n q^{2n^2}}{(-q;q^2)_{2n}}. \tag{1.4}$$

Andrews proved Ramanujan's identity [1] and also observed empirically [2, p. 710] that the coefficients of $g^s(q)$ have "a lengthy sign change pattern which alters fairly infrequently". Specifically, the coefficients alternate in sign except for sporadic pairs of consecutive coefficients which have the same sign. For example, the alternating pattern is respected up to q^{500} except for the pairs (14, 15), (49, 50), (102, 103), (175, 176), (268, 269), and (379, 380).

In this paper, we examine two signed q-series which are also closely related to $g(q)$ but whose sign patterns are much simpler — in fact, the coefficients will all be positive. Instead of the stacks and partition pairs mentioned above, our combinatorial framework is that of overpartitions [13]. Recall that an overpartition is a partition in which the first occurrence of a number may be overlined. For example, the 14 overpartitions of 4 are

$$\begin{gathered}(4),\quad (\overline{4}),\quad (3,1),\quad (\overline{3},1),\quad (3,\overline{1}),\quad (\overline{3},\overline{1}),\quad (2,2),\quad (\overline{2},2),\\ (2,1,1),\quad (\overline{2},1,1),\quad (2,\overline{1},1),\quad (\overline{2},\overline{1},1),\quad (1,1,1,1),\quad (\overline{1},1,1,1).\end{gathered} \tag{1.5}$$

Also recall that overpartitions may be represented by a Frobenius symbol, that is, a two-rowed array

$$\begin{pmatrix} a_1 & a_2 & \cdots & a_m \\ b_1 & b_2 & \cdots & b_m \end{pmatrix},$$

where the top row is a partition into distinct non-negative parts, the bottom row is an overpartition into non-negative parts, and $n = m + \sum a_i + \sum b_i$. With this representation, the 14 overpartitions of 4 are

$$\begin{gathered}\begin{pmatrix}3\\0\end{pmatrix},\ \begin{pmatrix}3\\\overline{0}\end{pmatrix},\ \begin{pmatrix}2\\1\end{pmatrix},\ \begin{pmatrix}2\\\overline{1}\end{pmatrix},\ \begin{pmatrix}1\\2\end{pmatrix},\ \begin{pmatrix}1\\\overline{2}\end{pmatrix},\ \begin{pmatrix}0\\3\end{pmatrix},\ \begin{pmatrix}0\\\overline{3}\end{pmatrix},\\ \begin{pmatrix}2&0\\0&0\end{pmatrix},\ \begin{pmatrix}2&0\\\overline{0}&0\end{pmatrix},\ \begin{pmatrix}1&0\\1&0\end{pmatrix},\ \begin{pmatrix}1&0\\\overline{1}&0\end{pmatrix},\ \begin{pmatrix}1&0\\1&\overline{0}\end{pmatrix},\ \begin{pmatrix}1&0\\\overline{1}&\overline{0}\end{pmatrix}.\end{gathered} \tag{1.6}$$

For more on the correspondence between overpartitions and their Frobenius representations, see [13].

It is readily seen that the q-series in (1.1) and (1.3) are generating functions for overpartitions where there are no non-overlined parts larger than the number of overlined parts. It is then natural to consider $\overline{p}(m,n)$, the number of overpartitions of n having m non-overlined parts larger than the number of overlined parts, and as a companion, $\overline{p}^u(m,n)$, the number of overpartitions of n having m overlined parts larger than the number of non-overlined parts. Using elementary combinatorial arguments we have the two-variable generating functions

$$F(z,q) := \sum_{n,m\geq 0} \overline{p}(m,n) z^m q^n = \sum_{n\geq 0} \frac{q^{n(n+1)/2}}{(q)_n^2 (zq^{n+1})_\infty}, \tag{1.7}$$

$$F^u(z,q) := \sum_{n,m\geq 0} \overline{p}^u(m,n) z^m q^n = \sum_{n\geq 0} \frac{q^n (-q)_n (-zq^{n+1})_\infty}{(q)_n}. \tag{1.8}$$

Note that $F(0,q) = g(q)$ is Ramanujan's q-series.

Our first pair of results concern the specializations

$$\begin{aligned}F(-1,q) &= \sum_{n\geq 0} \frac{q^{n(n+1)/2}}{(q)_n^2(-q^{n+1})_\infty}\\ &= 1 + 2q^2 + 2q^3 + 4q^4 + 4q^5 + 8q^6 + 10q^7 + 16q^8 + 20q^9 + 30q^{10} + \cdots\end{aligned}$$

and

$$F^u(-1,q) = \sum_{n\geq 0} \frac{q^n(-q)_n(q^{n+1})_\infty}{(q)_n}$$

$$= 1 + 2q^2 + 4q^3 + 6q^4 + 10q^5 + 16q^6 + 26q^7 + 40q^8 + 62q^9 + 92q^{10} + \cdots.$$

Observe that $F(-1,q)$ is the generating function for $\overline{p}_e(n) - \overline{p}_o(n)$, where $\overline{p}_e(n)$ (respectively, $\overline{p}_o(n)$) is the number of overpartitions of n such that there are an even (respectively, odd) number of non-overlined parts larger than the number of overlined parts, while $F^u(-1,q)$ is the generating function for $\overline{p}^u_e(n) - \overline{p}^u_o(n)$, where $\overline{p}^u_e(n)$ (respectively, $\overline{p}^u_o(n)$) is the number of overpartitions of n having an even (respectively, odd) number of overlined parts larger than the number of non-overlined parts.

Despite the signs in the generating functions, these overpartition differences both turn out to be positive. We exhibit this positivity via two combinatorial identities involving the Frobenius symbol of an overpartition. For some other identities involving these Frobenius symbols, see [14, 18, 19].

Theorem 1.1. *Let $a(n)$ denote the number of overpartitions of n whose Frobenius symbols have only odd parts in the top row. Then for all positive integers n we have*

$$\overline{p}_e(n) - \overline{p}_o(n) = a(n). \tag{1.9}$$

In particular, for all positive integers $n > 1$,

$$\overline{p}_e(n) > \overline{p}_o(n). \tag{1.10}$$

Theorem 1.2. *Let $b(n)$ denote the number of overpartitions of n whose Frobenius symbols have only positive parts in the top row. Then for all positive integers n we have*

$$\overline{p}^u_e(n) - \overline{p}^u_o(n) = b(n). \tag{1.11}$$

In particular, for all positive integers $n > 1$,

$$\overline{p}^u_e(n) > \overline{p}^u_o(n). \tag{1.12}$$

The overpartition identities are proved using elementary q-series transformations, and the positivity will follow from the transformed generating function. As an illustration, take $n = 4$. Using (1.5) and (1.6), we find that $\overline{p}_e(4) = 9, \overline{p}_o(4) = 5$, and $a(4) = 4$, while $\overline{p}^u_e(4) = 10, \overline{p}^u_o(4) = 4$, and $b(4) = 6$.

It turns out that $F(-1,q)$ and $F^u(-1,q)$ both have elegant expressions in terms of false theta functions. We state these together as one result.

Theorem 1.3. *We have*

$$\sum_{n\geq 0}(\overline{p}_e(n)-\overline{p}_o(n))q^n = \frac{1}{(q)_\infty}\sum_{n\geq 0} q^{n(3n+1)/2}(1-q^{2n+1}), \tag{1.13}$$

$$\sum_{n\geq 0}(\overline{p}_e^u(n)-\overline{p}_o^u(n))q^n = \frac{(-q)_\infty}{(q)_\infty}\left[1-2\sum_{n\geq 1} q^{n(3n-1)/2}(1-q^n)\right]. \tag{1.14}$$

Theorem 1.3 allows us to apply Ingham's Tauberian theorem to deduce asymptotic formulas for $\overline{p}_e(n)-\overline{p}_o(n)$ and $\overline{p}_e^u(n)-\overline{p}_o^u(n)$. We use the usual notation $p(n)$ for the number of partitions of n and $\overline{p}(n)$ for the number of overpartitions of n.

Theorem 1.4. *As* $n\to\infty$,

$$\overline{p}_e(n)-\overline{p}_o(n) \sim \frac{1}{6\sqrt{3n}}\exp\left(\pi\sqrt{\frac{2n}{3}}\right) \sim \frac{2}{3}p(n).$$

Theorem 1.5. *As* $n\to\infty$,

$$\overline{p}_e^u(n)-\overline{p}_o^u(n) \sim \frac{1}{3}\overline{p}(n).$$

These results are illustrated in Tables 1 and 2.

Note that the asymptotic behavior is quite different for the two difference functions. From work of Zuckerman [25], we have

$$\overline{p}(n) \sim \frac{1}{8n}\exp(\pi\sqrt{n}),$$

which implies that $\overline{p}_e(n)-\overline{p}_o(n)$ is small compared to $\overline{p}(n)$. On the other hand, $\overline{p}_e^u-\overline{p}_o^u(n)$ is approximately $\overline{p}(n)/3$.

Table 1. Comparison of $p(n)$ and $\overline{p}_e(n)-\overline{p}_o(n)$ for $n\leq 10{,}000$ (values rounded to four decimal places).

n	$p(n)$	$\overline{p}_e(n)-\overline{p}_o(n)$	$(\overline{p}_e(n)-\overline{p}_o(n))/p(n)$
100	1.906×10^{8}	1.288×10^{8}	0.6756
500	2.300×10^{21}	1.543×10^{21}	0.6708
1000	2.406×10^{31}	1.611×10^{31}	0.6696
5000	1.698×10^{74}	1.134×10^{74}	0.6680
10,000	3.617×10^{106}	2.415×10^{106}	0.6676

Table 2. Comparison of $\overline{p}(n)$ and $\overline{p}_e^u(n)-\overline{p}_o^u(n)$ for $n\leq 10{,}000$ (values rounded to four decimal places).

n	$\overline{p}(n)$	$\overline{p}_e^u(n)-\overline{p}_o^u(n)$	$(\overline{p}_e^u(n)-\overline{p}_o^u(n))/\overline{p}(n)$
100	5.329×10^{10}	1.895×10^{10}	0.3557
500	7.945×10^{26}	2.730×10^{26}	0.3436
1000	1.729×10^{39}	5.890×10^{38}	0.3406
5000	7.447×10^{91}	2.507×10^{91}	0.3366
10,000	3.413×10^{131}	1.146×10^{131}	0.3357

The rest of the paper is organized as follows. In Sec. 2, we prove the main results. In Sec. 3, we examine other classes of partitions using a similar weight on specific parts. By doing so, we generate several interesting new partition functions and find relations to Ramanujan's mock theta functions or false theta functions. In one case we prove an unexpected Ramanujan-type congruence modulo 5. We close in Sec. 4 with some remarks.

2. Proofs of Theorems 1.1–1.5

In this section, we prove the main results. We begin with Theorem 1.1.

Proof of Theorem 1.1. We require a transformation of Jackson [16, Eq. (III.4), Appendix III],

$$\sum_{n\geq 0}\frac{(a)_n(b)_n}{(q)_n(c)_n}z^n = \frac{(az)_\infty}{(z)_\infty}\sum_{n\geq 0}\frac{(a)_n(c/b)_n(-bz)^n q^{\binom{n}{2}}}{(q)_n(c)_n(az)_n}. \tag{2.1}$$

Using this with $z=-q/a$, $c=-b=q$, and $a\to\infty$ we have

$$F(-1,q) = \frac{1}{(-q)_\infty}\sum_{n\geq 0}\frac{(-q)_n q^{n(n+1)/2}}{(q)_n^2} = \sum_{n\geq 0}\frac{q^{n^2+n}(-1)_n}{(q)_n(q^2;q^2)_n}. \tag{2.2}$$

The nth summand of the last q-series generates a Frobenius symbol as follows. First, q^n counts the number of columns. Second, the term $q^{n^2}/(q^2;q^2)_n$ generates a partition into n distinct odd parts for the top row. Finally, the term $(-1)_n/(q)_n$ contributes an overpartition into n non-negative parts for the bottom row (as explained in [13]). Therefore,

$$F(-1,q) = \sum_{n\geq 0} a(n)q^n,$$

and (1.9) follows. The positivity in (1.10) is deduced from (2.2). □

We now turn to Theorem 1.2. The proof is similar, with Heine's transformation in place of Jackson's transformation.

Proof of Theorem 1.2. We require the third Heine transformation [16, Eq. (III.3), Appendix III],

$$\sum_{n\geq 0}\frac{(a)_n(b)_n}{(c)_n(q)_n}z^n = \frac{(abz/c)_\infty}{(z)_\infty}\sum_{n\geq 0}\frac{(c/a)_n(c/b)_n}{(c)_n(q)_n}(abz/c)^n \quad \text{for } |z|, |abz/c|<1. \tag{2.3}$$

Using this with $c=z=-b=q$ and $a=0$ we have

$$F^u(-1,q) = (q)_\infty\sum_{n\geq 0}\frac{q^n(-q)_n}{(q)_n^2} \tag{2.4}$$

$$= \sum_{n\geq 0}\frac{(-1)_n q^{n(n+3)/2}}{(q)_n^2}. \tag{2.5}$$

Here, the nth summand generates a Frobenius symbol much like before. The term q^n again counts the number of columns, while $(-1)_n/(q)_n$ generates an overpartition into non-negative parts in the bottom row. The remaining part, $q^{n(n+1)/2}/(q)_n$, contributes a partition into distinct positive parts in the top row. Therefore,

$$F^u(-1,q) = \sum_{n\geq 0} b(n)q^n,$$

and (1.11) follows. The positivity in (1.12) is deduced from (2.5). □

Next, we treat Theorem 1.3. The proof uses the Heine transformation and an identity from the lost notebook in the first case and an identity of Warnaar in the second case.

Proof of Theorem 1.3. We begin with the first Heine transformation [16, Eq. (III.1), Appendix III],

$$\sum_{n\geq 0} \frac{(a)_n(b)_n}{(q)_n(c)_n} z^n = \frac{(b)_\infty(az)_\infty}{(c)_\infty(z)_\infty} \sum_{n\geq 0} \frac{(c/b)_n(z)_n}{(q)_n(az)_n} b^n \quad \text{for } |q|, |z|, |b| < 1. \tag{2.6}$$

By setting $a = 1/z$, $c = -q$, $z = zq$, and $b \to 0$ in (2.6) and (2.3), we find the following two-variable version of Ramanujan's identity (1.2):

$$\begin{aligned} \frac{(q)_\infty}{(-q)_\infty(zq)_\infty} \sum_{n\geq 0} \frac{(zq)_n q^{n(n+1)/2}}{(q)_n^2} &= \sum_{n\geq 0} \frac{(1/z)_n(zq)^n}{(q^2;q^2)_n} \\ &= \frac{1}{(zq)_\infty} \sum_{n\geq 0} \frac{(-zq)_n(-1)^n q^{n(n+1)/2}}{(q^2;q^2)_n}. \end{aligned} \tag{2.7}$$

This implies that

$$F(z,q) = \frac{(-q)_\infty}{(q)_\infty(zq)_\infty} \sum_{n\geq 0} \frac{(-zq)_n(-1)^n q^{n(n+1)/2}}{(q^2;q^2)_n}.$$

From the lost notebook [3, Entry 9.4.2], we have the false theta identity

$$\sum_{n\geq 0} \frac{(-1)^n q^{n(n+1)/2}}{(-q)_n} = \sum_{n\geq 0} q^{n(3n+1)/2}(1-q^{2n+1}), \tag{2.8}$$

which gives

$$(q)_\infty F(-1,q) = \sum_{n\geq 0} \frac{(-1)^n q^{n(n+1)/2}}{(-q)_n} = \sum_{n\geq 0} q^{n(3n+1)/2}(1-q^{2n+1}), \tag{2.9}$$

and this is (1.13).

For (1.14) we simply use the case $a = 1$ of Warnaar's identity [23, p. 390],

$$\sum_{n\geq 0} \frac{(-aq)_n q^n}{(q)_n (a^2q)_n} = \frac{(-aq)_\infty}{(q)_\infty (a^2q)_\infty} \left[1 - (1+a) \sum_{n\geq 1} a^{3n-2} q^{n(3n-1)/2} (1 - aq^n) \right] \tag{2.10}$$

applied to (2.4). □

Before continuing, we give two corollaries of Theorem 1.3. First, using

$$\frac{1}{(q)_\infty} = \sum_{n\geq 0} p(n) q^n$$

and

$$\frac{(-q)_\infty}{(q)_\infty} = \sum_{n\geq 0} \overline{p}(n) q^n$$

together with (1.13) and (1.14), we obtain formulas for $\overline{p}_e(n) - \overline{p}_o(n)$ and $\overline{p}_e^u(n) - \overline{p}_o^u(n)$ in terms of $p(n)$ and $\overline{p}(n)$, respectively.

Corollary 2.1. *For all positive integers n we have*

$$\overline{p}_e(n) - \overline{p}_o(n) = p(n) - p(n-1) + p(n-2) - p(n-5) + \cdots, \tag{2.11}$$

and

$$\overline{p}_e^u(n) - \overline{p}_o^u(n) = \overline{p}(n) - 2\overline{p}(n-1) + 2\overline{p}(n-2) - 2\overline{p}(n-5) + \cdots. \tag{2.12}$$

Next, we give a characterization of $\overline{p}_e^u(n) - \overline{p}_o^u(n)$ modulo 4 which follows from (1.14).

Corollary 2.2. *For all positive integers $n \geq 1$,*

$$\overline{p}_e^u(n) - \overline{p}_o^u(n) \equiv \begin{cases} 2 \pmod 4 & \text{if } n \text{ is a square or generalized pentagonal number} \\ & \text{(but not both)}, \\ 0 \pmod 4 & \text{otherwise.} \end{cases}$$

Proof. Using the identity

$$\frac{(q)_\infty}{(-q)_\infty} = \sum_{n\in\mathbb{Z}} (-1)^n q^{n^2},$$

the generating function for overpartitions satisfies

$$\frac{(-q)_\infty}{(q)_\infty} = \frac{1}{1 + 2\sum_{n\geq 1}(-1)^n q^{n^2}} \equiv 1 + 2\sum_{n\geq 1} q^{n^2} \pmod 4.$$

Together with (1.14) we then have

$$\sum_{n\geq 0}(\overline{p}_e^u(n)-\overline{p}_o^u(n))q^n \equiv \left(1+2\sum_{n\geq 1}q^{n^2}\right)\left(1+2\sum_{n\geq 1}q^{n(3n-1)/2}(1+q^n)\right) \pmod 4,$$

and the result follows. □

Note that there exist numbers which are both squares and pentagonal numbers (see [22, A036353]).

We are now ready to prove Theorems 1.4 and 1.5. For these we use Ingham's Tauberian theorem [17, Theorem 1.1], as presented in [10].

Theorem 2.3 ([10, Theorem 1.1]). *Let $f(q)=\sum_{n\geq 0}a(n)q^n$ be a power series whose radius of convergence is equal to 1 and whose coefficients $a(n)$ are non-negative and weakly increasing. Suppose that for $A>0,\lambda,\alpha\in\mathbb{R}$,*

$$f(e^{-t})\sim \lambda t^{\alpha}e^{A/t} \quad \text{as } t\to 0^+, \quad f(e^{-z})\ll |z|^{\alpha}e^{A/|z|} \quad \text{as } z\to 0,$$

with $z=x+iy$ $(x>0, y\in\mathbb{R})$ in each region of the form $|y|\leq cx$ for $c>0$. Then

$$a(n)\sim \frac{\lambda}{2\sqrt{\pi}}\frac{A^{\frac{\alpha}{2}+\frac{1}{4}}}{n^{\frac{\alpha}{2}+\frac{3}{4}}}e^{2\sqrt{An}}$$

as $n\to\infty$.

Proof of Theorem 1.4. Let $q=e^{-z}$ and $f(z)=e^{-\frac{3}{2}z^2}$. By Euler–Maclaurin summation [10, Theorem 1.2],

$$\begin{aligned}
&\sum_{n\geq 0}q^{n(3n-1)/2}(1-q^{2n+1})\\
&\quad= e^{-z/24}\sum_{n\geq 0}[e^{-\frac{3}{2}\left(n+\frac{1}{6}\right)^2 z}-e^{-\frac{3}{2}\left(n+\frac{5}{6}\right)^2 z}]\\
&\quad= e^{-z/24}\sum_{n\geq 0}\left[f\left(\left(n+\frac{1}{6}\right)\sqrt{z}\right)-f\left(\left(n+\frac{5}{6}\right)\sqrt{z}\right)\right]\\
&\quad= e^{-z/24}\left[\frac{1}{\sqrt{z}}\int_0^{\infty}f(t)dt-B_1\left(\frac{1}{6}\right)-\frac{1}{\sqrt{z}}\int_0^{\infty}f(t)dt+B_1\left(\frac{5}{6}\right)+\mathcal{O}(z)\right]\\
&\quad= e^{-z/24}\left(\frac{2}{3}+\mathcal{O}(z)\right)
\end{aligned}$$

uniformly, as $z\to 0$ in $\{x+iy:|y|\leq cx\}$ for every fixed $c>0$, where $B_n(t)$ is the nth Bernoulli polynomial. Hence, we obtain, as $z\to 0$ in $\{x+iy:|y|\leq cx\}$ for every fixed $c>0$,

$$F(-1,e^{-z})\sim\sqrt{\frac{2z}{9\pi}}\exp\left(\frac{\pi^2}{6z}\right)$$

with the asymptotic

$$\log(q)_\infty \sim -\frac{\pi^2}{6z} + \frac{1}{2}\log\left(\frac{2\pi}{z}\right) + \frac{z}{24}.$$

To apply Ingham's Tauberian theorem, we now need to prove that $\overline{p}_e(n) - \overline{p}_o(n)$ is weakly increasing. To see this, note that (2.2) gives

$$\begin{aligned}(1-q)\sum_{n\geq 0}(\overline{p}_e(n) - \overline{p}_o(n))q^n &= (1-q)\sum_{n\geq 0}\frac{(-1)_n q^{n^2+n}}{(q)_n(q^2;q^2)_n}\\ &= (1-q) + \sum_{n\geq 1}\frac{(-1)_n q^{n^2+n}}{(q^2)_{n-1}(q^2;q^2)_n},\end{aligned}$$

which has non-negative coefficients for $n \geq 2$. Therefore, $\overline{p}_e(n) - \overline{p}_o(n)$ is weakly increasing for $n \geq 2$. By Theorem 2.3, we may now conclude that

$$\overline{p}_e(n) - \overline{p}_o(n) \sim \frac{1}{6\sqrt{3n}}\exp\left(\pi\sqrt{\frac{2n}{3}}\right)$$

as $n \to \infty$. □

Proof of Theorem 1.5. As in the proof of Theorem 1.4, we can obtain the asymptotic behavior of the generating function for $\overline{p}_e^u(n) - \overline{p}_o^u(n)$. Setting $q = e^{-z}$ and $f(z) = \exp(-\frac{3}{2}z^2)$, Euler–Maclaurin summation gives

$$\begin{aligned}\sum_{n\geq 1} q^{n(3n-1)/2}(1-q^{2n}) &= e^{-z/24}\sum_{n\geq 0}\left[f\left(\left(n+\frac{5}{6}\right)\sqrt{z}\right) - f\left(\left(n+\frac{7}{6}\right)\sqrt{z}\right)\right]\\ &= e^{-z/24}\left(\frac{1}{3} + O(z)\right)\end{aligned}$$

uniformly, as $z \to 0$ in $\{x+iy : |y| \leq cx\}$ for every fixed $c > 0$.

Hence, we have as $z \to 0$ in $\{x+iy : |y| \leq cx\}$ for every fixed $c > 0$,

$$\sum_{n\geq 0}(\overline{p}_e^u(n) - \overline{p}_o^u(n))q^n \sim \frac{1}{6}\sqrt{\frac{z}{\pi}}\exp\left(\frac{\pi^2}{4z}\right).$$

Since we can see that $\overline{p}_e^u(n) - \overline{p}_o^u(n)$ is weakly increasing for $n \geq 2$ from

$$(1-q)\sum_{n\geq 0}(\overline{p}_e^u(n) - \overline{p}_o^u(n))q^n = (1-q)\sum_{n\geq 0}\frac{(-1)_n q^{n(n+3)/2}}{(q)_n^2},$$

by applying Theorem 2.3, we have the asymptotic formula

$$\overline{p}_e^u(n) - \overline{p}_o^u(n) \sim \frac{1}{24n}\exp(\pi\sqrt{n})$$

as $n \to \infty$. □

3. Other Weighted Partition Functions

In this section, we replace overpartitions by other types of partitions but use a similar weight on specific types of parts. We touch on three different cases, the first of which is closely related to $F^u(-1,q)$.

3.1. *Bipartitions*

A bipartition π of n is a pair of partitions λ_r and λ_b where λ_r and λ_b are ordinary partitions and the sum of their parts is n. Let us say that the parts in λ_r are colored red and the parts in λ_b are colored blue. Define $b_e(n)$ (respectively, $b_o(n)$) to be the number of bipartitions of n which have an even (respectively, odd) number of red parts larger than the number of blue parts. Let $B(q)$ denote the generating function for $b_e(n) - b_o(n)$,

$$\begin{aligned} B(q) &:= \sum_{n\geq 0}(b_e(n) - b_o(n))q^n \\ &= 1 + 3q^2 + 4q^3 + 10q^4 + 14q^5 + 29q^6 + 44q^7 + 79q^8 + 120q^9 + 199q^{10} + \cdots . \end{aligned}$$

Elementary combinatorial arguments followed by an appeal to (2.4) give that

$$B(q) = \sum_{n\geq 0} \frac{q^n}{(q)_n^2(-q^{n+1})_\infty} = \frac{1}{(q^2;q^2)_\infty} F^u(-1,q),$$

from which we conclude the following.

Theorem 3.1. *For all positive integers $n > 1$,*

$$b_e(n) > b_o(n).$$

Moreover, we have the expression

$$\sum_{n\geq 0}(b_e(n) - b_o(n))q^n = \frac{1}{(q)_\infty^2}\left[1 - 2\sum_{n\geq 1} q^{n(3n-1)/2}(1-q^n)\right]. \tag{3.1}$$

Using "Ramanujan's method" [6, 7], the false theta identity (3.1) can be used to deduce the following, perhaps unexpected, Ramanujan-type congruence.

Theorem 3.2. *For all non-negative integers n, we have*

$$b_e(5n+4) - b_o(5n+4) \equiv 0 \pmod 5.$$

Proof. Recall Jacobi's identity

$$(q)_\infty^3 = \sum_{k\geq 0}(-1)^k(2k+1)q^{k(k+1)/2}.$$

Combined with (3.1), we have

$$B(q) \equiv \frac{1}{(q^5;q^5)_\infty}\left(\sum_{k\geq 0}(-1)^k(2k+1)q^{k(k+1)/2} - 2\sum_{\substack{k\geq 0\\ n\geq 1}}(2k+1)q^{k(k+1)/2+n(3n-1)/2}(1-q^n)\right) \pmod 5.$$

Since $k(k+1)/2 \not\equiv 4 \pmod 5$ the first term has no contribution modulo 5 to the coefficient of q^{5n+4} in $B(q)$. As for the second term, it is easy to check that if $k(k+1)/2 + n(3n-1)/2$ or $k(k+1)/2 + n(3n+1)/2$ is congruent to 4 modulo 5, then $k \equiv 2 \pmod 5$. In this case, the factor $(2k+1)$ ensures that there is again no contribution modulo 5 to the coefficient of q^{5n+4} in $B(q)$. □

3.2. *Partitions without repeated even parts I*

Let $ped(n)$ denote the number of partitions of n without repeated even parts. We define $ped_e(n)$ (respectively, $ped_o(n)$) to be the number of partitions counted by $ped(n)$ having an even (respectively, odd) number of odd parts larger than *twice* the number of even parts. Then we have

$$\begin{aligned} H(q) :&= \sum_{n\geq 0}(ped_e(n) - ped_o(n))q^n \\ &= \sum_{n\geq 0}\frac{q^{n^2+n}}{(q;q^2)_n(q^2;q^2)_n(-q^{2n+1};q^2)_\infty} \\ &= 1 - q + 2q^2 - q^3 + 4q^4 - 2q^5 + 7q^6 - 4q^7 + 10q^8 - 6q^9 + 17q^{10} - \cdots. \end{aligned} \tag{3.2}$$

While Ramanujan's series $g^s(q)$ is a reminder that looks can be deceiving, it turns out that the coefficients of $H(q)$ do indeed alternate as suggested by (3.2).

Theorem 3.3. *For all positive integers n we have*

$$(-1)^n(ped_e(n) - ped_o(n)) > 0.$$

Proof. We require the second Heine transformation [16, Eq. (III.2), Appendix III],

$$\sum_{n\geq 0}\frac{(a)_n(b)_n z^n}{(c)_n(q)_n} = \frac{(c/b)_\infty(bz)_\infty}{(c)_\infty(z)_\infty}\sum_{n\geq 0}\frac{(abz/c)_n(b)_n(c/b)^n}{(bz)_n(q)_n}. \tag{3.3}$$

With $q = q^2$, $a = -c = q$, $z = -q^2/b$, and $b \to \infty$ we find that

$$H(-q) = \frac{(-q^2;q^2)_\infty}{(q^2;q^4)_\infty} \sum_{n\geq 0} \frac{q^{n^2}}{(-q^2;q^2)_n}. \tag{3.4}$$

This clearly has positive coefficients. □

Recall Ramanujan's third-order mock theta function

$$\phi(q) = \sum_{n\geq 0} \frac{q^{n^2}}{(-q^2;q^2)_n}.$$

By (3.4), we have

$$H(-q) = \frac{(-q^2;q^2)_\infty}{(q^2;q^4)_\infty} \phi(q).$$

Such products of modular forms and mock theta functions are called *mixed* mock modular forms [20]. It is possible to obtain asymptotic formulas for the coefficients of such forms (see [9, 11], for example), though we shall not pursue this here.

We close this section with a combinatorial identity for the coefficients of $H(-q)$. Following the lead of Fine [15, Sec. 26] in his treatment of Ramanujan's mock theta function

$$\psi(q) = \sum_{n\geq 1} \frac{q^{n^2}}{(q;q^2)_n},$$

we say that a partition into odd parts is *without gaps* if all odd parts less than the largest occur. If λ is a partition and $k \geq 0$, we refer to the k smallest parts of λ as an *initial partition.*

Theorem 3.4. *Let $c(n)$ denote the number of partitions counted by $ped(n)$ such that odd parts occur an even number of times, except possibly for an initial partition into odd parts without gaps $\{1, 3, \ldots, 2k-1\}$, wherein all parts occur an odd number of times, and in which case even parts are all $> 2k$. Then*

$$c(n) = (-1)^n (ped_e(n) - ped_o(n)).$$

Proof. We have

$$\begin{aligned} H(-q) &= \frac{(-q^2;q^2)_\infty}{(q^2;q^4)_\infty} \sum_{n\geq 0} \frac{q^{n^2}}{(-q^2;q^2)_n} \\ &= \sum_{n\geq 0} \frac{q^{n^2}}{(q^2;q^4)_n} \times \frac{(-q^{2n+2};q^2)_\infty}{(q^{4n+2};q^4)_\infty}, \end{aligned}$$

and this is easily seen to be the generating function for $c(n)$. Specifically, the nth summand generates an initial partition into odd parts $\leq 2n-1$ without gaps (which is empty if $n = 0$), each part occurring an odd number of times, while the product in the numerator contributes the distinct even parts $\geq 2n+2$ and the product in the denominator contributes odd parts $\geq 2n+1$ occurring an even number of times. □

3.3. *Partitions without repeated even parts* II

In this section, we again consider partitions without repeated parts, but this time we reverse the roles played by the even and odd parts. Namely, let $ped_e^e(n)$ (respectively, $ped_o^e(n)$) be the number of partitions counted by $ped(n)$ having an even (respectively, odd) number of even parts larger than twice the number of odd parts. Then we have

$$\begin{aligned} H^e(q) &:= \sum_{n\geq 0}(ped_e^e(n) - ped_o^e(n))q^n \\ &= \sum_{n\geq 0} \frac{q^n(-q^2;q^2)_n(q^{2n+2};q^2)_\infty}{(q^2;q^2)_n} \\ &= 1 + q + 3q^3 + 2q^4 + 4q^5 + 7q^6 + 6q^7 + 12q^8 + 12q^9 + 21q^{10} + \cdots. \end{aligned} \tag{3.5}$$

We show that the coefficients are positive, as suggested by (3.5).

Theorem 3.5. *For all positive integers $n \neq 2$, we have*

$$ped_e^e(n) > ped_o^e(n).$$

Proof. By Jackson's transformation (2.1) with $q = q^2, z = q, c = -a = q^2$, and $b \to 0$ we have

$$\begin{aligned} H^e(q) &= (q^2;q^2)_\infty \sum_{n\geq 0} \frac{(-q^2;q^2)_n q^n}{(q^2;q^2)_n^2} \\ &= \frac{(q^2;q^2)_\infty}{(q;q^2)_\infty} \sum_{n\geq 0} \frac{q^{2n^2+n}(-q^2;q^2)_n(-q^{2n+3};q^2)_\infty}{(q^2;q^2)_n^2}. \end{aligned} \tag{3.6}$$

The positivity then follows after recalling Gauss' identity:

$$\frac{(q^2;q^2)_\infty}{(q;q^2)_\infty} = \sum_{n\geq 0} q^{n(n+1)/2}.$$

□

4. Concluding Remarks

The weighting procedure described in this paper can be applied to many other types of partitions. To give just one more example, let $p_e(n)$ (respectively, $p_o(n)$) be the number of partitions of n having an even (respectively, odd) number of even parts larger than twice of the number of odd parts. Then, we find that

$$\begin{aligned} \sum_{n\geq 0}(p_e(n) - p_o(n))q^n &= \sum_{n\geq 0} \frac{q^n}{(q^2;q^2)_n^2(-q^{2n+2};q^2)_\infty} \\ &= 1 + q + 3q^3 + 3q^4 + 5q^5 + 7q^6 + 9q^7 + 16q^8 + 18q^9 + \cdots. \end{aligned}$$

Comparing this with (3.6), we see that

$$\sum_{n\geq 0}(p_e(n)-p_o(n))q^n = (-q;q^2)_\infty \sum_{n\geq 0} \frac{q^{2n^2+n}(-q^2;q^2)_n(-q^{2n+3};q^2)_\infty}{(q^2;q^2)_n^2},$$

and we conclude that

$$p_e(n) > p_o(n)$$

for all positive integers $n \neq 2$.

Our positivity proofs have used elementary q-series transformations, but there are some situations which might be more subtle. For example, reversing the roles of even and odd parts in the previous paragraph, let $p'_e(n)$ (respectively, $p'_o(n)$) be the number of partitions of n having an even (respectively, odd) number of odd parts larger than twice of the number of even parts. Then

$$\begin{aligned}\sum_{n\geq 0}(p'_e(n)-p'_o(n))q^n &= \frac{1}{(-q;q^2)_\infty}\sum_{n\geq 0}\frac{q^{2n}(-q;q^2)_n}{(q)_{2n}} \\ &= 1-q+2q^2-q^3+5q^4-q^5+9q^6-q^7+16q^8 \\ &\quad +28q^{10}+4q^{11}+47q^{12}+11q^{12}+77q^{13}+26q^{15}+\cdots,\end{aligned}$$

leading one to suspect that $p'_e(n)-p'_o(n) > 0$ for all $n \neq 1,3,5,7$. This does not appear to follow easily from transformations used in this paper.

Instead of weighting according to the number of parts of some type that are larger than the number of parts of another type, one could also use the number of parts of some type that are *less than or equal to* the number of parts of another type. For instance, if $\overline{p}'_e(n)$ (respectively, $\overline{p}'_o(n)$) denotes the number of overpartitions of n having an even (respectively, odd) number of non-overlined parts less than or equal to the number of overlined parts, then

$$\sum_{n\geq 0}(\overline{p}'_e(n)-\overline{p}'_o(n))q^n = \sum_{n\geq 0}\frac{q^{n(n+1)/2}}{(q)_n(-q)_n(q^{n+1})_\infty} \tag{4.1}$$

$$= \frac{1}{(q)_\infty}\sum_{n\geq 0}\frac{q^{n(n+1)/2}}{(-q)_n}$$

$$= 1+2q+2q^2+6q^3+6q^4+12q^5+16q^6+26q^7+\cdots. \tag{4.2}$$

The positivity of $\overline{p}'_e(n)-\overline{p}'_o(n)$ follows immediately. Note that the sum in the middle line above is a famous q-series from Ramanujan's lost notebook [5]. For a different combinatorial interpretation of (4.1) and an asymptotic formula for the coefficients, see [21, Theorems 1.1 and 1.7].

The possibilities are many, and there are undoubtedly more interesting q-series waiting be discovered.

Acknowledgments

Byungchan Kim was supported by the Basic Science Research Program through the National Research Foundation of Korea (NRF) funded by the Ministry of Education (NRF-2016R1D1A1A09917344) and the Ministry of Science and ICT (NRF-2019R1F1A1043415). Eunmi Kim was supported by Basic Science Research Program through the National Research Foundation of Korea (NRF) funded by the Ministry of Education (NRF-2019R1A6A1A11051177) and the Ministry of Science and ICT (NRF-2018R1A2A1A05079095).

References

[1] G. E. Andrews, Ramanujan's "lost" notebook. IV. Stacks and alternating parity in partitions, *Adv. Math.* **53**(1) (1984) 55–74.

[2] G. E. Andrews, Unsolved problems: Questions and conjectures in partition theory, *Amer. Math. Monthly* **93**(9) (1986) 708–711.

[3] G. E. Andrews and B. C. Berndt, *Ramanujan's Lost Notebook. Part* I (Springer, New York, 2005).

[4] G. E. Andrews and B. C. Berndt, *Ramanujan's Lost Notebook. Part* II (Springer, New York, 2009).

[5] G. E. Andrews, F. J. Dyson and D. Hickerson, Partitions and indefinite quadratic forms, *Invent. Math.* **91**(3) (1988) 391–407.

[6] G. E. Andrews and R. Roy, Ramanujan's method in q-series congruences, *The Wilf Festschrift* (*Philadelphia, PA, 1996*), *Electron. J. Combin.* **4**(2) (1997), Research Paper 2.

[7] B. C. Berndt, C. Gugg and S. Kim, Ramanujan's elementary method in partition congruences, in *Partitions, q-Series, and Modular Forms*, Developments in Mathematics, Vol. 23 (Springer, New York, 2012), pp. 13–21.

[8] B. C. Berndt, B. Kim and A. J. Yee, Ramanujan's lost notebook: Combinatorial proofs of identities associated with Heine's transformation or partial theta functions, *J. Combin. Theory Ser. A* **117**(7) (2010) 857–973.

[9] K. Bringmann, J. Dousse, J. Lovejoy and K. Mahlburg, Overpartitions with restricted odd differences, *Electron. J. Combin.* **22**(3) (2015) Article No. 3.17.

[10] K. Bringmann, C. Jennings-Shaffer and K. Mahlburg, On a Tauberian theorem of Ingham and Euler–Maclaurin summation, preprint (2019); arXiv:1910.03036.

[11] K. Bringmann and K. Mahlburg, An extension of the Hardy–Ramanujan circle method and applications to partitions without sequences, *Amer. J. Math.* **133**(4) (2011) 1151–1178.

[12] K. Bringmann and K. Mahlburg, Asymptotic formulas for stacks and unimodal sequences, *J. Combin. Theory Ser. A* **126** (2014) 194–215.

[13] S. Corteel and J. Lovejoy, Overpartitions, *Trans. Amer. Math. Soc.* **356**(4) (2004) 1623–1635.

[14] S. Corteel, J. Lovejoy and O. Mallet, An extension to overpartitions of the Rogers–Ramanujan identities for even moduli, *J. Number Theory* **128**(6) (2008) 1602–1621.

[15] N. J. Fine, *Basic Hypergeometric Series and Applications*, Mathematical Surveys and Monographs, Vol. 27 (American Mathematical Society, Providence, RI, 1988).

[16] G. Gasper and M. Rahman, *Basic Hypergeometric Series*, Encyclopedia of Mathematics and its Applications, Vol. 96, 2nd edition (Cambridge University Press, Cambridge, 2004).

[17] A. Ingham, A Tauberian theorem for partitions, *Ann. Math.* **42** (1941) 1075–1090.

[18] J. Lovejoy, Overpartition theorems of the Rogers–Ramanujan type, *J. London Math. Soc. (2)* **69**(3) (2004) 562–574.

[19] J. Lovejoy, Rank and conjugation for the Frobenius representation of an overpartition, *Ann. Comb.* **9**(3) (2005) 321–334.

[20] J. Lovejoy and R. Osburn, Mixed mock modular q-series, *J. Indian Math. Soc. (N.S.)*, Special volume to commemorate the 125th birth anniversary of Srinivasa Ramanujan (2013) 45–61.

[21] H. T. Ngo and R. C. Rhoades, Integer partitions, probabilities and quantum modular forms, *Res. Math. Sci.* **4** (2017) Article No. 17.

[22] *The On-Line Encyclopedia of Integer Sequences*, https://oeis.org [accessed 22 October 2019].

[23] S. O. Warnaar, Partial theta functions. I. Beyond the lost notebook, *Proc. London Math. Soc. (3)* **87**(2) (2003) 363–395.

[24] G. N. Watson, The final problem: An account of the Mock theta functions, *J. London Math. Soc.* **11**(1) (1936) 55–80.

[25] H. S. Zuckerman, On the coefficients of certain modular forms belonging to subgroups of the modular group, *Trans. Amer. Math. Soc.* **45** (1939) 298–321.

© 2024 World Scientific Publishing Company
https://doi.org/10.1142/9789811277375_0026

Twin Nicomachean q-identities and conjectures for the associated discriminants, polynomials, and inequalities*

Seon-Hong Kim†

Department of Mathematics
Sookmyung Women's University, Seoul, 140-742 Korea
shkim17@sookmyung.ac.kr

Kenneth B. Stolarsky

Department of Mathematics
University of Illinois at Urbana-Champaign, Urbana, 61801 USA
stolarsky.ken@gmail.com

We insert additional variables into Warnaar's q-analogue of Nicomachus' identity and other related identities, and compute discriminants with respect to q. Factorization of these discriminants reveals pairs of partitions that conjecturally relate in the manner of Wheatstone. The factorization also yields, conjecturally, families of polynomials with relations to various Molien series, remarkable rational generating functions, and notable root distributions. For a q-analogue of Nicomachus' identity produced by Cigler, we provide proofs of the partition properties. We also state and in part prove tight inequalities for the elements of two interlacing sequences that led us to the "twin" of Warnaar's q-analogue.

Keywords: Discriminants; Molien series; Nicomachean identity; partitions; power means; q-analogues; squared triangular numbers; sums of cubes; zeros of polynomials.

Mathematics Subject Classification 2020: 11B65, 11P81, 11B83, 30C15

1. Introduction

The ancient Nicomachean identity

$$1^3 + 2^3 + 3^3 + \cdots + n^3 = (1 + 2 + 3 + \cdots + n)^2 = \left(\frac{n(n+1)}{2}\right)^2 \tag{1.1}$$

has an elegant q-analogue first found by Warnaar [12], namely

$$\sum_{k=1}^{n} \frac{q^{2(n-k)}}{q+1}\left(\frac{q^k - 1}{q-1}\right)^2\left(\frac{q^{2k}-1}{q-1}\right) = \frac{1}{(q+1)^2}\left(\frac{q^n - 1}{q-1}\frac{q^{n+1}-1}{q-1}\right)^2. \tag{1.2}$$

*This article originally appeared in *International Journal of Number Theory*, **17**(3), pp. 621–645 © 2021 World Scientific Publishing Company. DOI: 10.1142/S1793042120400175
†Seon-Hong Kim was supported by NRF of Korea.

The identity (1.1) yields nothing of significance in the $n \to \infty$ limit, while (1.2) for $|q| < 1$ yields

$$\begin{aligned} G(q) &= \frac{1}{(1-q)^2(1-q^2)^2} \\ &= 0 \cdot q^{-2} + 0 \cdot q^{-1} + 1 + 2q + 5q^2 + 8q^3 + 14q^4 + 20q^5 + \cdots . \end{aligned} \tag{1.3}$$

The sequence of sums of three consecutive coefficients here is

$$\{1, 3, 8, 15, 27, 42, 64, 90, \ldots\} \tag{1.4}$$

and it is an elementary exercise in "series sectioning" to show that the entries in odd places are the positive cubes in increasing order. Thus one might expect that (1.1) has a twin identity for the even places sequence

$$\{s(1), s(2), s(3), \ldots\} = \{3, 15, 42, 90, \ldots\}.$$

In fact

$$s(k) = \frac{1}{2}k(k+1)(2k+1) = 3\sum_{j=1}^{k} j^2$$

and the twin identity is

$$\sum_{k=1}^{n} s(k) = \frac{n(n+1)}{2} \cdot \frac{(n+1)(n+2)}{2}. \tag{1.5}$$

Here, rather than the square of a triangular number, we have the product of consecutive triangular numbers. Could the ancient Greeks have recorded (1.5) in a writing that is now lost? It is however the case that the present authors were unaware of its existence prior to Fall, 2019. This identity, a twin of (1.1), has a q-analogue that is a twin of Warnaar's (1.2), and that we prove in a similar way.

Proposition 1.1.

$$\sum_{k=1}^{n} \frac{q^{2(n-k)}}{q+1} \frac{q^k - 1}{q-1} \frac{q^{k+1} - 1}{q-1} \frac{q^{2k+1} - 1}{q-1} = \frac{(q^n - 1)(q^{n+1} - 1)^2(q^{n+2} - 1)}{(q-1)^2(q^2-1)^2}. \tag{1.6}$$

Proof. The identity

$$T(n) = \frac{(q^n - 1)(q^{n+1} - 1)(q^{2n+1} - 1)}{(q-1)^2(q^2-1)} + q^2 T(n-1),$$

where $T(n)$ is the right side of (1.6), shows that $T(n)$ is the last term of the sum on the left side (the $k = n$ term) of (1.6) plus the product of q^2 with $T(n-1)$. Upon iterating this formula $n - 1$ times, we obtain the left side of (1.6). □

The central theme of this paper is the introduction of discriminants into the study of such finite q-analogues. In the case of (1.6) this leads in particular to a

t-analogue of the generating function for certain Vassiliev invariants (see Sec. 6, especially Conjecture 6.7 and formulas (6.6), (6.8) and (6.9)).

Our main contribution is the construction of certain discriminants after inserting one or two additional variables into these twin Nicomachean identities. We believe our constructions are non-obvious and that they lead to remarkable conjectures concerning the "tame" polynomials that appear in the factorization of these discriminants (e.g., the polynomials generated by the generating functions of Conjecture 6.7). A standard reference for discriminants and their properties is [6].

We admit our inability to resolve the above conjectures. Each computer-generated factorization also has an identifiable "wild" polynomial factor whose coefficients become quite large as n increases. This greatly impedes our efforts at finding proofs. We do show in Sec. 10 that at least in one case computer data reveals that a certain sequence of wild polynomials has a somewhat comprehensible structure.

Looking ahead, we shall see in Sec. 4 examples of a discriminant selecting and summing certain infinite series, a phenomenon that is new to us. We shall also see discriminants pairing "interesting" partitions of certain integers N with additional but "still interesting" partitions of such N. For an example of the latter, consider the partition of the squared triangular number 10^2 given by the coefficients of

$$1 + 2^3 t + 3^3 t^2 + 4^3 t^3. \tag{1.7}$$

An appropriate discriminant pairs this with the partition given by the coefficients of

$$\begin{aligned} &1 + 3t + 7t^2 + 13t^3 + 13t^4 + 15t^5 + 13t^6 + 12t^7 + 9t^8 + 7t^9 \\ &\quad + 4t^{10} + 2t^{11} + t^{12}. \end{aligned} \tag{1.8}$$

The latter is a member of a family of polynomials that conjecturally have every root an algebraic unit, absolute value at $t = -1$ a perfect square, degree a multiple of 4, and exactly 1/4 of all roots inside of the unit circle. The polynomial of (1.8) also has the "Wheatstone property" with respect to (1.7): see Sec. 7.

We note that Schlosser [10] has taken (1.2) as a starting point and has then put a variety of similar finite q-analogues into the framework of basic hypergeometric series (A standard reference for such series is [5]). We also have (1.2) as our starting point, but go in a rather different direction. We also mention the work of Shanks on finite q-identities. His device is the introduction of certain ratios of determinants that occur in the theory of nonlinear convergence acceleration transformations. This is described in [2] (see also the references therein), a paper that also takes the step of putting the finite identities into the framework of basic hypergeometric series.

The variable t is introduced into (1.2) in Sec. 3. Section 4 analyzes the discriminant as a function of two variables u and t, while in Sec. 5 the one variable case (set $u = 1$) is pursued. The associated polynomials are discussed in Secs. 6 and 7. In particular, they are related to certain Molien series; see [11] for an introduction to Molien series. Generalizations are pursued in Sec. 8, and in Sec. 9 our procedure is applied to a rather different q-analogue of (1.1) founded by Cigler [3]. Here, we are able to prove the Wheatstone property. In Sec. 11, we examine how the summands in the twin Nicomachean identities are related to each other by certain inequalities. Here, we find a polynomial $U(k, q)$ that behaves in an unexpected way for $0 < q < 1$. In Sec. 12, we show that it makes sense to speak of "one-third of Nicomachus' identity". Here, we not only find polynomial sequences with conjecturally exactly 1/3 of their zeros small, but also prove that some associated polynomials have all their zeros on the unit circle. Finally, in Sec. 13, we provide infinitely many Warnaar-type q-analogues, together with their corresponding twins.

2. A Construction

Here, we introduce some auxiliary expressions.

Definition 2.1. Let (T, W) and (U, V) be, respectively, the four binomials

$$(-q^2 + t, -1 + q^2 t) \quad \text{and} \quad (-q + t, -1 + qt).$$

Definition 2.2. Call the polynomial $Q(q)$ a reversal of $P(q)$ if for some integer m,

$$q^m P\left(\frac{1}{q}\right) = \pm Q(q).$$

We observe that if Q is a reversal of P, then P is a reversal of Q. Since

$$q^2 T\left(\frac{1}{q}\right) = W(q) \quad \text{and} \quad qU\left(\frac{1}{q}\right) = V(q),$$

we see that W and V are, respectively, reversals of T and U.

Definition 2.3. Let $\{P_1, P_2, P_3, P_4\}$ be the four triple binomial products formed, respectively, by expanding

$$TUVW \cdot \left(\frac{1}{T}, \frac{1}{U}, \frac{1}{V}, \frac{1}{W}\right).$$

Thus,

$$\begin{aligned} P_1 &= (-q + t)(-1 + qt)(-1 + q^2 t), \\ P_2 &= (-q^2 + t)(-1 + q^2 t)(-1 + qt), \\ P_3 &= (-q^2 + t)(-1 + q^2 t)(-q + t), \\ P_4 &= (-q^2 + t)(-q + t)(-1 + qt). \end{aligned}$$

Also, set

$$M = (-1 + q)^3 (1 + q)(q + (q + 1)^2 t + qt^2).$$

We now introduce new variables t and u.

Definition 2.4. Let

$$L(n,q,t) = \sum_{k=1}^{n} \frac{q^{2(n-k)}}{1+q} \left(\frac{q^k-1}{q-1}\right)^2 \left(\frac{q^{2k}-1}{q-1}\right) t^k \tag{2.1}$$

and

$$L(n,t) = \lim_{q\to 1} L(n,q,t) = \sum_{k=1}^{n} k^3 t^k.$$

We now introduce N and R, objects central to this paper.

Definition 2.5. Let

$$\begin{aligned} N &= N(n,q,t,u) \\ &= t^n(-P_1 + 2q^n P_2 - 2q^{3n+1} P_3 + q^{4n+2} P_4) + uq^{2n} M, \end{aligned} \tag{2.2}$$

and define R by

$$N = (q-1)^3(q+1)R$$

so also

$$R = R(n,q,t,u).$$

For large n the expansion of the numerator of the right side of (1.2) reveals a lacunary polynomial. The above expressions were constructed after examining (2.1) for a sizeable n with this in mind.

We observe that each of P_1 and P_4 have degree 4 in q, so the second factor of the first term of (2.2) has degree $4n+6$ in q. Call it $C = C(n,q,t)$. Note that

$$q^{4n+6} C\left(n, \frac{1}{q}, t\right) = -C(n,q,t)$$

so as a polynomial in q, the C is self-reversal. Thanks to a similar property of the coefficient of u, we have

$$q^{4n+6} N\left(n, \frac{1}{q}, t, u\right) = -N(n,q,t,u)$$

so the same is true of N. One can also verify that R is a polynomial.

3. The Warnaar (q,t) Analogue

The following identity is not immediately obvious, but it is quickly verified by computer algebra.

Theorem 3.1.

$$L(n,q,t) = \frac{tN(n,q,t,1)}{(q-1)^3(q+1)TUVW}.$$

Remark 3.2. (1) We have

$$N(n,q,1,1) = (q-1)^3(q+1)(1-q^n)^2(1-q^{n+1})^2.$$

(2) The expression $L(n,q,t)$ "unifies" Warnaar's identity with the expression $L(n,t)$ which is a rather simple minded t-analogue of the sum of cubes. However, the connection is rather more than this. We shall see in Sec. 4, why we have earlier introduced the variable u into the N-function, even though it has no clear place in the expression $L(n,q,t)$ of Theorem 3.1. It allows a discriminant to sum infinite series.

4. The Discriminant of $R(n,q,t,u)$

Denote by $\Delta_q R$ the discriminant of R with respect to q. Computer factorization of $\Delta_q R$ for many values of n show that it has "tame" factors of interest, even though the presence of at least one "wild" factor has prevented us from even conjecturing a closed-form formula for $\Delta_q R$.

For $n = 5$, we have

$$\begin{aligned}\Delta_q R = 2^{20}t^{110}(t^6-u)^2((1-t^2)^2(1+3t^2+5t^4)+(1+t^2)(u-1))\\ ((1-t)^4(1+2^3t+3^3t^2+4^3t^3+5^3t^4)\\ +(1+4t+t^2)(u-1))\cdot S(t,u)^2,\end{aligned} \tag{4.1}$$

where $S(t,u)$ has 471 terms and very large coefficients. The following sums are well known:

$$\sum_{k=1}^{\infty}(2k-1)t^{2(k-1)} = \frac{1+t^2}{(1-t^2)^2} \tag{4.2}$$

and

$$\sum_{k=1}^{\infty}k^3t^{k-1} = \frac{1+4t+t^2}{(1-t)^4}. \tag{4.3}$$

Observe that all factors on the right of (4.2) and (4.3) are quite visible in $\Delta_q R$. In some sense, the $\Delta_q R$ senses that the finite sum $L(n,q,t)$ has embedded in it the initial segments of the series in (4.2) and (4.3), extrapolates them to infinity, sums them, and then somehow hides the extrapolation error in the $u-1$ factors.

The $\Delta_q R$ also seems to be "aware" of Wheatstone's famous proof of (1.1). He observed that

$$\sum_{j=1}^{m}(m^2-m+2j-1) = m^3$$

so, thanks to an easily established summand non-overlapping property, the sum of positive cubes is the sum of consecutive odd numbers up to n^2+n-1. Thus, the discriminant has summed two infinite series whose partial sums are intimately related.

5. The Discriminant of $R(n, q, t, 1)$

In this section, we let $u = 1$. In this case, we have further factorization and the $n = 5$ case becomes

$$\begin{aligned}\Delta_q R = {} & 2^{20} t^{110} (t-1)^{22} (1+t)^{10} (1-t+t^2)^2 (1+t+t^2)^4 \\ & (1+3t^2+5t^4)(1+2^3 t+3^3 t^2+4^3 t^3+5^3 t^4) \\ & F_5(t)^4 G_5(t)^4 H_5(t)^2.\end{aligned}$$

Here, F_5, G_5, H_5 are all polynomials of degree 16. However, H_5 is unwieldy, with large coefficients that can be either positive or negative, and we see no clear pattern in them. On the other hand, the $G_5(t)$ is a polynomial with constant term $n^2 = 5^2$ (we conjecture that this generalizes), and coefficients that are all positive and much smaller than those of $H_5(t)$. We are most interested in $F_5(t)$ which has even smaller coefficients, all positive, and many properties of interest (already foreshadowed in the discussion of (1.7) and (1.8) in the introduction for the case $n = 4$).

6. The $F_n(t)$ and Certain Generators

One could define $F_n(t)$ as the conjectural unique factor of $\Delta_q R(n, q, t, 1)$ that is raised to the fourth power and has both its constant term and leading coefficient equal to 1. Can one conjecture an exact formula for $F_n(t)$?

Definition 6.1. Let

$$\begin{aligned}a_0(q) &= (1+q)(1+q+q^2),\\ a_1(q) &= 1+2q-q^4+2n(1-q^2)(1+q+q^2),\\ a_3(q) &= 2q^2(1+q+q^2),\\ a_4(q) &= q^3(1+q)\end{aligned}$$

and

$$b_0(q) = (1-q)^3(1-q^2).$$

Let

$$\begin{aligned}A(n,q) &= a_0(q) - q^n a_1(q) - q^{3n} a_3(q) + q^{4n} a_4(q),\\ B(n,q) &= a_0(q) b_0(q),\end{aligned}$$

and

$$f(n,q) = \frac{A(n,q)}{B(n,q)}. \tag{6.1}$$

Theorem 6.2. *The rational function $f(n, q)$ is a polynomial of degree $4(n-1)$ and*

$$\lim_{q\to 1} f(n,q) = \left(\frac{n(n+1)}{2}\right)^2. \tag{6.2}$$

Proof. If $q^3 = 1$ then $q^4 = q$ and $q^{4n} = q^n$, so one easily sees that $A(n,q)$ is divisible by $1+q+q^2$. By direct computation $A(n,q)$ and its first three derivatives with respect to q are zero at $q = 1$, and the same holds for $A(n,q)$ and its first derivative at $q = -1$. Thus, $A(n,q)$ is a multiple of $B(n,q)$ and the degree of $f(n,q)$ is $4(n-1)$. The limit (6.2) is obtained by four applications of L'Hopital's rule (computer algebra advised for this). □

The following is an immediate consequence.

Corollary 6.3. *The first n terms of $f(n,q)$ coincide with the Molien series sequence generated by*

$$\frac{1}{(1-q)^4(1+q)}.$$

If $f(n,q)$ is reversed, then its first n terms coincide with the Molien series sequence generated by

$$\frac{1}{(1-q)^2(1-q^2)(1-q^3)}.$$

The above Corollary only involves looking at $a_0(q)$, $a_4(q)$ and $b_0(q)$. For more about these series, see OEIS, items A002623 and A000601. We add that the coefficient of q^{3n} is simply a shift of the sequence generated by $2G(x)$ where $G(x)$ is given by (1.3). The sequence that $G(x)$ generates is A006918. Among the coefficients of powers of q^n occurring in (6.1), that of q^n is the most intricate. It is a linear combination of the generator of A002620 and a function generating a sequence (possibly a Molien series sequence?) that is not listed in OEIS.

A variety of other identities for $f(n,q)$ can be read directly from its formula. For example,

$$f(2m,-1) = -m^2 \quad \text{and} \quad f(2m+1,-1) = (m+1)^2.$$

Theorem 6.4. *We have*

$$f(n,q) = q^{-n}L(n,q,q).$$

Proof. We can prove this by computer algebra. □

We now observe that $f(4,q)$ is the polynomial of (1.8) with t replaced by q. This will lead us to a conjecture that is fundamental to this investigation.

Definition 6.5. Let

$$G = G(j,t,z) = \frac{1+t(t+1)^2z+t^4z^2}{(1-z)(1-tz)(1-t^3z)^j(1-t^4z)}, \tag{6.3}$$

where $j \geq 1$ is an integer. Define polynomials $W(j,n,t)$ by

$$G(j,t,z) = 1 + \sum_{n=1}^{\infty} W(j,n,t)z^n. \tag{6.4}$$

Conjecture 6.6. *We have*

$$F_n(t) = f(n,t).$$

Moreover, the $F_n(t)$ *polynomials are essentially generated by* $G(2,t,z)$*. More precisely,*

$$t^{4(n-1)} F_n\left(\frac{1}{t}\right) = W(2, n-1, t).$$

To get a feeling for $G(2,t,z)$, set $t = 1$. One obtains (see (4.3))

$$\frac{1+4z+z^2}{(1-z)(1-z)^4} = \sum_{n=1}^{\infty}\left(\sum_{k=1}^{n} k^3\right) z^{n-1}$$

and the connection with cubes becomes explicit. It seems that the $W(j,n,t)$ polynomials are of independent interest for all integers $j \geq 1$ and we shall look at them further in Sec. 8. But what we have done so far all stems from the identity (1.2). What would our procedures yield for the twin identity (1.6)? If we introduce a t-variable and define an appropriate numerator R^* as before, we find that the analogue to the discriminant factorization (4.1) for $n = 5$ is

$$\begin{aligned}\Delta_q R^* = -2^4 3^{-4} t^{110}(t-1)^{20}(1+t)^6(1+t+t^2)^{12}(1-t^2+2t^3)^4 \\ (1+t+2t^2+2t^3+3t^4)(3+15t+42t^2+90t^3+165t^4) \\ (F_5^*(t))^4 (G_5^*(t))^4 (H_5^*(t))^2,\end{aligned} \tag{6.5}$$

where

$$\begin{aligned}F_5^*(t) = 3(1 + 2t + 4t^2 + 7t^3 + 11t^4 + 10t^5 + 11t^6 + 12t^7 + 10t^8 \\ + 9t^9 + 9t^{10} + 6t^{11} + 5t^{12} + 4t^{13} + 2t^{14} + t^{15} + t^{16})\end{aligned}$$

and $H_5^*(t)$ has degree 27 and very large coefficients. We note that the coefficients of the $3 + 15t + 42t^2 + \cdots$ polynomial in (6.5) are consecutive $s(k)$ numbers, and the discriminant is pairing them with the coefficients of F_5^*. In each case, the coefficient sum is 315. Calculation indicates that the $F_n^*(t)$ polynomials arising from (1.6) are of two different types depending upon the parity of n.

Conjecture 6.7. *The* ${F_n}^*(t)$ *polynomials are generated by rational functions in* t *and* z *with denominator*

$$(1-z)(1-t^2z)^2(1-t^6z)(1-t^8z). \tag{6.6}$$

For $n \geq 1$ *odd, we have polynomials of degree* $4(n-1)$ *and the numerator of their generator is*

$$1 + t(2+t^2)\frac{1-t^5}{1-t}z + t^4(1+2t^2)\frac{1-t^5}{1-t}z^2 + t^{11}z^3. \tag{6.7}$$

For $n \geq 2$ even, we have polynomials of degree $4n - 5$ and the numerator of their generator is

$$1 + t + t^3 + t^2(1 + t^2)\frac{1 - t^5}{1 - t}z + t^7(1 + t^2 + t^3)z^2. \tag{6.8}$$

For n odd and $t = 1$, the generator is

$$\frac{(1 + x)(1 + 14x + x^2)}{(1 - x)^5}$$

the generating function of A187756. For n even and $t = 1$, the generator is

$$\frac{(3 + x)(1 + 3x)}{(1 - x)^5} \tag{6.9}$$

the generating function of the Vassiliev invariants of fourth order for the torus knots. See A180324 and [1].

Remark 6.8. We do not understand why the fifth cyclotomic polynomial makes multiple appearances here. We do observe that when $t = -1$ the generators become $+(1 - x)^{-2}$ and $-(1 - x)^{-3}$, respectively.

7. Wheatstone Conjectures

In Sec. 4, we have described how Wheatstone proved (1.1) by breaking cubes into sums of odd numbers. Since the coefficients of the $f(n, t)$ add to the square of the nth triangular number by (6.2), it is natural to ask if they share this property of the distinct positive odd integers.

Definition 7.1. Given two sequences or multisets of positive integers S_1 and S_2, we say that S_2 has the Wheatstone property with respect to S_1 if S_2 can be partitioned into sub-multisets such that if each is replaced by its sum, the result is S_1. We say S_2 is Wheatstone with respect to S_1.

Clearly, if S_2 is Wheatstone with respect to S_1, the sum of the elements of S_2 equals the sum of the elements of S_1. Also, S_2 must have at least as many elements as S_1.

Conjecture 7.2. *The multiset of coefficients of $f(n, t)$, and of its conjectural equal $F_n(t)$, have the Wheatstone property with respect to a uniquely determined initial segment of the sequence of the cubes of the positive integers.*

For $F_3(t)$, we have the coefficients

$$\{1, 3, 7, 6, 7, 5, 4, 2, 1\}.$$

Once the obligatory $\{1\}$ is chosen for the first multisubset, any multisubset of the remaining elements with sum 27 (there are several) leaves behind numbers than sum to 8. For $F_n(t)$ with larger n one needs to choose multisubsets a bit more carefully, but it seems that the number of ways of constructing such a "Wheatstone disjoint union" increases. For $n = 5$, we have

$$1 = 1,$$
$$1 + 7 = 8,$$
$$2 + 3 + 22 = 27,$$
$$7 + 11 + 21 + 25 = 64,$$
$$4 + 13 + 14 + 19 + 23 + 25 + 27 = 125.$$

This is not unique. Since, e.g., $3+22 = 25$ and $7+11 = 4+14$ we easily find others.

Conjecture 7.3. *Let $E_n(q)$ be the polynomial on the right side of* (1.2). *The multiset of its coefficients satisfies the conclusion of Conjecture* 7.2.

Conjecture 7.3 has been verified for $n \leq 8$. One may create further conjectures of this type based on the coefficients of $L(1, q, q^m)$ for m a positive integer. We leave it to the reader to investigate the corresponding question for the twin identity (1.6).

Remark 7.4. As the title of [13] indicates, (1.1) can be viewed as being about arithmetic progressions. Good examples of sequences that are "almost" arithmetic progressions are Beatty sequences. For a paper about Beatty sequences with a Nicomachean theme, see [9].

8. The $W(j, n, t)$ Polynomials

Recall the polynomials defined by (6.3) and (6.4). Just as with $j = 2$, the $j = 1$ polynomials have an immediate connection with q-analogues of cubes.

Theorem 8.1. *Let*

$$a = a(n,t) = \frac{1-t^n}{1-t}, \quad b = b(n,t) = a(n,-t), \quad c = c(n,t) = \frac{1+t^n}{1-t^n},$$

and

$$P(n,t) = \begin{cases} a^3bc, & n \text{ even}, \\ a^3b, & n \text{ odd}. \end{cases}$$

Then every root of $W(1, n, t)$ has absolute value 1, and

$$W(1, n-1, t) = P(n, t).$$

Proof. We can prove this by computer algebra. □

The above contrasts considerably with the case $j = 2$ in which all polynomials seem to be irreducible.

Now, call roots of absolute value greater than 1 large roots and those with absolute value less than one small roots.

Conjecture 8.2. *Let $j \geq 2$. Then*

$$W(j,n,x) = 1 + 2x + 4x^2 + (j+5)x^3 + \cdots + (j+1)x^{4n-1} + x^{4n}$$

has exactly n large roots. In other words, exactly $1/4$ of the roots are large.

Conjecture 8.2 implies that the sum of all roots of $W(j,n,x)$ is $-(j+1)$.

Conjecture 8.3. *The sum of the large roots of $W(j,n,x)$ is $-j+\epsilon(n)$ where $\epsilon(n) \to 0$ as $n \to \infty$.*

The above conjecture asserts that most of the root sum, for n and j large, is due to the large roots, even though $3/4$ of the roots are (conjecturally) small. The property of having exactly $1/4$ of the roots either inside or outside of the unit circle seems robust in two different ways.

Conjecture 8.4. *For the polynomials generated by the quotient (6.7) divided by (6.6), one of the analogues of the reversal of W for the twin Nicomachean identity, exactly $1/4$ of the roots are small. Moreover, if we replace the numerator of (6.7) by 1 or do the same for the numerator of G, again exactly $1/4$ of the roots will be either all inside or all outside the unit circle.*

It is now tempting to (1) symmetrize the denominator of G by multiplying it with another $(1-tz)$ factor and to use 1 as the numerator, and (2) do the same for (6.6), introducing a further factor of $(1-t^6z)$ in the denominator.

Conjecture 8.5. *If*

$$\frac{1}{(1-z)(1-tz)^2(1-t^3z)^2(1-t^4z)} = 1 + \sum_{n=1}^{\infty} g_n(t)z^n, \tag{8.1}$$

then each self-reciprocal polynomial $g_n(t)$ has degree $4n$ with n large roots, n small roots, and $2n$ roots on the unit circle. For the polynomials constructed in this way from (6.6), the degree of the nth one is $8n$, and it has $2n$ small roots, $2n$ large roots, and $4n$ roots on the unit circle.

9. The Cigler-Wheatstone q-Analogue

Another paper that takes (1.2) in a direction different from Schlosser's is that of Cigler [3]. He produced a new q-analogue of (1.1) by taking Wheatstone's observation [13] that

$$\sum_{j=1}^{n}(n^2 - n + 2j - 1) = n^3, \tag{9.1}$$

replacing it by the q-analogue

$$\sum_{j=1}^{n} q^{n-j}\frac{1-q^{n^2-n+2j-1}}{1-q} = \frac{1-q^n}{1-q}\cdot\frac{1-q^{n^2}}{1-q}, \tag{9.2}$$

and following Wheatstone's procedure to produce his analogue, a result that does not obviously fit into the basic hypergeometric framework. Our interest here is in seeing how the discriminant process will take the partition of n^3 in (9.1) and produce a different partition.

Theorem 9.1. *If every summand on the left of* (9.2) *is multiplied by* t^j, *the sum over* j *becomes*

$$\frac{-tQ(q,t)}{(-1+q)(q-t)(-1+qt)},$$

where

$$\begin{aligned} Q(q,t) &= q^n - q^{n^2+1} + t(q^{n^2} - q^{n+1}) + t^n(q^{n^2+n+1}-1) \\ &\quad + t^{n+1}(q - q^{n^2+n}). \end{aligned}$$

Proof. We can prove this by computer algebra. □

Example 9.2. Let $n = 5$. Then $\Delta_q Q$, the discriminant of Q with respect to q is

$$\begin{aligned} \Delta_q Q &= t^{40}(1-t)^{10}(1+t)^2(1-t+t^2-t^3+t^4) \\ &\quad (21+23t+25t^2+27t^3+29t^4)^3 \\ &\quad (5(1+t+t^2+\cdots+t^{19}+t^{20}) + (4+3t^2+2t^4+t^6)(t^{21}+t^{22}))^4 \cdot I(t)^2, \end{aligned}$$

where $I(t)$ is a polynomial of degree 90 with large coefficients, many exceeding 10^{19}. Here,

$$21+23t+25t^2+27t^3+29t^4$$

is the t-analogue of the Wheatstone sum for 5^3, while the coefficients of the factor

$$5(1+t+t^2+\cdots+t^{19}+t^{20}) + (4+3t^2+2t^4+t^6)(t^{21}+t^{22})$$

before $I(t)^2$ is the new partitioning of 5^3.

Conjecture 9.3. *The above new partitioning of* n^3 *is given by the coefficients of the power series expansion of*

$$\frac{n}{1-t} - \frac{t^{n^2-n+1}(1-t^{2n})}{(1-t)(1-t^2)}. \tag{9.3}$$

That (9.3) actually approaches n^3 as $t \to 1$ does not seem at all obvious, but this can be shown by two applications of L'Hopital's rule. Now, by (9.1) we have

$$n^3 = \sum_{j=1}^{n}(n^2-n+1+2(j-1)) = n(n^2-n+1) + 2(1+2+\cdots+(n-1)).$$

What the discriminant tells us in this case is that n^2-n+1 copies of the integer n plus two copies of every smaller positive integer add to n^3. This immediately raises a question of a by now familiar type: can these numbers be batched so exactly the summands of (9.1) are recovered? For example, when $n=3$, we have

$$3^3 = 7+9+11 = (3+3+1)+(3+3+3)+(3+3+2+2+1).$$

Definition 9.4. For integers $n \geq 1$, let $\mathrm{Wh}(n)$ be the sequence of integers

$$\{n^2-n+2j-1 : 1 \leq j \leq n\}.$$

Let $J(n)$ be the multiset of n^2+n-1 integers consisting of

(1) n^2-n+1 copies of n,
(2) 2 copies of each of $\{1,2,\ldots,n-1\}$.

Theorem 9.5. *There exists a multiset partitioning of $J(n)$ into an ordered list of sub-multisets such that the sum of the elements in the jth term is*

$$n^2-n+2j-1, \quad 1 \leq j \leq n,$$

the jth element of $\mathrm{Wh}(n)$.

Example 9.6. Let $(m)_n$ denote the multiset consisting of m copies of n. Then for $n=6$ an example would be

$$\{\{(5)_6,1\},\{(5)_6,3\},\{(5)_6,5\},\{(6)_6,1\},\{(6)_6,3\},\{(4)_6,2,2,4,4,5\}\}$$

which collapses to $\{31,33,35,37,39,41\}$ upon addition (and which in turn collapses to 6^3 upon addition).

Proof of Theorem 9.5. We provide the proof only for n odd. A similar argument applies to the n even case. Write $n=2m+1$.
Step 1. From each of the numbers

$$n^2-n+2j-1, \quad 1 \leq j \leq m,$$

subtract $n-1$ copies of n. The remainders will be the

$$n^2-n+2j-1-n(n-1) = 2j-1, \quad 1 \leq j \leq m.$$

These are the consecutive odd integers from 1 to $2m-1$ $(=n-2)$.
Step 2. From each of the numbers

$$n^2-n+2j-1, \quad m+1 \leq j \leq 2m,$$

subtract n copies of n. The remainders will be the

$$n^2-n+2j-1-n^2 = -(2m+1)+2j-1 = 2(j-m-1), \quad m+1 \leq j \leq 2m.$$

These are the consecutive even integers from 0 to $2m-2$ $(=n-3)$.

The number of copies of n used above is

$$(n-1)m + n \cdot m = n^2 - n + 1 - \frac{n+1}{2} < n^2 - n + 1$$

and thus less than the number of copies of n in $J(n)$. Also, the remainders are exactly the distinct numbers

$$\{1, 2, 3, \ldots, n-2\}$$

all of which are less than n and belong to $J(n)$. Thus, the n's selected from $J(n)$ in the above two steps, together with the remainders listed there, form a proper subset of $J(n)$. Now,

$$n(n^2 - n + 1) + 2\sum_{j=1}^{n-1} j = \sum_{j=1}^{n} (n^2 - n + 2j - 1) = n^3.$$

Thus, the last term of the middle expression above, namely, $n^2 + n - 1$, must be the sum of all of the elements of $J(n)$ that were not among the subtracted n's nor among the resulting remainders. □

Remark 9.7. As the previous example indicates, for n even all of the remainders are odd, at most $n-1$, and no remainder occurs more than twice.

10. A "Wild" Polynomial of Interest

We now apply our discriminant procedure to a very simple q-analogue and show that the "wild" polynomial therein has a remarkable structure of a sort the authors have not previously seen. We begin with the identity

$$\sum_{k=1}^{n} \frac{1-q^k}{1-q} t^k = \frac{t(-1+q+t^n(1-q^{n+1}) + qt^{n+1}(q^n - 1))}{(1-q)(1-qt)(t-1)}$$

and take the discriminant of the second factor of the numerator with respect to q for $n = 13$. We obtain

$$(-1+t)^{28} t^{156} (13 + 12t + 11t^2 + \cdots + 2t^{11} + t^{12})^2$$
$$(1 + 2t + 3t^2 + \cdots + 12t^{11} + 13t^{12})^2 \cdot y(t),$$

so in this case the partition pairing is rather trivial. Here the "wild" $y(t)$ has degree $11 \cdot 12 = 132$ with large coefficients, many exceeding 10^{22}. Note that

$$11 \cdot 12 = (11 + 10 + 9 + \cdots + 2 + 1) + (1 + 2 + 3 + \cdots + 10 + 11). \qquad (10.1)$$

When the zeros of $y(t)$ are plotted we observe two cluster patterns. The first has its rightmost cluster in the first quadrant, a tight cluster of 11 zeros. Proceeding counterclockwise in a roughly circular contour centered at the origin, we find next a tight cluster of 10 zeros, then 9, and so on. Finally, we observe a single zero in the fourth quadrant. The second pattern is simply the complex conjugate of the first.

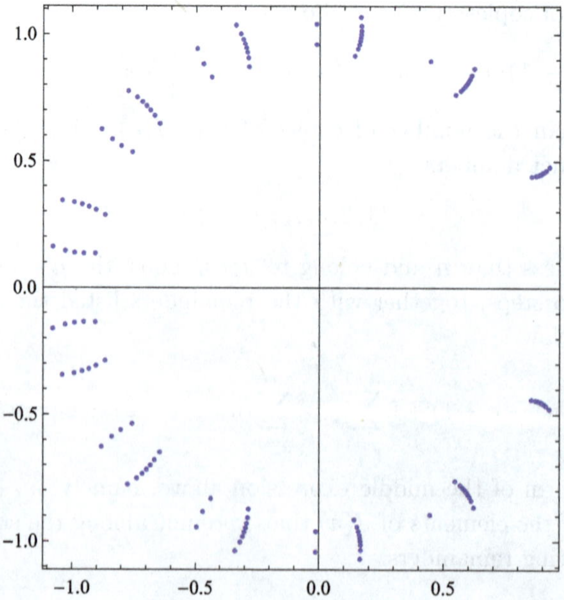

Fig. 1. The zeros of $y(t)$.

Thus we have a geometric realization of the identity (10.1) that also illustrates the fundamental theorem of algebra. We note that in each sizeable cluster the zeros seem to lie on arcs of rather small curvature.

11. Cubic Interlacing Inequalities

The sequence (1.4) displays the interlacing of the cubes with the $s(k)$. If we display the interlacing as

$$k^3 < s(k) = 3\sum_{j=1}^{k} j^2 < (k+1)^3$$

we see a sort of discrete version of

$$\int 3x^2\,dx = x^3 + C.$$

The interlacing here is special in that $s(k)$ is not merely between k^3 and $(k+1)^3$, but between their arithmetic and geometric means. This follows from the identity as follows.

Proposition 11.1. *For any positive integer k,*

$$s(k)^2 - k^3(k+1)^3 = \left(\frac{k(k+1)}{2}\right)^2 \tag{11.1}$$

and

$$s(k) + \frac{2k+1}{2} = \frac{k^3 + (k+1)^3}{2}.$$

However, the arithmetic mean upper bound is not very sharp. If written in the form

$$s(k)^{1/3} < \left(\frac{k^3 + (k+1)^3}{2}\right)^{1/3}$$

it suggests considering the class of pth power means

$$M_p(a,b) = \left(\frac{a^p + b^p}{2}\right)^{1/p}$$

and looking for an M_p upper bound with $p < 3$.

Conjecture 11.2. *Let*

$$E(k) = 648k^3(M_{1/3}(k, k+1) - s(k)^{1/3}).$$

Then

$$0 < E(k) < 1 \quad \textit{and} \quad \lim_{k\to\infty} E(k) = 1.$$

Observe that $0 < E(k) < 1$ is equivalent to

$$M_{1/3}(k, k+1) - \frac{1}{648k^3} < s(k)^{1/3} < M_{1/3}(k, k+1). \tag{11.2}$$

The $p = 1/3$ mean is discussed in [8] where it is stated that in physics it is known as the Lorenz mean. The fact that the $p = 1/3$ mean exceeds $s(k)^{1/3}$ is a consequence (Corollary 11.4) of the following result.

Theorem 11.3. *If $k > 0$ and $x > 0$, then*

$$\begin{aligned}\left(\frac{1}{2}k(k+x)(2k+x)\right)^{1/3} &\le \frac{(x+k) - k}{\log(x+k) - \log k} \le \left(\frac{k^{1/3} + (k+x)^{1/3}}{2}\right)^3 \\ &\le \frac{27}{8}\left(\frac{1}{k+x}\right)^2 \left(\frac{1}{k+x} + \frac{1}{2k+x}\right)^{-3}.\end{aligned} \tag{11.3}$$

with equality if and only if $x = 0$.

Proof. Denote the above inequality, in an obvious notation, by

$$E_1(x) \le E_2(x) \le E_3(x) \le E_4(x).$$

Here, $E_2(x) \le E_3(x)$ is shown in [8]. Next, $E_1(x) \le E_2(x)$ is equivalent to

$$L(x) = \log\frac{k+x}{k} \le \frac{x}{\left(\frac{1}{2}k(k+x)(2k+x)\right)^{1/3}} = R(x).$$

This holds with equality when $x = 0$ and its validity, with strict inequality for $x > 0$, follows from the differential identity

$$(R'(x))^3 - (L'(x))^3 = \frac{x^4(3k+x)(3k+2x)}{27k(k+x)^4(2k+x)^4} > 0.$$

Similar considerations establish $E_3(x) \leq E_4(x)$. □

It follows from Theorem 11.3 by taking $x = 1$ that the 1/3 power mean is an upper bound for $s(k)^{1/3}$.

Corollary 11.4.

$$s(k)^{1/3} < M_{1/3}(k, k+1).$$

Example 11.5. Take $k = 10$ in (11.3). Then the differences $E_{i+1} - E_i$ for $1 \leq i \leq 3$ are

$$\{1.2016\cdots \times 10^{-6}, 1.3357\cdots \times 10^{-7}, 0.0003308\cdots\}.$$

Conjecture 11.6. *If k is large then the continued fraction of $E((3k+1)^3)$ is*

$$\{0, 1, B_3(k), 2, 2, 3, 14, B_8(k), \ldots\},$$

where $B_3(k) = 6k(1+3k+3k^2)$, $3196 < B_3(k)/B_8(k) < 3197$, and $B_3(k)/B_8(k) \to 3196$ as $k \to \infty$.

The geometric lower bound (11.1) is, from the q-analogue point of view, rather a "cliffhanger". Designate the q-analogues of k^3 and $s(k)$ taken from (1.2) and (1.6) by

$$\mathrm{wcube}(k,q) = q^{2n-2k}\frac{(q^k-1)^2}{(q-1)^3}\frac{q^{2k}-1}{q+1}$$

and

$$s(k,q) = q^{2(n-k)}\frac{(1-q^k)(1-q^{k+1})(1-q^{2k+1})}{(1-q)^3(1+q)}.$$

Then

$$s(k,q)^2 - \mathrm{wcube}(k,q)\mathrm{wcube}(k+1,q)$$
$$= \frac{q^{4(n-k)-2}}{(1+q)^2}\left(\frac{q^k-1}{q-1}\right)^2\left(\frac{q^{k+1}-1}{q-1}\right)^2 U(k,q), \tag{11.4}$$

where

$$U(k,q) = \frac{T(k,q)}{q-1}$$

with

$$T(k,q) = 1 + q - q^{2k} - q^{2k+1} - 2q^{2k+2} + q^{4k+2} + q^{4k+3}.$$

Observe that (11.4) is another q-analogue of the squared kth triangular number.

Now (11.1) requires that $U(k,q) \to 1$ as $q \to 1$. But a plot of $U(k,q)$ for $0 < q < 9/10$ with (e.g.,) $k = 10$ shows the graph of an always negative function becoming more negative at an increasing rate. That the limit is 1 is, however, an easy consequence of L'Hopital's rule. Thus, this polynomial must suddenly increase rapidly in value for some q very close to 1. For sizeable k, where does this sudden change occur? Make the change of variable from q to b given by

$$q = 1 - \frac{1}{bk^2}.$$

Then, by multiple applications of the binomial expansion,

$$U(k,q) = 1 - \frac{8}{b} - \frac{2}{bk^2} - \frac{10}{bk} + \frac{16}{b^2k} + O\left(\frac{1}{b^2k^2}\right).$$

Thus, for k large, there is a transition to small in absolute value negative numbers when q is near $1 - 1/8k^2$, and for b large the value is indeed close to 1. For k large the roots of $T(k,\, q) = 0$ have a distribution of interest with conjecturally exactly $2k$ large roots, but this is beyond the scope of this paper.

Remark 11.7. The Nicomachean identity (1.1) has already inspired work on associated inequalities. See [7].

12. One Third of Nicomachus' Identity

We say that the binomial coefficient identity

$$\sum_{m=1}^{n} \binom{m+1}{3} = \binom{n+2}{4} \tag{12.1}$$

is one-third of Nicomachus' identity. The justification for this is the identity

$$\sum_{m=1}^{n} \left(\frac{2}{3}m\left(m^2 + \frac{1}{2}\right) + \frac{1}{3}t(m-1)m(m+1)\right)$$
$$= \left(\frac{n(n+1)}{2}\right)^2 + 2\binom{n+2}{4}(t-1). \tag{12.2}$$

Here, $t = 1$ is (1.1) while if both sides are divided by $2t$, then the limit as t goes to infinity is (12.1). For a more detailed justification of this terminology, see the analysis of the function $A = A(a,x)$ that is introduced after Conjecture 12.10.

The above suggests applying our discriminant procedure to

$$\sum_{k=1}^{n} q^{2(n-k)} \frac{1-q^{k-1}}{1-q} \frac{1-q^k}{1-q} \frac{1-q^{k+1}}{1-q} t^k.$$

The factorization of the resulting discriminants for positive integers n produces among other things two sequences of polynomials whose coefficients are positive integers, that provide, conjecturally, partitions of $\binom{n+2}{4}$ that are alternative to those of (12.1).

Definition 12.1. Define polynomials $J_n(t)$ by

$$\frac{1}{(1-x)(1-tx)^2(1-t^2x)(1-t^3x)} = \sum_{n=0}^{\infty} J_n(t)x^n, \tag{12.3}$$

so $J_0(t) = 1$ and $J_1(t) = 1 + 2t + t^2 + t^3$.

We shall see that $J_n(t)$ has degree $3n$.

Lemma 12.2. *If $r_{3k}(t)$ and $s_{2j}(t)$ are polynomials of exact degrees $3k$ and $2j$, respectively, and*

$$\left(\sum_{k=0}^{\infty} r_{3k}(t)x^k\right)\left(\sum_{j=0}^{\infty} s_{2j}(t)x^j\right) = \sum_{m=0}^{\infty} u_m(t)t^m,$$

then $u_m(t)$ has exact degree $3m$.

Proof. If k, j and m are non-negative integers constrained by $k + j = m$, then

$$\max(3k + 2j) = 3m.$$

The result follows by Cauchy multiplication rule. □

To see that $J_n(t)$ has exact degree $3n$, observe that

$$\frac{1}{(1-tx)^2(1-t^2x)} = \sum_{j=0}^{\infty} t^j x^j \sum_{h=1}^{j+1} ht^{j+1-h}$$

and

$$\frac{1}{(1-x)(1-t^3x)} = \sum_{k=0}^{\infty}\left(\sum_{h=0}^{k} t^{3h}\right) x^k.$$

Conjecture 12.3. *One of the polynomial sequences produced by our discriminant process is $J_n(t)$, $n = 0, 1, 2, \ldots$. It has exactly n small roots, so exactly $1/3$ of its zeros are small. The coefficients of the lower order terms of the $J_n(t)$ converge to the Molien series A000601. The coefficients of the lower order terms of the reversed polynomial $t^{3n}J_n(1/t)$ converge to the Molien series A266769.*

For the other sequence, we shall only conjecture that it provides yet another alternative to the partition described in (12.1), that all of its roots are large, and that its reversal provides a sequence of coefficients converging to the sequence A028289.

We note that the absolute value of $J_n(-1)$ is always a triangular number, and that $J_{3n-1}(t)$, $n \geq 1$, is always divisible by $1 + t + t^2$. Despite the appearance of this factor $1/3$ of the time, computer graphics indicate that of all regions near the unit circle, the density of zeros of $J_n(t)$ is most sparse in the vicinity of the three cube roots of unity.

Definition 12.4. Set

$$J_n^*(t) = \frac{J_{3n-1}(t)}{1+t+t^2}.$$

Conjecture 12.5.

$$\text{Resultant}(1+t+t^2, J_n(t)) = \begin{cases} \left(\frac{n}{3}+1\right)^2, & n \equiv 0 \bmod 3, \\ \left(\frac{n-1}{3}+1\right)^2, & n \equiv 1 \bmod 3, \\ 0, & n \equiv 2 \bmod 3. \end{cases}$$

The $J_n^*(t)$ polynomials take us back to (1.1) and squared triangular numbers.

Conjecture 12.6.

$$\text{Resultant}(1+t+t^2, J_n^*(t)) = \sum_{k=1}^{n} k^3 = \left(\frac{n(n+1)}{2}\right)^2.$$

Now, symmetrize the generator on the left of (12.3) by putting another factor of $1-t^2x$ into the denominator. Call the nth polynomial produced by this symmetrized generator $j_n(t)$. Thus, $j_0(t) = 1$ and $j_1(t) = 1+2t+2t^2+t^3$.

Conjecture 12.7. *The polynomial $j_n(t)$ is a self-reciprocal polynomial of degree $3n$ that has exactly n zeros on the unit circle if n is even, and exactly $n+2$ zeros on the unit circle if n is odd.*

Definition 12.8. Let S and T be finite sets of distinct points on the unit circle. Assume that $\text{Im}(z) > 0$ for all z in $S \cup T$. Also, assume

$$\text{card}(S) = \text{card}(T) = m,$$

and let

$$S^* = S \cup \{-1\} \cup \{+1\}$$

so $\text{card}(S^*) = m+2$. Number the elements of S^* as

$$\{s_1, s_2, s_3, \ldots, s_{m+2}\}$$

so that

$$\text{Re}(s_i) < \text{Re}(s_{i+1}), \quad 1 \le i \le m+1.$$

Then we say T "almost interlaces" S if each element of T is between two elements of S^* on the unit circle, and each open arc (s_i, s_{i+1}) contains at most one element of T.

It is easily seen and important to note that if T almost interlaces S, then there is at least one open arc

$$I_i = (s_i, s_{i+1})$$

that contains no element of T. We can now make precise the manner in which the small zeros of $j_n(t)$ interlace the zeros of that polynomial that lie on the unit circle.

Conjecture 12.9. *Let S be the zeros of $j_n(t)$, where $n \geq 2$ is even, that lie on the unit circle and have positive imaginary parts. Then project each small zero of that polynomial having positive imaginary part onto the unit circle along a ray emanating from the origin. Let T be the resulting subset of the unit circle. Then S and T have the same cardinality $n/2$ and T almost interlaces S. Moreover, there is an open interval I_i that contains no point of T, such that its closure contains the cube root of unity $(-1 + i\sqrt{3})/2$.*

Next, define $j_n^*(t)$ by $j_n^*(t) = j(t)$ if $1 + t + t^2$ and $j_n(t)$ are relatively prime, and by $j_n^*(t) = j_n(t)/(1 + t + t^2)$ otherwise.

Conjecture 12.10. *Write $n = 3m + r$ where m and r are positive integers and $0 \leq r \leq 2$. Then*

$$\text{Resultant}(1 + t + t^2, j_n^*(t)) = \begin{cases} (m+1)^2, & r = 0, \\ \left(\dfrac{(m+1)(m+2)}{2}\right)^2, & r = 1 \text{ or } 2. \end{cases}$$

To explain where the notion of "one third of Nicomachus' identity" came from, we shall now "trisect" individual cubes. The cubes are generated by the right side

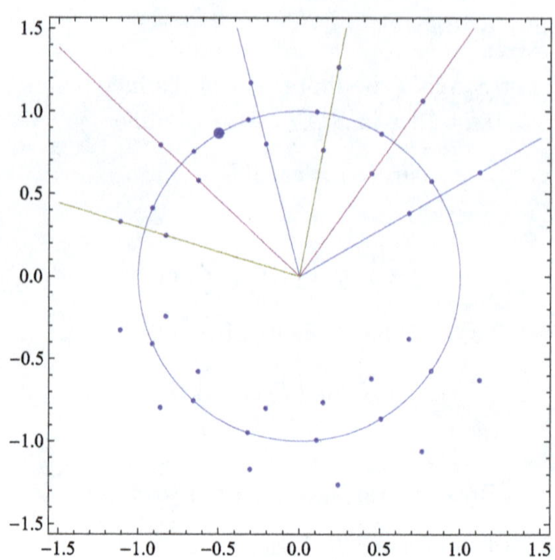

Fig. 2. The zeros of $j_{12}(t)$ and $(-1 + i\sqrt{3})/2$.

of (4.3). We bring in a new parameter a and think of (4.3) as the $a = 1$ case of

$$A = A(a,x) = \frac{1+(a+1)^2x + a^2x^2}{(1-x)^2(1-a^2x)^2}.$$

The expansion of A yields polynomials whose coefficients represent partitions of cubes. Now, A can be sectioned as

$$\begin{aligned} A(a,x) &= \frac{(1+x)(1+a^2x)}{(1-x)^2(1-a^2x)^2} + \frac{2ax}{(1-x)^2(1-a^2x)^2} \\ &= A_1(a,x) + A_2(a,x). \end{aligned}$$

Note that

$$A_2(a,x) = 2\sum_{n=1}^{\infty}\sum_{j=1}^{n} j(n-j)a^{2j-1}x^{n-1}. \tag{12.4}$$

We now observe that the coefficient of x^3 in A is

$$7 + 6a + 15a^2 + 8a^3 + 15a^4 + 6a^5 + 7a^6$$

while that of x^3 in $A_2(a,x)$ is

$$6a + 8a^3 + 6a^5.$$

At $a = 1$ these have the values 64 and 20, respectively, with ratio 5/16. Further such calculations indicate that $A_2(a,x)$ is "roughly 1/3 of A". Since

$$A_2(1,x) = \frac{2x}{(1-x)^4} = 2(x + 4x^2 + 10x^3 + 20x^4 + 35x^5 + \cdots)$$

the origin of the binomial coefficients that featured in (12.1) and (12.2) becomes apparent.

Now, replace a by z and write

$$A_2(z,x) = \sum_{n=1}^{\infty} p_n(z)x^{n-1}.$$

Theorem 12.11. *Except for $z = 0$, all zeros of $p_n(z)$ lie on the unit circle.*

Proof. Since $z^{2j-1} = z \cdot z^{2(j-1)}$, it suffices by (12.4) to show that

$$f(n,z) = \sum_{j=1}^{n} j(n-j)z^{j-1}$$

has all zeros on the unit circle. Now,

$$\begin{aligned} f(n,z) &= \frac{1-n+(n+1)z-(n+1)z^n+(n-1)z^{n+1}}{(z-1)^3} \\ &= \frac{2}{z-1}z^{\frac{n+1}{2}}\frac{d}{dz}\left(\frac{1-z^n}{1-z}\cdot\frac{1}{z^{(n-1)/2}}\right). \end{aligned}$$

Now, if any polynomial $g = g(z)$ has all of its zeros on the unit circle, then so does the rational function

$$\frac{d}{dz}\left(\frac{g(z)}{z^{d/2}}\right)$$

where d is the degree of g. This is [4, Lemma 4.2.1] with the j of that paper set equal to 1. The result follows. □

13. A Remark

There is much more to be done here, since there are in fact (1) infinitely many Nicomachus' identities and (2) infinitely many twin Nicomachus' identities. Let $T(n)$ be the nth triangular number and j a positive integer. For (1), we have

$$\sum_{k=1}^{n} k(k+j)(k+2j) = T(n)T(n+2j)$$

and for (2), we have

$$\sum_{k=1}^{n} \frac{1}{2}k(k+j)(2k+j) = T(n)T(n+j).$$

In this vein, we have infinitely many Warnaar's identities

$$\sum_{k=1}^{n} \frac{q^{2(n-k)}}{q+1}\frac{q^k-1}{q-1}\frac{q^{k+j}-1}{q-1}\frac{q^{2k+j}-1}{q-1} = \frac{(q^n-1)(q^{n+1}-1)(q^{j+n}-1)(q^{j+n+1}-1)}{(q-1)^4(q+1)^2} \tag{13.1}$$

and our corresponding twin identities are

$$\sum_{k=1}^{n} \frac{q^{2(n-k)}}{q+1}\frac{q^k-1}{q-1}\frac{q^{k+j+1}-1}{q-1}\frac{q^{2k+j+1}-1}{q-1} = \frac{(q^n-1)(q^{n+1}-1)(q^{j+n+1}-1)(q^{j+n+2}-1)}{(q-1)^4(q+1)^2}. \tag{13.2}$$

When $j = 0$, (13.1) and (13.2) become (1.2) and (1.6), respectively.

Acknowledgments

Seon-Hong Kim was supported by NRF of Korea by Basic Science Research Program through the National Research Foundation of Korea(NRF) funded by the Ministry of Education (2018R1D1A1B07045435).

References

[1] S. V. Allënov, Explicit formulas for Vassiliev invariants of the fourth order for knots, *J. Math. Sci. (N.Y.)* **157** (2009) 413–423.

[2] G. E. Andrews, I. P. Goulden and D. M. Jackson, Shanks' convergence acceleration transform, Padé approximants and partitions, *J. Combin. Theory Ser. A* **43** (1986) 70–84.

[3] J. Cigler, Sum of cubes: Old proofs suggest new q-analogues, preprint (2014).

[4] D. W. Farmer and R. C. Rhoades, Differentiation evens out zero spacings, *Trans. Amer. Math. Soc.* **357** (2005) 3789–3811.

[5] G. Gasper and M. Rahman, *Basic Hypergeometric Series*, Encyclopedia of Mathematics and its Applications, Vol. 35. (Cambridge University Press, Cambridge, 1990).

[6] I. M. Gelfand, M. M. Kapranov and A. V. Zelevinsky, *Discriminants, Resultants and Multidimensional Determinants*, American Mathematical Society (Birkhäuser, Boston, 1994).

[7] M. S. Klamkin and D. J. Newman, Inequalities and identities for sums and integrals, *Amer. Math. Monthly* **83** (1975) 26–30.

[8] T. P. Lin, The power mean and the logarithmic mean, *Amer. Math. Monthly* **81** (1974) 879–883.

[9] F. Luca, G. Polanco and W. Zudilin, A variation on the theme of Nicomachus, *Bull. Aust. Math. Soc.* **97** (2018) 367–373.

[10] M. Schlosser, q-analogues of the sums of consecutive integers, squares, cubes, quarts and quints, *Electron. J. Combin.* **11** (2004) #R71.

[11] N. J. A. Sloane, Error-correcting codes and invariant theory: new applications of a nineteenth-century technique, *Amer. Math. Monthly* **84** (1977) 82–107.

[12] S. O. Warnaar, On the q-analogue of the sum of cubes, *Electron. J. Combin.* **11** (2004) #N13.

[13] C. Wheatstone, On the formation of powers from arithmetical progressions, *Proc. Roy. Soc. London* **7** (1854) 145–151.

© 2024 World Scientific Publishing Company
https://doi.org/10.1142/9789811277375_0027

A combinatorial proof of a recurrence relation for the sum of divisors function*

Sun Kim

Department of Mathematics, and Institute of Pure and Applied Mathematics, Jeonbuk National University, Jeonju 54896, Republic of Korea
ssunny8079@hanmail.net

Dedicated to Bruce C. Berndt on his 80th birthday

We give a combinatorial proof of a generalization of an identity involving the sum of divisors function $\sigma(n)$ and the partition function $p(n)$, which is a companion of Euler's recurrence formula for $\sigma(n)$.

Keywords: Euler's recurrence formula; Euler's pentagonal number theorem; sum of the divisors function; theta functions.

Mathematics Subject Classification 2020: 05A17, 11P81

1. Introduction

One of Euler's famous identities is a recurrence formula for the sum of divisors function

$$\sigma(n) := \sum_{d|n} d.$$

Theorem 1.1. *For every $n > 0$, we have*

$$\sum_{k=-\infty}^{\infty} (-1)^k \sigma\left(n - \frac{k(3k+1)}{2}\right) = \begin{cases} (-1)^{k-1} n & \text{if} \quad n = \frac{k(3k+1)}{2}, k \in \mathbb{Z} \\ 0 & \text{otherwise.} \end{cases} \tag{1.1}$$

Euler [3, p. 234] derived the above result using logarithmic differentiation of Euler's pentagonal number theorem

$$F(x) := \prod_{k=1}^{\infty} (1 - x^k) = \sum_{k=-\infty}^{\infty} (-1)^k x^{\frac{k(3k+1)}{2}}. \tag{1.2}$$

*This article originally appeared in *International Journal of Number Theory*, **17**(3), pp. 647–653 © 2021 World Scientific Publishing Company. DOI: 10.1142/S1793042120400084

The proof is elementary. We begin with

$$xF'(x) = F(x)\frac{xF'(x)}{F(x)} = -\sum_{k=-\infty}^{\infty}(-1)^k x^{\frac{k(3k+1)}{2}}\sum_{k=1}^{\infty}\frac{kx^k}{1-x^k}$$
$$= -\sum_{k=-\infty}^{\infty}(-1)^k x^{\frac{k(3k+1)}{2}}\sum_{j=1}^{\infty}\sigma(j)x^j.$$

On the other hand,

$$xF'(x) = \sum_{k=-\infty}^{\infty}(-1)^k\frac{k(3k+1)}{2}x^{\frac{k(3k+1)}{2}}.$$

By equating coefficients of x^n, we obtain Theorem 1.1.

Note that from (1.2), we can easily derive Euler's recurrence formula for the partition function $p(n)$ as follows:

$$\sum_{k=-\infty}^{\infty}(-1)^k p\left(n-\frac{k(3k+1)}{2}\right) = 0. \tag{1.3}$$

Thus, we can view (1.1) as an analogue of (1.3) for $\sigma(n)$.

Considering $xF'(x)/F(x)$ instead of $xF'(x)$, we can deduce an identity, which is a companion of the identity (1.1). From

$$-\sum_{j=1}^{\infty}\sigma(j)x^j = \prod_{k=1}^{\infty}(1-x^k)^{-1}\sum_{k=-\infty}^{\infty}(-1)^k\frac{k(3k+1)}{2}x^{\frac{k(3k+1)}{2}},$$

we obtain a formula for $\sigma(n)$ in terms of $p(n)$.

Theorem 1.2. *Let $n \geq 1$. Then,*

$$-\sigma(n) = \sum_{k=-\infty}^{\infty}(-1)^k\frac{k(3k+1)}{2}p\Big(n-\frac{k(3k+1)}{2}\Big).$$

Even if it is not hard to derive Theorems 1.1 and 1.2 analytically, it is very intriguing to find combinatorial proofs of them. A combinatorial proof for Theorem 1.1, which is based on a double counting argument, can be found in [7, pp. 182–183] and [6, p. 53]. However, there is no known combinatorial proof for Theorem 1.2.

In this paper, we generalize Theorem 1.2 and also give a combinatorial proof of the generalization. We can obtain this generalization by applying the previous argument to the general theta function, which is defined by

$$f(a,b) := \sum_{n=-\infty}^{\infty}a^{\frac{n(n+1)}{2}}b^{\frac{n(n-1)}{2}} = (-a;ab)_\infty(-b;ab)_\infty(ab;ab)_\infty, \tag{1.4}$$

where $|ab| < 1$ and $(z;q)_\infty = \Pi_{i=0}^{\infty}(1-aq^n)$. The latter equality is called Jacobi's triple product identity [1, p. 21].

We remark that a generalization of Theorem 1.1 can also be found along the same lines. Moreover, the combinatorial argument from [7, pp. 182–183] can be applied to the generalization with a little modification. For the detailed proof, the reader is directed to [4].

Lastly, we also show that the quintuple product identity can be employed to yield similar sorts of identities and their combinatorial proofs can be given in a similar fashion.

2. A Generalization of Theorem 1.2 and Its Combinatorial Proof

Theorem 2.1. *Let n and m be positive integers, and put $n+m=l$. Then, for $N \geq 1$, we have*

$$-\sigma_{n,m}(N) = \sum_{k=-\infty}^{\infty} (-1)^k \frac{k^2(n+m)+k(n-m)}{2} p_{n,m}\left(N - \frac{k^2(n+m)+k(n-m)}{2}\right), \tag{2.1}$$

where $p_{n,m}(N)$ is the number of partitions of N into parts congruent to n, m or l $\pmod{l}$ and

$$\sigma_{n,m}(k) = \sum_{\substack{d|k \\ d \equiv n,m,l \pmod{l}}} d.$$

Note that Theorem 2.1 with $n=1$, $m=2$ reduces to Theorem 1.2.

We first deduce Theorem 2.1 using logarithmic differentiation of $f(-q^n, -q^m)$. By (1.4), we have

$$\begin{aligned} f(-q^n, -q^m) &= \prod_{k=0}^{\infty} (1-q^{n+kl})(1-q^{m+kl})(1-q^{(k+1)l}) \\ &= \sum_{k=-\infty}^{\infty} (-1)^k q^{\frac{k^2(n+m)+k(n-m)}{2}}. \end{aligned} \tag{2.2}$$

Consider

$$\begin{aligned} -q\frac{\frac{d}{dq}f(-q^n,-q^m)}{f(-q^n,-q^m)} &= \sum_{k=0}^{\infty} \frac{(n+kl)q^{n+kl}}{1-q^{n+kl}} + \sum_{k=0}^{\infty} \frac{(m+kl)q^{m+kl}}{1-q^{m+kl}} + \sum_{k=0}^{\infty} \frac{(k+1)lq^{(k+1)l}}{1-q^{(k+1)l}} \\ &= \sum_{k=1}^{\infty} \sigma_{n,m}(k) q^k. \end{aligned}$$

On the other hand, we also have

$$-q\frac{\frac{d}{dq}f(-q^n,-q^m)}{f(-q^n,-q^m)} = \frac{-1}{f(-q^n,-q^m)} \sum_{k=-\infty}^{\infty} (-1)^k \frac{k^2(n+m)+k(n-m)}{2} q^{\frac{k^2(n+m)+k(n-m)}{2}}.$$

Thus, we obtain

$$-\sum_{k=1}^{\infty}\sigma_{n,m}(k)q^k = \frac{1}{f(-q^n,-q^m)}\sum_{k=-\infty}^{\infty}(-1)^k\frac{k^2(n+m)+k(n-m)}{2}q^{\frac{k^2(n+m)\pm k(n-m)}{2}},$$

which implies Theorem 2.1.

A combinatorial proof of Theorem 2.1. Define a set

$$A_{n,m}(N) = \{(\pi,\lambda) : |\pi| + |\lambda| = N, \pi \in D_{n,m}, \lambda \in P_{n,m}\},$$

where $|\pi|$ is the sum of parts of π, $D_{n,m}$ is the set of partitions into distinct parts congruent to n, m or $l \pmod{l}$, and $P_{n,m}$ is the set of partitions into parts congruent to n, m or $l \pmod{l}$. We also let

$$B_{n,m}(N) = \sum_{(\pi,\lambda)\in A_{n,m}(N)}(-1)^{\ell(\pi)}|\pi|,$$

where $\ell(\pi)$ is the number of parts of π. We show that $B_{n,m}(N)$ is equal to both sides of (2.1). The involution of the Jacobi triple product identity [8] implies a bijective proof of (2.2). Thus, bijectively, we have

$$\begin{aligned}B_{n,m}(N) &= \sum_{(\pi,\lambda)\in A_{n,m}(N)}(-1)^{\ell(\pi)}|\pi| = \sum_{b=0}^{N}(N-b)p_{n,m}(b)\sum_{\pi\in D_{n,m}(N-b)}(-1)^{\ell(\pi)}\\ &= \sum_{k=-\infty}^{\infty}(-1)^k\frac{k^2(n+m)+k(n-m)}{2}p_{n,m}\left(N-\frac{k^2(n+m)+k(n-m)}{2}\right).\end{aligned}$$

Next, we first show that the number of pairs $(\pi,\lambda)\in A_{n,m}(N)$ with $\ell(\pi)$ even is equal to the number of those with $\ell(\pi)$ odd. Let $s(\pi)$ be the smallest part of a partition π and say $s(\pi)=\infty$ if $\pi=\emptyset$. If $s(\pi)\le s(\lambda)$, then move $s(\pi)$ to the partition λ. And, if $s(\pi)>s(\lambda)$, then move $s(\lambda)$ to the partition π. Obviously, this map is an involution, so we complete the proof of our claim. By this involution, we have

$$\begin{aligned}B_{n,m}(N) &= \sum_{(\pi,\lambda)\in A_{n,m}(N)}(-1)^{\ell(\pi)}|\pi| = \sum_{s(\pi)\le s(\lambda)}(-1)^{\ell(\pi)}|\pi| + \sum_{s(\pi)>s(\lambda)}(-1)^{\ell(\pi)}|\pi|\\ &= \sum_{a=1}^{N}\sum_{\substack{s(\pi)=a\\ s(\pi)\le s(\lambda)}}(-1)^{\ell(\pi)}|\pi| + \sum_{a=1}^{N}\sum_{\substack{s(\lambda)=a\\ s(\pi)>s(\lambda)}}(-1)^{\ell(\pi)}|\pi|\\ &= \sum_{a=1}^{N}\sum_{\substack{s(\pi)=a\\ s(\pi)\le s(\lambda)}}(-1)^{\ell(\pi)}(|\pi|-(|\pi|-a))\\ &= \sum_{a=1}^{N}a\sum_{\substack{s(\pi)=a\\ s(\pi)\le s(\lambda)}}(-1)^{\ell(\pi)}.\end{aligned}$$

It suffices to show that

$$\sum_{\substack{s(\pi)=a \\ s(\pi)\leq s(\lambda)}} (-1)^{\ell(\pi)} = \begin{cases} -1, & \text{if } a \mid N, \\ 0, & \text{otherwise.} \end{cases}$$

Let $L(\pi)$ be the largest part of a partition π and say $L(\pi) = 0$ if $\pi = \emptyset$. Consider a pair $(\pi, \lambda) \in A_{n,m}(N)$ with $s(\pi) = a \leq s(\lambda)$. Let μ be the partition consisting of the parts of π that are greater than $s(\pi) = a$. If $L(\mu) \geq L(\lambda)$ and $\mu \neq \emptyset$, then move the largest part of μ to the partition λ. And, if $L(\mu) < L(\lambda)$, except in the case when $L(\mu) = 0 < L(\lambda) = a$, then move the largest part of λ to the partition μ. Say that we obtain the new partition pair (π', λ'). Note that $s(\pi') = a \leq s(\lambda')$, and $\ell(\pi)$ and $\ell(\pi')$ have different parity. But, the above map fails when $\pi = a$ (i.e. $\mu = \emptyset$) and $\lambda = \emptyset$, or $\pi = a$ and $\lambda = a + \cdots + a$. Thus, we have $a \mid N$. Therefore,

$$\sum_{\substack{s(\pi)=a \\ s(\pi)\leq s(\lambda)}} (-1)^{\ell(\pi)} = \begin{cases} -1, & \text{if } a \mid N, \\ 0, & \text{otherwise,} \end{cases}$$

whence

$$B_{n,m}(N) = \sum_{(\pi,\lambda)\in A_{n,m}(N)} (-1)^{\ell(\pi)} |\pi| = - \sum_{\substack{a \mid N \\ a \equiv n,m,L \pmod{L}}} a = -\sigma_{n,m}(N).$$

□

This completes the proof of (2.1).

3. Further Results

Let k, l be positive integers. By the quintuple product identity [2, 5],

$$\begin{aligned} F(q) &:= \prod_{n=0}^{\infty} (1 - q^{\frac{k+l}{2}+kn})(1 - q^{k(n+1)})(1 - q^{\frac{k-l}{2}+kn})(1 - q^{l+2kn})(1 - q^{2k-l+2kn}) \\ &= \sum_{n=-\infty}^{\infty} q^{\frac{3k}{2}n^2 + \frac{2k-3l}{2}n} - q^{\frac{3k}{2}n^2 + \frac{4k-3l}{2}n + \frac{k-l}{2}}. \end{aligned}$$

Repeating the same argument in Sec. 2, we have

$$-q \frac{\frac{d}{dq} F(q)}{F(q)} = \sum_{n=1}^{\infty} \epsilon_{k.l}(n) q^n, \tag{3.1}$$

where

$$\epsilon_{k,l}(n) = \sum_{\substack{d \mid n \\ d \equiv \frac{k+l}{2}, \frac{k-l}{2}, 0 \pmod{k}}} d + \sum_{\substack{d \mid n \\ d \equiv l, 2k-l \pmod{2k}}} d.$$

Also,

$$q\frac{d}{dq}F(q) = \sum_{n=-\infty}^{\infty}\left(\frac{3k}{2}n^2 + \frac{2k-3l}{2}n\right)q^{\frac{3k}{2}n^2+\frac{2k-3l}{2}n}$$
$$-\left(\frac{3k}{2}n^2 + \frac{4k-3l}{2}n + \frac{k-l}{2}\right)q^{\frac{3k}{2}n^2+\frac{4k-3l}{2}n+\frac{k-l}{2}}. \tag{3.2}$$

Equating the two expressions for $q\frac{dF(q)}{dq}$ from (3.1) and (3.2), we conclude that

$$\left(\sum_{n=-\infty}^{\infty} q^{\frac{3k}{2}n^2+\frac{4k-3l}{2}n+\frac{k-l}{2}} - q^{\frac{3k}{2}n^2+\frac{2k-3l}{2}n}\right)\sum_{n=1}^{\infty}\epsilon_{k,l}(n)q^n$$
$$= \sum_{n=-\infty}^{\infty}\left(\frac{3k}{2}n^2 + \frac{2k-3l}{2}n\right)q^{\frac{3k}{2}n^2+\frac{2k-3l}{2}n}$$
$$-\left(\frac{3k}{2}n^2 + \frac{4k-3l}{2}n + \frac{k-l}{2}\right)q^{\frac{3k}{2}n^2+\frac{4k-3l}{2}n+\frac{k-l}{2}}. \tag{3.3}$$

The identity (3.3) has the following arithmetic interpretation.

Theorem 3.1. *For $N \geq 1$,*

$$\sum_{j+\frac{3k}{2}n^2+\frac{4k-3l}{2}n+\frac{k-l}{2}=N}\epsilon_{k,l}(j) - \sum_{j+\frac{3k}{2}n^2+\frac{2k-3l}{2}n=N}\epsilon_{k,l}(j)$$
$$= \begin{cases} N, & \text{if } N = \frac{3k}{2}n^2 + \frac{2k-3l}{2}n, \\ -N, & \text{if } N = \frac{3k}{2}n^2 + \frac{4k-3l}{2}n + \frac{k-l}{2}, \\ 0, & \text{otherwise.} \end{cases}$$

Next, considering two different representations of $qF(q)\frac{dF(q)}{dq}$ from (3.1) and (3.2), we obtain the following identity.

Theorem 3.2. *For $N \geq 1$,*

$$\epsilon_{k,l}(N) = \sum_{n=-\infty}^{\infty}\left(\frac{3k}{2}n^2 + \frac{4k-3l}{2}n + \frac{k-l}{2}\right)\rho_{k,l}\left(N - \frac{3k}{2}n^2 - \frac{4k-3l}{2}n - \frac{k-l}{2}\right)$$
$$-\sum_{n=-\infty}^{\infty}\left(\frac{3k}{2}n^2 + \frac{2k-3l}{2}n\right)\rho_{k,l}\left(N - \frac{3k}{2}n^2 - \frac{2k-3l}{2}n\right),$$

where $\rho_{k,l}(n)$ is the number of partitions of n into parts congruent to $\frac{k+l}{2}, \frac{k-l}{2}, 0 \pmod{k}$ or $l, 2k-l \pmod{2k}$.

Remark 3.1. Using the bijective proof of the quintuple product identity [5], we can give combinatorial proofs of Theorems 3.1 and 3.2, which are similar to those for Theorems 1.1 and 2.1. We omit the details.

Acknowledgments

The author would like to thank Bruce C. Berndt for very helpful comments and encouragement. This work was partially supported by the National Research Foundation of Korea(NRF) Grant funded by the Korean Government(MSIP) (NRF-2016R1A5A1008055).

References

[1] G. E. Andrews, *The Theory of Partitions*, (Addison–Wesley, Reading, MA, 1976); reissued by (Cambridge University Press, Cambridge, 1998).
[2] S. Cooper, The quintuple product identity, *Int. J. Number Theory*, **2** (2006) 115–161.
[3] L. Euler, Commentationes arithmeticae collectae, T. I et II, (Petropoli, 1849).
[4] S. Kim, Doctoral dissertation at U. (Illinois at Urbana-Champaign, 2009).
[5] S. Kim, A bijective proof of the quintuple product identity, *Int. J. Number Theory*, **6**(2) (2010) 247–256.
[6] I. Pak, Partition bijections, a survey, *Ramanujan J.* **12** (2006) 5–75.
[7] B. A. Venkov, *Elementary Number Theory*, (Wolters-Noordhoff Publishing, Groningen, 1970).
[8] J. Zolnowsky, A direct combinatorial proof of the Jacobi identity, *Discrete Math.* **9** (1974) 293–298.

© 2024 World Scientific Publishing Company
https://doi.org/10.1142/9789811277375_0028

(k, ℓ)-Regular partitions

Louis W. Kolitsch
Department of Mathematics and Statistics, University of Tennessee at Martin, Martin, TN 38238, USA
lkolitsc@utm.edu

Dedicated to Bruce Berndt in celebration of his 80th birthday

In this paper, (k, ℓ)-regular partitions will be introduced. Interesting properties of these partitions for specific choices of k and ℓ will be explored.

Keywords: Regular partitions; Rogers–Ramanujan functions; Ramanujan-type congruences.

Mathematics Subject Classification 2070: 11P81

1. Introduction

A k-regular partition of n is a partition of n that contains no part that is a multiple of k. These partitions have been extensively studied. Before introducing the (k, ℓ)-regular partitions, let us review some of the properties of k-regular partitions.

Theorem 1.1. *If we let $b_k(n)$ be the number of k-regular partitions of n, then*

$$\sum_{n=0}^{\infty} b_k(n)q^n = \frac{(q^k;q^k)_\infty}{(q;q)_\infty},$$

where $(a;b)_\infty = \prod_{i=0}^{\infty}(1-ab^i)$.

The generating function in Theorem 1.1 can be rewritten as

$$\frac{(q^k;q^k)_\infty}{(q;q)_\infty} = \prod_{i=1}^{\infty}(1+q^i+q^{2i}+\cdots+q^{(k-1)i}).$$

This gives us the following theorem.

Theorem 1.2. *The number of k-regular partitions of n is the same as the number of partitions of n where each part occurs fewer than k times.*

From [3], we have the following congruence properties.

Theorem 1.3. *For $n \geq 0$,*

$$b_5(20n+5) \equiv 0 \pmod{2},$$

$$b_5(20n+13) \equiv 0 \pmod{2},$$

$$b_{13}(9n+7) \equiv 0 \pmod{3},$$

$$b_{13}(27n+22) \equiv 0 \pmod{3},$$

$$b_{13}(81n+67) \equiv 0 \pmod{3}.$$

From [7], we list the following members of an infinite family of congruences modulo powers of 5 for the 5-regular partitions.

Theorem 1.4. *For $n \geq 0$,*

$$b_5(5n+4) \equiv 0 \pmod{5},$$

$$b_5(25n+4) \equiv 0 \pmod{5},$$

$$b_5(125n+104) \equiv 0 \pmod{25}.$$

In the next section, we will define (k,ℓ)-regular partitions and show that they satisfy combinatorial properties that are very similar to the properties of k-regular partitions.

2. Definition and Interpretation of (k,ℓ)-Regular Partitions

For positive integers k and ℓ with $k < \ell$, the (k,ℓ)-regular partitions of n are defined to be the partitions of n that contain no part that is a multiple of k or ℓ. The number of (k,ℓ)-regular partitions of n will be denoted by $b_{k,\ell}(n)$. These should not be confused with the (k,ℓ)-regular bipartitions of n that appear in the literature [6]. The (k,ℓ)-regular bipartitions are pairs of partitions S and T where S is a k-regular partition, T is an ℓ-regular partition, and the sum of the parts in $S \cup T$ is n. The number of (k,ℓ)-regular bipartitions of n is denoted by $B_{k,\ell}(n)$.

The generating function for the (k,ℓ)-regular partitions is given in the following theorem.

Theorem 2.1. *If we let $m = lcm(k,\ell)$, then*

$$\sum_{n=0}^{\infty} b_{k,\ell}(n)q^n = \frac{(q^k;q^k)_\infty (q^\ell;q^\ell)_\infty}{(q;q)_\infty (q^m;q^m)_\infty}.$$

When k and ℓ are relatively prime, this generating function can be rewritten to obtain the following result.

Theorem 2.2. *Suppose k and ℓ are relatively prime.*

(i) *The number of (k,ℓ)-regular partitions of n is the same as the number of partitions of n where each part is congruent to $1, 2, \ldots,$ or $k-1$ modulo k and occurs fewer than ℓ times.*

(ii) *The number of (k,ℓ)-regular partitions of n is the same as the number of partitions of n where each part is congruent to $1, 2, \ldots,$ or $\ell-1$ modulo ℓ and occurs fewer than k times.*

Proof. When k and ℓ are relatively prime, $m = lcm(k,\ell) = k\ell$. Case (i) now follows immediately by looking at the generating function for the second class of partitions and rewriting it to get the generating function for the (k,ℓ)-regular partitions:

$$\begin{aligned}
&\prod_{i=1}^{k-1}\prod_{j=0}^{\infty}(1+q^{i+kj}+q^{2(i+kj)}+\cdots+q^{(\ell-1)(i+kj)})\\
&\quad=\prod_{i=1}^{k-1}\prod_{j=0}^{\infty}\frac{1-q^{\ell(i+kj)}}{1-q^{i+kj}}\\
&\quad=\prod_{j=0}^{\infty}\frac{(1-q^{k(j+1)})(1-q^{\ell(k(j+1))})\left(\prod_{i=1}^{k-1}(1-q^{\ell(i+kj)})\right)}{(1-q^{k(j+1)})(1-q^{\ell(k(j+1))})\left(\prod_{i=1}^{k-1}(1-q^{(i+kj)})\right)}\\
&\quad=\frac{(q^k;q^k)_\infty(q^\ell;q^\ell)_\infty}{(q;q)_\infty(q^{k\ell};q^{k\ell})_\infty}.
\end{aligned}$$

Case (ii) follows in a similar manner. □

As an immediate corollary, we have the following result.

Corollary 2.3. *If k and ℓ are relatively prime, then the number of partitions of n where each part is congruent to $1, 2, \ldots,$ or $k-1$ modulo k and occurs fewer than ℓ times is equal to the number of partitions of n where each part is congruent to $1, 2, \ldots,$ or $\ell-1$ modulo ℓ and occurs fewer than k times.*

3. Congruences Modulo 2 for (2, 5)- and (3, 4)-Regular Partitions

Paralleling the congruence results for 5-regular partitions, we can get the following congruence for $(2,5)$-regular partitions.

Theorem 3.1. *For $n \geq 0$,*

$$b_{2,5}(4n+3) \equiv 0 \pmod{2}.$$

Proof. This theorem comes from using the identity

$$\frac{(q^5;q^5)_\infty}{(q;q)_\infty}=\frac{(q^8;q^8)_\infty(q^{20};q^{20})_\infty^2}{(q^2;q^2)_\infty^2(q^{40};q^{40})_\infty}+q\frac{(q^4;q^4)_\infty^3(q^{10};q^{10})_\infty(q^{40};q^{40})_\infty}{(q^2;q^2)_\infty^3(q^8;q^8)_\infty(q^{20};q^{20})_\infty}$$

from [4] to rewrite the generating function for $(2,5)$-regular partitions as

$$\frac{(q^8;q^8)_\infty(q^{20};q^{20})_\infty^2}{(q^2;q^2)_\infty(q^{10};q^{10})_\infty(q^{40};q^{40})_\infty}+q\frac{(q^4;q^4)_\infty^3(q^{40};q^{40})_\infty}{(q^2;q^2)_\infty^2(q^8;q^8)_\infty(q^{20};q^{20})_\infty}.$$

Substituting q^2 for q in line (20) in [5], we have

$$\frac{1}{(q^2;q^2)_\infty^2} = \frac{(q^{16};q^{16})_\infty^5}{(q^4;q^4)_\infty^5(q^{32};q^{32})_\infty^2} + \frac{2q^2(q^8;q^8)_\infty^2(q^{32};q^{32})_\infty^2}{(q^4;q^4)_\infty^5(q^{16};q^{16})_\infty}$$

and using this fact, the generating function for $(2,5)$-regular partitions can be rewritten as

$$\frac{(q^8;q^8)_\infty(q^{20};q^{20})_\infty^2}{(q^2;q^2)_\infty(q^{10};q^{10})_\infty(q^{40};q^{40})_\infty} + q\frac{(q^{16};q^{16})_\infty^5(q^{40};q^{40})_\infty}{(q^4;q^4)_\infty^2(q^8;q^8)_\infty(q^{20};q^{20})_\infty(q^{32};q^{32})_\infty^2}$$
$$+ 2q^3\frac{(q^8;q^8)_\infty(q^{32};q^{32})_\infty^2(q^{40};q^{40})_\infty}{(q^4;q^4)_\infty^2(q^{16};q^{16})_\infty(q^{20};q^{20})_\infty}. \tag{3.1}$$

The factor of 2 in the last expression in (3.1) gives the desired result. □

The following theorem gives a congruence modulo 2 for $(3,4)$-regular partitions.

Theorem 3.2. *For* $n \geq 0$,

$$b_{3,4}(3n+2) \equiv 0 \pmod 2.$$

Proof. This theorem comes from using the identity

$$\frac{(q^4;q^4)_\infty}{(q;q)_\infty} = \frac{(q^{12};q^{12})_\infty(q^{18};q^{18})_\infty^4}{(q^3;q^3)_\infty^3(q^{36};q^{36})_\infty^2} + q\frac{(q^6;q^6)_\infty^2(q^9;q^9)_\infty^3(q^{36};q^{36})_\infty}{(q^3;q^3)_\infty^4(q^{18};q^{18})_\infty^2}$$
$$+ 2q^2\frac{(q^6;q^6)_\infty(q^{18};q^{18})_\infty(q^{36};q^{36})_\infty}{(q^3;q^3)_\infty^3}$$

from [5] to rewrite the generating function for $(3,4)$-regular partitions as

$$\sum_{n=0}^{\infty} b_{3,4}(n)q^n = \frac{(q^{18};q^{18})_\infty^4}{(q^3;q^3)_\infty^2(q^{36};q^{36})_\infty^2} + q\frac{(q^6;q^6)_\infty^2(q^9;q^9)_\infty^3(q^{36};q^{36})_\infty}{(q^3;q^3)_\infty^3(q^{12};q^{12})_\infty(q^{18};q^{18})_\infty^2}$$
$$+ 2q^2\frac{(q^6;q^6)_\infty(q^{18};q^{18})_\infty(q^{36};q^{36})_\infty}{(q^3;q^3)_\infty^2(q^{12};q^{12})_\infty}. \tag{3.2}$$

The factor of 2 in the last expression in (3.2) gives the desired result. □

In the next section three congruences for $(3,5)$-regular partitions modulo 5 will be presented.

4. Congruences Modulo 5 for (3, 5)-Regular Partitions

Theorem 4.1. *For* $n \geq 0$,

(i) $b_{3,5}(25n+12) \equiv 0 \pmod 5$,

(ii) $b_{3,5}(25n+17) \equiv 0 \pmod 5$,

(iii) $b_{3,5}(25n+22) \equiv 0 \pmod 5$.

To prove this theorem we will derive the generating function for $b_{3,5}(5k+2)$, i.e.

$$\sum_{k=0}^{\infty} b_{3,5}(5k+2)q^k,$$

and then show that the coefficients of q^{5n+j} for $j = 2, 3, 4$ are congruent to 0 modulo 5. To accomplish this, we need a series of lemmas and a theorem involving the following functions:

$$H(q) = \frac{1}{(q^2; q^5)_\infty (q^3; q^5)_\infty} \quad \text{and} \quad G(q) = \frac{1}{(q; q^5)_\infty (q^4; q^5)_\infty}$$

associated with the Rogers–Ramanujan identities. In these lemmas, we let

$$x_k = \frac{G(q^k)}{H(q^k)}.$$

The results in the following seven lemmas are presented in [1] as Eqs. (2.10), (2.11), (2.12), (3.11), (3.12), (3.13), and (3.14), respectively.

Lemma 4.2.

$$x_1^5 - 11q - \frac{q^2}{x_1^5} = \frac{(q; q)_\infty^6}{(q^5; q^5)_\infty^6}.$$

Lemma 4.3.

$$\frac{(q; q)_\infty}{(q^5; q^5)_\infty} = \frac{(q^{25}; q^{25})_\infty}{(q^5; q^5)_\infty} \left(x_5 - q - \frac{q^2}{x_5} \right).$$

Lemma 4.4.

$$\begin{aligned} \frac{(q^5; q^5)_\infty}{(q; q)_\infty} = \frac{(q^{25}; q^{25})_\infty^5}{(q^5; q^5)_\infty^5} \Bigg(&\left(x_5^4 - \frac{3q^5}{x_5} \right) + q\left(x_5^3 + \frac{2q^5}{x_5^2} \right) \\ &+ q^2 \left(2x_5^2 - \frac{q^5}{x_5^3} \right) + q^3 \left(3x_5 + \frac{q^5}{x_5^4} \right) + 5q^4 \Bigg). \end{aligned}$$

Lemma 4.5.

$$\begin{aligned} A(q) &= x_1^4 x_3^2 + \frac{q^4 x_1^2}{x_3^4} + \frac{x_3^4}{x_1^2} + \frac{q^4}{x_1^4 x_3^2} \\ &= 5q^2 - 18q^4 \frac{(q^{15}; q^{15})_\infty^5 (q; q)_\infty}{(q^5; q^5)_\infty^5 (q^3; q^3)_\infty} + 2\frac{(q^5; q^5)_\infty^5 (q^3; q^3)_\infty}{(q^{15}; q^{15})_\infty^5 (q; q)_\infty}. \end{aligned}$$

Lemma 4.6.

$$\begin{aligned} B(q) &= \frac{q^2 x_1^3}{x_3} - x_1 x_3^3 - \frac{q^4}{x_1 x_3^3} + \frac{q^2 x_3}{x_1^3} \\ &= \frac{13q^2 - A(q)}{2}. \end{aligned}$$

Lemma 4.7.

$$C(q) = x_1^2 x_3 + \frac{q^2 x_1}{x_3^2} - \frac{x_3^2}{x_1} - \frac{q^2}{x_1^2 x_3} = 3q.$$

Lemma 4.8.

$$D(q) = x_1^3 x_3^4 - \frac{q^6}{x_1^3 x_3^4} + \frac{q^2 x_3^3}{x_1^4} - \frac{q^4 x_1^4}{x_3^3}$$

$$= -\frac{51}{2}q^3 + \frac{3}{2}qA(q) + q^2\frac{(q;q)_\infty^6}{(q^5;q^5)_\infty^6} + \frac{(q^3;q^3)_\infty^6}{(q^{15};q^{15})_\infty^6} + 22q^3.$$

The following lemma follows from Lemma 4.2 by calculating the product of $x_1^5 - 11q - \frac{q^2}{x_1^5}$ and $x_3^5 - 11q^3 - \frac{q^6}{x_3^5}$.

Lemma 4.9.

$$x_1^5 x_3^5 - \frac{q^6 x_1^5}{x_3^5} - \frac{q^2 x_3^5}{x_1^5} + \frac{q^8}{x_1^5 x_3^5}$$

$$= \frac{(q;q)_\infty^6 (q^3;q^3)_\infty^6}{(q^5;q^5)_\infty^6 (q^{15};q^{15})_\infty^6} + 11q^3\frac{(q;q)_\infty^6}{(q^5;q^5)_\infty^6} + 11q\frac{(q^3;q^3)_\infty^6}{(q^{15};q^{15})_\infty^6} + 121q^4.$$

The following lemma comes from adding identities (iv) and (v) from [2, Theorem 5.1].

Lemma 4.10.

$$\frac{q^2 x_3}{x_1^3} + \frac{q^2 x_1^3}{x_3} + \frac{q^4}{x_1 x_3^3} + x_1 x_3^3 = 9q^4\frac{(q^{15};q^{15})_\infty^5 (q;q)_\infty}{(q^5;q^5)_\infty^5 (q^3;q^3)_\infty} + \frac{(q^5;q^5)_\infty^5 (q^3;q^3)_\infty}{(q^{15};q^{15})_\infty^5 (q;q)_\infty}.$$

The following lemma comes from rewriting the left-hand side of (i) from [2, Theorem 5.1] using the functions G and H, multiplying the equation by $\frac{1}{H(q)^3 H(q^3)}$, and using the fact that $\frac{1}{G(q)H(q)} = \frac{(q;q)_\infty}{(q^5;q^5)_\infty}$.

Lemma 4.11.

$$\frac{1}{H(q)^3 G(q^3)} - \frac{1}{G(q)^3 H(q^3)} = \frac{3q(q;q)_\infty^2 (q^{15};q^{15})_\infty^2}{(q^5;q^5)_\infty^4}.$$

We will now use Lemmas 4.2–4.10 to establish the following theorem.

Theorem 4.12.

$$1 = G(q)^2 H(q) G(q^3)^3 - q^2 G(q) H(q)^2 H(q^3)^3$$
$$- qG(q)^3 G(q^3) H(q^3)^2 - qH(q)^3 G(q^3)^2 H(q^3).$$

Proof. If we define

$$F(q) = G(q)^2 H(q) G(q^3)^3 - q^2 G(q) H(q)^2 H(q^3)^3$$
$$- qG(q)^3 G(q^3) H(q^3)^2 - qH(q)^3 G(q^3)^2 H(q^3),$$

it is easy to see that

$$1 = \frac{F(q)}{G(q)^2H(q)G(q^3)^3} + \frac{q^2}{x_1x_3^3} + \frac{qx_1}{x_3^2} + \frac{q}{x_1^2x_3},$$

$$1 = \frac{-F(q)}{q^2G(q)H(q)^2H(q^3)^3} + \frac{x_1x_3^3}{q^2} - \frac{x_1^2x_3}{q} - \frac{x_3^2}{qx_1},$$

$$1 = \frac{-F(q)}{qG(q)^3G(q^3)H(q^3)^2} + \frac{x_3^2}{qx_1} - \frac{q}{x_1^2x_3} - \frac{x_3}{x_1^3},$$

$$1 = \frac{-F(q)}{qH(q)^3G(q^3)^2H(q^3)} + \frac{x_1^2x_3}{q} - \frac{qx_1}{x_3^2} - \frac{x_1^3}{x_3}.$$

Combining these four equations, adding the first two and subtracting the last two, we obtain

$$\frac{F(q)^2}{G(q)^3H(q)^3G(q^3)^3H(q^3)^3} = \left(\frac{q^2x_3}{x_1^3} + \frac{q^2x_1^3}{x_3} + \frac{q^4}{x_1x_3^3} + x_1x_3^3 + 2\left(\frac{q^3}{x_1^2x_3} - qx_1^2x_3 + \frac{q^3x_1}{x_3^2} - \frac{qx_3^2}{x_1}\right)\right).$$

Squaring both sides of this equation we obtain

$$\begin{aligned}
&\frac{F(q)^4}{G(q)^6H(q)^6G(q^3)^6H(q^3)^6}\\
&= \left(\frac{q^2x_3}{x_1^3} + \frac{q^2x_1^3}{x_3} + \frac{q^4}{x_1x_3^3} + x_1x_3^3\right)^2\\
&\quad + 4\left(\frac{q^2x_3}{x_1^3} + \frac{q^2x_1^3}{x_3} + \frac{q^4}{x_1x_3^3} + x_1x_3^3\right)\left(\frac{q^3}{x_1^2x_3} - qx_1^2x_3 + \frac{q^3x_1}{x_3^2} - \frac{qx_3^2}{x_1}\right)\\
&\quad + 4\left(\frac{q^3}{x_1^2x_3} - qx_1^2x_3 + \frac{q^3x_1}{x_3^2} - \frac{qx_3^2}{x_1}\right)^2\\
&= \left(9q^4\frac{(q^{15};q^{15})_\infty^5(q;q)_\infty}{(q^5;q^5)_\infty^5(q^3;q^3)_\infty} + \frac{(q^5;q^5)_\infty^5(q^3;q^3)_\infty}{(q^{15};q^{15})_\infty^5(q;q)_\infty}\right)^2\\
&\quad + 4\left(\frac{q^5}{x_1^5} - q^3x_1^5 + \frac{q^7}{x_3^5} - qx_3^5 - qD(q)\right) + 4(2q^2A(q) - 17q^4)\\
&= 81q^8\frac{(q^{15};q^{15})_\infty^{10}(q;q)_\infty^2}{(q^5;q^5)_\infty^{10}(q^3;q^3)_\infty^2} + \frac{(q^5;q^5)_\infty^{10}(q^3;q^3)_\infty^2}{(q^{15};q^{15})_\infty^{10}(q;q)_\infty^2} - 36q^6\frac{(q^{15};q^{15})_\infty^5(q;q)_\infty}{(q^5;q^5)_\infty^5(q^3;q^3)_\infty}\\
&\quad + 4q^2\frac{(q^5;q^5)_\infty^5(q^3;q^3)_\infty}{(q^{15};q^{15})_\infty^5(q;q)_\infty} - 8q^3\frac{(q;q)_\infty^6}{(q^5;q^5)_\infty^6} - 8q\frac{(q^3;q^3)_\infty^6}{(q^{15};q^{15})_\infty^6} - 114q^4. \qquad (4.1)
\end{aligned}$$

To complete the proof, we will now show that the right-hand side of (4.1) is equal to

$$\frac{1}{G(q)^6H(q)^6G(q^3)^6H(q^3)^6} = \frac{(q;q)_\infty^6(q^3;q^3)_\infty^6}{(q^5;q^5)_\infty^6(q^{15};q^{15})_\infty^6},$$

which forces $F(q)^4 = 1$. A quick check of the constant term in the series expansion of $F(q)$ yields the desired result of $F(q) = 1$.

We begin by looking at $C(q) \cdot D(q)$ in two different ways. On the one hand,

$$\begin{aligned} C(q) \cdot D(q) &= 3q\Big(-\frac{51}{2}q^3 + \frac{3}{2}qA(q) + q^2 \frac{(q;q)_\infty^6}{(q^5;q^5)_\infty^6} + \frac{(q^3;q^3)_\infty^6}{(q^{15};q^{15})_\infty^6} + 22q^3 \Big) \\ &= 12q^4 - 81q^6 \frac{(q^{15};q^{15})_\infty^5 (q;q)_\infty}{(q^5;q^5)_\infty^5 (q^3;q^3)_\infty} + 9q^2 \frac{(q^5;q^5)_\infty^5 (q^3;q^3)_\infty}{(q^{15};q^{15})_\infty^5 (q;q)_\infty} \\ &\quad + 3q^3 \frac{(q;q)_\infty^6}{(q^5;q^5)_\infty^6} + 3q \frac{(q^3;q^3)_\infty^6}{(q^{15};q^{15})_\infty^6}. \end{aligned}$$

We also have

$$\begin{aligned} C(q) \cdot D(q) &= \left(x_1^2 x_3 + \frac{q^2 x_1}{x_3^2} - \frac{x_3^2}{x_1} - \frac{q^2}{x_1^2 x_3} \right) \left(x_1^3 x_3^4 - \frac{q^6}{x_1^3 x_3^4} + \frac{q^2 x_3^3}{x_1^4} - \frac{q^4 x_1^4}{x_3^3} \right) \\ &= \left(x_1^5 x_3^5 - \frac{q^6 x_1^5}{x_3^5} - \frac{q^2 x_3^5}{x_1^5} + \frac{q^8}{x_1^5 x_3^5} \right) + \left(\frac{q^4 x_3}{x_1^3} + \frac{q^4 x_1^3}{x_3} - \frac{q^6}{x_1 x_3^3} - q^2 x_1 x_3^3 \right) \\ &\quad + \left(q^2 x_1^4 x_3^2 + \frac{q^6 x_1^2}{x_3^4} + \frac{q^2 x_3^4}{x_1^2} + \frac{q^6}{x_1^4 x_3^2} \right) - \left(\frac{q^4 x_1^6}{x_3^2} + \frac{q^8}{x_1^2 x_3^6} + x_1^2 x_3^6 + \frac{q^4 x_3^2}{x_1^6} \right) \\ &= \left(\frac{(q;q)_\infty^6 (q^3;q^3)_\infty^6}{(q^5;q^5)_\infty^6 (q^{15};q^{15})_\infty^6} + 11q^3 \frac{(q;q)_\infty^6}{(q^5;q^5)_\infty^6} + 11q \frac{(q^3;q^3)_\infty^6}{(q^{15};q^{15})_\infty^6} + 121q^4 \right) \\ &\quad + q^2 B(q) + q^2 A(q) - (B(q)^2 + 2q^2 A(q) - 4q^4). \end{aligned}$$

Substituting for $A(q)$ and $B(q)$ we find that

$$\begin{aligned} C(q) \cdot D(q) = {} & \frac{(q;q)_\infty^6 (q^3;q^3)_\infty^6}{(q^5;q^5)_\infty^6 (q^{15};q^{15})_\infty^6} + 11q^3 \frac{(q;q)_\infty^6}{(q^5;q^5)_\infty^6} + 11q \frac{(q^3;q^3)_\infty^6}{(q^{15};q^{15})_\infty^6} + 126q^4 \\ & - 45q^6 \frac{(q^{15};q^{15})_\infty^5 (q;q)_\infty}{(q^5;q^5)_\infty^5 (q^3;q^3)_\infty} + 5q^2 \frac{(q^5;q^5)_\infty^5 (q^3;q^3)_\infty}{(q^{15};q^{15})_\infty^5 (q;q)_\infty} \\ & - 81q^8 \frac{(q^{15};q^{15})_\infty^{10} (q;q)_\infty^2}{(q^5;q^5)_\infty^{10} (q^3;q^3)_\infty^2} - \frac{(q^5;q^5)_\infty^{10} (q^3;q^3)_\infty^2}{(q^{15};q^{15})_\infty^{10} (q;q)_\infty^2}. \end{aligned}$$

Equating the two expressions for $C(q) \cdot D(q)$ and rewriting the equation, we get the desired result. □

To prove Theorem 4.1, we also need the following corollary of Theorem 4.12.

Corollary 4.13.

$$\frac{(q^{15};q^{15})_\infty}{(q^3;q^3)_\infty} \left(\frac{x_3^2}{x_1} - q - \frac{q^2}{x_1^2 x_3} - \frac{q x_3}{x_1^3} \right) = (q;q^5)_\infty^3 (q^4;q^5)_\infty^3 (q^6;q^{15})_\infty (q^9;q^{15})_\infty.$$

Proof. When written in terms of the functions G and H, the corollary simply states that

$$G(q^3)H(q^3)\left(\frac{H(q)G(q^3)^2}{G(q)H(q^3)^2} - q - \frac{q^2H(q)^2H(q^3)}{G(q)^2G(q^3)} - \frac{qH(q)^3G(q^3)}{G(q)^3H(q^3)}\right) = \frac{1}{G(q)^3H(q^3)},$$

which follows immediately from Theorem 4.12. □

We are now ready to present the proof of Theorem 4.1, parts (i) and (iii).

Proof. Using Lemmas 4.3 and 4.4, we have

$$\begin{aligned}\sum_{i=0}^{\infty} b_{3,5}(i)q^i &= \frac{(q^3;q^3)_\infty(q^5;q^5)_\infty}{(q;q)_\infty(q^{15};q^{15})_\infty} \\ &= \left(\frac{(q^5;q^5)_\infty}{(q;q)_\infty}\right)\left(\frac{(q^3;q^3)_\infty}{(q^{15};q^{15})_\infty}\right) \\ &= \frac{(q^{25};q^{25})_\infty^5}{(q^5;q^5)_\infty^5}\left(\left(x_5^4 - \frac{3q^5}{x_5}\right) + q\left(x_5^3 + \frac{2q^5}{x_5^2}\right) + q^2\left(2x_5^2 - \frac{q^5}{x_5^3}\right)\right. \\ &\quad \left. + q^3\left(3x_5 + \frac{q^5}{x_5^4}\right) + 5q^4\right) \cdot \frac{(q^{75};q^{75})_\infty}{(q^{15};q^{15})_\infty}\left(x_{15} - q^3 - \frac{q^6}{x_{15}}\right).\end{aligned}$$

Expanding the products on the right-hand side and selecting those terms that involve powers of q of the form $5k+2$, we get

$$\begin{aligned}&\sum_{k=0}^{\infty} b_{3,5}(5k+2)q^{5k+2} \\ &\quad = \frac{(q^{75};q^{75})_\infty(q^{25};q^{25})_\infty^5}{(q^{15};q^{15})_\infty(q^5;q^5)_\infty^5}\left(2q^2x_5^2x_{15} - \frac{2q^{12}}{x_5^2x_{15}} - q^7\left(\frac{x_{15}}{x_5^3} + \frac{x_5^3}{x_{15}} + 5\right)\right).\end{aligned}$$

Dividing by q^2 and replacing q by $q^{1/5}$, we obtain

$$\sum_{k=0}^{\infty} b_{3,5}(5k+2)q^k = \frac{(q^{15};q^{15})_\infty(q^5;q^5)_\infty^5}{(q^3;q^3)_\infty(q;q)_\infty^5}\left(2x_1^2x_3 - \frac{2q^2}{x_1^2x_3} - \frac{qx_3}{x_1^3} - \frac{qx_1^3}{x_3} - 5q\right).$$

Using Lemma 4.7 and Corollary 4.13, we can rewrite the right-hand side of the previous equation as

$$\begin{aligned}&\frac{(q^{15};q^{15})_\infty(q^5;q^5)_\infty^5}{(q^3;q^3)_\infty(q;q)_\infty^5}\left(2x_1^2x_3 - \frac{2q^2}{x_1^2x_3} - \frac{qx_3}{x_1^3} - \frac{qx_1^3}{x_3} - 2q - x_1^2x_3 - \frac{q^2x_1}{x_3^2} + \frac{x_3^2}{x_1} + \frac{q^2}{x_1^2x_3}\right) \\ &\quad = \frac{(q^{15};q^{15})_\infty(q^5;q^5)_\infty^5}{(q^3;q^3)_\infty(q;q)_\infty^5}\left(x_1^2x_3 - \frac{q^2}{x_1^2x_3} - \frac{qx_3}{x_1^3} - \frac{qx_1^3}{x_3} - 2q - \frac{q^2x_1}{x_3^2} + \frac{x_3^2}{x_1}\right) \\ &\quad = \frac{(q^{15};q^{15})_\infty(q^5;q^5)_\infty^5}{(q^3;q^3)_\infty(q;q)_\infty^5}\left(\frac{x_3^2}{x_1} - q - \frac{q^2}{x_1^2x_3} - \frac{qx_3}{x_1^3}\right)\left(1 + \frac{x_1^3}{x_3}\right)\end{aligned}$$

$$= \frac{(q^5;q^5)^5_\infty}{(q;q)^5_\infty}\left((q;q^5)^3_\infty(q^4;q^5)^3_\infty(q^6;q^{15})_\infty(q^9;q^{15})_\infty\right)\left(1+\frac{x_1^3}{x_3}\right)$$

$$= \frac{(q^5;q^5)^5_\infty}{(q;q)^5_\infty}\Big((q;q^5)^3_\infty(q^4;q^5)^3_\infty(q^6;q^{15})_\infty(q^9;q^{15})_\infty$$

$$+ (q^2;q^5)^3_\infty(q^3;q^5)^3_\infty(q^3;q^{15})_\infty(q^{12};q^{15})_\infty\Big)$$

$$= \frac{(q^5;q^5)^2_\infty}{(q;q)^5_\infty(q^{15};q^{15})_\infty}\left(\left(\sum_{j=-\infty}^{\infty}(-1)^j q^{\frac{5j^2+3j}{2}}\right)^3\left(\sum_{j=-\infty}^{\infty}(-1)^j q^{\frac{15j^2+3j}{2}}\right)\right.$$

$$\left. + \left(\sum_{j=-\infty}^{\infty}(-1)^j q^{\frac{5j^2+j}{2}}\right)^3\left(\sum_{j=-\infty}^{\infty}(-1)^j q^{\frac{15j^2+9j}{2}}\right)\right).$$

The last equality above comes from applying Jacobi's triple product identity to each of the following infinite products:

$$(q;q^5)_\infty(q^4;q^5)_\infty, (q^6;q^{15})_\infty(q^9;q^{15})_\infty, (q^2;q^5)_\infty(q^3;q^5)_\infty, (q^3;q^{15})_\infty(q^{12};q^{15})_\infty.$$

We will now show that the coefficients on the terms of the form q^{5n+2} and q^{5n+4} in

$$P(q) = \left(\sum_{j=-\infty}^{\infty}(-1)^j q^{\frac{5j^2+3j}{2}}\right)^3\left(\sum_{j=-\infty}^{\infty}(-1)^j q^{\frac{15j^2+3j}{2}}\right)$$

$$+ \left(\sum_{j=-\infty}^{\infty}(-1)^j q^{\frac{5j^2+j}{2}}\right)^3\left(\sum_{j=-\infty}^{\infty}(-1)^j q^{\frac{15j^2+9j}{2}}\right)$$

are all zeros. Since

$$\frac{(q^5;q^5)^2_\infty}{(q;q)^5_\infty(q^{15};q^{15})_\infty} \equiv \frac{(q^5;q^5)_\infty}{(q^{15};q^{15})_\infty} \pmod 5$$

and only involves powers of q that are multiples of 5, this will complete the proof of Theorem 4.1, parts (i) and (iii).

We will begin with an ordered 4-tuple of integers $V = (r, s, t, u)$ such that

$$\frac{5r^2+r}{2} + \frac{5s^2+s}{2} + \frac{5t^2+t}{2} + \frac{15u^2+9u}{2} = 5n+2,$$

and note that in $P(q)$ the term q^{5n+2} determined by this 4-tuple is counted with weight $(-1)^{r+s+t+u}$. Since

$$r+s+t-u \equiv 4 \pmod 5,$$

the transformation μ defined by

$$\mu(V) = \bigg(\frac{1}{5}(-2r + 3s + 3t - 3u - 2), \frac{1}{5}(3r - 2s + 3t - 3u - 2),$$
$$\frac{1}{5}(3r + 3s - 2t - 3u - 2), \frac{1}{5}(r + s + t + 4u + 1)\bigg)$$

maps V to a 4-tuple of integers (R, S, T, U) such that

$$\frac{5R^2 + 3R}{2} + \frac{5S^2 + 3S}{2} + \frac{5t^2 + 3T}{2} + \frac{15U^2 + 3U}{2} = 5n + 2$$

and the corresponding weight is given by $(-1)^{R+S+T+U} = (-1)^{r+s+t-u+1}$. If the original weight is 1, then this new weight is -1 and vice versa. Since the transformation μ is invertible, we conclude that the coefficient on q^{5n+2} in $P(q)$ is zero.

Similarly, if $W = (r, s, t, u)$ is a 4-tuple of integers such that

$$\frac{5r^2 + r}{2} + \frac{5s^2 + s}{2} + \frac{5t^2 + t}{2} + \frac{15u^2 + 9u}{2} = 5n + 4,$$

then we will have

$$r + s + t - u \equiv 3 \pmod 5$$

and we can define the transformation β by

$$\beta(W) = \bigg(\frac{1}{5}(2r - 3s - 3t + 3u - 1), \frac{1}{5}(-3r + 2s - 3t + 3u - 1),$$
$$\frac{1}{5}(-3r - 3s + 2t + 3u - 1), \frac{1}{5}(-r - s - t - 4u - 2)\bigg),$$

which yields a 4-tuple of integers (R, S, T, U) with

$$\frac{5R^2 + 3R}{2} + \frac{5S^2 + 3S}{2} + \frac{5T^2 + 3T}{2} + \frac{15U^2 + 3U}{2} = 5n + 4$$

and the weight has the opposite sign. Since the transformation β is invertible, we conclude that the coefficient on q^{5n+4} in $P(q)$ is zero. □

Unfortunately, the coefficients on the terms of the form q^{5n+3} in $P(q)$ are not all zero which means an alternate way to prove part (ii) of Theorem 4.1 is needed. We will first show that the coefficients on the terms of the form q^{5n+3} in $P(q)$ can be obtained from the coefficients on the terms of the form q^{5n+3} in $P(q)^2$ and then show that the coefficients on the terms of the form q^{5n+3} in $P(q)^2$ are all multiples of 5.

Since $P(q)$ contains no terms with the power on q congruent to 1 or 4 modulo 5, we can write

$$P(q) = X + qY + q^3 Z,$$

where X, Y, and Z are each power series that only contain terms where the powers on q are multiples of 5. If we square this expression for $P(q)$ and look for terms where the power on q is congruent to 3 modulo 5, we find that these terms are $2q^3XZ$. Thus, the terms of the form q^{5n+3} in $P(q)$ are given by $\frac{1}{2X}$ times the terms of the form q^{5n+3} in $P(q)^2$. Since X only contains terms where the powers on q are multiples of 5, this means that the coefficients on the terms of the form q^{5n+3} in $P(q)$ are congruent to 0 modulo 5 if the coefficients on the terms of the form q^{5n+3} in $P(q)^2$ are congruent to 0 modulo 5 (which we will now establish).

Using Lemma 4.11 and the fact that $\dfrac{1}{G(q)H(q)} = \dfrac{(q;q)_\infty}{(q^5;q^5)_\infty}$, we have

$$\begin{aligned}
P(q)^2 &= \left(\frac{1}{H(q)^3G(q^3)} + \frac{1}{G(q)^3H(q^3)}\right)^2 \\
&= \left(\frac{1}{H(q)^3G(q^3)} - \frac{1}{G(q)^3H(q^3)}\right)^2 + \frac{4}{G(q)^3H(q)^3G(q^3)H(q^3)} \\
&= \frac{9q^2(q;q)_\infty^4(q^{15};q^{15})_\infty^4}{(q^5;q^5)_\infty^8} + \frac{4(q;q)_\infty^3(q^3;q^3)_\infty}{(q^5;q^5)_\infty^3(q^{15};q^{15})_\infty}.
\end{aligned}$$

Now, using Lemma 4.4, we see

$$\begin{aligned}
P(q)^2 &= \frac{9q^2(q^{25};q^{25})_\infty^4\left(x_5 - q - \frac{q^2}{x_5}\right)^4(q^{15};q^{15})_\infty^4}{(q^5;q^5)_\infty^8} \\
&\quad + \frac{4(q^{25};q^{25})_\infty^3\left(x_5 - q - \frac{q^2}{x_5}\right)^3(q^{75};q^{75})_\infty\left(x_{15} - q^3 - \frac{q^6}{x_{15}}\right)}{(q^5;q^5)_\infty^3(q^{15};q^{15})_\infty}.
\end{aligned}$$

Expanding the right-hand side of this equation, we find that the terms of the form q^{5n+3} in $P(q)^2$ are given by

$$\begin{aligned}
&\frac{9(q^{25};q^{25})^4(q^{15};q^{15})_\infty^4}{(q^5;q^5)_\infty^8}\left(\frac{2q^8}{x_5^2} - 4q^3x_5^3\right) \\
&\quad + \frac{4(q^{25};q^{25})^3(q^{75};q^{75})_\infty}{(q^5;q^5)_\infty^3(q^{15};q^{15})_\infty}\left(\frac{3q^8}{x_5^2} - q^3x_5^3 + 5q^3x_{15}\right).
\end{aligned}$$

Modulo 5 this is equivalent to

$$\begin{aligned}
&\frac{4(q^{25};q^{25})^4(q^{75};q^{75})_\infty}{(q^5;q^5)_\infty^3(q^{15};q^{15})_\infty(q^{25};q^{25})_\infty}\left(\frac{2q^8}{x_5} - 4q^3x_5^3\right) \\
&\quad + \frac{4(q^{25};q^{25})^3(q^{75};q^{75})_\infty}{(q^5;q^5)_\infty^3(q^{15};q^{15})_\infty}\left(\frac{3q^8}{x_5^2} - q^3x_5^3\right),
\end{aligned}$$

which clearly simplifies to an expression that is congruent to 0 modulo 5. This completes the proof of Theorem 4, part (*ii*).

5. Concluding Remarks

If we rewrite the equations,

$$\frac{5r^2 + r}{2} + \frac{5s^2 + s}{2} + \frac{5t^2 + t}{2} + \frac{15u^2 + 9u}{2} = 5n + 2$$

and

$$\frac{5R^2 + 3R}{2} + \frac{5S^2 + 3S}{2} + \frac{5t^2 + 3T}{2} + \frac{15U^2 + 3U}{2} = 5n + 2,$$

as

$$(10r + 1)^2 + (10s + 1)^2 + (10t + 1)^2 + 3(10u + 3)^2 = 200n + 110$$

and

$$(10R + 3)^2 + (10S + 3)^2 + (10T + 3)^2 + 3(10U + 1)^2 = 200n + 110,$$

we obtain the following corollary.

Corollary 5.1. *Let n be a nonnegative integer and let r, s, t, and u be integers. The number of representations of $200n + 110$ in the form $(10r + 1)^2 + (10s + 1)^2 + (10t + 1)^2 + 3(10u + 3)^2$ is the same as the number of representations of $200n + 110$ in the form $(10r + 3)^2 + (10s + 3)^2 + (10t + 3)^2 + 3(10u + 1)^2$.*

If we rewrite the equations,

$$\frac{5r^2 + r}{2} + \frac{5s^2 + s}{2} + \frac{5t^2 + t}{2} + \frac{15u^2 + 9u}{2} = 5n + 4$$

and

$$\frac{5R^2 + 3R}{2} + \frac{5S^2 + 3S}{2} + \frac{5t^2 + 3T}{2} + \frac{15U^2 + 3U}{2} = 5n + 4,$$

as

$$(10r + 1)^2 + (10s + 1)^2 + (10t + 1)^2 + 3(10u + 3)^2 = 200n + 190$$

and

$$(10R + 3)^2 + (10S + 3)^2 + (10T + 3)^2 + 3(10U + 1)^2 = 200n + 190,$$

we obtain the following corollary.

Corollary 5.2. *Let n be a nonnegative integer and let r, s, t, and u be integers. The number of representations of $200n + 190$ in the form $(10r + 1)^2 + (10s + 1)^2 + (10t + 1)^2 + 3(10u + 3)^2$ is the same as the number of representations of $200n + 190$ in the form $(10r + 3)^2 + (10s + 3)^2 + (10t + 3)^2 + 3(10u + 1)^2$.*

Acknowledgment

I thank the referees for their helpful suggestions and corrections.

References

[1] Z. Ahmed, N. Baruah and M. Dastidar, New congruences modulo 5 for the number of 2-color partitions, *J. Number Theory* **157** (2015) 184–198.

[2] C. Gugg, Modular equations for cubes of the Rogers–Ramanujan and Ramanujan–Gőllnitz–Gordon functions and their associated continued fractions, *J. Number Theory* **132** (2012) 1519–1553.

[3] N. Calkin, N. Drake, K. James, S. Law, P. Lee, D. Penniston and J. Radder, Divisibility properties of the 5-regular and 13-regular partition functions, *Integers* **8** (2008), A60, 10pp.

[4] M. Hirschhorn and J. Sellers, Elementary proofs of parity results for 5-regular partitions, *Bull. Aust. Math. Soc.* **81** (2010) 58–63.

[5] M. S. Mahadeva Naika and S. Shivaprasada Nayaka, Congruences for (2,3)-regular partition with designated summands, *Note Mat.* **36**(2) (2016) 99–123.

[6] L. Wang, Arithmetic properties of (k, ℓ)-regular bipartitions, *Bull. Aust. Math. Soc.* **95** (2017) 353–364.

[7] L. Wang, Congruences for 5-regular partitions modulo powers of 5, *Ramanujan J.* **44**(2) (2017) 343–358.

© 2024 World Scientific Publishing Company
https://doi.org/10.1142/9789811277375_0029

A combinatorial construction for two formulas in Slater's list*

Kağan Kurşungöz
Faculty of Engineering and Natural Sciences
Sabancı University, İstanbul, Turkey
kursungoz@sabanciuniv.edu

We set up a combinatorial framework for inclusion-exclusion on the partitions into distinct parts to obtain an alternative generating function of partitions into distinct and non-consecutive parts. In connection with Rogers–Ramanujan identities, the generating function yields two formulas in Slater's list. The same formulas were constructed by Hirschhorn. Similar formulas were obtained by Bringmann, Mahlburg and Nataraj. We also use staircases to give alternative triple series for partitions into d-distinct parts for any $d \geq 2$.

Keywords: Integer partition; partition generating function; Rogers–Ramanujan identities; Slater's list.

Mathematics Subject Classification 2020: 05A17, 05A15, 11P84

1. Introduction

Number 19 in Slater's list [6] is

$$(-q;q)_\infty \sum_{n\geq 0} \frac{(-1)^j q^{3j^2}}{(q^2;q^2)_j(-q;q)_{2j}} = \frac{1}{(q;q^5)_\infty(q^4;q^5)_\infty}. \tag{1.1}$$

Here, we use the q-Pochhammer symbols [3]

$$(a;q)_n = \prod_{j=1}^{n}(1-aq^{j-1})$$

for $j \in \mathbb{N} \cup \{\infty\}$.

The companion to (1.1) in [6] is number 15:

$$(-q;q)_\infty \sum_{n\geq 0} \frac{(-1)^j q^{3j^2-2j}}{(q^2;q^2)_j(-q;q)_{2j}} = \frac{1}{(q^2;q^5)_\infty(q^3;q^5)_\infty}. \tag{1.2}$$

*This article originally appeared in *International Journal of Number Theory*, **17**(3), pp. 655–663
© 2021 World Scientific Publishing Company. DOI: 10.1142/S1793042120400114

In his Ph.D. thesis [4, Chap. 5], Hirschhorn gave a combinatorial construction of (1.1) and

$$(-q;q)_\infty \sum_{n\geq 0} \frac{(-1)^j q^{3j^2+2j}}{(q^2;q^2)_j(-q;q)_{2j+1}} = \frac{1}{(q;q^5)_\infty(q^4;q^5)_\infty}. \tag{1.3}$$

He showed the equivalence of (1.2) and (1.3), as well. He elaborated on the *length of runs* (called sequences in [4]) in a partition into distinct parts.

In this paper, we define certain moves on partitions into distinct parts, and apply inclusion-exclusion on the number of runs of length at least two to obtain the formulas (1.1) and (1.3). By adding staircases, we show that it is possible to get alternative triple series as generating functions of partitions into d-distinct parts.

Section 2 has the necessary definitions and q-series formulas we will use. In Sec. 3, we give the main construction. Section 4 discusses how to enhance the construction and obtain the desired q-series identities.

It should be also noted that Bringmann *et al.* [2] found a generating function for partitions into distinct parts without k-sequences (runs of length k) for $k \geq 2$:

$$\mathcal{C}_k(x;q) = \sum_{j,r\geq 0} \frac{(-1)^j x^{kj+r} q^{\frac{(r+kj)(r+kj+1)}{2}+k\frac{j(j-1)}{2}}}{(q^k;q^k)_j(q;q)_r}. \tag{1.4}$$

Should we expand this as a double power series in x and q, the exponent of q is the number being partitioned, and the exponent of x is the number of parts. Obviously, $\mathcal{C}_2(1;q)$ is identical to (1.1), and $\mathcal{C}_2(q;q)$ to (1.2). The combinatorial construction of (1.4) in [2] is reminiscent of Hirshhorn's construction of (1.1), but they have an alternative proof by solving q-difference equations, asymptotic formulas for the enumerants they consider, as well as a fairly extensive list of references.

2. Definitions and Auxiliary Formulas

A partition of a positive integer n is a finite non-decreasing sequence of non-negative integers whose sum is n [1]

$$\lambda : \lambda_1 + \lambda_2 + \cdots + \lambda_l = n,$$

where $0 \leq \lambda_1 \leq \lambda_2 \leq \cdots \leq \lambda_l$. The number being partitioned, n, is called the *weight* of the partition, denoted by $|\lambda|$. The number of parts, l, is called the length of the partition, denoted by $l(\lambda)$. Allowing zeros in a partition does not change the weight, but it changes the length.

Given a positive integer d, a partition (into positive parts) is called d-distinct if $\lambda_j - \lambda_{j-1} \geq d$ for $j = 2, 3, \ldots, l$. 1-distinct partitions are just partitions into distinct parts. 2-distinct partitions are partitions into distinct and non-consecutive parts.

In a partition into distinct parts, a maximal streak of consecutive parts is called a *run*. For example, the partition $1, 2, 3, 5, 7, 8, 9$ has three runs: $\{1, 2, 3\}, \{5\}, \{7, 8, 9\}$.

Definition 2.1. A *raft* in a partition into distinct parts is a pair of consecutive parts $[k, k+1]$ such that $k+2$ is not a part.

In other words, a raft is the largest two parts in a run with two or more parts. Please observe the maximum number of rafts in a partition into distinct parts is the number of runs with two or more parts. Not all largest pairs in runs need to be designated as rafts. For instance, the partition $1, 2, 3, 5, 7, 8, 9$ could have

(1) no designated rafts as $1, 2, 3, 5, 7, 8, 9$,
(2) one designated raft as $1, [2, 3], 5, 7, 8, 9$ or $1, 2, 3, 5, 7, [8, 9]$,
(3) or two designated rafts as $1, [2, 3], 5, 7, [8, 9]$.

In the last instance, the rafts $[2, 3]$ and $[8, 9]$ can be compared in the obvious way. The former will be called the smaller, and the latter the larger. Because there are no other designated rafts between them, those two rafts will be called *successive* rafts.

For convenience, we indicate the designated rafts by square brackets around them.

If the next smallest part after the raft $[k, k+1]$ is at least $k+4$, that is, if the way is *clear ahead*, then the raft can move forward as follows:

$$\begin{array}{c}(\text{parts } \leq k-1), [k, k+1], (\text{parts } \geq k+4) \\ \Big\downarrow \text{ one forward move} \\ (\text{parts } \leq k-1), [k+1, k+2], (\text{parts } \geq k+4).\end{array} \tag{2.1}$$

Instead, if there is a run of length $s \geq 1$ containing $k+3$ with no designated raft at the end, then the raft moves forward as follows:

$$\begin{array}{c}(\text{parts } \leq k-1), [k, k+1], k+3, k+4, \ldots, k+s+2, (\text{parts } \geq k+s+4) \\ \Big\downarrow \text{ one forward move} \\ (\text{parts } \leq k-1), [k+1, k+2], k+3, k+4, \ldots, k+s+2, (\text{parts } \geq k+s+4) \\ \Big\downarrow \text{ rearranging} \\ (\text{parts } \leq k-1), k+1, k+2, \ldots [k+s+1, k+s+2], (\text{parts } \geq k+s+4).\end{array} \tag{2.2}$$

With or without an obstacle, a forward move increases the weight of the partition by two. The rearrangement does not change the weight. Backward moves as inverses of forward moves are defined analogously.

The careful reader will have noticed that Definition 2.1 stipulates that rafts do not *collide* nor are they *docked to the same platform*. That is, there can be at most one designated raft in a run. For instance,

$$\ldots, [k, k+1], [k+2, k+3], \ldots$$

is not an admissible configuration. We will have more to say about this requirement after the proof of Theorem 3.4.

Below are some formulas from [3] that we are going to use in the proofs.

$$(-z;q)_\infty = \sum_{n\geq 0} \frac{q^{n(n-1)/2} z^n}{(q;q)_n} \tag{2.3}$$

$$\frac{(az;q)_\infty}{(z;q)_\infty} = \sum_{n\geq 0} \frac{(a;q)_n z^n}{(q;q)_n} \tag{2.4}$$

$$\sum_{n\geq 0} \frac{(a;q)_n (b;q)_n (c/ab)^n}{(q;q)_n (c;q)_n} = \frac{(c/a;q)_\infty (c/b;q)_\infty}{(c;q)_\infty (c/ab;q)_\infty}. \tag{2.5}$$

3. Main Results

The proof of the following lemma is straightforward, and left to the reader.

Lemma 3.1. *Let $[k, k+1]$ and $[l, l+1]$ be successive rafts in a given partition into distinct parts where $k < l$.*

(1) *When conditions exist, a forward move on $[l, l+1]$ allows at least one forward move on $[k, k+1]$.*
(2) *When conditions exist, a backward move on $[k, k+1]$ allows at least on backward move on $[l, l+1]$.*

Lemma 3.2. *Let β be a minimal partition into distinct parts having exactly k designated rafts for $k \geq 1$. That is, no further backward moves are possible on any of the designated rafts of β. Then, a generating function of such β's is*

$$\sum_{\beta} q^{|\beta|} = \sum_{m\geq 0} q^{\binom{3k+m}{2} - 3\binom{k}{2}} \begin{bmatrix} m+k-1 \\ k-1 \end{bmatrix}_{q^{-1}} (-q^{3k+m+1};q)_\infty. \tag{3.1}$$

Proof. Let $[r_1, r_1+1]$ be the smallest raft in β. Then, β must contain the parts $1, 2, \ldots, r_1 - 1$. Any missing part among those would have allowed a backward move on $[r_1, r_1+1]$. In addition, $r_1 + 1$ cannot be a part in β.

Induction on k gives β as

$$\begin{aligned} \beta = {} & 1, 2, \ldots, r_1 - 1, [r_1, r_1+1], \\ & r_1+3, r_1+4, \ldots, r_2 - 1, [r_2, r_2+1], \\ & \vdots \\ & r_{k-1}+3, r_{k-1}+4, \ldots, r_k - 1, [r_k, r_k+1] \\ & (\text{parts } \geq r_k + 3 \text{ containing no rafts}). \end{aligned}$$

Here, $r_j \geq r_{j-1} + 3$ for $j = 2, 3, \ldots, k$, so $r_k \geq 3k - 2$.

In order to have k rafts among the parts $1, 2, \ldots, r_k + 1$, we need $k - 1$ missing parts. The missing parts of the displayed β above are $r_1 + 2,\ r_2 + 2, \ldots, r_{k-1} + 2$. Since $r_{j-1} + 2 \leq r_j - 1$, the missing parts are pairwise at least three apart.

For a moment, we focus on these missing parts. To generate them, we start with $3, 6, \ldots, 3k - 3$, and add μ_1 to $3k-3$, μ_2 to $3k-6$, etc. $r_{k-1} = 3k-3+\mu_1 \leq r_k - 1$, so $\mu_1 \leq r_k - 3k - 2$. To retain the difference conditions between μ's, we must have $\mu_1 \geq \mu_2 \geq \cdots \geq \mu_{k-1} \geq 0$. Thus, μ's form a restricted partition into at most $k - 1$ parts, all of which are at most $r_k - 3k + 2$. As such, the μ's are generated by the Gaussian polynomial $\begin{bmatrix} r_k - 3k + 2 + k - 1 \\ k - 1 \end{bmatrix}$. Thus, the parts of the complementary part of β are generated by

$$q^{3\binom{k}{2}} \begin{bmatrix} r_k - 2k + 1 \\ k - 1 \end{bmatrix}. \tag{3.2}$$

To obtain the generating function of the portion of β containing parts that are at most $r_k + 1$, we need to extract the missing parts from among $1,\ 2, \ldots, r_k + 1$. This amounts to replacing q by q^{-1} in (3.2), and multiplying by $q^{\binom{r_k+2}{2}}$:

$$q^{\binom{r_k+2}{2} - 3\binom{k}{2}} \begin{bmatrix} r_k - 2k + 1 \\ k - 1 \end{bmatrix}_{q^{-1}}. \tag{3.3}$$

The parts of β that are at least r_k+3 merely form a partition into distinct parts. They are generated by $(-q^{r_k+3}; q)_\infty$, to be multiplied by (3.3). Then, summing over $r_k \geq 3k - 2$ and the change of parameter $r_k = m + 3k - 2$ will conclude the proof. □

Remark. If a partition into distinct parts has no designated rafts, it is generated by $(-q; q)_\infty$. This is not the $k = 0$ case of Lemma 3.2, so it has to be handled separately.

Corollary 3.3. *Let λ be a partition into distinct parts having exactly k designated rafts for $k \geq 1$. A generating function for such λ is*

$$\sum_\lambda q^{|\lambda|} = \sum_{m \geq 0} q^{\binom{3k+m}{2} - 3\binom{k}{2}} \begin{bmatrix} m + k - 1 \\ k - 1 \end{bmatrix}_{q^{-1}} \frac{(-q^{3k+m+1}; q)_\infty}{(q^2; q^2)_k}. \tag{3.4}$$

Proof. The only difference of this generating function from (3.1) is the factor $1/(q^2; q^2)_k$.

This factor generates a partition into k even parts (allowing zeros) $\eta_1 \geq \eta_2 \geq \cdots \geq \eta_k \geq 0$. Beginning with a minimal partition β into distinct parts with k designated rafts, we move the largest raft forward $\eta_1/2$ times, the next largest $\eta_2/2$ etc. in this order. Thanks to Lemma 3.1 and the comparison between η's, all moves are admissible. This will give us a λ as described in the statement.

Conversely, given such λ, we perform $\eta_k/2$ backward moves on the smallest raft so that no further backward moves on it are possible. This uniquely determines even η_k. Then, we perform $\eta_{k-1}/2$ backward moves on the next smallest raft so that no further backward moves on it are possible. This uniquely determines even η_{k-1}. Because of Lemma 3.1, $\eta_{k-1} \geq \eta_k$. We continue with the next smallest raft etc. to eventually obtain the partition η's into k even parts (allowing zeros).

The fact that forward and backward moves are inverses of each other, and that we perform the moves in the exact reverse order finishes the proof. □

Theorem 3.4.

$$(-q;q)_\infty + \sum_{k\geq 1}(-1)^k \sum_{m\geq 0} q^{\binom{3k+m}{2}-3\binom{k}{2}} \begin{bmatrix} m+k-1 \\ k-1 \end{bmatrix}_{q^{-1}} \frac{(-q^{3k+m+1};q)_\infty}{(q^2;q^2)_k}, \tag{3.5}$$

generates partitions into distinct parts in which there can be no designated rafts.

Remark. If we cannot designate a raft in a partition into distinct parts, then that partition has no runs. In particular, it does not have any consecutive parts. Thanks to the first Rogers–Ramanujan Identity, we have

$$(-q;q)_\infty + \sum_{k\geq 1}(-1)^k \sum_{m\geq 0} q^{\binom{3k+m}{2}-3\binom{k}{2}} \begin{bmatrix} m+k-1 \\ k-1 \end{bmatrix}_{q^{-1}} \frac{(-q^{3k+m+1};q)_\infty}{(q^2;q^2)_k}$$
$$= \frac{1}{(q;q^5)_\infty (q^4;q^5)_\infty}. \tag{3.6}$$

Proof of Theorem 3.4. This is a standard inclusion-exclusion argument. The first term $(-q;q)_\infty$ generates partitions into distinct parts, which we will interpret as partitions with no designated rafts. These are weighted by $(-1)^0$. The double sum generates partitions into distinct parts, weighted by $(-1)^k$, where $k \geq 1$ is the number of designated rafts in them.

The number of runs in a partition into distinct parts, say K, is the maximum possible number of designated rafts the partition can have. If $K = 0$, then the partition can have no designated rafts. If $K > 0$, then we can designate k rafts in $\binom{K}{k}$ different ways, each of which is weighted by $(-1)^k$. In total, the partition is counted

$$\sum_{k=0}^{K} \binom{K}{k}(-1)^k = (1-1)^K = 0$$

times, yielding the proof. □

If we allowed configurations such as $[k, k+1]$, $[k+2, k+3]$, then, the partition $1, 2, 3, 4$ would have been counted $+1$ times as $1, 2, 3, 4$; -1 times as $1, 2, [3, 4]$; and $+1$ times as $[1, 2]$, $[3, 4]$; a total of $+1$ times. Yet, it should have been annihilated by inclusion-exclusion. If we tried to solve this problem, insisting that rafts could be adjacent, then there will be other problems either making the moves or raft designation ambiguous, or making the generating functions unnecessarily complicated.

Proof of (1.1). We will manipulate the left-hand side of (3.5). In

$$(-q;q)_\infty + \sum_{k\geq 1}(-1)^k \sum_{m\geq 0} q^{\binom{3k+m}{2}-3\binom{k}{2}} \begin{bmatrix} m+k-1 \\ k-1 \end{bmatrix}_{q^{-1}} \frac{(-q^{3k+m+1};q)_\infty}{(q^2;q^2)_k},$$

utilize

$$\begin{bmatrix} m+k-1 \\ k-1 \end{bmatrix}_{q^{-1}} = q^{-m(k-1)} \begin{bmatrix} m+k-1 \\ k-1 \end{bmatrix} = q^{-m(k-1)} \frac{(q^k;q)_m}{(q;q)_m},$$

as well as

$$(-q^{3k+m+1};q)_\infty = \frac{(-q;q)_\infty}{(-q;q)_{3k+m}},$$

to obtain

$$(-q;q)_\infty + (-q;q)_\infty \sum_{k\geq 1} \frac{(-1)^k q^{3k^2}}{(q^2;q^2)_k} \sum_{m\geq 0} \frac{q^{\binom{m}{2}+2km+m}(q^k;q)_m}{(q;q)_m(-q^{3k+1};q)_m}.$$

$$= (-q;q)_\infty + (-q;q)_\infty \sum_{k\geq 1} \frac{(-1)^k q^{3k^2}}{(q^2;q^2)_k(-q;q)_{3k}}$$

$$\times \lim_{b\to 0} \sum_{m\geq 0} \frac{(-1/b;q)_m (q^k;q)_m}{(q;q)_m(-q^{3k+1};q)_m} (bq^{2k+1})^m.$$

Using the q-Gauss summation (2.5), this transforms to

$$= (-q;q)_\infty + (-q;q)_\infty \sum_{k\geq 1} \frac{(-1)^k q^{3k^2}}{(q^2;q^2)_k(-q;q)_{3k}} \lim_{b\to 0} \frac{(bq^{3k+1};q)_\infty(-q^{2k+1};q)_\infty}{(-q^{3k+1};q)_\infty(bq^{2k+1};q)_\infty}$$

$$= (-q;q)_\infty + (-q;q)_\infty \sum_{k\geq 1} \frac{(-1)^k q^{3k^2}}{(q^2;q^2)_k(-q;q)_{3k}} \frac{(-q^{2k+1};q)_\infty}{(-q^{3k+1};q)_\infty}$$

$$= (-q;q)_\infty + (-q;q)_\infty \sum_{k\geq 1} \frac{(-1)^k q^{3k^2}}{(q^2;q^2)_k(-q;q)_{2k}}$$

$$= (-q;q)_\infty \sum_{k\geq 0} \frac{(-1)^k q^{3k^2}}{(q^2;q^2)_k(-q;q)_{2k}}.$$

Theorem 3.4 finishes the proof. □

4. Shifts and Staircases

It is possible to keep track of the number of parts in the processes described in Sec. 3. The proofs of the respective results apply *mutatis mutandis*. We record the intermediate formulas here for convenience.

Let β be a minimal partition described in Lemma 3.2. Then,

$$\sum_{\beta} q^{|\beta|} x^{l(\beta)} = \sum_{m \geq 0} q^{\binom{3k+m}{2} - 3\binom{k}{2}} x^{2k+m} \begin{bmatrix} m+k-1 \\ k-1 \end{bmatrix}_{q^{-1}} (-xq^{3k+m+1}; q)_\infty.$$

Let λ be a partition described in Corollary 3.3. Then,

$$\sum_{\lambda} q^{|\lambda|} x^{l(\lambda)} = \sum_{m \geq 0} q^{\binom{3k+m}{2} - 3\binom{k}{2}} x^{2k+m} \begin{bmatrix} m+k-1 \\ k-1 \end{bmatrix}_{q^{-1}} \frac{(-xq^{3k+m+1}; q)_\infty}{(q^2; q^2)_k}.$$

Upon accounting for the number of parts, (3.6) becomes

$$(-xq; q)_\infty + \sum_{k \geq 1} (-1)^k \sum_{m \geq 0} q^{\binom{3k+m}{2} - 3\binom{k}{2}} x^{2k+m} \begin{bmatrix} m+k-1 \\ k-1 \end{bmatrix}_{q^{-1}} \frac{(-xq^{3k+m+1}; q)_\infty}{(q^2; q^2)_k}$$
$$= \sum_{n \geq 0} \frac{q^{n^2} x^n}{(q; q)_n}.$$

Then, we can go through the same line of reasoning as in the proof of Theorem 3.4 to obtain

$$(-xq; q)_\infty \sum_{k \geq 0} \frac{(-1)^k q^{3k^2} x^{2k}}{(q^2; q^2)_k (-xq; q)_{2k}} = \sum_{n \geq 0} \frac{q^{n^2} x^n}{(q; q)_n}. \tag{4.1}$$

In (4.1), if we substitute $x = q$ and appeal to the second Rogers–Ramanujan identity [5], we arrive at (1.3).

If we expand the left-hand side of (4.1) as a power series in x, it will be possible to insert staircases into partitions. Using (2.3) and (2.4), a generating function of partitions into parts with pairwise difference at least two is

$$\sum_{m,n,k \geq 0} \frac{q^{\binom{n+1}{2}}}{(q; q)_n} \cdot \frac{(-1)^k q^{3k^2}}{(q^2; q^2)_k} \cdot \frac{(q^{2k}; q)_m (-q)^m}{(q; q)_m} \cdot x^{n+2k+m}.$$

To make the pairwise difference at least $(2+d)$, we insert the staircase

$$0 + d + 2d + \cdots + (l(\lambda) - 1)d$$

to each partition λ. In the generating function as a power series in x, It amounts to replacing x^N by $x^N q^{d\binom{N}{2}}$. This operation yields the generating function

$$\sum_{m,n,k \geq 0} \frac{q^{\binom{n+1}{2} + d\binom{n}{2}}}{(q; q)_n} \cdot \frac{(-1)^k q^{3k^2 + d\binom{2k}{2}}}{(q^2; q^2)_k} \cdot \frac{(q^{2k}; q)_m q^{d\binom{m}{2}} (-q)^m}{(q; q)_m} \cdots$$
$$\times x^{n+2k+m} q^{2dnk + dnm + 2dkm} \tag{4.2}$$

for partitions into parts with pairwise difference at least $(2+d)$. Unfortunately, and as to be expected, unless $d = 0$, series-product identities do not seem to help reduce the triple sum in (4.2) into a single sum, or even a double sum.

Acknowledgments

This paper is dedicated to Bruce C. Berndt for his 80th birthday. We thank George E. Andrews and Michael D. Hirshhorn for their guidance and help. We also thank the anonymous referees for their helpful comments, and for pointing out [2].

References

[1] G. E. Andrews, *The Theory of Partitions* (Cambridge University Press, 1984).
[2] K. Bringmann, K. Mahlburg and K. Nataraj, Distinct parts partitions without sequences, *Electron. J. Combin.* **22**(3) (2015) p. 3.3.
[3] G. Gasper and M. Rahman, *Basic Hypergeometric Series*, 2nd edn., Encyclopedia of Mathematics and its Applications, Vol. 96 (Cambridge University Press, 2004).
[4] M. D. Hirschhorn, Developments in the theory of partitions, Ph.D. thesis, University of New South Wales (1979).
[5] S. Ramanujan and L. J. Rogers, Proof of certain identities in combinatory analysis, *Proc. Cambridge Phil. Soc.* **19** (1919) 211–216.
[6] L. J. Slater, Further identities of the Rogers-Ramanujan Type, *Proc. London Math. Soc. Ser. 2* **54** (1952) 147–167.

© 2024 World Scientific Publishing Company
https://doi.org/10.1142/9789811277375_0030

Some observations on Lambert series, vanishing coefficients and dissections of infinite products and series

James Mc Laughlin
Mathematics Department, West Chester University,
25 University Avenue, West Chester, PA 19383, USA
jmclaughlin2@wcupa.edu

Andrews and Bressoud, Alladi and Gordon, and others, have proven, in a number of papers, that the coefficients in various arithmetic progressions in the series expansions of certain infinite q-products vanish. In the present paper, it is shown that these results follow automatically (simply by specializing parameters) in an identity derived from a special case of Ramanujan's ${}_1\psi_1$ identity.

Likewise, a number of authors have proven results about the m-dissections of certain infinite q-products using various methods. It is shown that many of these m-dissections also follow automatically (again simply by specializing parameters) from this same identity alluded to above.

Two identities that may be considered as extensions of two identities of Ramanujan are also derived.

It is also shown how applying similar ideas to certain other Lambert series gives rise to some rather curious q-series identities, such as, for any positive integer m,

$$\frac{\left(q, q, a, \frac{q}{a}, \frac{bq}{d}, \frac{dq}{b}, \frac{aq}{bd}, \frac{bdq}{a}; q\right)_\infty}{\left(b, \frac{q}{b}, d, \frac{q}{d}, \frac{a}{b}, \frac{bq}{a}, \frac{a}{d}, \frac{dq}{a}; q\right)_\infty} = \sum_{r=0}^{m-1} q^r \frac{\left(q^m, q^m, aq^{2r}, \frac{q^{m-2r}}{a}, \frac{bq^m}{d}, \frac{dq^m}{b}, \frac{aq^m}{bd}, \frac{bdq^m}{a}; q^m\right)_\infty}{\left(bq^r, \frac{q^{m-r}}{b}, dq^r, \frac{q^{m-r}}{d}, \frac{aq^r}{b}, \frac{bq^{m-r}}{a}, \frac{aq^r}{d}, \frac{dq^{m-r}}{a}; q^m\right)_\infty}$$

and

$$(aq;q)_\infty \sum_{n=1}^{\infty} \frac{na^n q^n}{(q;q)_n} = \sum_{r=1}^{m} (aq^r;q^m)_\infty \sum_{n=1}^{\infty} \frac{na^n q^{nr}}{(q^m;q^m)_n}.$$

Applications to the Fine function $F(a, b; t)$ are also considered.

Keywords: q-Series; infinite products; infinite q-products; vanishing coefficients; m-dissections; Ramanujan's ${}_1\psi_1$ identity.

Mathematics Subject Classification 2020: 11B65, 33D15

1. Introduction

The origins of this paper lie in the entirely elementary observation that if

$$f(a,b,c,q) := \sum_{n=-\infty}^{\infty} \frac{ab^n}{1-cq^n}, \tag{1.1}$$

and m is any positive integer, then after replacing n with $nm+r$ and rearranging into arithmetic progressions modulo m,

$$f(a,b,c,q) = \sum_{r=0}^{m-1} f(ab^r, b^m, cq^r, q^m). \tag{1.2}$$

A similar statement can be made for a function

$$f^*(a,b,c,q) := \sum_{n=0}^{\infty} \frac{ab^n}{1-cq^n}. \tag{1.3}$$

The usefulness of this derives from the fact if $f(a,b,c,q)$ has some other representation as a basic hypergeometric series or infinite q-product, say $f(a,b,c,q) = g(a,b,c,q)$, then for any positive integer m,

$$g(a,b,c,q) = \sum_{r=0}^{m-1} g(ab^r, b^m, cq^r, q^m). \tag{1.4}$$

A similar statement may be made about $f^*(a,b,c,q)$.

As will be shown, this will lead to proofs of various results on vanishing coefficients, m-dissections of various infinite products, and will also result in a number of new, somewhat unusual q-series expansions involving an integer parameter m. In the case of the results on vanishing coefficients and dissection of infinite products, the results will follow immediately upon specialization of parameters in a general identity, in contrast to previous proofs of these results, which employed various manipulations of infinite series.

Some examples of existing results proved by the methods of the present paper are given below. The first two use a general identity derived from Ramanujan's ${}_1\psi_1$ summation formula (valid for $|q|<1$, $|b/a|<|z|<1$),

$$\sum_{n=-\infty}^{\infty} \frac{(a;q)_n z^n}{(b;q)_n} = \frac{(b/a,q,az,q/az;q)_\infty}{(q/a,b,z,b/az;q)_\infty}. \tag{1.5}$$

The first example is a general result on vanishing coefficients.

Theorem 1.1 ([17]). *Let $k>1$, $m>1$ be positive integers. Let $r=sm+t$, for some integers s and t, where $0\le s<k$, $1\le t<m$ and r and k are relatively prime. Let*

$$\frac{(q^{r-tk}, q^{mk-(r-tk)}; q^{mk})_\infty}{(q^r, q^{mk-r}; q^{mk})_\infty} =: \sum_{n=0}^{\infty} c_n q^n, \tag{1.6}$$

then c_{kn-rs} is always zero.

The second is an example of the application of the general identity to m-dissections of infinite products, where the 3-dissection of the infinite product on the left below is given.

Corollary 1.2 ([14, Theorem 3.2]). *If* $|q| < 1$, *then*

$$\frac{(q^7, q^5; q^{12})_\infty}{(q, q^{11}; q^{12})_\infty} = \frac{(q^6, q^{30}; q^{36})_\infty}{(q^{12}, q^{12}, q^{24}, q^{24}; q^{36})_\infty} \times \left(\frac{(q^9, q^{18}, q^{18}, q^{27}; q^{36})_\infty}{(q^3, q^{33}; q^{36})_\infty} + q \frac{(q^6, q^{15}, q^{21}, q^{30}; q^{36})_\infty}{(q^3, q^{33}; q^{36})_\infty} + q^2 (q^{18}, q^{18}; q^{36})_\infty \right). \tag{1.7}$$

Iteration of the main identity leads to infinite product identities such as

$$\frac{(q^5, q^5; q^5)_\infty}{(q, q^4; q^5)_\infty} = \frac{1}{1-q^2} \prod_{k=1}^{\infty} \left(1 + q^{2^{k-1}} \frac{(q^2, q^{2^k 5-2}, q^{2^{k-1}7+2}, q^{2^{k-1}3-2}; q^{2^k 5})_\infty}{(q^{2^{k-1}5+2}, q^{2^{k-1}5-2}, q^{2^k+2}, q^{2^{k+2}-2}; q^{2^k 5})_\infty} \right). \tag{1.8}$$

After applying similar ideas to Bailey's ${}_6\psi_6$ summation formula, the following identity is a consequence.

Theorem 1.3. *If* $|q| < 1$ *and* m *is a positive integer, then*

$$\frac{\left(q, q, a, \frac{q}{a}, \frac{bq}{d}, \frac{dq}{b}, \frac{aq}{bd}, \frac{bdq}{a}; q\right)_\infty}{\left(b, \frac{q}{b}, d, \frac{q}{d}, \frac{a}{b}, \frac{bq}{a}, \frac{a}{d}, \frac{dq}{a}; q\right)_\infty} = \sum_{r=0}^{m-1} q^r \frac{\left(q^m, q^m, aq^{2r}, \frac{q^{m-2r}}{a}, \frac{bq^m}{d}, \frac{dq^m}{b}, \frac{aq^m}{bd}, \frac{bdq^m}{a}; q^m\right)_\infty}{\left(bq^r, \frac{q^{m-r}}{b}, dq^r, \frac{q^{m-r}}{d}, \frac{aq^r}{b}, \frac{bq^{m-r}}{a}, \frac{aq^r}{d}, \frac{dq^{m-r}}{a}; q^m\right)_\infty}. \tag{1.9}$$

Finally, examples of applying similar ideas to other identities leads to the following somewhat unusual expansions involving the integer parameter m.

Corollary 1.4. *Let* m *be a positive integer.*

(i) *If* $|q| < 1$, *then*

$$\frac{1}{(aq; q)_\infty} \sum_{n=1}^{\infty} \frac{n(-a)^n q^{n(n+1)/2}}{(q; q)_n} = \sum_{r=1}^{m} \frac{1}{(aq^r; q^m)_\infty} \sum_{n=1}^{\infty} \frac{n(-a)^n q^{mn(n-1)/2+nr}}{(q^m; q^m)_n}. \tag{1.10}$$

(ii) *If* $|q|, |aq| < 1$, *then*

$$(aq; q)_\infty \sum_{n=1}^{\infty} \frac{na^n q^n}{(q; q)_n} = \sum_{r=1}^{m} (aq^r; q^m)_\infty \sum_{n=1}^{\infty} \frac{na^n q^{nr}}{(q^m; q^m)_n}. \tag{1.11}$$

2. A Special Case Ramanujan's ${}_1\psi_1$ Summation Formula

The main tool used is an identity that follows from a special case Ramanujan's ${}_1\psi_1$ summation formula (1.5), after specializing one of the parameters. Observe that the right-hand side of (2.1) provides the p-dissection of the left-hand side, where by "p-dissection" here we mean in terms of powers of z, rather than powers of q.

Proposition 2.1. *If p is a positive integer and $|q|<1$, then*

$$\frac{(q,q,az,q/(az);q)_\infty}{(a,q/a,z,q/z;q)_\infty}=\sum_{j=0}^{p-1} z^j\frac{(q^p,q^p,aq^jz^p,q^{p-j}/(az^p);q^p)_\infty}{(aq^j,q^{p-j}/a,z^p,q^p/z^p;q^p)_\infty}. \tag{2.1}$$

Proof. Set $b=aq$ in (1.5) and divide both sides by $1-a$ to get

$$\begin{aligned}\frac{(q,q,az,q/(az);q)_\infty}{(a,q/a,z,q/z;q)_\infty}&=\sum_{n=-\infty}^{\infty}\frac{z^n}{1-aq^n}\\&=\sum_{j=0}^{p-1}\sum_{n=-\infty}^{\infty}\frac{z^{np+j}}{1-aq^{np+j}}\\&=\sum_{j=0}^{p-1}z^j\sum_{n=-\infty}^{\infty}\frac{(z^p)^n}{1-(aq^j)(q^p)^n},\end{aligned} \tag{2.2}$$

and (2.1) for $|q|<|z|<1$ follows immediately upon employing the first equality in (2.2), with q replaced with q^p, a with aq^r, and z with z^p to each of the inner sums in the last expression. The full result follows by analytic continuation. □

The identity (2.1) can be slightly rearranged and iterated to give the following identity.

Corollary 2.2. *If m and p are positive integers, then*

$$\begin{aligned}&\frac{(q,q,az,q/(az);q)_\infty}{(a,q/a,z,q/z;q)_\infty}\\&\quad=\frac{(q^{p^m},q^{p^m},az^{p^m},q^{p^m}/(az^{p^m});q^{p^m})_\infty}{(a,q^{p^m}/a,z^{p^m},q^{p^m}/z^{p^m};q^{p^m})_\infty}\\&\quad\times\prod_{k=1}^{m}\left(1+\sum_{j=1}^{p-1}z^{jp^{k-1}}\frac{(a,q^{p^k}/a,aq^{jp^{k-1}}z^{p^k},q^{p^k-jp^{k-1}}/(az^{p^k});q^{p^k})_\infty}{(aq^{jp^{k-1}},q^{p^k-jp^{k-1}}/a,az^{p^k},q^{p^k}/(az^{p^k});q^{p^k})_\infty}\right).\end{aligned} \tag{2.3}$$

Proof. Rewrite (2.1) in the form

$$\begin{aligned}&\frac{(q,q,az,q/(az);q)_\infty}{(a,q/a,z,q/z;q)_\infty}\\&\quad=\frac{(q^p,q^p,az^p,q^p/(az^p);q^p)_\infty}{(a,q^p/a,z^p,q^p/z^p;q^p)_\infty}\sum_{j=0}^{p-1}z^j\frac{(a,q^p/a,aq^jz^p,q^{p-j}/(az^p);q^p)_\infty}{(aq^j,q^{p-j}/a,az^p,q^p/(az^p);q^p)_\infty}.\end{aligned} \tag{2.4}$$

Now iterate, noting that the infinite product before the sum on the right-hand side of (2.4) has the same form as the infinite product on the left-hand side, with the replacements $q \to q^p$ and $z \to z^p$. Note also that the $j = 0$ term in the sum on the right simplifies to 1. $\square$

The limit as $m \to \infty$ of the individual terms on the right-hand side of (2.4) will not exist for all values of z, but will for some values of z. We next give an example of an identity where these limits do exist, derived by making the substitutions $q \to q^5$, $z \to q$, $a \to q^2$ in (2.4) and setting $p = 2$.

Example 1. If m is a positive integer, then

$$\frac{(q^5, q^5; q^5)_\infty}{(q, q^4; q^5)_\infty} = \frac{(q^{2^m 5}, q^{2^m 5}, q^{2^m+2}, q^{2^{m+2}-2}; q^{2^m 5})_\infty}{(q^2, q^{2^m 5-2}, q^{2^m}, q^{2^{m+2}}; q^{2^m 5})_\infty} \times \prod_{k=1}^{m} \left(1 + q^{2^{k-1}} \frac{(q^2, q^{2^k 5-2}, q^{2^{k-1}7+2}, q^{2^{k-1}3-2}; q^{2^k 5})_\infty}{(q^{2^{k-1}5+2}, q^{2^{k-1}5-2}, q^{2^k+2}, q^{2^{k+2}-2}; q^{2^k 5})_\infty}\right). \tag{2.5}$$

Upon letting $m \to \infty$, it follows that

$$\frac{(q^5, q^5; q^5)_\infty}{(q, q^4; q^5)_\infty} = \frac{1}{1-q^2} \prod_{k=1}^{\infty} \left(1 + q^{2^{k-1}} \frac{(q^2, q^{2^k 5-2}, q^{2^{k-1}7+2}, q^{2^{k-1}3-2}; q^{2^k 5})_\infty}{(q^{2^{k-1}5+2}, q^{2^{k-1}5-2}, q^{2^k+2}, q^{2^{k+2}-2}; q^{2^k 5})_\infty}\right). \tag{2.6}$$

3. Extensions of Two Identities of Ramanujan

Before considering vanishing coefficients and m-dissections, we consider two other quite general identities, which may be regarded as extensions of two identities of Ramanujan, as will be shown below.

Corollary 3.1. *Let m be a positive integer, and let $\omega = \exp(2\pi i/m)$ be a primitive m root of unity. If $a, z \neq 0$ and $|q| < 1$, then*

$$\sum_{j=0}^{m-1} \frac{\left(q, q, \omega^j az, \dfrac{q}{\omega^j az}; q\right)_\infty}{\left(a, \dfrac{q}{a}, \omega^j z, \dfrac{q}{\omega^j z}; q\right)_\infty} = m \frac{\left(q^m, q^m, az^m, \dfrac{q^m}{az^m}; q^m\right)_\infty}{\left(a, \dfrac{q^m}{a}, z^m, \dfrac{q^m}{z^m}; q^m\right)_\infty}, \tag{3.1}$$

$$\sum_{j=0}^{m-1} \omega^j \frac{\left(q, q, \omega^j az, \dfrac{q}{\omega^j az}; q\right)_\infty}{\left(\omega^j a, \dfrac{q}{\omega^j a}, z, \dfrac{q}{z}; q\right)_\infty} = ma^{m-1} \frac{\left(q^m, q^m, a^m zq^{m-1}, \dfrac{q}{a^m z}; q^m\right)_\infty}{\left(a^m, \dfrac{q^m}{a^m}, zq^{m-1}, \dfrac{q}{z}; q^m\right)_\infty}. \tag{3.2}$$

Proof. For (3.1), take the first equality (2.2) and for each j, $0 \leq j \leq m-1$, replace z with $\omega^j z$. Add the resulting m identities together, so that the resulting left-hand side is the left-hand side of (3.1), and the right-hand side becomes

$$\sum_{n=-\infty}^{\infty} \frac{z^n}{1-aq^n} \sum_{j=0}^{m-1} (\omega^n)^j.$$

That this equals the right-hand side of (3.1) follows from the fact that

$$\sum_{j=0}^{m-1} (\omega^n)^j = \begin{cases} m, & m|n, \\ 0, & \text{otherwise.} \end{cases}$$

This proves (3.1) for $|q| < |z| < 1$, and the full result follows by analytic continuation.

For (3.2), similarly take the first equality (2.2) and for each j, $0 \leq j \leq m-1$, replace a with $\omega^j a$. Multiply the jth resulting identity by ω^j and add all m identities together, so that the resulting left-hand side is the left-hand side of (3.2), and the right-hand side becomes

$$\sum_{n=-\infty}^{\infty} z^n \sum_{j=0}^{m-1} \frac{\omega^j}{1-\omega^j aq^n}.$$

Now,

$$\begin{aligned} \sum_{j=0}^{m-1} \frac{\omega^j}{1-\omega^j aq^n} &= -\sum_{j=0}^{m-1} \frac{1}{q^n} \frac{d}{da} \ln(1-\omega^j aq^n) \\ &= -\frac{1}{q^n} \frac{d}{da} \ln \left(\prod_{j=0}^{m-1} (1-\omega^j aq^n) \right) \\ &= -\frac{1}{q^n} \frac{d}{da} \ln \left(1-a^m q^{mn}\right) \\ &= \frac{ma^{m-1} q^{(m-1)n}}{1-a^m q^{mn}}. \end{aligned}$$

Hence the right-hand side is

$$ma^{m-1} \sum_{n=-\infty}^{\infty} \frac{(zq^{m-1})^n}{1-a^m (q^m)^n},$$

and (3.2) follows for $|q| < |z| < 1$ after one further application of (2.2) (with q replaced with q^m, a replaced with a^m, and z replaced with zq^{m-1}). Once again, the full result follows by analytic continuation. □

Note that if z is replaced with z/a in the $m = 2$ cases of (3.1) and (3.2), then after some elementary q-product manipulations the following pair of identities appear:

$$\left(z, \frac{q}{z}, -a, -\frac{q}{a}, q, q; q\right)_\infty + \left(-z, -\frac{q}{z}, a, \frac{q}{a}, q, q; q\right)_\infty$$

$$= 2\left(\frac{zq}{a}, \frac{aq}{z}, az, \frac{q^2}{az}, q^2, q^2; q^2\right)_\infty, \tag{3.3}$$

$$\left(z, \frac{q}{z}, -a, -\frac{q}{a}, q, q; q\right)_\infty - \left(-z, -\frac{q}{z}, a, \frac{q}{a}, q, q; q\right)_\infty$$

$$= 2a\left(\frac{z}{a}, \frac{aq^2}{z}, azq, \frac{q}{az}, q^2, q^2; q^2\right)_\infty. \tag{3.4}$$

These are the two identities in **Entry 29** in Chap. 16 of Ramanujan's second notebook (see [5, p. 45]), so that in a sense (3.1) and (3.2) may be regarded as extensions of Ramanujan's two identities.

4. Vanishing Coefficients

As remarked above, the right-hand side of (2.1) provides a p-dissection of the left side in powers of z, in that each term in the sum on the right-hand side contains powers of z lying in a single arithmetic progression modulo p. For the purposes of giving simple proofs of some of the results stated in the introduction, the parameter a is now specialized so that the coefficients of all powers of z in one arithmetic progression vanish.

Corollary 4.1. *Let s be an integer, $0 \le s \le p-1$. If the sequence $\{c_n(q)\}_{n=-\infty}^{\infty}$ is defined by*

$$\frac{(q, q, q^{p-s}z^{1-p}, q^{s+1-p}z^{p-1}; q)_\infty}{(q^{p-s}z^{-p}, q^{s+1-p}z^p, z, q/z; q)_\infty} = \sum_{n=-\infty}^{\infty} c_n(q) z^n. \tag{4.1}$$

Then $c_{pn+s}(q) = 0$.

Proof. Set $a = q^{p-s}z^{-p}$ in (2.1), so that the left-hand side of (2.1) becomes the left-hand side of (4.1). The jth term, for $0 \le j \le p-1$, on the right-hand side of (2.1) becomes

$$z^j \frac{(q^p, q^p, q^{p-s+j}, q^{s-j}; q^p)_\infty}{(q^{p-s+j}z^{-p}, q^{s-j}z^p, z^p, q^p/z^p; q^p)_\infty},$$

which vanishes when $j = s$, giving the result. □

There are a number of results in the literature about coefficients in the series expansion of various infinite q-products vanishing in certain arithmetic progressions, and also some results giving the m-dissection of various infinite products for some small positive integers $m > 1$ (more details of some of these results are given below). These results have been proved in a number of ways, often involving various q-series identities and manipulation of infinite series, and in some cases using expansions of

Hardy–Ramanujan–Rademacher type developed from the infinite products. In the present paper it is shown that many of these follow automatically from various forms of (2.1) (such as (4.1)), upon simply specializing some of the parameters (without any need for further manipulation of series or appeal to other q-series identities).

We first review some of the history of the results on vanishing coefficients (previous results on m-dissections will be described later, when discussing the results in the present paper on the topic).

In [20], Richmond and Szekeres showed that if the sequence $\{c_m\}$ is defined by

$$F(q) := \frac{(q^3, q^5; q^8)_\infty}{(q, q^7; q^8)_\infty} =: \sum_{m=0}^{\infty} c_m q^m, \tag{4.2}$$

then c_{4n+3} is always zero. Likewise, they showed that d_{4n+2} is zero for all n, if the sequence $\{d_m\}$ is defined by

$$\frac{1}{F(q)} =: \sum_{m=0}^{\infty} d_m q^m.$$

Their results were derived from Hardy–Ramanujan–Rademacher expansions they developed of the infinite products. They also conjectured that if

$$G(q) := \frac{(q^5, q^7; q^{12})_\infty}{(q, q^{11}; q^{12})_\infty} =: \sum_{m=0}^{\infty} a_m q^m,$$

then a_{6n+5} is always zero, and if

$$\frac{1}{G(q)} =: \sum_{m=0}^{\infty} b_m q^m,$$

then b_{6n+3} is always zero.

In [2], Andrews and Bressoud proved the following theorem, from which the results of Richmond and Szekeres followed as special cases.

Theorem 4.2. *If $1 \leq r < k$ are relatively prime integers of opposite parity and*

$$\frac{(q^r, q^{2k-r}; q^{2k})_\infty}{(q^{k-r}, q^{k+r}; q^{2k})_\infty} =: \sum_{n=0}^{\infty} \phi_n q^n, \tag{4.3}$$

then $\phi_{kn+r(k-r+1)/2}$ is always zero.

Andrews and Bressoud used Ramanujan's ${}_1\psi_1$ summation formula (1.5) to derive the result in Theorem 4.2, after replacing q with q^{2k}, specializing a, b and z and employing some q-series manipulations. The two results proved by Richmond and Szekeres, and the two results conjectured by them, follow from the cases $(k, r) = (4, 3)$, $(4, 1)$, $(6, 5)$ and $(6, 1)$, respectively.

Alladi and Gordon [1] prove a generalization of Theorem 4.2.

Theorem 4.3. *Let $1 < m < k$ and let $(s, km) = 1$ with $1 \leq s < mk$. Let $r^* = (k-1)s$ and $r \equiv r^* \pmod{mk}$, with $1 \leq r < mk$.*

Put $r' = \lceil \frac{r^*}{mk} \rceil \pmod{k}$ *with* $1 \leq r' < k$. *Write*

$$\frac{(q^r, q^{mk-r}; q^{mk})_\infty}{(q^s, q^{mk-s}; q^{mk})_\infty} = \sum_{n=0}^{\infty} a_n q^n.$$

Then $a_n = 0$ *for* $n \equiv rr' \pmod{k}$.

Alladi and Gordon [1] also prove a companion result to that in their Theorem 4.3.

Theorem 4.4. *Let* m, k, s, r^*, r, r' *be as in Theorem* 4.4 *with* k *odd. Write*

$$\frac{(q^r, q^{mk-r}; q^{mk})_\infty}{(-q^s, -q^{mk-s}; q^{mk})_\infty} = \sum_{n=0}^{\infty} a'_n q^n.$$

Then $a'_n = 0$ *for* $n \equiv rr' \pmod{k}$.

In [17], the present author proved the following results.

Theorem 4.5. *Let* $k > 1$, $m > 1$ *be positive integers. Let* $r = sm + t$, *for some integers* s *and* t, *where* $0 \leq s < k$, $1 \leq t < m$ *and* r *and* k *are relatively prime. Let*

$$\frac{(q^{r-tk}, q^{mk-(r-tk)}; q^{mk})_\infty}{(q^r, q^{mk-r}; q^{mk})_\infty} =: \sum_{n=0}^{\infty} c_n q^n, \tag{4.4}$$

then c_{kn-rs} *is always zero.*

Theorem 4.6. *Let* $k > 1$, $m > 1$ *be positive integers, with* k *odd. Let* $r = sm + t$, *for some integers* s *and* t, *where* $0 \leq s < k$, $1 \leq t < m$ *and* r *and* k *are relatively prime. Let*

$$\frac{(q^{r-tk}, q^{mk-(r-tk)}; q^{mk})_\infty}{(-q^r, -q^{mk-r}; q^{mk})_\infty} =: \sum_{n=0}^{\infty} d_n q^n, \tag{4.5}$$

then d_{kn-rs} *is always zero.*

The proofs of these theorems in that paper also employed the Ramanujan's ${}_1\psi_1$ summation formula (1.5), but in the present paper it is used in a different way to give an m-dissection of a special case of the product on the right-hand side of (1.5), from which these results follow automatically.

Theorems 4.5 and 4.6 now follow in a straightforward manner from Corollary 4.1.

Corollary 4.7. *Theorems* 4.5 *and* 4.6 *are true.*

Proof. For Theorem 4.5, replace q with q^{mk} and set $p = k$ in Corollary 4.1. Set $z = q^{mk-r}$, where r is as in Theorem 4.5 ($1 \leq r < mk$, $\gcd(r, k) = 1$, so that $r = sm + t$, for some integers s and t, where $0 \leq s < k$, $1 \leq t < m$). Let this particular value of s be the s in Corollary 4.1. Then

$$\frac{(q^{km}, q^{km}, q^{r-k(r-ms)}, q^{mk-(r-k(r-ms))}; q^{km})_\infty}{(q^{k(r-ms)}, q^{km-k(r-ms)}, q^r, q^{km-r}; q^{km})_\infty}$$

$$= \sum_{j=0}^{k-1} q^{j(km-r)} \frac{\left(q^{k^2 m}, q^{k^2 m}, q^{km(j+k-s)}, q^{km(s-j)}; q^{k^2 m}\right)_\infty}{\left(q^{k(km-r)}, q^{kr}, q^{k(jm-sm+r)}, q^{k(-jm+km+sm-r)}; q^{k^2 m}\right)_\infty}. \tag{4.6}$$

After multiplying both sides by the reciprocal of

$$\frac{(q^{km}, q^{km}; q^{km})_\infty}{(q^{k(r-ms)}, q^{km-k(r-ms)}; q^{km})_\infty}$$

and noting that $r - ms = t$, the left-hand side of (4.6) becomes the left-hand side of (4.4). In the new right-hand side of (4.6), the q-products are such that all powers of q are multiples of k, and since r is invertible modulo k (recall $\gcd(r, k) = 1$), each term of the sum contains all powers of q in a single arithmetic progression modulo k. Theorem 4.5 follows, since the term corresponding to $j = s$ is identically zero.

The proof of Theorem 4.6 is similar, the only real difference being to set $z = -q^{mk-r}$. □

In [13], Hirschhorn gave the first examples of a new class of infinite q-products which also have the property that the coefficients in the series expansions vanish in one or more arithmetic progressions.

Specifically, he showed that if the sequences $\{a_n\}$ and $\{b_n\}$ are defined by

$$\sum_{n=0}^{\infty} a_n q^n := (-q, -q^4; q^5)_\infty (q, q^9; q^{10})^3_\infty,$$

$$\sum_{n=0}^{\infty} b_n q^n := (-q^2, -q^3; q^5)_\infty (q^3, q^7; q^{10})^3_\infty,$$

then $a_{5n+2} = a_{5n+4} = b_{5n+1} = b_{5n+4} = 0$.

Similar results were proven by Baruah and Kaur [4], Tang [22, 23] and in [24] by Tang and Xia. A natural question to ask is: Can any of these results be proved by the present methods? Unfortunately this does not appear to this author to be possible, as the types of infinite products dealt with in the cited papers have a form that is too different from that of the infinite products in the present paper. However, the infinite products in the papers cited above are examined in another recent paper [18] by the present author.

5. m-Dissections

In none of the papers referenced above did the authors give the relevant dissection of the infinite products, i.e. closed formulae for the arithmetic progressions in which the coefficients do not vanish. These dissections may be found elsewhere, for example the 4-dissection of the product (4.2) was given by Hirschhorn in [10] and Chen and Huang in [8], using different methods (and both methods differ from the method in the present paper). In those papers, producing the dissection required some effort, using various q-series manipulations and the Jacobi triple product identity.

As an illustration of the present method, where the dissections are produced automatically, their result is reproduced in the next corollary.

Corollary 5.1. *The 4-dissection of the infinite product (4.2) is given by*

$$\frac{(q^3,q^5;q^8)_\infty}{(q,q^7;q^8)_\infty}=\frac{(q^{12},q^{12},q^{20},q^{20};q^{32})_\infty}{(q^8,q^{16},q^{16},q^{24};q^{32})_\infty}+q\frac{(q^4,q^{12},q^{20},q^{28};q^{32})_\infty}{(q^8,q^8,q^{24},q^{24};q^{32})_\infty}+q^2\frac{(q^4,q^{12},q^{20},q^{28};q^{32})_\infty}{(q^8,q^{16},q^{16},q^{24};q^{32})_\infty}. \tag{5.1}$$

Proof. In (4.6), set $k=4$, $m=2$, $r=7$ and $s=3$. Multiply both sides by

$$\frac{(q^4,q^4;q^8)_\infty}{(q^8,q^8;q^8)_\infty}$$

and the result follows upon rewriting the products $(q^4;q^8)_\infty$ and $(q^8;q^8)_\infty$ in terms of the base q^{32}. □

We next consider the limitations of the expansion provided by (2.1) in producing m-dissections. In this identity, if q is replaced with q^t and then the substitutions $z=q^r$ and $a=q^s$ are made (some integers r and s satisfying $1\le r,s<t$), the identity

$$\frac{(q^t,q^t,q^{r+s},q^{t-r-s};q^t)_\infty}{(q^s,q^{t-s},q^r,q^{t-r};q^t)_\infty}=\sum_{j=0}^{p-1}q^{jr}\frac{(q^{pt},q^{pt},q^{pr+s+jt},q^{(p-j)t-pr-s};q^{pt})_\infty}{(q^{jt+s},q^{(p-j)t-s},q^{pr},q^{(t-r)p};q^{pt})_\infty} \tag{5.2}$$

is produced. If it is desired that the right-hand side represents a p-dissection of the left-hand side, then it is necessary that

(a) all of the powers of q in each of the infinite products on the right-hand side must be multiples of p;
(b) The integer r must satisfy $\gcd(r,p)=1$.

Hence $p|(sj+t)$, $0\le j\le p-1$, and hence $p|s$ and $p|t$, or $s=pu$ and $t=pv$ for some integers u and v.

For example, setting $t=12$, $s=6$ and $r=1$ in (5.2) leads to

$$\frac{(q^{12},q^{12},q^7,q^5;q^{12})_\infty}{(q^6,q^6,q,q^{11};q^{12})_\infty}=\sum_{j=0}^{p-1}q^j\frac{(q^{12p},q^{12p},q^{12j+p+6},q^{12(p-j)-p-6};q^{12p})_\infty}{(q^{12j+6},q^{12(p-j)-6},q^p,q^{11p};q^{12p})_\infty}, \tag{5.3}$$

and letting $p=2,3$ and 6 immediately produces, in turn, the 2-, 3- and 6-dissections of

$$\frac{(q^7,q^5;q^{12})_\infty}{(q,q^{11};q^{12})_\infty} \tag{5.4}$$

proven by Lin [14], with no manipulation of infinite series or the use of other theta function identities being necessary (Lin also gave the 4-dissection of (5.4), but the method described here does not provide that). We give the 3-dissection as an example.

Corollary 5.2 ([14, Theorem 3.2]). *The 3-dissection of* (5.4) *is given by*

$$\frac{(q^7,q^5;q^{12})_\infty}{(q,q^{11};q^{12})_\infty} = \frac{(q^6,q^{30};q^{36})_\infty}{(q^{12},q^{12},q^{24},q^{24};q^{36})_\infty} \times \left(\frac{(q^9,q^{18},q^{18},q^{27};q^{36})_\infty}{(q^3,q^{33};q^{36})_\infty} + q\frac{(q^6,q^{15},q^{21},q^{30};q^{36})_\infty}{(q^3,q^{33};q^{36})_\infty} + q^2(q^{18},q^{18};q^{36})_\infty \right). \tag{5.5}$$

Proof. Let $p = 3$ in (5.3), and the result follows after a little algebra and some elementary q-product manipulations. □

Remarks. (1) Lin [14] also gave the 2-, 3-, 4- and 6-dissections of the reciprocal of the infinite product (5.4), and the 2-, 3- and 6-dissections may be similarly produced by first setting $t = 12$, $s = 6$ and $r = 5$ in (5.2), and then setting $p = 2, 3$ and 6 in turn in the resulting identity.

(2) The reciprocal of the infinite product (5.4) is also equal to a special case of a quite general continued fraction recorded by Ramunajan in his second notebook (see [19], [5, Entry 12, p. 24]),

$$\frac{1}{1-ab} + \frac{(a-bq)(b-aq)}{(1-ab)(1+q^2)} + \frac{(a-bq^3)(b-aq^3)}{(1-ab)(1+q^4)} + \cdots = \frac{(a^2q^3,b^2q^3;q^4)_\infty}{(a^2q,b^2q;q^4)_\infty}, \tag{5.6}$$

after replacing q with q^3, then setting $a = q^2$ and $b = q$, and finally multiplying both sides by $1 - q$. This special case was further investigated by Naika *et al.* in [15], where explicit evaluations, reciprocity theorems and integral representations were developed.

The formula (5.2) and the requirements that $p|t$ and $p|s$ also demonstrate some of the limitations of this method of producing m-dissections. For example, it cannot produce any dissections of products that arise from setting $t = 5$, and hence does not lead to any results on vanishing coefficients (if they existed) in these cases. More generally, the formula (5.2) will not give dissections of any product that arises when $t = p$, p a prime.

The formula (5.2) in some cases may be partially successful at recovering results proved elsewhere by other methods. For example, Hirschhorn and Roselin in [11] consider Ramanujan's cubic continued fraction

$$RC(q) := \frac{1}{1} + \frac{q+q^2}{1} + \frac{q^2+q^4}{1} + \frac{q^3+q^6}{1} + \cdots = \frac{(q,q^5;q^6)_\infty}{(q^3,q^3;q^6)_\infty}, \tag{5.7}$$

where the 2-, 3-, 4- and 6-dissections of $RC(q)$ and its reciprocal were given. The present method gives the 2-dissections of these, but not the 3-, 4- and 6-dissections.

Corollary 5.3. *The 2-dissections of $RC(q)$ and $1/RC(q)$ are given by*

$$RC(q) = \frac{(q^4,q^8;q^{12})_\infty}{(q^6;q^{12})^2_\infty}\left(\frac{(q^4,q^8;q^{12})_\infty}{(q^2,q^{10};q^{12})_\infty} + q\frac{(q^2,q^{10};q^{12})_\infty}{(q^4,q^8;q^{12})_\infty}\right),$$

$$\frac{1}{RC(q)} = \frac{1}{(q^6;q^{12})^4_\infty}\left((q^4,q^8;q^{12})^2_\infty - q(q^2,q^{10};q^{12})^2_\infty\right).$$

Proof. For $RC(q)$, in (5.2) set $t=6$, $s=2$, $r=3$ and $p=2$, and for $1/RC(q)$ set $t=6$, $s=2$, $r=1$ and $p=2$. The results follow, as above, after some elementary algebra and q-series manipulations. □

Remarks. (1) These 2-dissections were represented differently by the authors in [11], where they were given in terms of infinite products of the form $(q^k;q^k)_\infty$.

(2) Recall that the Borwein cubic function $a(q)$ (see [7]) is defined by

$$a(q) = \sum_{m,n=-\infty}^{\infty} q^{m^2+mn+n^2}.$$

We note in passing that it was shown in [6] that

$$a(q) - a(q^2) = 6q\frac{(q,q^5,q^6,q^6;q^6)_\infty}{(q^2,q^3,q^3,q^4;q^6)_\infty}, \tag{5.8}$$

where the infinite product on the right-hand side is the same as that on the left-hand side of (5.2), after setting $t=6$, $s=2$, $r=3$, the same choices that were made in deriving the 2-dissection of $RC(q)$.

6. An Identity that Follows from Bailey's ${}_6\psi_6$ Summation Formula

The next identity, which we believe to be a new observation, is somewhat similar to that of (2.1), but unfortunately does not appear to have any applications in terms of m-dissections or vanishing coefficients. However, we include it as we believe it deserves to be known, and it may have applications which the author does not presently see.

Theorem 6.1. *If $|q|<1$ and m is a positive integer, then*

$$\frac{\left(q,q,a,\frac{q}{a},\frac{bq}{d},\frac{dq}{b},\frac{aq}{bd},\frac{bdq}{a};q\right)_\infty}{\left(b,\frac{q}{b},d,\frac{q}{d},\frac{a}{b},\frac{bq}{a},\frac{a}{d},\frac{dq}{a};q\right)_\infty}$$
$$= \sum_{r=0}^{m-1} q^r \frac{\left(q^m,q^m,aq^{2r},\frac{q^{m-2r}}{a},\frac{bq^m}{d},\frac{dq^m}{b},\frac{aq^m}{bd},\frac{bdq^m}{a};q^m\right)_\infty}{\left(bq^r,\frac{q^{m-r}}{b},dq^r,\frac{q^{m-r}}{d},\frac{aq^r}{b},\frac{bq^{m-r}}{a},\frac{aq^r}{d},\frac{dq^{m-r}}{a};q^m\right)_\infty}. \tag{6.1}$$

Proof. In Bailey's [3] ${}_6\psi_6$ identity (which holds for $|qa^2/(bcde)| < 1$)

$$\sum_{n=-\infty}^{\infty} \frac{(q\sqrt{a}, -q\sqrt{a}, b, c, d, e; q)_n}{(\sqrt{a}, -\sqrt{a}, aq/b, aq/c, aq/d, aq/e; q)_n} \left(\frac{qa^2}{bcde}\right)^n$$

$$= \frac{(aq, aq/bc, aq/bd, aq/be, aq/cd, aq/ce, aq/de, q, q/a; q)_\infty}{(aq/b, aq/c, aq/d, aq/e, q/b, q/c, q/d, q/e, qa^2/bcde; q)_\infty}, \tag{6.2}$$

set $c = a/b$, $e = a/d$, and multiply both sides of the resulting identity by

$$\frac{(1-a)}{(1-b)(1-d)(1-a/b)(1-a/d)}$$

to get

$$\sum_{n=-\infty}^{\infty} \frac{(1-aq^{2n})q^n}{(1-bq^n)(1-dq^n)(1-aq^n/b)(1-aq^n/d)}$$

$$= \frac{\left(q, q, a, \frac{q}{a}, \frac{bq}{d}, \frac{dq}{b}, \frac{aq}{bd}, \frac{bdq}{a}; q\right)_\infty}{\left(b, \frac{q}{b}, d, \frac{q}{d}, \frac{a}{b}, \frac{bq}{a}, \frac{a}{d}, \frac{dq}{a}; q\right)_\infty}. \tag{6.3}$$

The result (6.1) now follows, after splitting the sum on the left of (6.3) into sums of terms in the same arithmetic progression modulo m ($n \to mn + r$, $-\infty < n < \infty$, $0 \le r \le m-1$) and using (6.3) on each of these sums (with the replacements $q \to q^m$, $a \to aq^{2r}$, $b \to bq^r$ and $d \to dq^r$). □

One might ask if is possible to specialize the parameters in (6.1) to derive an interesting zero sum. The obvious choice, setting $a = 1$, does not lead to anything interesting, as the non-zero terms in the sum on the right cancel in pairs. However, another choice does lead to something less trivial.

Corollary 6.2. *If $|q| < 1$ and m is a positive integer, then*

$$\sum_{r=0}^{m} \frac{\left(adq^{2r}, \dfrac{q^m}{adq^{2r}}; q^m\right)_\infty q^r}{\left(aq^r, \dfrac{q^m}{aq^r}, aq^{r-1}, \dfrac{q^m}{aq^{r-1}}, dq^r, \dfrac{q^m}{dq^r}, dq^{r+1}, \dfrac{q^m}{dq^{r+1}}; q^m\right)_\infty} = 0 \tag{6.4}$$

Proof. In (6.1), set $b = dq$, then replace a with ad, and finally cancel any q-products that are independent of the summation variable r. □

7. Miscellaneous Identities

In this section, we consider several other functions, each of which has a number of different series representations, one of which is in terms of Lambert series. The Lambert series representation in each case is such that when the series is "sectioned", so that when terms in each arithmetic progression modulo m ($m > 1$

a positive integer) are grouped together, each of the resulting new Lambert series is of the same form as the original, albeit with shifted parameters. This in turn permits each of the other series representations to be similarly expressed as a sum of m series of the same type. To illustrate this, we first recall some other results of the present author (see [16]). Those results are collected and slightly reformulated to make easier to express the new results.

Recall that a *WP-Bailey pair* is a pair of sequences $(\alpha_n(a,k,q),\ \beta_n(a,k,q))$ satisfying $\alpha_0(a,k,q) = \beta_0(a,k,q) = 1$, and for $n > 0$,

$$\beta_n(a,k,q) = \sum_{j=0}^{n} \frac{(k/a;q)_{n-j}(k;q)_{n+j}}{(q;q)_{n-j}(aq;q)_{n+j}} \alpha_j(a,k,q). \tag{7.1}$$

If $k = 0$, then the pair of sequences $(\alpha_n(a,q), \beta_n(a,q))$ is called a *Bailey pair with respect to a*.

Theorem 7.1. *Let* $(\alpha_n(a,k,q),\ \beta_n(a,k,q))$ *be a WP-Bailey pair and let* $f_i(a,k,z,q)$, $i = 1,2,3$ *be defined as follows:*

$$f_1(a,k,z,q) = \sum_{n=1}^{\infty} \left(\frac{kq^n}{1-kq^n} + \frac{q^n a/z}{1-q^n a/z} - \frac{aq^n}{1-aq^n} - \frac{q^n k/z}{1-q^n k/z} \right),$$

$$f_2(a,k,z,q) = \sum_{n=1}^{\infty} \frac{(1-kq^{2n})(z,k/a;q)_n}{(1-kq^n)(qk/z,qa;q)_n(1-q^n)} \left(\frac{qa}{z}\right)^n,$$

$$\begin{aligned} f_3(a,k,z,q) = {} & \sum_{n=1}^{\infty} \frac{(q\sqrt{k},-q\sqrt{k},z;q)_n(q;q)_{n-1}}{(\sqrt{k},-\sqrt{k},qk,qk/z;q)_n} \left(\frac{qa}{z}\right)^n \beta_n(a,k,q) \\ & - \sum_{n=1}^{\infty} \frac{(z;q)_n(q;q)_{n-1}}{(qa,qa/z;q)_n} \left(\frac{qa}{z}\right)^n \alpha_n(a,k,q). \end{aligned} \tag{7.2}$$

Then for all quadruples of values (a,k,z,q) *for which all three functions are defined, we have*

$$f_1(a,k,z,q) = f_2(a,k,z,q) = f_3(a,k,z,q). \tag{7.3}$$

The point here is that it follows easily from the definition of $f_1(a,k,z,q)$ (upon splitting the series into arithmetic progressions modulo m (after the replacement $n \to mn - r$, $n \geq 1$, $0 \leq r \leq m-1$), that if $m > 1$ is an integer, then

$$f_1(a,k,z,q) = \sum_{r=0}^{m-1} f_1(aq^{-r}, kq^{-r}, z, q^m), \tag{7.4}$$

and thus that (7.4) also holds if $f_1(a,k,z,q)$ is replaced with $f_2(a,k,z,q)$ or $f_3(a,k,z,q)$.

We ignore the choices for a WP-Bailey pair in $f_3(a,k,z,q)$, and instead use $f_2(a,k,z,q)$ to illustrate what has been said with the following somewhat unusual identity (which follows after the replacement $r \to m - r$).

Corollary 7.2. *If $|qa/z| < 1$, then for each positive integer m,*

$$\sum_{n=1}^{\infty} \frac{(1-kq^{2n})(z,k/a;q)_n}{(1-kq^n)(aq,kq/z;q)_n(1-q^n)} \left(\frac{qa}{z}\right)^n$$

$$= \sum_{r=1}^{m}\sum_{n=1}^{\infty} \frac{\left(1-kq^{m(2n-1)+r}\right)\left(z,\frac{k}{a};q^m\right)_n}{\left(1-kq^{m(n-1)+r}\right)\left(aq^r,\frac{kq^r}{z};q^m\right)_n(1-q^{mn})} \left(\frac{q^r a}{z}\right)^n. \qquad (7.5)$$

Upon letting $k \to 0$ in Theorem 7.1, then the next corollary follows.

Corollary 7.3. *Let $(\alpha_n(a,q), \beta_n(a,q))$ be a Bailey pair with respect to a, and let $g_i(a,z,q)$, $i = 1,2,3,4$, be defined by*

$$g_1(a,z,q) = \sum_{n=1}^{\infty}\left(\frac{aq^n/z}{1-aq^n/z} - \frac{aq^n}{1-aq^n}\right),$$

$$g_2(a,z,q) = \sum_{n=1}^{\infty} \frac{(z;q)_n}{(qa;q)_n(1-q^n)}\left(\frac{qa}{z}\right)^n$$

$$g_3(a,z,q) = \sum_{n=1}^{\infty}(z;q)_n(q;q)_{n-1}\left(\frac{qa}{z}\right)^n \beta_n(a,q)$$

$$- \sum_{n=1}^{\infty} \frac{(z;q)_n(q;q)_{n-1}}{(qa,qa/z;q)_n}\left(\frac{qa}{z}\right)^n \alpha_n(a,q),$$

$$g_4(a,z,q) = -\sum_{n=1}^{\infty} \frac{(1-aq^{2n})(z;q)_n q^{n(n+1)/2}}{(1-aq^n)(qa/z;q)_n(1-q^n)}\left(\frac{-a}{z}\right)^n. \qquad (7.6)$$

Then, for any triple of values for a, z and q for which all of the series converge, it holds that

$$g_1(a,z,q) = g_2(a,z,q) = g_3(a,z,q) = g_4(a,z,q). \qquad (7.7)$$

Note that $g_4(a,z,q)$ is a special case of $g_3(a,z,q)$, derived by inserting the "unit" Bailey pair

$$\alpha_n(a,q) = \frac{(q\sqrt{a},-q\sqrt{a},a;q)_n}{(\sqrt{a},-\sqrt{a},q;q)_n}(-1)^n q^{n(n-1)/2},$$

$$\beta_n(a,q) = \begin{cases} 1, & n=0, \\ 0, & n>1. \end{cases}$$

By a similar argument to that above, it easily follows that if m is any positive integer, then for $1 \le i \le 4$,

$$g_i(a,z,q) = \sum_{r=0}^{m-1} g_i(aq^{-r},z,q^m). \qquad (7.8)$$

We defer giving a specific example until after stating the special case of Corollary 7.3 derived by letting $z \to \infty$.

Corollary 7.4. *Let $(\alpha_n(a,q), \beta_n(a,q))$ be a Bailey pair with respect to a, and let $h_i(a,q)$, $i = 1, \ldots, 6$ be defined by*

$$\begin{aligned}
h_1(a,q) &= \sum_{n=1}^{\infty} \frac{aq^n}{1-aq^n},\\
h_2(a,q) &= -\sum_{n=1}^{\infty} \frac{q^{n(n+1)/2}(-a)^n}{(qa;q)_n(1-q^n)}\\
h_3(a,q) &= -\sum_{n=1}^{\infty} (q;q)_{n-1}(-a)^n q^{n(n+1)/2}\beta_n(a,q)\\
&\quad + \sum_{n=1}^{\infty} \frac{(q;q)_{n-1}(-a)^n q^{n(n+1)/2}}{(qa;q)_n}\alpha_n(a,q),\\
h_4(a,q) &= \sum_{n=1}^{\infty} \frac{(1-aq^{2n})q^{n^2}a^n}{(1-aq^n)(1-q^n)},\\
h_5(a,q) &= \frac{-1}{(aq;q)_\infty} \sum_{n=1}^{\infty} \frac{n(-a)^n q^{n(n+1)/2}}{(q;q)_n},\\
h_6(a,q) &= (aq;q)_\infty \sum_{n=1}^{\infty} \frac{na^n q^n}{(q;q)_n}.
\end{aligned} \tag{7.9}$$

Then, for any pair of values for a and q for which all of the series converge, it holds that

$$h_1(a,q) = h_2(a,q) = h_3(a,q) = h_4(a,q) = h_5(a,q) = h_6(a,q). \tag{7.10}$$

Remark. The equivalence of $h_5(a,q)$ and $h_6(a,q)$ to $h_1(a,q), \ldots, h_4(a,q)$ is shown in [16].

Once again, by reasoning as above, it is easy to show that if m is any positive integer, then for $1 \leq i \leq 6$,

$$h_i(a,q) = \sum_{r=0}^{m-1} h_i(aq^{-r}, q^m). \tag{7.11}$$

The following pair of somewhat curious identities follow from $h_5(a,q)$ and $h_6(a,q)$ (after making the replacement $r \to m - r$ on the left-hand side).

Corollary 7.5. *Let m be a positive integer.*

(i) *If $|q| < 1$, then*

$$\begin{aligned}
&\frac{1}{(aq;q)_\infty} \sum_{n=1}^{\infty} \frac{n(-a)^n q^{n(n+1)/2}}{(q;q)_n}\\
&= \sum_{r=1}^{m} \frac{1}{(aq^r;q^m)_\infty} \sum_{n=1}^{\infty} \frac{n(-a)^n q^{mn(n-1)/2+nr}}{(q^m;q^m)_n}.
\end{aligned} \tag{7.12}$$

If $|aq| < 1$, *then*

$$(aq;q)_\infty \sum_{n=1}^{\infty} \frac{na^n q^n}{(q;q)_n} = \sum_{r=1}^{m} (aq^r;q^m)_\infty \sum_{n=1}^{\infty} \frac{na^n q^{nr}}{(q^m;q^m)_n}. \tag{7.13}$$

8. Further Applications — The Fine Function $F(a,b;t)$

The ideas in the present paper may be applied in a number of other situations. One of these is the collection of identities that exist involving the function

$$F(a,b;t) := \sum_{n=0}^{\infty} \frac{(aq;q)_n}{(bq;q)_n} t^n, \tag{8.1}$$

which was investigated in [9] by Fine. One important identity involving the function $F(a,b;t)$ that he found was the *Rogers-Fine identity*, discovered previously by Rogers [21]:

$$\sum_{n=0}^{\infty} \frac{(\alpha;q)_n}{(\beta;q)_n} \tau^n = \sum_{n=0}^{\infty} \frac{(1-\alpha\tau q^{2n})(\alpha,\alpha\tau q/\beta;q)_n \beta^n \tau^n q^{n^2-n}}{(\beta;q)_n(\tau;q)_{n+1}}. \tag{8.2}$$

Note that the left-hand side of (8.2) is $F(\alpha/q,\beta/q;\tau)$. Fine derived several other identities involving the function on the left-hand side of (8.2) (for simplicity in the present paper, results are stated in terms of this series, rather than $F(a,b;t)$), including

$$\sum_{n=0}^{\infty} \frac{(\alpha;q)_n}{(\beta;q)_n} \tau^n = \sum_{n=0}^{\infty} \frac{(\beta/\alpha;q)_n(-\alpha t)^n q^{n(n-1)/2}}{(\beta;q)_n(t;q)_{n+1}}. \tag{8.3}$$

From the perspective of the present paper, the special cases of these identities that are of interest arise from setting $\beta = \alpha q$ and dividing both sides in each case by $1-\alpha$. This leads to

$$\begin{aligned} F_1(\alpha,\tau,q) := \sum_{n=0}^{\infty} \frac{\tau^n}{1-\alpha q^n} &= \sum_{n=0}^{\infty} \frac{(1-\alpha\tau q^{2n})\alpha^n\tau^n q^{n^2}}{(1-\alpha q^n)(1-\tau q^n)} =: F_2(\alpha,\tau,q) \\ &= \sum_{n=0}^{\infty} \frac{(q;q)_n(-\alpha t)^n q^{n(n-1)/2}}{(\alpha;q)_{n+1}(t;q)_{n+1}} =: F_3(\alpha,\tau,q). \end{aligned} \tag{8.4}$$

The point in stating these special cases is that, by similar reasoning to that employed earlier, it follows that if m is any positive integer, then

$$F_i(\alpha,\tau,q) = \sum_{r=0}^{m-1} t^r F_i(\alpha q^r, \tau^m, q^m), \quad i = 1,2,3. \tag{8.5}$$

A similar result may be derived from any other identity for the function $F(a,b;t)$.

We leave it to the reader to possibly find some useful applications of these facts.

References

[1] K. Alladi and B. Gordon, Vanishing coefficients in the expansion of products of Rogers-Ramanujan type., *Proc. Rademacher Centenary Conference*, eds. G. E. Andrews and D. M. Bressoud, Contemp. Math., Vol. 166, (Amer. Math. Soc., 1994), pp. 129–139.

[2] G. E. Andrews and D. M. Bressoud, Vanishing coefficients in infinite product expansions, *J. Austral. Math. Soc. Ser. A* **27**(2) (1979) 199–202.

[3] W. N. Bailey, Series of hypergeometric type which are infinite in both directions, *Quart. J. Math. (Oxford)* **7** (1936) 105–115.

[4] N. D. Baruah and M. Kaur, Some results on vanishing coefficients in infinite product expansions, *Ramanujan J.* **53**(3) (2020) 551–568.

[5] B. C. Berndt, *Ramanujan's Notebooks. Part III* (Springer-Verlag, New York, 1991).

[6] B. C. Berndt, *Ramanujan's Notebooks. Part V* (Springer-Verlag, New York, 1998).

[7] J. M. Borwein and P. B. Borwein, A cubic counterpart of Jacobi's identity and the AGM, *Trans. Amer. Math. Soc.* **323**(2) (1991) 691–701.

[8] S.-D. Chen and S.-S. Huang, On the series expansion of the Göllnitz–Gordon continued fraction, *Int. J. Number Theory* **1**(1) (2005) 53–63.

[9] N. J. Fine, *Basic Hypergeometric Series and Applications*, Mathematical Surveys and Monographs, Vol. 27 (American Mathematical Society, Providence, RI, 1988).

[10] M. D. Hirschhorn, On the expansion of a continued fraction of Gordon, *Ramanujan J.* **5**(4) (2001) 369–375.

[11] M. D. Hirschhorn and Roselin, On the 2-, 3-, 4- and 6-dissections of Ramanujan's cubic continued fraction and its reciprocal, in *Ramanujan Rediscovered*, Ramanujan Math. Soc. Lect. Notes Ser., Vol. 14 (Ramanujan Math. Soc., Mysore, 2010), pp. 125–138.

[12] M. D. Hirschhorn, On the 2- and 4-dissections of the Rogers–Ramanujan functions, *Ramanujan J.* **40**(2) (2016) 227–235.

[13] M. D. Hirschhorn, Two remarkable q-series expansions, *Ramanujan J.* **49**(2) (2019) 451–463.

[14] B. L. S. Lin, On the expansion of a continued fraction of order twelve, *Int. J. Number Theory* **9**(8) (2013) 2019–2031.

[15] M. S. Mahadeva Naika, B. N. Dharmendra and K. Shivashankara, A continued fraction of order twelve, *Cent. Eur. J. Math.* **6**(3) (2008) 393–404.

[16] J. Mc Laughlin, Some new transformations for Bailey pairs and WP-Bailey pairs, *Cent. Eur. J. Math.* **8**(3) (2010) 474–487.

[17] J. Mc Laughlin, Further results on vanishing coefficients in infinite product expansions, *J. Aust. Math. Soc.* **98**(1) (2015) 69–77.

[18] J. Mc Laughlin, New infinite q-product expansions with vanishing coefficients, *Ramanujan J.* **55**(2) (2021) 733–760.

[19] S. Ramanujan, *Notebooks* (Tata Institute of Fundamental Research, Bombay, 1957).

[20] B. Richmond and G. Szekeres, The Taylor coefficients of certain infinite products, *Acta Sci. Math.* (*Szeged*) **40**(3–4) (1978) 347–369.

[21] L. J. Rogers, On two theorems of combinatory analysis and some allied identities, *Proc. London Math. Soc (2).* **16** (1917), 315–336.

[22] D. Tang, Vanishing coefficients in four quotients of infinite product expansions, *Bull. Aust. Math. Soc.* **100** (2019) 216–224.

[23] D. Tang, Vanishing coefficients in some q-series expansions, *Int. J. Number Theory* **15**(4) (2019) 763–773.

[24] D. Tang and E. X. W. Xia, Several q-series related to Ramanujan's theta functions, *Ramanujan J.* **53** (2020) 705–724.

© 2024 World Scientific Publishing Company
https://doi.org/10.1142/9789811277375_0031

On the sum of parts with multiplicity at least 2 in all the partitions of n*

Mircea Merca

Academy of Romanian Scientists
Ilfov 3, Sector 5, Bucharest, Romania
mircea.merca@profinfo.edu.ro

Ae Ja Yee

Department of Mathematics
The Pennsylvania State University
University Park, PA 16802, USA
yee@psu.edu

Dedicated to Bruce Berndt on the occasion of his 80th birthday

In this paper, we investigate the sum of distinct parts that appear at least 2 times in all the partitions of n providing new combinatorial interpretations for this sum. A connection with subsets of $\{1, 2, \ldots, n\}$ is given in this context. We provide two different proofs of our results: analytic and combinatorial. In addition, considering the multiplicity of parts in a partition, we provide a combinatorial proof of the truncated pentagonal number theorem of Andrews and Merca.

Keywords: Integer partitions; generating functions; partition identities; truncated partition theorems.

Mathematics Subject Classification 2020: 11P83, 05A17, 05A19

1. Introduction

A partition of a positive integer n is a weakly decreasing sequence of positive integers whose sum is n. The positive integers in the sequence are called parts [1]. For a partition λ of n, we denote it by $\lambda \vdash n$. Also, we denote by $p(n)$ the number of the partitions of n. For example, the following are the partitions of 5:

$$5, \quad 4+1, \quad 3+2, \quad 3+1+1, \quad 2+2+1, \quad 2+1+1+1, \quad 1+1+1+1+1, \tag{1.1}$$

so $p(5) = 7$. For convenience, we define $p(0) = 1$.

*This article originally appeared in *International Journal of Number Theory*, **17**(3), pp. 665–681 © 2021 World Scientific Publishing Company. DOI: 10.1142/S1793042120400205

In the theory of partitions, besides the partition function $p(n)$, studies on partitions with various statistics can be found in the literature. For instance, Wilf investigated the average number of different sizes of parts in a partition of n [7]; Knopfmacher and Mays studied the sum of different part sizes in partitions [5].

In this paper, we consider the following counting functions.

Definition 1.1. For a positive integer n, we define

(i) $a(n)$ to be the sum of parts counted without multiplicity in all the partitions of n, i.e.

$$a(n) := \sum_{\lambda \vdash n} \text{different parts of } \lambda,$$

(ii) $a_e(n)$ to be the sum of even parts counted without multiplicity in all the partitions of n, i.e.

$$a_e(n) := \sum_{\lambda \vdash n} \text{different even parts of } \lambda,$$

(iii) $a_o(n)$ to be the sum of odd parts counted without multiplicity in all the partitions of n, i.e.

$$a_o(n) := \sum_{\lambda \vdash n} \text{different odd parts of } \lambda.$$

It is clear that $a(n) = a_e(n) + a_o(n)$. For example, we see from (1.1) that

$$\begin{aligned} a_o(5) &= 5 + 1 + 3 + (3 + 1) + 1 + 1 + 1 = 16, \\ a_e(5) &= 0 + 4 + 2 + 0 + 2 + 2 + 0 = 10, \\ a(5) &= 26. \end{aligned}$$

The first of our result is the generating functions for $a(n), a_e(n)$ and $a_o(n)$.

Theorem 1.2. *We have*

$$\begin{aligned} \sum_{n=1}^{\infty} a(n)q^n &= \frac{1}{(q;q)_\infty} \cdot \frac{q}{(1-q)^2}, \\ \sum_{n=1}^{\infty} a_e(n)q^n &= \frac{1}{(q;q)_\infty} \cdot \frac{2q^2}{(1-q^2)^2}, \\ \sum_{n=1}^{\infty} a_o(n)q^n &= \frac{1}{(q;q)_\infty} \cdot \frac{q(1+q^2)}{(1-q^2)^2}, \end{aligned}$$

where

$$(a;q)_\infty = \prod_{n=0}^{\infty} (1 - aq^n) \quad \textit{for } |q| < 1.$$

Furthermore, these three functions $a_o(n)$, $a_e(n)$, and $a(n)$ are closely related to another counting function $b(n)$, which is the sum of distinct parts that appear at least 2 times in all the partitions of n.

Theorem 1.3. *For all $n \geq 1$,*

(a) $a_e(n) = 2b(n)$;
(b) $a_o(n) = b(n+1) + b(n-1)$;
(c) $a(n) = b(n+1) + 2b(n) + b(n-1)$.

We note that the sequence $b(n)$ can be found in the On-Line Encyclopedia of Integer Sequence [6, A103650]. Also on the page for $A103650$, we find without proof its generating function:

$$\sum_{n=0}^{\infty} b(n)q^n = \frac{1}{(q;q)_\infty} \cdot \frac{q^2}{(1-q^2)^2}. \tag{1.2}$$

In [2], while investigating the truncated Euler's pentagonal number theorem, Andrews and Merca introduced the partition function $M_k(n)$, which counts the number of partitions of n where k is the least positive integer that is not a part and there are more parts $>k$ than there are parts $<k$. For instance, we have $M_3(18) = 3$ because the three partitions in question are

$$5+5+5+2+1, \quad 6+5+4+2+1, \quad 7+4+4+2+1.$$

We have the following result.

Theorem 1.4. *For any positive integers k and n,*

$$(-1)^{k-1}\left(\sum_{j=-(k-1)}^{k} (-1)^j b(n-j(3j-1)/2) - \frac{1+(-1)^n}{2}\cdot\frac{n}{2}\right)$$
$$= \sum_{j=1}^{\lfloor n/2 \rfloor} jM_k(n-2j).$$

An immediate consequence of this theorem is the following infinite family of linear inequalities.

Corollary 1.5. *For any positive integers k and n,*

$$(-1)^{k-1}\left(\sum_{j=-(k-1)}^{k} (-1)^j b(n-j(3j-1)/2) - \frac{1+(-1)^n}{2}\cdot\frac{n}{2}\right) \geq 0,$$

with strict inequality if $n > k(3k+1)/2$.

When $k \to \infty$, Theorem 1.4 yields the following corollary.

Corollary 1.6. *For any positive integers k and n,*

$$\sum_{j=-\infty}^{\infty} (-1)^j b(n - j(3j-1)/2) = \frac{1+(-1)^n}{2} \cdot \frac{n}{2}.$$

The result in the corollary above is reminiscent of the recurrence relation of Euler's partition function $p(n)$.

In [3], Andrews and Merca considered an analogue of the truncated pentagonal number theorem for the function $p_{od}(n)$, the number of partitions of n in which odd parts are not repeated, and they introduced the function $MP_k(n)$, the number of partitions of n in which the first part larger than $2k-1$ is odd and appears exactly k times, and all other odd parts appear at most once. For example, $MP_2(19) = 10$, and the partitions in question are

$$9+9+1, \quad 9+5+5, \quad 8+5+5+1, \quad 7+7+3+2, \quad 7+7+2+2+1,$$
$$7+5+5+2, \quad 6+5+5+3, \quad 6+5+5+2+1, \quad 5+5+3+2+2+2,$$
$$5+5+2+2+2+2+1.$$

We have the following result connecting the sequences $b(n)$ and $MP_k(n)$.

Theorem 1.7. *For any positive integers k and n,*

$$(-1)^{k-1}\left(\sum_{j=0}^{2k-1}(-1)^j b(n - j(j+1)/2) - \frac{1+(-1)^n}{2}\cdot c\left(\frac{n}{2}\right)\right)$$
$$= \sum_{j=1}^{\lfloor n/2 \rfloor} c(j) MP_k(n-2j),$$

where $c(n)$ is the number of subsets S of $\{1,2,\ldots,n\}$ which contain a number that is greater than the sum of the other numbers in S.

There is a substantial amount of numerical evidence to conjecture the following inequality.

Conjecture 1.8. *For $n > 0$,*

$$\sum_{j=1}^{\lfloor n/2 \rfloor} c(j) MP_k(n-2j) \leq \sum_{j=1}^{\lfloor n/2 \rfloor} j M_k(n-2j).$$

The organization of this paper is as follows. We will first prove Theorems 1.2 and 1.3 in Sec. 2. In Sec. 3, we will provide a proof of Theorem 1.4. In Sec. 4, we will provide a proof of Theorem 1.7. In Sec. 5, Theorems 1.2, 1.3, 1.4, and 1.7 will be proved combinatorially. In addition, we will provide a combinatorial proof of the truncated pentagonal number theorem of Andrews and Merca [2] in Sec. 5.

2. Proofs of Theorems 1.2 and 1.3

We first sketch the proof of the generating function for $b(n)$ in (1.2). Note that the generating function for partitions where the power of z keeps track of parts with multiplicity >1 is

$$\prod_{j=1}^{\infty}(1+q^j+z^j(q^{2j}+q^{3j}+q^{4j}+\cdots)) = \prod_{j=1}^{\infty}\left(1+q^j+z^j\frac{q^{2j}}{1-q^j}\right)$$
$$= \frac{1}{(q;q)_\infty}\prod_{j=1}^{\infty}(1+(z^j-1)q^{2j}).$$

Taking the derivative with respect to z, and setting z equal to 1, we obtain the expression of the generating function for $b(n)$:

$$\sum_{n=1}^{\infty} b(n)q^n = \frac{1}{(q;q)_\infty}\sum_{j=1}^{\infty} jq^{2j} = \frac{1}{(q;q)_\infty}\cdot\frac{q^2}{(1-q^2)^2}.$$

The generating function for $a(n)$ can be obtained in a similar way:

$$\sum_{n=1}^{\infty} a(n)q^n = \frac{d}{dz}\prod_{j=1}^{\infty}(1+z^j(q^j+q^{2j}+q^{3j}+\cdots))|_{z=1}$$
$$= \frac{d}{dz}\prod_{j=1}^{\infty}\left(1+z^j\frac{q^j}{1-q^j}\right)\Bigg|_{z=1}$$
$$= \frac{d}{dz}\frac{1}{(q;q)_\infty}\prod_{j=1}^{\infty}(1+(z^j-1)q^j)|_{z=1}$$
$$= \frac{1}{(q;q)_\infty}\sum_{j=1}^{\infty} jq^j$$
$$= \frac{1}{(q;q)_\infty}\cdot\frac{q}{(1-q)^2}.$$

The generating function for $a_e(n)$ can be obtained in the same way. Also, for the generating function for $a_o(n)$, we can use the fact that $a_o(n) = a(n) - a_e(n)$. Thus, we will omit the details. This completes the proof of Theorem 1.2.

Theorem 1.3 immediately follows from Theorem 1.2 and the generating function for $b(n)$ in (1.2).

3. Proof of Theorem 1.4

In [2], Andrews and Merca considered Euler's pentagonal number theorem

$$(q;q)_\infty = \sum_{n=-\infty}^{\infty}(-1)^n q^{n(3n-1)/2},$$

and they proved the following truncated form: For any $k \geq 1$,

$$\frac{(-1)^{k-1}}{(q;q)_\infty} \sum_{n=-(k-1)}^{k} (-1)^n q^{n(3n-1)/2} = (-1)^{k-1} + \sum_{n=k}^{\infty} \frac{q^{\binom{k}{2}+(k+1)n}}{(q;q)_n} \begin{bmatrix} n-1 \\ k-1 \end{bmatrix}, \tag{3.1}$$

where $(a;q)_n = (a;q)_\infty/(aq^n;q)_\infty$, and

$$\begin{bmatrix} n \\ k \end{bmatrix} = \begin{cases} \dfrac{(q;q)_n}{(q;q)_k (q;q)_{n-k}}, & \text{if } 0 \leq k \leq n, \\ 0, & \text{otherwise.} \end{cases}$$

We note that the series on the right-hand side of (3.1) is the generating function for $M_k(n)$, i.e.

$$\sum_{n=0}^{\infty} M_k(n) q^n = \sum_{n=k}^{\infty} \frac{q^{\binom{k}{2}+(k+1)n}}{(q;q)_n} \begin{bmatrix} n-1 \\ k-1 \end{bmatrix}. \tag{3.2}$$

Multiplying both sides of (3.1) by

$$\sum_{n=0}^{\infty} n q^{2n} = \frac{q^2}{(1-q^2)^2},$$

we obtain

$$(-1)^{k-1} \left(\left(\sum_{n=1}^{\infty} b(n) q^n \right) \left(\sum_{n=-(k-1)}^{k} (-1)^n q^{n(3n-1)/2} \right) - \sum_{n=0}^{\infty} n q^{2n} \right)$$
$$= \left(\sum_{n=0}^{\infty} n q^{2n} \right) \left(\sum_{n=0}^{\infty} M_k(n) q^n \right),$$

and Theorem 1.4 follows easily considering Cauchy's multiplication of two power series.

4. Connections with Subsets of $\{1, 2, \ldots, n\}$

Recall the definition of $c(n)$ from Theorem 1.7. It is easy to see that $c(n)$ equals the number of subsets of $\{1, 2, \ldots, n\}$ whose largest element is greater than the sum of the other numbers. For example, $c(4) = 11$, since the subsets of $\{1, 2, 3, 4\}$ that contain a number greater than the sum of the other numbers in the set are

$$\{1\}, \{2\}, \{3\}, \{4\}, \{1,2\}, \{1,3\}, \{1,4\}, \{2,3\}, \{2,4\}, \{3,4\}, \{1,2,4\}.$$

This sequence can be found in the On-Line Encyclopedia of Integer Sequence [6, A095944], where its generating function is given without proof. The idea of the proof belongs to Helenius [6, A095944].

Theorem 4.1. *We have*

$$\sum_{n=1}^{\infty} c(n) q^n = \frac{q}{(1-q)^2} \cdot (-q;q)_\infty.$$

Proof. For a subset S of $\{1, 2, \ldots, n\}$, let k be the largest element in S and m be the sum of the remaining elements in S. Then, the remaining elements form a partition of m into distinct parts. Thus, for given k, m with $k > m$,

$$c(n) = \sum_{k=1}^{n} \sum_{m=0}^{k-1} Q(m) = \sum_{m=0}^{n-1} \sum_{k=m+1}^{n} Q(m) = \sum_{m=0}^{n-1} (n-m)Q(m),$$

where $Q(m)$ denotes the number of partitions of m into distinct parts. Hence,

$$\begin{aligned}\sum_{n=1}^{\infty} c(n)q^n &= \sum_{n=1}^{\infty} \sum_{m=0}^{n-1} (n-m)Q(m)q^n \\ &= \sum_{m=0}^{\infty} Q(m)q^m \sum_{n=1}^{\infty} nq^n \\ &= \frac{q}{(1-q)^2} \sum_{m=0}^{\infty} Q(m)q^m \\ &= \frac{q}{(1-q)^2} \cdot (-q;q)_\infty,\end{aligned}$$

where the last equality follows from

$$\sum_{n=0}^{\infty} Q(n)q^n = (-q;q)_\infty.$$

This completes the proof. □

From this proof, we see that $c(n)$ also counts the number of overpartitions in which non-overlined parts are distinct, only one overlined part is allowed, and the sum of the parts does not exceed n.

We are now ready to prove Theorem 1.7.

4.1. *Proof of Theorem* 1.7

The following theta identity is often attributed to Gauss [1, p. 23, Eq. (2.2.13)]:

$$\sum_{n=0}^{\infty} (-q)^{n(n+1)/2} = \frac{(q^2;q^2)_\infty}{(-q;q^2)_\infty}.$$

Andrews and Merca [3] considered this identity and proved the following truncated form:

$$\begin{aligned}&\frac{(-q;q^2)_\infty}{(q^2;q^2)_\infty} \sum_{j=0}^{2k-1} (-q)^{j(j+1)/2} \\ &= 1 + (-1)^{k-1} \frac{(-q;q^2)_k}{(q^2;q^2)_{k-1}} \sum_{j=0}^{\infty} \frac{q^{k(2k+2j+1)}(-q^{2k+2j+3};q^2)_\infty}{(q^{2k+2j+2};q^2)_\infty}.\end{aligned} \tag{4.1}$$

We note that the series on the right-hand side is the generating function for $MP_k(n)$:

$$\sum_{n=0}^{\infty} MP_k(n)q^n = \frac{(-q;q^2)_k}{(q^2;q^2)_{k-1}} \sum_{j=0}^{\infty} \frac{q^{k(2k+2j+1)}(-q^{2k+2j+3};q^2)_\infty}{(q^{2k+2j+2};q^2)_\infty}. \quad (4.2)$$

Multiplying both sides of (4.1) by

$$\frac{q^2}{(1-q^2)^2} \cdot (-q^2;q^2)_\infty,$$

we deduce the following identity

$$(-1)^{k-1}\left(\frac{1}{(q;q)_\infty}\frac{q^2}{(1-q^2)^2}\sum_{j=0}^{2k-1}(-q)^{j(j+1)/2} - \frac{q^2}{(1-q^2)^2}\frac{1}{(q^2;q^4)_\infty}\right)$$
$$= \frac{q^2}{(1-q^2)^2}(-q^2;q^2)_\infty \frac{(-q;q^2)_k}{(q^2;q^2)_{k-1}} \sum_{j=0}^{\infty} \frac{q^{k(2k+2j+1)}(-q^{2k+2j+3};q^2)_\infty}{(q^{2k+2j+2};q^2)_\infty}.$$

This identity can be rewritten as

$$(-1)^{k-1}\left(\left(\sum_{n=0}^{\infty} b(n)q^n\right)\left(\sum_{n=0}^{2k-1}(-q)^{n(n+1)/2}\right) - \sum_{n=0}^{\infty} c(n)q^{2n}\right)$$
$$= \left(\sum_{n=0}^{\infty} c(n)q^{2n}\right)\left(\sum_{n=0}^{\infty} MP_k(n)q^n\right),$$

where we have invoked (1.2), Theorems 4.1 and (4.2). The proof follows easily considering Cauchy's multiplication of two power series.

As a consequence of Theorem 1.7, we obtain the following infinite family of linear inequalities.

Corollary 4.2. *Let k and n be positive integers. Then*

$$(-1)^{k-1}\left(\sum_{j=0}^{2k-1}(-1)^j b(n-j(j+1)/2) - \frac{1+(-1)^n}{2}\cdot c\left(\frac{n}{2}\right)\right) \geq 0,$$

with strict inequality if $n \geq k(2k+1)$.

For example:

$$b(n) - b(n-1) \geq \frac{1+(-1)^n}{2}\cdot c\left(\frac{n}{2}\right),$$

$$b(n) - b(n-1) - b(n-3) + b(n-6) \leq \frac{1+(-1)^n}{2}\cdot c\left(\frac{n}{2}\right).$$

When $k \to \infty$, Theorem 1.7 yields the following result.

Corollary 4.3. *Let n be a positive integer. Then*

$$\sum_{j=0}^{\infty}(-1)^j b(n-j(j+1)/2) = \frac{1+(-1)^n}{2}\cdot c\left(\frac{n}{2}\right).$$

From this corollary, we deduce the following decomposition of $c(n)$:

$$c(n) = \sum_{j=0}^{\infty}(-1)^j b(2n - j(j+1)/2).$$

5. Combinatorial Proofs

5.1. *Proof of* $a_e(n) = 2b(n)$

For a partition $\lambda \vdash n$, let $2a$ be an even part of λ. We split the even part $2a$ into two a's keeping the remaining parts unchanged. We call the resulting partition μ. Then, it is clear that $\mu \vdash n$, and the multiplicity of a in μ is at least 2. Clearly, we can recover λ from μ by merging two copies of a in μ to get one $2a$. Thus,

$$\begin{aligned} a_e(n) &= \sum_{\lambda \vdash n} \text{ different even parts of } \lambda \\ &= 2\sum_{\mu \vdash n} \text{ different parts of } \mu \text{ with multiplicity } > 1 \\ &= 2b(n). \end{aligned}$$

Example 5.1. $n = 5$. The following are partitions of 5:

$$5, \quad 4+1, \quad 3+2, \quad 3+1+1, \quad 2+2+1, \quad 2+1+1+1, \quad 1+1+1+1+1.$$

Below listed are λ's with at least one even part and the corresponding μ's. In each λ, an even part is circled and the parts corresponding to the circled part are underlined in μ.

λ	μ
$\textcircled{4}+1$	$\underline{2+2}+1$
$3+\textcircled{2}$	$3+\underline{1+1}$
$\textcircled{2}+2+1$	$2+\underline{1+1}+1$
$\textcircled{2}+1+1+1$	$\underline{1+1}+1+1+1$

5.2. *Proof of* $a_o(n) = b(n+1) + b(n-1)$

For a partition $\lambda \vdash n$, let $2a-1$ be an odd part of λ. We remove the part $2a-1$ and add two a's keeping the remaining parts unchanged. We call the resulting partition μ. Similarly, we remove $2a-1$ and add two $(a-1)$'s keeping the remaining parts of λ unchanged. We call the resulting partition ν. Then, it is clear that $\mu \vdash (n+1)$ and $\nu \vdash (n-1)$. Also, we can easily recover λ from μ by merging two copies of a in μ and then subtracting 1 from it to get one $2a-1$. Thus,

$$\begin{aligned} a_o(n) &= \sum_{\lambda \vdash n} \text{different odd parts of } \lambda \\ &= \sum_{\mu \vdash n+1} \text{different parts of } \mu \text{ with multiplicity } > 1 \\ &\quad + \sum_{\nu \vdash n-1} \text{different parts of } \nu \text{ with multiplicity } > 1 \\ &= b(n+1) + b(n-1). \end{aligned}$$

Example 5.2. $n = 5$. The following are partitions of 5:

$$5, \quad 4+1, \quad 3+2, \quad 3+1+1, \quad 2+2+1, \quad 2+1+1+1, \quad 1+1+1+1+1.$$

In what follows, listed are λ's with at least one odd part and the corresponding μ's and ν's. In each λ, an odd part is circled and the parts corresponding to the circled parts are underlined in μ's and ν's, respectively. For convenience, 0's are allowed as parts in ν's.

$\lambda \vdash 5$	$\mu \vdash 6$	$\nu \vdash 4$
$\textcircled{5}$	$\underline{3+3}$	$\underline{2+2}$
$4+\textcircled{1}$	$4+\underline{1+1}$	$4+\underline{0+0}$
$\textcircled{3}+2$	$\underline{2+2}+2$	$2+\underline{1+1}$
$\textcircled{3}+1+1$	$\underline{2+2}+1+1$	$\underline{1+1}+1+1$
$3+\textcircled{1}+1$	$3+\underline{1+1}+1$	$3+1+\underline{0+0}$
$2+2+\textcircled{1}$	$2+2+\underline{1+1}$	$2+2+\underline{0+0}$
$2+\textcircled{1}+1+1$	$2+\underline{1+1}+1+1$	$2+1+1+\underline{0+0}$
$\textcircled{1}+1+1+1+1$	$\underline{1+1}+1+1+1+1$	$1+1+1+1+\underline{0+0}$

5.3. *Proof of the generating function for $a_e(n)$*

The generating functions for $a(n)$ and $a_e(n)$ can be proved combinatorially and the proofs are very similar. Since we will need $a_e(n)$ for Theorem 1.4, we give the proof of $a_e(n)$ only.

Let us recall the generating function for $a_e(n)$ from Theorem 1.2:

$$\sum_{n=1}^{\infty} a_e(n)q^n = \frac{1}{(q;q)_\infty} \cdot \frac{2q^2}{(1-q^2)^2}. \tag{5.1}$$

Let $\overline{A}_e(n)$ be the set of colored overpartitions of n in which there exists exactly one overlined part, which is even, and at most one non-overlined even part may be colored blue. For example, the following are in $\overline{A}_e(6)$:

$$\overline{6}, \overline{4}+2, \overline{4}+2_b, 4_b+\overline{2}, 4+\overline{2}, \overline{4}+1+1, 3+\overline{2}+1, \overline{2}+2+2,$$
$$\overline{2}+2_b+2, \overline{2}+2+1+1, \overline{2}+2_b+1+1, \overline{2}+1+1+1+1,$$

where the subscript b is used for blue parts.

Then, it is clear that

$$\sum_{n=1}^{\infty} |\overline{A}_e(n)| q^n = \frac{1}{(q;q)_\infty} \cdot \frac{q^2}{(1-q^2)} \cdot \frac{1}{(1-q^2)}, \tag{5.2}$$

where the infinite product, $q^2/(1-q^2)$, and $1/(1-q^2)$ on the right-hand side account for non-colored and non-overlined parts, even overlined parts, and blue colored parts, respectively. Thus, (5.1) is equivalent to

$$a_e(n) = 2|\overline{A}_e(n)|.$$

To prove this, we will first define a subset $\overline{P}_e(n)$ of overpartitions of n such that

$$a_e(n) = \sum_{\lambda \in \overline{P}_e(n)} \text{overlined part of } \lambda, \tag{5.3}$$

and then we set up a surjection from $\overline{A}_e(n)$ to $\overline{P}_e(n)$, which will lead us to

$$\frac{1}{2} \sum_{\lambda \in \overline{P}_e(n)} \text{overlined part of } \lambda = |\overline{A}_e(n)|. \tag{5.4}$$

Let $\overline{P}_e(n)$ be the set of overpartitions of n with exactly one overlined part, which is even. For example,

$$\overline{P}_e(6) = \{\overline{6}, \overline{4}+2, 4+\overline{2}, \overline{4}+1+1, 3+\overline{2}+1,$$
$$\overline{2}+2+2, \overline{2}+2+1+1, \overline{2}+1+1+1+1\}.$$

Suppose that an ordinary partition π of n has exactly m different even parts, say

$$\alpha_1 > \alpha_2 > \cdots > \alpha_m.$$

We create m copies of π and overline α_j in the jth copy of π. Since each copy of π has exactly one even overlined part, it is in $\overline{P}_e(n)$. Furthermore, the overlined parts are different, so the m copies of π are all different overpartitions in $\overline{P}_e(n)$.

Conversely, let λ be an overpartition in $\overline{P}_e(n)$. Suppose that $\overline{\beta}$ is the overlined part and there are $m-1$ different even parts in λ not equal to β, i.e.

$$\beta_1 > \beta_2 > \cdots > \beta_{m-1}.$$

Then, by replacing $\overline{\beta}$ by β and β_j by $\overline{\beta}_j$, we get a different overpartition in $\overline{P}_e(n)$. Also, if we replace the overlined parts by non-overlined parts in these overpartitions, we will get the same ordinary partition.

Therefore, there is a one-to-one correspondence between a partition with m different even parts and a set of m overpartitions in $\overline{P}_e(n)$. This proves the identity in (5.3).

We now set up a surjection from $\overline{A}_e(n)$ to $\overline{P}_e(n)$. For a colored overpartition μ in $\overline{A}_e(n)$, if there is a blue part, we merge it with the overlined even part keeping the resulting part overlined. For example, for $\mu = 4 + \overline{2}$, we merge 4 and $\overline{2}$ to get $\overline{6}$. Since the color is dropped while the overline is kept, the resulting partition belongs to $\overline{P}_e(n)$. This map is certainly surjective since $\overline{P}_e(n)$ is a subset of $\overline{A}_e(n)$ and if there is no blue part in μ, no parts are merged and μ is mapped to itself. We now determine the number of pre-images of $\lambda \in \overline{P}_e(n)$. If $\overline{2a}$ is the overlined part of λ, then there are exactly a ways of merging an overlined part and a blue part to get $2a$, namely $\overline{2x}$ and $2y$ for $1 \leq x \leq a$ and $y = a - x$. Thus, for each partition of $\overline{P}_e(n)$, there are as many pre-images as a half the overlined part. This proves the identity in (5.4).

By (5.2)–(5.4), we see that

$$\sum_{n\geq 0} a_e(n)q^n = \frac{1}{(q;q)_\infty} \cdot \frac{2q^2}{(1-q^2)} \cdot \frac{1}{(1-q^2)}$$

as desired.

5.4. *Proof of the truncated pentagonal number theorem*

In this section, we provide a combinatorial proof of the truncated pentagonal number theorem of Andrews and Merca from [2], namely the identity in (3.1), which is equivalent to

$$(-1)^{k-1}\sum_{j=0}^{k-1}(-1)^j(p(n-j(3j+1)/2) - p(n-(j+1)(3j+2)/2)) = M_k(n). \quad (5.5)$$

For $k = 1$, it is clear that both sides generate partitions into parts >1, i.e.

$$p(n) - p(n-1) = M_1(n). \quad (5.6)$$

Recall the generating function of $M_k(n)$ for $k \geq 1$ in (3.2). Upon some simplifications, the right-hand side of (3.2) may be rewritten

$$\sum_{n\geq 0} \frac{q^{\binom{k}{2}+(k+1)k+(k+1)n}}{(q;q)_{k-1}(q;q)_n(1-q^{n+k})},$$

and we define the coefficient of q^n in this series by $m_k(n)$, i.e.

$$\sum_{n\geq 0} m_k(n)q^n := \sum_{n\geq 0} \frac{q^{\binom{k}{2}+(k+1)k+(k+1)n}}{(q;q)_{k-1}(q;q)_n(1-q^{n+k})}. \quad (5.7)$$

Since

$$\sum_{n\geq 0}\frac{q^{\binom{k}{2}+(k+1)k+(k+1)n}}{(q;q)_{k-1}(q;q)_n(1-q^{n+k})}$$

$$=\frac{q^{\binom{k}{2}+(k+1)k}}{(q;q)_{k-1}}\left(\sum_{n\geq 0}\frac{q^{(k+1)n}}{(q;q)_n}+\frac{q^k}{(1-q^k)}+\sum_{n\geq 1}\frac{q^{(k+2)n+k}}{(q;q)_n(1-q^{n+k})}\right),$$

we can easily see that $m_k(n)$ counts the number of partitions of n satisfying the following conditions:

(i) $f_i \geq 1$ for $i = 1, \ldots, k-1$;
(ii) $f_{k+1} \geq k$;
(iii) if $f_k \geq 1$, then $f_{k+1} = k$ and $f_k \leq (x - k - 1)$, where x is the smallest part $> k+1$. Here, if there is no part $> k+1$, we take $x = \infty$.

Throughout this paper, f_i denotes the multiplicity of the part i in a partition. By (5.7), we also see that

$$\sum_{n\geq 0} m_{k+1}(n)q^n = \sum_{n\geq 0}\frac{q^{\binom{k+1}{2}+(k+2)(k+1)+(k+2)n}}{(q;q)_k(q;q)_n(1-q^{n+k+1})}$$

$$=\sum_{n\geq 1}\frac{q^{\binom{k}{2}+(k+1)k+(k+2)n+2k}}{(q;q)_k(q;q)_{n-1}(1-q^{n+k})}$$

$$=\frac{q^{\binom{k}{2}+(k+1)k}}{(q;q)_{k-1}}\sum_{n\geq 1}\frac{q^{(k+2)n+2k}}{(1-q^k)(q;q)_{n-1}(1-q^{n+k})}.$$

The last summation formula gives a different combinatorial interpretation of $m_{k+1}(n)$, namely it counts the number of partitions of n satisfying the following conditions:

(i) $f_i \geq 1$ for $i = 1, \ldots, k-1$;
(ii) $f_{k+1} = k$;
(iii) $f_k \geq 1$ and there exists a part $> k+1$ such that $f_k > (x - k - 1) \geq 1$, where x is the smallest part $> k+1$.

Thus, it follows from the conditions of $m_k(n)$ and $m_{k+1}(n)$ that for $k \geq 1$, $m_k(n) + m_{k+1}(n)$ counts the number of partitions of n satisfying the following conditions:

(i) $f_i \geq 1$ for $i = 1, \ldots, k-1$;
(ii) $f_{k+1} \geq k$;
(iii) if $f_k \geq 1$, then $f_{k+1} = k$.

Since $q^{\binom{k}{2}+(k+1)k}/(q;q)_\infty$ generates partitions satisfying the first two conditions above and $q^{\binom{k}{2}+(k+1)k+k+(k+1)}/(q;q)_\infty$ generates partitions satisfying the first two conditions plus $f_k \geq 1$ and $f_{k+1} > k$, we see that

$$\sum_{n\geq 0}(m_k(n)+m_{k+1}(n))q^n = \frac{q^{\binom{k}{2}+(k+1)k}(1-q^{2k+1})}{(q;q)_\infty}.$$

Equivalently,

$$m_k(n)+m_{k+1}(n) = p(n-k(3k+1)/2) - p(n-(k+1)(3k+2)/2). \tag{5.8}$$

Thus, to prove the truncated pentagonal number theorem combinatorially, we only need to prove combinatorially that for $k \geq 1$ and $n \geq 0$,

$$M_k(n) = m_k(n). \tag{5.9}$$

Let us compare the summands of the right-hand sides of their generating functions in (3.2) and (5.7), i.e.

$$\frac{q^{\binom{k}{2}+(k+1)k+(k+1)n}}{(q;q)_{n+k-1}(1-q^{n+k})}\begin{bmatrix} n+k-1 \\ k-1 \end{bmatrix} = \frac{q^{\binom{k}{2}+(k+1)k+(k+1)n}}{(q;q)_{k-1}(q;q)_n(1-q^{n+k})}.$$

In [1], there exists a combinatorial proof of

$$\frac{1}{(q;q)_{n+k-1}}\begin{bmatrix} n+k-1 \\ k-1 \end{bmatrix} = \frac{1}{(q;q)_{k-1}(q;q)_n}.$$

Therefore, by (5.6), (5.8), and (5.9), we have proved (5.5) combinatorially.

5.5. *Proof of Theorem 1.4*

By Theorem 1.3(a) and (5.3), we see that Theorem 1.4 is equivalent to

$$(-1)^{k-1}\left(\sum_{j=-(k-1)}^{k}(-1)^j\left(\sum_{\lambda\in\overline{P}_e(n-j(3j-1)/2)}\text{overlined part of }\lambda\right)\right.$$
$$\left.-\frac{1+(-1)^n}{2}\cdot n\right) = \sum_{j=1}^{\lfloor n/2\rfloor} 2jM_k(n-2j).$$

For an overpartition λ in $\overline{P}_e(n)$, by separating the overlined part of λ, we can make a pair of an ordinary partition μ and a partition ν of only one single part which is even. For instance, $\lambda = \overline{2}+2+1+1$ corresponds to $(\mu,\nu) = (2+1+1, 2)$. Thus,

$$\sum_{\lambda\in\overline{P}_e(n)}\text{overlined part of }\lambda = \sum_{m=1}^{\lfloor n/2\rfloor} 2m \sum_{\mu\vdash(n-2m)} 1 = \sum_{m=1}^{\lfloor n/2\rfloor} 2m\cdot p(n-2m),$$

from which it follows that

$$
\begin{aligned}
&(-1)^{k-1} \sum_{j=-(k-1)}^{k} (-1)^j \sum_{\lambda \in \overline{P}_e(n-j(3j-1)/2)} \text{overlined part of } \lambda \\
&= (-1)^{k-1} \sum_{j=-(k-1)}^{k} (-1)^j \sum_{m=1}^{\lfloor (n-j(3j-1)/2)/2 \rfloor} 2m \cdot p(n - j(3j-1)/2 - 2m) \\
&= \sum_{m=1}^{\lfloor n/2 \rfloor} 2m \left((-1)^{k-1} \sum_{j=-(k-1)}^{k} (-1)^j p(n-2m-j(3j-1)/2) \right) \\
&= \sum_{m=1}^{\lfloor n/2 \rfloor} 2m \cdot M_k(n-2m) + (-1)^{k-1} \frac{1+(-1)^n}{2} n, \qquad (5.10)
\end{aligned}
$$

where the second equality is just a rearrangement of the terms in the double summations, assuming that $p(N) = 0$ if $N < 0$, and the last equality follows from the truncated pentagonal number theorem.

Here, we note that there is the correction factor $(-1)^{k-1}n$ for n even. This is because if n is even and $m = n/2$, without this correction factor, we have

$$(-1)^{k-1} np(0) = nM_k(0).$$

However, $p(0) = 1$ and $M_k(0) = 0$. Thus, the correction factor is necessary. Since the last equality can be proved combinatorially due to the proof given in Sec. 5.4, this completes our combinatorial proof.

5.6. *Proof of Theorem* 1.7

By Theorem 1.3(a), (5.3), and (5.4), we see that Theorem 1.7 is equivalent to

$$
\begin{aligned}
&(-1)^{k-1} \left(\sum_{j=0}^{2k-1} (-1)^j \left(\sum_{\lambda \in \overline{A}_e(n-j(j+1)/2)} 1 \right) - \frac{(1+(-1)^n)}{2} \cdot c\left(\frac{n}{2}\right) \right) \\
&\quad = \sum_{j=1}^{\lfloor n/2 \rfloor} c(j) MP_k(n-2j).
\end{aligned}
$$

For a colored overpartition λ in $\overline{A}_e(n)$, let f_i be the multiplicity of the non-colored and non-overlined part of size i in λ. We define

$$
\begin{aligned}
\mu &= (1^{f_1 - 2\lfloor f_1/2 \rfloor}, 2^{f_2}, 3^{f_3 - 2\lfloor f_3/2 \rfloor}, 4^{f_4}, 5^{f_5 - 2\lfloor f_5/2 \rfloor}, \ldots), \\
\nu &= (\overline{x}, y, 1^{\lfloor f_1/2 \rfloor}, 3^{\lfloor f_3/2 \rfloor}, 5^{f\lfloor f_5/2 \rfloor}, \ldots),
\end{aligned}
$$

where $\overline{2x}$ is the overlined part of λ and $2y$ is the colored part if exists. Then, it is clear that

$$|\lambda| = |\mu| + 2|\nu|.$$

Recall the Sylvester map between partitions into odd parts and partitions into distinct parts, namely the bijection for the following Euler's partition theorem [1]:

$$\frac{1}{(q;q^2)_\infty} = (-q;q)_\infty.$$

Finally, we remove y from ν, and then we apply the Sylvester map to the remaining non-overlined odd parts to obtain non-overlined distinct parts. This process creates an overpartition ρ in which there exist exactly one overlined part and distinct non-overline parts such that $|\rho| \leq |\nu|$. It follows from the remark after the proof of Theorem 4.1 that for a given $|\nu|$, the number of such ρ's is $c(|\nu|)$. Also, all the odd parts of μ are distinct, so μ is counted by $p_{od}(|\mu|)$. Thus,

$$\sum_{\lambda \in \overline{A}_e(n)} 1 = \sum_{m=1}^{\lfloor n/2 \rfloor} c(m) \cdot p_{od}(n-2m),$$

from which it follows that

$$\begin{aligned}
&(-1)^{k-1} \sum_{j=0}^{2k-1} (-1)^j \left(\sum_{\lambda \in \overline{A}_e(n-j(j+1)/2)} 1 \right) \\
&= (-1)^{k-1} \sum_{j=0}^{2k-1} (-1)^j \sum_{m=1}^{\lfloor (n-j(j+1)/2)/2 \rfloor} c(m) \cdot p_{od}(n - j(j+1)/2 - 2m) \\
&= \sum_{m=1}^{\lfloor n/2 \rfloor} c(m) \left((-1)^{k-1} \sum_{j=0}^{2k-1} (-1)^j p_{od}(n-2m-j(j+1)/2) \right) \\
&= \sum_{m=1}^{\lfloor n/2 \rfloor} c(m) MP_k(n-2m) + (-1)^{k-1} \frac{(1+(-1)^n)}{2} \cdot c\left(\frac{n}{2}\right),
\end{aligned}$$

where the second equality is just a rearrangement of the terms in the double summations, assuming that $p_{od}(N) = 0$ if $N < 0$, and the last equality follows from (4.1) and (4.2), i.e. the truncated theta series identity of Andrews and Merca [3].

As seen in the previous section, we note that there is a correction factor, i.e. $(-1)^{k-1}c(n/2)$ for n even. This is because $p_{od}(0) = 1$ but $MP_k(0) = 0$.

In [4], there is given a combinatorial proof of the truncated theta series identity of Andrews and Merca i.e.

$$(-1)^{k-1} \sum_{j=0}^{2k-1} (-1)^j p_{od}(n - j(j+1)/2) = MP_k(n).$$

Therefore, this completes our proof.

Acknowledgments

The authors would like to thank the anonymous referee for his/her valuable comments. The second author was partially supported by a grant (#633963) from the Simons Foundation.

References

[1] G. E. Andrews, *The Theory of Partitions*, Cambridge Mathematical Library (Cambridge University Press, Cambridge, 1998), reprint of the 1976 original.

[2] G. E. Andrews and M. Merca, The truncated pentagonal number theorem, *J. Combin. Theory Ser. A* **119** (2012) 1639–1643.

[3] G. E. Andrews and M. Merca, Truncated theta series and a problem of Guo and Zeng, *J. Combin. Theory Ser. A* **154** (2018) 610–619.

[4] C. Ballantine, M. Merca, D. Passary and A. J. Yee, Combinatorial proofs of two truncated theta series theorems, *J. Combin. Theory Ser. A* **160** (2018) 168–185.

[5] A. Knopfmacher and M. E. Mays, The sum of distinct parts in compositions and partitions, *Bull. Inst. Combin. Appl.* **25** (1999) 66–78.

[6] N. J. A. Sloane, The On-Line Encyclopedia of Integer Sequences, Published electronically at http://oeis.org.

[7] H. Wilf, Three problems in combinatorial asymptotics, *J. Combin. Theory Ser. A* **35** (1983) 199–207.

© 2024 World Scientific Publishing Company
https://doi.org/10.1142/9789811277375_0032

Mock modular Eisenstein series with Nebentypus*

Michael H. Mertens
Department of Mathematical Sciences, University of Liverpool
Liverpool L69 7ZL, United Kingdom
m.h.mertens@liverpool.ac.uk

Ken Ono†
Department of Mathematics, University of Virginia
Charlottesville, VA 22904, USA
ken.ono691@gmail.com

Larry Rolen
Department of Mathematics, 1420 Stevenson Center
Vanderbilt University, Nashville, TN 37240, USA
larry.rolen@vanderbilt.edu

In Celebration of Bruce Berndt's 80th Birthday.

By the theory of Eisenstein series, generating functions of various divisor functions arise as modular forms. It is natural to ask whether further divisor functions arise systematically in the theory of mock modular forms. We establish, using the method of Zagier and Zwegers on holomorphic projection, that this is indeed the case for certain (twisted) "small divisors" summatory functions $\sigma_\psi^{\mathrm{sm}}(n)$. More precisely, in terms of the weight 2 quasimodular Eisenstein series $E_2(\tau)$ and a generic Shimura theta function $\theta_\psi(\tau)$, we show that there is a constant α_ψ for which

$$\mathcal{E}_\psi^+(\tau) := \alpha_\psi \cdot \frac{E_2(\tau)}{\theta_\psi(\tau)} + \frac{1}{\theta_\psi(\tau)} \sum_{n=1}^{\infty} \sigma_\psi^{\mathrm{sm}}(n) q^n$$

is a half integral weight (polar) mock modular form. These include generating functions for combinatorial objects such as the Andrews *spt*-function and the "consecutive parts" partition function. Finally, in analogy with Serre's result that the weight 2 Eisenstein

*This article originally appeared in *International Journal of Number Theory*, **17**(3), pp. 683–697
© 2021 World Scientific Publishing Company. DOI: 10.1142/S179304212040028X
†Corresponding author.

series is a p-adic modular form, we show that these forms possess canonical congruences with modular forms.

Keywords: Mock modular forms; Eisenstein series.

Mathematics Subject Classification 2020: 11F03, 11F37, 11F30

1. Introduction and Statement of Results

In the theory of mock theta functions and its applications to combinatorics as developed by Andrews, Hickerson, Watson and many others, various formulas for q-series representations have played an important role. For instance, the generating function $R(\zeta;q)$ for partitions organized by their ranks is given by

$$R(\zeta;q) := \sum_{\substack{n\geq 0\\ m\in\mathbb{Z}}} N(m,n)\zeta^m q^n = \sum_{n\geq 0} \frac{q^{n^2}}{(\zeta q;q)_n(\zeta^{-1}q;q)_n} = \frac{1-\zeta}{(q;q)_\infty}\sum_{n\in\mathbb{Z}} \frac{(-1)^n q^{\frac{n(3n+1)}{2}}}{1-\zeta q^n},$$

where $N(m,n)$ is the number of partitions of n of rank m and $(a;q)_n := \prod_{j=0}^{n-1}(1-aq^j)$ is the usual q-Pochhammer symbol. The function $R(\zeta;q)$, when ζ is specialized to any root of unity times a certain rational power of q, is (up to a certain rational power of q) a mock modular form on a congruence subgroup; see [10, 23, 37] for discussion and applications. For instance, Ramanujan's original example of the third-order mock theta function $f(q)$ is

$$f(q) := R(-1:q) = \sum_{n\geq 0} \frac{q^{n^2}}{(-q;q)_n^2},$$

which shows that the Fourier coefficients of $f(q)$ enumerate the number of partitions of n with even rank minus the number of partitions of n with odd rank.

Many identities involving partition ranks are subtle and require delicate q-series techniques to prove. Thus, it is interesting to ask for further constructions of mock theta functions that will likely play a future role in the study of ranks and related partition functions. Here we take inspiration from the fact that many classical modular forms are generated by Eisenstein series which, thanks to their representations as Lambert series, have nice q-expansions in terms of divisor power sums. Such representations have long been a staple in the theory of modular forms and their applications.

To review one classical situation, we first recall some notation. The vth power divisor functions $\sigma_v(n) := \sum_{d|n} d^v$, for odd $v \geq 3$, arise naturally in the theory of modular forms. Indeed, their generating functions are essentially the classical Eisenstein series $G_k(\tau)$, defined for positive even integers k by

$$G_k(\tau) := \frac{1}{2}\zeta(1-k) + \sum_{n=1}^{\infty} \sigma_{k-1}(n)q^n,$$

where $q := e^{2\pi i\tau}$. In particular, if $k \geq 4$, then $G_k(\tau)$ is a weight k holomorphic modular form for the full modular group $\mathrm{SL}_2(\mathbb{Z})$.

Hecke's well-known theory [20] of modular forms includes many further divisor functions. For example, for a primitive Dirichlet character χ modulo N, consider the complementary divisor functions

$$\sigma_{v,\chi}(n) := \sum_{d|n} \chi(d)d^v \quad \text{and} \quad \sigma'_{v,\chi}(n) := \sum_{d|n} \chi(n/d)d^v.$$

If $k > 2$ and $(-1)^k = \chi(-1)$, then Hecke's work implies that the functions

$$G_{k,\chi}(\tau) := \frac{1}{2}L(1-k,\chi) + \sum_{n=1}^{\infty} \sigma_{k-1,\chi}(n)q^n \quad \text{and} \quad \widehat{G}_{k,\chi}(\tau) := \sum_{n=1}^{\infty} \sigma'_{k-1,\chi}(n)q^n,$$

where $L(s,\chi)$ denotes the standard Dirichlet L-function of χ, are holomorphic weight k modular forms on $\Gamma_0(N)$ with Nebentypus χ.

The weight $k = 2$ case is particularly interesting. Hecke obtained a modification of $G_2(\tau)$ which establishes that

$$E_2^*(\tau) := 1 - 24\sum_{n=1}^{\infty} \sigma_1(n)q^n - \frac{3}{\pi\Im(\tau)}$$

is a weight 2 harmonic Maaß form. Therefore, we may view $E_2(\tau) := 1 - 24\sum_{n=1}^{\infty}\sigma_1(n)q^n$ as a "mock modular form", the holomorphic part of the non-holomorphic modular form $E_2^*(\tau)$ (see for instance [8, Sec. 6.1] for more details). In view of this example, it is natural to ask whether there are further divisor functions which arise naturally in the theory of mock modular forms. We show that this is indeed the case for the generating functions of certain twisted "small divisors" summatory functions. In particular, we shall associate these functions with Shimura's weight 1/2 and 3/2 theta functions.

To make this precise, suppose that $\psi \pmod{f_\psi}$ is a primitive Dirichlet character with conductor $f_\psi > 1$. We then define $\lambda_\psi := (1-\psi(-1))/2$, so that $\lambda_\psi = 0$ (respectively, $\lambda_\psi = 1$) if ψ is even (respectively, odd). The corresponding Shimura theta function [34] of weight $\lambda_\psi + \frac{1}{2}$ on $\Gamma_0(4f_\psi^2)$ with Nebentypus ψ is defined as

$$\theta_\psi(\tau) := \begin{cases} \displaystyle\sum_{n=1}^{\infty} \psi(n)q^{n^2} & \text{if } \lambda_\psi = 0, \\ \displaystyle\sum_{n=1}^{\infty} \psi(n)nq^{n^2} & \text{if } \lambda_\psi = 1. \end{cases} \tag{1.1}$$

It is a well-known fact that θ_ψ is a modular form, more precisely we have $\theta_\psi(\tau) \in M_{\frac{1}{2}}(\Gamma_0(4f_\psi^2),\psi)$ if $\lambda_\psi = 0$, respectively, $\theta_\psi \in S_{\frac{3}{2}}(\Gamma_0(4f_\psi^2),\psi\chi_{-4})$ with χ_{-4} denoting the nontrivial character modulo 4 if $\lambda_\psi = 1$.

We define a twisted "small divisors" summatory function for any nontrivial Dirichlet character ψ. To ease notation, if n is a positive integer, then we define its small divisors by

$$D_n := \{d\,|\,n : 1 \le d \le n/d \text{ and } d \equiv n/d \pmod 2\}. \tag{1.2}$$

Then, the ψ-twisted small divisors summatory function is defined by

$$\sigma_\psi^{\mathrm{sm}}(n) := \sum_{d \in D_n} \psi\left(\frac{(n/d)^2 - d^2}{4}\right) d. \tag{1.3}$$

We let $\mathcal{E}_\psi^+(\tau)$ denote its generating function divided by $\theta_\psi(\tau)$, as in the following expression:

$$\mathcal{E}_\psi^+(\tau) := \alpha_\psi \cdot \frac{E_2(\tau)}{\theta_\psi(\tau)} + \frac{1}{\theta_\psi(\tau)} \cdot \sum_{n=1}^{\infty} \sigma_\psi^{\mathrm{sm}}(n) q^n, \tag{1.4}$$

where α_ψ is a suitable, explicitly computable constant (see Sec. 2.2), depending only on ψ.

It turns out that $\mathcal{E}_\psi^+(\tau)$ is a polar mock modular form, the holomorphic part of a weight $\frac{3}{2} - \lambda_\psi$ *polar harmonic Maaß form*. Recall that polar harmonic Maaß forms have holomorphic parts which are permitted to have poles in $\mathbb{H}$ (see [8, Chap. 13]). Its nonholomorphic part is given by the normalized period integral

$$\mathcal{E}_\psi^-(\tau) := (-1)^{\lambda_\psi} \frac{(2\pi)^{\lambda_\psi - 1/2} i}{8\Gamma(1/2 + \lambda_\psi)} \cdot \int_{-\overline{\tau}}^{i\infty} \frac{\theta_{\overline{\psi}}(z)}{(-i(z+\tau))^{3/2-\lambda_\psi}} dz. \tag{1.5}$$

Using the constructive method of Zagier and Zwegers (for example, see [4, 9, 22]), where harmonic Maaß forms are constructed via holomorphic projection, we obtain the following theorem.

Theorem 1.1. *Assuming the notation above, we have that* $\mathcal{E}_\psi(\tau) := \mathcal{E}_\psi^+(\tau) + \mathcal{E}_\psi^-(\tau)$ *is a weight* $\frac{3}{2} - \lambda_\psi$ *polar harmonic weak Maaß form on* $\Gamma_0(4f_\psi^2)$ *with Nebentypus* $\overline{\psi} \cdot \chi_{-4}^{\lambda_\psi}$.

Remark 1.2. The harmonic Maaß forms $\mathcal{E}_\psi(\tau)$ are, up to a constant factor, preimages of $\theta_\psi(\tau)$ under the Bruinier–Funke ξ-operator (see [8, Sec. 5.2]). In other words, Shimura's theta functions are the *shadows* of these canonical polar harmonic Maaß forms.

Remark 1.3. It is clear from the definition that the "mock Eisenstein series" $\mathcal{E}_\psi^+(\tau)$ can only have poles in the zeros of $\theta_\psi(\tau)$.

Since the space $M_2^!(\Gamma_0(4f_\psi))$ of weakly holomorphic modular forms of weight 2 is infinite-dimensional and θ_ψ has only finitely many poles in a fundamental domain of $\Gamma_0(4f_\psi))$, it is possible to construct a form $G_\psi \in M_2^!(\Gamma_0(4f\psi))$ such that G_ψ/θ_ψ has the same principal part at each zero of θ_ψ as $\mathcal{E}_\psi$, wherefore the difference $\mathcal{E}_\psi - G_\psi/\theta_\psi$ is a harmonic Maaß form without any poles in the upper half-plane.

Remark 1.4. An elementary computation reveals that the q-series generating function of $\sigma_\psi^{\mathrm{sm}}(n)$ (i.e. without the $1/\theta_\psi(\tau)$ factor) can be given in terms of derivatives of Appell–Lerch sums as studied by Zwegers [38, 39]. For a special case of this, see for instance [27, Lemma 2]. The forms obtained in Theorem 1.1 have the pleasing property that both parts depend on ψ in a simple way which is analogous to the standard theory of holomorphic modular forms with Nebentypus.

A natural question arises as to which of the functions $\mathcal{E}_\psi^+(\tau)$ from Theorem 1.1 are actually mock modular forms, that is, which do not actually have poles on the upper half-plane. For example, this situation occurs whenever $\theta_\psi(\tau)$ has no zeros in the upper half-plane. Due to the infinite product expansion of the Dedekind eta function $\eta(\tau) := q^{\frac{1}{24}} \prod_{n\geq 1}(1-q^n)$, eta-quotients automatically have no zeros on the upper half-plane. In fact, Rouse and Webb showed in [32] that the *only* modular forms on $\Gamma_0(N)$ with integer Fourier coefficients and no zeros on the upper half-plane are eta quotients. The classification of such theta functions was completed by Mersmann [26] and Lemke Oliver in [24].

Theorem 1.5. *The only nontrivial primitive characters ψ where $\theta_\psi(\tau)$ is an eta-quotient (in which case $\mathcal{E}_\psi^+(\tau)$ is a mock modular form), are when ψ is in the set of Kronecker characters $\chi_D := (\frac{D}{\cdot})$ where D is in the following set of fundamental discriminants*:

$$\{-8, -4, -3, 2, 12, 24\}.$$

Several of the mock Eisenstein series in Theorem 1.5 occur naturally in the theory of partitions. Here we point two particularly well-known examples. We let $spt(n)$ denote the Andrews *smallest parts function* [3], which counts the number of smallest parts among partitions of n, whose mock modular properties were first exhibited and exploited by Bringmann [7] and by Folsom and the second author in [17]. Furthermore, we will denote by $cpt(n)$ the total number of parts in all partitions of n into *consecutive integers.* Connections between odd divisors of integers and partitions into consecutive parts have also been noted previously (see works of Hirschhorn and Hirschhorn [21] as well as Sylvester and Franklin [36]).

We summarize these combinatorial examples which arise out of Theorem 1.5 in the following:

Corollary 1.6. *The following are true*:

(i) *The functions*

$$\mathcal{E}_{\chi_{12}}^+(\tau) = -2q^{-1}\sum_{n\geq 1} spt(n)q^{24n} + \frac{1}{6\theta_{\chi_{12}}(\tau)}E_2(24\tau)$$

and

$$\mathcal{E}_{\chi_{24}}^+(\tau) = -2q^{-1}\sum_{n\geq 1}(-1)^n spt(n)q^{24n} - \frac{1}{24\theta_{\chi_{24}}(\tau)}E_2(24\tau)$$

are mock modular forms of weight $3/2$.

(ii) *The function*

$$\theta_{\chi_2}(\tau)\cdot\mathcal{E}_{\chi_2}^+(\tau) = 2\sum_{n\geq 1}(-1)^n cpt(n)q^{8n} + \frac{1}{16}E_2(8\tau)$$

is a mixed mock modular form, i.e. the product of a mock modular form and a modular form (see for instance [8, *Sec.* 13.2]) *of weight* 2.

Note that in slight abuse of notation, we have not inserted $E_2(\tau)$ in the above corollary, but rather $E_2(24\tau)$, respectively, $E_2(8\tau)$ because thus the coefficients remain supported on the same arithmetic progression as indicated in the given sums and the transformation behavior of the completed function can be improved. For instance, the function $\theta_{\chi_{12}}(\tau/24)\mathcal{E}_{\chi_{12}}(\tau/24)$ turns out to be modular of weight 2 for the full modular group $\mathrm{SL}_2(\mathbb{Z})$.

Remark 1.7. There are other prominent examples which have appeared in the literature outside of combinatorial number theory. For instance, in work related to quantum black holes, we find

$$\theta_{\chi_{-4}}(\tau)\cdot\mathcal{E}^+_{\chi_{-4}}\left(\frac{\tau}{8}\right) = \sum_{\substack{r>s>0\\ r-s\equiv 1 \pmod 2}} (-1)^r s q^{\frac{rs}{2}} =: F_2^{(2)}(\tau),$$

where $F_2^{(2)}(\tau)$ is the series which Dabholkar *et al.* computed using holomorphic projection (cf. [13, Sec. 7.1, Example 4]). They use this form to construct the mock modular form

$$h^{(2)}(\tau) := \frac{24F_2^{(2)}(\tau) - E_2(\tau)}{\eta(\tau)^3},$$

which has shadow $\eta^3(\tau)$, and was first described in works of Eguchi–Taormina [16] and Eguchi–Ooguri–Taormina–Yang [15] in relation to the elliptic genus. It has since played an important role in Mathieu Moonshine [12, 14] and in relation to SO(3) Donaldson invariants of $\mathbb{C}P^2$ [25].

Although the Eisenstein series $E_2(\tau)$ is not modular, Serre [33] proved that it is a p-adic modular form for every prime p. Namely, for every prime power p^t he identified $E_2(\tau) \pmod{p^t}$ as the reduction of a genuine holomorphic modular form. Moreover, he found that such sequences of modular forms, as $t \to +\infty$, have weights that p-adically converge to 2.

In view of Theorem 1.1, it is natural to ask whether the mock modular Eisentstein series $\mathcal{E}^+_\psi(\tau)$ possess similar p-adic properties. In analogy with Serre's construction, we indeed show that these mock modular Eisenstein series satisfy analogous congruences with genuine meromorphic modular forms modulo arbitrary prime powers. Under repeated iteration of the U-operator

$$\left(\sum_{n\gg-\infty} a(n)q^n\right)\Bigg| U(p) := \sum_{n\gg-\infty} a(pn)q^n, \tag{1.6}$$

we obtain the following result.

Theorem 1.8. *Let $p > 3$ be a prime and $a, b \in \mathbb{N}$. Then there is a weight 2 meromorphic modular form $F_\psi^{(a,b)}(\tau)$ for the group $\Gamma_0(4f_\psi^2 p^m)$, with $m = \max(2a - b, 1 - 2v_p(f_\psi))$, with trivial Nebentypus such that we have the congruence*

$$(\theta_\psi(p^{2a}\tau)\mathcal{E}^+_\psi(\tau))\,|\,U(p^b) \equiv F_\psi^{(a,b)}(\tau) \pmod{p^{\min(a,b)}}.$$

Remark 1.9. In the special case of $spt(n)$, these p-adic properties have been studied previously. Andrews found [3] that the spt-function satisfies the following Ramanujan-type congruences:

$$spt(5n+4) \equiv 0 \pmod{5}, \quad spt(7n+5) \equiv 0 \pmod{7},$$
$$spt(13n+6) \equiv 0 \pmod{13},$$

and the second author and Folsom in [17] obtained many further congruences, see also [31]. Garvan [18] and later Ahlgren and Kim [2] used the modularity properties to study the context of Andrews' congruences. They showed that Andrews' congruences are "explained" by the triviality of the space of cusp forms of level one on and weights $5+1, 7+1, 13+1$, and surmised the general conjecture that no further congruences of this type exist for primes $\ell \geq 11$. As for prime power congruences, results close to p-adic modularity along the lines of Theorem 1.8 have been obtained in [6].

The paper is organized as follows. In Sec. 2.1, we will review the theory of holomorphic projection and establish some preliminary results, which we will exploit in Sec. 2.2 to prove our main results on the construction of the polar mock modular Eisenstein series. We then derive the examples, including $spt(n)$ and $cpt(n)$, where we obtain mock modular forms in Sec. 3. We conclude in Sec. 4 with the proof of our p-adic congruence results.

2. Modularity Properties of New Divisor Function Generating Functions

Here we prove Theorem 1.1. To this end we first recall the important work of Zwegers, which we then manipulate to obtain the desired generating functions.

2.1. *Holomorphic projection*

Holomorphic projection is now an important tool in the study of non-holomorphic modular forms, in particular harmonic Maaß forms. It was first introduced by Sturm [35] and further developed by Gross and Zagier in their seminal work on derivatives of L-functions [19]. For some applications in the context of mock modular forms and harmonic Maaß forms, the reader may consult for instance [1, 4, 9, 22, 29, 30], among several others. For background on the subject of harmonic Maaß forms, we refer the reader to [8].

Let us first recall the definition of holomorphic projection and explicit expressions for holomorphic projection of the product of a harmonic Maaß form with a holomorphic modular form. These can be found, e.g., in [22], where a vector-valued version of the same question was considered, or in [29], where Rankin–Cohen brackets of such functions were examined.

The usual holomorphic projection operator is defined for a (not necessarily holomorphic) modular form Φ, growing at most polynomially approaching the cusps, with Fourier expansion

$$\Phi(\tau) = \sum_{n\in\mathbb{Z}} c(n,y)q^n, \quad y := \Im(\tau), \tag{2.1}$$

of weight $k \geq 2$ for some subgroup $\Gamma \leq \mathrm{SL}_2(\mathbb{Z})$ as

$$\pi_{hol}(\Phi)(\tau) := c_0 + \sum_{n=1}^{\infty} c_n q^n$$

with

$$c_n := \frac{(4\pi n)^{k-1}}{\Gamma(k-1)} \int_0^\infty c(n,y)e^{-4\pi n y} y^{k-2} dy, \quad (n \geq 1), \tag{2.2}$$

provided that the integral converges, and where c_0 is an appropriate constant term. This operator maps into the space of holomorphic modular forms for $k > 2$ and to the space of holomorphic quasimodular forms for $k = 2$. It also behaves like a projection in that it acts as the identity on holomorphic functions, and it preserves the Petersson inner product, i.e. for every cusp form $g \in S_k(\Gamma)$, we have

$$\langle \Phi, g\rangle = \langle \pi_{hol}(\Phi), g\rangle.$$

This is in fact the origin of the operator's definition.

We mention for the sake of completeness that in many applications, the growth conditions on the function Φ are too restrictive. If instead of moderate growth towards the cusps, one requires that at each cusp $\mathfrak{a}$ there is a polynomial $H_{\mathfrak{a}} \in \mathbb{C}[X]$ such that

$$\Phi - H_{\mathfrak{a}}(q^{-1/h}),$$

where h denotes the width of the cusp $\mathfrak{a}$, has moderate growth approaching $\mathfrak{a}$, and requires that the coefficients of Φ satisfy $c(n,y) = O(y^{2-k})$ as $y \to 0$ uniformly for all $n > 0$, one obtains a *regularized* holomorphic projection operator

$$\pi_{hol}^{reg}(\Phi)(\tau) := H_{i\infty}(q^{-1}) + \sum_{n=1}^{\infty} \tilde{c}_n q^n$$

with

$$\tilde{c}_n := \frac{(4\pi n)^{k-1}}{\Gamma(k-1)} \lim_{s\to 0} \int_0^\infty c(n,y)e^{-4\pi n y} y^{k-2-s} dy, \quad (n \geq 1).$$

For further details, the reader may consult for instance [30].

It is a basic fact (see for instance [8, Sec. 4.2]) that a harmonic Maaß form $f \in H_k(\Gamma)$ of weight $k \neq 1$ naturally splits into a holomorphic and a non-holomorphic part,

$$f(\tau) = f^+(\tau) + f^-(\tau)$$

with

$$f^+(\tau) = \sum_{n \gg -\infty} c_f^+(n) q^n \quad \text{and}$$
$$f^-(\tau) = \frac{(4\pi y)^{1-k}}{1-k} \overline{c_f^-(0)} + \sum_{n=1}^{\infty} \overline{c_f^-(n)} n^{k-1} \Gamma(1-k; 4\pi n y) q^{-n}. \tag{2.3}$$

With this, one obtains the following result which is a special case of [29, Theorem 4.6].

Proposition 2.1. *Let $f \in H_k(\Gamma)$ be a harmonic Maaß form of weight $k \in \frac{1}{2}\mathbb{Z}\backslash\{1\}$ with an expansion as in (2.3) with $c_f^-(0) = 0$ and $g \in M_{2-k}(\Gamma)$ with Fourier expansion*

$$g(\tau) = \sum_{n=1}^{\infty} b(n) q^n.$$

Then the regularized holomorphic projection $\pi_{hol}^{reg}(f \cdot g)$ is well-defined and we have

$$\pi_{hol}^{reg}(f \cdot g)(\tau) = f^+(\tau) g(\tau) + \sum_{r=1}^{\infty} c(r) q^r$$

where

$$c(r) = \frac{\Gamma(2-k)}{k-1} \sum_{m=1}^{\infty} \overline{c_f^-(m)} b(m+r)[(r+m)^{k-1} - m^{k-1}].$$

It is sometimes convenient to define holomorphic projection for merely translation-invariant rather than modular functions as done for instance in [9, Sec. 3]. The results there will be the key to the proof of Theorem 1.1 and are summarized in the following lemma (cf. [9, Lemmas 3.1–3.3]).

Lemma 2.2. *Let $\Phi : \mathbb{H} \to \mathbb{C}$ be a translation invariant function with Fourier expansion as in (2.1) and $\kappa > 1$. Then the following are true:*

(1) *We have that*

$$\pi_{hol} f(\tau) = \frac{(\kappa-1)(2i)^\kappa}{4\pi} \int_{\mathbb{H}} \frac{f(x+iy) y^\kappa}{(\tau - x + iy)^\kappa} \frac{dx dy}{y^2}, \tag{2.4}$$

provided that both expressions are well-defined.

(2) *If $f : \mathbb{H} \to \mathbb{C}$ is holomorphic and translation invariant so that the right-hand side of (2.4) converges absolutely, we have for either of the possible expressions of π_{hol} that $\pi_{hol} f = f$.*

(3) *If the right-hand side of (2.4) converges absolutely, we have for all $\gamma \in SL_2(\mathbb{Z})$ and $\tau \in \mathbb{H}$ that*

$$\pi_{hol}(f|_\kappa \gamma)(\tau) = (\pi_{hol} f)|_\kappa \gamma(\tau).$$

Remark 2.3. We point out that for instance Proposition 2.1 holds for translation-invariant functions, provided that their Fourier expansions have the correct form.

From this, one can derive the following result from [9, Proposition 3.5].

Proposition 2.4. *Let* $f \colon \mathbb{H} \to \mathbb{C}$ *be a translation invariant function such that* $|f(\tau)|\Im(\tau)^r$ *is bounded on* $\mathbb{H}$ *for some* $r > 0$*. If the weight* k *holomorphic projection of* f *vanishes identically for some* $k > r+1$ *and* $\xi_k f$ *is modular of weight* $2-k$ *for some subgroup* $\Gamma \leq SL_2(\mathbb{Z})$*, then* f *is modular of weight* k *for* Γ*.*

2.2. *Proof of Theorem* 1.1

Proof of Theorem 1.1. It is clear from the definition that $\mathcal{E}_\psi$ is annihilated by the weight $1/2 + \lambda_\psi$ hyperbolic Laplacian away from the singularities, which lie precisely in the zeros of θ_ψ. Since the Fourier coefficients of $\theta_\psi \mathcal{E}_\psi^+$ at ∞ grow at most polynomially, wherefore the growth of the function towards any cusp can be at most polynomially and thus $\mathcal{E}_\psi$ itself can only grow at most linearly exponentially towards any cusp. It follows from the known transformation properties of θ_ψ under the full modular group that $\mathcal{E}_\psi^-$ also grows at most polynomially towards the cusps, so that $\mathcal{E}_\psi$ has the correct growth properties towards the cusps for a harmonic Maaß form. We are therefore left with showing modularity. The expression for $\pi_{hol}(\theta_\psi \mathcal{E}_\psi)$ in Lemma 2.2 (1) is easily seen to be well-defined, because of the polynomial growth towards the cusps, as discussed above. A standard computation further reveals that

$$\mathcal{E}_\psi^- = \frac{1/2 - \lambda_\psi}{2\Gamma(1/2 + \lambda_\psi)} \sum_{n=1}^{\infty} \overline{\psi(n)} n^{1-2\lambda_\psi} \Gamma(\lambda_\psi - 1/2; 4\pi n^2 y) q^{-n^2}.$$

Under the assumption that the constant α_ψ in the definition of $\mathcal{E}_\psi^+$ is zero, we obtain from Proposition 2.1 (see also the remark following Lemma 2.2) that

$$\pi_{hol}(\theta_\psi \mathcal{E}_\psi) = \sum_{n=1}^{\infty} \sigma_\psi^{\mathrm{sm}}(n) q^n - \sum_{m > r \geq 1} \psi(mr)(m-r) q^{m^2 - r^2},$$

which is easily seen to be identically zero. Furthermore, we have that

$$\xi_2(\theta_\psi \mathcal{E}_\psi)(\tau) = C y^{\lambda_\psi + 1/2} |\theta_\psi(\tau)|^2$$

for some constant C is clearly modular of weight 0 for the group $\Gamma_0(4f_\psi^2)$ with Nebentypus $\chi_{-4}^{\lambda_\psi}$. However, Proposition 2.4 does not immediately apply since the required growth conditions are not satisfied. However, by writing the generating function $\sum_{n=1}^{\infty} \sigma_\psi^{\mathrm{sm}}(n) q^n$ in terms of derivatives of Appell-Lerch sums (cf. the third remark following Theorem 1.1 and [28, Chap. IV]), and using the known modular properties of these functions (see for instance [39]), one can compute the asymptotic of this function as τ approaches any rational number. We find that there is a linear combination $F(\tau)$ of (modular) Eisenstein series for $\Gamma_0(4f_\psi^2)$ of weight 2 as well as the quasimodular Eisenstein series E_2, such that the function

$$\sum_{n=1}^{\infty} \sigma_{\psi}^{\mathrm{sm}}(n)q^n + \theta_{\psi}(\tau)\mathcal{E}^-(\tau) + F(\tau) + \alpha_{\psi}E_2(\tau)$$

behaves like a cusp form when τ approaches a rational number. We leave the details to the reader. Therefore, this function satisfies the hypotheses of Proposition 2.4. Adding modular Eisenstein series does not affect modularity, so we may deduce that

$$\sum_{n=1}^{\infty} \sigma_{\psi}^{\mathrm{sm}}(n)q^n + \theta_{\psi}(\tau)\mathcal{E}^-(\tau) + \alpha_{\psi}E_2(\tau)$$

is modular. This concludes the proof. □

Remark 2.5. A very similar complication for holomorphic projection in weight 2 is dealt with in an analogous way in the seminal work of Gross and Zagier [19, Proposition 6.2], and has continued to arise in subsequent papers involving holomorphic projection. It would be very interesting to find an exact (or more conceptual) formula for α_{ψ} in all of these situations, rather than having to rely on the solvability of a system of linear equations in modular forms.

Remark 2.6. As an alternative to the proof given above, one might also define a modified regularized holomorphic projection operator for translation invariant functions by the constant term in the Laurent expansion around $s = 0$ of

$$\frac{(\kappa-1)(2i)^{\kappa+2s}}{4\pi} \lim_{T\to\infty} \int_{\mathbb{H}_T} \frac{f(x+iy)y^{k+2s}}{(\tau - x + iy)^{\kappa+2s}} \frac{dxdy}{y^2}.$$

This however does not in general produce holomorphic functions, but might add powers of $\Im(\tau)$, just like when one applies Hecke's trick to find the transformation properties of E_2. Other than that, Lemma 2.2 essentially still holds. The power of $\Im(\tau)$ introduced by the regularization may be replaced by a multiple of E_2, again verifying the claim.

Remark 2.7. The argument that enables us to use holomorphic projection in the above setup using Appell–Lerch sums would also suffice to establish modularity, without relying on holomorphic projection at all.

It seems conceivable that one can derive asymptotics for Appell–Lerch sums near cusps more directly. For example, a straightforward modification of [11, Lemma 4.3] shows that derivatives of Appell–Lerch sums have the right order of magnitude approaching cusps of $\Gamma_0(4N)$, so a more careful analysis should yield the correct asymptotic expansion.

3. Proof of Theorem 1.5 and Corollary 1.6

The proof of Theorem 1.5 follows directly from the characterized list of theta functions which are also eta quotients in [24], as per the discussion preceding the theorem.

Proof of Corollary 1.6. First, note that in part (i), the character χ_{24} is $\left(\frac{2}{n}\right)$ times the first character χ_{12}. Thus, switching from the first example to the second twists the theta function in the denominator of (1.4) by $\left(\frac{2}{n}\right)$, and a similar check can be performed on the character in the sum in (1.3). Thus, it is readily checkable that the extra twist by $(-1)^n$ occurs in this case.

It remains to check that the *spt* and *cpt* functions arise out of these constructions. For both, the mock modularity properties are already known, at least implicitly in the literature. For instance, the mock modularity of the generating function of *spt*(n) is described in [17], and our Theorem 1.5 establishes the mock modularity of our series $\mathcal{E}^+_{\chi_{12}}(\tau)$. Similarly, for *cpt*$(n)$, the generating function can be shown elementary means to have an Appell–Lerch-type generating function

$$\sum_{n\geq 1} cpt(n)q^n = \sum_{n\geq 1} \frac{nq^{\frac{n(n+1)}{2}}}{1-q^n}$$

(see the proof of Joerg Arndt in the comments on the OEIS sequence A204217). This is readily expressible as a theta function times a Taylor coefficient in an elliptic variable of Zwegers' μ-function [38], which by the general theory there is a weight 3/2 mock modular form. Normalizing by dividing by the theta function, the identity for *cpt*(n) is thus reduced to checking that two weight 3/2 mock modular forms in known spaces are equal. Thus, to prove the claims, one first needs to check that they live in the same space of mock modular forms, that is, have the same weight, level, Nebentypus, shadow, and that their principal parts at the cusps agree.

Let us focus on the case of *spt*, it works in much the same way for *cpt*. It has been established in [7] that

$$F(\tau) := q^{-1}\sum_{n=1}^{\infty} spt(n)q^{24n} - \frac{1}{12}\frac{E_2(24\tau)}{\eta(24\tau)} - \frac{i}{4\pi\sqrt{2}}\int_{-\bar{\tau}}^{i\infty} \frac{\eta(24z)}{(i(z+\tau))^{3/2}}dz$$

is a harmonic Maaß form for $\Gamma_0(576)$ with Nebentypus $\chi_{12} = \left(\frac{12}{\bullet}\right)$. By direct comparison, using the famous fact that $\theta_{\chi_{12}}(\tau) = \eta(24\tau)$, we therefore see from Theorem 1.5 that $\mathcal{E}_{\chi_{12}} - 2F$ is a weakly holomorphic modular form of weight 3/2 for $\Gamma_0(576)$ with Nebentypus χ_{12}. After multiplying it by a suitable cusp form (as it turns out $\eta(24\tau)$ suffices), we are left with checking that a holomorphic modular form is zero, which can be done checking only finitely many coefficients. $\square$

4. Proof of Theorem 1.8

We now turn to the behavior of the generating functions modulo powers of primes.

Proof of Theorem 1.8. As remarked after the statement of Theorem 1.1, there is a weakly holomorphic modular form $G_\psi \in M_2^!(\Gamma_0(4f_\psi^2))$ such that $\mathcal{E} - G_\psi/\theta_\psi$ has no poles in the upper half-plane, so we may and will assume this from now on. By Proposition 2.1, we find that

$$\pi_{hol}^{reg}(\theta_\psi(p^{2a}\tau)\mathcal{E}_\psi(\tau)) = \theta_\psi(p^{2a}\tau)\mathcal{E}_\psi^+(\tau) + \sum_{r=1}^{\infty}\left(\sum_{\substack{(p^a m)^2-n^2=r\\ m,n\geq 0}} \psi(mn)(p^a m-n)\right) q^r \tag{4.1}$$

is a weakly holomorphic quasimodular form. This is congruent to a weakly holomorphic modular form $F \in M_2^!(\Gamma_0(4f_\psi^2 p^{2a}))$ because the Eisenstein series E_2 is congruent to the holomorphic modular form $E_2(\tau) - NE_2(N\tau) \in M_2(\Gamma_0(N))$ for any $N > 1$.

Since $(p^a m)^2 - n^2 = (p^a m - n)(p^a m + n)$ factors, we can rewrite the inner sum in the second term in (4.1) as a sum over small divisors of r,

$$\sum_{\substack{(p^a m)^2-n^2=r\\ m,n\geq 0}} \chi(mn)(p^a m-n) = \sum_{\substack{d|r, d\leq r/d\\ d+r/d\equiv 0 \pmod{2p^a}\\ d-r/d\equiv 0 \pmod 2}} \chi\left(\frac{(r/d)^2-d^2}{4p^a}\right) d, \tag{4.2}$$

as in the proof of Theorem 1.1. We now apply the operator $U(p^b)$ to (4.1). This corresponds to replacing r by $p^b r$ in (4.2), and then the condition $d + p^b r/d \equiv 0 \pmod{2p^a}$ forces $d \equiv 0 \pmod{p^{\min(a,b)}}$, so that we find that

$$\pi_{hol}^{reg}(\theta_\psi(p^{2a}\tau)\mathcal{E}_\psi(\tau)) \,|\, U(p^b) \equiv (\theta_\psi(p^{2a}\tau)\mathcal{E}_\psi^+(\tau)) \,|\, U(p^b) \equiv F \,|\, U(p^b) \pmod{p^{\min(a,b)}}.$$

By [5, Lemmas 6 and 7], $F \,|\, U(p^b)$ is modular for $\Gamma_0(4f_\psi^2 p^m)$ for $m = \max(2a - b, 1 - 2v_p(f_\psi))$, as stated in the Theorem. This completes the proof. □

Acknowledgments

The authors thank Scott Ahlgren for noting an error in an earlier version of this paper. We also thank the anonymous referee for various suggestions that helped improving the paper. Ken Ono thanks the support of the NSF and the Asa Griggs Candler Fund and the Thomas Jefferson Fund.

References

[1] S. Ahlgren and N. Andersen, Euler-like recurrences for smallest parts functions, *Ramanujan J.* **36**(1–2) (2015) 237–248.

[2] S. Ahlgren and B. Kim, Congruences for a mock modular form on and the smallest parts function, *J. Number Theory* **189** (2018) 81–89.

[3] G. E. Andrews, The number of smallest parts in the partitions of n, *J. Reine Angew. Math.* **624** (2008) 133–142.

[4] G. E. Andrews, R. C. Rhoades and S. P. Zwegers, Modularity of the concave composition generating function, *Algebra Number Theory* **7**(9) (2013) 2103–2139.

[5] A. O. L. Atkin and J. Lehner, Hecke operators on $\Gamma_0(m)$, *Math. Ann.* **185** (1970) 134–160.

[6] E. Belmont, H. Lee, A. Musat and S. Trebat-Leder, ℓ-adic properties of partition functions, *Monatsh. Math.* **173**(1) (2014) 1–34.

[7] K. Bringmann, On the explicit construction of higher deformations of partition statistics, *Duke Math J.* **144** (2007) 195–233.

[8] K. Bringmann, A. Folsom, K. Ono and L. Rolen, *Harmonic Maass Forms and Mock Modular Forms: Theory and Applications* (American Mathematical Society, Providence, 2018).

[9] K. Bringmann, B. Kane and S. Zwegers, On a completed generating function of locally harmonic Maass forms, *Compos. Math.* **150**(5) (2014) 749–762.

[10] K. Bringmann and K. Ono, Dyson's ranks and Maass forms, *Ann. Math.* **171** (2010) 419–449.

[11] K. Bringmann and L. Rolen, Radial limits of mock theta functions, *Res. Math. Sci.* **1** (2015), Article ID: 17, 18 pp.

[12] M. Cheng, $K3$ surfaces, $N = 4$ Dyons, and the Mathieu group M_{24}, *Commun. Number Theory Phys.* **4** (2010) 623–657.

[13] A. Dabholkar, S. Murthy and D. Zagier, Quantum Black Holes, Wall Crossing, and Mock Modular Forms, to appear in Cambridge Monographs in Mathematical Physics, 151 p.

[14] T. Eguchi, H. Ooguri and Y. Tachikawa, Notes on the $K3$ surface and the Mathieu group M_{24}, *Exp. Math.* **20**(1) (2011) 91–96.

[15] T. Eguchi, H. Ooguri, S.-K. Yang and A. Taormina, Superconformal algebras and string compactification on manifolds with $SU(N)$ Holonomy, *Nucl. Phys. B* **315** (1989) 193.

[16] T. Eguchi and A. Taormina, On the unitary representation of $N = 2$ and $N = 4$ superconformal algebras, *Phys. Lett. B* **210** (1988) 125.

[17] A. Folsom and K. Ono, The *spt*-function of Andrews, *Proc. Natl. Acad. Sci. USA* **105**(51) (2008) 20152–20156.

[18] F. Garvan, Congruences for Andrews' *spt*-function modulo powers of 5, 7, and 13, *Trans. Amer. Math. Soc.* **364**(9) (2012) 4847–4873.

[19] B. H. Gross and D. B. Zagier, Heegner points and derivatives of L-series, *Invent. Math.* **84** (1986) 225–320.

[20] E. Hecke, Theorie der Eisensteinschen Reihen höherer Stufe und ihre Anwendung auf Funktionentheorie und Arithmetik, *Abh. Math. Sem. Hamburg. Univ.* **5** (1927) 199–224; E. Hecke, *Mathematische Werke* (Vandenhoek & Ruprecht Göttingen, 1959), pp. 461–486.

[21] M. D. Hirschhorn and P. M. Hirschhorn, Partitions into consecutive parts, *Math. Mag.* **78**(5) (2005) 396–397.

[22] Ö. Imamoḡlu, M. Raum and O. Richter, Holomorphic projections and Ramanujan's mock theta functions, *Proc. Natl. Acad. Sci. USA* **111**(11) (2014) 3961–3967.

[23] S.-Y. Kang, Mock Jacobi forms in basic hypergeometric series, *Compos. Math.* **145**(3) (2009) 553–565.

[24] R. Lemke Oliver, Eta-quotients and theta functions, *Adv. Math.* **241** (2013) 1–17.

[25] A. Malmendier and K. Ono, SO(3)-Donaldson invariants of $\mathbb{C}P^2$ and mock theta functions, *Geom. Topol.* **16** (2012) 1767–1833.

[26] G. Mersmann, Holomorphe η-produkte und nichtverschwindende ganze modulformen für $\Gamma_0(n)$, Diploma Thesis, University of Bonn (1991).

[27] M. H. Mertens, Mock modular forms and class number relations, *Res. Math. Sci.* **1** (2014), Article ID: 6, 16 pp.

[28] M. H. Mertens, Mock modular forms and class numbers of quadratic forms, Ph.D. thesis, Universität zu Köln (2014).

[29] M. H. Mertens, Eichler–Selberg type identities for mixed mock modular forms, *Adv. Math.* **301** (2016) 359–382.
[30] M. H. Mertens and K. Ono, Special values of shifted convolution Dirichlet series, *Mathematika* **62**(1) (2016) 47–66.
[31] K. Ono, Congruences for the Andrews *spt* function, *Proc. Natl. Acad. Sci. USA* **108**(2) (2011) 473–476.
[32] J. Rouse and J. J. Webb, Spaces of modular forms spanned by eta-quotients, *Adv. Math.* **272** (2015) 200–224.
[33] J.-P. Serre, Formes modulaires et fonctions zêta p-adiques, (French) *Modular Functions of One Variable*, III, Springer Lecture Notes in Mathematics, Vol. 350 (Proceedings International Summer School, University Antwerp, 1972, 1973), pp. 191–268.
[34] G. Shimura, On modular forms of half-integral weight, *Ann. Math.* **97** (1973) 440–481.
[35] J. Sturm, Projections of C^∞ automorphic forms, *Bull. Amer. Math. Soc.* (*N.S.*) **2**(3) (1980) 435–439.
[36] J. J. Sylvester and F. Franklin, A constructive theory of partitions, arranged in three acts, an interact and an exodion, *Amer J. Math.* **5** (1882) 251–330.
[37] D. Zagier, Ramanujan's mock theta functions and their applications [d'après Zwegers and Bringmann-Ono], *Séminaire Bourbaki, 60ème année, 2007–2008, no. 986, Astérisque*, Vol. 326 (Societe Mathematique de France, 2009), pp. 143–164.
[38] S. Zwegers, Mock Theta Functions, Ph.D. thesis, Universiteit Utrecht (2002).
[39] S. Zwegers, Multivariable Appell functions and nonholomorphic Jacobi forms, *Res. Math. Sci.* **6** (2019), Article ID: 16, 15 pp.

© 2024 World Scientific Publishing Company
https://doi.org/10.1142/9789811277375_0033

The largest size of an $(s, s+1)$-core partition with parts of the same parity*

Hayan Nam
Department of Mathematics
Duksung Women's University
Seoul 01370, Republic of Korea
hnam@duksung.ac.kr

Myungjun Yu†
Center for Mathematical Challenges
Korea Institute for Advanced Study
Seoul 02455, Republic of Korea
mjyu.math@gmail.com

Finding the largest size of a partition under certain restrictions has been an interesting subject to study. For example, it is proved by Olsson and Stanton that for two coprime integers s and t, the largest size of an (s,t)-core partition is $(s^2-1)(t^2-1)/24$. Xiong found a formula for the largest size of a $(t, mt+1)$-core partitions with distinct parts. In this paper, we find an explicit formula for the largest size of an $(s, s+1)$-core partition such that all parts are odd (or even).

Keywords: Core partitions; partitions; order ideal; largest size.

Mathematics Subject Classification 2020: 11P81, 06A07

1. Introduction

Let $\mathbb{N}$ denote the set of nonnegative integers. Let $\lambda = (\lambda_\ell, \lambda_{\ell-1}, \ldots, \lambda_1)$ be a partition such that $0 < \lambda_\ell \le \lambda_{\ell-1} \le \cdots \le \lambda_1$. We define the *size* of a partition λ, denoted by $|\lambda|$, to be $\sum_{i=1}^{\ell} \lambda_i$. The *Ferrers diagram* of a partition λ is a collection of squares whose ith row consists of λ_i squares. Each square in a Ferrers diagram can be assigned by the unique number called the *hook length*, which is the sum of the number of squares on its right, the number of squares below it, and 1 for itself. The *beta set* β_λ of λ is defined by the set of hook lengths of the squares on the first

*This article originally appeared in *International Journal of Number Theory*, **17**(3), pp. 699–712 © 2021 World Scientific Publishing Company. DOI: 10.1142/S179304212040014X
†Corresponding author.

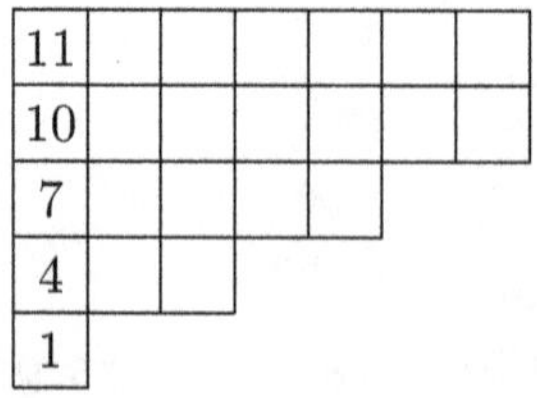

Fig. 1. The Ferrers diagram of $\lambda = (1, 3, 5, 7, 7)$.

column of the Ferrers diagram of λ. Then by definition of the beta set, we have

$$\beta_\lambda = \{\lambda_\ell, \lambda_{\ell-1} + 1, \ldots, \lambda_{\ell-i} + i, \ldots, \lambda_1 + \ell - 1\}.$$

For example, Fig. 1 shows the Ferrers diagram of a partition $\lambda = (1, 3, 5, 7, 7)$ and its beta set $\beta_\lambda = \{1, 4, 7, 10, 11\}$.

For a positive number t, we say that a partition λ is a t-core partition (or simply t-core) if none of the hook lengths of λ are divisible by t. We say λ is $(a_1, a_2, \ldots, a_n)$-core partition if it is simultaneously a_i-core for all $1 \leq i \leq n$.

The enumeration of simultaneous core partitions has been an interesting topic to mathematicians. For examples, Straub [8] computed the number of $(s, ds - 1)$-core partitions into distinct parts. Kravitz [6] found a formula for the number of $(s, s + r)$-core partitions with d-distinct parts. Xiong [9] obtained the recurrence relation for the number of simultaneous $(t, t + 1, \ldots, t + p)$-cores. Baek and the authors [3] found an expression for the number of $(a_1, a_2, \ldots, a_m)$-core partitions.

On the other hand, people have been also studying the average size and the largest size of a simultaneous core partition under various conditions. Let s and t be coprime positive integers. Johnson [4] proved Armstrong's conjecture which asserts that the average size of a simultaneous (s, t)-core partition is $(s + t + 1)$ $(s - 1)(t - 1)/24$. Olsson and Stanton [7] found that the largest size of an (s, t)-core partition is $(s^2 - 1)(t^2 - 1)/24$. Xiong [10] evaluated the largest size of a $(t, mt + 1)$-core partitions with distinct parts. Yang *et al.* [11] proved a conjecture on the largest size of an $(s, s + 1, s + 2)$ core partition. Baek and the authors [3] computed the largest size of a self-conjugate $(s, s + 1, s + 2)$-core partition.

In this paper, we find a formula for the largest size of an $(s, s + 1)$-core partition such that all parts are odd (or even) as follows.

Theorem 1.1. *We have*

$$\max(\{|\lambda| : \lambda \textit{ is an } (s, s+1)\textit{-core and all its parts are odd}\})$$
$$= \begin{cases} \dfrac{1}{96}(s^4 + 4s^3 + 26s^2 + 44s - 75) & \textit{if } s \textit{ is odd,} \\[2ex] \dfrac{1}{96}(s^4 + 4s^3 + 20s^2 + 32s - 96) & \textit{if } s \textit{ is even.} \end{cases}$$

$$\max(\{|\lambda| : \lambda \textit{ is an } (s, s+1)\textit{-core and all its parts are even}\})$$
$$= \begin{cases} \dfrac{1}{96}(s^4 + 4s^3 + 2s^2 - 4s - 3) & \textit{if } s \textit{ is odd,} \\ \dfrac{1}{96}(s^4 + 4s^3 - 4s^2 - 16s) & \textit{if } s \textit{ is even.} \end{cases}$$

The main ingredient of the proof of the above theorem is a bijection between core partitions and the order ideals given by their beta sets. Using this bijection, it boils down to finding an order ideal with the largest possible "size" in the sense of Definition 2.11 (see Remark 2.12). This method has been already used in the literature such as [1, 2, 8, 11].

In Sec. 2, we review the bijection between core partitions and order ideals. Also, we define various objects to be used in the proof of Theorem 1.1. Via the bijection in Sec. 2, we may think of the order ideal corresponding to the $(s, s+1)$-core partition with parts of the same parity having the largest possible size. In Sec. 3, this order ideal is shown to satisfy several properties (Propositions 3.4, 3.5 and 3.10) that determine it uniquely with a precise description.

2. A Bijection Between Core Partitions and Order Ideals

Throughout this paper, we fix an integer $s > 1$.

Definition 2.1. Let $\mathcal{C}_{s,s+1}$ be the set of $(s, s+1)$-core partitions. Define

$$\mathcal{C}^{\mathrm{O}}_{s,s+1} := \{\lambda \in \mathcal{C}_{s,s+1} : \text{all parts of } \lambda \text{ are odd}\},$$
$$\mathcal{C}^{\mathrm{E}}_{s,s+1} := \{\lambda \in \mathcal{C}_{s,s+1} : \text{all parts of } \lambda \text{ are even}\}.$$

Remark 2.2. Johnson [5] recently studied the cardinalities of $\mathcal{C}^{\mathrm{O}}_{s,s+1}$ and $\mathcal{C}^{\mathrm{E}}_{s,s+1}$.

Definition 2.3. Let

$$\mathbb{P}_{s,s+1} = \mathbb{N} \backslash \{n \in \mathbb{N} : n = as + b(s+1) \text{ for some } a, b \in \mathbb{N}\}$$

be a poset whose partial order $\preceq$ is given by requiring that $x \in \mathbb{P}_{s,s+1}$ covers $y \in \mathbb{P}_{s,s+1}$ if and only if $x - y$ is s or $s+1$. An *order ideal* I of $\mathbb{P}_{s,s+1}$ is a subset of $\mathbb{P}_{s,s+1}$ such that $x \preceq y$ and $y \in I$ imply $x \in I$. Let $\mathcal{O}_{s,s+1}$ denote the set of order ideals of $\mathbb{P}_{s,s+1}$.

Remark 2.4. We may visualize $\mathbb{P}_{s,s+1}$ by means of a diagram (often referred to as the Hasse diagram) depicted in Fig. 2. In this figure, note that the partial order $\preceq$ is defined in a way that an element covers two elements right below (SW and SE directions) it. It is well known that $\max(\mathbb{P}_{s,s+1}) = s^2 - s - 1$: more generally, the *Frobenius number* of two coprime positive integers m and n (i.e. the largest number which cannot be represented as a linear combination of m and n with nonnegative integer coefficients) is $mn - m - n$.

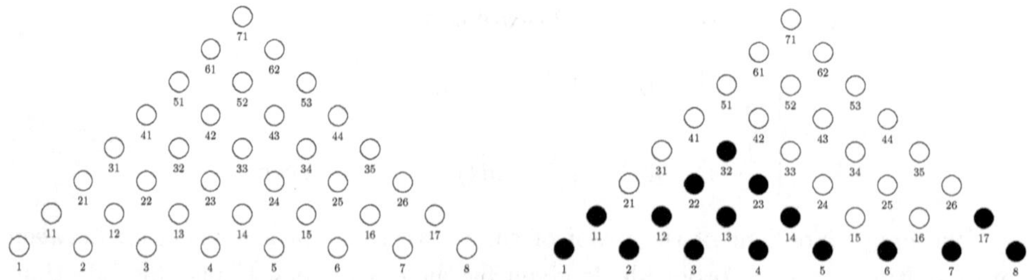

Fig. 2. The poset $\mathbb{P}_{9,10}$ and an order ideal whose elements are colored black.

Recall that a partition $\lambda = (\lambda_\ell, \lambda_{\ell-1}, \ldots, \lambda_1)$ such that $\lambda_\ell \le \lambda_{\ell-1} \le \cdots \le \lambda_1$ has its beta set $\beta_\lambda = \{\lambda_\ell, \lambda_{\ell-1}+1, \ldots, \lambda_{\ell-i}+i, \ldots, \lambda_1+\ell-1\}$.

Lemma 2.5. *There is a bijection*

$$\varphi: \mathcal{C}_{s,s+1} \xrightarrow{\sim} \mathcal{O}_{s,s+1}$$

sending a partition λ to its beta set β_λ.

Proof. See [1]. □

Definition 2.6. For $1 \le i \le s-1$, define

$$R_i := \{n \in \mathbb{N} : 1 + (s+1)(i-1) \le n \le s - 1 + s(i-1)\},$$

which is the set of integers on the ith layer of $\mathbb{P}_{s,s+1}$. Let

$$l_i := s - 1 + s(i-1)$$

denote the largest integer in R_i.

Remark 2.7. We give a justification of why the set R_i is equal to the set of integers on the ith layer of $\mathbb{P}_{s,s+1}$. Recall that $\max(\mathbb{P}_{s,s+1}) = s^2 - s - 1$. Note that

$$\mathbb{P}_{s,s+1} = \{1 \le n \le s^2 - s - 1 : n \ne as + c \text{ for } a \ge 1 \text{ and } 0 \le c \le a\}. \quad (2.1)$$

The first layer of $\mathbb{P}_{s,s+1}$ consists of the elements that cover no element in $\mathbb{P}_{s,s+1}$. Therefore, $R_1 = \{1, 2, \ldots, s-1\}$ is equal to the set of integers on the first layer of $\mathbb{P}_{s,s+1}$. Let z_i be an element in the set of integers on the ith layer of $\mathbb{P}_{s,s+1}$. Then there is a sequence $z_1, z_2, \ldots, z_i$ such that

- $z_1 \in R_1$ and
- z_{j+1} covers z_j for $1 \le j \le i-1$.

Then one can show inductively that R_i is indeed the set of integers on the ith layer of $\mathbb{P}_{s,s+1}$ by using (2.1) and the definition of the partial order of $\mathbb{P}_{s,s+1}$: x covers y if and only if $x - y = s$ or $s+1$.

Definition 2.8. Let $A = \{a_1, a_2, \ldots, a_n\}$ be a finite set of integers such that $a_i < a_{i+1}$ for all $1 \leq i \leq n-1$. We say A is *parity-alternating* if $a_{i+1} - a_i$ is odd for all $1 \leq i \leq n-1$. We say A is an OE set if

- A is parity-alternating and
- $a_1 = \min(A)$ is odd.

Similarly, we say A is an EO set if

- A is parity-alternating and
- $a_1 = \min(A)$ is even.

Here, O and E stand for odd and even, respectively. If $I \in \mathcal{O}_{s,s+1}$, we call I an OE order ideal (respectively, EO order ideal) if I is an OE set (respectively, EO set).

Definition 2.9. We define

$$\mathcal{O}^{\mathrm{OE}}_{s,s+1} := \{I \in \mathcal{O}_{s,s+1} : I \text{ is an OE set}\},$$

$$\mathcal{O}^{\mathrm{EO}}_{s,s+1} := \{I \in \mathcal{O}_{s,s+1} : I \text{ is an EO set}\}.$$

Recall the map $\varphi : \mathcal{C}_{s,s+1} \xrightarrow{\sim} \mathcal{O}_{s,s+1}$ in Lemma 2.5.

Lemma 2.10. *The restrictions of the map φ yield the following two bijections:*

$$\mathcal{C}^{\mathrm{O}}_{s,s+1} \xrightarrow{\sim} \mathcal{O}^{\mathrm{OE}}_{s,s+1},$$

$$\mathcal{C}^{\mathrm{E}}_{s,s+1} \xrightarrow{\sim} \mathcal{O}^{\mathrm{EO}}_{s,s+1}.$$

Proof. Recall that φ sends $\lambda = (\lambda_\ell, \lambda_{\ell-1}, \ldots, \lambda_1)$ to $\beta_\lambda = \{\lambda_\ell, \lambda_{\ell-1}+1, \ldots, \lambda_1 + \ell - 1\}$. Note that all λ_i's have the same parity if and only if β_λ is parity-alternating. Then the lemma as follows. □

Definition 2.11. Let I be an order ideal of $\mathbb{P}_{s,s+1}$. Define a *size* of I by

$$\mathfrak{s}(I) := \sum_{n \in I} n - \sum_{j=0}^{|I|-1} j.$$

Remark 2.12. In fact, $\mathfrak{s}(I)$ is the same as the size of the corresponding partition of I via the bijection in Lemma 2.5.

Definition 2.13. Let $I \in \mathcal{O}_{s,s+1}$. Define $I_i \subseteq I$ by

$$I_i := I \cap R_i.$$

In other words, $I_i \subseteq I$ is the set of integers on the ith layer of I. By definition, $I = \cup_{i \geq 1} I_i$.

For example, if $s = 9$ and I is as in Fig. 2, then $I_2 = \{11, 12, 13, 14, 17\}$ and $I_3 = \{22, 23\}$.

3. $(s, s+1)$-Core Partitions with Parts of the Same Parity

Let I and J be order ideals of $\mathbb{P}_{s,s+1}$.

Definition 3.1. We say J is OE-largest if

- $J \in \mathcal{O}^{\mathrm{OE}}_{s,s+1}$ and
- $\mathfrak{s}(J) = \max(\{\mathfrak{s}(I) : I \in \mathcal{O}^{\mathrm{OE}}_{s,s+1}\})$.

Similarly, we say J is EO-largest if

- $J \in \mathcal{O}^{\mathrm{EO}}_{s,s+1}$ and
- $\mathfrak{s}(J) = \max(\{\mathfrak{s}(I) : I \in \mathcal{O}^{\mathrm{EO}}_{s,s+1}\})$.

Remark 3.2. To prove our main theorem (Corollary 3.11), we only need to compute $\mathfrak{s}(J)$ when J is EO-largest or OE-largest by Lemma 2.10 and Remark 2.12. In Propositions 3.4, 3.5 and 3.10, we prove that an EO-largest order ideal and an OE-largest order ideal satisfy certain properties that actually determine them uniquely.

Definition 3.3. For an integer $i \geq 1$, we say I_i is *connected* if either $I_i = \emptyset$ or I_i consists of consecutive integers. We say I is *connected* if I_i is connected for every $i \geq 1$.

For example, if I is an order ideal in Fig. 2, then I_1, I_3, I_4 are connected whereas I_2 is not.

Proposition 3.4. *If J is* OE*-largest or* EO*-largest, then J is connected.*

Proof. Suppose by contradiction that k is the smallest positive integer such that J_k is not connected. Then for any $i < k$, we have J_i is connected. Also J_k can be expressed uniquely as a disjoint union of "connected components" $K_1, K_2, \ldots, K_a$ $(a \geq 2)$ such that

(1) Each K_i is nonempty and K_i consists of consecutive integers and
(2) $\min(K_{i+1}) - \max(K_i) > 1$ for all $1 \leq i \leq a-1$.

Define the "closure" of K_1 in J by

$$\overline{K_1} := K_1 \cup \left\{ y \in \bigcup_{i>k} J_i : y \succeq x \text{ for some } x \in K_1 \right\} \subset J.$$

Let $\min(K_2) - \max(K_1) = m > 1$. Note that for any c such that $\max(K_1) < c < \min(K_2)$, we have $c \notin J$. Since J is parity-alternating, the parity of $\min(K_2)$ and $\max(K_1)$ are different, so m is odd $(m \geq 3)$. Now let $J = \{a_1, a_2, \ldots, a_t\}$ where $a_1 < a_2 < \cdots < a_t$. Define

$$a_i' := \begin{cases} a_i + m - 1 & \text{if } a_i \in \overline{K_1}, \\ a_i & \text{if } a_i \notin \overline{K_1}. \end{cases}$$

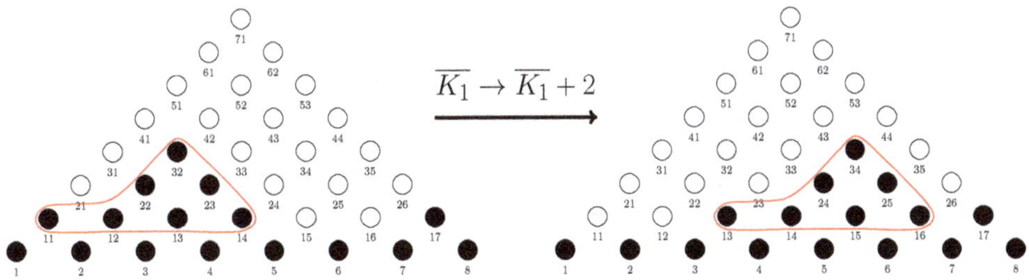

Fig. 3. A "connectification" process in the proof of Proposition 3.4.

Let $J' = \{a'_1, a'_2, \ldots, a'_t\}$. Then $a'_1 < a'_2 < \cdots < a'_t$, and $a_i \equiv a'_i \pmod 2$ for all i. Thus J' is parity-alternating. Examples of J and J' are illustrated in Fig. 3 on the left side and on the right side, respectively. We claim that J' is an order ideal. We need to show

$$\text{if } a'_i \notin J'_1 \quad \text{then } a'_i - (s+1), a'_i - s \in J'.$$

If $k = 1$ or $a_i \notin K_1$, this is clear by the fact that J is an order ideal. Suppose that $k \geq 2$ and $a_i \in K_1$. Then $a_i - (s+1) \in J_{k-1}$. Also, we have $\min(K_2) - s \in J_{k-1}$ and $\min(K_2) - a_i \geq m$. Note that

$$a_i - (s+1) < a'_i - (s+1) < a'_i - s < \min(K_2) - s.$$

Since J_{k-1} is connected, it follows that

$$a'_i - (s+1), a'_i - s \in J_{k-1} \subset J',$$

so J' is an order ideal. Moreover, we have $\min(J) = a_1 \equiv a'_1 = \min(J') \pmod 2$. It follows that

- $J \in \mathcal{O}^{\text{EO}}_{s,s+1} \Rightarrow J' \in \mathcal{O}^{\text{EO}}_{s,s+1}$ and
- $J \in \mathcal{O}^{\text{OE}}_{s,s+1} \Rightarrow J' \in \mathcal{O}^{\text{OE}}_{s,s+1}$.

Clearly $\mathfrak{s}(J') > \mathfrak{s}(J)$, which is a contradiction. □

Recall that $l_i = \max(R_i)$ (Definition 2.6).

Proposition 3.5. *Suppose that J is* OE*-largest or* EO*-largest. If $J_i \neq \emptyset$, then $l_i \in J_i$.*

Proof. By Proposition 3.4, we know J_i is connected for all $i \geq 1$. By contradiction, suppose that k is the smallest integer such that J_k is nonempty and $l_k \notin J_k$. Define J'_i so that

$$J'_i = \begin{cases} J_i & \text{if } i < k, \\ (J_k + 1) \cup \{\min(J_k)\} & \text{if } i = k, \\ J_i + 1 & \text{if } i > k, \end{cases}$$

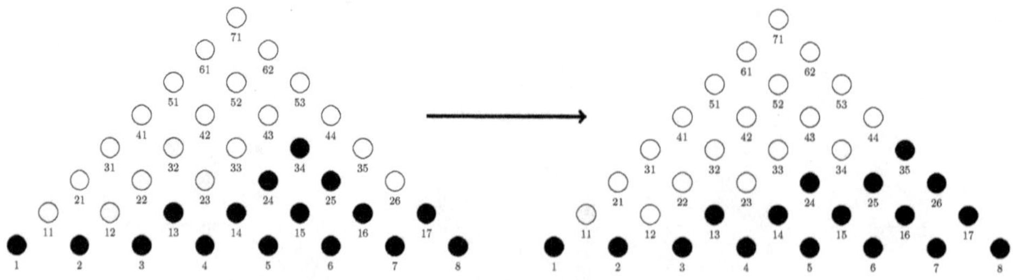

Fig. 4. A process in the proof of Proposition 3.5.

where $J_i + 1 := \{m+1 : m \in J_i\}$. One can check that $J' := \cup_{i \geq 1} J_i'$ is an order ideal by recalling the following facts:

- J is an order ideal.
- $l_{k-1} \in J_{k-1}$.
- J_{k-1} is connected.

See Fig. 4 for a visualization. One can see that $J \in \mathcal{O}^{\mathrm{OE}}_{s,s+1}$ (respectively, $J \in \mathcal{O}^{\mathrm{EO}}_{s,s+1}$) if and only if $J' \in \mathcal{O}^{\mathrm{OE}}_{s,s+1}$ (respectively, $J' \in \mathcal{O}^{\mathrm{EO}}_{s,s+1}$). By Definition 2.11, we have

$$\mathfrak{s}(J') - \mathfrak{s}(J) = \max(J) + 1 - |J| > 0,$$

which is a contradiction. □

Definition 3.6. Let I be an order ideal of $\mathbb{P}_{s,s+1}$. Define

$$I^i := \bigcup_{k=1}^{i} I_k.$$

Definition 3.7. Let $I \in \mathcal{O}^{\mathrm{OE}}_{s,s+1}$ (respectively, $I \in \mathcal{O}^{\mathrm{EO}}_{s,s+1}$). We say I is i-packed if there is no $J \in \mathcal{O}^{\mathrm{OE}}_{s,s+1}$ (respectively, $J \in \mathcal{O}^{\mathrm{EO}}_{s,s+1}$) such that $I^i \subsetneq J^i$.

Remark 3.8. Roughly speaking, I is i-packed if I^i contains "as many elements as possible". For example, if I is 1-packed, then $I^1 = I_1 = \{1, 2, \ldots, s-1\}$ or $\{2, 3, \ldots, s-1\}$ depending on whether $I \in \mathcal{O}^{\mathrm{OE}}_{s,s+1}$ or $I \in \mathcal{O}^{\mathrm{EO}}_{s,s+1}$. If I is 2-packed and $I \in \mathcal{O}^{\mathrm{OE}}_{s,s+1}$, then

$$I^2 = \{n \in \mathbb{N} : 1 \leq n \leq s-1\} \cup \{n \in \mathbb{N} : s+2 \leq n \leq 2s-1\}.$$

If I is 2-packed and $I \in \mathcal{O}^{\mathrm{EO}}_{s,s+1}$, then

$$I^2 = \{n \in \mathbb{N} : 2 \leq n \leq s-1\} \cup \{n \in \mathbb{N} : s+4 \leq n \leq 2s-1\} \quad \text{or}$$

$$I^2 = \{n \in \mathbb{N} : 2 \leq n \leq s-2\} \cup \{n \in \mathbb{N} : s+3 \leq n \leq 2s-2\}.$$

To show the latter statement, note that $2 \in I$ since I is 2-packed, and proceed with considering two different cases: $s+3 \notin I$ or $s+3 \in I$. See Fig. 5, for examples of 2-packed order ideals in $\mathcal{O}^{\mathrm{EO}}_{s,s+1}$.

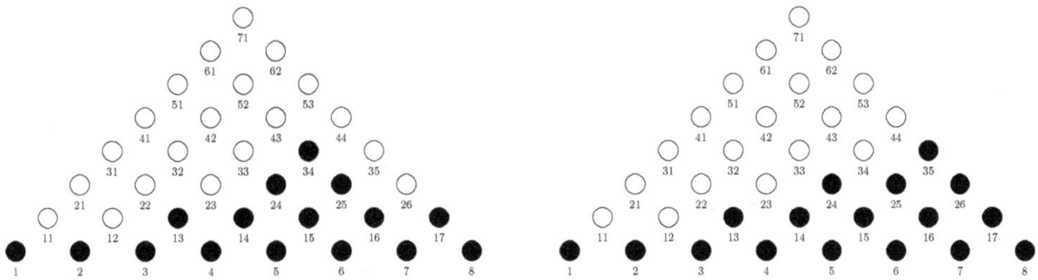

Fig. 5. Examples of 2-packed order ideals in $\mathcal{O}^{\mathrm{EO}}_{s,s+1}$.

Lemma 3.9. *Suppose that $I \in \mathcal{O}^{\mathrm{OE}}_{s,s+1}$ (respectively, $I \in \mathcal{O}^{\mathrm{EO}}_{s,s+1}$) is connected and $l_i \in I_i$ if $I_i \neq \emptyset$. Suppose that I is $(k-1)$-packed but not k-packed. Let*

$$A = \begin{cases} \{l_k\} & \text{if } I_k = \emptyset \text{ and } s \text{ is odd,} \\ \{l_k - 1\} & \text{if } I_k = \emptyset \text{ and } s \text{ is even,} \\ \{n \in \mathbb{N} : \min(I_k) - 2 \leq n \leq l_k\} & \text{if } I_k \neq \emptyset. \end{cases}$$

Then $I^{k-1} \cup A \in \mathcal{O}^{\mathrm{OE}}_{s,s+1}$ (respectively, $I^{k-1} \cup A \in \mathcal{O}^{\mathrm{EO}}_{s,s+1}$).

Proof. Assume that $I \in \mathcal{O}^{\mathrm{OE}}_{s,s+1}$ (in the case $I \in \mathcal{O}^{\mathrm{EO}}_{s,s+1}$, the proof is identical to what follows if $\mathcal{O}^{\mathrm{OE}}_{s,s+1}$ is replaced by $\mathcal{O}^{\mathrm{EO}}_{s,s+1}$). Since I is not k-packed, there exists an order ideal $J \in \mathcal{O}^{\mathrm{OE}}_{s,s+1}$ such that $I^k \subsetneq J^k$, so in particular, $I^{k-1} \subseteq J^{k-1}$. Since I is $(k-1)$-packed, this implies that $I^{k-1} = J^{k-1}$, so $I_k \subsetneq J_k$. By definition of order ideal, $\min(J_k) - (s+1) \in J_{k-1} = I_{k-1}$. By the hypotheses on I, we have

$$\{n \in \mathbb{N} : \min(J_k) - (s+1) \leq n \leq l_{k-1}\} \subseteq I_{k-1}. \tag{3.1}$$

Suppose first that $I_k = \emptyset$. Then J_k is nonempty. Clearly $\min(J_k) - l_{k-1} \equiv 1 \pmod 2$ by the parity-alternating condition of J. Then noting that $l_k - l_{k-1} = s$, we see that

- s is odd $\Rightarrow \min(J_k) \equiv l_k \pmod 2$ and $\min(J_k) \leq l_k$,
- s is even $\Rightarrow \min(J_k) \equiv l_k - 1 \pmod 2$ and $\min(J_k) \leq l_k - 1$.

Therefore, $I^{k-1} \cup A$ is parity-alternating. It follows from (3.1) that $I^{k-1} \cup A$ is an order ideal. This completes the proof when $I_k = \emptyset$.

Now suppose that $I_k \neq \emptyset$. By the parity-alternating condition of I and J, we have $\min(J_k) \equiv \min(I_k) \pmod 2$. Note also that $\max(J_k) \leq \max(I_k)$ because $l_i \in I_i$. Therefore, by $I_k \subsetneq J_k$, we have $\min(J_k) \leq \min(I_k) - 2$. Again it follows from (3.1) that $I^{k-1} \cup A$ is an order ideal and it is clear that $I^{k-1} \cup A$ is parity-alternating, so, we are done. □

Proposition 3.10. *Suppose that J is* OE*-largest or* EO*-largest. Then J is i-packed for all $i \geq 1$.*

Proof. We assume that J is OE-largest as it follows similarly in the EO-largest case. By Propositions 3.4 and 3.5, we know that J is connected and $l_i \in J_i$ if J_i is nonempty. Let us assume that k is the smallest integer such that J is not k-packed and we will derive a contradiction. Then J is $(k-1)$-packed.

Suppose first that $J_k = \emptyset$. Then $J = J^{k-1}$. Define

$$J'_i = \begin{cases} J_i & \text{if } i < k, \\ \{l_k\} & \text{if } i = k \text{ and } s \text{ is odd}, \\ \{l_k - 1\} & \text{if } i = k \text{ and } s \text{ is even}, \\ \emptyset & \text{if } i > k. \end{cases}$$

By Lemma 3.9,

$$J' = \bigcup_{i \geq 1}^{k} J'_i \in \mathcal{O}^{\mathrm{OE}}_{s,s+1}.$$

It follows from Definition 2.11 that

$$\mathfrak{s}(J') - \mathfrak{s}(J) = c - |J| > 0,$$

where $c = l_k$ or $c = l_k - 1$ depending on whether s is odd or even. This is a contradiction to the fact that J is OE-largest.

Now, we assume that $J_k \neq \emptyset$. Define

$$J''_i = \begin{cases} J_i & \text{if } i < k, \\ (J_k - 2) \cup \{l_k - 1\} & \text{if } i = k, \\ J_{i-1} + s - 1 & \text{if } i > k, \end{cases}$$

where $J_i + a = \{m + a : m \in J_i\}$ for any integer a.

We claim that

$$J'' := \bigcup_{i \geq 1} J''_i$$

is parity-alternating. Since J is parity-alternating, it follows that

$$\bigcup_{i<k} J_i \quad \text{and} \quad \bigcup_{i>k} (J_{i-1} + s - 1)$$

are also parity-alternating. Moreover $(J_k - 2) \cup \{l_k - 1\}$ is parity-alternating because it consists of consecutive integers. Therefore, to prove that J'' is parity-alternating, it remains to show that

$$\begin{aligned} \max(J_{k-1}) &\not\equiv \min(J_k - 2) \pmod 2 \quad \text{and} \\ l_k - 1 &\not\equiv \min(J_k + s - 1) \pmod 2. \end{aligned}$$

The former follows from the fact that J is parity-alternating. The latter follows from $\min(J_k) \not\equiv \max(J_{k-1}) = l_{k-1} \pmod 2$ and $l_k - l_{k-1} = s$ (recall Definition 2.6). Thus J'' is parity-alternating.

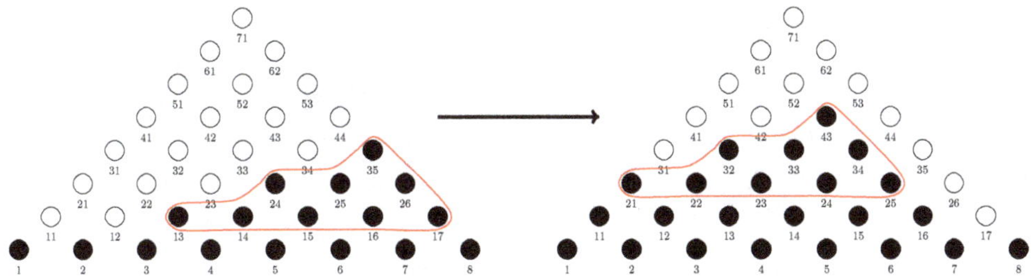

Fig. 6. A process in the proof of Proposition 3.10.

To see J'' is an order ideal, one need to show for $x \in J''$ that $x-(s+1)$ and $x-s$ (when positive) are in J''. This is actually proved for $x \in J''^k$ in Lemma 3.9. If $x \in J''_{k+1}$, then $x-(s+1), x-s \in (J_k - 2) \cup (J_k - 1)$, so they are in J''_k. If $x \in J''_i$ for $i > k+1$, then $x-(s-1) \in J_{i-1}$, so $x-(s-1)-(s+1), x-(s-1)-s \in J_{i-2}$ by the fact that J is an order ideal. Equivalently, $x-s, x-(s+1) \in J''_{i-1}$. Therefore, it follows that $J'' \in \mathcal{O}^{\mathrm{OE}}_{s,s+1}$. We depict the construction of J'' in Fig. 6. Now, we have by Definition 2.11 that

$$\mathfrak{s}(J'') - \mathfrak{s}(J) = \sum_{i \in J \setminus J^{k-1}} (s-1) + \sum_{i \in J''_k} i - \sum_{i=0}^{|J''_k|-1} (|J| + i). \tag{3.2}$$

Clearly, we have

$$\sum_{i \in J''_k} i > \sum_{i=0}^{|J''_k|-1} (|J^{k-1}| + i). \tag{3.3}$$

Combining (3.2) and (3.3), we obtain

$$\mathfrak{s}(J'') - \mathfrak{s}(J) > (s - 1 - |J''_k|)(|J| - |J^{k-1}|) \geq 0,$$

which is a contradiction. □

The following is merely a restatement of Theorem 1.1.

Corollary 3.11. *We have*

$$\max(\{|\lambda| : \lambda \in \mathcal{C}^{\mathrm{O}}_{s,s+1}\}) = \begin{cases} \dfrac{1}{96}(s^4 + 4s^3 + 26s^2 + 44s - 75) & \text{if } s \text{ is odd,} \\[2ex] \dfrac{1}{96}(s^4 + 4s^3 + 20s^2 + 32s - 96) & \text{if } s \text{ is even.} \end{cases}$$

$$\max(\{|\lambda| : \lambda \in \mathcal{C}^{\mathrm{E}}_{s,s+1}\}) = \begin{cases} \dfrac{1}{96}(s^4 + 4s^3 + 2s^2 - 4s - 3) & \text{if } s \text{ is odd,} \\[2ex] \dfrac{1}{96}(s^4 + 4s^3 - 4s^2 - 16s) & \text{if } s \text{ is even.} \end{cases}$$

Proof. For positive integers $a \leq b$, let

$$[a,b] := \{n \in \mathbb{N} : a \leq n \leq b\}.$$

Let $\mathcal{I}$ be OE-largest and $\mathcal{J}$ be EO-largest. Note that by Lemma 2.10 and Remark 2.12, we only need to compute $\mathfrak{s}(\mathcal{I})$ and $\mathfrak{s}(\mathcal{J})$. It follows from Propositions 3.4, 3.5 and 3.10 that

$$\mathcal{I}_i = \begin{cases} [1, s-1] & \text{if } i = 1, \\ [(i-1)(s+2), is-1] & \text{if } 2 \leq i \leq \lfloor (s+1)/2 \rfloor, \\ \emptyset & \text{if } i > \lfloor (s+1)/2 \rfloor. \end{cases}$$

We have

$$|\mathcal{I}| = s - 1 + \sum_{i=2}^{\lfloor (s+1)/2 \rfloor} (s - 2i + 2),$$

and

$$\mathfrak{s}(\mathcal{I}) = \sum_{i=1}^{s-1} i + \sum_{i=2}^{\lfloor (s+1)/2 \rfloor} ((2i-1)s + 2i - 3)(s - 2i + 2)/2 - \sum_{i=0}^{|\mathcal{I}|-1} i.$$

Then it is straightforward to show that

$$\mathfrak{s}(\mathcal{I}) = \begin{cases} \frac{1}{96}(s^4 + 4s^3 + 26s^2 + 44s - 75) & \text{if } s \text{ is odd}, \\ \frac{1}{96}(s^4 + 4s^3 + 20s^2 + 32s - 96) & \text{if } s \text{ is even}. \end{cases}$$

Now, we compute $\mathfrak{s}(\mathcal{J})$. By Propositions 3.4, 3.5 and 3.10, we get

$$\mathcal{J}_i = \begin{cases} [2 + (i-1)(s+2), is-1] & \text{if } 1 \leq i \leq \lfloor (s-1)/2 \rfloor, \\ \emptyset & \text{if } i > \lfloor (s-1)/2 \rfloor. \end{cases}$$

Then, we have

$$|\mathcal{J}| = \sum_{i=1}^{\lfloor (s-1)/2 \rfloor} (s - 2i),$$

and

$$\mathfrak{s}(\mathcal{J}) = \sum_{i=1}^{\lfloor (s-1)/2 \rfloor} \left(\sum_{k=is+2i-s}^{is-1} k \right) - \sum_{i=0}^{|\mathcal{J}|-1} i.$$

Again, it is straightforward to check that

$$\mathfrak{s}(\mathcal{J}) = \begin{cases} \frac{1}{96}(s^4 + 4s^3 + 2s^2 - 4s - 3) & \text{if } s \text{ is odd}, \\ \frac{1}{96}(s^4 + 4s^3 - 4s^2 - 16s) & \text{if } s \text{ is even}. \end{cases}$$

□

Remark 3.12. By using the map in Lemma 2.5, we can recover the $(s, s+1)$-core partition with only odd (even) parts that has the largest size. Let λ be a partition. For every positive integer k, we write $n_\lambda(k)$ for the number of parts of λ that are equal to k. Let θ (respectively, δ) be the largest $(s, s+1)$-core partition with only odd (respectively, even) parts. Then by the descriptions of $\mathcal{I}_i$ and $\mathcal{J}_i$ in Corollary 3.11 and the bijection in Lemma 2.5, we have

$$n_\theta(k) = \begin{cases} s-1 & \text{if } k = 1, \\ s-2i+2 & \text{if } k = 1+i(i-1) \text{ for } 2 \le i \le \lfloor (s+1)/2 \rfloor, \\ 0 & \text{otherwise,} \end{cases}$$

and

$$n_\delta(k) = \begin{cases} s-2i & \text{if } k = i(i+1) \text{ for } 1 \le i \le \lfloor (s-1)/2 \rfloor, \\ 0 & \text{otherwise.} \end{cases}$$

Remark 3.13. In general, one can ask for the largest size of an $(a_1, a_2, \ldots, a_n)$-core partition with parts of the same parity. In the case that $a_i = s+i-1$, we expect some modification of our proof would give a formula for the largest size of such a partition.

Acknowledgments

We are grateful to Armin Straub for throwing an interesting question that led to this project. We thank the anonymous referee for comments to improve the exposition of this paper. The second author was supported by a KIAS Individual Grant (SP075201) via the Center for Mathematical Challenges at Korea Institute for Advanced Study.

References

[1] J. Anderson, Partitions which are simultaneously t_1- and t_2-core, *Discrete Math.* **248**(1–3) (2002) 237–243.

[2] J. Baek, H. Nam and M. Yu, A bijective proof of Amdeberhan's conjecture on the number of $(s, s+2)$-core partitions with distinct parts, *Discrete Math.* **341**(5) (2018) 1294–1300.

[3] J. Baek, H. Nam and M. Yu, Johnson's bijections and their application to counting simultaneous core partitions, *Eur. J. Combin.* **75** (2019) 43–54.

[4] P. Johnson, Lattice points and simultaneous core partitions, *Electron. J. Combin.* **25**(3) (2018) 19.

[5] P. Johnson, Simultaneous cores with restrictions and a question of Zaleski and Zeilberger, preprint (2018), arXiv:1802.09621.

[6] N. Kravitz, On the number of simultaneous core partitions with d-distinct parts, *Discrete Math.* **342**(12) (2019) 111592.

[7] J. R. B. Olsson and D. Stanton, Block inclusions and cores of partitions, *Aequationes Math.* **74**(1–2) (2007) 90–110.

[8] A. Straub, Core partitions into distinct parts and an analog of Euler's theorem, *Eur. J. Combin.* **57** (2016) 40–49.

[9] H. Xiong, The number of simultaneous core partitions, preprint (2014), arXiv:1409.7038.
[10] H. Xiong On the largest sizes of certain simultaneous core partitions with distinct parts, *Eur. J. Combin.* **71** (2018) 33–42.
[11] J. Y. X. Yang, M. X. X. Zhong and R. D. P. Zhou, On the enumeration of $(s, s+1, s+2)$-core partitions, *Eur. J. Combin.* **49** (2015) 203–217.

OPEN ACCESS
© 2024 The Author(s)
https://doi.org/10.1142/9789811277375_0034

Holonomic relations for modular functions and forms: First guess, then prove*

Peter Paule[†] and Cristian-Silviu Radu[‡]

Research Institute for Symbolic Computation (RISC)
Johannes Kepler University, A-4040 Linz, Austria
[†]*Peter.Paule@risc.uni-linz.ac.at*
[‡]*Silviu.Radu@risc.uni-linz.ac.at*

Dedicated to Bruce Berndt at the occasion of his 80th birthday.

One major theme of this paper concerns the expansion of modular forms and functions in terms of fractional (Puiseux) series. This theme is connected with another major theme, holonomic functions and sequences. With particular attention to algorithmic aspects, we study various connections between these two worlds. Applications concern partition congruences, Fricke–Klein relations, irrationality proofs *a la* Beukers, or approximations to pi studied by Ramanujan and the Borweins. As a major ingredient to a "first guess, then prove" strategy, a new algorithm for proving differential equations for modular forms is introduced.

Keywords: Modular functions; modular forms; holonomic differential equations; holonomic functions and sequences; Fricke–Klein relations; q-series; partition congruences.

Mathematics Subject Classification 2020: 05A30, 11F03, 68W30, 11F33, 11P83

1. Introduction

The study of holonomic functions and sequences satisfying linear differential and difference equations, respectively, with polynomial coefficients has roots tracing back (at least) to the time of Gauß. More recently, such objects have become fundamental ingredients for the modern theory of enumerative combinatorics, see for example, the work of Stanley [35]. Concerning the promotion of the algorithmic relevance of these objects, a major role was played by Zeilberger's pioneering "holonomic systems approach" to special functions identities [41].

*This Open Access article originally appeared in *International Journal of Number Theory*, **17**(3), pp. 713–759 © 2021 The Author(s). DOI: 10.1142/S1793042120400278. It is distributed under the terms of the Creative Commons Attribution 4.0 (CC BY) License which permits use, distribution and reproduction in any medium, provided the original work is properly cited.

The ubiquitous nature of holonomic functions and sequences is documented by the numerous examples given in the literature, for instance, in the magnum opus [35].[1] However, when browsing through Ramanujan's Notebooks [8–12] and Lost Notebooks [3–7], one finds it surprisingly difficult to identify holonomic sequences or functions — apart from classical hypergeometric series having hypergeometric summands which, by definition, satisfy a linear recurrence, with polynomial coefficients, of order 1. To some extent this phenomenon can be explained as follows: Many of Ramanujan's entries are related to partition numbers $p(n)$ or variations of such. As a matter of fact, proved in Proposition 3.1, $(p(n))_{n\geq 0}$ is not holonomic — despite being one of the most prominent combinatorial and number theoretical sequences!

One objective of this paper is to cure Ramanujan's work from this "non-holonomy syndrome." This will be done by considering holonomic differential equations and algebraic relations, satisfied by modular functions and forms which are associated to the given non-holonomic sequences, respectively, functions. Our study is driven by algorithmic consideration; applications concern results related to Ramanujan's work, partition congruences, and special functions relations of Fricke–Klein type which, for instance, arise in approximations for π and in Beukers' modular form version of Apèry's proof of $\zeta(3) \notin \mathbb{Q}$. In all these examples, the computer algebra package `GeneratingFunctions`, written by Christian Mallinger in Mathematica [26][2], is playing a crucial role.[3]

Another major aspect of this paper concerns differential equations involving modular forms and functions. In general, any modular form satisfies a *nonlinear* third-order differential equation with constant coefficients, see, for instance [40, Proposition 16]. But in many contexts relevant to Ramanujan's work, modular form theory guarantees the existence of *holonomic* differential equations. Zagier [40, Proposition 21] introduces to this fact as follows: "... it is at the heart of the original discovery of modular forms by Gauss and of the later work of Fricke and Klein and others, and appears in modern literature as the theory of Picard–Fuchs differential equations or of the Gauss–Manin connection — but it is not nearly as well known as it ought to be." The story connected to this quote is continued in Sec. 6. As a contribution to this topical area, in the spirit of the "first guess, then prove" paradigm,[4] in this paper together with [31] we present an algorithm which automatically proves a claim stating that a given modular form of positive weight satisfies a differential equation which is linear and with coefficients being polynomials in a given modular function.

[1] In [35], holonomic sequences are called P-recursive (polynomially recursive), holonomic functions are called D-finite (differentiably finite).

[2] This package is freely available upon password request by email to the first named author.

[3] A much more powerful multivariate package `Holonomic Functions` has been developed by Koutschan [23] also in the Mathematica system.

[4] This paradigm, e.g., was popularized by George Polya in many of his writings.

Our paper is structured as follows. To make the exposition as self-contained as possible, in Sec. 2 we present the most important basic facts on modular functions and forms needed. Section 3 does the same for univariate holonomic functions and sequences, an illustrative example is taken from Ramanujan's "Quarterly Reports". Section 4 introduces to one of the major themes, the expansion of modular forms and functions in terms of fractional (Puiseux) series. In Sec. 5, we present case studies dealing with relations of Fricke–Klein type, irrationality proofs based on modular forms *a la* Beukers, and approximations to pi which were first discovered by Ramanujan and then extensively studied by Jonathan and Peter Borwein. In Sec. 6, again with focus on an algorithmic setting, we discuss aspects of a classical connection of modular forms with differential equations. This connection traces back to Gauß and was popularized prominently by Zagier. In addition, we present a brief account[5] of a new algorithm for proving differential equations for modular forms. Beginning with Sec. 7, we try to make the following point: connecting Puiseux expansions for modular functions with holonomic differential equations is of interest for the theory of partition congruences. After an introductory example inspired by Ramanujan's $11 \mid p(11n+6)$ observation, in Sec. 8 we discuss algebraic relations between modular functions and corresponding issues of computational relevance. In Sec. 9, the focus is shifted from algebraic relations to differential equations involving modular functions. Choosing Andrews' 2-colored Frobenius partitions, Sec. 10 presents another case study to illustrate aspects treated in Secs. 8 and 9. Linear differential equations with algebraic function coefficients can be transformed into holonomic differential equations. Appendix A contains a proof of this fact;[6] it also comments very shortly on zero-recognition of meromorphic functions on Riemann surfaces.

2. Conventions and Basic Facts: Modular Functions and Forms

Conventions used throughout this paper: N denotes a positive integer,

$$\mathbb{H} := \{z \in \mathbb{C} : \operatorname{Im}(z) > 0\}, \quad \hat{\mathbb{C}} := \mathbb{C} \cup \{\infty\}, \quad \text{and} \quad \hat{\mathbb{Q}} := \mathbb{Q} \cup \{\infty\}.$$

The ring of formal power series over the complex numbers, respectively, over the integers $\mathbb{Z}$, is denoted by $\mathbb{C}[[z]]$, respectively, $\mathbb{Z}[[z]]$. The ring of univariate polynomials with complex coefficients is denoted by $\mathbb{C}[X]$, and by $\mathbb{C}[X,Y]$ in the bivariate case.

The group $\mathrm{SL}_2(\mathbb{Z}) = \{\left(\begin{smallmatrix} a & b \\ c & d\end{smallmatrix}\right) \in \mathbb{Z}^{2\times 2} : ad - bc = 1\}$ acts on elements τ from the upper half complex plane $\mathbb{H}$ by

$$\begin{pmatrix} a & b \\ c & d \end{pmatrix} \tau := \frac{a\tau + b}{c\tau + d}.$$

[5]Full details are given in [31].

[6]Proposition 6.2.

Congruence subgroups Γ are subgroups of $\mathrm{SL}_2(\mathbb{Z})$ which are specified by congruence conditions on their entries.[7] The so-called principal congruence subgroup is defined as

$$\Gamma(N) := \left\{ \begin{pmatrix} a & b \\ c & d \end{pmatrix} \in \mathrm{SL}_2(\mathbb{Z}) : \begin{pmatrix} a & b \\ c & d \end{pmatrix} \equiv \begin{pmatrix} 1 & 0 \\ 0 & 1 \end{pmatrix} \pmod N \right\}.$$

A congruence subgroup Γ is called to be of level N, if $\Gamma(N) \subseteq \Gamma$ and N is minimal with this property. Besides $\Gamma(N)$, we need the congruence subgroups

$$\Gamma_0(N) := \left\{ \begin{pmatrix} a & b \\ c & d \end{pmatrix} \in \mathrm{SL}_2(\mathbb{Z}) : \begin{pmatrix} a & b \\ c & d \end{pmatrix} \equiv \begin{pmatrix} \star & \star \\ 0 & \star \end{pmatrix} \pmod N \right\}$$

and

$$\Gamma_1(N) := \left\{ \begin{pmatrix} a & b \\ c & d \end{pmatrix} \in \mathrm{SL}_2(\mathbb{Z}) : \begin{pmatrix} a & b \\ c & d \end{pmatrix} \equiv \begin{pmatrix} 1 & \star \\ 0 & 1 \end{pmatrix} \pmod N \right\}.$$

There are many refinements, for instance, for positive integers N, M, P such that $M \mid NP$:

$$\Gamma(N,M,P) := \left\{ \begin{pmatrix} a & b \\ c & d \end{pmatrix} \in \mathrm{SL}_2(\mathbb{Z}) : a \equiv d \equiv 1 \pmod M, \right.$$
$$\left. b \equiv 0 \pmod P, c \equiv 0 \pmod N \right\}.$$

In Sec. 6.3, we will need $\Gamma(2,4,2)$ together with properties which we retrieve by using the computer algebra system Magma.[8] See also Examples 2.6, 4.2 and Sec. 5.3.

2.1. *Modular forms*

For a function $f : \mathbb{H} \to \hat{\mathbb{C}}$, $k \in \mathbb{Z}_{\geq 0}$, and $\gamma = \left(\begin{smallmatrix} a & b \\ c & d \end{smallmatrix}\right) \in \mathrm{SL}_2(\mathbb{Z})$ we define the weight k slash-action as usual by

$$(f|_k\gamma)(\tau) := (c\tau + d)^{-k} f(\gamma\tau), \quad \tau \in \mathbb{H}.$$

Also as usual, we define (e.g., [28, Definition 1.8]): Let $f : \mathbb{H} \to \hat{\mathbb{C}}$ be meromorphic, $k \in \mathbb{Z}_{\geq 0}$, and Γ a congruence subgroup. Then f is called a meromorphic *modular form* of weight k for Γ, if for all $\gamma \in \Gamma$,

$$(f|_k\gamma)(\tau) = f(\tau), \quad \tau \in \mathbb{H}, \tag{2.1}$$

and if for each $\gamma_0 \in \mathrm{SL}_2(\mathbb{Z})$ there exists $N_0 = N_0(\gamma_0) \in \mathbb{Z}_{>0}$ and $n_0 = n_0(\gamma_0) \in \mathbb{Z}$ together with a Fourier expansion

$$(f|_k\gamma_0)(\tau) = \sum_{n \geq n_0} a_{\gamma_0}(n) q_{N_0}^n, \tag{2.2}$$

where $q_{N_0} := e^{2\pi i\tau/N_0}$ and $a_{\gamma_0}(n_0) \neq 0$.

[7]More precisely, such Γ is a discrete subgroup of $\mathrm{SL}_2(\mathbb{R})$ which is commensurable with $\mathrm{SL}_2(\mathbb{Z})$ (i.e. $\Gamma \cap \mathrm{SL}_2(\mathbb{Z})$ has finite index in Γ and $\mathrm{SL}_2(\mathbb{Z})$) and $\Gamma(N) \subseteq \Gamma$ for some N.

[8]In the Magma system, this group is specified as `CongruenceSubgroup([N,M,P]);` see http://magma.maths.usyd.edu.au/magma/handbook/text/1589.

Modular forms with weight $k = 0$ are called *modular functions*; in this case we write $f \mid \gamma_0$ instead of $f|_0\gamma_0$. Also if $k = 0$, the integer $n_0 = n_0(\gamma_0)$ is called the order of $f \mid \gamma_0$ at infinity; notation: $n_0 = \operatorname{ord}(f \mid \gamma_0)$.

One can extend the action of $\mathrm{SL}_2(\mathbb{Z})$ on $\mathbb{H}$ to an action on $\hat{\mathbb{H}} := \mathbb{H} \cup \hat{\mathbb{Q}}$ by defining $\gamma\infty = a/c$ for $\gamma = \left(\begin{smallmatrix} a & b \\ c & d \end{smallmatrix}\right)$. If $\gamma_0\infty = \gamma_1\infty = a/c$ for $\gamma_j \in \mathrm{SL}_2(\mathbb{Z})$, then $n_0 = \operatorname{ord}(f \mid \gamma_0) = \operatorname{ord}(f \mid \gamma_1)$. If $\gamma_0 = \left(\begin{smallmatrix} 1 & 0 \\ 0 & 1 \end{smallmatrix}\right) \in \mathrm{SL}_2(\mathbb{Z})$, then we write in short $\operatorname{ord} f := \operatorname{ord}(f \mid \gamma_0)$ for the order of f at infinity.

2.2. *Modular functions*

For algorithmic zero recognition, modular functions to us are most important. For those owing to $k = 0$, we have the invariance $f(\gamma\tau) = f(\tau)$ for all $\gamma \in \Gamma$, and the expansions (2.2) then allow to extend f meromorphically to all the points $a/c \in \hat{\mathbb{Q}}$ where $a/0 := \infty$. To this end, the first step is to extend the action of $\mathrm{SL}_2(\mathbb{Z})$ on $\mathbb{H}$ to an action on $\hat{\mathbb{H}} := \mathbb{H} \cup \hat{\mathbb{Q}}$ as already mentioned. Notation: for any congruence subgroup Γ the Γ-orbit of $\tau \in \hat{\mathbb{H}}$ is written as $[\tau]_\Gamma := \{\gamma\tau : \gamma \in \Gamma\}$. We will write $[\tau]$ instead of $[\tau]_\Gamma$, if the subgroup Γ is clear from the context.

After this first step, the meromorphic extension of a modular function f from $\mathbb{H}$ to a function on $\hat{\mathbb{H}}$ is done by choosing $\gamma_0 = \left(\begin{smallmatrix} a & b \\ c & d \end{smallmatrix}\right) \in \mathrm{SL}_2(\mathbb{Z})$ such that $\gamma_0\infty = a/c$, and one defines

$$f(a/c) := (f \mid \gamma_0)(\infty) := \begin{cases} \infty & \text{if } n_0 < 0, \\ a_{\gamma_0}(0) & \text{if } n_0 = 0 \\ 0 & \text{if } n_0 > 0 \end{cases}.$$

A straightforward verification shows that this definition is independent from the choice of γ_0. This means, if one chooses another $\gamma_1 \in \mathrm{SL}_2(\mathbb{Z})$ such that $\gamma_1\infty = a/c$, one would obtain the same value for $f(a/c)$. Moreover, $f(a/c) = f(\gamma\frac{a}{c})$ for any $\gamma \in \Gamma$. Together with (2.1), this means, a modular function is constant on all the Γ-orbits $[\tau]_\Gamma$. The set of all such orbits, denoted by $X(\Gamma)$, can be equipped with the structure of a compact Riemann surface. Hence a modular function f with respect to Γ can be interpreted as a function $\hat{f} : X(\Gamma) \to \hat{\mathbb{C}}$; in fact, such $\hat{f}$ will be meromorphic.

Orbits $[\tau]_\Gamma$, where $\tau = a/c \in \hat{\mathbb{Q}}$, are called cusps. Congruence subgroups have only a finite index in $\mathrm{SL}_2(\mathbb{Z})$, hence $\hat{\mathbb{Q}} = \{\gamma\infty : \gamma \in \mathrm{SL}_2(\mathbb{Z})\} = \mathrm{SL}_2(\mathbb{Z})\infty = \cup_j \Gamma\gamma_j\infty = \cup_j[\gamma_j\infty]_\Gamma$ is a disjoint union of *finitely many* cusps. Fourier expansions as in (2.2) are called expansions of f at the cusp $[a/c]$, or simply at a/c.

Special attention is given to the case $a/c = \infty$ (i.e. $c = 0$). In this case, we can exploit the fact that each congruence subgroup contains a translation matrix $\left(\begin{smallmatrix} \pm 1 & N_0 \\ 0 & \pm 1 \end{smallmatrix}\right)$ with $N_0 \in \mathbb{Z}_{>0}$ minimal; this means, a modular function f satisfying (2.1) has minimal period N_0. Suppose for all $\tau \in \mathbb{H}$ with $\mathrm{Im}(\tau)$ sufficiently large, the

Fourier expansion of f at infinity[9] is:

$$f(\tau) = \sum_{n \geq n_0} a(n) q^{n/N_0}. \tag{2.3}$$

In various contexts, we need to put emphasis on representing f in the Fourier variable q; in such cases we write

$$\tilde{f}(q) := \sum_{n \geq n_0} a(n) q^{n/N_0} = f(\tau) \quad \text{for } q = e^{2\pi i \tau} \text{ sufficiently small.} \tag{2.4}$$

Remark 2.1. As mentioned above, $\hat{\mathbb{Q}}$ is a disjoint union of finitely many cusps. As a consequence, if f is a non-constant modular function holomorphic on $\mathbb{H}$ and with $f(a/c) = \infty$ for some $a/c \in \hat{\mathbb{Q}}$, then $f(r) = \infty$ for an infinite set of values $r \in \hat{\mathbb{Q}}$. In view of (2.4) this means, $\tilde{f}(z) = \infty$ for infinitely many z on the complex unit circle $|q| = 1$.

2.3. *Zero recognition of modular functions*

For zero recognition of a modular function f, one exploits its extension to a meromorphic function $\hat{f} : X(\Gamma) \to \hat{\mathbb{C}}$ on the compact Riemann surface $X(\Gamma)$. Namely, if such functions are non-constant, they have the property that

$$\text{number of poles of } \hat{f} = \text{number of zeros of } \hat{f}, \tag{2.5}$$

counting multiplicities; for further details, see Lemma A.2.

Remark 2.2. Particularly straightforward applications of this fact are with regard to modular functions having a possible pole only at infinity:

$$\begin{aligned} M^{\infty}(\Gamma) := \{ f : \mathbb{H} \cup \hat{\mathbb{Q}} \to \hat{\mathbb{C}} : \hat{f} \text{ has poles only at } \infty, \text{ and} \\ f \text{ is a modular function for the congruence subgroup } \Gamma \}. \end{aligned} \tag{2.6}$$

We note that $M^{\infty}(\Gamma)$ is a $\mathbb{C}$-algebra; i.e. a ring with multiplication by scalars from $\mathbb{C}$. To prove equality $f_1 = f_2$ of two modular functions in $M^{\infty}(\Gamma)$, it is sufficient to inspect the q-expansion at infinity,

$$f_1 - f_2 = \frac{c_{-m_0}}{q^{m_0/N_0}} + \cdots + \frac{c_{-1}}{q^{1/N_0}} + c_0 + c_1 q^{1/N_0} + \cdots,$$

and to verify that

$$c_{-m_0} = \cdots = c_{-1} = c_0 = 0.$$

Note that in this case $f_1 - f_2 \in M^{\infty}(\Gamma)$, respectively, $\hat{f}_1 - \hat{f}_2 : X(\Gamma) \to \hat{\mathbb{C}}$, has no pole but a zero at infinity; hence according to (2.5) it must be a constant.

[9] In other words, at the point $\infty \in \hat{\mathbb{Q}}$, respectively, at the cusp $[\infty]$.

Finally, for the field of modular functions *meromorphic* on $\mathbb{H}$, we write:

$$M(\Gamma) := \{f : \mathbb{H} \cup \hat{\mathbb{Q}} \to \hat{\mathbb{C}} : f \text{ is meromorphic on } \mathbb{H}, \text{ and}$$
$$f \text{ is a modular function for the congruence subgroup } \Gamma\}; \tag{2.7}$$

for the $\mathbb{C}$-vector space of modular forms of weight $k \geq 1$ for Γ we write $M_k(\Gamma)$.

2.4. *Examples of modular forms and functions*

Example 2.3. Consider the Dedekind eta function,

$$\eta(\tau) := e^{\pi i \tau/12} \prod_{j=1}^{\infty}(1 - e^{2\pi i \tau j}) = q^{1/24} \prod_{j=1}^{\infty}(1 - q^j), \quad \tau \in \mathbb{H}.$$

For $\gamma = \left(\begin{smallmatrix} a & b \\ c & d \end{smallmatrix}\right) \in \mathrm{SL}_2(\mathbb{Z})$ it transforms as

$$\eta\left(\frac{a\tau + b}{c\tau + d}\right) = v(\gamma)(c\tau + d)^{1/2}\eta(\tau), \tag{2.8}$$

where for the root one takes the principal branch[10] and with $v(\gamma)$ being some 24th root of unity depending on γ. For details on Δ and $v(\gamma)$ see, e.g., [16, Theorem 5.8.1]. As a consequence of (2.8), the Delta function

$$\Delta(\tau) := \eta(\tau)^{24}, \quad \tau \in \mathbb{H},$$

is a modular form of weight 12 for $\mathrm{SL}_2(\mathbb{Z})$; i.e. $\Delta \in M_{12}(\mathrm{SL}_2(\mathbb{Z}))$.

Example 2.4. In Sec. 5.1, we will need the normalized Eisenstein series

$$E_4(\tau) = 1 + 240 \sum_{n=1}^{\infty} \sum_{d \mid n} d^3 q^n = 1 + 240q + 2160q^2 + \cdots, \quad q = e^{2\pi i \tau} \quad \text{with } \tau \in \mathbb{H}.$$

As shown in [16, Sec. 5.2], $E_4 \in M_4(\mathrm{SL}_2(\mathbb{Z}))$.

Example 2.5. The most important modular function is the Klein j function, also called modular invariant

$$j(\tau) = \frac{E_4(\tau)^3}{\Delta(\tau)} = \frac{1}{q} + 744 + 196884q + 21493760q^2 + \cdots, \quad q = e^{2\pi i \tau} \quad \text{with } \tau \in \mathbb{H}.$$

Being the quotient of two modular forms in $M_{12}(\mathrm{SL}_2(\mathbb{Z}))$, one has that $j \in M(\Gamma(1))$. Since j has no pole in $\mathbb{H}$, $j \in M^{\infty}(\Gamma(1))$; see [16, Sec. 5.7].

Example 2.6. Consider the Jacobi theta functions with $q = e^{2\pi i \tau}$,

$$\theta_2(\tau) = \sum_{n \in \mathbb{Z}+1/2} q^{n^2/2} = 2\frac{\eta(2\tau)^2}{\eta(\tau)}, \quad \tau \in \mathbb{H}, \tag{2.9}$$

[10] i.e. if $z = re^{i\varphi}$, $r \in \mathbb{R}_{\geq 0}$ and $-\pi < \varphi \leq \pi$, then $z^{1/2} = \sqrt{r}e^{i\varphi/2}$.

and

$$\theta_3(\tau) = \sum_{n\in\mathbb{Z}} q^{n^2/2} = 1 + 2\sum_{n=1}^{\infty} q^{n^2/2} = \frac{\eta(\tau)^5}{\eta(\tau/2)^2\eta(2\tau)^2}, \quad \tau \in \mathbb{H}. \tag{2.10}$$

One has, as a consequence of (2.8), that for all $\gamma \in \Gamma(2,4,4)$,

$$\theta_2(\gamma\tau)^2 = (c\tau + d)\,\theta_2(\tau)^2$$

and for all $\gamma \in \Gamma(2,4,2)$,

$$\theta_3(\gamma\tau)^2 = (c\tau + d)\theta_3(\tau)^2.$$

In other words, θ_2^2 and θ_3^2 are modular forms of weight 1 for $\Gamma(2,4,4) \subseteq \Gamma(2,4,2)$.[11]

3. Univariate Holonomic Functions and Sequences

An introduction to univariate holonomic theory and related algorithms is given in [22].[12] In this section, we recall only the most fundamental holonomic notions. To illustrate the main concepts, we apply computer algebra to an example arising in the work of Ramanujan. Further details and proofs of the presented holonomic yoga, can be found in [22]; related theoretical extensions and applications in enumerative combinatorics can be found in [35].

A sequence $(a(n))_{n\geq 0}$ is holonomic[13] $:\Leftrightarrow$ there exist polynomials $p_j(X)$, not all zero, such that

$$p_d(n)a(n+d) + \cdots + p_0(n)a(n) = 0, \quad n \geq 0. \tag{3.1}$$

In the given context, we restrict to complex sequences; as a consequence, the $p_j(X)$ are from $\mathbb{C}[X]$, the ring of polynomials with coefficients in $\mathbb{C}$. The same restriction to $\mathbb{C}$ applies to functions.

Let $y(z) = \sum_{n=0}^{\infty} a(n)z^n$ be a function analytic at 0 or, alternatively, a formal power series with complex coefficients: $y(x)$ is holonomic[14] $:\Leftrightarrow$ there exist polynomials $P_j(X) \in \mathbb{C}[X]$, not all zero, such that

$$P_m(z)y^{(m)}(z) + \cdots + P_0(z)y(z) = 0. \tag{3.2}$$

The following fact, despite easy to prove, is of fundamental importance.

Proposition 3.1. *If $y(z)$ is a formal power series or a function analytic at* 0:

$$y(z) = \sum_{n=0}^{\infty} a(n)z^n \textit{ is holonomic } \Leftrightarrow (a(n))_{n\geq 0} \textit{ is holonomic.}$$

We will see many applications below. But first we apply Proposition 3.1 to prove

[11] Recalling definition (2.3), note that $\theta_2(\tau)$ and $\theta_3(\tau)$ have period $N_0 = 2$. — In addition, we note that allowing multiplier systems in the definition of modular forms, θ_2 and θ_3 are modular forms of weight 1/2 for a bigger congruence subgroup than $\Gamma(2,4,2)$; see, e.g., [16].

[12] The book by Kauers and Paule [22] is intended as a computer algebra supplement to "Concrete Mathematics" [21].

[13] P-recursive (polynomially recursive) in [35].

[14] D-finite (differentiably finite) in [35].

Corollary 3.1. *The sequence of partition numbers $(p(n))_{n\geq 0}$ is not holonomic.*

Proof. Suppose it is holonomic. Then Proposition 3.1 implies that $y(z) := \sum_{n=0}^{\infty} p(n)z^n$ is holonomic and would satisfy a differential equation as in (3.2) with $P_m(z) \neq 0$, say. As a polynomial, $P_m(z)$ has only finitely many zeros; thus any analytic solution to (3.2) can have only finitely many singularities on its circle of convergence. This gives a contradiction to $y(z) = \prod_{j=1}^{\infty}(1 - z^j)^{-1}$ having infinitely many singularities on the complex unit circle. □

For the same reason, modular forms and modular functions cannot be holonomic. We restrict to prove an important special case explicitly:

Corollary 3.2. *The q-expansion $\hat{f}(q)$ of any non-constant function $f \in M^{\infty}(\Gamma)$, Γ a congruence subgroup, is not holonomic.*

Proof. The statement follows from the fact that $\tilde{f}(z) = \infty$ for infinitely many $z = e^{2\pi i \tau}$ on the complex unit circle; see Remark 2.1. □

Example 3.2. For fixed $n \in \mathbb{Z}_{\geq 1}$ consider the power series

$$y(z) := \sum_{k=0}^{\infty} a_n(k) z^k \quad \text{where } a_n(k) := n\frac{(n+2k-1)!}{(n+k)!k!}(-1)^k, \tag{3.3}$$

which arose in the work of Ramanujan. Before we point to references, we demonstrate how $y(z)$ can be studied by following the "holonomic paradigm." To this end, as announced in the introduction, we use a RISC software package written in the Mathematica system. After putting the package in a directory where we open a Mathematica session, we read it in as follows[15]:

```
In[1]:= << RISC`GeneratingFunctions`
```

```
Package GeneratingFunctions version 0.8
written by Christian Mallinger © RISC-JKU
```

As a hypergeometric sequence $(a_n(k))_{k\geq 0}$ is holonomic. This means, the quotient of two consecutive terms is a rational function

$$\frac{a_n(k+1)}{a_n(k)} = -\frac{(2k+n)(2k+n+1)}{(k+1)(k+n+1)},$$

in other words, it satisfies recurrence of order 1.

[15]The package is freely available at https://combinatorics.risc.jku.at/software upon password request via email to the first named author.

Hence, by Proposition 3.1, $y(z)$ is holonomic and we use the package to pass from the order 1 recurrence as input, to a differential equation for $y(z)$:

```
In[2]:= de = RE2DE[{(k + 1)(k + n + 1)a[k + 1] + (2k + n)(2k + n + 1)a[k] == 0,
                a[0] == 1}, a[k], y[z]]
Out[2]= {(n + n^2)y[z] + (1 + n + 6z + 4nz)y'[z] + (z + 4z^2)y''[z] = 0, y[0] = 1, y'[0] = -n}.
```

We use a command from the Mathematica system to solve it:

```
In[3]:= DSolve[de, y[z], z, Assumptions → (n ∈ Integers && n > 0)]
    DSolve : For some branches of the general solution, the given boundary
        conditions lead to an empty solution.
```

From the system's reply, we extract that the built-in solver[16] cannot handle a generic positive integer n in this differential equation, so we ask the system to solve special instances of it:

```
In[4]:= {DSolve[de /.n → 1, y[z], z], DSolve[de /.n → 2, y[z], z]}
Out[4]= {{{y[z] → (-1 + Sqrt[1 + 4z])/(2z)}}, {{y[z]- → (1 + 2z - Sqrt[1 + 4z])/(2z^2)}}}
```

We could continue with $n = 3, 4$, and so on, to collect more data for guessing a closed form for $y(z)$. However, in our context it is more instructive to consider an alternative way to proceed. The output expressions in `Out[4]` already suggest that $y(x)$ might be an algebraic function (respectively, power series); this means, a function (respectively, power series) which satisfies a polynomial relation with polynomial coefficients. Connecting to holonomic objects, we have the

Proposition 3.3. *Any algebraic function (respectively, power series) $y(z)$ is holonomic. More concretely, if $P(z, y(z)) = 0$ for a polynomial $P(X,Y) \in \mathbb{C}[X][Y]$ of degree $m = \deg_Y P(X,Y)$ in Y, then $y(z)$ satisfies a homogeneous differential equation of order m as in (3.2), or an inhomogeneous differential equation of order $m-1$ of the form*

$$Q_{m-1}(z)y^{(m-1)}(z) + \cdots + Q_0(z)y(z) = Q(z) \tag{3.4}$$

with polynomials $Q_j(X), Q(X) \in \mathbb{C}[X]$.

Proof. The proof in [35, Theorem 6.4.6] works the same for our case where $P(X,Y)$ is not necessarily irreducible. □

Example 3.4 (Example 3.2 continued). To obtain a closed form for $y(z)$, our strategy is to apply *computer-supported guessing* successively for $n = 1$,2, etc. The procedure to do so, `GuessAE`, is part of the `GeneratingFunctions` package. As input we take the first 15 coefficients $a_n(0), \ldots, a_n(14)$ successively fixing $n = 1$, $n = 2$, and $n = 3$. As it turns out, these 15 initial values are sufficient to automatically

[16] Using Mathematica V11.3.0.

guess an algebraic equation in each instance:

```
In[5]:= aList[n_] := Table[n (n + 2k − 1)!/((n + k)!k!) (−1)^k, {k, 0, 14}];
In[6]:= GuessAE[aList[1], y[z]]
Out[6]= {{−1 + y[z] + zy[z]^2 = 0, y[0] = 1}, ogf}
In[7]:= GuessAE[aList[2], y[z]]
Out[7]= {{1 + (−1 − 2z)y[z] + z^2y[z]^2 = 0, y[0] = 1}, ogf}
In[8]:= GuessAE[aList[3], y[z]]
Out[8]= {{−1 + (1 + 3z)y[z] + z^3y[z]^2 = 0, y[0] = 1}, ogf[17]}
```

From this data, after solving the quadratic equations and fixing the sign of the solution, the following conjecture becomes evident:

$$y(z) = \left(\frac{2}{1+\sqrt{4z+1}}\right)^n, \quad n \geq 1. \tag{3.5}$$

Remarks 3.3 (on Example 3.2 and Example 3.4). (i) Using computer-supported holonomic guessing, we derived the closed form representation (3.5) as a *conjecture*. On the other hand, by holonomic methods we computed (and thus proved!) the differential equation in `Out[2]` which is satisfied by $y(z)$ for generic $n \in \mathbb{Z}_{>0}$.[18] Hence the task to prove (3.5) is trivialized: just take the differential equation and verify that it is satisfied by the right-hand side of (3.5).

(ii) With the `GeneratingFunctions` package, one can automatically obtain from any specific algebraic equation the corresponding differential equation; for example, starting with the relation in `Out[8]` gives,

```
In[9]:= AE2DE[{−1 + (1 + 3z)y[z] + z^3y[z]^2 == 0, y[0] == 1}, y[z]]
Out[9]= 3 − 3(1 + 4z + 2z^2)y[z] − (z + 5z^2 + 4z^3)y'[z] = 0
```

By differentiation we turn this inhomogeneous equation into a homogeneous one:

```
In[10]:= D[3 − 3(1 + 4z + 2z^2)y[z] − (z + 5z^2 + 4z^3)y'[z], z] //Simplify
Out[10]= −(1 + z)(12y[z] + 2(2 + 9z)y'[z] + z(1 + 4z)y''[z])
```

Up to the factor $-(1+z)$, this is the case $n = 3$ of `Out[2]`.

(iii) Identity (3.5) stems from the first of Ramanujan's "Quarterly Reports"; it is entry (1.12) in [8]. As explained in [8], this identity arose in the context of Ramanujan's "Master Theorem"; Bruce Berndt presents a wonderful account of this theorem, and also tells the story of Ramanujan's reports.

[17] "ogf" means, one has to take $y(z)$ as an "ordinary generating function" $y(z) = \sum a(k)x^k$, in contrast to an interpretation as an "exponential generating function (egf)."

[18] In fact, this differential equation is satisfied by $y(z)$ also if n is a nonzero real number. Thus, defining the $a_n(k)$ for such n (using the gamma function instead of factorials), the identity (3.5) extends to real n.

(iv) Consider the first 15 coefficients $a_1(0), \ldots, a_1(14)$ where we fix $n = 1$:

```
In[11]:= aList[1]

Out[11]= {1, −1, 2, −5, 14, −42, 132, −429, 1430, −4862, 16796, −58786, 208012, −742900, 2674440}
```

These are the first 15 Catalan numbers $C_n = \frac{1}{n+1}\binom{2n}{n}$ with alternating signs. Indeed, consulting Sloane's OEIS at http://oeis.org (entry A00976, and also entries like A090749) one finds that Ramanujan's numbers for fixed $n \geq 1$ are forming the nth diagonals of the table of the celebrated Ballot numbers $B(\ell, k) := \frac{\ell-k}{\ell+k}\binom{\ell+k}{\ell}$, where $\ell > 0$ and $0 \leq k \leq \ell$:

$$|a_n(k)| = B(n+k, k), \quad n \geq 1, k \geq 0.$$

The Catalan numbers sit on the main diagonal $(B(1+k, k))_{k \geq 0}$.

4. Puiseux Expansions

In this section, we present the basic mechanism of Puiseux expansion which will be used to connect holonomic functions and sequences with modular functions and forms.

As a major theme, we will study the following setting. Given a Laurent series

$$g = \sum_{k=M}^{\infty} g(k)x^k, \quad M \in \mathbb{Z} \text{ fixed },$$

and a power series

$$h = x^m(1 + \underbrace{h_1 x + h_2 x^2 + \cdots}_{=:\psi = \psi(x)}), m \in \mathbb{Z}_{\geq 1} \text{ fixed},$$

find $(c(k))_{k \geq M}$ such that

$$g = \sum_{k=M}^{\infty} c(k) h^{\frac{k}{m}}, \tag{4.1}$$

where to define $h^{\frac{k}{m}} := (h^{\frac{1}{m}})^k$ we choose the branch

$$h^{\frac{1}{m}} := x(1+\psi)^{\frac{1}{m}} := \sum_{n=0}^{\infty} \binom{1/m}{n} \psi^n.$$

Expansions as in (4.1) are called Puiseux series or fractional series. Such expansions always exist, for formal power series as well as for complex functions g and h being meromorphic and holomorphic, respectively, at 0. The coefficients can be computed as follows.

Lemma 4.1. *Given g and h as above. Then there exists a sequence $(c(k))_{k \geq M}$ of complex numbers such that*

$$g = \sum_{k=M}^{\infty} c(k) h^{\frac{k}{m}}.$$

Moreover, the $c(k)$ are uniquely determined by

$$c(k) := \textit{coefficient of } x^k \textit{ in } \sum_{j=M}^{\infty} g(j)W(x)^j, \quad k \geq M,$$

where W is the compositorial inverse of $U(x) := h^{\frac{1}{m}}(x) = x(1 + h_1 x + \cdots)^{\frac{1}{m}}$; i.e.

$$W(U(x)) = U(W(x)) = x.$$

Proof.

$$\sum_{k=M}^{\infty} c(k) h^{\frac{k}{m}}(x) = \sum_{k=M}^{\infty} (h^{\frac{1}{m}}(x))^k \left(\text{coefficient of } z^k \text{ in } \sum_{j=M}^{\infty} g(j) W(z)^j \right)$$

$$= \sum_{j=M}^{\infty} g(j) W(h^{\frac{1}{m}}(x))^j = \sum_{j=M}^{\infty} g(j) x^j = g. \qquad \square$$

To illustrate this kind of Puiseux expansion, we take a classical example from Zagier's classical exposition [40].

Example 4.2. Recall the modular forms $\theta_2^2 \in M_1(\Gamma(2,4,4))$ and $\theta_3^2 \in M_1(\Gamma(2,4,2))$ from Example 2.6. Let $x := e^{i\pi\tau} = q^{1/2}$. As in [40, p. 63] choose

$$g = \theta_3(\tau)^2 = 1 + 4x + 4x^2 + \cdots \quad \text{and}$$

$$h = \frac{\theta_2(\tau)^4}{2^4 \theta_3(\tau)^4} = x(1 - 8x + 44x^2 - \cdots);$$

i.e. $m = 1$. Here $h(\tau) = \lambda(\tau)/16$ is the normalized modular lambda function. Applying (2.8) reveals that $\lambda \in M(\Gamma(2,4,2))$.[19] Using a computer algebra system like Mathematica, according to Lemma 4.1 the coefficients $c(k)$ can be computed as follows:

```
In[12]:= g = 1 + 4x + 4x^2 + 4x^4 + 8x^5 + O[x]^6;

In[13]:= U = x(1 - 8x + 44x^2 - 192x^3 + 718x^4 + O[x]^5);

In[14]:= W = InverseSeries[U]

Out[14]= x + 8x^2 + 84x^3 + 992x^4 + O[x]^5

In[15]:= ComposeSeries[g, W]

Out[15]= 1 + 4x + 36x^2 + 400x^3 + 4900x^4 + O[x]^5
```

Interpreting the Mathematica output, we obtained

$$\theta_3(\tau)^2 = \sum_{k=0}^{\infty} c(k) \left(\frac{\lambda(\tau)}{16}\right)^k = 1 + 4\frac{\lambda(\tau)}{16} + 36\frac{\lambda(\tau)^2}{16^2} + 400\frac{\lambda(\tau)^3}{16^3} + 4900\frac{\lambda(\tau)^4}{16^4} + \cdots,$$

[19] The product forms (2.9) and (2.10) of θ_2^2 and θ_3^2 imply that they are holomorphic in $\mathbb{H}$.

With the same ease, one can compute sufficiently many further values of the $c(k)$ for additional information. For example, using the first 12 values, we can automatically guess a holonomic recurrence for the $c(k)$:

```
In[16]:= cList = {1, 4, 36, 400, 4900, 63504, 853776, 11778624, 165636900, 2363904400,
                  34134779536, 497634306624};

In[17]:= GuessRE[cList, c[k]]

Out[17]= {{-4(1 + 2k)^2 c[k] + (1 + k)^2 c[1 + k] = 0, c[0] = 1}, ogf}
```

The output recurrence tells that

$$c(k) = \frac{(1/2)_k(1/2)_k}{k!2} 16^k = \binom{2k}{k}^2, \quad k \geq 0, \tag{4.2}$$

where $(a)_k := a(a+1)\cdots(a+k-1)$. Summarizing, on the level of generating functions we have derived the conjecture that[20]

$$\theta_3(\tau)^2 = \sum_{k=0}^{\infty} \binom{2k}{k}^2 \left(\frac{\lambda(\tau)}{16}\right)^k = {}_2F_1\left(\begin{matrix}\frac{1}{2}\ \frac{1}{2}\\ 1\end{matrix}; \lambda(\tau)\right). \tag{4.3}$$

Remark 4.3. In view of Proposition 3.1, an alternative way to describe the $c(k)$ is to derive a holonomic differential equation for $y(z) := \sum_{k=0}^{\infty} c(k)z^k$. Using the `GeneratingFunctions` package, this can be done as follows[21]:

```
In[18]:= GuessDE[cList, y[z]]

Out[18]= {{4y[z] + (-1 + 32z)y'[z] + (-z + 16z^2)y''[z] = 0, y[0] = 1, y'[0] = 4}, ogf}
```

Comparing the guessed differential equation to the classical hypergeometric differential equation $z(z-1)y''(z) + ((a+b+1)z - c)y'(z) + aby(z) = 0$, which has $y(z) = {}_2F_1(a, b; c; z)$ as a solution, again produces the generating function identity (4.3).

Using computer algebra, we derived relation (4.3) as a *conjecture.* In Sec. 6, we explain how such identities, once guessed, can be *proved* routinely in computer-assisted fashion. As a concrete example, in Sec. 6.3 we present a computer-supported proof of the equation in `Out[18]` which, relating back to $h = h(\tau)$ and $g = g(\tau) = y(h(\tau))$, reads as

$$(16h^2 - h)\frac{d^2y(h)}{dh^2} + (32h-1)\frac{dy(h)}{dh} + 4y(h) = 0 \quad \text{with } y(z) := \sum_{k=0}^{\infty} c(k)z^k. \tag{4.4}$$

[20] In [40, (73)] the binomial coefficient $\binom{2n}{n}$ should be replaced by its square; i.e. $\binom{2n}{n}^2$.
[21] Another option is to transform the recurrence in `Out[17]` to the desired differential equation by the command `RE2DE[Out[17][[1]],c[k],y[x]]`.

Remark 4.4. Using the chain rule

$$g'(\tau) = \left.\frac{dy(z)}{dz}\right|_{z=h(\tau)} h'(\tau) \quad \text{and}$$

$$g''(\tau) = \left.\frac{d^2y(z)}{dz^2}\right|_{z=h(\tau)} (h'(\tau))^2 + \left.\frac{dy(z)}{dz}\right|_{z=h(\tau)} h''(\tau),$$

the differential equation (4.4) translates into the differential equation:

$$\frac{h(-1+16h)}{(h')^2}g'' + \frac{hh'' - 16h^2h'' - (h')^2 + 32h(h')^2}{(h')^3}g' + 4g = 0.$$

Despite being linear in the $g^{(k)}$, the coefficients are not modular forms anymore.[22] Nevertheless, as explained in Sec. 6, one can overcome this problem by a different processing of the holonomic version (4.4).

5. Case Studies: Puiseux Expansions of Modular Forms

In this section, we present further examples to illustrate computational aspects.

5.1. *Fricke–Klein relations*

We begin by considering the second classical example given in [40, p. 63] where, taking $q = e^{2\pi i\tau}$, Zagier chooses

$$g = E_4(\tau) = 1 + 240q + 2160q^2 + \cdots \quad \text{and} \quad h = \frac{1}{j(\tau)} = q(1 - 744q + \cdots).$$

In the setting of Lemma 4.1 this means, $m = 1$ again. Here $E_4(\tau) \in M_4(\mathrm{SL}_2(\mathbb{Z}))$ is the normalized Eisenstein series from Example 2.4 and $j \in M^\infty(\mathrm{SL}_2(\mathbb{Z})))$ the modular invariant from Example 2.5. Again, we determine $c(k)$ such that

$$E_4(\tau) = \sum_{k=0}^{\infty} c(k) \left(\frac{1}{j(\tau)}\right)^k.$$

As above, the coefficients can be computed as follows:

```
In[19]:= g = 1 + 240q + 2160q^2 + 6720q^3 + 17520q^4 + O[q]^5;

In[20]:= U = q(1 − 744q + 356652q^2 − 140361152q^3 + 49336682190q^4 + O[q]^5);

In[21]:= W = InverseSeries[U]

Out[21]= q + 744q^2 + 750420q^3 + 872769632q^4 + 1102652742882q^5 + O[q]^6

In[22]:= ComposeSeries[g, W]

Out[22]= 1 + 240q + 180720q^2 + 183321600q^3 + 213917180400q^4 + O[q]^5
```

This means, we obtained

$$E_4(\tau) = 1 + 240\frac{1}{j(\tau)} + 180720\frac{1}{j(\tau)^2} + 183321600\frac{1}{j(\tau)^3} + \cdots$$

[22] As the anonymous referee pointed out, they are quasi-modular forms.

As in our first example, we compute further coefficients to produce a guess on their structure. To this end, we input the coefficients from $c(0)$ to $c(20)$ collected in the list `cList`:

```
In[23]:= cList = {1, 240, 180720, 183321600, 213917180400, ...};

In[24]:= GuessRE[cList, c[k]]

Out[24]= {{36864(1 + k)(1 + 3k)(2 + 3k)(4 + 3k)(5 + 3k)c[k] − 24(3 + 2k)(4 + 3k)(5 + 3k)(21 + 24k + 8k^2)c[1 +
         k] + (2 + k)^5 c[2 + k] == 0, c[0] == 1, c[1] == 240}, ogf}
```

Now, if we would proceed as in Example 4.2, we would need the first 56 coefficients $c(k)$ to guess a differential equation. An algorithmic version of Proposition 3.1 allows us to directly transform the recurrence `Out[24]` into a differential equation:

```
In[25]:= RE2DE[%24[[1]], c[k], f[x]]

Out[25]= {240(−1 + 6144x)f[x] + (1 − 30000x + 81100800x^2)f'[x] + 15(x − 9864x^2 + 17252352x^3)f''[x] + 5(5x^2 −
         29616x^3 + 38486016x^4)f^(3)[x] + 10(x^3 − 4320x^4 + 4478976x^5)f^(4) + (x^4 − 3456x^5 + 2985984x^6)f^(5) =
         0, f[0] = 1, f'[0] = 240, f''[0] = 361440, f^(3)[0] = 1099929600, f^(3)[0] = 513401232960
0}
```

Without giving this differential equation explicitly, Zagier makes the following comment on this example: "Since $g := E_4$ is a modular form of weight 4, it should satisfy a fifth-order linear differential equation with respect to $h(\tau) := 1/j(\tau)$, but by the third proof above[23] one should even have that the fourth root of E_4 satisfies a second-order differential equation, and indeed one finds

$$\sqrt[4]{E_4(\tau)} = {}_2F_1\left(\begin{matrix}\frac{1}{12}\ \frac{5}{12}\\ 1\end{matrix}; \frac{12^3}{j(\tau)}\right) = 1 + \frac{1\cdot 5}{1\cdot 1}\frac{12}{j(\tau)} + \frac{1\cdot 5\cdot 13\cdot 17}{1\cdot 1\cdot 2\cdot 2}\frac{12^2}{j(\tau)^2} + \cdots \tag{5.1}$$

a classical identity which can be found in the works of Fricke and Klein."

Needless to say, that, as above, the corresponding differential equation in a computer-assisted way can be derived (as a guess) and proved (as the example Sec. 6.3) without any effort. Instead of going through this, we do a square root variation of the problem and determine $c(k)$ such that

$$\sqrt[2]{E_4(\tau)} = \sum_{k=0}^{\infty} c(k)\left(\frac{1}{j(\tau)}\right)^k.$$

Again, the coefficients can be computed as follows:

```
In[26]:= g = Series[1 + 240 Sum[DivisorSigma[3, n]q^n, {n, 1, 55}], {q, 0, 55}];

In[27]:= gRoot2 = Series[g^(1/2), {q, 0, 55}];
         A test whether the square root is built correctly:

In[28]:= gRoot2^2 − g
```

[23]We come back to Zagier's proofs in Sec. 6 in which we discuss proving.

```
Out[28]= O[q]^56
In[29]:= Del = Series[q Product[(1 - q^n)^24, {n, 1, 55}], q, 0, 55];
In[30]:= j = Series[g^3/Del, {q, 0, 55}];
In[31]:= h = Series[1/j, {q, 0, 55}];
In[32]:= U = h; W = InverseSeries[U];
In[33]:= comp = ComposeSeries[gRoot2, W];
In[34]:= cRoot2List11 = CoefficientList[Normal[Series[comp, {q, 0, 11}]], q];
In[35]:= cRoot2List11
Out[35]= {1, 120, 83160, 81681600, 93699005400, 117386113965120, 155667030019300800,
    214804163196079142400, 305240072216678400087000, 443655767845074392936328000,
    656486312795713480715743268160, 985646873056680684690542988249600}
In[36]:= GuessRE[cRoot2List11, c[k]]
Out[36]= {{-24(1 + 2k)(1 + 6k)(5 + 6k)c[k] + (1 + k)^3 c[1 + k] = 0, c[0] = 1}, ogf}
```

The output recurrence `Out[36]` was guessed by using only the first 12 coefficients, it means that the $c(k)$ form a hypergeometric sequence more concretely:

$$\sqrt[2]{E_4(\tau)} = {}_3F_2\left(\begin{matrix}\frac{1}{6} & \frac{1}{2} & \frac{5}{6} \\ 1 & 1 & \end{matrix}; \frac{12^3}{j(\tau)}\right) = \sum_{n=0}^{\infty} \frac{(6n)!}{(3n)!(n!)^3} \frac{1}{j(\tau)^n}. \tag{5.2}$$

As hypergeometric series, identities (5.1) and (5.2) are related by Clausen's formula, for further details, including relevance of (5.2) to Calabi–Yau varieties and string theory, see [25].

5.2. *Relations connected to irrationality proofs*

The next example is taken from Beukers' modular-form-based proof [14] for the irrationality of $\zeta(3)$. Following Beukers, respectively, Zagier [40, pp. 63–64], choose

$$g = \prod_{n=1}^{\infty} \frac{(1-q^{2n})^7(1-q^{3n})^7}{(1-q^n)^5(1-q^{6n})^5} = 1 + 5q + 13q^3 + 23q^3 + 29q^4 + 30q^5 + 31q^6 + \cdots$$

and

$$\begin{aligned} h &= q \prod_{n=1}^{\infty} \frac{(1-q^n)^{12}(1-q^{6n})^{12}}{(1-q^{2n})^{12}(1-q^{3n})^{12}} \\ &= q(1 - 12q + 66q^2 - 220q^3 + 495q^4 - 804q^5 + \cdots); \end{aligned}$$

again $q = e^{2\pi i \tau}$. Relating to the setting of Lemma 4.1, $m = 1$, and we determine $c(k)$ such that

$$\prod_{n=1}^{\infty} \frac{(1-q^{2n})^7(1-q^{3n})^7}{(1-q^n)^5(1-q^{6n})^5} = \sum_{k=0}^{\infty} c(k) q^k \prod_{n=1}^{\infty} \frac{(1-q^n)^{12k}(1-q^{6n})^{12k}}{(1-q^{2n})^{12k}(1-q^{3n})^{12k}}.$$

As above, the coefficients $c(k)$ can be computed as follows:

In[37]:= $\mathbf{g = q\prod_{n=1}^{25}\frac{(1-q^{2n})^7(1-q^{3n})^7}{(1-q^n)^5(1-q^{6n})^5};\quad h = q\prod_{n=1}^{25}\frac{(1-q^n)^{12}(1-q^{6n})^{12}}{(1-q^{2n})^{12}(1-q^{3n})^{12}};}$

In[38]:= **U = h; W = InverseSeries[U];**

In[39]:= **comp = ComposeSeries[g, W]; Series[comp, q, 0, 6]**

Out[39]= $1 + 5q + 73q^2 + 1445q^3 + 33001q^4 + 819005q^5 + 21460825q^6 + O[q]^7$

This means

$$g = \sum_{k=0}^{\infty} c(k)h^k = 1 + 5h + 73h^2 + 1445h^3 + 33001h^4 + 819005h^5 + \cdots$$

To gain further insight into the structure of the $c(k)$, we proceed as above

In[40]:= **cList20 = CoefficientList[Normal[Series[comp, {q, 0, 20}]], q];**

In[41]:= **GuessRE[cList20, c[k]]**

Out[41]= $\{\{(1+k)^3c[k] - (3+2k)(39+51k+17k^2)c[1+k] + (2+k)^3c[2+k] == 0, c[0] == 1, c[1] == 5\}, \text{ogf}\}$

This guess, recurrence `Out[41]`, was computed by using only the first 21 coefficients. The sequence $(c(k))_{k\geq 0}$ can be defined as a definite hypergeometric sum

$$c(k) = \sum_{\ell=0}^{k} \binom{k}{\ell}^2 \binom{k+\ell}{\ell}^2, \quad k \geq 0,$$

which satisfies the same recurrence with the same initial conditions, and which can be proven automatically by using implementations of Zeilberger's holonomic systems approach [41]. This sequence is the celebrated Apéry sequence which, as beautifully described in [36], has played a key role in Apéry's proof of $\zeta(3) \notin \mathbb{Q}$.

According to Proposition 3.1 the recurrence in `Out[41]` translates into a differential equation for the generating function $y(x) = \sum_{k\geq 0} c(k)x^k$. E.g., with the `GeneratingFunctions` package, one computes:

In[42]:= **RE2DE[%41[[1]], c[k], y[x]]**

Out[42]= $\{(-5+x)y[x] + (1-112x+7x^2)y'[x] + 3(x-51x^2+2x^3)y''[x] + (x^2-34x^3+x^4)y^{(3)} = 0,$
$y[0] = 1, y'[0] == 5, y''[0] == 146\}$

Owing to Proposition 6.1, discussed in the next section, the order 3 of the differential equation is no surprise: Using the modular transformation property of $\eta(\tau) = q^{1/24}\prod_{n=1}^{\infty}(1-q^n)$, one can show that g is a modular form of weight 2 for $\Gamma_1(6)$; moreover, h is a Hauptmodul[24] for $\Gamma_1(6)$, and $X(\Gamma_1(6))$ has genus 0. This means, when taking $\sqrt{g}$ instead of g, one obtains a differential equation of order 2:

In[43]:= **g = Series[g^{1/2}, q, 0, 25]; Series[g, q, 0, 6]**

[24] I.e. each function in $M(\Gamma_1(6))$ can be represented as a rational function in h.

Out[43]= $1+\frac{5q}{2}+\frac{27q^2}{8}+\frac{49q^3}{16}+\frac{147q^4}{128}+\frac{459q^5}{256}+\frac{2511q^6}{1024}+O[q]^7$

In[44]:= **U = h; W = InverseSeries[U]; comp = ComposeSeries[g, W];**

In[45]:= **Series[comp, q, 0, 6]**

Out[45]= $1+\frac{5q}{2}+\frac{267q^2}{8}+\frac{10225q^3}{16}+\frac{1836275q^4}{128}+\frac{90191115q^5}{256}+\frac{9386644575q^6}{1024}+O[q]^7$

In[46]:= **cList11 = CoefficientList[Normal[Series[comp, q, 0, 11]], q]; GuessRE[cList11, c[k]]**

Out[46]= $\{\{(1+2k)^2c[k]-2(107+170k+68k^2)c[1+k]+4(2+k)^2c[2+k]=0, c[0]=1, c[1]=5/2\}, \text{ogf}\}$

In[47]:= **RE2DE[%[[1]], c[k], y[x]]**

Out[47]= $\{(-10+x)y[x]+4(1-51x+2x^2)y'[x]+4(x-34x^2+x^3)y''[x]=0, y[0]=1, y'[0]=5/2\}$

In [13], Beukers gives further details about how the differential equations in `Out[42]` and `Out[47]` are related.

5.3. *Relations for approximations to* π

Relation (5.2) can also be found as entry (c) of Theorem 5.7 in the famous monograph [15] by Jon and Borwein. This entry is listed in the company of 11 identities [15, Theorems 5.6 and 5.7] of the same flavor. Besides other applications, the Borweins used these identities to derive and explain identities which Ramanujan [33] gave (without too many details) to establish formulas to approximate π, respectively, $1/\pi$.

The 12 identities listed in [15, Theorems 5.6 and 5.7] can be written in our format where g is a modular form of some weight and h is a Hauptmodul. For this translation, one only needs the formulas

$$K(k(\tau)) = \frac{\pi}{2}\theta_3(\tau)^2 \quad \text{where } k(\tau)^2 = \lambda(\tau),$$

for the complete elliptic integral of the first kind $K(k(\tau))$ with modulus $k(\tau)$. In [15], the 12 identities were discovered (and proved) by using tools from hypergeometric functions like Kummer's identity or Clausen's product formula, and by "piecing together quadratic and cubic transformations given in [18, Sec. 2.1]."

Knowing that θ_3^2 is a modular form in $M_1(\Gamma(2,4,2)$ and $\lambda \in M(\Gamma(2,4,2))$, in our setting the proofs of all these 12 identities become routine and can be carried out with full computer-support. We illustrate this by choosing as a concrete example [15, Theorem 5.6(ai)],

$$\frac{2}{\pi}K = {}_2F_1\left(\begin{matrix}\frac{1}{4} & \frac{1}{4} \\ & 1\end{matrix}; \; 4k^2(1-k^2)\right). \tag{5.3}$$

(i) *Computer-supported discovery of* (5.3). In our setting, this relation can be discovered as follows: for

$$h := 4\lambda(1-\lambda)/64 = x - 24x^2 + \cdots \quad \text{with } x = e^{\pi i\tau},$$

we determine $c(k)$ such that

$$g := \theta_3(\tau)^2 = \sum_{k=0}^{\infty} c(k)h(\tau)^k.$$

```
In[48]:= g = Series[EllipticTheta[3, 0, x]^2, {x, 0, 9}]
Out[48]= 1 + 4x + 4x^2 + 4x^4 + 8x^5 + 4x^8 + 4x^9 + O[x]^10
In[49]:= λ = Series[EllipticTheta[2, 0, x]^4 / EllipticTheta[3, 0, x]^4, {x, 0, 9}];
In[50]:= h = Series[1/64 4λ(1 − λ), {x, 0, 8}]
Out[50]= x − 24x^2 + 300x^3 − 2624x^4 + 18126x^5 − 105504x^6 + 538296x^7 − 2471424x^8 + O[x]^9
In[51]:= U = h; W = InverseSeries[U]; comp = ComposeSeries[g, W];
In[52]:= cList = CoefficientList[Normal[Series[comp, {x, 0, 8}]], x]
Out[52]= {1, 4, 100, 3600, 152100, 7033104, 344622096, 17582760000, 924193822500}
In[53]:= GuessRE[cList, c[k]]
Out[53]= {{−4(1 + 4k)^2 c[k] + (1 + k)^2 c[1 + k] == 0, c[0] == 1}, ogf}
```

This means, the program guesses that

$$c(k) = \frac{(1/4)_k(1/4)_k}{(1)_k}\frac{4^{3k}}{k!},\ k \geq 0,$$

which confirms (5.3).

(ii) *Computer-supported proof of* (5.3). The first step is to *guess* a holonomic differential equation satisfied by $y(z) = \sum_{k=0}^{\infty} c(k)z^k$. For various purposes, including order estimation of the desired differential equations, modular function/form knowledge is useful. In our case, $X(\Gamma(2,4,2))$ has genus 0. In addition, g is a modular form of weight 1 not only for $\Gamma(2,4,2)$, but also for the larger group

$$\Gamma := \Gamma(2,4,2) \cup \begin{pmatrix} 0 & -1 \\ 1 & 0 \end{pmatrix} \Gamma(2,4,2); \tag{5.4}$$

moreover, h is a Hauptmodul for this group.[25] As a consequence, according to Proposition 6.1 there exists a differential equation of order 2:

```
In[54]:= RE2DE[{−4(1 + 4k)^2 c[k] + (1 + k)^2 c[1 + k] == 0, c[0] == 1}, c[k], y[z]]
Out[54]= {−4y[z] − (−1 + 96z)y'[z] − (−z + 64z^2)y''[z] = 0, y[0] = 1, y'[0] = 4}
```

As a consequence, in order to prove (5.3) we have to prove that

$$4y(h) + (-1 + 96h)\frac{dy(h)}{dh} + (-h + 64h^2)\frac{d^2y(h)}{dh^2} = 0 \quad \text{with } y(z) := \sum_{k=0}^{\infty} c(k)z^k, \tag{5.5}$$

[25] We are grateful to the anonymous referee for pointing to this larger group.

and

$$y(h)|_{h=0} = 1 \quad \text{and} \quad \left(\frac{d}{dh}y\right)(h)|_{h=0} = 4, \tag{5.6}$$

where, as above, $g = y(h) = 1 + 4h + 100h^2 + 3600h^3 + 152100h^4 + \cdots$, and $h = x - 24x^2 + 300x^3 - 2624x^4 + 18126x^5 + \cdots$ with $x = e^{\pi i \tau}$.

The verification of (5.6) is immediate from the series expansion of y in powers of h. The "computer-proof" of (5.5) works analogously to that of (4.4) given in Sec. 6.3.

6. Proving Holonomic Differential Equations for Modular Forms

Using computer algebra, we derived relations like (5.3) as *conjectures*. In this section, we explain how such identities, once guessed, can be *proved* routinely in computer-assisted fashion. As pointed out in Sec. 5.3, such proofs carry out the equivalent task to show that the given modular form g and the given modular function h satisfy an associated holonomic differential equation for example, the differential equation (5.5) is associated to (5.3).

6.1. *Holonomic DEs for modular forms*

In several examples, we showed how these associated holonomic differential equations can be derived, as a *guess*, in computer-supported fashion. Modular form theory guarantees the existence of such differential equations. Zagier [40, Proposition 21] introduces to this fact as follows: "... it is at the heart of the original discovery of modular forms by Gauss and of the later work of Fricke and Klein and others, and appears in modern literature as the theory of Picard–Fuchs differential equations or of the Gauss–Manin connection — but it is not nearly as well known as it ought to be. Here is a precise statement:"

Proposition 6.1. *Let $g(\tau)$ be a modular form of weight $k > 0$ and $h(\tau)$ a modular function, both with respect to the congruence subgroup Γ. Express $g(\tau)$ locally as $y(h(\tau))$. Then the function $y(h)$ satisfies a linear differential equation of order $k+1$ with algebraic coefficients, or with polynomial coefficients if the compact Riemann surface $X(\Gamma)$ has genus* 0 *and* $\mathrm{ord}(h(\tau)) = 1$.[26]

After stating this proposition, Zagier [40, p. 21] continues: "This proposition is perhaps the single most important source of applications of modular forms in other branches of mathematics, so with no apology we sketch three different proofs, ..."

Zagier's third proof is constructive; i.e. given g and h, it constructs the associated differential equation. Along this line, the following observation is highly relevant for applying holonomic proving strategies[aa]:

[26] In view of Lemma 4.1, $\mathrm{ord}\, h = 1$ means $m = 1$.
[aa] We were not able to find this proposition in the literature.

Proposition 6.2. *In the setting of Proposition* 6.1, *the function* $y(h)$ *satisfies a linear differential equation with rational coefficients also when the genus of* $X(\Gamma)$ *is nonzero or when* $\mathrm{ord}(h(\tau)) > 1$. — *In these cases, the order of the differential equation in general will be larger than* $k+1$.

Proof. See Appendix A.1. □

Another major aspect we want to stress is that, as an alternative to constructing the associated holonomic differential equation, one can follow the "first guess, then prove" strategy. This means, using software like the `GeneratingFunctions` package, one first guesses the holonomic equation, and then proves it by using the following algorithm.

6.2. *An algorithm to prove holonomic DEs for modular forms*

We describe an algorithm, ModFormDE, which solves the following problem.

GIVEN a modular form $g \in M_k(\Gamma)$ with weight $k \in \mathbb{Z}_{\geq 1}$ for the congruence subgroup Γ, and a modular function $h \in M(\Gamma)$; moreover, suppose g has a local expansion of the form

$$g(\tau) = y(h(\tau)) \quad \text{where } y(z) := \sum_{k=0}^{\infty} c(k) z^k.$$

PROVE that $y(h)$ satisfies a holonomic differential equation of the form

$$P_m(h)y^{(m)}(h) + P_{m-1}(h)y^{(m-1)}(h) + \cdots + P_0(h)y(h) = 0, \tag{6.1}$$

where $P_j(X)$ are given polynomials in $\mathbb{C}[X]$ with $P_m(X) \neq 0$.

Note that $y^{(k)}(h) := \frac{dy}{dz}(z)|_{z=h}$.

Our algorithm ModFormDE is based on work of Yang [38]. So, before describing the steps of ModFormDE, we recall notation and notions used in [38].

First, we recall the differential operators used by Yang:

- for functions $\varphi(\tau)$ defined on $\mathbb{H}$ having q-expansions[bb] $\tilde{\varphi}(q) = \sum_{k \geq n_0} a(k) q^k$,

$$D_q\varphi = D_q\varphi(\tau) := \frac{1}{2\pi i}\frac{d\varphi}{d\tau}(\tau) = \frac{1}{2\pi i}\varphi'(\tau) = q\,\tilde{\varphi}'(q);$$

- for functions $\psi = \psi(z)$ being analytic in z

$$D_h\psi := h\frac{d\psi}{dz}(h).$$

Note that owing to $q = e^{2\pi i \tau}$, $D_q = \frac{1}{2\pi i}\frac{d}{d\tau}$; also, $D_h y = hy'(h)$.

[bb]As in [38], in this general description we assume $q = e^{2\pi i \tau}$. But the setting without any difficulties generalizes from q to $x = q^{1/N_0} = e^{2\pi i \tau / N_0}$; see, e.g., Sec. 6.3 or [31].

As in [38], define the fundamental functions

$$G_1 := \frac{D_q h}{h} = \frac{1}{2\pi i}\frac{h'}{h} \quad \text{and} \quad G_2 := \frac{D_q g}{g} = \frac{1}{2\pi i}\frac{g'}{g}; \tag{6.2}$$

note that $h' = h'(\tau) = \frac{dh(\tau)}{d\tau}$ and $g' = g'(\tau) = \frac{dg(\tau)}{d\tau}$.

Because of

$$g'(\tau) = \frac{d}{d\tau}y(h(\tau)) = \frac{dy}{dz}(h(\tau))\frac{dh}{d\tau}(\tau) = y'(h(\tau))h'(\tau),$$

we have

$$D_h y(h) = hy'(h) = h\frac{g'}{h'} = g\frac{G_2}{G_1}. \tag{6.3}$$

Similarly, one has

$$h^2 y''(h) = D_h^2 y - D_h y. \tag{6.4}$$

Yang [38, p. 9] computed

$$D_h^2 y = (-p_2)\cdot g + (-p_1)\cdot g\frac{G_2}{G_1} + \left(1 - \frac{1}{k}\right)\cdot g\frac{G_2^2}{G_1^2}, \tag{6.5}$$

where the $p_j \in M(\Gamma)$ are modular functions[cc] defined as

$$p_1 = p_1(h) := \frac{D_q G_1 - 2G_1 G_2/k}{G_1^2} \quad \text{and} \quad p_2 = p_2(h) := -\frac{D_q G_2 - G_2^2/k}{G_1^2}.$$

Yang [38, p. 10] also computed that

$$D_h^3 y = \alpha_3 \cdot g\frac{G_2^3}{G_1^3} + \alpha_2 \cdot g\frac{G_2^2}{G_1^2} + \alpha_1 \cdot g\frac{G_2}{G_1} + \alpha_0 \cdot g,$$

with α_j being polynomials in h, p_1, p_2, p_1', and p_2'; and this pattern continues. Define the polynomial ring

$$R := \mathbb{C}\left[h, p_1, p_2, \frac{dp_1}{dh}, \frac{dp_2}{dh}, \ldots, \frac{d^m p_1}{dh^m}, \frac{d^m p_2}{dh^m}, \ldots\right].$$

Note that the elements of this ring are modular functions in $M(\Gamma)$.

Yang showed that any expression of the form,

$$Y := Q_m(h)D_h^m y + Q_{m-1}(h)D_h^{m-1} y + \cdots + Q_0(h)y, \tag{6.6}$$

with polynomials $Q_j(X) \in \mathbb{C}[X]$, can be written into "Yang form" as

$$Y = \alpha_m \cdot g\frac{G_2^m}{G_1^m} + \alpha_{m-1} \cdot g\frac{G_2^{m-1}}{G_1^{m-1}} + \cdots + \alpha_0 \cdot g \quad \text{with } \alpha_j \in R. \tag{6.7}$$

In [31], we show that these α_j are uniquely determined.

[cc]This is proved in [38, Lemma 1]. Hence the p_j are also algebraic functions in $h \in M(\Gamma)$; see the remark after Theorem 8.1.

Our algorithm ModFormDE consists of the following steps:

Step 0. Rewrite the left side of (6.1) into the form (6.6).

Step 1. Transform the resulting expression into Yang form (6.7).

Step 2. Owing to the uniqueness of the coefficients in (6.7), the proof of (6.1) finally is reduced to prove that

$$\alpha_m = 0, \alpha_{m-1} = 0, \dots, \alpha_0 = 0. \tag{6.8}$$

Since the α_j are modular functions, this task, owing to (2.5), reduces to determine upper bounds for the number of poles of each α_j. Such bounds are given explicitly in [31]; because of $\alpha_j \in R$, it is sufficient to provide such bounds for h and for the $\frac{d^n p_j}{dh^n}$, $n \geq 0$. The zero test is completed by expanding the q-series expansions of the α_j at infinity[dd] to sufficiently high powers of q such that the number of possible zeros exceeds the corresponding bounds for the poles.

6.3. *The ModFormDE algorithm on an example*

To illustrate the behavior of the ModFormDE algorithm, we choose the task to prove the validity of (4.4). For a better matching to Yang [38, Example 1], we choose $h := \lambda$, instead of $h = \lambda/16$ as in the setting of Example 4.2. As a consequence, the differential equation (4.4) changes into the equivalent form

$$h(h-1)\frac{d^2y}{dz^2}(h) + (2h-1)\frac{dy}{dh}(h) + \frac{1}{4}y(h) = 0$$

$$\text{with } y(z) := \sum_{k=0}^{\infty} \frac{c(k)}{16^k} z^k, \quad \text{and where } g = y(h). \tag{6.9}$$

Step 0. Use (6.3) and (6.4) to rewrite the left side of the differential equation in (6.9) into the form (6.6):

$$Y := (h-1)D_h^2 y + hD_h y + \frac{h}{4}y. \tag{6.10}$$

Step 1. By using (6.3) and (6.5), we transform (6.10) into Yang form (6.7)[ee]:

$$Y = \alpha_1 \cdot g\frac{G_2}{G_1} + \alpha_0 \cdot g \quad \text{with } \alpha_1 := -p_1 h + p_1 + h \in M(\Gamma(2,4,2)), \tag{6.11}$$

$$\alpha_0 := -p_2 h + p_2 + \frac{h}{4} \in M(\Gamma(2,4,2)).$$

Step 2. The proof of (6.9) is reduced to zero recognition of the modular functions α_1 and α_0.

[dd] As mentioned, if required we can work also with x-expansions at infinity where $x = q^{1/N_0}$ for fixed $N_0 \in \mathbb{Z}_{\geq 1}$.

[ee] It is important to note that in this application q is replaced by $x := q^{1/2}$ compare the remark given in the first paragraph of Example 4.2.

Step 2a. α_1 is a modular function for $\Gamma(2,4,2)$. Hence, the task to prove $\alpha_1 = 0$, owing to (2.5), reduces to determine upper bounds for the total number poles of α_1, respectively, $\hat{\alpha_1}$. Obviously,

$$\begin{aligned}\text{no. of poles}(\alpha_1) &= \text{no. of poles}(p_1(1-h)+h)\\ &\leq \text{no. of poles}(h) + \text{no. of poles}(p_1) + \text{no. of poles}(h). \qquad (6.12)\end{aligned}$$

To proceed, one needs further information about $\Gamma(2,4,2)$, respectively, $X(\Gamma(2,4,2))$. For instance, it has three cusps at 0, 1, and ∞. To obtain such information, one can run systems like Magma, for example:

```
> G:=CongruenceSubgroup([2,4,2]);
> Cusps(G);
[
    oo,
    0,
    1
]
> Widths(G);
[ 2, 2, 2 ]
```

Inspecting the x-expansions ($x = e^{\pi i\tau}$) at these cusps, one sees that h, respectively $\hat{h}$, has its only pole at 1 with multiplicity 1. To bound the number of poles of p_1, in view of $g \in M_1(\Gamma(2,4,2))$, i.e. weight $k = 1$, we apply from [31, Theorem 5.3]:

$$\text{no. of poles}(p_1) \leq 8\text{no. of poles}(h) + 3\text{NofPoles}(g^2) + (2k+4)(g_{\Gamma(2,4,2)} - 1).$$

Here $g_{\Gamma(2,4,2)} = 0$, the genus of $X(\Gamma(2,4,2)$ in addition, according to the definition given in [31], one has $\text{NofPoles}(g^2) = 2$.

Remark 6.3. Our definition [31, Definition 6.12] of $\text{NofPoles}(g^2)$ uses a notion of "order of modular forms" for even weight; see [31, Definition 6.9]. This order is coinciding with the order of a differential on a compact Riemann surface, although *a priori* there is no direct connection between these two notions.

Hence, we obtain

$$\text{no. of poles}(p_1) \leq 8 + 3 \cdot 2 + 6(-1) = 8;$$

and, turning back to (6.12),

$$\text{no. of poles}(\alpha_1) \leq 1 + 8 + 1 = 10.$$

This means, to prove $\alpha_1 = 0$, it is sufficient to verify that the first 10 coefficients in the x-expansion of α_1 at ∞ are zero.[ff]

[ff]Recall that $x = q^{1/2}$.

Step 2b. The task to prove $\alpha_0 = 0$ works analogously to Step 2a. In this case, from [31, Theorem 5.3] gives

$$\text{no. of poles}(p_2) \leq 22.$$

This means, to prove $\alpha_0 = 0$, it is sufficient to verify that the first 24 coefficients in the x-expansion of α_0 at ∞ are zero.

Note, this algorithmic proving strategy carries over to the general case (6.1). One can establish explicit formulas for the bounds on the number of poles of p_1 and p_2, and of the derivatives involved. A full account of all these details is given in [31]. In addition, in a forthcoming version of this paper we present an improved degree estimation. For example, in this setting, for p_1 and p_2 as above

$$\text{no. of poles}(p_1) \leq 2 \quad \text{and} \quad \text{no. of poles}(p_2) \leq 12. \tag{6.13}$$

7. Puiseux Expansions of Modular Functions

Given a modular function h, in the previous sections we discussed the case when its "Puiseux mate" g is a *modular form* with positive integer weight. From this section on, we consider what happens when the weight of g is zero, i.e. when both g and h are modular functions. Again we will follow a "first guess, then prove" strategy and we shall see that holonomic guessing will work essentially the same. On the other hand, in contrast to the case of modular forms, for modular *functions* g, proving will be a more elementary routine step, owing to classical theory.

All the modular functions in this section and in the following sections are for congruence subgroups $\Gamma_0(N)$. Consequently, the abbreviations

$$M(N) := M(\Gamma_0(N)) \quad \text{and} \quad M^\infty(N) := M^\infty(\Gamma_0(N))$$

will be convenient.

To illustrate the basic features and the potential for applications in the field of partition congruences, we again consider a concrete example.

Define

$$f_2(\tau) := \frac{\eta(\tau)\eta(2\tau)^3}{\eta(11\tau)^3\eta(22\tau)} = q^{-2}\prod_{k=1}^{\infty}\frac{(1-q^k)(1-q^{2k})^3}{(1-q^{11k})^3(1-q^{22k})} \in M(22),$$

$$f_3(\tau) := \frac{\eta(\tau)^3\eta(2\tau)}{\eta(11\tau)\eta(22\tau)^3} = q^{-3}\prod_{k=1}^{\infty}\frac{(1-q^k)^3(1-q^{2k})}{(1-q^{11k})(1-q^{22k})^3} \in M(22).$$

Owing to Newman's Lemma [28, Theorem 1.64] it is straightforward to show that these eta-quotients are in $M(22)$. In [30], we derived and proved the relation

$$\frac{1}{q^4}\prod_{j=1}^{\infty}\frac{(1-q^j)^{12}}{(1-q^{11j})^{11}}\sum_{n=0}^{\infty}p(11n+6)q^n = 11F_2^2 + 11^2F_3 + 11^3F_2 + 11^4, \tag{7.1}$$

for the generating function of the partition numbers $p(11n+6)$ with

$$F_2(\tau) := f_2(\tau) - (U_2 f_3)(\tau) = q^{-2} + 2q^{-1} - 12 + 5q + 8q^2 + \cdots \in M^\infty(11),$$

$$F_3(\tau) := f_3(\tau) - 4(U_2 f_2)(\tau) = q^{-3} - 3q^{-2} - 5q^{-1} + 24 - \cdots \in M^\infty(11),$$

where

$$U_2 \sum_{n=N}^{\infty} a(n)q^n := \sum_{2n\geq N}^{\infty} a(2n)q^n.$$

As the main problem in this section, we consider Puiseux expansion of the left side of (7.1). Concretely, we consider the following task: Given

$$\begin{aligned} g(\tau) &:= \frac{1}{q^4}\prod_{j=1}^{\infty}\frac{(1-q^j)^{12}}{(1-q^{11j})^{11}}\sum_{n=0}^{\infty}p(11n+6)q^n \in M^\infty(11) \\ &= \frac{11}{q^4} + \frac{165}{q^3} + \frac{748}{q^2} + \frac{1639}{q} + 3553 + 4136q + O(q^2) \end{aligned} \tag{7.2}$$

and

$$\begin{aligned} f(\tau) &:= F_2(\tau) = f_2(\tau) - (U_2 f_3)(\tau) \in M^\infty(11) \\ &= \frac{1}{q^2} + \frac{2}{q} - 12 + 5q + 8q^2 + O(q^3), \end{aligned} \tag{7.3}$$

determine $c(k)$ such that

$$g = \sum_{k=-4}^{\infty} c(k)\left(\frac{1}{f}\right)^{\frac{k}{2}} = \sum_{k=-4}^{\infty} c(k)\left(\frac{1}{F_2}\right)^{\frac{k}{2}}. \tag{7.4}$$

Using the `GeneratingFunctions` package, according to Lemma 4.1 with $h := (\frac{1}{f})^{1/2}$ and $m = 2$, the coefficients $c(k)$ can be computed as follows:

In[55]:= $\mathbf{g = \frac{11}{q^4} + \frac{165}{q^3} + \frac{748}{q^2} + \frac{1639}{q} + 3553 + 4136q + 6347q^2 + 3586q^3 + 7414q^4 + O[q]^5;}$

The first eight coefficients of g:

In[56]:= **gg[n_] := Coefficient[Normal[g], q, n]**

In[57]:= **Table[gg[k], k, −4, 4]**

Out[57]= {11, 165, 748, 1639, 3553, 4136, 6347, 3586, 7414}

In[58]:= $\mathbf{f = \frac{1}{q^2} + \frac{2}{q} - 12 + 5q + 8q^2 + q^3 + 7q^4 - 11q^5 + 10q^6 + O[q]^7;}$

In[59]:= $\mathbf{h = Series[q\left(\frac{1}{q^2 f}\right)^{1/2}, \{q, 0, 6\}]}$

Out[59]= $q - q^2 + \frac{15q^3}{2} - 23q^4 + \frac{855q^5}{8} - \frac{3481q^6}{8} + O[q]^7$

In[60]:= **U = h; W = InverseSeries[U]; ComposeSeries[g, W]**

Out[60]= $\frac{11}{q^4} + \frac{121}{q^3} + \frac{605}{q^2} + \frac{5687}{2q} + \frac{14641}{2} + \frac{84095q}{8} + O[q]^2$

Alternatively, we can express the answer in powers of h:

In[61]:= **Clear[h]**

In[62]:= $\mathbf{Normal[Sum[gg[k]W^k, \{k, -4, 3\}]] + O[h]^2\ /.\ q \to h}$

Out[62]= $\frac{11}{h^4} + \frac{121}{h^3} + \frac{605}{h^2} + \frac{5687}{2h} + \frac{14641}{2} + \frac{84095h}{8} + O[h]^2$

This means, we obtained

$$g = \sum_{k=-4}^{\infty} c(k) \left(\frac{1}{f}\right)^{\frac{k}{2}} = \sum_{k=-4}^{\infty} c(k) h^k = \frac{11}{h^4} + \frac{121}{h^3} + \frac{605}{h^2} + \frac{5687}{2h} + \frac{14641}{2} + \cdots$$

Using additional terms in the q-expansions of f and g, one computes in the same fashion that for $h := (\frac{1}{f})^{1/2}$:

$$h^4 g = 11 + 11^2 h + 11^2 \cdot 5h^2 + \frac{11^2 \cdot 47}{2} h^3 + \frac{11^4}{2} h^4 + \frac{11^2 \cdot 5 \cdot 139}{8} h^5)$$
$$- \frac{11^2 \cdot 17 \cdot 199}{16} h^7 + \frac{11^2 \cdot 3 \cdot 50993}{128} h^9 - \frac{11^2 \cdot 67 \cdot 107 \cdot 347}{256} h^{11} + \cdots. \quad (7.5)$$

We note explicitly that all the coefficients are quotients with integers as numerators, and denominators being powers of 2. This is owing to

$$(1+\psi)^{\frac{1}{2}} = \sum_{n=0}^{\infty} \binom{1/2}{n} \psi^n \quad \text{and} \quad \binom{1/2}{n+1} = \frac{1}{2} \frac{(-1)^n}{4^n} \frac{1}{n+1} \binom{2n}{n},$$

together with the following lemma for formal power series which is folklore.

Lemma 7.1. *Suppose* $\psi = 1 + f_1 x + f_2 x^2 + \cdots \in \mathbb{Z}[[x]]$. *Then*

$$\frac{1}{\psi} = 1 - f_1 x + (f_1^2 - f_2) x^2 + \cdots \in \mathbb{Z}[[x]]$$

and

$$\phi = x - f_1 x^2 + (2f_1^2 - f_2) x^3 + \cdots \in \mathbb{Z}[[x]],$$

where ϕ *is the compositorial inverse of* $x\psi$, *i.e.* $\phi(x\psi(x)) = \phi(x)\psi(\phi(x)) = x$.

In the next step, we take the first 24 coefficients in the expansion (7.5) of $y(h) := h^4 g$ and use the `GeneratingFunctions` package to guess an algebraic equation[gg]:

```
In[63]:= cList = {11, 121, 605, 5687/2, 14641/2, 84095/8, 0, -(409343/16), 0, 18510459/128,
    0, -(301004803/256), 0, 12660067739/1024, 0, -(323203962851/2048),
    0, 75351596221803/32768, 0, -(2379607141538539/65536), 0,
    157195485892282497/262144, 0, -(5317572127702768593/524288)}

In[64]:= Clear[h]; GuessAE[cList, y[h], 2, 6]

Out[64]= {{y[h]^2 - 11(1331h^4 + 110h^2 + 2)y[h] - 121(14641h^6 + 1331h^4 + 11h^2 - 1) = 0,
    y[0] = 11}, ogf}
```

With $y(h) := h^4 g = \frac{1}{f^2} g$, the algebraic relation in `Out[64]` translates into an algebraic relation involving f and g which, owing to its importance, we state it as a separate proposition.

[gg]The addition of 2 and 6 in the procedure call `In[64]` asks the program to guess an algebraic relation of degree 2 with polynomial coefficients of degree maximally 6.

Proposition 7.2. *The modular functions f and g in $M^\infty(11)$ as given in* (7.3) *and* (7.2) *satisfy*

$$g^2 - 11(11^3 + 11 \cdot 10f + 2f^2)g - 11^2(11^4 f + 11^3 f^2 + 11f^3 - f^4) = 0. \quad (7.6)$$

Proof. Knowing from [30] that $f, g \in M^\infty(11)$, according to Remark 2.2, to prove (7.6) it suffices to show that the principal part and the constant term in the q-expansion of the left side of (7.6) are equal to zero. □

Remark 7.3. To verify this by actual computation, it turns out that the precision we entered for f and g in our Mathematica session suffices:

```
In[65]:= Y^2 − 11(11^3 + 110X + 2X^2)Y − 11^2(11^4X + 11^3X^2 + 11X^3 − X^4) /.{Y → g, X → f}
Out[65]= O[q]^1
```

8. Existence of Algebraic Relations

Holonomic guessing can be used to set up algebraic relations like (7.6). As explained in Sec. 7, *proving* such relations between two modular *functions* in $M^\infty(N)$ amounts to straightforward computational verification. In this section, we discuss why such algebraic relations exist, state some facts of computational relevance, and connect to Puiseux's theorem for solving algebraic equations.

In [29, Theorem 7.1] we proved the following theorem:

Theorem 8.1. *Given a compact Riemann surface S together with non-constant functions f and g being meromorphic on S. Suppose f has m poles, counted with their multiplicity. If for any $p \in S$ the condition,*

$$p \text{ a pole of } g \Rightarrow p \text{ a pole of } f, \quad (8.1)$$

holds, then there exist POLYNOMIALS $c_1, \ldots, c_m \in \mathbb{C}[X]$ *such that*

$$g^m + c_1(f)g^{m-1} + \cdots + c_m(f) = 0. \quad (8.2)$$

Without the pole condition (8.1), this theorem holds with the c_j being *rational functions* in $\mathbb{C}(X)$. In this version the statement is folklore; see for instance [20, §8.2 and §8.3]. Our version, where condition (8.1) forces the c_j to be *polynomials*, is algorithmically relevant, but it seems to be less known. A special instance of Theorem 8.1 where f is allowed to have poles only at one point of S is used in [39].

In the classical context, i.e. without the pole condition (8.1), another useful result is proved in [39, Lemma 1]:

Lemma 8.2. *Let f and g be as in Theorem* 8.1. *Suppose f and g have m and M poles, respectively, counted with their multiplicity. If* $\gcd(m, M) = 1$ *then there exists a polynomial,*

$$c(X, Y) = Y^m + c_1(X)Y^{m-1} + \cdots + c_m(X) \in \mathbb{C}(X)[Y], \quad (8.3)$$

which is irreducible over $\mathbb{C}(X)$, *and where the* $c_j(X) \in \mathbb{C}(X)$ *are rational functions such that* $c(f, g) = 0$.

Now we are ready to state and prove a general theorem which explains the existence of algebraic relations such as (7.6).

Theorem 8.3. *Let* f *and* g *be modular functions from* $M^\infty(N)$ *with* $\operatorname{ord} f = -m \leq -1$ *and* $\operatorname{ord} g = -M \leq -1$. *Then there exist polynomials* $c_j(X) \in \mathbb{C}[X]$ *such that*

$$g^m + c_1(f)g^{m-1} + \cdots + c_m(f) = 0.$$

Moreover, if

$$c(X,Y) := Y^m + c_1(X)Y^{m-1} + \cdots + c_m(X) \in \mathbb{C}[X][Y] = \mathbb{C}[X,Y] \tag{8.4}$$

is irreducible with $\deg c_m(X) = M$, *then*

$$\deg c_j(X) \leq M - 1, \quad j = 1, \ldots, m-1. \tag{8.5}$$

The condition

$$\gcd(m, M) = 1, \tag{8.6}$$

is sufficient, but not necessary, to imply that $c(X,Y)$ *is irreducible with degree* $\deg c_m(X) = M$.

Proof. The existence of $c(X,Y) \in \mathbb{C}[X,Y]$ such that $c(f,g) = 0$ is implied by Theorem 8.1 with $S = \mathbb{C} \cup \{\infty\}$, the Riemann sphere. Note that both f and g have their only pole at ∞ with multiplicities m and M, respectively.

To prove the second part of the theorem, consider the polynomial

$$d(X,Y) := X^M + d_1(Y)X^{M-1} + \cdots + d_M(Y) \in \mathbb{C}[Y][X] = \mathbb{C}[X,Y]$$

with $d(f,g) = 0$; $d(X,Y)$ exists in this form by applying the same argument as for $c(X,Y)$. The polynomials $c(X,Y), d(X,Y) \in \mathbb{C}[X,Y]$ cannot be relatively prime. Suppose they are, then owing to $c(f,g) = 0 = d(f,g)$ there are infinitely many common roots which contradicts the fact that the set of common roots of two bivariate polynomials has to be finite, see, e.g., [1, Exercise 1.3.8]. As a consequence, the irreducible $c(X,Y)$ has to be a factor of $d(X,Y)$, i.e.

$$d(X,Y) = (a_0(Y)X^L + a_1(Y)X^{L-1} + \cdots + a_L(Y))c(X,Y)$$

for some polynomials $a_j \in \mathbb{C}[Y]$. Now the assumption $\deg c_m(X) = M$ implies $L = 0$ and $a_0(Y) = a_0 \in \mathbb{C}$, and comparing the coefficients of X^iY^j of both sides of $d(X,Y) = a_0 \cdot c(X,Y)$ proves (8.5).

The third part of the theorem is an immediate consequence of Lemma 8.2 which implies that both $c(X,Y)$ and $d(X,Y)$ are irreducible. Hence, as in part two

$$d(X,Y) = a_0 \cdot c(X,Y) \quad \text{for some } a_0 \in \mathbb{C}.$$

Applying coefficient comparison gives $\deg c_m(X) = M$.

Finally, in Example 8.4 we show that the condition (8.6) is not necessary for $c(X,Y)$ to be irreducible with $\deg c_m(X) = M$. □

Example 8.4. Consider the relation (7.6) with modular functions $f, g \in M^\infty(11)$ as in (7.3) and (7.2), where the orders $\operatorname{ord} f = -2$ and $\operatorname{ord} g = -4$ are not relatively prime. Nevertheless, using computer algebra one can check that the corresponding polynomial

$$\begin{aligned} c(X,Y) &= Y^2 + c_1(X)Y + c_2(X) \\ &= Y^2 - 11(11^3 + 11\cdot 10X + 2X^2)Y - 11^2(11^4X + 11^3X^2 + 11X^3 - X^4), \end{aligned}$$

such that $c(f,g) = 0$ as in (7.6), is irreducible in $\mathbb{C}[X,Y]$ with $\deg c_2(X) = 4 = M$.

Recall that, in view of the Laurent series

$$g = \sum_{k=-4}^{\infty} c(k)\left(\frac{1}{f}\right)^{\frac{k}{2}} = \sum_{k=-4}^{\infty} c(k)h^k \quad \text{with } h := \left(\frac{1}{f}\right)^{1/2}, \tag{8.7}$$

we set $y(h) := h^4 g$ to obtain a power series expansion

$$y(h) = \sum_{k=0}^{\infty} c(k-4)h^k = 11 + 11^2h + 11^2\cdot 5h^2 + \frac{11^2\cdot 47}{2}h^3 + \cdots,$$

where the coefficients $c(k)$ form a holonomic sequence with coefficients as in (7.5). Using the `GeneratingFunctions` package, we algorithmically *guessed* the algebraic relation `Out[64]`

$$y(h)^2 - 11(11^3h^4 + 11\cdot 10h^2 + 2)y(h) - 11^2(11^4h^6 + 11^3h^4 + 11h^2 - 1) = 0. \tag{8.8}$$

Noting that $y(h) := h^4 g$ with $h := (\frac{1}{f})^{1/2}$, this relation is equivalent to (7.6)

$$g^2 - 11(11^3 + 11\cdot 10f + 2f^2)g - 11^2(11^4f + 11^3f^2 + 11f^3 - f^4) = 0.$$

We want to stress that this equivalence is no surprise in view of "Puiseux's theorem"; see, e.g., [35, Theorem 6.1.5, Proposition 6.1.6, Corollary 6.1.7].

Theorem 8.5 (Puiseux's Theorem). *Let $P(X,Y) \in \mathbb{C}[X][Y]$ be an irreducible polynomial in Y with $\deg_Y P(X,Y) = m$. Then there exist positive integers $d_1,\ldots,d_r$ such that $d_1 + \cdots + d_r = m$, and for each $j \in \{1,\ldots,r\}$ there exist Laurent series*

$$y_{j,\ell}(X) = \sum_{k=M_j}^{\infty} a_{j,\ell}(k)e^{\frac{2\pi i k\ell}{d_j}}X^k \text{ with } a_{j,\ell}(k) \in \mathbb{C} \text{ such that } P(X^{d_j}, y_{j,\ell}(X)) = 0,$$

for $\ell \in \{0,\ldots,d_j - 1\}$.

After the substitution $X \to X^{1/d_j}$, the Laurent series $y_{j,\ell}(X)$ in the theorem become objects of the form $y_{j,\ell}(X^{1/d_j}) = \sum_{k=M_j}^{\infty} \tilde{a}_{j,\ell}(k)X^{k/d_j}$; i.e. Puiseux series (also called fractional series) which satisfy $P(X, y_{j,\ell}(X^{1/d_j})) = 0$.

Remark 8.6. The classic version of "Puiseux's theorem" traces back to Newton. For proofs in the setting of complex analysis see, for example [19, Chap. 7] or

[24, Chap. 6.3]. For computational aspects see, for instance, [22, Theorem 6.2] or [37, Chap. 4.3].

Also in our context Puiseux's theorem can be useful for computational reasons: it implies lower bounds supplementing the general degree bound given in Theorem 8.3(8.5). We restrict to state a special case:

Corollary 8.1. *Let $P(X,Y)$ be as in Puiseux's Theorem* 8.5 *with $r = 1$ (i.e. $d_1 = m$) and with coefficient polynomials $P_n(X) \in \mathbb{C}[X]$ such that*

$$P(X,Y) = P_m(X)Y^m + P_{m-1}(X)Y^{m-1} + \cdots + P_0(X).$$

Suppose $d = \deg P_m(X)$, then for $\ell = 1, \ldots, m$ and M_1 as in Theorem 8.5:

$$\text{lowest power of } X \text{ in } P_{m-\ell}(X) \geq \max\left\{d + \ell\frac{M_1}{m}, 0\right\}. \tag{8.9}$$

Proof. The statement follows directly by comparing coefficients of Y^k in

$$\begin{aligned}P(X,Y) &= P_m(X)(Y - y_1(X^{1/m}))\cdots(Y - y_m(X^{1/m}))\\ &= P_m(X)Y^m - P_m(X)(y_1(X^{1/m}) + \cdots + y_m(X^{1/m})Y^{m-1} + \cdots\\ &\quad + (-1)^m P_m(X)y_1(X^{1/m})\cdots y_m(X^{1/m}),\end{aligned}$$

where the $y_\ell(X)$ are the $y_{1,\ell}(X)$ in Theorem 8.5. To this end, note that the fractional series formed by the elementary symmetric functions have to be polynomials. □

Example 8.7. Consider

$$\begin{aligned}P(X,Y) := {}& X^4Y^2 - 11X^2(2 + 11\cdot 10X + 11^3X^2)Y\\ &+ 11^2(1 - 11X - 11^3X^2 - 11^4X^3),\end{aligned} \tag{8.10}$$

i.e. $P(1/f, g) = 0$ with f as in (7.3) and g as in (7.2) with expansion (7.4). In other words, the relation $P(1/f, g) = 0$ is nothing but (7.6), produced from the equivalent relation `Out[64]` by setting $y(h) := h^4 g$ with $h^2 = 1/f$. In the light of of Puiseux's Theorems 8.5 and (8.7), this means

$$\begin{aligned}P(X,Y) &= X^4(Y - y_1(X^{1/2}))(Y - y_2(X^{1/2}))\\ &= X^4Y^2 + P_1(X)Y + P_0(X),\end{aligned}$$

with

$$y_1(X^{1/2}) := \sum_{k=-4}^{\infty} c(k)X^{k/2} \quad \text{and} \quad y_2(X^{1/2}) := \sum_{k=-4}^{\infty} (-1)^k c(k)X^{k/2}.$$

The coefficient polynomials $P_0(X)$ and $P_1(X)$ can be recovered from truncated versions y_j of the $y_j(X^{1/2})$ as follows. For instance, set

$$y_1 := \frac{11}{X^2} + \frac{121}{X^{3/2}} + \frac{605}{X} + \frac{5687}{2X^{1/2}} + \frac{14641}{2} + \frac{84095X^{1/2}}{8} - \frac{409343X^{3/2}}{16} + \frac{18510459X^{5/2}}{128}$$

and

$$y_2 := \frac{11}{X^2} - \frac{121}{X^{3/2}} + \frac{605}{X} - \frac{5687}{2X^{1/2}} + \frac{14641}{2} - \frac{84095X^{1/2}}{8} + \frac{409343X^{3/2}}{16} - \frac{18510459X^{5/2}}{128}.$$

Then

$$y_1 + y_2 = \frac{11}{X^2}(2 + 110X + 1331X^2)$$

and

$$y_1y_2 = \frac{11^2}{X^4}(1 - 11X - 1331X^2 - 14641X^3) - \frac{36421581163X^2}{128} - \frac{1891760432903X^3}{512} + \cdots .$$

Finally, in view of Proposition 8.1, we note that with $P_2(X) := X^4$:

$$d = \deg P_2(X) = 4,$$

and one has for the lowest powers of X in $P_1(X)$, respectively, $P_0(X)$:

$$2 = \text{lowest power of } X \text{ in } P_1(X) \geq 4 + \frac{-4}{2} = 2$$

and

$$0 = \text{lowest power of } X \text{ in } P_0(X) \geq 4 - 4 = 0;$$

this confirms the estimates given by (8.9).

9. Holonomic Differential Equations for Modular Functions

In this section, we continue to study the situation where g is a modular *function* as h. But, in contrast to algebraic relations as considered in Sec. 8, we now put our focus on *differential equations.* Again we see that the underlying mathematics is somewhat simpler than that for modular *forms* g. Nevertheless, applications of this setting still seem to be of some interest. In Secs. 9 and 10, we illustrate this aspect by examples related to congruences for partition numbers.

Suppose we are interested in $\sum_{n=0}^{\infty} p(11n+6)q^n$ with coefficients being the numbers of partitions of $11n + 6$. Using Smoot's implementation [34] of Radu's

Ramanujan-Kolberg algorithm, we compute a q-product, $q = e^{2\pi i\tau}$, such that when multiplied to this generating function

$$g(\tau) := \frac{1}{q^4}\prod_{j=1}^{\infty}\frac{(1-q^j)^{12}}{(1-q^{11j})^{11}}\sum_{n=0}^{\infty}p(11n+6)q^n \in M^{\infty}(11). \tag{9.1}$$

Given $f = \frac{1}{q^2}+\frac{2}{q}-12+5q+8q^2+\cdots \in M^{\infty}(11)$ as in (7.3), in Sec. 7 we considered the task to determine $c(k)$ such that

$$g = \sum_{k=-4}^{\infty} c(k)\left(\frac{1}{f}\right)^{\frac{k}{2}}.$$

Using holonomic guessing and modular function proving, we derived and proved the algebraic relation (7.6) between f and g, for convenience we display (7.6) again:

$$g^2 = 11(11^3 + 11\cdot 10f + 2f^2)g + 11^2(11^4 f + 11^3 f^2 + 11f^3 - f^4). \tag{7.6}$$

Connecting this fact with the q-expansion (9.1), one obtains a proof of a classical congruence of Ramanujan [32]:

Theorem 9.1. *For non-negative integer n*

$$p(11n+6) \equiv 0 \pmod{11}. \tag{9.2}$$

Proof. By (7.6) each coefficient of g^2 has 11 as a factor. Hence 11 divides each coefficient of g, and thus $11 \mid p(11n+6)$. □

For computational purposes, it is convenient to set $y(h) := h^4 g$ with $h := (\frac{1}{f})^{1/2}$, and to rewrite (7.6) into the equivalent form (8.8), which we repeat to display for the reader's convenience:

$$y(h)^2 - 11(11^3h^4 + 11\cdot 10h^2 + 2)y(h) - 11^2(11^4h^6 + 11^3h^4 + 11h^2 - 1) = 0. \tag{8.8}$$

This means

$$y(h) = \sum_{k=0}^{\infty} c(k-4)h^k = 11 + 11^2h + 11^2\cdot 5h^2 + \frac{11^2\cdot 47}{2}h^3 + \cdots \tag{9.3}$$

is an algebraic power series and, according to Proposition 3.3, the coefficients $c(k)$ constitute a holonomic sequence.

After these preliminary remarks, we finally turn to the aspect of *differential equations*. To this end, using the `GeneratingFunctions` package, we transform (8.8) into a holonomic differential equation:

In[66]:= $\mathbf{AE2DE[y[h]^2 - 11(11^3h^4 + 110h^2 + 2)y[h] - 121(11^4h^6 + 11^3h^4 + 11h^2 - 1) == 0, y[h]]}$

Out[66]= $-44 - 1716h^2 - 7986h^4 + 351384h^6 + 3543122h^8 + 4(1 + 94h^2 + 2178h^4 + 14641h^6)y[h]$
$\quad - (4h + 188h^3 + 2904h^5 + 14641h^7)y'[h] = 0$

Note that the differential equation `Out[66]` is valid for $y(h)$ as in (9.3), where h can be an indeterminate or a complex number sufficiently close to 0, depending on a formal power series or analytic function context.

We also note that, according to Proposition 3.3, the existence of `Out[66]` in the form of an *inhomogeneous* differential equation of order 1 is no surprise, owing to the degree 2 of the algebraic relation.

In order to learn more about the $c(k)$, we set for convenience

$$d(k) := c(k-4), \quad k \geq 0, \tag{9.4}$$

and use the `GeneratingFunctions` package to transform (8.8), via the differential equation `Out[66]`, into a recurrence:

```
In[67]:= DE2RE[%66, y[h], d[k]]

Out[67]= {-2357947691(-4+k)(-2+k) d[k] - 701538156(-2+k)k d[2+k] - 161051(-176+666k+443k^2) d[4+
         k] - 2662(-9104-1218k+545k^2) d[6+k] + 242(60032+17226k+1075k^2) d[8+k] + 12(151726+33975k+
         1827k^2) d[10+k] + 8(10+k)(1087+86k) d[12+k] + 8(13+k)(14+k) d[14+k] = 0, d[0] = 11, d[2] =
         605, d[3] = (47d[1])/2, d[4] = 14641/2, d[5] = (695d[1])/8, d[6] = 0, d[7] = -((3383d[1])/16), d[8] =
         0, d[9] = (152979d[1])/128, d[10] = 0, d[11] = -((2487643d[1])/256),
          d[12] = 0}
```

The output also contains initial conditions, one observes that $c[1]$ can be freely chosen and the recurrence is still valid. In order to match the initial conditions with the corresponding values of $c(k)$, respectively, with $y(0) = c(-4) = 11$ and $y'(0) = 11^2 = c(-3)$, we choose `d[1]` $:= c(-3) = 11^2$.

An alternative way to produce a recurrence for the $d(k)$ is by computer-supported *guessing*. Again we use the `GeneratingFunctions` package, as input we take the first 35 coefficients $d(0) := c(-4), \ldots, d(34) := c(30)$:

```
In[68]:= dList35 = {11, 121, 605, 5687/2, 14641/2, 84095/8, 0, -409343/16, 0, 18510459/128,
          0, -301004803/256, 0, 12660067739/1024, 0, -323203962851/2048, 0,
          75351596221803/32768, 0, -2379607141538539/65536, 0,
          157195485892282497/262144, 0, -5317572127702768593/524288, 0,
          728392438513303248223/4194304, 0, -25092691314419458322263/8388608, 0,
          1733298832317545489307443/33554432, 0,
          - 59906702629110941247099115/67108864, 0,
          3311691446840933891508912442 7/2147483648, 0}

In[69]:= GuessRE[cList35, d[k], 6, 3]

Out[69]= {{14641(-4+k)(-2+k)k d[k] + 2904(-2+k)(-1+k)k d[2+k] +
          188(-2+k)k(2+k) d[4+k] + 4(-2+k)k(5+k) d[6+k] = 0,
          d[0] = 11, d[1] = 121, d[2] = 605, d[3] = 5687/2, d[4] = 14641/2,
          d[5] = 84095/8, d[6] = 0, d[7] = -(409343/16), d[8] = 0}, ogf}
```

With the procedure call in `In[69]` we asked the program to guess a recurrence of order 6 with polynomial coefficients of degree maximally 3, the output `Out[69]` shows that such a recurrence indeed exists.

The following remark is elementary, but important for practical computations: As we have seen in `Out[69]`, when using the *inhomogeneous* differential equation `Out[66]`, one obtains a recurrence which overshoots the order significantly, 14 instead of 6. The explanation for this increase of the order is caused by the presence of the powers h^2 up to h^8 in the inhomogeneous polynomial part. Using as input the *homogeneous* version of `Out[66]`, one also obtains a valid recurrence which is produced by comparing coefficients of h^k, and this comparison is correct from $k = 9$ on:

```
In[70]:= DE2RE[−4(1+94x^2+2178x^4+14641x^6)y[x]+(4x+188x^3+2904x^5+14641x^7)y'[x] ==
         0, y[x], d[k]]

Out[70]= 14641(−4 + k)d[k] + 2904(−1 + k)d[2 + k] + 188(2 + k)d[4 + k] + 4(5 + k)d[6 + k] = 0
```

This is, up to a common multiplicative factor, the guessed recurrence `Out[66]` but this time we *proved* this relation by an algorithmic version of the fundamental holonomic conversion, Proposition 3.1.

To summarize, we rewrite `Out[70]` as

$$\begin{aligned} d(k) = \frac{-1}{4(k-1)}&(11^4(k-10)d(k-6) + 2^3 3 \cdot 11^2(k-7)d(k-4) \\ &+ 2^2 47(k-4)d(k-2)), \end{aligned}$$

and apply the relabeling (9.4):

Proposition 9.2. *Given f as in (7.3) and g as in (9.1). Then the uniquely determined coefficients $c(k)$ such that*

$$g = \sum_{k=-4}^{\infty} c(k) \left(\frac{1}{f}\right)^{\frac{k}{2}}, \tag{9.5}$$

for $k \geq 5$ satisfy the recurrence

$$c(k) = \frac{-1}{4(k+3)}(11^4(k-6)c(k-6) + 2^3 3 \cdot 11^2(k-3)c(k-4) + 2^2 47kc(k-2)) \tag{9.6}$$

with initial values

$$\begin{aligned} &c(-4) = 11, \quad c(-3) = 121, \quad c(-2) = 605, \quad c(-1) = 5687/2, \quad c(0) = 14641/2, \\ &c(1) = 84095/8, \quad c(2) = 0, \quad c(3) = -(409343/16), \quad c(4) = 0. \end{aligned}$$

Proposition 6.1 is fundamental when working with modular forms g of positive weight.[hh] We conclude this section with a version for the case when the weight is zero, i.e. when g is a modular function.

[hh]See the Zagier quote in Sec. 6.1.

Proposition 9.3. *Let f and g be modular functions from $M^\infty(N)$ with* $\operatorname{ord} f = -m \leq -1$, $\operatorname{ord} g = -M \leq -1$, *and $m \leq M$. For $h := (\frac{1}{f})^{1/m}$ express $g(\tau)$ locally as $y(h(\tau))$, i.e.*

$$g = y(h) = \sum_{k=-M}^{\infty} c(k) h^k. \tag{9.7}$$

Then $y(h)$ satisfies a homogeneous holonomic differential equation of order m,

$$P_m(h)\left(\frac{dy}{dh}\right)^m + \cdots + P_0(h)y = 0,$$

or an inhomogeneous differential equation of order $m-1$ of the form

$$Q_{m-1}(h)\left(\frac{dy}{dh}\right)^{m-1} + \cdots + Q_0(h)y = Q(h), \tag{9.8}$$

where $P_i(X)$, $Q_j(X)$, and $Q(X)$ are polynomials in $\mathbb{C}[X]$.

Proof. By Theorem 8.3 there exists a polynomial

$$P(X,Y) := Y^m + c_1(X)Y^{m-1} + \cdots + c_m(X) \in \mathbb{C}[X][Y] = \mathbb{C}[X,Y]$$

such that $P(f,g) = 0$. Equivalently we have

$$0 = P(1/h^m, g) = P(1/h^m, y(h)) = y(h)^m + c_1(1/h^m)y(h)^{m-1} + \cdots + c_m(1/h^m).$$

This means, $y(h)$ is an algebraic function satisfying a relation $0 = Q(h, y(h)) := P(1/h^m, y(h))$ where $Q(X,Y) \in \mathbb{C}(X)[X]$ is a polynomial with degree $m = \deg_Y Q(X,Y)$ in Y. The rest follows from Proposition 3.3. □

Remark 9.4. As noted in connection with Lemma 7.1, all the coefficients $c(k)$ in (9.5) are quotients with integers as numerators, and denominators being powers of 2, i.e. for each fixed k they have the form

$$c(k) = \frac{\text{integer}}{\text{power of } 2}. \tag{9.9}$$

In addition, the $c(k)$ inherit property (9.2) of the $p(11n+6)$ in the following sense: each numerator in the quotient representation (9.9) is divisible by 11. We proved this fact as a consequence of the algebraic relation (7.6), respectively (8.8). In the q-series interpretation,[ii] relation (9.7) reads as

$$\tilde{g}(q) = \frac{1}{q^4}\prod_{j=1}^{\infty}\frac{(1-q^j)^{12}}{(1-q^{11j})^{11}}\sum_{n=0}^{\infty} p(11n+6)q^n = \sum_{k=-4}^{\infty} c(k)\tilde{h}(q)^k. \tag{9.10}$$

Consequently, the 11 divisibility of the $c(k)$, by coefficient comparison of q-powers, implies $11 \mid p(11n+6)$, $n \geq 0$. In other words, Ramanujan's congruence (9.2) involving a *non-holonomic* sequence $p(11n+6)$ is implied by the analogous divisibility property of a *holonomic* sequence $c(k)$!

[ii]For the notation $\tilde{g}$, see (2.4)

10. Case Study: A Partition Congruence by Andrews

In this section, we present another example to illustrate the relevance of holonomic differential equations to partition congruences.

Consider the number $c\phi_2(n)$ of 2-colored Frobenius partitions of n introduced by Andrews [2], their generating function is

$$\sum_{n=0}^{\infty} c\phi_2(n)q^n = \prod_{j=1}^{\infty} \frac{(1-q^{2j})^5}{(1-q^j)^4(1-q^{4j})^2}. \tag{10.1}$$

Picking the subsequence $(c\phi_2(5n+3))_{n\geq 0}$, as for (9.1) we use Smoot's package [34] to compute a q-product $G(\tau)$, $q = e^{2\pi i\tau}$, which, when multiplied with the generating function, gives a modular function

$$g(\tau) := G(\tau)\sum_{n=0}^{\infty} c\phi_2(5n+3)q^n \in M^{\infty}(20). \tag{10.2}$$

Matching the product representation in (10.1) with modular function facts for eta-quotients, $N = 20$ for the ambient space $M^{\infty}(N)$ is a natural choice. Using the command

```
RKDelta[20, 20, {-4, 5, -2, 0, 0, 0}, 5, 3],
```

Smoot's package returns

$$G(\tau) = \frac{1}{q^9}\prod_{j=1}^{\infty} \frac{(1-q^j)^{14}(1-q^{4j})^7(1-q^{10j})^{10}}{(1-q^{2j})^{11}(1-q^{5j})^2(1-q^{20j})^{17}}.$$

The genus of $X(\Gamma_0(20))$ is 1, hence by Weierstraß' gap theorem[jj] there exists a modular function $F \in M^{\infty}(20)$ such that $\operatorname{ord} f = -2$. Such h can also be computed using Smoot's package:

$$f = \frac{1}{q^2}\prod_{j=1}^{\infty} \frac{(1-q^{2j})^3(1-q^{10j})}{(1-q^{4j})(1-q^{20j})^3} \in M^{\infty}(20). \tag{10.3}$$

Remark 10.1. In general, such q-product (respectivley, eta-quotient) generators of modular function spaces $M^{\infty}(N)$ — provided they exist for a particular choice of N — are not uniquely determined. Algorithmically, they are members of the solution space of a system of linear Diophantine equations and inequalities. Another element of the solution space for the given problem, for instance, is

$$f^{\star} = \frac{1}{q^2}\prod_{j=1}^{\infty} \frac{(1-q^{4j})^4(1-q^{10j})^2}{(1-q^{2j})^2(1-q^{20j})^4} \in M^{\infty}(20).$$

These two modular functions are related

$$f = f^{\star} + 5, \tag{10.4}$$

[jj]See, for instance [30, Theorem 12.2].

which owing to the fact that $f, f^\star \in M^\infty(20)$ can be easily verified by comparing the coefficients of q^{-2}, q^{-1}, and q^0 in their q-expansions. After the replacement $q^2 \to q$, relation (10.4) is Theorem 1.6.1(ii),

$$\psi^2(q) - 5\, q\psi^2(q^5) = (q;q)_\infty^2 \frac{\chi(-q)}{\chi(-q^5)} = q^{-1/4}\frac{\eta(\tau)^3\eta(10\tau)}{\eta(2\tau)\eta(5\tau)},$$

in Andrews and Berndt [3], see also (10.7) in Cooper's monograph [17].

Again we use the `GeneratingFunctions` package to determine $c(k)$ such that

$$g = \sum_{k=-9}^{\infty} c(k)\left(\frac{1}{f}\right)^{\frac{k}{2}};$$

i.e. according to Lemma 4.1 with $h := (\frac{1}{f})^{1/2}$ and $m = 2$, the coefficients $c(k)$ can be computed as follows:

In[71]:= $\mathbf{g = \frac{20}{q^9} + \frac{165}{q^8} + \frac{270}{q^7} + \frac{1190}{q^6} + \frac{1100}{q^5} + \frac{3680}{q^4} + \frac{2990}{q^3} + \frac{8220}{q^2} + \frac{6520}{q} + 16825 + O[q]^1;}$

The first eight coefficients of g:

In[72]:= **gg[n_] := Coefficient[Normal[g], q, n]**

In[73]:= **Table[gg[k], k, −9, 0]**

Out[73]= {20, 165, 270, 1190, 1100, 3680, 2990, 8220, 6520, 16825}

In[74]:= $\mathbf{f = \frac{1}{q^2} - 3 + q^2 + 2q^4 + 2q^6 - 2q^8 - q^{10} - 4q^{14} - 2q^{16} + 5q^{18} + O[q]^{19};}$

In[75]:= $\mathbf{h = Series[q\left(\frac{1}{q^2 f}\right)^{1/2}, \{q, 0, 14\}]}$

Out[75]= $q + \frac{3q^3}{2} + \frac{23q^5}{8} + \frac{83q^7}{16} + \frac{1099q^9}{128} + \frac{3637q^{11}}{256} + \frac{23091q^{13}}{1024} + O[q]^{14}$

In[76]:= **U = h; W = InverseSeries[U]; ComposeSeries[g, W]**

Out[76]= $\frac{20}{q^9} + \frac{165}{q^8} + \frac{540}{q^7} + \frac{3170}{q^6} + \frac{10525}{2q^5} + \frac{22640}{q^4} + \frac{23615}{q^3} + \frac{71400}{q^2} + \frac{1508395}{32q} + 84000 + O[q]^1$

To match our problem setting, we express the answer in powers of h:

In[77]:= **Clear[h]**

In[78]:= $\mathbf{Normal[Sum[gg[k]W^k, \{k, -9, 0\}]] + O[h]^1\ /.\ q \to h}$

Out[78]= $\frac{20}{h^9} + \frac{165}{h^8} + \frac{540}{h^7} + \frac{3170}{h^6} + \frac{10525}{2h^5} + \frac{22640}{h^4} + \frac{23615}{h^3} + \frac{71400}{h^2} + \frac{1508395}{32h} + 84000 + O[h]^1$

This means, we obtained

$$g = \sum_{k=-4}^{\infty} c(k)\left(\frac{1}{f}\right)^{\frac{k}{2}} = \sum_{k=-9}^{\infty} c(k)h^k$$

$$= \frac{20}{h^9} + \frac{165}{h^8} + \frac{540}{h^7} + \frac{3170}{h^6} + \frac{10525}{2h^5} + \cdots$$

In the next step, we guess an algebraic relation between f and g. To this end, we put the first 60 coefficients $c(-9), c(-10), \ldots, c(50)$ into a list:

```
In[79]:= cList = {20, 165, 540, 3170, 10525/2, 22640, 23615, 71400, 1508395/32, 84000,
          792795/32, 0, -(2775035/256), 0, 856325/64, 0, ...,
          - (49862066867855050577211121507905/2251799813685248), 0,
          68861170372827393869036145946955/2251799813685248, 0}
In[80]:= AlgRel = GuessAE[cList, Y[h], 2, 18][[1, 1]]
Out[80]= -25(-16 + 225h^2 + 21760h^4 + 435680h^6 + 4480640h^8 + 27700736h^10 + 107417600h^12 + 257024000h^14 +
          348160000h^16 + 204800000h^18) + 10(33h + 634h^3 + 4528h^5 + 14280h^7 + 16800h^9)Y[h] - Y[h]^2 = 0
```

The program has guessed the algebraic relation for

$$Y(h) := \sum_{k=0}^{\infty} c(k-9)h^k = h^9 \sum_{k=-9}^{\infty} c(k)h^k = h^9 g, \tag{10.5}$$

hence to relate to g, we substitute accordingly:

```
In[81]:= AlgRel /.Y[h] → h^9 g
Out[81]= -g^2 h^18 + 10gh^9(33h + 634h^3 + 4528h^5 + 14280h^7 + 16800h^9) - 25(-16 + 225h^2 + 21760h^4 + 435680h^6 +
          4480640h^8 + 27700736h^10 + 107417600h^12 + 257024000h^14 + 348160000h^16 + 204800000h^18) = 0
```

Finally, in `Out[81]` we substitute $h = (\frac{1}{f})^{1/2}$ to obtain, after factoring the coefficient polynomials, the following result.

Proposition 10.2. *The modular functions f and g in $M^\infty(20)$ as given in* (10.3) *and* (10.2) *satisfy*

$$\begin{aligned} g^2 &- 10(f+4)(4200 + 2520f + 502f^2 + 33f^3)g \\ &+ 25(f+4)^4(800000 + 560000f + 144000f^2 + 15600f^3 + 481f^4 - 16f^5) = 0. \end{aligned} \tag{10.6}$$

Proof. Knowing that $f, g \in M^\infty(20)$, to prove (10.6) it suffices to show that the principal part and the constant term in the q-expansion of the left side of (10.6) are equal to zero:

```
In[82]:= f = 1/q^2 - 3 + q^2 + 2q^4 + 2q^6 - 2q^8 - q^10 - 4q^14 - 2q^16 + 5q^18 + O[q]^19;
In[83]:= g = 20/q^9 + 165/q^8 + 270/q^7 + 1190/q^6 + 1100/q^5 + 3680/q^4 + 2990/q^3 + 8220/q^2 + 6520/q + 16825 + 11500q +
          28760q^2 + 17340q^3 + 36820q^4 + 22050q^5 + 40760q^6 + 19680q^7 + 36345q^8 + 9000q^9 + O[q]^10;
In[84]:= g^2 - 10(f + 4)(4200 + 2520f + 502f^2 + 33f^3)g + 25(f + 4)^4(800000 + 560000f + 144000f^2 +
          15600f^3 + 481f^4 - 16f^5)
Out[84]= O[q]^1
```

□

Relation (10.6) implies a result which has been proven already by Andrews [2, Corollary 10.1]:

Corollary 10.1. *For non-negative integer n*

$$c\phi_2(5n+3) \equiv 0 \pmod 5. \tag{10.7}$$

Proof. By (10.6) each coefficient of g^2 has 5 as a factor. Hence 5 divides each coefficient of g, and thus $5 \mid c\phi_2(5n+3)$. □

Remark 10.3. Another direct consequence of relation (10.6) is $g^2 \equiv f^8 \pmod 2$, hence $g \equiv f^4 \pmod 2$ or equivalently

$$g \equiv \frac{1}{q^8} \prod_{j=1}^{\infty} \frac{(1-q^{2j})^{12}(1-q^{10j})^4}{(1-q^{4j})^4(1-q^{20j})^{12}} \pmod 2.$$

Taking the q-product factor $G(\tau)$ of $g(\tau)$ from (10.2) into account, one can reduce this further. By applying freshman's dream, $(1+x)^p \equiv 1 + x^p \pmod p$ with p a prime, one, for instance, obtains

$$\sum_{n=0}^{\infty} c\phi_2(5n+3)q^n \equiv q \prod_{j=1}^{\infty} \frac{(1-q^{10j})^5}{(1-q^{4j})^3} \pmod 2. \tag{10.8}$$

In other words, the q-series expansion of the q-product on the right side describes the parity sequence of the partition numbers $c\phi_2(5n+3)$.

Finally, we return to the aspect of differential and recurrence equations. First, using the `GeneratingFunctions` package, we convert relation `AlgRel` in `Out[80]`, which is equivalent to (10.6), into a holonomic differential equation:

```
In[85]:= AE2DE[AlgRel[[1]], Y[h]]
```

Notice that here h is taken as an indeterminate.

```
Out[85]= 330 + 12255h^2 + 205730h^4 + 2029080h^6 + 12774880h^8 + 52211200h^10 + 134438400h^12 + 198400000h^14 +
   128000000h^16 + (108h + 2807h^3 + 28810h^5 + 147660h^7 + 380600h^9 + 396000h^11)Y[h] - (2 + 67h^2 + 895h^4 +
   6246h^6 + 24340h^8 + 50600h^10 + 44000h^12)Y'[h] = 0
```

To obtain a recurrence for the coefficients $d(n) := c(n-9)$ of $Y(h)$, we input, analogously to `Out[70]`, the *homogeneous* version of the differential equation in `Out[85]`:

```
In[86]:= DE2RE[(108h + 2807h^3 + 28810h^5 + 147660h^7 + 380600h^9 + 396000h^11)Y[h]
     - (2 + 67h^2 + 895h^4 + 6246h^6 + 24340h^8 + 50600h^10 + 44000h^12)Y'[h] = 0, d[n]]

Out[86]= -44000(-9 + n)d[n] - 2200(-127 + 23n)d[2 + n] - 20(-2515 + 1217n)d[4 + n] - 2(4333 + 3123n)d[6 +
   n] + (-4353 - 895n)d[8 + n] + (-562 - 67n)d[10 + n] - 2(12 + n)d[12 + n] = 0
```

As with `Out[70]`, the way we derived this recursion also *proves* it. Analogously to Proposition 9.2, we summarize what one obtains after relabeling and using $d(n) = c(n-9)$:

Proposition 10.4. *Given f as in (10.3) and g as in (10.2). Then the uniquely determined coefficients $c(k)$ such that*

$$g = \sum_{n=-9}^{\infty} c(n) \left(\frac{1}{f}\right)^{\frac{n}{2}}, \tag{10.9}$$

for $n \geq 9$ satisfy the holonomic recurrence

$$\begin{aligned} c(n) = \frac{-1}{2(n+9)} & (2^5 5^3 \cdot 11(n-12)\, c(n-12) + 2^3 5^2 \cdot 11(23n-196)\, c(n-10) \\ & + 2^2\, 5(1217n-6166)\, c(n-8) + 2(3123n-5036)\, c(n-6) \\ & + (895n+1668)\, c(n-4) + (67n+361)\, c(n-2)), \end{aligned} \tag{10.10}$$

with initial values

$$\begin{aligned} & c(-9) = 20,\ c(-8) = 165,\ c(-7) = 540,\ c(-6) = 3170,\ c(-5) = 10525/2, \\ & c(-4) = 22640,\ c(-3) = 23615,\ c(-2) = 71400,\ c(-1) = 1508395/32, \\ & c(0) = 84000,\ c(1) = 792795/32,\ c(2) = 0,\ c(3) = -(2775035/256),\ c(4) = 0, \\ & c(5) = 856325/64,\ c(6) = 0,\ c(7) = -(246729945/8192),\ \textit{and}\ c(8) = 0. \end{aligned}$$

11. Conclusion

In this paper, we attempt to connect two worlds which, at the first glance, look very different: the "web of modularity", i.e. modular forms and functions, with the universe of holonomic functions and sequences. It is our hope to inspire subsequent investigations in this direction.

To point to one of the more general aspects: in the given context, we see quite some application potential of the "first guess, then prove" strategy. For instance, it led us to develop, on the "shoulders" of Yang [38], a new algorithm, ModFormDE, to verify holonomic differential equations involving modular forms; see Sec. 6 and, for full details [31]. Concerning the computational complexity of this approach, we remark that the holonomic procedure to *guess* the differential equation only requires to solve a system of linear equations. The *proving* step by algorithm ModFormDE boils down to zero-testing of the coefficients α_j in (6.7). Each of these tests needs linear time in the number of coefficients of the corresponding q-expansions, respectively, x-expansions with $x = q^{1/N_0}$, at infinity of the particular α_j.

We restrict to mention only one more concrete aspect for further exploration. Choosing modular functions g and h, as explained, one can relate a non-holonomic sequence with a holonomic sequence, for example

$$(p(11n+6))_{n\geq 0} \leftrightarrow (c(k))_{k\geq 0},$$

where $c(k)$ is defined by the holonomic recurrence (9.6). As remarked at the end of Sec. 9, Ramanujan's observation $11 \mid p(11n+6)$ is implied by the corresponding divisibility property of the holonomic sequence $(c(k))_{k\geq 0}$. Nevertheless, in the examples we considered (e.g., the sequence defined in (10.10) is another such instance) we found it a non-trivial task to prove the corresponding arithmetic property of the $c(k)$ directly from the defining holonomic recurrence.

Acknowledgments

Our sincerest thanks go to the anonymous referee for reading the paper so carefully, his/her comments helped to improve our paper significantly. — In October 2019, while working on parts of this paper, Peter Paule enjoyed the overwhelming hospitality of Bill Chen and his team at the Center for Applied Mathematics, Tianjin University. Peter Paule and Cristian-Silviu Radu were supported by grant SFB F50-06 of the Austrian Science Fund (FWF).

Appendix A. Proof of Proposition 6.2 and Zero-Recognition of Meromorphic Functions on Riemann Surfaces

A.1. *Proof of Proposition* 6.2

Proposition 6.2 is an immediate consequence of the following, more general Proposition A.1 which we state and prove in the setting of the ring of formal power series over the complex numbers, denoted by $\mathbb{C}[[z]]$. The ring of polynomials with complex coefficients is denoted by $\mathbb{C}[z]$; its quotient field, the rational functions, by $\mathbb{C}(z)$. This notation is extended to multivariate polynomials, $\mathbb{C}[z_1, \ldots, z_n]$, and to rational functions, $\mathbb{C}(z_1, \ldots, z_n)$.

Proposition A.1. *Let* $y(z) = \sum_{k=0}^{\infty} c(k)z^k \in \mathbb{C}[[z]]$ *be such that*

$$A_\ell(z)y^{(\ell)}(z) + \cdots + A_0(z) = 0, \tag{A.1}$$

where the $A_j(z) \in \mathbb{C}[[z]]$ *are algebraic Puiseux series, not all zero. Then* $y(z)$ *is holonomic series; i.e. there exist polynomials* $P_k(z) \in \mathbb{C}[z]$, *not all zero, such that*

$$P_d(z)y^{(d)}(z) + \cdots + P_0(z) = 0. \tag{A.2}$$

Proof. By definition, each $A_j(z)$ satisfies an algebraic equation $P_j(z, A_j(z)) = 0$ with $P_j(X, Y) \in \mathbb{C}(X)[Y]$. As a consequence, for each j the derivative $A_j'(z)$ is a rational function in z and $A_j(z)$, i.e.

$$A_j'(z) = \frac{dA_j(z)}{dz} \in \mathbb{C}(z, A_j(z)).$$

This property extends to the higher derivatives $A_j^{(i)}(z)$. Hence the rational function field

$$K := \mathbb{C}(z)(A_0(z), \ldots, A_\ell(z), A_0'(z), \ldots, A_\ell'(z), A_0''(z), \ldots, A_\ell''(z), \ldots),$$

generated by z and all possible derivatives $A_j^{(i)}(z)$, is nothing but the field generated by z and the $A_j(z)$, i.e.

$$K = \mathbb{C}(z)(A_0(z), \ldots, A_\ell(z)). \tag{A.3}$$

Again using the fact that $A_j(z)$ are algebraic, one sees that K is finitely generated as a $\mathbb{C}(z)$-module

$$K = \{\beta_1(z)b_1(z) + \cdots + \beta_r(z)b_r(z) : \beta_j(z) \in \mathbb{C}(z)\}. \tag{A.4}$$

As a concrete choice of the linear generators $b_j(z)$, for instance, one can take all the power products

$$A_0(z)^{a_0} \cdots A_\ell(z)^{a_\ell} \quad \text{with } a_0 < \deg_Y P_0(X,Y), \ldots, a_\ell < \deg_Y P_\ell(X,Y).$$

The next observation is that for $k \geq \ell$:

$$y^{(k)}(z) = c_{k,1}(z)y^{(\ell-1)}(z) + \cdots + c_{k,\ell}(z) \quad \text{for some } c_{k,1}(z), \ldots, c_{k,\ell}(z) \in K. \tag{A.5}$$

This can be seen by the following steps which can be applied successively to prove the statement by mathematical induction. Assuming $A_\ell(z) \neq 0$, rewrite (A.1) as

$$y^{(\ell)}(z) = -\frac{A_{\ell-1}(z)}{A_\ell(z)}y^{(\ell-1)}(z) - \cdots - \frac{A_0(z)}{A_\ell(z)}, \tag{A.6}$$

which by (A.3) is of the required form. Taking the derivative, again by (A.3), gives

$$y^{(\ell+1)}(z) = d_\ell(z)y^{(\ell)}(z) + \cdots + d_0(z) \quad \text{for some } d_j(z) \in K.$$

Now reduce with (A.6) to arrive at the required form.

Fact (A.5) prepares us to define another set of linear generators by

$$\{B_1, \ldots, B_{r\ell}\} := \{b_i(z)\, y^{(j)}(z) : 1 \leq i \leq r, 0 \leq j \leq \ell - 1\}. \tag{A.7}$$

Using this set and setting $d := r\ell$, fact (A.5) translates into a matrix relation

$$\begin{bmatrix} y(z) \\ y'(z) \\ \cdots \\ y^{(d)}(z) \end{bmatrix} = \begin{bmatrix} c_{0,1}(z) & c_{0,2}(z) & \cdots & c_{0,d}(z) \\ c_{1,1}(z) & c_{1,2}(z) & \cdots & c_{1,d}(z) \\ \cdots & \cdots & \cdots & \cdots \\ c_{d,1}(z) & c_{d,2}(z) & \cdots & c_{d,d}(z) \end{bmatrix} \begin{bmatrix} B_1 \\ B_2 \\ \cdots \\ B_d \end{bmatrix}, \tag{A.8}$$

where the $c_{i,j}$ are rational functions in $\mathbb{C}(z)$. Denote the $(d+1) \times d$ matrix by C. Its transpose C^t is a $d \times (d+1)$ matrix, hence there exists a nonzero vector $x = (x_0(z), \ldots, x_d(z)) \in \mathbb{C}(z)^{d+1}$ such that $C^t x^t = 0$. Equivalently, $xC = 0$, and as a consequence of applying this to both sides of (A.8),

$$x_0(z)y(z) + x_1(z)y'(z) + \cdots + x_d(z)y^{(d)}(z) = 0,$$

where the rational functions $x_j(z) \in \mathbb{C}(z)$ are not all zero. Multiplying $x_j(z)$ with a common denominator completes the proof of Proposition A.1. $\square$

A.2. *Zero recognition of meromorphic functions*

The basic fundamental fact we use for zero recognition of meromorphic functions on Riemann surfaces is:

Lemma A.2. *Let f be a non-constant meromorphic function on a compact Riemann surface X. Then*

$$\sum_{x \in X} \operatorname{ord}_x f = 0.$$

Proof. See, for instance, [27, Proposition 4.12]. □

In other words,

$$\text{number of poles of } f = \text{number of zeros of } f,$$

counting multiplicities.

Here $\operatorname{ord}_{x_0} f$ is defined as follows: Suppose $f(x) = \sum_{n \geq m} c_n(\varphi(x) - \varphi(x_0))^n$, $c_m \neq 0$, is the local Laurent expansion of g at x_0 using the local coordinate chart $\varphi : U_0 \to \mathbb{C}$ which homeomorphically maps a neighborhood U_0 of $x_0 \in X$ to an open set $V_0 \subseteq \mathbb{C}$. Then $\operatorname{ord}_{x_0} f := m$.

References

[1] W. W. Adams and P. Loustaunau, *An Introduction to Gröbner Bases*, Graduate Studies in Mathematics, Vol. 3 (AMS, 1994).

[2] G. E. Andrews, Generalized Frobenius Partitions, *Mem. Amer. Math. Soc.* **49**(301) (1984).

[3] G. E. Andrews and B. C. Berndt, *Ramanujan's Lost Notebook: Part I* (Springer, 2005).

[4] G. E. Andrews and B. C. Berndt, *Ramanujan's Lost Notebook: Part II* (Springer, 2009).

[5] G. E. Andrews and B. C. Berndt, *Ramanujan's Lost Notebook: Part III* (Springer, 2012).

[6] G. E. Andrews and B. C. Berndt, *Ramanujan's Lost Notebook: Part IV* (Springer, 2013).

[7] G. E. Andrews and B. C. Berndt, *Ramanujan's Lost Notebook: Part V* (Springer, 2018).

[8] B. C. Berndt, *Ramanujan's Notebooks: Part I* (Springer, 1985).

[9] B. C. Berndt, *Ramanujan's Notebooks: Part II* (Springer, 1989).

[10] B. C. Berndt, *Ramanujan's Notebooks: Part III* (Springer, 1991).

[11] B. C. Berndt, *Ramanujan's Notebooks: Part IV* (Springer, 1994).

[12] B. C. Berndt, *Ramanujan's Notebooks: Part V* (Springer, 1998).

[13] F. Beukers, Another congruence for the Apéry numbers, *J. Number Theory* **25** (1987) 201–210.

[14] F. Beukers, Irrationality proofs using modular forms, *Astérisque* **147–148** (1987) 271–283.

[15] J. M. Borwein and P. B. Borwein, *Pi and the AGM* (John Wiley & Sons, 1987).

[16] H. Cohen and F. Strömberg, *Modular Forms: A Classical Approach*, Graduate Studies in Mathematics, Vol. 179 (AMS, 2017).

[17] S. Cooper, *Ramanujan's Theta Functions* (Springer, 2017).
[18] A. Erdély *et al.*, *Higher Transcendental Functions*, Vols. 1–3 (McGraw-Hill, 1953–1955).
[19] G. Fischer, *Plane Algebraic Curves*, Student Mathematical Library, Vol. 15 (AMS, 2001).
[20] O. Forster, *Lectures on Riemann Surfaces* (Springer, 1981).
[21] R. L. Graham, D. E. Knuth and O. Patashnik, *Concrete Mathematics — A Foundation for Computer Science*, 2nd edn. (Addison-Wesley, 1994).
[22] M. Kauers and P. Paule, *The Concrete Tetrahedron* (Springer, 2011).
[23] C. Koutschan, *HolonomicFunctions* (*User's Guide*). RISC Technical Report 10-01, J. Kepler University (2010), https://risc.jku.at/research_topic/computer-algebra-for-combinatorics/#publications.
[24] K. Lamotke, *Riemannsche Flächen* (Springer, 2009).
[25] B. H. Lian and S.-T. Yau, Integrality of certain exponential series, in *Lecture Notes in Algebra and Geometry, Proceedings of the International Conference on Algebra and Geometry*, M.-C. Kang, ed. (Int. Press Cambridge, 1998), pp. 215–227. Appeared 1995 as Mirror maps, modular relations and hypergeometric series I, arXiv:hep-th/9507151.
[26] C. Mallinger, Algorithmic Manipulations and Transformations of Univariate Holonomic Functions and Sequences (Diploma Thesis). RISC, Johannes Kepler University (1996), https://risc.jku.at/research_topic/computer-algebra-for-combinatorics/#publications.
[27] R. Miranda, *Algebraic Curves and Riemann Surfaces*, Graduate Studies in Mathematics, Vol. 5 (AMS, 1995).
[28] K. Ono, *The Web of Modularity: Arithmetic of the Coefficients of Modular Forms and q-series*, CBMS Regional Conference Series in Mathematics, Vol. 102 (AMS, CBMS, 2004).
[29] P. Paule and C.-S. Radu, A unified algorithmic framework for Ramanujan's congruences modulo powers of 5, 7, and 11 submitted (2018), https://risc.jku.at/research_topic/computer-algebra-for-combinatorics/#publications.
[30] P. Paule and C.-S. Radu, A proof of the Weierstraß gap theorem not using the Riemann-Roch formula, *Ann. Combin.* **23** (2019) 963–1007.
[31] P. Paule and C.-S. Radu, An algorithm to prove holonomic differential equations for modular forms, submitted (2020), https://risc.jku.at/research_topic/computer-algebra-for-combinatorics/#publications.
[32] S. Ramanujan, Some properties of $p(n)$, the number of partitions of n, *Proc. Cambridge Philos. Soc.* **19** (1919) 207–210.
[33] S. Ramanujan, Modular equations and approximations to π, *Quart. J. Math.* **45** (1914) 350–372.
[34] N. A. Smoot, On the computation of identities relating partition numbers in arithmetic progressions with eta quotients: An implementation of Radu's algorithm, *J. Symb. Comput.* to appear (2020), https://arxiv.org/abs/2004.03948.
[35] R. P. Stanley, *Enumerative Combinatorics*, Vol. 2 (Cambridge University Press, 1999).
[36] A. J. van der Poorten, A proof that Euler missed ... Apérys proof of the irrationality of $\zeta(3)$, *Math. Intelligencer* **1** (1978) 195–203.
[37] R. J. Walker, *Algebraic Curves* (Dover, 1962).
[38] Y. Yang, On differential equations satisfied by modular forms, *Math. Z* **246** (2004) 1–19.

[39] Y. Yang, Defining equations of modular curves, *Adv. Math.* **204**(2) (2006) 481–508.

[40] D. Zagier, Elliptic Modular Forms and their Applications, in *The 1-2-3 of Modular Forms*, K. Ranestad, ed. (Universitext, Springer, 2008), pp. 1–103.

[41] D. Zeilberger, A holonomic systems approach to special functions identities, *J. Comput. Appl. Math.* **32** (1990) 321–368.

© 2024 World Scientific Publishing Company
https://doi.org/10.1142/9789811277375_0035

Equal sums of two cubes of quadratic forms*

Bruce Reznick
Department of Mathematics
University of Illinois at Urbana-Champaign
Urbana, IL 61801, USA
reznick@illinois.edu

This paper is dedicated to my friend and colleague Bruce Berndt on his 80th birthday.

We give a complete description of all solutions to the equation $f_1^3 + f_2^3 = f_3^3 + f_4^3$ for quadratic forms $f_j \in \mathbb{C}[x, y]$ and show how Ramanujan's example can be extended to three equal sums of pairs of cubes. We also give a complete census in counting the number of ways a sextic $p \in \mathbb{C}[x, y]$ can be written as a sum of two cubes. The extreme example is $p(x, y) = xy(x^4 - y^4)$, which has six such representations.

Keywords: Cubic forms; sextic forms; diophantine equations.

Mathematics Subject Classification 2020: 11E76, 14M99, 11D41, 11D45

1. Introduction

In 1913, Ramanujan [11; 12, p. 326] (see [2, p. 56; 6, p. 201]) posed to the *Journal of the Indian Mathematical Society* the following question: "Shew that

$$(6x^2 - 4xy + 4y^2)^3 = (3x^2 + 5xy - 5y^2)^3 + (4x^2 - 4xy + 6y^2)^3 + (5x^2 - 5xy - 3y^2)^3, \quad (1.1)$$

and find other quadratic expressions satisfying similar relations." Write (1.1) as $R_1^3(x, y) = R_2^3(x, y) + R_3^3(x, y) + R_4^3(x, y)$ for short.

In 1914, Narayanan [10] replaced the integers in (1.1) with the variables ℓ, m, n, p and solved the resulting equations; namely, $m^3 + n^3 = p^3 - \ell^3 = mp^2 + n\ell^2$,

*This article originally appeared in *International Journal of Number Theory*, **17**(3), pp. 761–786 © 2021 World Scientific Publishing Company. DOI: 10.1142/S1793042120400308

over $\mathbb{R}$.

$$(\ell x^2 - nxy + ny^2)^3 = (px^2 + mxy - my^2)^3 + (nx^2 - nxy + \ell y^2)^3 + (mx^2 - mxy - py^2)^3;$$
$$\ell = \lambda(\lambda^3+1), \quad m = 2\lambda^3 - 1, \quad n = \lambda(\lambda^3 - 2), \quad p = \lambda^3 + 1. \tag{1.2}$$

Write (1.2) as $N_{1,\lambda}^3(x,y) = N_{2,\lambda}^3(x,y) + N_{3,\lambda}^3(x,y) + N_{4,\lambda}^3(x,y)$, and note $N_{j,2} = 3R_j$.

Equation (1.1) can be rewritten as two equal sums of two cubes in three different ways, and in two of the three ways, there is a third equal sum of two cubes. First,

$$\begin{aligned}(4x^2 - 4xy + 6y^2)^3 &+ (5x^2 - 5xy - 3y^2)^3 \\ &= (6x^2 - 4xy + 4y^2)^3 - (3x^2 + 5xy - 5y^2)^3 \\ &= (6x^2 - 8xy + 6y^2)^3 - (3x^2 - 11xy + 3y^2)^3 \\ &= 63(x^2 + xy + y^2)(3x^2 - 3xy + y^2)(x^2 - 3xy + 3y^2).\end{aligned} \tag{1.3}$$

We also have

$$\begin{aligned}(6x^2 - 4xy + 4y^2)^3 &- (5x^2 - 5xy - 3y^2)^3 \\ &= (4x^2 - 4xy + 6y^2)^3 + (3x^2 + 5xy - 5y^2)^3 \\ &= \left(\frac{94}{21}x^2 - \frac{8}{21}xy + \frac{94}{21}y^2\right)^3 + \left(\frac{23}{21}x^2 - \frac{199}{21}xy + \frac{23}{21}y^2\right)^3 \\ &= (13x^2 - 23xy + 13y^2)(7x^2 + xy + y^2)(x^2 + xy + 7y^2),\end{aligned} \tag{1.4}$$

and

$$\begin{aligned}(6x^2 - 4xy + 4y^2)^3 &- (4x^2 - 4xy + 6y^2)^3 \\ &= (3x^2 + 5xy - 5y^2)^3 + (5x^2 - 5xy - 3y^2)^3 \\ &= 8(x - y)(x + y)(x^2 - xy + y^2)(19x^2 - 11xy + 19y^2).\end{aligned} \tag{1.5}$$

It can be shown that there is no third representation in (1.5). Furthermore, (1.4) follows from (1.3) (with the rows permuted) upon making the unimodular linear change of variables: $(x,y) \to (\frac{5x-2y}{\sqrt{21}}, \frac{3x+3y}{\sqrt{21}})$.

Comparable versions of these properties apply to the Narayanan formulas (see (1.14)). More to the point, up to transposition of terms, changes of variable and taking $\lambda \in \mathbb{C}$, we shall show that (1.2) *completely* describes the solution in binary quadratic forms $f_j = f_j(x,y) \in \mathbb{C}[x,y]$ to

$$p = f_1^3 + f_2^3 = f_3^3 + f_4^3. \tag{1.6}$$

Our analysis comes from looking at the equation in quadratic forms over $\mathbb{C}$ and studying the properties of the common sum p.

We begin with some notations, following those in [15]. For $m \ge 3$, let $\zeta_m = e^{\frac{2\pi i}{m}}$ and $\omega = \zeta_3$. Two forms in $\mathbb{C}[x,y]$ are *distinct* if they are not proportional. The identity (1.6) is *honest* if the f_j's are pairwise distinct. A *flip* of (1.6) is either of the two equivalent identities

$$p_1 = f_1^3 - f_3^3 = -f_2^3 + f_4^3, \quad p_2 = f_1^3 - f_4^3 = -f_2^3 + f_3^3. \tag{1.7}$$

There seems to be no obvious way of deriving p_1 or p_2 from p in (1.7). If (1.6) holds, we say that the family $\mathcal{F} = \{\{f_1, f_2\}, \{f_3, f_4\}\}$ *represents* p, with the understanding that two families $\mathcal{F}$ and $\mathcal{G}$ are identified if $\{\{f_1^3, f_2^3\}, \{f_3^3, f_4^3\}\} = \{\{g_1^3, g_2^3\}, \{g_3^3, g_4^3\}\}$; we do not care about the order of the summands, or powers of ω multiplying the quadratics. For a sextic form $p \in \mathbb{C}[x, y]$, we define $N(p)$ to be the number of pairwise-nonsimilar families $\mathcal{F}$ representing p.

If $M(x, y) = (\alpha x + \beta y, \gamma x + \delta y)$ is an invertible linear change of variables (or *linear change* for short), and $f \in \mathbb{C}[x, y]$ is a form, define $f \circ M$ by $(f \circ M)(x, y) = f(\alpha x + \beta y, \gamma x + \delta y)$. A *scaling* is a linear change in which $\beta = \gamma = 0$. If $\deg f = d$, and $\delta = \alpha$ in a scaling M, then $f \circ M = \alpha^d f$. If M is a linear change, and $g = f \circ M$, then f and g are *similar*, and if $\mathcal{G} = \mathcal{F} \circ M$, the $\mathcal{F}$ and $\mathcal{G}$ will also be called similar.

More generally, suppose the equation

$$p = f_1^3 + f_2^3 \tag{1.8}$$

holds. If M is a linear change, then (1.8) implies that $p \circ M = (f_1 \circ M)^3 + (f_2 \circ M)^3$. It may happen that $p = p \circ M$, but that $\{(f_1 \circ M)^3, (f_2 \circ M)^3\} \neq \{f_1^3, f_2^3\}$: this seems to be the inherent mechanism behind multiple representations.

The following underlying identity is central to our analysis. For $\alpha \in \mathbb{C}$,

$$(\alpha x^2 - xy + \alpha y^2)^3 + \alpha(-x^2 + \alpha xy - y^2)^3 = (\alpha^2 - 1)(\alpha x^3 + y^3)(x^3 + \alpha y^3). \tag{1.9}$$

(This can easily be verified by setting $v = x^2 + y^2$ and $w = xy$ and noting that $v^3 - 3vw^2 = x^6 + y^6$.) Observe that the sum is a quadratic in $\{x^3, y^3\}$, and so if $(x, y) \mapsto (\omega x, \omega^2 y)$, then the sum is unchanged, although the summands are changed. Writing $\alpha = \lambda^3$, we can bring in the outside coefficient and obtain

$$\begin{aligned}
&(\lambda^3 x^2 - xy + \lambda^3 y^2)^3 + (-\lambda x^2 + \lambda^4 xy - \lambda y^2)^3 \\
&= (\lambda^3 \omega^2 x^2 - xy + \lambda^3 \omega y^2)^3 + (-\lambda \omega^2 x^2 + \lambda^4 xy - \lambda \omega y^2)^3 \\
&= (\lambda^3 \omega x^2 - xy + \lambda^3 \omega^2 y^2)^3 + (-\lambda \omega x^2 + \lambda^4 xy - \lambda \omega^2 y^2)^3 \\
&= p_{1,\lambda}(x, y) := (\lambda^6 - 1)(\lambda^3 x^3 + y^3)(x^3 + \lambda^3 y^3).
\end{aligned} \tag{1.10}$$

Write the summands in (1.10) as:

$$\begin{aligned}
&F_{1,\lambda}(x, y) = \lambda^3 x^2 - xy + \lambda^3 y^2, \quad F_{2,\lambda}(x, y) = -\lambda x^2 + \lambda^4 xy - \lambda y^2, \\
&F_{3,\lambda}(x, y) = F_{1,\lambda}(\omega x, \omega^2 y), \quad F_{4,\lambda}(x, y) = F_{2,\lambda}(\omega x, \omega^2 y), \\
&F_{5,\lambda}(x, y) = F_{1,\lambda}(\omega^2 x, \omega y), \quad F_{6,\lambda}(x, y) = F_{2,\lambda}(\omega^2 x, \omega y).
\end{aligned} \tag{1.11}$$

If $\lambda = 0$ or $\lambda^6 = 1$, then the identities of (1.10) are not honest, so we shall assume that $\lambda(\lambda^6 - 1) \neq 0$. Let $\mathcal{F}_{1,\lambda} = \{\{F_{1,\lambda}, F_{2,\lambda}\}, \{F_{3,\lambda}, F_{4,\lambda}\}\}$, $\mathcal{F}_{2,\lambda} = \{\{F_{3,\lambda}, F_{4,\lambda}\}, \{F_{5,\lambda}, F_{6,\lambda}\}\}$ and $\mathcal{F}_{3,\lambda} = \{\{F_{5,\lambda}, F_{6,\lambda}\}, \{F_{1,\lambda}, F_{2,\lambda}\}\}$. Observe that under the scaling $(x,y) \mapsto (\omega x, \omega^2 y)$, $\mathcal{F}_{1,\lambda} \mapsto \mathcal{F}_{2,\lambda} \mapsto \mathcal{F}_{3,\lambda} \mapsto \mathcal{F}_{1,\lambda}$. Thus, the three sets of equations $F_{1,\lambda}^3 + F_{2,\lambda}^3 = F_{3,\lambda}^3 + F_{4,\lambda}^3$, $F_{1,\lambda}^3 + F_{2,\lambda}^3 = F_{5,\lambda}^3 + F_{6,\lambda}^3$, and $F_{3,\lambda}^3 + F_{4,\lambda}^3 = F_{5,\lambda}^3 + F_{6,\lambda}^3$ are similar to each other. The "cleanest" versions of the flips come from $\mathcal{F}_{2,\lambda}$

$$\begin{aligned} F_{4,\lambda}^3(x,y) - F_{5,\lambda}^3(x,y) &= -F_{3,\lambda}^3(x,y) + F_{6,\lambda}^3(x,y) \\ &= p_{2,\lambda}(x,y) \\ &:= ((1+\lambda^6)x^3 + 3\lambda^3 x^2 y - \lambda^3 y^3) \\ &\quad \times (-\lambda^3 x^3 + 3\lambda^3 x y^2 + (1+\lambda^6)y^3) : \qquad (1.12) \\ F_{4,\lambda}^3(x,y) - F_{6,\lambda}^3(x,y) &= F_{5,\lambda}^3(x,y) - F_{3,\lambda}^3(x,y) \\ &= p_{3,\lambda}(x,y) \\ &:= 3\sqrt{-3}\, xy(x-y)(x+y)(\lambda^3 x + y)(x + \lambda^3 y). \qquad (1.13) \end{aligned}$$

We now present some symmetries of (1.10). Since $F_{j,-\lambda}(x,y) = -F_{j,\lambda}(x,-y)$, $\mathcal{F}_{j,-\lambda}$ is similar to $\mathcal{F}_{j,\lambda}$. Further, $F_{1,\lambda^{-1}} = -\lambda^{-4} F_{2,\lambda}$ and $F_{2,\lambda^{-1}} = -\lambda^{-4} F_{1,\lambda}$, etc., so $\mathcal{F}_{j,\lambda^{-1}}$ is similar to $\mathcal{F}_{j,\lambda}$. Under the unimodular linear change

$$(x,y) \mapsto \frac{1}{\sqrt{1-\lambda^6}}(\lambda^3 x + y, -(x + \lambda^3 y)),$$

the system of identities

$$F_{1,\lambda}^3(x,y) + F_{2,\lambda}^3(x,y) = F_{3,\lambda}^3(x,y) + F_{4,\lambda}^3(x,y) = F_{5,\lambda}^3(x,y) + F_{6,\lambda}^3(x,y)$$

becomes

$$\begin{aligned} F_{7,\lambda}^3(x,y) + F_{8,\lambda}^3(x,y) &= -F_{3,\lambda}^3(x,y) + F_{6,\lambda}^3(x,y) = -F_{5,\lambda}^3(x,y) + F_{4,\lambda}^3(x,y); \\ F_7(x,y) &= \frac{1}{1-\lambda^6}((2\lambda^3+\lambda^9)x^2 + (1+5\lambda^6)xy + (2\lambda^3+\lambda^9)y^2), \\ F_8(x,y) &= -\frac{\lambda}{1-\lambda^6}((1+2\lambda^6)x^2 + (5\lambda^3+\lambda^9)xy + (1+2\lambda^6)y^2). \end{aligned}$$

Of course, $p_1(x,y) \mapsto p_2(x,y)$ under this linear change. This means that each $\mathcal{F}_{j,\lambda}$ is similar to one of its flips.

If we make the linear change $(x,y) \mapsto (x + \omega^2 y, x + \omega y)$ into (1.10), we obtain an enhanced version of (1.2), with a third sum:

$$\begin{aligned} N_{4,\lambda}^3(x,y) + N_{3,\lambda}^3(x,y) &= -N_{2,\lambda}^3(x,y) + N_{1,\lambda}^3(x,y) \\ &= (-px^2 + (m+2p)xy - py^2)^3 \\ &\quad + (\ell x^2 + (n-2\ell)xy + \ell y^2)^3. \qquad (1.14) \end{aligned}$$

Upon continuing with the linear change which takes (1.10) into (1.12), we get a flipped version of (1.2) and another third equal sum, but with denominators. A slightly different linear change gives a simple version in $\mathbb{Q}(\lambda)[x,y]$: under $(x,y) \mapsto (x-\sqrt{-3}\,y, x+\sqrt{-3}\,y)$, and multiplication by -1, (1.10) becomes

$$\begin{aligned}((1-2\lambda^3)x^2 &+ 3(1+2\lambda^3)y^2)^3 + (\lambda(2-\lambda^3)x^2 - 3\lambda(2+\lambda^3)y^2)^3\\ &= ((1+\lambda^3)x^2 + 6\lambda^3xy + 3(1-\lambda^3)y^2)^3\\ &\quad + (-\lambda(1+\lambda^3)x^2 - 6\lambda xy + 3\lambda(1-\lambda^3)y^2)^3\\ &= ((1+\lambda^3)x^2 - 6\lambda^3xy + 3(1-\lambda^3)y^2)^3\\ &\quad + (-\lambda(1+\lambda^3)x^2 + 6\lambda xy + 3\lambda(1-\lambda^3)y^2)^3.\end{aligned}$$

It is also worth noting that under the linear change $(x,y) \mapsto (x+\tau y, -i(\tau x - y))$, $\tau = \sqrt{1-\lambda^6} - i\lambda^3$, (which is invertible provided $\lambda^6 \neq 1$), (1.13) becomes an equation of the shape $(ax^2+bxy+ay^2)^3 + (ax^2-bxy+ay^2)^3 = (rx^2+sy^2)^3 + (sx^2+ry^2)^3$, and $p_{3,\lambda}$ becomes a multiple of $x^6 + (4\lambda^6-1)x^4y^2 + (4\lambda^6-1)x^2y^4 + y^6$. This phenomenon is explored in Theorem 3.1.

This paper has two parts. The main result of the first part is the following theorem.

Theorem 1.1. *Every honest identity* (1.6) *for binary sextics is similar to some* $\mathcal{F}_{2,\lambda}$ *with* $\lambda(\lambda^6-1) \neq 0$, *up to a possible flip.*

There is a crucial intermediate step in the proof of Theorem 1.1. Any four binary quadratic forms are linearly dependent, and a given dependence is not affected by a linear change. We shall say that an honest (1.6) is an identity of *Type*(T) for $T \in \mathbb{C}$ if, perhaps after a flip, the following two equations hold:

$$f_1^3 + f_2^3 = f_3^3 + f_4^3, \quad f_1 + f_2 = T(f_3 + f_4). \tag{1.15}$$

We show (see Lemma 4.2) that $T(T^3-1) \neq 0$ in an honest family of Type(T). Of course, the same equation is both Type(T) and Type(T^{-1}), and factors of ω^k do not matter.

The following identities show that (1.2) and (1.9) are both of Type(λ^2):

$$N_{2,\lambda}^3 + N_{4,\lambda}^3 = N_{1,\lambda}^3 - N_{3,\lambda}^3, \quad N_{2,\lambda} + N_{4,\lambda} = \lambda^2(N_{1,\lambda} - N_{3,\lambda});$$

$$F_{5,\lambda}^3 - F_{3,\lambda}^3 = F_{4,\lambda}^3 - F_{6,\lambda}^3, \quad F_{5,\lambda} - F_{3,\lambda} = \lambda^2(F_{4,\lambda} - F_{6,\lambda}).$$

We prove Theorem 1.1 in two stages. After a few technical lemmas, we show that after a linear change, for any honest solution (1.6), f_1 and f_2 are both even and that f_3 and f_4 are not (see Corollary 2.5). We then determine all honest (1.6) in which f_3, f_4 are not both even, but $f_3^3 + f_4^3$ is (see Theorems 3.1 and 3.2) and show that they must be of Type(T) for some T. (Geometrically, this says that any quadratic curve which lies on the surface $z_1^3 + z_2^3 + z_3^3 + z_4^3 = 0$ must in fact lie on the intersection of the surface with a hyperplane $z_i + z_j + T(z_k + z_\ell) = 0$ for some permutation of the indices.) We finally show that any two honest solutions of (1.6) of Type(T) are similar (Theorem 4.3), and are similar to (1.10) (or (1.2)) with

$T = \lambda^2$. Sándor [18] proved in 1996, using different methods, that any solution to (1.6) must be Type(T) for some T, and that two such solutions in $\mathbb{Q}[x, y]$ with the same T must be similar.

We also explore solutions to (1.6) with $f_j \in \mathbb{Q}[x, y]$. If such an equation has type $T = \lambda^2$, then it is clear from linear algebra that $T \in \mathbb{Q}$. If $\lambda \in \mathbb{Q}$, then (1.14) lies in $\mathbb{Q}[x, y]$. The preprint version of this paper contained a conjecture that this condition is necessary, as well as sufficient. The referee pointed out that this was not correct, with suggestions to the correct result and its proof. We show at the end of the paper (Theorem 7.4) that the solutions in $\mathbb{Q}[x, y]$ occur precisely for T which are $a^2 + 3b^2$ for $a, b \in \mathbb{Q}$. Sándor had also proved this, in a different way, in [18].

In the second part of the paper, we give a complete description of $N(p)$, the number of different ways that a binary sextic form is a sum of two cubes. A key result (see Theorem 5.1) is that a form p (of degree $3k$) is a sum of two cubes if and only if $p = h_1 h_2 h_3$ where the h_j's are distinct, but linearly dependent. There are two important families of sextics: for $t \in \mathbb{C}$, let

$$A_t(x, y) := x^6 + tx^4y^2 + tx^2y^4 + y^6, \quad B_t(x, y) := x^6 + tx^3y^3 + y^6. \tag{1.16}$$

Observe that $p_{1,\lambda} = \lambda^3(\lambda^6 - 1)B_{\lambda^3+\lambda^{-3}}$, and as we have seen, $p_{3,\lambda}$ is similar to $A_{4\lambda^6-1}$. Every A_t and B_t is thus similar to $p_{1,\lambda}$ or $p_{3,\lambda}$ for λ with $\lambda(1 - \lambda^6) \neq 0$ except for $A_{-1}, A_3, B_{\pm 2}$.

We give a census of $N(p)$ for binary sextics: (i) a binary sextic p is a sum of two cubes (that is, $N(p) \geq 1$) if and only if $p = \ell^3 q$, where ℓ is linear and q is a square-free cubic or p is similar to $q(x^2, y^2)$, where q is a square-free cubic (see Theorem 5.3); (ii) a binary sextic p has $N(p) = 2$ if and only if p is similar to A_t for $t \in \mathbb{C}$, except that $N(A_3) = 0$, $N(A_{-1}) = 1$, $N(A_0) = N(A_{15}) = 4$ and $N(A_{-5}) = 6$ (see Theorem 5.4); (iii) a binary sextic p has $N(p) = 3$ if and only if p is similar to B_t for $t \in \mathbb{C}$, except that $N(B_{\pm 2}) = 0$, $N(B_0) = 4$ and $N(B_{\pm 5\sqrt{-2}}) = 6$, (see Theorem 5.5); (iv) up to similarity, there are two sextics with $N(p) > 3$:

$$Q_1(x, y) = x^6 + y^6 \quad \text{or} \quad Q_2(x, y) = xy(x^4 - y^4). \tag{1.17}$$

To be specific, Q_1 is similar to A_0, A_{15}, B_0 and $N(Q_1) = 4$ and Q_2 is similar to A_{-5} and $B_{\pm 5\sqrt{-2}}$ and $N(Q_2) = 6$ (see Theorem 5.6). Section 6 gives some extra attention to the representations of Q_1, Q_2 and their similarities.

In the final section, we give some different directions that this study might go. We show that the classical Euler–Binet parametrization of equal sums of two cubes over $\mathbb{Q}$ is also valid over $\mathbb{C}(x_1, \ldots, x_n)$ (see Theorem 7.1 and Corollary 7.2). We apply the usual "point addition" of points on the curve $x_1^3 + x_2^3 = x_3^3 + x_4^3 = A$ to show that $(F_{1,\lambda}, F_{2,\lambda})$"+"$(F_{3,\lambda}, F_{4,\lambda}) = (F_{5,\lambda}, F_{6,\lambda})$ (see Theorem 7.3); the denominators disappear. We show, separately, that a flip of the Euler–Binet parametrization can be added to find a third representation as a sum of cubes of polynomials (see (7.14).) Finally, we present a few previous results from the huge literature, including those of Sándor [18], which when combined with Euler–Binet, allows us a quick proof of

Theorem 7.4. Besides [18], we have not found a systematic analysis of (1.6) over $\mathbb{C}[x, y]$, and we have not found (1.9) or, indeed, any three-fold identities.

2. Preliminary Lemmas

We begin with several old simple lemmas, giving proofs for completeness. The first is a special case of, for example, [15, Theorem 1.1].

Lemma 2.1. *If* $\{\alpha_k x + \beta_k y\}, 1 \le k \le r \le 4$, *are pairwise distinct linear forms, then* $\{(\alpha_k x + \beta_k y)^3\}$ *is linearly independent. In particular, if* (1.6) *holds and* $\{h_j\}$ *is honest, then it cannot be the case that each* h_j *is even.*

Proof. If $r < 4$, add more distinct linear forms to assume that $r = 4$. The matrix of $\{(\alpha_j x + \beta_j y)^3\}$ with respect to the basis $\{\binom{3}{i} x^{3-i} y^i\}$ is $[\alpha_j^{3-i} \beta_j^i]$, which is Vandermonde, with determinant $\prod_{1 \le j < k \le 4} (\alpha_j \beta_k - \alpha_k \beta_j)$. This determinant is nonzero because each pair of linear forms is distinct.

Suppose p is a cubic form and

$$p(x, y) = (\alpha_1 x + \beta_1 y)^3 + (\alpha_2 x + \beta_2 y)^3 = (\alpha_3 x + \beta_3 y)^3 + (\alpha_4 x + \beta_4 y)^3. \tag{2.1}$$

Then $0 = p - p$ gives a formal linear dependence of four cubics, which must result from pairwise cancellation; that is, the original representations were the same.

Finally, by comparing coefficients, the equation

$$(\alpha_1 x^2 + \beta_1 y)^3 + (\alpha_2 x^2 + \beta_2 y^2)^3 = (\alpha_3 x^2 + \beta_3 y^2)^3 + (\alpha_4 x^2 + \beta_4 y^2)^3$$

implies (2.1), and so cannot happen in an honest family. □

Lemma 2.2. *Suppose* $g_1, g_2 \in \mathbb{C}[x_1, \dots, x_n]$ *are distinct forms. Then for* $d \ge 2$, *the set* $\{g_1^{d-k} g_2^k : 0 \le k \le d\}$ *is linearly independent.*

Proof. Suppose $\sum_{k=0}^{d} \lambda_k g_1^{d-k} g_2^k = 0$ for a nonzero choice of $\{\lambda_k\}$. Then

$$\sum_{k=0}^{d} \lambda_k x^{d-k} y^k = \prod_{j=1}^{d} (\alpha_j x + \beta_j y) \implies \prod_{j=1}^{d} (\alpha_j g_1 + \beta_j g_2) = 0;$$

thus $\alpha_j g_1 + \beta_j g_2 = 0$ for some j, violating the distinctness hypothesis. □

We need an old fact about simultaneous diagonalization; there does not seem to be a standard easy-to-find modern proof, a different proof is shown in [15, Theorem 3.2].

Theorem 2.3. *If* $f_1(x, y)$ *and* $f_2(x, y)$ *are relatively prime quadratic forms, then there is a linear change* M *so that* $f_1 \circ M$ *and* $f_2 \circ M$ *are both even.*

Proof. We may assume $rank(f_1) \ge rank(f_2) \ge 1$, and after a preliminary linear change, take $f_1(x, y) = x^2$ or $x^2 + y^2$. In the first case, $rank(f_2) = 1$, so $f_2 = \ell^2$ for a linear ℓ which can become y after a linear change, so $(f_1, f_2) \to (x^2, y^2)$. Otherwise, we have $f_1(x,y) = x^2 + y^2$ and $f_2(x,y) = ax^2 + 2bxy + cy^2$. Since f_1 and f_2 are relatively prime, $x \pm iy$ is not a factor of f_2 and so $a \pm 2bi - c \neq 0$.

The quadratic $\lambda f_1 + f_2$ has discriminant

$$\Delta(\lambda) = 4(\lambda + a)(\lambda + c) - (2b)^2 = 4(\lambda^2 + (a+c)\lambda + (ac - b^2));$$

$$\mathrm{Disc}(\Delta(\lambda)) = 16((a+c)^2 - 4(ac - b^2)) = 16(a + 2bi - c)(a - 2bi - c) \neq 0.$$

Thus, there exist $\lambda_1 \neq \lambda_2$ so that each quadratic $\lambda_j f_1 + f_2$ is perfect square; that is, $\lambda_j f_1 + f_2 = \ell_j^2$. This implies that both f_1 and f_2 are linear combinations of ℓ_1^2, ℓ_2^2. A linear change taking $(\ell_1, \ell_2) \mapsto (x, y)$ completes the diagonalization. □

In order to apply Theorem 2.3, we need a small technical lemma.

Lemma 2.4. *Suppose $p = f_1^3 + f_2^3 = f_3^3 + f_4^3$ for quadratic f_1, f_2 and f_1 and f_2 have a nontrivial common factor. Then $\{f_1^3, f_2^3\} = \{f_3^3, f_4^3\}$. Thus in any honest instance of* (1.6), *the f_j's are pairwise relatively prime.*

Proof. Suppose ℓ is a linear form and $f_1 = \ell\ell_1$ and $f_2 = \ell\ell_2$. Then

$$\ell^3 \mid f_3^3 + f_4^3 = (f_3 + f_4)(f_3 + \omega f_4)(f_3 + \omega^2 f_4).$$

Since the three factors on the right are quadratic, ℓ must divide at least two of them; it follows that ℓ divides both f_3 and f_4. By writing $f_3 = \ell\ell_3$ and $f_4 = \ell\ell_4$, we see that $\ell_1^3 + \ell_2^3 = \ell_3^3 + \ell_4^3$, and since the original equation was honest, the ℓ_j's are pairwise distinct. This is impossible by Lemma 2.1. □

Putting the results of this section together, we have the following corollary.

Corollary 2.5. *If an honest* (1.6) *holds, then after a linear change, f_1 and f_2 are even, (and hence so is p), but f_3 and f_4 are not both even; thus*

$$(ax^2 + bxy + cy^2)^3 + (dx^2 + exy + fy^2)^3 \tag{2.2}$$

is even, where $(b, e) \neq (0, 0)$.

3. Even Sums of the Cubes of Non-Even Quadratic Forms

Our goal in this section is to show that every quadratic solution to (1.6) is a family of Type(T) for some T.

How can it happen that $f_3^3 + f_4^3$ is even when at least one of $\{f_3, f_4\}$ is not even? An obvious case is

$$f_3(x,y) = ax^2 + bxy + cy^2, \quad f_4(x,y) = ax^2 - bxy + cy^2, \tag{3.1}$$

which, as in [15], we call the *tame* case; otherwise we are in the *wild* case. If $a = 0$, then it follows from (3.1) that y divides f_3 and f_4, and by Lemma 2.4, this cannot happen, so $a \neq 0$. Similarly, $c \neq 0$. Thus, we may scale x and y and assume that f_3, f_4 are $x^2 \pm \gamma xy + y^2$ for some $\gamma \neq 0$.

Theorem 3.1. *The tame case occurs in a family of Type*$((1 + \frac{3}{4}\gamma^2)^{1/3})$.

Proof. Observe that

$$\begin{aligned}(x^2 + \gamma xy + y^2)^3 &+ (x^2 - \gamma xy + y^2)^3 \\ &= 2(x^6 + 3(1+\gamma^2)x^4y^2 + 3(1+\gamma^2)x^2y^4 + y^6) \\ &= 2A_{3(1+\gamma^2)}(x, y). \qquad (3.2)\end{aligned}$$

Let $\{f_3(x,y), f_4(x,y)\} = \{x^2 \pm \gamma xy + y^2\}$. Honesty requires $\gamma \neq 0$. By hypothesis, $2A_{3(1+\gamma^2)}$ is a sum of cubes of two even quadratics in a unique way by Lemma 2.1.

Note that (3.2) implies that

$$\begin{aligned}2(x^6 + 3(1+\gamma^2)x^4y^2 + 3(1+\gamma^2)x^2y^4 + y^6) &= (r_\gamma x^2 + s_\gamma y^2)^3 + (s_\gamma x^2 + r_\gamma y^2)^3 \\ \iff r_\gamma^3 + s_\gamma^3 = 2, \quad 3r_\gamma^2 s_\gamma + 3r_\gamma s_\gamma^2 &= 3r_\gamma s_\gamma(r_\gamma + s_\gamma) = 6(1+\gamma^2) \\ \implies (r_\gamma + s_\gamma)^3 = 8 + 6\gamma^2. \qquad (3.3)\end{aligned}$$

Observe that if $\gamma^2 = -\frac{4}{3}$, then $0 = (r_\gamma + s_\gamma)^3$, so $s_\gamma = -r_\gamma$ and $r_\gamma^3 + s_\gamma^3 = 0$, so we take $\gamma^2 \neq -\frac{4}{3}$. Up to $(r_\gamma, s_\gamma) \mapsto \omega^k(r_\gamma, s_\gamma)$ and a choice of cube root,

$$r_\gamma + s_\gamma = (8 + 6\gamma^2)^{1/3} \neq 0 \implies r_\gamma s_\gamma = \frac{2(1+\gamma^2)}{(8+6\gamma^2)^{1/3}},$$

and so r_γ and s_γ are the roots of the quadratic equation

$$X^2 - (8 + 6\gamma^2)^{1/3} X + \frac{2(1+\gamma^2)}{(8+6\gamma^2)^{1/3}} = 0.$$

Let $\{f_1(x,y), f_2(x,y)\} = \{r_\gamma x^2 + s_\gamma y^2, s_\gamma x^2 + r_\gamma y^2\}$. Since $(r_\gamma - s_\gamma)^2 = (r_\gamma + s_\gamma)^2 - 4r_\gamma s_\gamma = -\frac{2\gamma^2}{(8+6\gamma^2)^{1/3}} \neq 0$, these roots are distinct, and since $(r_\gamma + s_\gamma)(f_3 + f_4) = 2(f_1 + f_2)$, the equation $f_1^3 + f_2^3 = f_3^3 + f_4^3$ is a Type$(\frac{r_\gamma + s_\gamma}{2})$ family. □

Theorem 3.2. *If*

$$p(x,y) = f_1^3(x,y) + f_2^3(x,y) := (ax^2 + bxy + cy^2)^3 + (dx^2 + exy + fy^2)^3 \qquad (3.4)$$

is even and a sum of two even cubes $f_3^3(x,y) + f_4^3(x,y)$, $(b,e) \neq (0,0)$, *and* $(d,e,f) \neq \omega^k(a,-b,c)$, *then a flip of* $f_1^3 + f_2^3 = f_3^3 + f_4^3$ *is a Type*(T) *family for some* T *and* p *has a third representation as a sum of two cubes.*

Proof. By considering the coefficients of x^5y, x^3y^3, xy^5 in (3.4), we need to solve

$$3a^2b + 3d^2e = 6abc + b^3 + 6def + e^3 = 3bc^2 + 3ef^2 = 0. \qquad (3.5)$$

If $a = 0$ in (3.5), then $d^2e = 0$. If $d = 0$, then $a = d = 0$ implies a common factor in the quadratics, violating Lemma 2.4. Hence $a = e = 0$, so $b^3 = 0$ and $b = e = 0$.

These contradictions imply that $a \neq 0$; similar arguments show that $cef \neq 0$ and now, if $b = 0$, then $d^2e = 0$ and $e \neq 0$ imply $d = 0$, so $b \neq 0$ after all. Similarly $e \neq 0$. Thus all variables in (3.5) are nonzero.

By a scaling of (x, y), we may assume $a = c = 1$, so

$$p(x,y) = (x^2 + bxy + y^2)^3 + (dx^2 + exy + fy^2)^3 \tag{3.6}$$

is even, and (3.5) becomes

$$3b + 3d^2e = 6b + b^3 + 6def + e^3 = 3b + 3f^2e = 0. \tag{3.7}$$

It follows that $b = -d^2e$ and $f^2 = d^2$; the remaining equation becomes

$$0 = -6d^2e - d^6e^3 + 6def + e^3 = e^3(1 - d^6) + 6de(f - d). \tag{3.8}$$

If $f = d$ in (3.8), then $d^6 = 1$, so up to a power of ω, $d \in \{1, -1\}$. If $d = 1$, then $e = -b, f = 1$ implies that (3.6) is tame; if $d = -1$, then $e = -b, f = -1$ implies that $p = 0$. In the remaining case, $f = -d$ and $e^2(1 - d^6) = 12d^2$, so $e = \pm\frac{2\sqrt{3}d}{\sqrt{1-d^6}}$, $d^6 \neq 1$. By taking $y \mapsto -y$ if necessary, we may choose one square root and rewrite (3.6) as

$$p(x,y) = \left(x^2 - \frac{2\sqrt{3}d^3}{\sqrt{1-d^6}}xy + y^2\right)^3 + \left(dx^2 + \frac{2\sqrt{3}d}{\sqrt{1-d^6}}xy - dy^2\right)^3. \tag{3.9}$$

Write (3.9) as $p = f_1^3 + f_2^3$. Pull d^3 out of the second factor and let $r = d^3$. A computation shows that

$$p(x,y) = (1+r)x^6 + \frac{3(1+10r+r^2)}{1-r}x^4y^2 + \frac{3(1-10r+r^2)}{1+r}x^2y^4 + (1-r)y^6.$$

We use the Sylvester algorithm (see [14, Theorem 2.1]) to write p as a sum of two cubes of even quadratics. In this way, and omitting details, we find that

$$\begin{aligned} p(x,y) = r&\left(-\frac{2+3r+r^2}{1-r^2}\cdot x^2 + \frac{2-3r+r^2}{1-r^2}\cdot y^2\right)^3 \\ &+\left(\frac{1+3r+2r^2}{1-r^2}\cdot x^2 + \frac{1-3r+2r^2}{1-r^2}\cdot y^2\right)^3. \end{aligned} \tag{3.10}$$

Write (3.10) as $f_3^3 + f_4^3$, and restore $r = d^3$, so we now have

$$\begin{aligned} f_1(x,y) &= x^2 - \frac{2\sqrt{3}d^3}{\sqrt{1-d^6}}xy + y^2, \quad f_2(x,y) = dx^2 + \frac{2\sqrt{3}d}{\sqrt{1-d^6}}xy - dy^2, \\ f_3(x,y) &= -\frac{d(2+3d^3+d^6)}{1-d^6}\cdot x^2 + \frac{d(2-3d^3+d^6)}{1-d^6}\cdot y^2, \\ f_4(x,y) &= \frac{1+3d^3+2d^6}{1-d^6}\cdot x^2 + \frac{1-3d^3+2d^6}{1-d^6}\cdot y^2. \end{aligned} \tag{3.11}$$

Putting this together, (3.9)–(3.11) imply that

$$f_1^3(x,y) - f_4^3(x,y) = f_3^3(x,y) - f_2^3(x,y),$$
$$f_1(x,y) + d^2 f_2(x,y) = d^2 f_3(x,y) + f_4(x,y) \quad (= (1+d^3)x^2 + (1-d^3)y^2).$$

Thus, the wild case flips into a Type(d^2) family. Since p is even, and f_3, f_4 are not, we also have $p = f_5^3 + f_6^3$ where $f_5(x,y) = f_3(x,-y)$ and $f_6(x,y) = f_4(x,-y)$. □

4. Equations of Type(T)

In this section we completely describe the solutions to (1.6) of Type(T). We begin with a probably familiar result from Diophantine analysis.

Proposition 4.1. *Suppose $\Phi(u,v) = au^2 + 2buv + cv^2$ is a rank two quadratic form in $\mathbb{C}[x,y]$. Then any two honest solutions $(p_i, q_i, r_i), i = 1, 2$, in binary quadratic forms to the following equation are similar.*

$$\Phi(p,q) = r^2. \tag{4.1}$$

Proof. Write (4.1) as $(a_{11}p + a_{12}q)(a_{21}p + a_{22}q) = r^2$, where the factors on the left are distinct. Since $\gcd(p,q) = 1$, $\gcd(a_{11}p + a_{12}q, a_{21}p + a_{22}q) = 1$ as well. It follows by unique factorization that $(a_{11}p + a_{12}q, a_{21}p + a_{22}q, r) = (g^2, h^2, gh)$, for suitable distinct linear forms g, h. Let $[b_{ij}] = [a_{ij}]^{-1}$. Then

$$(p,q,r) = (b_{11}g^2 + b_{12}h^2, b_{21}g^2 + b_{22}h^2, gh).$$

In particular, (p_j, q_j, r_j) comes from (g_j, h_j), and the linear change M taking the honest pairs of linear forms (g_1, h_1) into (g_2, h_2) will take (p_1, q_1, r_1) into (p_2, q_2, r_2). □

Lemma 4.2. *If (1.6) is honest and a Type(T) family, then $T(T^3 - 1) \neq 0$.*

Proof. If $T = 0$, then $f_2 = -f_1$, violating honesty. Suppose $T^3 = 1$, so $T = \omega^k$. Then by $(f_3, f_4) \mapsto \omega^k(f_3, f_4)$ we may assume that $f_1 + f_2 = f_3 + f_4$. In this case, we have

$$(f_1 + f_2)^2 - \frac{f_1^3 + f_2^3}{f_1 + f_2} = (f_3 + f_4)^2 - \frac{f_3^3 + f_4^3}{f_3 + f_4} \implies f_1 f_2 = f_3 f_4. \tag{4.2}$$

This implies that $\{f_1, f_2\} = \{f_3, f_4\}$, again violating honesty. □

Theorem 4.3. *Suppose $\{f_1, f_2, f_3, f_4\}$ is an honest Type(T) family; specifically*

$$f_1^3 + f_2^3 = f_3^3 + f_4^3, \quad f_1 + f_2 = T(f_3 + f_4), \quad T(T^3 - 1) \neq 0, \tag{4.3}$$

and let $T = \lambda^2$. Then there is a linear change M so that $\{f_1 \circ M, f_2 \circ M\} = \{F_{3,\lambda}, -F_{5,\lambda}\}$ and $\{f_3 \circ M, f_4 \circ M\} = \{-F_{4,\lambda}, F_{6,\lambda}\}$.

Proof. As in (4.2), after dividing the equations in (4.3) we obtain

$$f_1^2 - f_1 f_2 + f_2^2 = T^{-1}(f_3^2 - f_3 f_4 + f_4^2). \tag{4.4}$$

It follows that

$$\begin{aligned} 3f_1f_2 &= (f_1+f_2)^2 - (f_1^2 - f_1f_2 + f_2^2) \\ &= (T^2 - T^{-1})f_3^2 + (2T^2 + T^{-1})f_3f_4 + (T^2 - T^{-1})f_4^2. \end{aligned} \tag{4.5}$$

But f_1 and f_2 are quadratic forms, and also the roots of the quadratic

$$\begin{aligned} &(X - f_1)(X - f_2) \\ &\quad = X^2 - T(f_3+f_4)X + \frac{1}{3}((T^2 - T^{-1})(f_3^2 + f_4^2) + (2T^2 + T^{-1})f_3f_4) \\ &\quad \implies \{f_1, f_2\} = \left\{\frac{T}{2}(f_3+f_4) \pm \frac{1}{2}\sqrt{\Delta}\right\}; \end{aligned}$$

$$\Delta = \frac{1}{3T}((4 - T^3)f_3^2 - (4 + 2T^3)f_3f_4 + (4 - T^3)f_4^2) = (f_2 - f_1)^2. \tag{4.6}$$

Consider now the quadratic form Φ, which has rank 2 if $T^3 \neq 1$.

$$\Phi(u,v) = \frac{1}{3T}((4 - T^3)u^2 - (4 + 2T^3)uv + (4 - T^3)v^2).$$

We have seen that $\Phi(f_3, f_4) = (f_2 - f_1)^2$. It may be checked that

$$\Phi(-F_{4,\lambda}, F_{6,\lambda}) = (\lambda^3x^2 + 2xy + \lambda^3y^2)^2. \tag{4.7}$$

Thus by Proposition 4.1, there is a linear change M so that $f_3 = -F_{4,\lambda} \circ M$ and $f_4 = F_{6,\lambda} \circ M$. It is routine to check that the quadratic equation (4.6) then solves to give $\{f_1 \circ M, f_2 \circ M\} = \{F_{3,\lambda}, -F_{5,\lambda}\}$. □

Proof of Theorem 1.1. Combine Theorems 3.1, 3.2 and 4.3. □

The historical motivation for the study of (1.6) was to find parametrizations of equal sums of pairs of rational cubes, so there is a special interest in solutions to (1.6) in which $f_j \in \mathbb{Q}[x,y]$. Since every solution to (1.6) is a Type(T) family for $T = \lambda^2$, we know that if $T = a^2, a \in \mathbb{Q}$, then there is a Type($T$) family in $\mathbb{Q}[x,y]$. In fact, there is such a family if and only if $T = a^2 + 3b^2$ for $a, b \in \mathbb{Q}$. We are indebted to the referee for suggesting this result and giving strong directions towards its proof, using [18]. For expository reasons, we defer Theorem 7.4 to the end of the paper.

Finally, a 1595 identity of Vieta (see [16]) becomes a version of (1.6) upon clearing denominators

$$(x(x^3 - y^3))^3 + (y(x^3 - y^3))^3 = (x(x^3 + 2y^3))^3 + (-y(2x^3 + y^3))^3;$$

the four quartics above are linearly independent. It seems unlikely that the methods of this paper are helpful when f_j in (1.6) have degree greater than two.

5. How Many Ways is a Sextic a Sum of Two Cubes?

We turn to a more general question. Lundqvist *et al.* proved in [9] that every binary sextic in $\mathbb{C}[x,y]$ can be written in infinitely many different ways as a sum of three cubes of quadratic forms. It is natural to wonder which binary sextics can be written as a sum of two cubes, and in how many ways.

We need some more general notation: for distinct forms $F, G \in \mathbb{C}[x_1, \dots, x_n]$, write $X = \langle F, G\rangle$ for the linear subspace $\{c_1F + c_2G\}$, and write $X^3 = \langle F^3, F^2G, FG^2, G^3\rangle$; X^3 is the set of all $h(F, G)$ for binary cubic forms h.

Theorem 5.1. *A form $p \in \mathbb{C}[x_1, \dots, x_n]$ of degree $3r$ can be written as $p = f_1^3 + f_2^3$ for distinct forms f_i of degree r if and only if it has a factorization $p = g_1g_2g_3$ in which the g_k's are distinct but linearly dependent and $\langle f_1, f_2\rangle = \langle g_1, g_2, g_3\rangle$. If p belongs to m different subspaces $\langle F_j, G_j\rangle^3$ as above, then $N(p) \le m$. If p is not divisible by the square of a form of degree r, then $N(p) = m$.*

Proof. In one direction,

$$p = f_1^3 + f_2^3 \implies p = (f_1 + f_2)(f_1 + \omega f_2)(f_1 + \omega^2 f_2) := g_1g_2g_3. \tag{5.1}$$

If any two of the g_i's are proportional in (5.1), then so are f_1 and f_2, and p is a cube contrary to hypothesis. For dependence, $g_j \in \langle f_1, f_2\rangle$, also, $g_1 + \omega g_2 + \omega^2 g_3 = 0$.

Conversely, if $P = g_1g_2g_3$ and g_1 and g_2 are distinct with $g_3 \in X = \langle g_1, g_2\rangle$, there exist $\alpha, \beta \ne 0$ so that $g_3 = \alpha g_1 + \beta g_2$. The sum of two cubes follows from an old formula (recall that $\omega - \omega^2 = \sqrt{-3}$):

$$\begin{aligned} p = g_1g_2g_3 &= g_1g_2(\alpha g_1 + \beta g_2) \\ &= \frac{1}{3\sqrt{-3}\alpha\beta} \cdot ((\omega\alpha g_1 - \beta g_2)^3 + (-\alpha g_1 + \omega\beta g_2)^3). \end{aligned} \tag{5.2}$$

Suppose p had two different expressions as a sum of two cubes of forms in $\langle f_1, f_2\rangle$

$$p = (c_{1,1}f_1 + c_{2,1}f_2)^3 + (c_{3,1}f_1 + c_{4,1}f_2)^3 = (c_{1,2}f_1 + c_{2,2}f_2)^3 + (c_{3,2}f_1 + c_{4,2}f_2)^3.$$

Then by the linear independence of $\{f_1^{3-k}f_2^k\}$ from Lemma 2.2, it follows that

$$(c_{1,1}x + c_{2,1}y)^3 + (c_{3,1}x + c_{4,1}y)^3 = (c_{1,2}x + c_{2,2}y)^3 + (c_{3,2}x + c_{4,2}y)^3,$$

which contradicts Lemma 2.1.

Thus, every representation of $p = f_1^3 + f_2^3$ identifies the subspace $\langle f_1, f_2\rangle^3$. Conversely, if $p \in \langle f_1, f_2\rangle^3$, then there is a cubic form h so that $p = h(f_1, f_2)$ and

$$h(x, y) = \sum_{j=1}^{2}(\alpha_j x + \beta_j y)^3 \implies p(x, y) = \sum_{j=1}^{2}(\alpha_j f_1 + \beta_j f_2)^3.$$

If $p \in \langle f_1, f_2\rangle^3$, then p is a sum of two cubes, unless h is a cube (and hence so is p), or $h(x, y) = (\alpha_1 x + \beta_1 y)^2(\alpha_2 x + \beta y)$, so p is divisible by $(\alpha_1 f_1 + \beta_1 f_2)^2$. □

Our study of sextics relies critically on the behavior of cubics as a sum of cubes. An important corollary was known in the 19th century (see also e.g., [14, Theorem 5.2]). A binary cubic q is *square-free* if it is a product of three pairwise distinct linear factors.

Proposition 5.2. *If p is a binary cubic which is not the cube of a linear form, then $p = \ell_1^3 + \ell_2^3$ for distinct linear forms ℓ_j if and only if p it is square-free, and this representation is unique,*

Proof. In the general case, $f = \ell_1\ell_2\ell_3$ is a product of three distinct linear forms; any three such forms are linearly dependent. The other cases are $f = \ell^3$ and $f = \ell_1^2\ell_2$, and the necessary factorization is impossible. □

For Theorems 5.4–5.6, recall (1.16) and (1.17).

Theorem 5.3. *A binary sextic $p(x, y)$ is an honest sum of two cubes ($N(p) \ge 1$) if and only if one of the two conditions hold:* (i) *$p = \ell^3 q$, where ℓ is linear form and q is a square-free cubic; or* (ii) *p is similar to $q(x^2, y^2)$, where q is a square-free cubic, so p is similar to an even binary sextic.*

Theorem 5.4. *A binary sextic p has $N(p) = 2$ if and only if p is similar to A_t for $t \in \mathbb{C}$, with the following exceptional values: $N(A_3) = 0$, $N(A_{-1}) = 1$, $N(A_0) = N(A_{15}) = 4$ and $N(A_{-5}) = 6$.*

Theorem 5.5. *A binary sextic p has $N(p) = 3$ if and only if p is similar to B_t for $t \in \mathbb{C}$, except that $N(B_{\pm 2}) = 0$, $N(B_0) = 4$ and $N(B_{\pm 5\sqrt{-2}}) = 6$.*

Theorem 5.6. *The binary sextics p with $N(p) > 3$ are similar to Q_1 or Q_2 : $N(Q_1) = 4$ and $N(Q_2) = 6$; Q_1 is similar to A_0, A_{15}, B_0; Q_2 is similar to A_{-5} and $B_{\pm 5\sqrt{-2}}$.*

Proof of Theorem 5.3. Suppose $p = f_1^3 + f_2^3$ is a binary sextic with $N(p) \ge 1$. If f_1 and f_2 are not distinct, then p is a cube, so f_1 and f_2 are distinct. If $\gcd(f_1, f_2) = \ell$ is linear, then $f_1 = \ell\ell_1$ and $f_2 = \ell\ell_2$, where ℓ_1 and ℓ_2 are distinct. Thus, $p = \ell^3(\ell_1^3 + \ell_2^3)$ satisfies (i). If f_1 and f_2 are relatively prime, then by Theorem 2.3, we may make a linear change M so that both $f_1 \circ M$ and $f_2 \circ M$ are even; that is, there exist distinct linear forms ℓ_j so that $(f_j \circ M)(x, y) = \ell_j(x^2, y^2)$; now let $q = \ell_1^3 + \ell_2^3$; this is (ii). □

Theorem 5.7. *If p is a binary sextic with a square factor, then $N(p) \le 1$.*

Proof. Suppose $\ell^k \,|\, p$ for a linear factor ℓ, where $k \ge 2$. Suppose $k \ge 3$ and $p = f_1^3 + f_2^3$ for quadratic forms f_1, f_2. Then as in Lemma 2.4, ℓ must divide at least two of $\{f_1 + \omega^k f_2\}$, and so $\ell \,|\, f_1, f_2$, so p has no other representation as a sum of two cubes.

Now suppose $k = 2$, and after a linear change, take $\ell = y$, so that for some $c_j \in \mathbb{C}$,

$$p(x, y) = \lambda y^2(x + c_1 y)(x + c_2 y)(x + c_3 y)(x + c_4 y).$$

To apply Theorem 5.2, we need to write $p = g_1 g_2 g_3$ for linearly dependent factors. If y divides two of the g_j's, it must divide the third, which is impossible, hence we may assume that $g_1 = y^2$. If $N(p) \geq 2$, then after reindexing if necessary, each of these two different sets is dependent:

$$\{y^2, (x + c_1 y)(x + c_2 y), (x + c_3 y)(x + c_4 y)\},$$
$$\{y^2, (x + c_1 y)(x + c_3 y), (x + c_2 y)(x + c_4 y)\}.$$

But dependence implies that $c_1 + c_2 = c_3 + c_4$ and $c_1 + c_3 = c_2 + c_4$, so $c_3 = c_2$ and $c_4 = c_1$ and $(x + c_1 y)(x + c_2 y) = (x + c_3 y)(x + c_4 y)$, so the factors are not distinct. $\square$

We isolate those exceptional cases in Theorems 5.4 and 5.5 with square factors.

Theorem 5.8. *We have* $N(A_3) = 0$, $N(B_{\pm 2}) = 0$, *and* $N(A_{-1}) = 1$.

Proof. By the first argument of the proof of Theorem 5.7, since $A_3(x, y) = (x^2 + y^2)^3$, in any representation $A_3 = f_1^3 + f_2^3$, both f_1 and f_2 are multiples of $x^2 + y^2$, so that they are not distinct. This also follows from Liouville's solution for Fermat's last theorem in polynomials (see [17, pp. 263–265] for a proof).

We have seen that if ℓ^2 (but not ℓ^3) divides a sextic p and p has a factorization that partitions into three dependent factors, then one of those factors must be ℓ^2. The only feasible partitions for $B_{\pm 2}(x, y) = (x^3 \pm y^3)^2$ are $\{(x \pm y)^2, (x \pm \omega y)^2, (x \pm \omega^2 y)^2\}$, which are linearly independent; thus $N(B_{\pm 2}) = 0$.

Finally, consider A_{-1}, which factors as $(x - y)^2(x + y)^2(x^2 + y^2)$. Each of the two squares must be a factor, and $\{(x-y)^2, (x+y)^2, x^2+y^2\} \subset \langle x^2+y^2, xy\rangle$. There is a representation for $2A_{-1}$ in (3.2) with $\gamma = \sqrt{-4/3}$. Thus $N(A_{-1}) = 1$. $\square$

It is worth mentioning that $A_{-1}(x, y) = (x^2 - y^2)^2(x^2 + y^2)$, so $A_{-1}(x, y) = q_1(x^2, y^2)$, where $q_1(x, y) = (x - y)^2(x + y)$ is not square-free. But $\tilde{A}_{-1}(x, y) := A_{-1}(x + y, x - y) = 32x^4y^2 + 32x^2y^4 = q_2(x^2, y^2)$, where $q_2(x, y) = 32xy(x + y)$ *is* square-free. Although A_{-1} and $\tilde{A}_{-1}$ are similar, q_1 and q_2 are not.

Now suppose that $N(p) \geq 2$. By Theorem 1.1, we know that after a linear change, p appears as the common sum in (1.10), (1.12) or (1.13), and in the first two cases, $N(p) \geq 3$. Since (1.12) is a linear change of (1.10), we may ignore it. We now apply Theorem 5.1 to $p_{3,\lambda}$ and to $p_{1,\lambda}$, which have already been conveniently split into six linear factors. There are 15 ways to divide six factors into three unordered pairs.

Proof of Theorems 5.4–5.6. Up to a constant which can be ignored, we have $p_{3,\lambda}(x, y) = xy(x - y)(x + y)(\alpha x + y)(x + \alpha y)$, where $\alpha = \lambda^3 \notin \{0, -1, 1\}$, which cause repeated factors. It is not hard to check the 15 possibilities, and we suppress the details. In two cases, the factors are always dependent

$$\{x(\alpha x + y), y(x + \alpha y), (x + y)(x - y)\} = \langle \alpha x^2 + xy, xy + \alpha y^2\rangle,$$
$$\{x(x + \alpha y), y(\alpha x + y), (x + y)(x - y)\} = \langle x^2 + \alpha xy, \alpha xy + y^2\rangle.$$

There are two cases when there are multiple dependencies. If $\alpha \in \{\pm 2, \pm\frac{1}{2}\}$, there are two additional cases of dependency, and if $\alpha = \pm i$, there are four additional cases. Thus, $N(p_{3,\lambda}) = 2$ for $\lambda(1-\lambda^6) \neq 0$ unless $\alpha \in \{\pm 2, \pm\frac{1}{2}, \pm i\}$.

If $\alpha = \lambda^3 = \pm i$, then up to powers of ω, $\lambda^2 = -1$. In the language of Theorem 3.1, $\frac{r_\gamma + s_\gamma}{2} = \lambda^2 \implies r_\gamma + s_\gamma = -2 = (8+6\gamma^2)^{1/3} \implies 3(1+\gamma^2) = -5$, so $p_{3,\pm i}$ is similar to A_5. If $\alpha = \pm 2, \pm\frac{1}{2}$, then $\lambda^2 = 2^{\pm 2/3}$ and $r_\gamma + s_\gamma = 2^{1/3}, 2^{5/3} \implies 8 + 6\gamma^2 = 2, 32 \implies 3(1+\gamma^2) = 0, 15$, so $p_{3,\lambda}$ is similar to $A_0 = Q_1$ or A_{15}.

Up to a constant,

$$p_{1,\lambda}(x,y) = (\lambda x + y)(\lambda x + \omega y)(\lambda x + \omega^2 y)(x + \lambda y)(x + \lambda\omega y)(x + \lambda\omega^2 y).$$

As we would hope, there are three cases in which the factors are always dependent:

$$\begin{aligned}&\{(\lambda x + y)(x+\lambda y), (\lambda x + \omega y)(x + \lambda\omega^2 y), (\lambda x + \omega^2 y)(x + \lambda\omega y)\},\\ &\{(\lambda x + y)(x+\lambda\omega y), (\lambda x + \omega y)(x + \lambda y), (\lambda x + \omega^2 y)(x + \lambda\omega^2 y)\},\\ &\{(\lambda x + y)(x+\lambda\omega^2 y), (\lambda x + \omega y)(x + \lambda\omega y), (\lambda x + \omega^2 y)(x + \lambda y)\};\end{aligned} \tag{5.3}$$

the subspaces are $\langle x^2 + \omega^k y^2, xy\rangle$. There are a few cases with multiple dependencies: when $\lambda = \pm i$ (so $\alpha = \pm i$), there is one extra case. In this case, $p_{1,\pm i}(x,y) = \pm 2i Q_1(x,y)$. The other cases in which a dependency occurs are when $\lambda^4 + 4\lambda^2 + 1 = 0$, up to $\lambda \mapsto \omega^k\lambda$. For example, suppose

$$\begin{aligned}&\{(\lambda x + y)(x + \lambda y), (x + \lambda\omega y)(x + \lambda\omega^2 y), (\lambda x + \omega y)(\lambda x + \omega^2 y)\}\\ &\quad = \{\lambda x^2 + (\lambda^2+1)xy + \lambda y^2, x^2 - \lambda xy + \lambda y^2, \lambda^2 x^2 - \lambda xy + y^2\}\end{aligned}$$

is linearly dependent. This happens if and only if

$$\begin{vmatrix} \lambda & \lambda^2+1 & \lambda \\ 1 & -\lambda & \lambda^2 \\ \lambda^2 & -\lambda & 1 \end{vmatrix} = (\lambda^2 - 1)(\lambda^4 + 4\lambda^2 + 1) = 0.$$

In computations that Ramanujan could probably do in his sleep,

$$\begin{aligned}&\lambda^4 + 4\lambda^2 + 1 = 0 \implies \lambda^2 = -2 \pm \sqrt{3} \implies \lambda = \pm_1 \left(\frac{\sqrt{6} \pm_2 \sqrt{2}}{2}\right) i\\ &\qquad \implies \lambda^3 + \lambda^{-3} = \pm 5\sqrt{-2}.\end{aligned} \tag{5.4}$$

Since B_t and B_{-t} are similar via $y \mapsto -y$, we focus on $B_{5\sqrt{-2}}$. Let $\eta = \frac{\sqrt{6}+\sqrt{2}}{2}$, so $\lambda = \eta i$ is a root. We have a linear change with bizarre coefficients:

$$B_{5\sqrt{-2}}(\zeta_8^2 \eta x + \zeta_8 y, x + \zeta_8^3 \eta y) = 54\zeta_8^3 \eta^3 Q_2(x,y), \tag{5.5}$$

showing that $B_{5\sqrt{-2}}$ is similar to Q_2. We give a geometric explanation for (5.5) in the next section.

□

The instance of (1.6) with the simplest coefficients is probably

$$\begin{aligned}(x^2+xy-y^2)^3+(x^2-xy-y^2)^3 &= 2(x^2)^3+2(-y^2)^3 = 2x^6-2y^6\\ &= (\omega x^2+xy-\omega^2y^2)^3+(\omega x^2-xy-\omega^2y^2)^3\\ &= (\omega^2x^2+xy-\omega y^2)^3+(\omega^2x^2-xy-\omega y^2)^3. \end{aligned} \tag{5.6}$$

With $(x,y)\mapsto(x+y,x-y)$, (5.6) is due to Girardin in 1910 (see [4, p. 550]; the earliest exact version of (5.6) we have found is by Elkies in 1995 (see [3, p. 542]). Observe that (5.6) is simply (1.9) with $\lambda=i$ and $y\mapsto iy$, and it also a scaling of Q_1. (We have $2x^6-2y^6=Q_1(rx,sy)$ if $r^6=2, s^6=-2$.) Unsurprisingly, since $\lambda=i$, a flip of (5.6) is similar to Q_2:

$$(\omega x^2+xy-\omega^2y^2)^3-(\omega^2x^2+xy-\omega y^2)^3=-3\sqrt{-3}(x^5y-xy^5). \tag{5.7}$$

Finally, we remark that while (5.6) is presented as a Type(-1) family, we have

$$(x^2+xy-y^2)+(x^2-xy-y^2)=2^{-1/3}(2^{1/3}x^2+(-2^{1/3}y^2)),$$

which gives a Type($2^{2/3}$) family from (3.2), with $y\mapsto iy$. Thus the Type parameter may vary when more than three representations occur.

6. More on the Extra Representations

As we saw in the last section, there are two special cases of sextics with more than three representations and we treat them separately. First, note that

$$Q_1(x,y)=x^6+y^6=A_0(x,y)=B_0(x,y);\quad A_{15}(x,y)=\frac{1}{2}A_0(x+y,x-y).$$

For purposes of analyzing the factorizations, we note that with $\lambda=i$, it is easier to use powers of $\nu:=\zeta_{12}$:

$$Q_1(x,y)=(x-\nu y)(x-\nu^3y)(x-\nu^5y)(x-\nu^7y)(x-\mu^9y)(x-\nu^{11}y).$$

Keeping in mind that $i=\nu^3,\omega=\nu^4$, and rearranging (5.3) a bit, we have that the three-dependent factorizations of Q_1 are as follows:

$$\{(x+\nu y)(x+\nu^{11}y),(x+\nu^3y)(x+\nu^9y),(x+\nu^5y)(x+\nu^7y)\},$$

$$\{(x+\nu y)(x+\nu^3y),(x+\nu^7y)(x+\nu^9y),(x+\nu^5y)(x+\nu^{11}y)\},$$

$$\{(x+\nu y)(x+\nu^7y),(x+\nu^3y)(x+\nu^5y),(x+\nu^9y)(x+\nu^{11}y)\}.$$

These live in $\langle x^2+y^2,xy\rangle,\langle x^2+y^2,\omega xy\rangle,\langle x^2+\omega^2y^2,xy\rangle$ respectively. The fourth dependent factorization is

$$\{(x+\nu y)(x+\nu^7y),(x+\nu^3y)(x+\nu^9y),(x+\nu^5y)(x+\nu^{11}y)\}\subseteq\langle x^2,y^2\rangle.$$

The best way of visualizing the four equal pairs of sums seems to be (5.6).

The other case is somewhat more mysterious. Since $Q_2(x,y)=xy(x^4-y^4)$, it is simple to work out all 15 factorizations into three quadratics. The following six are dependent:

$$\{xy,(x+y)(x+iy),(x-y)(x-iy)\} \subseteq \langle x^2+iy^2, xy\rangle,$$
$$\{xy,(x+y)(x-iy),(x-y)(x+iy)\} \subseteq \langle x^2-iy^2, xy\rangle,$$
$$\{x(x+y),y(x-y),(x+iy)(x-iy)\} \subseteq \langle x^2+xy, x^2+y^2\rangle,$$
$$\{x(x+iy),y(x-iy),(x+y)(x-y)\} \subseteq \langle x^2+ixy, x^2-y^2\rangle,$$
$$\{x(x-y),y(x+y),(x+iy)(x-iy)\} \subseteq \langle x^2-xy, x^2+y^2\rangle,$$
$$\{x(x-iy),y(x+iy),(x+y)(x-y)\} \subseteq \langle x^2-ixy, x^2-y^2\rangle.$$

We could simply write Q_2 explicitly as an element in $\langle F,G\rangle^3$ in these six cases. It is more interesting to derive them from earlier work; see (6.1), (6.3), (6.4) below.

First, observe that $r_{3,i}(x,y) = r_{3,-i}(x,y) = -3\sqrt{-3}Q_2(x,y) = (\sqrt{-3})^3Q_2(x,y)$. One would think that this gives four representations of Q_2, coming from (1.13); however the representation for $\lambda = -i$ is a permutation of that from $\lambda = i$, and there are only two distinct ones:

$$\begin{aligned} -3\sqrt{-3}Q_2(x,y) &= (\nu^5x^2+xy+\nu y^2)^3 + (\nu^7x^2-xy+\nu^{11}y^2)^3, \\ 3\sqrt{-3}Q_2(x,y) &= (\nu^{11}x^2+xy+\nu^7y^2)^3 + (\nu x^2-xy+\nu^5y^2)^3. \end{aligned} \tag{6.1}$$

These come from $\langle x^2-ixy, x^2-y^2\rangle$ and $\langle x^2+ixy, x^2-y^2\rangle$, respectively. However, $Q_2(x,y) = -iQ_2(x,iy)$, so

$$Q_2(x,y) = f_1(x,y)^3 + f_2(x,y)^3 \implies Q_2(x,y) = (if_1(x,iy))^3 + (if_2(x,iy))^3. \tag{6.2}$$

In this way, we immediately obtain two more representations:

$$\begin{aligned} -3\sqrt{-3}Q_2(x,y) &= (\nu^{10}x^2+xy+\nu^8y^2)^3 + (\nu^8x^2-xy+\nu^{10}y^2)^3, \\ -3\sqrt{-3}Q_2(x,y) &= (\nu^4x^2+xy+\nu^2y^2)^3 + (\nu^2x^2-xy+\nu^4y^2)^3. \end{aligned} \tag{6.3}$$

These are in $\langle x^2+xy, x^2+y^2\rangle$ and $\langle x^2-xy, x^2+y^2\rangle$, as one would expect; (6.2) simply permutes the equations, and we get no more. Since $\nu^4 = \omega$ and $\nu^2 = -\omega^2$, the second equation in (6.3) recovers (5.7).

We have $Q_2(\frac{x+y}{\sqrt{2}}, \frac{x-y}{\sqrt{2}}) = Q_2(x,y)$, so after some simplification, we obtain the last two representations of Q_2

$$\begin{aligned} 6\sqrt{-6}Q_2(x,y) &= (\zeta_8^5x^2+\sqrt{6}xy+\zeta_8^7y^2)^3 + (\zeta_8x^2+\sqrt{6}xy+\zeta_8^3y^2)^3. \\ 6\sqrt{-6}Q_2(x,y) &= (\zeta_8^7x^2-\sqrt{-6}xy+\zeta_8^5y^2)^3 + (\zeta_8^3x^2-\sqrt{-6}xy+\zeta_8y^2)^3. \end{aligned} \tag{6.4}$$

These are in $\langle x^2+iy^2, xy\rangle$ and $\langle x^2-iy^2, xy\rangle$. Although it might seem daunting to consider checking whether any two of these six expressions are similar, the fact that they live in different subspaces shows that this is impossible.

Finally, we discuss the connection of Q_2 and $B_{5\sqrt{-2}}$. To do so, we need an old idea of Felix Klein; see also [15, p. 731]. Associate to each nonzero linear form

$\ell(x,y) = sx - ty$ the image of $t/s \in \mathbb{C}^*$ on the unit sphere S^2 under the Riemann map and vice versa. (Assign $\ell(x,y) = y$ to ∞ and $(0,0,1)$.) The *Klein set* of $p(x,y) = \prod_{j=1}^{k}(s_j x - t_j y)$ is the image of the k points t_j/s_j on S^2 under the Riemann map. Every rotational symmetry of the Klein set of p has an interpretation as a symmetry of p under a linear change.

There are two particularly symmetric six-point sets on S^2. One is a hexagon along a great circle, say the equator. Note that $Q_1(x,y) = x^6 + y^6 = \prod_{j=0}^{5} (x + \zeta_{12}^{2j+1} y)$ has such a hexagon as its Klein set. The other natural choice is the vertex set of a regular octahedron, and the Klein set of Q_2 is $\{\pm e_k\}$:

$$Q_2(x,y) = xy(x-y)(x+y)(x-iy)(x+iy) = xy(x^4 - y^4).$$

One may rotate an octahedron so that the top and bottom are antipodal triangular faces parallel to the equator. One set of coordinates of the vertices is:

$$\left\{ \pm\left(\frac{2}{\sqrt{6}}, 0, \frac{\sqrt{2}}{\sqrt{6}}\right), \pm_1 \left(\frac{-1}{\sqrt{6}}, \pm_2 \frac{\sqrt{3}}{\sqrt{6}}, \frac{\sqrt{2}}{\sqrt{6}}\right) \right\}. \tag{6.5}$$

The cubic which corresponds to the triangle in the northern hemisphere is

$$(x - \lambda_0 y)(x - \omega\lambda_0 y)(x - \omega^2\lambda_0 y) = x^3 - \frac{5 + 3\sqrt{3}}{\sqrt{2}} y^3, \quad \lambda_0 = \frac{\sqrt{6} + \sqrt{2}}{2}.$$

Similarly, the cubic for the southern hemisphere is

$$(x + \lambda_0^{-1} y)(x + \omega\lambda_0^{-1} y)(x + \omega^2\lambda_0^{-1} y) = x^3 + \frac{5 + 3\sqrt{3}}{\sqrt{2}} y^3.$$

Multiplying these together, we get another Klein polynomial for the octahedron:

$$\begin{aligned} \tilde{Q}_2(x,y) &= x^6 - 5\sqrt{2}x^3y^3 - y^6 \\ &\implies \tilde{Q}_2(x,iy) = x^6 + 5\sqrt{2}ix^3y^3 + y^6 = B_{5\sqrt{-2}}(x,y). \end{aligned}$$

The rotation relating $\{\pm e_k\}$ into (6.5) inspired the coefficients of (5.5).

There are, in general, $\frac{(3r)!}{3!(r!)^3}$ ways to arrange the $3r$ linear factors of a form p into three factors of degree r, and by Theorem 5.1, this gives an upper bound on the number of ways to write p as a sum of two cubes. It would be interesting to know how the actual bound grows for p. The natural analogues of Q_1, Q_2 are $x^{3r} + y^{3r}$ and $xy(x^{3r-2} - y^{3r-2})$.

7. Other Approaches to Sums of Two Cubes

The proof of the Euler–Binet parametrization of all solutions, found for example in [6, pp. 199–201], can easily be adapted to fields of characteristic zero. For our purposes, we look at rational functions over $\mathbb{C}$.

Theorem 7.1 (Euler–Binet). *Suppose $F = \mathbb{C}(x_1, \ldots, x_n)$ and suppose*

$$p = f_1^3 + f_2^3 = f_3^3 + f_4^3. \tag{7.1}$$

for pairwise distinct $f_1, f_2, f_3, f_4 \in F$*. Then there exists* $\mu, a, b \in F$ *so that*

$$\begin{aligned} f_1 &= \mu(1-(a-3b)(a^2+3b^2)), \quad f_2 = \mu((a+3b)(a^2+3b^2)-1), \\ f_3 &= \mu((a+3b)-(a^2+3b^2)^2), \quad f_4 = \mu((a^2+3b^2)^2-(a-3b)). \end{aligned} \tag{7.2}$$

Conversely, if f_1, f_2, f_3, f_4 *are given by* (7.2) *in terms of* μ, a, b*, then*

$$p = f_1^3 + f_2^3 = f_3^3 + f_4^3 = 18\mu^3 b(a^2+3b^2)(1-(a+b)^3-(a-b)^3+(a^2+3b^2)^3)). \tag{7.3}$$

Proof. Define g_i's by

$$f_1 = g_1 + g_2, \quad f_2 = g_1 - g_2, \quad f_3 = g_3 + g_4, \quad f_4 = g_3 - g_4, \tag{7.4}$$

so that (7.1) becomes

$$p = 2g_1(g_1^2 + 3g_2^2) = 2g_3(g_3^2 + 3g_4^2).$$

Since $p \neq 0$, $g_1^2 + 3g_2^2 \neq 0$ as well, and we may define

$$a = \frac{g_1g_3 + 3g_2g_4}{g_1^2 + 3g_2^2}, \quad b = \frac{g_1g_4 - g_3g_2}{g_1^2 + 3g_2^2}. \tag{7.5}$$

Observe that

$$ag_1 - 3bg_2 = g_3; \quad bg_1 + ag_2 = g_4; \quad a^2 + 3b^2 = \frac{g_3^2 + 3g_4^2}{g_1^2 + 3g_2^2} = \frac{g_1}{g_3}. \tag{7.6}$$

(In the original derivation, taken over $\mathbb{Q}$, (a, b) are defined by $a \pm b\sqrt{-3} = \frac{g_3 \pm g_4\sqrt{-3}}{g_1 \pm g_2\sqrt{-3}}$, which is unambiguous. We cannot do this here, because some coefficient of g_j might involve $\sqrt{-3}$, but (7.5) recaptures the essence.) Now let

$$\begin{aligned} &c = a(a^2+3b^2) - 1, \quad d = 3b(a^2+3b^2) \\ &\implies cg_1 - dg_2 = (a^2+3b^2)(ag_1 - 3bg_2) - g_1 = (a^2+3b^2)g_3 - g_1 = 0, \end{aligned} \tag{7.7}$$

so $cg_1 = dg_2$. Suppose $c = d = 0$. Looking at $d = 0$, $a^2 + 3b^2 = 0$ implies $c = -1$, so $b = 0$, and $ag_1 = g_3$, and $ag_2 = g_4$ by (7.6), so that $af_1 = f_3$ and $af_2 = f_4$ implying that (7.1) is not honest. Thus $(c, d) \neq (0, 0)$, and we write (g_1, g_2) with $\mu \in F$ as

$$g_1 = \mu d = 3\mu b(a^2+3b^2), \quad g_2 = \mu c = \mu(a(a^2+3b^2) - 1). \tag{7.8}$$

Now solve for g_3 and g_4 from (7.6):

$$g_3 = ag_1 - 3bg_2 = 3\mu b, \quad g_4 = bg_1 + ag_2 = \mu((a^2+3b^2)^2 - a). \tag{7.9}$$

Plug back into (7.4) and (7.5) to get (7.2). □

Corollary 7.2. *Suppose* f_1, f_2, f_3, f_4 *are forms of degree* k *satisfying* (7.1)*. Then up to a possible common factor, there exist forms* p, q, r *of degree* $\leq 2k$ *so that*

$$\begin{aligned} f_1 &= r(r^3-(p-3q)(p^2+3q^2)), \quad f_2 = r((p+3q)(p^2+3q^2)-r^3), \\ f_3 &= r^3(p+3q)-(p^2+3q^2)^2, \quad f_4 = (p^2+3q^2)^2 - r^3(p-3q). \end{aligned} \tag{7.10}$$

Proof. Define f_1, f_2, f_3, f_4 as above, and define a and b via (7.5) as rational functions with a common denominator, subject to possible cancellation:

$$a = \frac{p(x,y)}{r(x,y)}, \quad b = \frac{q(x,y)}{r(x,y)}. \tag{7.11}$$

The expressions for f_3, f_4 have a formal denominator of r^4, so we take $\mu(x,y) = r^4(x,y)$, with the understanding that cancellation may occur. By substituting (7.11) into (7.2), we obtain (7.10). □

Applying this to the quadruple $(f_1, f_2, f_3, f_4) = (F_{6,\lambda}, -F_{4,\lambda}, F_{3,\lambda}, -F_{5,\lambda})$, there is much cancellation and

$$a = -\frac{x^2+y^2}{2\lambda\, xy}, \quad b = \frac{-i(x^2-y^2)}{2\sqrt{3}\lambda\, xy}, \quad \mu = r^4 xy, \tag{7.12}$$

so that p and q are quadratic, and r is linear. Other choices for the f_j's lead to p, q, r of higher degree. There are $3^4 \cdot 4! = 1944$ ways to arrange the f_i's, counting cube roots of unity, and we cannot assert that a simpler set of parameters does not exist. In the famous Ramanujan case of $12^3 + 1^3 = 10^3 + 9^3 = 1729$, the integral version of (7.2) comes from $(a, b, \mu) = (\frac{10}{19}, \frac{7}{19}, -\frac{361}{42})$, but permuting 9 and 10 means that we need denominators of 266 and 333. On the other hand, the same identity flipped as $10^3 + (-1)^3 = (-9)^3 + 12^3$ comes from $(a, b, \mu) = (-\frac{3}{2}, \frac{1}{2}, 1)$. We note for later use that $5^3 + 4^3 = (-3)^3 + 6^3$ comes from $(a, b, \mu) = (0, 1, \frac{1}{2})$.

The other standard approach to equal sums of cubes arises from point-addition on the curve $X^3 + Y^3 = A$; see e.g., [19]. Assuming that $(X, Y) = (X_1, Y_1), (X_2, Y_2)$ lie on this curve, the cubic equation $(tX_1 + (1-t)Y_1)^3 + (tX_2 + (1-t)Y_2)^3 = A$ has two solutions $t = 0, 1$, and so the third may be computed; after simplification,

$$\begin{aligned} X_3 &= \frac{A(X_1 - X_2) + Y_1Y_2(X_2Y_1 - X_1Y_2)}{(X_1^2X_2 + Y_1^2Y_2) - (X_1X_2^2 + Y_1Y_2^2)}, \\ Y_3 &= \frac{A(Y_1 - Y_2) + X_1X_2(X_1Y_2 - X_2Y_1)}{(X_1^2X_2 + Y_1^2Y_2) - (X_1X_2^2 + Y_1Y_2^2)}. \end{aligned} \tag{7.13}$$

This computation (usually done over $\mathbb{Q}$), is still valid when X_i, Y_i are polynomials. Of course, the denominator means that the new solution is usually composed of rational functions. Somewhat astonishingly, (7.13) is applicable to (1.10), and we present a theorem whose only proof is direct computation.

Theorem 7.3. *If we take* $(X_1, Y_1) = (F_{1,\lambda}, F_{2,\lambda})$, $(X_2, Y_2) = (F_{3,\lambda}, F_{4,\lambda})$ *and* $A = p_{1,\lambda}(x, y)$ *in* (7.13), *then* $(X_3, Y_3) = (F_{5,\lambda}, F_{6,\lambda})$.

More generally, if we take the parametrizations from (7.2) to add (f_1, f_2) and (f_3, f_4), we obtain denominators. But if we add $(f_1, -f_4)$ and $(f_3, -f_2)$, which come from the flip $f_1^3 - f_4^3 = (-f_2)^3 + f_3^3$, we obtain a third polynomial solution which is apparently new.

$$f_1^3 - f_4^3 = -f_2^3 + f_3^3 = (\mu(1 + 2a(a^2 + 3b^2)))^3 - (\mu(2a + (a^2 + 3b^2)^2))^3. \tag{7.14}$$

A few caveats: even though (7.2) is a complete parametrization of solutions to two equal sums of two cubes; (7.14) is *not* a complete parametrization of solutions to three equal sums of two cubes. An extremely tedious application of Theorem 5.1 to the three flips of (7.2) shows that this is the only bonus representation.

As is the case with $\mathbb{Q}$, there can be arbitrarily large sets of equal pairs of sums of two cubes. For example, Rouse and the author give in [16] the complete (infinite) solution to the solution over rational functions of:

$$x^3 + y^3 = \left(\frac{p(x,y)}{r(x,y)}\right)^3 + \left(\frac{q(x,y)}{r(x,y)}\right)^3, \quad p, q, r \in \mathbb{C}[x,y].$$

for rational functions $(p/r, q/r)$. Clearing the denominator in any finite family of sums $x^3 + y^3 = (\frac{p_i}{r_i})^3 + (\frac{q_i}{r_i})^3, 1 \le i \le N$, gives a set of N equal sums.

We may also take an invariant-theory approach to $N(p) \ge 1$. In any sum of two cubes of quadratic forms:

$$\sum_{j=1}^{2} (\alpha_{j0}x^2 + \alpha_{j1}xy + \alpha_{j2}y^2)^3 = \sum_{k=0}^{6} c_k x^{6-k} y^k,$$

the seven c_k's are cubic polynomials in the six $\alpha'_{j\ell}s$, and since $7 > 6$, we know that the c_k's must be algebraically dependent. There are $\binom{n+6}{6}$ monomials in the c_j's of degree n; these are forms of degree $3n$ in the $\alpha'_{j\ell}s$, which comprise a vector space of dimension $\binom{3n+5}{5}$. Eventually, $\binom{n+6}{6} > \binom{3n+5}{5}$, so there must be dependence at some degree n. Unfortunately, the smallest n for which this happens is $n = 1442$.

We can be less brute-force and apply Theorem 5.1. Suppose our given cubic p is a sum of two cubes, factor it and expand it in the usual way. Write p as

$$\sum_{k=0}^{6} c_k x^{6-k} y^k = c_0 \left(x^6 + \sum_{k=1}^{6} e_k x^{6-k} y^k \right) = c_0 \prod_{j=1}^{6} (x + r_j y),$$

where the e_k's are the elementary symmetric functions in the r_j's. As noted earlier, there are 15 ways to divide the 6 r_j's into three pairs of roots, and the condition that the quadratic factors be dependent is equivalent to the vanishing of

$$H(r) := \prod_{\ell=1}^{15} \begin{vmatrix} 1 & 1 & 1 \\ r_{\sigma_\ell(1)} + r_{\sigma_\ell(2)} & r_{\sigma_\ell(3)} + r_{\sigma_\ell(4)} & r_{\sigma_\ell(5)} + r_{\sigma_\ell(6)} \\ r_{\sigma_\ell(1)} r_{\sigma_\ell(2)} & r_{\sigma_\ell(3)} r_{\sigma_\ell(4)} & r_{\sigma_\ell(5)} r_{\sigma_\ell(6)} \end{vmatrix},$$

where the product is taken over a suitable subset of S_6. (Of course, $H(r) = 0$ even if the factors are dependent, so this is a necessary but not sufficient condition.) Mathematica can compute $H(r)$ without too much difficulty, and in a few hours transform it into a symmetric function in the e_k's of degree 15. Now write $e_k = c_k/c_0$, make the substitution and multiply by c_0^{15} to get the relation. It has 1360 terms and is *isobaric* in the old sense: each monomial $\prod c_k^{m_k}$ in the product has $\sum m_k = 15, \sum k m_k = 45$. It seems likely that this is the skew invariant called I_{15} in the old literature. For more information, see [5], especially §143, §244 and Examples 20 and 21 in pp. 315–316. The original discovery is attributed there to Joubert.

Finally, here are some of the quadratic parametrizations of (1.6) which can be found in the literature. The earliest one found in [4, p. 554] was in J. R. Young's 1816 book *Algebra*, in S. Ward's edition of 1832, and in 1895, by the self-taught mathematician Artemas Martin (see [1]) in a journal he wrote, edited and typeset

$$(x^2+16xy-21y^2)^3+(-x^2+16xy+21y^2)^3+(2x^2-4xy+42y^2)^3$$
$$=(2x^2+4xy+42y^2)^3. \tag{7.15}$$

This is a Type(4) family. In fact, Young presented a one-parameter family of such solutions, of Type(n^2), which homogenizes to

$$(nx^2-6nxy+3(n^7-n)y^2)^3+(-x^2+6n^3xy+3(n^6-1)y^2)^3$$
$$=(nx^2+6nxy+3(n^7-n)y^2)^3+(-x^2-6n^3xy+3(n^6-1)y^2)^3. \tag{7.16}$$

By Theorem 1.1, these are similar to the Narayanan solutions from a century later, and since their sum is an even polynomial, there is not a third representation.

Hirschhorn has written several papers which explore Ramanujan's approach to (1.6) and related questions. In [7], he conjectured that an "amazing" identity of Ramanujan in his "Lost Notebook" could be proved via the Type(4) identity

$$(x^2+7xy-9y^2)^3+(2x^2-4xy+12y^2)^3=(2x^2+10y^2)^3+(x^2-9xy-y^2)^3, \tag{7.17}$$

and in [8, p. 388], he derived this as a special case of a more general formula, which homogenizes to the Type(n^2) identity:

$$(3x^2+6n^3xy+(1-n^6)y^2)^3+(3nx^2-6nxy+(n^7-n)y^2)^3$$
$$=(3x^2-6n^3xy+(1-n^6)y^2)^3+(3nx^2+6nxy+(n^7-n)y^2)^3. \tag{7.18}$$

Sándor [18] gave a beautiful solution to (1.6) as a conditional polynomial identity. (In 1873, Korneck [4, p. 556] (see [18, p. 122]) gave a similar family of identities.) Suppose $(w_1,w_2,w_3,w_4)\in\mathbb{C}^4$ satisfy

$$w_1^3+w_2^3=w_3^3+w_4^3, \tag{7.19}$$

and define $h_i=h_i(x,y)$ by

$$\begin{aligned}
h_1&=w_1(w_1+w_2)x^2-(w_3^2-w_4^2)xy-w_2(w_3+w_4)y^2,\\
h_2&=w_2(w_1+w_2)x^2+(w_3^2-w_4^2)xy-w_1(w_3+w_4)y^2,\\
h_3&=w_3(w_1+w_2)x^2-(w_1^2-w_2^2)xy-w_4(w_3+w_4)y^2,\\
h_4&=w_4(w_1+w_2)x^2+(w_1^2-w_2^2)xy-w_3(w_3+w_4)y^2.
\end{aligned} \tag{7.20}$$

Then

$$h_1^3+h_2^3-h_3^3-h_4^3=(w_1^3+w_2^3-w_3^3-w_4^3)((w_1+w_2)x^2+(w_3+w_4)y^2)^3=0 \tag{7.21}$$

and

$$\frac{h_1 + h_2}{h_3 + h_4} = \frac{w_1 + w_2}{w_3 + w_4}; \tag{7.22}$$

that is, $\{h_j(x, y)\}$ forms a Type($\frac{w_1+w_2}{w_3+w_4}$) family.

Observe that in any solution to (1.6), the coefficients of x^2, say (w_i) up to multiple, must satisfy (7.19). Sándor, in [18] divides into two cases, depending on whether (7.19) has distinct or repeated entries. First assume that the cubes do not pair off. If $(w_i) \subset \mathbb{C}^4$ satisfies (7.19), then for every choice of x, y, $(h_i(x, y))$ will also satisfy (7.19) (Theorem 1). If $w_i, x, y \in \mathbb{Z}$, and (7.20) holds then (7.21) and (7.22) will hold as well (Theorem 2), and conversely, if $w_i, h_i \in \mathbb{Z}^4$ are solutions to (7.19) and satisfy (7.22), then there exist $x, y \in \mathbb{Z}$ so that $h_i = h_i(x, y)$, up to multiple, in the sense of (7.20) (Theorem 3). Theorem 4 states, in effect, that the quotients that may arise in (7.22) are those that can be represented by $a^2 + 3b^2$ over $\mathbb{Q}$. Similar results hold if the (w_i) pair off, either as coefficients of x^2 or y^2 in (7.20) (Theorems 5–7).

Finally, we return to the question of which values of T arise for solutions in $\mathbb{Q}[x, y]$. Again, we thank the referee for pointing in this direction.

Theorem 7.4. *In any solution to* (1.6) *with* $f_j \in \mathbb{Q}[x, y]$ *of Type*(T), *we have* $T = a^2 + 3b^2$ *with* $a, b \in \mathbb{Q}$. *Conversely, if* $T = a^2 + 3b^2$ *with* $a, b \in \mathbb{Q}$, *then there is a solution to* (1.6) *of Type*(T) *with* $f_j \in \mathbb{Q}[x, y]$.

Proof. By Theorem 4.3 and (4.4), for any type T family, we have

$$T = \frac{f_3^2 - f_3 f_4 + f_4^2}{f_1^2 - f_1 f_2 + f_2^2} = \frac{(2f_3 - f_4)^2 + 3f_4^2}{(2f_1 - f_2)^2 + 3f_2^2}. \tag{7.23}$$

Suppose $f_j \in \mathbb{Q}[x, y]$. Since $\gcd(f_1, f_2) = 1$, the denominator is never 0, and upon evaluating the right-hand side at $(x, y) = (1, 0)$, say, we have for certain rationals c_j:

$$T = \frac{c_1^2 + 3c_2^2}{c_3^2 + 3c_4^2} = \left(\frac{c_1 c_3 + 3c_2 c_4}{c_3^2 + 3c_4^2}\right)^2 + 3\left(\frac{c_1 c_4 - c_2 c_3}{c_3^2 + 3c_4^2}\right)^2. \tag{7.24}$$

Conversely, suppose $T = a^2 + 3b^2$ with $a, b \in \mathbb{Q}$. Define $f_i \in \mathbb{C}$ by (7.2) with $\mu \neq 0$. Then by (7.4) and (7.6)

$$f_1^3 + f_2^3 = f_3^3 + f_4^3; \quad \frac{f_1 + f_2}{f_3 + f_4} = \frac{2g_1}{2g_3} = a^2 + 3b^2.$$

Now let $w_i = f_i$ and apply the construction from (7.20) to obtain a Type(T) family with coefficients in $\mathbb{Q}[x, y]$. □

We remark that a (complicated) Type($a^2 + 3b^2$) family on $\mathbb{Q}[x, y]$ also comes from an appropriate scaling of $\mathcal{F}_{2,\sqrt{a^2+3b^2}}$ under the linear change

$$(x, y) \mapsto \left(-\sqrt{a^2 + 3b^2}x + \frac{a + \sqrt{-3}b}{\sqrt{a^2 + 3b^2}}y, (a + \sqrt{-3}\, b)x - y\right).$$

In the simplest non-square case, $T = a^2 + 3b^2 = 3$. Putting $(a, b, \mu) = (0, 1, \frac{1}{2})$ in (7.2) gives $5^3 + 4^3 = (-3)^3 + 6^3$, and putting $(w_1, w_2, w_3, w_4) = (5, 4, -3, 6)$ into (7.20) yields four quadratics, so that (up to a factor of 3),

$$h_1(x, y) = 15x^2 + 9xy - 4y^2, \quad h_2(x, y) = 12x^2 - 9xy - 5y^2,$$

$$h_3(x, y) = -9x^2 - 3xy - 6y^2, \quad h_4(x, y) = 18x^2 + 3xy + 3y^2.$$

Finally, since $h_1^3 + h_2^3 = h_3^3 + h_4^3$ and $h_1 + h_2 = 3(h_3 + h_4)$, the two flips have each have a third representation; namely,

$$h_1^3 - h_4^3 = h_3^3 - h_2^3 = \left(\frac{3}{2}x^2 - 18xy + \frac{1}{2}y^2\right)^3 - \left(\frac{27}{2}x^2 - 6xy + \frac{9}{2}y^2\right)^3;$$

$$h_1^3 - h_3^3 = h_4^3 - h_2^3 = \left(\frac{216}{13}x^2 + \frac{165}{13}xy + \frac{72}{13}y^2\right)^3 - \left(\frac{102}{13}x^2 + \frac{261}{13}xy + \frac{34}{13}y^2\right)^3.$$

Acknowledgments

This project began 20 years ago when Bruce Berndt gave a seminar at Illinois about (1.1) and (1.2). The author foolishly believed that an algebraic approach would easily lead to all solutions, and posted a proof-free online set of notes [13] in 2000. Eventually, it has produced this paper and an earlier companion paper studying higher powers, [15]. He wishes to thank his present and former colleagues Michael Bennett, Bruce Berndt, Nigel Boston, Dan Grayson and Jeremy Rouse for helpful conversations, and Andrew Bremner, Noam Elkies and Michael Hirschhorn for encouraging and sending useful e-mails over the years. He also wants to thank the referee for a careful and sympathetic reading of the paper, as well as suggesting the key to the proof of Theorem 7.4. The author was supported by Simons Collaboration Grant 280987.

References

[1] P. R. Allaire and A. Cupillari, Artemas Martin: An amateur mathematician of the nineteenth century and his contribution to mathematics, *College Math. J.* **31** (2000), 22–34. MR1747645 (2001a:01022).

[2] B. C. Berndt, *Ramanujan's* Notebooks–Part IV (Springer-Verlag, New York, 1994). MR1261634 (95e:11028).

[3] H. Darmon and A. Granville, On the equations $z^m = F(x, y)$ and $Ax^p + By^q = Cz^r$, *Bull. London Math. Soc.* **27** (1995), 513–543. MR1348707 (96e:11042).

[4] L. E. Dickson, *History of the Theory of Numbers, Vol. II: Diophantine Analysis* (Chelsea, New York, 1966). MR0245500 (39 #6807b).

[5] E. B. Elliott, *The Algebra of Quantics*, 2nd edn. (Oxford: At the Clarendon Press, 1913).

[6] G. H. Hardy and E. M. Wright, *An introduction to the Theory of Numbers*, 5th edn (Oxford University Press, Oxford, 1979). MR0568909 (81i:10002).

[7] M. D. Hirschhorn, An amazing identity of Ramanujan, *Math. Mag.* **68** (1995) 199–201. MR1335148 (96f:11044).

[8] M. D. Hirschhorn, Two Diophantine equations studied by Ramanujan, *Math. Mag.* **71** (1998) 386–388. MR1573361.

[9] S. Lundqvist, A. Oneto, B. Reznick and B. Shapiro, On generic and maximal k-ranks of binary forms, *J. Pure Appl. Algebra* **223** (2019) 2062–2079. MR3906540.

[10] S. Narayanan, Solution to Problem 441, *J. Indian Math. Soc.* **VI** (1914) 226–227.

[11] S. Ramanujan, Problem 441, *J. Indian Math. Soc.* **V** (1913) 29.

[12] S. Ramanujan, *Collected Papers of Srinivasa Ramanujan* (AMS Chelsea, Providence, 2000). MR2280843 (2008b:11002).

[13] B. Reznick, https://faculty.math.illinois.edu/∼ reznick/92100notes.pdf.

[14] B. Reznick, On the length of forms, *Quadratic and Higher Degree Forms* 207–232, Developments in Mathematics, Vol. 31 (Springer, New York, 2013), pp. 207–232. MR3156559.

[15] B. Reznick, Linear dependent powers of binary quadratic forms, *Pac. J. Math.* **303** (2019) 729–755.

[16] B. Reznick and J. Rouse, On the sums of two cubes, *Int. J. Number Theory* **7** (2011) 1863–1882, http://arxiv.org/pdf/1012.5801, MR2854220.

[17] P. Ribenboim, *13 Lectures on Fermat's Last Theorem* (Springer-Verlag, New York–Heidelberg, 1979). MR0551363 (81f:10023).

[18] C. Sándor, On the equation $a^3+b^3+c^3=d^3$, *Period. Math. Hung.* **33** (1996) 121–134. MR1610133 (99c:11031).

[19] J. H. Silverman, Taxicabs and sums of two cubes, *Amer. Math. Monthly* **100** (1993) 331–340. MR1209462 (93m:11025).

© 2024 World Scientific Publishing Company
https://doi.org/10.1142/9789811277375_0036

Log-concavity results for a biparametric and an elliptic extension of the q-binomial coefficients*

Michael J. Schlosser[†] and Koushik Senapati[‡]
Fakultät für Mathematik, Universität Wien
Oskar-Morgenstern-Platz 1, A-1090 Vienna, Austria
[†]*michael.schlosser@univie.ac.at*
[‡]*koushik.bapan.19@gmail.com*

Ali K. Uncu
Johann Radon Institute for Computational and Applied Mathematics
Austrian Academy of Sciences
Altenbergerstraße 69, A-4040 Linz, Austria
akuncu@risc.jku.at

We establish discrete and continuous log-concavity results for a biparametric extension of the q-numbers and of the q-binomial coefficients. By using classical results for the Jacobi theta function we are able to lift some of our log-concavity results to the elliptic setting. One of our main ingredients is a putatively new lemma involving a multiplicative analogue of Turán's inequality.

Keywords: Log-concavity; q-series; binomial coefficients; theta functions; elliptic functions; Turán's inequality.

Mathematics Subject Classification 2020: 05A20, 05A10, 05A30, 11F27, 26D20, 33E05

1. Introduction

In this paper, we extend some well-known properties of the q-numbers and q-binomial coefficients (a.k.a. *Gaußian binomial coefficients*) to settings involving extra parameters. In particular, we establish discrete and continuous log-concavity results for certain uniparametric, biparametric, and even elliptic extensions of the q-numbers and q-binomial coefficients.

Given any complex $q \neq 1$, the q-analogue of a complex number x is defined by

$$[x]_q := \frac{1-q^x}{1-q}.$$

We refer to $[x]_q$ as a *q-number* (or *basic number*). The q-numbers play an important role in the theory of integer partitions (see Andrews' book [1] and

*This article originally appeared in *International Journal of Number Theory*, **17**(3), pp. 787–804
© 2021 World Scientific Publishing Company. DOI: 10.1142/S1793042120400187

historical references cited in there). One can recover x from $[x]_q$ by letting q tend to 1. Log-concavity results for q-binomial coefficients were given by Butler [2], Krattenthaler [5], and Sagan [7]. (In this connection it should be mentioned that the q-log-concavity or weighted log-concavity considered in the literature implies log-concavity if the weights are non-negative.) Recently Kalmykov and Karp [4] established log-concavity results (or equivalently, Turán-type inequalities) for specific basic hypergeometric series.

For $a, b, q \in \mathbb{C}$ we define the $a, b; q$-extension of a complex number x as follows:

$$[x]_{a,b;q} := \frac{(1-q^x)(1-aq^x)(1-bq)(1-aq/b)}{(1-q)(1-aq)(1-bq^x)(1-aq^x/b)}, \tag{1.1}$$

where the variables are chosen such that none of the denominator factors vanish. (This definition corresponds to that of the $a, b; q$-numbers considered by Schlosser and Yoo in [11–13, 15]. In the latter b has to be replaced by bq^{-1} to match the definition used in (1.1).) Letting $a \to 0$ followed by $b \to 0$ (or $b \to 0$ followed by $a \to \infty$), the $a, b; q$-numbers reduce to the basic numbers.

In Sec. 2, we list some elementary properties of the $a, b; q$-numbers and explain the various notions of log-concavity we are concerned about in this paper. Section 3 deals with results about the log-concavity of the $a, b; q$-numbers. The lemma proved in that section involves a multiplicative analogue of Turán's inequality and plays a key role in proving results involving $a, b; q$-numbers and $a, b; q$-binomial coefficients, and in proving results in the (more general) elliptic setting. Section 4 is devoted to log-concavity results for uniparametric and biparametric extensions of the q-binomial coefficient. The $a, b; q$-numbers can be further extended to the elliptic numbers that appeared in [11–13, 15], and are the contents of study in Secs. 5 and 6. Our elliptic numbers are indeed elliptic functions (i.e. they are meromorphic and doubly periodic); they are expressed as certain ratios of theta functions. Accordingly, the analysis in Secs. 5 and 6 involves some machinery from the theory of Jacobi theta functions (or, equivalently, of the Weierstraß sigma function) which is classical but not so well known in the community of q-series, which is the reason why we cover this material in separate sections of our paper. Finally, in Sec. 7, we conclude with an outlook of further open problems.

2. Preliminaries

It is a matter of simple algebra to verify for arbitrary x and y the following addition formula for the $a, b; q$-numbers defined in (1.1):

$$[x]_{a,b;q} + W_{a,b;q}(x)[y-x]_{aq^{2x},bq^x;q} = [y]_{a,b;q}, \tag{2.1a}$$

where $W_{a,b;q}(x)$ is the $a, b; q$-*weight*, defined by

$$W_{a,b;q}(x) = \frac{(1-aq^{1+2x})(1-b)(1-bq)(1-a/b)(1-aq/b)}{(1-aq)(1-bq^x)(1-bq^{1+x})(1-aq^x/b)(1-aq^{1+x}/b)} q^x. \tag{2.1b}$$

Now, if we impose $0 < q < 1$ and $0 < a < b < 1$ (in particular, all variables are presumed to be real), it is easy to see that for any real $x > 0$ we have

$$[x]_{a,b;q} > 0 \quad \text{and} \quad W_{a,b;q}(x) > 0, \tag{2.2}$$

as all the factors appearing in the respective quotients are manifestly positive.

It is also easy to observe the following three properties of the $a, b; q$-numbers and the associated $a, b; q$-weights:

$$[0]_{a,b;q} = 0 \quad \text{and} \quad W_{a,b;q}(0) = 1, \tag{2.3a}$$

$$[x]_{a,b;q} \geq [y]_{a,b;q} \quad \text{for } 0 < q < 1, \ x \geq y \geq 0, \quad \text{and} \quad 0 < a < b < 1, \tag{2.3b}$$

$$[x]_{a,b;q} = -W_{a,b;q}(x)\,[-x]_{aq^{2x},bq^x;q}. \tag{2.3c}$$

The relation (2.3b) follows from the addition formula in (2.1), with x and y having been interchanged, as for $x > y > 0$ the difference $[x]_{a,b;q} - [y]_{a,b;q}$ is $W_{a,b;q}(y)[x-y]_{aq^{2y},bq^y;q}$ which is positive by (2.2). While we could use the above relation (2.3c) to deal with the $a, b; q$-numbers of negative argument, in this paper we shall restrict our attention to the case that the arguments are non-negative real numbers.

There are two intermediate extensions from the basic numbers to the $a, b; q$-numbers. These two intermediate extensions correspond to the limits $b \to 0$, and to $a \to 0$, in the $a, b; q$-numbers, respectively. Specifically, one can let $b \to 0$ (or $b \to \infty$) in (1.1), by which one obtains the *$a; q$-numbers* (studied in [12, 14, 15]):

$$[x]_{a;q} := [x]_{a,0;q} = \frac{(1-q^x)(1-aq^x)}{(1-q)(1-aq)} q^{1-x}. \tag{2.4}$$

These do not only generalize the standard q-numbers $[x]_q$ obtained by letting $a \to 0$ in (2.4), but also the *quantum* numbers $\langle x \rangle_q := (q^x - q^{-x})/(q - q^{-1})$ (which frequently appear in physical models), obtained by letting $a \to -1$ in (2.4).

One can also let $a \to 0$ (or $a \to \infty$) in (1.1) and arrive at the *$(b; q)$-numbers*

$$[x]_{(b;q)} := [x]_{0,b;q} = \frac{(1-q^x)(1-bq)}{(1-q)(1-bq^x)}. \tag{2.5}$$

We decided to put parentheses in "$(b; q)$-numbers" but none in "$a; q$-numbers" to distinguish them in notation, thus to avoid confusion. (For instance, we have $[x]_{(0;q)} = [x]_q$ but $[x]_{0;q} = [x]_{q^{-1}}$.)

In terms of standard terminology for basic hypergeometric series (cf. [3]), the basic hypergeometric expression on the right-hand side of (2.4) is *well-poised* and that on the right-hand side of (2.5) is *balanced*. It should come to no surprise that the right-hand side of (1.1) is well-poised and balanced (while the corresponding expression for the weight in (2.1b) is even *very-well-poised* and balanced).

We now explain different notions of log-concavity.

Definition 1. A sequence of real numbers $(a_k)_{k=0}^{\infty}$ (indexed by non-negative integers) is called *log-concave* if

$$a_k^2 \geq a_{k+1}a_{k-1}, \tag{2.6}$$

for all $k \geq 1$. Similarly, one calls a sequence $(a_k)_{k=0}^{\infty}$ *strongly log-concave* if

$$a_k a_l \geq a_{k+1} a_{l-1} \tag{2.7}$$

for all positive integers k and l with $k \geq l$.

It is clear that if the a_k are all positive (or all negative), log-concavity implies strong log-concavity since (2.6) then is equivalent to

$$\frac{a_k}{a_{k+1}} \geq \frac{a_{k-1}}{a_k},$$

which can be iterated to establish (2.7).

We will also use the notions of log-concavity and strong log-concavity in the continuous setting.

Definition 2. A function $a(x)$ depending on a non-negative real variable x is called continuously log-concave if

$$a(x)^2 \geq a(x+r)a(x-r) \tag{2.8}$$

for all $x \geq r \geq 0$, and continuously strongly log-concave if

$$a(x)a(y) \geq a(x+r)a(y-r) \tag{2.9}$$

for all real $x \geq y \geq r \geq 0$.

Again it is easy to see that if $a(x) > 0$ for all $x \geq 0$ (or $a(x) < 0$ for all $x \geq 0$) log-concavity implies strong log-concavity since the down-shift by r of the arguments, $a(x)/a(x+r) \geq a(x-r)/a(x)$, can be iterated with an additional down-shift by $x-y$ to establish (2.9).

For a non-vanishing positive (or non-vanishing negative) function $a(x)$, $x \geq 0$, one can equivalently express the continuous strong log-concavity as

$$\frac{a(x+r)a(y-r)}{a(x)a(y)} \leq 1, \tag{2.10}$$

where $x \geq y \geq r \geq 0$. Similar reformulations can be applied to the other notions of log-concavity considered above and we will be using them when convenient.

Example 3. Maybe the most trivial example of a continuously strong log-concave function is the identity on $[0, \infty)$. Indeed, assuming $x \geq y \geq r > 0$ (the case $r = 0$ of (2.9) is trivial),

$$\frac{(x+r)(y-r)}{xy} < 1 \tag{2.11}$$

of course holds, since $xy - (x+r)(y-r) = r(r+x-y) > 0$. This simple fact is already responsible for the continuous strong log-concavity of the *continuous binomial coefficients* (which were recently studied by Salwinski [8] who proved identities satisfied by them, among them also a continuous binomial theorem), defined by

$$\binom{x}{k} = \frac{\Gamma(1+x)}{\Gamma(1+k)\Gamma(1+x-k)} \quad \text{for } x, k \in \mathbb{C},\ x \notin -1, -2, \ldots. \tag{2.12}$$

By virtue of Euler's product formula for the gamma function

$$\Gamma(1+x) = \prod_{j=1}^{\infty} \frac{j^{1-x}(1+j)^x}{x+j} \qquad \text{for } x \in \mathbb{C},\ x \notin -1, -2, \ldots, \tag{2.13}$$

we may rewrite Eq. (2.12) in the following convenient product form:

$$\binom{x}{k} = \prod_{j=1}^{\infty} \frac{(k+j)(x-k+j)}{j(x+j)} \qquad \text{for } x, k \in \mathbb{C},\ x \notin -1, -2, \ldots. \tag{2.14}$$

It is now easy to deduce the following result: *For any real x, y, k, l, r satisfying $x \geq y$, $k \geq l \geq r \geq 0$, and $y - l \geq x - k$, we have the continuous strong log-concavity*

$$\binom{x}{k}\binom{y}{l} \geq \binom{x}{k+r}\binom{y}{l-r}. \tag{2.15}$$

Proof. The $r = 0$ case is trivial, so assume $r > 0$. After canceling common factors we see that

$$\frac{\binom{x}{k+r}\binom{y}{l-r}}{\binom{x}{k}\binom{y}{l}} = \prod_{j=1}^{\infty} \frac{(k+r+j)(x-k-r+j)(l-r+j)(y-l+r+j)}{(k+j)(x-k+j)(l+j)(y-l+j)} < 1,$$

since by taking different instances of (2.11), we have

$$\frac{(k+r+j)(l-r+j)}{(k+j)(l+j)} < 1 \quad \text{and} \quad \frac{(y-l+r+j)(x-k-r+j)}{(y-l+j)(x-k+j)} < 1$$

for each $j \geq 1$, which establishes the claim. □

3. Log-Concavity of $a, b; q$-Numbers

To deal with the continuous log-concavity of the $a; q$-, the $(b; q)$- and the $a, b; q$-numbers we will make use of the following elementary result.

Proposition 4. *For $0 < q < 1$ and $0 \leq \nu < 1$ the function $x \mapsto (1 - \nu q^x)$, $x \geq 0$ is continuously strongly log-concave.*

Proof. Assume that $x \geq y \geq r \geq 0$. Using the definition (2.9) directly and rearranging the terms yields

$$(1-\nu q^x)(1-\nu q^y) - (1-\nu q^{x+r})(1-\nu q^{y-r}) = \nu q^{y-r}(1-q^r)(1-q^{x-y+r}) \geq 0.$$

□

In particular, we can utilize Proposition 4 to establish the following result involving the $a; q$-numbers.

Theorem 5. *For $0 < q < 1$ and $0 < a < 1$ the $a; q$-numbers $[x]_{a;q}$ are continuously strongly log-concave.*

Proof. Let $x \geq y \geq r \geq 0$. Using (2.10) we have to show that the fraction

$$\frac{[x+r]_{a;q}[y-r]_{a;q}}{[x]_{a;q}[y]_{a;q}} = \frac{(1-q^{x+r})(1-q^{y-r})}{(1-q^x)(1-q^y)} \frac{(1-aq^{x+r})(1-aq^{y-r})}{(1-aq^x)(1-aq^y)}$$

is less or equal to 1. Now, the two fractions on the right-hand side are both non-negative and less than or equal to 1 by virtue of Proposition 4 (with $\nu = 1$ and $\nu = a$), and (2.10). Thus their product is less than or equal to 1. □

The same argument is not applicable to the $(b;q)$-numbers $[x]_{(b;q)}$, nevertheless we can show that the $(b;q)$-numbers possess the continuously strong log-concavity property by elementary algebra.

Theorem 6. *For $0 < q < 1$ and $0 < b < 1$ the $(b;q)$-numbers $[x]_{(b;q)}$ are continuously strongly log-concave.*

Proof. Let $x \geq y \geq r \geq 0$. By direct computation, we see that

$$[x]_{(b;q)}[y]_{(b;q)} - [x+r]_{(b;q)}[y-r]_{(b;q)} = \frac{(1-q^r)(1-q^{x-y+r})(1-b)(1-bq)^2(1-bq^{x+y})}{(1-q)^2(1-bq^x)(1-bq^{x+r})(1-bq^y)(1-bq^{y-r})}q^{y-r}.$$

All the factors appearing in the fraction on the right-hand side are non-negative. Thus, the whole product is non-negative, so (2.9) holds. □

We now move our attention to the $a, b; q$-numbers. Before we show the continuous strong log-concavity of these numbers according to Definition 2, we show that these numbers satisfy an $a, b; q$-version of continuous strong log-concavity where certain shifts of the variables a and b occur.

Theorem 7. *For all real x, y, a, b, q satisfying $0 < q < 1$, $0 < a < b < 1$, and $x \geq y \geq r \geq 0$, the $a, b; q$-numbers $[x]_{a,b;q}$ satisfy the inequality*

$$[x]_{aq^{2r},bq^r;q}[y]_{a,b;q} \geq [x+r]_{a,b;q}[y-r]_{aq^{2r},bq^r;q}.$$

Proof. Using (2.1), we have

$$\begin{aligned}
&[x]_{aq^{2r},bq^r;q}[y]_{a,b;q} - [x+r]_{a,b;q}[y-r]_{aq^{2r},bq^r;q}\\
&\quad = [x]_{aq^{2r},bq^r;q}([r]_{a,b;q} + W_{a,b;q}(r)[y-r]_{aq^{2r},bq^r;q})\\
&\qquad - ([r]_{a,b;q} + W_{a,b;q}(r)[x]_{aq^{2r},bq^r;q})[y-r]_{aq^{2r},bq^r;q}\\
&\quad = [r]_{a,b;q}([x]_{aq^{2r},bq^r;q} - [y-r]_{aq^{2r},bq^r;q})\\
&\quad = [r]_{a,b;q}W_{aq^{2r},bq^r;q}(y-r)[x-y+r]_{aq^{2y},bq^y;q}.
\end{aligned}$$

The non-negativity of the last expression is clear from (2.2). □

We can actually avoid using shifts of the variables a and b and show (pure) continuous strong log-concavity of the $a, b; q$-numbers.

Theorem 8. *The $a, b; q$-numbers $[x]_{a,b;q}$ are continuously strongly log-concave. In particular, for all real x, y, a, b, q satisfying $0 < q < 1$, $0 < a < b < 1$, and*

$x \geq y \geq r \geq 0$, *we have*

$$[x]_{a,b;q}[y]_{a,b;q} \geq [x+r]_{a,b;q}[y-r]_{a,b;q}.$$

Theorem 8 is a direct consequence of a general lemma, see Lemma 9, which involves a multiplicative analogue of Turán's inequality. In its formulation, the role of the variable δ, which is assumed to be a fixed non-negative real number, at first may seem to be enigmatic. Indeed, for the sole purpose of proving Theorem 8, the case $\delta = 0$ would completely suffice. However, we shall also employ the lemma when dealing with theta functions in the proofs of Theorems 14 and 15 and there we require δ to be a specific positive real number.

Lemma 9. *Let* $0 \leq \delta < \lambda$. *Let* f *be a continuous positive real function on the interval* $[\delta, \lambda]$, *twice differentiable with values of first and second derivatives being negative on the interval* (δ, λ). *Let* $\delta < a \leq b < \lambda$. *Then we have the inequality*

$$\frac{f(\lambda)f(a)}{f(b)f(\lambda a/b)} \leq 1. \tag{3.1}$$

Proof. Without loss of generality, assume that $b \geq \lambda a/b$; if not, one can switch b and $\lambda a/b$. Let $x = f(\lambda a/b) - f(\lambda)$, $y = f(b) - f(\lambda a/b)$, and $z = f(a) - f(b)$. Using these new variables the claimed inequality (3.1) is equivalent to

$$f(\lambda)(z-x) - x(x+y) \leq 0.$$

The $z \leq x$ case is clear (since $x \geq 0$ and $x + y > 0$) while the case $z > x$ is actually vacuous. Indeed, $z > x$, i.e. $f(a) - f(b) > f(\lambda a/b) - f(\lambda)$, would be equivalent to

$$\frac{f(a) - f(\lambda a/b)}{\lambda - b} > \frac{f(b) - f(\lambda)}{\lambda - b},$$

which again is equivalent to

$$\frac{a}{b} \cdot \frac{f(\frac{\lambda a}{b}) - f(a)}{\frac{\lambda a}{b} - a} < \frac{f(\lambda) - f(b)}{\lambda - b}. \tag{3.2}$$

Recall that $a < \lambda a/b \leq b < \lambda$. By two applications of the mean value theorem there exist $c \in (a, \lambda a/b)$ and $d \in (b, \lambda)$ (in particular, $c < d$ must hold), such that

$$f'(c) = \frac{f(\frac{\lambda a}{b}) - f(a)}{\frac{\lambda a}{b} - a} \qquad \text{and} \qquad f'(d) = \frac{f(\lambda) - f(b)}{\lambda - b}.$$

Since, by assumption, the first derivatives of f are negative in (δ, λ), we obtain from (3.2) the following string of inequalities:

$$\frac{a}{b} f'(c) < f'(d) < 0.$$

If we subtract $f'(c)$ from both sides of the first inequality, and subsequently divide by $d - c$, we obtain

$$\left(\frac{a}{b} - 1\right) \frac{f'(c)}{d-c} < \frac{f'(d) - f'(c)}{d-c}.$$

Now, since $\frac{a}{b} - 1 < 0$ and $f'(c) < 0$ (by assumption), the product on the left-hand side of the inequality is positive. But the expression on the right-hand side is negative, as by the mean value theorem it is equal to the second derivative of f at some point in (c, d), which by assumption must be negative. This is a contradiction, thus the lemma is proved. □

Now, we are ready to prove Theorem 8.

Proof of Theorem 8. By the remarks following Definition 2 it suffices to show the continuous log-concavity of the $a, b; q$-numbers (1.1), i.e.

$$\frac{[x+r]_{a,b;q}[x-r]_{a,b;q}}{[x]_{a,b;q}^2} \leq 1. \tag{3.3}$$

Using the definition of the $a, b; q$-numbers (1.1), after some cancellations, we see that the left-hand side of (3.3) is

$$\frac{(1-q^{x+r})(1-q^{x-r})}{(1-q^x)^2} \frac{(1-aq^{x+r})(1-aq^{x-r})}{(1-aq^x)^2} \times \frac{(1-bq^x)^2}{(1-bq^{x+r})(1-bq^{x-r})} \frac{(1-aq^x/b)^2}{(1-aq^{x+r}/b)(1-aq^{x-r}/b)}. \tag{3.4}$$

For a fixed pair of real numbers $x \geq r > 0$, let

$$f(u) := f_{q,x,r}(u) = \frac{(1-uq^{x+r})(1-uq^{x-r})}{(1-uq^x)^2} \quad \text{for } u \in [0,1]. \tag{3.5}$$

We have

$$f'(u) = -\frac{(1-q^r)^2(1+uq^x)}{(1-uq^x)^3} q^{x-r} < 0,$$

and

$$f''(u) = -\frac{2(1-q^r)^2(2+uq^x)}{(1-uq^x)^4} q^{2x-r} < 0,$$

for any $u \in (0,1)$, which allows us to apply Lemma 9 with f as defined in (3.5), with $\delta = 0$ and $\lambda = 1$. This immediately establishes that the expression in (3.4) is ≤ 1. □

4. $a, b; q$-Binomial Coefficients and Log-Concavity

In this section, we recall the definition of the $a, b; q$-analogue of binomial coefficients (that first appeared in work of Schlosser [9] in the context of enumeration of weighted lattice paths and subsequently were studied in further work of Schlosser [10] in connection with a non-commutative binomial theorem, and in joint work of the Schlosser and Yoo [11, 12]; in particular, these papers contain their recurrence relations and combinatorial interpretations, also for the more general $a, b; q, p$- or elliptic binomial coefficients, which include the $a, b; q$-binomial coefficients obtained by letting $p \to 0$).

For the following definition we restrict the base q to satisfy $0 < |q| < 1$ (while for studying inequalities we further assume q to be real and satisfy $0 < q < 1$). For parameter $a \in \mathbb{C}$ and lower index $k \in \mathbb{C}$, the *q-shifted factorial* is defined as

$$(a;q)_k := \frac{(a;q)_\infty}{(aq^k;q)_\infty}, \quad \text{where } (a;q)_\infty = \prod_{j\geq 0}(1-aq^j),$$

cf. [3]. (For $a = q^{-n}$ with integers $n \geq k \geq 0$ the above definition for $(a;q)_k$ involves a pole which however can be removed.)

For brevity, we use the following compact notation for products of q-shifted factorials:

$$(a_1, \ldots, a_m; q)_k := (a_1;q)_k \cdots (a_m;q)_k, \quad \text{where } k \in \mathbb{C} \cup \{\infty\}.$$

For $x, k, a, b, q \in \mathbb{C}$, $0 < |q| < 1$, we define the $a, b; q$-binomial coefficient as

$$\begin{aligned}\begin{bmatrix} x \\ k \end{bmatrix}_{a,b;q} &:= \frac{(q^{1+k}, aq^{1+k}, bq^{1+k}, aq^{1-k}/b;q)_{x-k}}{(q, aq, bq^{1+2k}, aq/b;q)_{x-k}} \\ &= \frac{(q^{1+x-k}, aq^{1+x-k}, bq^{1+k}, aq^{1-k}/b;q)_k}{(q, aq, bq^{1+x}, aq^{1+x-2k}/b;q)_k}. \end{aligned} \tag{4.1}$$

(Note that the $a, b; q$-binomial coefficients are *not* symmetric in k and $x-k$, contrary to the symmetry for the ordinary and basic binomial coefficients.) If we assume x and k to be non-negative integers, and replace b by bq on the right-hand side, then definition (4.1) matches that of the $a, b; q$-binomial coefficient as used in [9–13, 15].

If we let $k = 1$ and divide b by q, the $a, b; q$-binomial coefficient reduces to the $a, b; q$-number defined in (1.1):

$$\begin{bmatrix} x \\ 1 \end{bmatrix}_{a,b/q;q} = [x]_{a,b;q}.$$

We define the $a;q$ and $(b;q)$ analogues of binomials by taking limits $b \to \infty$ and $a \to 0$ of (4.1), respectively. Then

$$\begin{bmatrix} x \\ k \end{bmatrix}_{a;q} := \frac{(q^{1+k}, aq^{1+k};q)_{x-k}}{(q, aq;q)_{x-k}} q^{k(k-x)} \quad \text{and} \quad \begin{bmatrix} x \\ k \end{bmatrix}_{(b;q)} := \frac{(q^{1+k}, bq^{1+k};q)_{x-k}}{(q, bq^{1+2k};q)_{x-k}}. \tag{4.2}$$

(While the $a;q$-binomial coefficients are symmetric in k and $x-k$, the $(b;q)$-binomial coefficients are not). It is clear that

$$\begin{bmatrix} x \\ 1 \end{bmatrix}_{a;q} = [x]_{a;q} \quad \text{and} \quad \begin{bmatrix} x \\ 1 \end{bmatrix}_{(b/q;q)} = [x]_{(b;q)},$$

with the $a;q$- and $(b;q)$-numbers defined in (2.4) and (2.5), respectively.

Parallel to Sec. 3, we start our log-concavity discussion with the $a;q$-binomial coefficients. For these we have continuous strong log-concavity with respect to the *lower* parameter of the binomial coefficient.

Theorem 10. *For any real a,x,y,k,l,r,q satisfying $0<q<1$, $0\le a<1$, $x\ge y$, $k\ge l\ge r\ge 0$, and $y-l\ge x-k$, we have the continuous strong log-concavity*

$$\begin{bmatrix} x \\ k \end{bmatrix}_{a;q}\begin{bmatrix} y \\ l \end{bmatrix}_{a;q} \ge \begin{bmatrix} x \\ k+r \end{bmatrix}_{a;q}\begin{bmatrix} y \\ l-r \end{bmatrix}_{a;q}. \tag{4.3}$$

Proof. After cancellations, we see that

$$\frac{\begin{bmatrix} x \\ k+r \end{bmatrix}_{a;q}\begin{bmatrix} y \\ l-r \end{bmatrix}_{a;q}}{\begin{bmatrix} x \\ k \end{bmatrix}_{a;q}\begin{bmatrix} y \\ l \end{bmatrix}_{a;q}} = q^{2r(r+k-l)}\frac{(q^{1+x-k-r},aq^{1+x-k-r},q^{1+k+r},aq^{1+k+r};q)_\infty}{(q^{1+x-k},aq^{1+x-k},q^{1+k},aq^{1+k};q)_\infty}$$
$$\times\frac{(q^{1+y-l+r},aq^{1+y-l+r},q^{1+l-r},aq^{1+l-r};q)_\infty}{(q^{1+y-l},aq^{1+y-l},q^{1+l},aq^{1+l};q)_\infty}.$$

Now, $q^{2r(r+l-l)}\le 1$ and by taking different instances of Proposition 4, we have

$$\frac{(1-\nu q^{k+r})(1-\nu q^{l-r})}{(1-\nu q^{k})(1-\nu q^{l})}\le 1 \quad\text{and}\quad \frac{(1-\nu q^{y-l+r})(1-\nu q^{x-k-r})}{(1-\nu q^{y-l})(1-\nu q^{x-k})}\le 1, \tag{4.4}$$

for $\nu=q^j$ and $\nu=aq^j$, where $j=1,2,\ldots$. Thus, the infinite product is also ≤ 1 and the theorem follows. □

We note that the $(b;q)$-binomial coefficients and $a,b;q$-binomial coefficients do not appear to be log-concave with respect to their lower parameter.

We now turn back to the $a,b;q$-binomial coefficients. We are able to prove the discrete strong log-concavity of $\begin{bmatrix} x \\ k \end{bmatrix}_{a,b;q}$ (here, x may be real but k should be a non-negative integer) with respect to the upper parameter x but require that the range of the variables a,b depends on the lower parameter k.

Theorem 11. *For any real a,b,x,y,q and non-negative integer k satisfying $0<q<1$, $0<a\le bq^k<1$, and $x\ge y\ge 1$, with the difference $x-y$ being a non-negative integer, we have the discrete strong log-concavity*

$$\begin{bmatrix} x \\ k \end{bmatrix}_{a,b;q}\begin{bmatrix} y \\ k \end{bmatrix}_{a,b;q} \ge \begin{bmatrix} x+1 \\ k \end{bmatrix}_{a,b;q}\begin{bmatrix} y-1 \\ k \end{bmatrix}_{a,b;q}. \tag{4.5}$$

Proof. All we need to show is that

$$\frac{\begin{bmatrix} x+1 \\ k \end{bmatrix}_{a,b;q}\begin{bmatrix} x-1 \\ k \end{bmatrix}_{a,b;q}}{\begin{bmatrix} x \\ k \end{bmatrix}_{a,b;q}^2}\le 1,$$

as then the iteration of the inequality

$$\frac{\begin{bmatrix} x \\ k \end{bmatrix}_{a,b;q}}{\begin{bmatrix} x+1 \\ k \end{bmatrix}_{a,b;q}}\ge\frac{\begin{bmatrix} x-1 \\ k \end{bmatrix}_{a,b;q}}{\begin{bmatrix} x \\ k \end{bmatrix}_{a,b;q}}$$

leads to (4.5). After canceling common factors, we obtain

$$\frac{\begin{bmatrix} x+1 \\ k \end{bmatrix}_{a,b;q} \begin{bmatrix} x-1 \\ k \end{bmatrix}_{a,b;q}}{\begin{bmatrix} x \\ k \end{bmatrix}_{a,b;q}^2} = \frac{g(1)g(a)}{g(bq^k)g(a/bq^k)}, \tag{4.6}$$

where

$$g(u) := g_{q,x,k}(x) = \frac{(1-uq^{x-k})(1-uq^{x+1})}{(1-uq^{x-k+1})(1-uq^{x})}.$$

Observe that

$$g_{q,x,k}(u) = \prod_{i=1}^{k} f_{q,x-k+i,1}(u),$$

where $f_{q,x,1}(u)$ is as defined in (3.5). Therefore, the right-hand side of (4.6) can be rewritten as

$$\frac{\begin{bmatrix} x+1 \\ k \end{bmatrix}_{a,b;q} \begin{bmatrix} x-1 \\ k-1 \end{bmatrix}_{a,b;q}}{\begin{bmatrix} x \\ k \end{bmatrix}_{a,b;q}^2} = \prod_{j=1}^{k} \frac{f_{q,x-k+j,1}(1) f_{q,x-k+j,1}(a)}{f_{q,x-k+j,1}(bq^k) f_{q,x-k+j,1}(a/bq^k)}.$$

The k terms of the product are each less than or equal to 1 by the $\delta = 0$, $\lambda = 1$, and $b \mapsto bq^k$ case of Lemma 3.1 with $f(u)$ as defined in (3.5) but with b replaced by bq^k. □

This is consistent with the log-concavity of the numbers $[n]_{a,b;q}$ that we proved in Theorem 8; that is namely just the $k = 1$ and $b \mapsto b/q$ case of Theorem 11.

5. Elliptic Numbers and Elliptic Binomial Coefficients

We start with explaining some standard notions about theta functions and elliptic functions. A function g of a complex variable z is *elliptic* if it is meromorphic and doubly periodic. It follows from the theory of abelian functions (cf. [16]) that there exists a non-negative integer s and complex numbers $a_1, \dots, a_s$, $b_1, \dots, b_s$, c, and p with $0 < |p| < 1$ such that

$$g(z) = c\,\frac{\theta(a_1 z, \dots, a_s z; p)}{\theta(b_1 z, \dots, b_s z; p)}, \quad \text{where } a_1 \cdots a_s = b_1 \cdots b_s \neq 0.$$

Here, $\theta(x_1, \dots, x_s; p) := \theta(x_1; p) \cdots \theta(x_s; p)$, where, for $x \neq 0$ and $|p| < 1$,

$$\theta(x;p) = (x, p/x; p)_\infty$$

is the *modified Jacobi theta function* (which in short we refer to as *theta function*) of argument x and *nome* p. It is clear that for $p = 0$ we have $\theta(x; 0) = (1 - x)$. So, the theta function can be considered to be a p-extension of a linear factor. The following properties hold for the theta function:

$$\theta(x;p) = -x\theta(1/x;p), \tag{5.1a}$$

$$\theta(px;p) = -\frac{1}{x}\theta(x;p), \tag{5.1b}$$

and the *addition formula*

$$\theta(xy, x/y, ut, u/t; p) - \theta(xt, x/t, uy, u/y; p) = \frac{u}{y}\theta(yt, y/t, xu, x/u; p). \tag{5.2}$$

The two identities in (5.1), which can be referred to as *inversion formula* and *quasi-periodicity*, are readily shown from the definition of the theta function, while the theta function addition formula (originally due to Weierstraß, see [17, Example 5, p. 451]) in (5.2) is not obvious but crucial in the theory of elliptic hypergeometric functions. (See e.g., Rosengren's lecture notes [6, Sec. 1.4] for the standard derivation of (5.2) involving complex analysis.)

We recall further classical facts about theta functions which we need. In particular, we would like to have formulas involving their first and second derivatives, so we can apply Lemma 9 with f involving a quotient of theta functions, rather than a quotient of linear factors as in (3.5). This will enable us to extend Theorems 8 and 11 to the elliptic setting.

The modified Jacobi theta function is a special case of a *sigma function.* According to [17, §21.43, p. 473] the function $\sigma(z) = \sigma(z \mid \omega_1, \omega_2)$ formed with the two periods $2\omega_1$, $2\omega_2$, is expressible in the form

$$\begin{aligned}\sigma(z \mid \omega_1, \omega_2) &= \frac{2\omega_1}{\pi}\exp\left(\frac{\eta z^2}{2\omega_1}\right)\sin\left(\frac{\pi z}{2\omega_1}\right)\prod_{n=1}^{\infty}\frac{\left(1 - 2\tilde{p}^{2n}\cos\dfrac{\pi z}{\omega_1} + \tilde{p}^{4n}\right)}{(1-\tilde{p}^{2n})^2},\\ &= \omega_1\, e^{\eta z^2/2\omega_1}(e^{\pi \mathrm{i}/2\omega_1} - e^{-\pi \mathrm{i}/2\omega_1})\frac{(\tilde{p}^2 e^{\pi \mathrm{i} z/\omega_1}, \tilde{p}^2 e^{-\pi \mathrm{i} z/\omega_1}; \tilde{p}^2)_\infty}{\pi \mathrm{i}\,(\tilde{p}^2;\tilde{p}^2)_\infty^2}\\ &= \frac{\mathrm{i}\,\omega_1 e^{(\eta z^2 - \pi \mathrm{i})/2\omega_1}\,\theta(e^{\pi \mathrm{i}/2\omega_1}; \tilde{p}^2)}{\pi\,(\tilde{p}^2;\tilde{p}^2)_\infty^2},\end{aligned} \tag{5.3a}$$

where

$$\eta = \frac{\pi^2}{12\omega_1}\left(1 - 24\sum_{n=1}^{\infty}\frac{\tilde{p}^{2n}}{(1-\tilde{p}^{2n})^2}\right) \quad \text{and} \quad \tilde{p} = e^{\frac{\pi \mathrm{i}\omega_2}{\omega_1}}. \tag{5.3b}$$

Here, $\mathrm{i}^2 = -1$. From (5.2) it is immediate that the sigma function satisfies the addition formula (cf. [17, Example 5, p. 451])

$$\begin{aligned}&\sigma(x+y)\sigma(x-y)\sigma(u+t)\sigma(u-t) - \sigma(x+t)\sigma(x-t)\sigma(u+y)\sigma(u-y)\\ &\quad = \sigma(y+t)\sigma(y-t)\sigma(x+u)\sigma(x-u).\end{aligned} \tag{5.4}$$

Now, let $\zeta(z) := \frac{\mathrm{d}}{\mathrm{d}z}\log\sigma(z)$ and $\wp(z) := -\frac{\mathrm{d}}{\mathrm{d}z}\zeta(z)$ (called the *Weierstraß ζ-function* and *Weierstraß $\wp$-function*, respectively; the latter is actually an elliptic function). Differentiation of both sides of (5.4) with respect to x, followed by division by $\sigma(x+y)\sigma(x-y)\sigma(u+t)\sigma(u-t)$ and putting $u = x$, gives (cf. [17, Example 38, p. 461])

$$\zeta(x+y) + \zeta(x-y) - \zeta(x+t) - \zeta(x-t) = \frac{\sigma'(0)\sigma(2x)\sigma(y+t)\sigma(y-t)}{\sigma(x+y)\sigma(x-y)\sigma(x+t)\sigma(x-t)}. \tag{5.5}$$

Now, it is not difficult to show directly, using (5.3), that $\sigma'(0) = 1$. After making the simultaneous substitutions $(x, y, t) \mapsto ((u+v)/2, (u-v)/2, -w-(u+v)/2)$ in (5.5) we obtain

$$\zeta(u) + \zeta(v) + \zeta(w) - \zeta(u+v+w) = \frac{\sigma(u+v)\sigma(u+w)\sigma(v+w)}{\sigma(u)\sigma(v)\sigma(w)\sigma(u+v+w)}, \tag{5.6}$$

which after the simultaneous substitutions $(u, v, w) \mapsto (2u, 2v, -u-v)$ reduces to

$$\zeta(2u) + \zeta(2v) - 2\zeta(u+v) = \frac{\sigma(2u+2v)\sigma^2(u-v)}{\sigma(2u)\sigma(2v)\sigma^2(u+v)}. \tag{5.7}$$

Finally, by differentiating (5.6) with respect to x and applying (5.6) to the result, one obtains, after another substitution of variables (cf. [17, Example 1, p. 451])

$$\wp(v) - \wp(u) = \frac{\sigma(u-v)\sigma(u+v)}{\sigma^2(u)\sigma^2(v)}. \tag{5.8}$$

This important formula can be used to prove the addition formula in (5.4) without knowing the definition of the Weierstraß $\wp$-function. While we will actually not use this function and the formula in (5.8) in this paper, we nevertheless include this very classical material to make our survey in this section more complete.

In the following, we will choose as half-periods $\omega_1 = \frac{1}{2}$, $\omega_2 = \frac{\tau}{2}$ (with $\tau \in \mathrm{i}\mathbb{R}^+$), and put $p = \tilde{p}^2$, so $p = e^{2\pi \mathrm{i}\tau}$. (Notice that the condition on τ guarantees that $0 < |p| < 1$.) With these specializations we have

$$\sigma(z) = \sigma\left(z \,\middle|\, \frac{1}{2}, \frac{\tau}{2}\right) = \frac{\mathrm{i}\, e^{\eta z^2 - \pi \mathrm{i} z}\, \theta(e^{2\pi \mathrm{i} z}; p)}{2\pi\, (p;p)_\infty^2}, \tag{5.9a}$$

where

$$\eta = \frac{\pi^2}{6}\left(1 - 24 \sum_{n=1}^{\infty} \frac{p^n}{(1-p^n)^2}\right). \tag{5.9b}$$

We are now ready to define our elliptic numbers and elliptic binomial coefficients. For $a, b, q, p \in \mathbb{C}$ with $|p| < 1$ we define the elliptic (or $a, b; q, p$-)extension of a number x (which does not need to be an integer) as follows:

$$[x]_{a,b;q,p} = \frac{\theta(q^x, aq^x, bq, aq/b; p)}{\theta(q, aq, bq^x, aq^x/b; p)}. \tag{5.10}$$

(This definition corresponds to that for the elliptic numbers considered in [11–14] subject to the substitution $b \mapsto bq^{-1}$.)

From (5.2) we have for all x and y the following addition formula for the elliptic numbers:

$$[x]_{a,b;q,p} + W_{a,b;q,p}(x)[y-x]_{aq^{2x}, bq^x; q,p} = [y]_{a,b;q,p}, \tag{5.11a}$$

where $W_{a,b;q,p}(x)$ is the *elliptic-weight*, defined by

$$W_{a,b;q,p}(x) = \frac{\theta(aq^{1+2x}, b, bq, a/b, aq/b; p)}{\theta(aq, bq^x, bq^{1+x}, aq^x/b, aq^{1+x}/b; p)} q^x. \tag{5.11b}$$

It is clear that the properties (2.2) and (2.3) readily extend to the elliptic level.

For parameter $a \in \mathbb{C}$, base $0 < |q| < 1$, nome $|p| < 1$ and non-negative integer k, the *theta* (or q,p-)*shifted factorial* is defined as

$$(a; q, p)_k := \theta(a, aq, \dots, aq^{k-1}; p),$$

cf. [3, Chap. 11]. For brevity, we use the following compact notation for products of theta shifted factorials:

$$(a_1, \dots, a_m; q, p)_k := (a_1; q, p)_k \cdots (a_m; q, p)_k, \quad \text{where } k \in \mathbb{N}_0.$$

Further, for $x, a, b, q, p \in \mathbb{C}$ with $0 < |q| < 1$, $|p| < 1$, and a non-negative integer k we define the elliptic (or $a, b; q, p$-)binomial coefficient as

$$\begin{bmatrix} x \\ k \end{bmatrix}_{a,b;q,p} := \frac{(q^{1+x-k}, aq^{1+x-k}, bq^{1+k}, aq^{1-k}/b; q, p)_k}{(q, aq, bq^{1+x}, aq^{1+x-2k}/b; q, p)_k}. \tag{5.12}$$

If we assume x and k being non-negative integers, and replace b by bq on the right-hand side, the definition in (5.12) matches that of the elliptic binomial coefficient as used in [9, 11, 12]. Clearly, $\left[\begin{smallmatrix} x \\ 1 \end{smallmatrix}\right]_{a,b/q;q,p} = [x]_{a,b;q,p}$.

6. Log-Concavity of Elliptic Numbers and Elliptic Binomial Coefficients

Just as the $a, b; q$-numbers satisfy an $a, b; q$-version of continuous strong log-concavity where certain shifts of the variables a and b occur (see Theorem 7), the elliptic numbers satisfy the following result.

Theorem 12. *Let q, p, x, y, r, a, b be real numbers satisfying $0 < q < 1$, $0 < p < 1$, $x \geq y \geq r \geq 0$, and $0 < a < b < 1$. Then the elliptic numbers satisfy the inequality*

$$[x]_{aq^{2r}, bq^r; q,p} [y]_{a,b;q,p} \geq [x + r]_{a,b;q,p} [y - r]_{aq^{2r}, bq^r; q,p}.$$

Proof. The proof is just like that of Theorem 7, but we use (5.11) instead of (2.1). □

In the remaining section, we will consider log-concavity results where the variables a and b do *not* shift. Our results crucially depend on the following proposition.

Proposition 13. *Let q, p, x, r be real numbers satisfying $0 < q < 1$, $x \geq r > 0$ and $0 < p < q^{2r}$, and let the function f be defined by*

$$f(u) := f_{q,p,x,r}(u) = \frac{\theta(uq^{x+r}, uq^{x-r}; p)}{\theta^2(uq^x; p)} \quad \text{for } u \in [\delta, \lambda], \tag{6.1}$$

where $\delta = pq^{-x-r}$ and $\lambda = q^{r-x}$. Then f is continuous and positive on the interval $[\delta, \lambda]$, with first and second derivatives being negative on the interval (δ, λ).

Proof. The continuity and positivity is clear. To compute the derivatives on (δ, λ), we rewrite $f(u)$ in terms of the sigma function. Let $p = e^{2\pi i \tau}$, $q = e^{2\pi i \nu}$, and $u=e^{2\pi\mathrm{i}z}$, where $\tau,\nu,z\in\mathrm{i}\mathbb{R}$. We have $\mathrm{d}u = 2\pi\mathrm{i}\,e^{2\pi\mathrm{i}z}\mathrm{d}z$, or equivalently, $\frac{\mathrm{d}}{\mathrm{d}z}=2\pi\mathrm{i}u\,\frac{\mathrm{d}}{\mathrm{d}u}$, and, according to (5.9a),

$$\begin{aligned} f(e^{2\pi\mathrm{i}z}) &= \frac{\theta(e^{2\pi\mathrm{i}((x+r)\nu+z)},e^{2\pi\mathrm{i}((x-r)\nu+z)};p)}{\theta^2(e^{2\pi\mathrm{i}(x\nu+z)};p)} \\ &= e^{-2\eta r^2\nu^2}\,\frac{\sigma(z+(x+r)\nu)\sigma(z+(x-r)\nu)}{\sigma^2(z+x\nu)}. \end{aligned}$$

Thus, the first derivative of f is

$$\begin{aligned} \frac{\mathrm{d}}{\mathrm{d}u}f(u) &= \frac{e^{-2\pi\mathrm{i}z}}{2\pi\mathrm{i}}\,\frac{\mathrm{d}}{\mathrm{d}z}f(e^{2\pi\mathrm{i}z}) \\ &= \frac{e^{-2\pi\mathrm{i}z-2\eta r^2\nu^2}}{2\pi\mathrm{i}}\,\frac{\mathrm{d}}{\mathrm{d}z}\,\frac{\sigma(z+(x+r)\nu)\sigma(z+(x-r)\nu)}{\sigma^2(z+x\nu)} \\ &= \frac{e^{-2\pi\mathrm{i}z-2\eta r^2\nu^2}}{2\pi\mathrm{i}\,\sigma^3(z+x\nu)}[\sigma(z+x\nu)\sigma(z+(x+r)\nu)\sigma'(z+(x-r)\nu) \\ &\quad +\sigma(z+x\nu)\sigma'(z+(x+r)\nu)\sigma(z+(x-r)\nu) \\ &\quad -2\sigma'(z+x\nu)\sigma(z+(x+r)\nu)\sigma(z+(x-r)\nu)] \\ &= \frac{e^{-2\pi\mathrm{i}z}}{2\pi\mathrm{i}}f(e^{2\pi\mathrm{i}z})[\zeta(z+(x+r)\nu)+\zeta(z+(x-r)\nu)-2\zeta(z+x\nu)] \\ &= \frac{e^{-2\pi\mathrm{i}z-2\eta r^2\nu^2}}{2\pi\mathrm{i}}\,\frac{\sigma(2(z+x\nu))\sigma^2(r\nu)}{\sigma^4(z+x\nu)} = -q^{x-r}\,\frac{(p;p)_\infty^2\,\theta^2(q^r;p)\theta(u^2q^{2x};p)}{\theta^4(uq^x;p)}, \end{aligned}$$

where we applied (5.7) in the penultimate equality, and used (5.9a) to rewrite the expression in terms of the modified theta function in the last equality. The final expression obtained for the first derivative of $f(u)$ is clearly negative. (The minus sign in front of the product stems from collecting all the factors i from the applications of (5.9a); indeed, $\mathrm{i}^3/\mathrm{i}^5=-1$.)

We still need to compute the second derivative of f and show that it is also negative. We have

$$\begin{aligned} \frac{\mathrm{d}^2}{\mathrm{d}u^2}f(u) &= -q^{x-r}(p;p)_\infty^2\,\theta^2(q^r;p)\,\frac{\mathrm{d}}{\mathrm{d}u}\,\frac{\theta(u^2q^{2x};p)}{\theta^4(uq^x;p)} \\ &= -q^{x-r}\,\frac{(p;p)_\infty^2\,\theta^2(q^r;p)\theta(u^2q^{2x};p)}{\theta^4(uq^x;p)} \\ &\quad\times\left(\frac{\mathrm{d}}{\mathrm{d}u}\log\theta(u^2q^{2x};p)-4\frac{\mathrm{d}}{\mathrm{d}u}\log\theta(uq^x;p)\right). \end{aligned}$$

All we need to do is to show that

$$\frac{\mathrm{d}}{\mathrm{d}u}\log\theta(u^2q^{2x};p) > 4\frac{\mathrm{d}}{\mathrm{d}u}\log\theta(uq^x;p).$$

Since the logarithmic derivative of a product is the sum of the logarithmic derivatives, we thus need to show

$$\sum_{j=0}^{\infty}\frac{\mathrm{d}}{\mathrm{d}u}\log[(1-p^ju^2q^{2x})+(1-p^{j+1}u^{-2}q^{-2x})]$$

$$>4\sum_{j=0}^{\infty}\frac{\mathrm{d}}{\mathrm{d}u}\log[(1-p^juq^{x})+(1-p^{j+1}u^{-1}q^{-x})].$$

The inequality actually holds term-wise, for each j. Indeed, comparison of the logarithmic derivatives for fixed j amounts to

$$\frac{-2p^juq^{2x}}{1-p^ju^2q^{2x}}+\frac{4p^{j+1}u^{-3}q^{-2x}}{1-p^{j+1}u^{-2}q^{-2x}}>\frac{-4p^jq^{x}}{1-p^juq^{x}}+\frac{4p^{j+1}u^{-2}q^{-x}}{1-p^{j+1}u^{-1}q^{-x}},$$

which is easy to verify, as already the first summand on the left-hand side is greater than the first summand on the right-hand side, which analogously holds for the second summands on both sides. □

Having set up all the ingredients, the results now immediately extend from the $a,b;q,p$-case to the elliptic case.

Theorem 14. *The elliptic numbers $[x]_{a,b;q,p}$ are continuously strongly log-concave. In particular, for all real numbers q,p,x,y,r,a,b satisfying $0<q<1$, $0<p<q^{2r}$, $x\geq y\geq r\geq 0$, and $pq^{-x-r}<a<b<1$ we have*

$$[x]_{a,b;q,p}[y]_{a,b;q,p}\geq[x+r]_{a,b;q,p}[y-r]_{a,b;q,p}.$$

Proof. The proof follows the lines of that of the proof of Theorem 8 but we use, instead of (3.5), $f(u)$ as defined in (6.1), where $u\in[\delta,1]$ with $\delta=pq^{-x-r}$. As a consequence of Proposition 13, f satisfies on $[\delta,1]$ (which is contained in $[\delta,\lambda]$ with $\lambda=q^{r-x}$) the necessary requirements for application of Lemma 9, which we invoke with $\lambda=1$. □

Theorem 15. *For any real q,p,a,b,x,y and non-negative integer k satisfying $0<q<1$, $0<p<q^2$, $pq^{-x-1}<a\leq bq^k<1$, and $x\geq y\geq 1$, with the difference $x-y$ being a non-negative integer, we have the discrete strong log-concavity of elliptic binomial coefficients*

$$\begin{bmatrix}x\\k\end{bmatrix}_{a,b;q,p}\begin{bmatrix}y\\k\end{bmatrix}_{a,b;q,p}\geq\begin{bmatrix}x+1\\k\end{bmatrix}_{a,b;q,p}\begin{bmatrix}y-1\\k\end{bmatrix}_{a,b;q,p}.$$

Proof. The proof follows the lines of that of the proof of Theorem 11 but we use, instead of (3.5), $f(u)$ as defined in (6.1), where $u\in[\delta,1]$ with $\delta=pq^{-x-1}$, for application of Lemma 9. □

7. Conclusion

We have shown that fundamental log-concavity results which are known to hold for (the q-)numbers and (q-)binomial coefficients can be extended to the $a, b; q$-case and even to the elliptic case. Further closely related questions remain open, for instance about the log-concavity of $a, b; q$- or elliptic extensions of those other sequences considered by Sagan in [7] or of the $a, b; q$- or elliptic rook numbers (considered by the Schlosser and Yoo [12, 13]).

Acknowledgments

This paper is dedicated to Bruce Berndt, on the occasion of his 80th birthday. All three authors thank the FWF Austrian Science Fund for generous support. In particular, the first author was partially supported by grants F50-08 and P32305, the second author was fully supported by grants F50-08 and F50-10, and the third author was fully supported by grants F50-07, F50-09, and F50-11.

References

[1] G. E. Andrews, *The Theory of Partitions* (Cambridge University Press, Cambridge, 1998).

[2] L. M. Butler, The q-log concavity of q-binomial coefficients, *J. Combin. Theory Ser. A* **54**(1) (1990) 54–63.

[3] G. Gasper and M. Rahman, *Basic Hypergeometric Series*, 2nd edn., Encyclopedia of Mathematics and Its Applications, Vol. 96 (Cambridge University Press, Cambridge, 2004).

[4] S. I. Kalmykov and D. B. Karp, Inequalities for series in q-shifted factorials and q-gamma functions, *J. Math. Anal. Appl.* **460**(1) (2018) 332–351.

[5] C. Krattenthaler, On the q-log-concavity of Gaussian binomial coefficients, *Monatsh. Math.* **107** (1989) 333–339.

[6] H. Rosengren, Elliptic hypergeometric functions, in *Lectures on Orthogonal Polynomials and Special Functions*, eds. H. S. Cohl and M. E. H. Ismail (Cambridge University Press, 2000), https://www.cambridge.org/core/books/lectures-on-orthogonal-polynomials-and-special-functions/DC6C460E9C89A9AD8669B7BE82C6D980.

[7] B. E. Sagan, Inductive proofs of q-log concavity, *Discrete Math.* **99**(1–3) (1992) 289–306.

[8] D. Salwinski, The continuous binomial coefficient: An elementary approach, *Amer. Math. Monthly* **125**(3) (2018) 231–244.

[9] M. J. Schlosser, Elliptic enumeration of nonintersecting lattice paths, *J. Combin. Theory Ser. A* **114**(3) (2007) 505–521.

[10] M. J. Schlosser, A noncommutative weight-dependent generalization of the binomial theorem, *Sém. Lothar. Combin.* **B81j** (2020) 24 pp.

[11] M. J. Schlosser and M. Yoo, An elliptic extension of the general product formula for augmented rook boards, *European J. Combin.* **58** (2016) 247–266.

[12] M. J. Schlosser and M. Yoo, Elliptic rook and file numbers, *Electron. J. Combin.* **24**(1) (2017) Article ID #P1.31, 47 pp.

[13] M. J. Schlosser and M. Yoo, Elliptic extensions of the alpha-parameter model and the rook model for matchings, *Adv. Appl. Math.* **84** (2017) 8–33.

[14] M. J. Schlosser and M. Yoo, Basic hypergeometric summations from rook theory, in *Analytic Number Theory, Modular Forms and q-Hypergeometric Series: In Honor of Krishna Alladi's 60th Birthday, University of Florida, Gainesville, March 2016*, eds. G. E. Andrews and F. Garvan, Springer Proceedings in Mathematics & Statistics, Vol. 221 (Springer, 2017), pp. 677–692

[15] M. J. Schlosser and M. Yoo, Weight-dependent commutation relations and combinatorial identities, *Discrete Math.* **341** (2018) 2308–2325.

[16] H. Weber, *Elliptische Functionen und Algebraische Zahlen* (Vieweg-Verlag, Braunschweig, 1897).

[17] E. T. Whittaker and G. N. Watson, *A Course of Modern Analysis*, 4th edn. (Cambridge University Press, Cambridge, 1962).

© 2024 World Scientific Publishing Company
https://doi.org/10.1142/9789811277375_0037

Analysis and combinatorics of partition zeta functions*

Robert Schneider

Department of Mathematics
University of Georgia
Athens, Georgia 30602, USA
Robert.Schneider@uga.edu

Andrew V. Sills†

Department of Mathematical Sciences
Georgia Southern University
Statesboro, Georgia 30458, USA
ASills@georgiasouthern.edu

Dedicated to Bruce C. Berndt on the occasion of his 80th birthday.

We examine "partition zeta functions" analogous to the Riemann zeta function but summed over subsets of integer partitions. We prove an explicit formula for a family of partition zeta functions already shown to have nice properties — those summed over partitions of fixed length — which yields complete information about analytic continuation, poles and trivial roots of the zeta functions in the family. Then we present a combinatorial proof of the explicit formula, which shows it to be a zeta function analog of MacMahon's partial fraction decomposition of the generating function for partitions of fixed length.

Keywords: Integer partitions; zeta functions; partition zeta functions; MacMahon decomposition.

Mathematics Subject Classification 2020: 11P83, 11M32, 05A17

1. Introduction: Partition Zeta Functions

Here, we study an interesting class of objects dwelling at the intersection of partition theory and the theory of L-functions. Let $\mathcal{P}$ denote the set of integer partitions (see e.g., [1, 2]), with $\mathcal{P}_S$ being partitions whose parts all belong to a subset $S \subseteq \mathbb{N}$ of natural numbers. Let $\lambda = (\lambda_1, \lambda_2, \lambda_3, \ldots, \lambda_r)$ denote a generic partition, $\lambda_1 \geq$

*This article originally appeared in *International Journal of Number Theory*, **17**(3), pp. 805–814 © 2021 World Scientific Publishing Company. DOI: 10.1142/S1793042120400023
†Corresponding author.

$\lambda_2 \geq \cdots \geq \lambda_r \geq 1$, with $\emptyset$ the empty partition. Let $|\lambda|$ denote the sum of parts or *size* of λ, with $|\emptyset| := 0$. Let $N(\lambda)$ denote the product of the parts or *norm* of the partition, with $N(\emptyset) := 1$. Let $\ell(\lambda) := r$ denote the number of parts or *length* of a partition, with $\ell(\emptyset) := 0$.

Then in analogy to the Riemann zeta function $\zeta(s) := \sum_{n=1}^{\infty} n^{-s}$, convergent for $\mathrm{Re}(s) > 1$, we give the definition of a *partition zeta function* as studied in [5, 8, 9].

Definition 1.1. For a proper subset $\mathcal{P}' \subset \mathcal{P}$ and a value $s \in \mathbb{C}$ for which the series converges, we define a *partition zeta function* to be the following sum over partitions in $\mathcal{P}'$:

$$\zeta_{\mathcal{P}'}(s) := \sum_{\lambda \in \mathcal{P}'} N(\lambda)^{-s}. \tag{1.1}$$

We note that $\zeta_{\mathcal{P}}(s)$ itself diverges[a]: partitions of the shape $\lambda = (1, 1, 1, \ldots, 1)$ contribute infinitely many 1's to the sum. If $\mathcal{P}'$ is of the class $\mathcal{P}_S$ of partitions into elements from some subset $S \subsetneq \mathbb{N}$ where $1 \notin S$, then the associated zeta function has an Euler product as well

$$\zeta_{\mathcal{P}_S}(s) = \prod_{n \in S} (1 - n^{-s})^{-1}, \tag{1.2}$$

and standard techniques for working with products give other variants.

Of course, in this setting, the Riemann zeta function $\zeta(s)$ represents the case $\zeta_{\mathcal{P}_{\mathbb{P}}}(s)$ (sum over partitions into prime parts). These combinatorial zeta functions (as well as related *partition Dirichlet series*) form a highly general class of objects, yet they share many structural laws, which specialize to well-known classical zeta function (and Dirichlet series) identities, as well as more exotic non-classical cases such as the following, proved in [8], for partitions into even parts, distinct parts and parts not equal to 1, respectively,

$$\zeta_{\mathcal{P}_{\text{even}}}(2) = \frac{\pi}{2}, \quad \zeta_{\mathcal{P}_{\text{dist}}}(2) = \frac{\sinh \pi}{\pi}, \quad \zeta_{\mathcal{P}_{\neq 1}}(3) = \frac{3\pi}{\cosh\left(\frac{1}{2}\pi\sqrt{3}\right)}. \tag{1.3}$$

Since the choice of subset $\mathcal{P}' \subsetneq \mathcal{P}$ as well as domain of $s \in \mathbb{C}$ determine a given partition zeta function, each pair of choices yields distinctive yet disparate evaluations and analytic properties.

Then identifying interesting non-classical cases seems like something of a needle-in-a- haystack type of undertaking. One wonders: are there non-classical families of partition zeta functions having such nice properties that they truly parallel, for instance, Euler's evaluations of $\zeta(s)$ at positive even arguments

$$\zeta(2m) = \pi^{2m} \times \text{rational number?}$$

[a]Similarly if the maximum number of 1's appearing in partitions in $\mathcal{P}'$ is unrestricted.

2. A Nice Family of Partition Zeta Functions

A particularly nice class of zeta functions is defined in [8], that generalizes $\zeta(s)$ in a more fruitful direction than just writing $\zeta(s) = \zeta_{\mathcal{P}_{\mathbb{P}}}(s)$.

Definition 2.1. For $\mathrm{Re}(s) > 1$, we define the family of zeta sums taken over all partitions of fixed length $k \geq 0$

$$\zeta_{\mathcal{P}}(\{s\}^k) := \sum_{\ell(\lambda)=k} \frac{1}{N(\lambda)^s},$$

with $\zeta_{\mathcal{P}}(\{s\}^0) := N(\emptyset)^{-s} = 1$.

The $k = 1$ case is $\zeta(s)$. For $|z| < 1$, we have the obvious generating function:

$$\prod_{n=1}^{\infty}(1 - zn^{-s})^{-1} = \sum_{k=0}^{\infty} \zeta_{\mathcal{P}}(\{s\}^k) z^k. \tag{2.1}$$

The first author proves in [8] that for argument $s = 2$ and $k \geq 1$, these partition zeta functions are actually rational multiples of Euler's zeta values (and of π^{2k}).

Theorem 2.2 (Schneider). *For $k \geq 1$, we have*

$$\zeta_{\mathcal{P}}(\{2\}^k) = \frac{2^{2k-1} - 1}{2^{2k-2}} \zeta(2k).$$

We note that setting $k = 0$ in Proposition 2.2 suggests formally that $\zeta(0) = \frac{2^{-2}}{2^{-1}-1}\zeta_{\mathcal{P}}(\{2\}^0) = -1/2$, which is the correct value for $\zeta(0)$ obtained through analytic continuation.

In [5], Ono *et al.* prove other facts about partition zeta functions, including special cases yielding analytic continuation (restricted, however, to the right half of the complex plane) and a farther-reaching follow-up to the preceding theorem.

Theorem 2.3 (Ono–Rolen–Schneider). *For $m \geq 1, k \geq 1$, we have*

$$\zeta_{\mathcal{P}}(\{2m\}^k) = \pi^{2mk} \times \textit{rational number}.$$

So these zeta sums over partitions of fixed length do indeed form a family like Euler's zeta values; but zeta values are only the beginning of the story of the zeta function, as Riemann displayed in a brilliant sequel [7] to Euler's work. Then it is natural to ask about analytic properties of the general case $\zeta_{\mathcal{P}}(\{s\}^k), \mathrm{Re}(s) > 1$. Here, we prove a closed formula for the partition zeta function for each $k \geq 0$, that speaks to this question.

Let $\lambda \vdash n$ mean that λ is a partition of size $n \geq 0$, and let $m_j = m_j(\lambda) \geq 0$ denote the *multiplicity* (i.e. number of occurrences) of the integer $j \geq 1$ as a part of partition λ.

Theorem 2.4. *For $\mathrm{Re}(s) > 1, k \geq 0$, we have*

$$\zeta_{\mathcal{P}}(\{s\}^k) = \sum_{\lambda \vdash k} \frac{\zeta(s)^{m_1}\zeta(2s)^{m_2}\zeta(3s)^{m_3}\cdots\zeta(ks)^{m_k}}{N(\lambda)\, m_1!\, m_2!\, m_3! \cdots m_k!},$$

where the sum on the right-hand side is taken over the partitions of size $k \geq 0$.

To prove the theorem, we need the Maclaurin expansion of the natural logarithm for $|x| < 1$

$$-\log(1-x) = \sum_{j=1}^{\infty} \frac{x^j}{j}, \tag{2.2}$$

as well as a partition version of *Faà di Bruno's formula.*

Theorem 2.5 (Faà di Bruno). *Take* $\exp(t) := e^t$ *and* $a_j, x \in \mathbb{C}$ *such that* $\sum_{j=1}^{\infty} a_j x^j$ *converges. Then we have*

$$\exp\left(\sum_{j=1}^{\infty} a_j x^j\right) = \sum_{\lambda \in \mathcal{P}} x^{|\lambda|} \frac{a_1^{m_1} a_2^{m_2} a_3^{m_3} \cdots}{m_1!\, m_2!\, m_3! \cdots} = \sum_{k=0}^{\infty} x^k \sum_{\lambda \vdash k} \frac{a_1^{m_1} a_2^{m_2} a_3^{m_3} \cdots\ a_k^{m_k}}{m_1!\, m_2!\, m_3! \cdots m_k!}. \tag{2.3}$$

Proof of Theorem 2.4. For $|z| < 1, \mathrm{Re}(s) > 1$, by (2.2) we rewrite the product side of generating function (2.1) as

$$\prod_{n=1}^{\infty} \exp(-\log(1 - zn^{-s})) = \exp\left(\sum_{n=1}^{\infty}\sum_{j=1}^{\infty} \frac{z^j}{n^{js} j}\right) = \exp\left(\sum_{j=1}^{\infty} \frac{\zeta(js)}{j} z^j\right). \tag{2.4}$$

Now, setting $x = z$ and $a_j = \zeta(js)/j$ in (2.3) and noting $N(\lambda) = \prod_{j \geq 1} j^{m_j}$ for each partition λ, comparing the coëfficient of z^k with the right-hand side of (2.1) gives the theorem. □

Theorem 2.4 yields much information about the analytic properties of $\zeta_{\mathcal{P}}(\{s\}^k)$.

Corollary 2.6. *The partition zeta function* $\zeta_{\mathcal{P}}(\{s\}^k)$:

(1) *Inherits analytic continuation from* $\zeta(s)$, *to the entire complex plane minus poles*;
(2) *Has poles at* $s = 1, 1/2, 1/3, 1/4, \ldots, 1/k$ *with the order of* $s = 1/j$ *being* $\lfloor k/j \rfloor$;
(3) *Has trivial roots at* $s = -2, -4, -6, -8$, *etc.*;
(4) *Does not have roots at the nontrivial roots of* $\zeta(s)$.

Proof. All four items of the corollary follow immediately from well-known analytic properties of the Riemann zeta function. That $\zeta_{\mathcal{P}}(\{s\}^k)$ inherits analytic continuation from the finite combination of $\zeta(js)$ factors is obvious, and the pole at $s = 1/j$ comes from the pole of $\zeta(js)$ at $js = 1$ with order of the pole coming from the partition of k with the maximum possible number $\lfloor k/j \rfloor$ of j's. For $s \in -2\mathbb{N}$, all the $\zeta(js)$ vanish trivially, giving trivial roots of the partition zeta function; but on the assumption no single value s can serve as a nontrivial root for all the $\zeta(js), 1 \leq j \leq k$, simultaneously, the partition zeta function does not entirely vanish at any nontrivial root of $\zeta(s)$. □

3. Combinatorial Approach

By contrast with the analytic proof of the previous section, in this section, we give a combinatorial proof of Theorem 2.4. We take an almost identical approach to the second author's work [10] generating partitions of fixed length and allowing the symmetric group to act on these partitions to create *integer compositions.*

Let $\mathcal{C}_k$ denote the set of k-tuples of positive integers $c = (c_1, c_2, \ldots, c_k)$. These are *integer compositions of length k* (or "k-compositions"). As with partitions, the c_j in $c = (c_1, \ldots, c_k)$ are the *parts* of c, and the sum of parts $|c|$ is the *size* of c.

We proceed by example. Recall that $\zeta_{\mathcal{P}}(\{s\}^0) = 1$ and the $k = 1$ case is simply

$$\zeta(s) = \sum_{n\geq 1} n^{-s} = \zeta_{\mathcal{P}}(\{s\}^1).$$

The $k = 2$ case of Theorem 2.4 may be proved combinatorially as follows:

$$\begin{aligned}
\sum_{\lambda\vdash 2} \frac{\zeta(s)^{m_1}\zeta(2s)^{m_2}}{N(\lambda)m_1!m_2!} &= \frac{1}{2}\zeta(s)^2 + \frac{1}{2}\zeta(2s) \\
&= \frac{1}{2}\sum_{n_1\geq 1} n_1^{-s} \sum_{n_2\geq 1} n_2^{-s} + \frac{1}{2}\sum_{n\geq 1}(n^2)^{-s} \\
&= \frac{1}{2}\sum_{(n_1,n_2)\in\mathcal{C}_2} \frac{1}{(n_1n_2)^s} + \frac{1}{2}\sum_{\substack{(n_1,n_2)\in\mathcal{C}_2\\ n_1=n_2}} \frac{1}{(n_1n_2)^s} \\
&= \frac{1}{2}\left(\sum_{\substack{(n_1,n_2)\in\mathcal{C}_2\\ n_1\neq n_2}} \frac{1}{(n_1n_2)^s} + \sum_{\substack{(n_1,n_2)\in\mathcal{C}_2\\ n_1=n_2}} \frac{1}{(n_1n_2)^s}\right) \\
&\quad + \frac{1}{2}\sum_{\substack{(n_1,n_2)\in\mathcal{C}_2\\ n_1=n_2}} \frac{1}{(n_1n_2)^s} \\
&= \frac{1}{2}\sum_{\substack{(n_1,n_2)\in\mathcal{C}_2\\ n_1\neq n_2}} \frac{1}{(n_1n_2)^s} + \left(\frac{1}{2}+\frac{1}{2}\right)\sum_{\substack{(n_1,n_2)\in\mathcal{C}_2\\ n_1=n_2}} \frac{1}{(n_1n_2)^s} \\
&= \frac{1}{2}\sum_{\substack{(n_1,n_2)\in\mathcal{C}_2\\ n_1\lessgtr n_2}} \frac{1}{(n_1n_2)^s} + \sum_{\substack{(n_1,n_2)\in\mathcal{C}_2\\ n_1=n_2}} \frac{1}{(n_1n_2)^s} \\
&= \sum_{n_1>n_2\geq 1} \frac{1}{(n_1n_2)^s} + \sum_{n_1=n_2\geq 1} \frac{1}{(n_1n_2)^s} \\
&= \sum_{\substack{\pi\in\mathcal{P}\\ \ell(\pi)=2}} \frac{1}{N(\pi)^s} = \zeta_{\mathcal{P}}(\{s\}^2).
\end{aligned}$$

The details for the case $k = 3$ are as follows:

$$\sum_{\lambda\vdash 3} \frac{\zeta(s)^{m_1}\zeta(2s)^{m_2}\zeta(3s)^{m_3}}{N(\lambda)m_1!m_2!m_3!} = \frac{1}{6}(\zeta(s))^3 + \frac{1}{2}\zeta(2s)\zeta(s) + \frac{1}{3}\zeta(3s)$$

$$= \frac{1}{6}\sum_{n_1\geq 1} n_1^{-s} \sum_{n_2\geq 1} n_2^{-s} \sum_{n_3\geq 1} n_3^{-s} + \frac{1}{2}\sum_{n_1\geq 1}(n_1^2)^{-s}\sum_{n_2\geq 1} n_2^{-s} + \frac{1}{3}\sum_{n\geq 1}(n^3)^{-s}$$

$$= \left[\frac{1}{6}\sum_{(n_1,n_2,n_3)\in\mathcal{C}_3} + \left(\frac{1}{6}\sum_{\substack{(n_1,n_2,n_3)\in\mathcal{C}_3\\ n_1=n_2}} + \frac{1}{6}\sum_{\substack{(n_1,n_2,n_3)\in\mathcal{C}_3\\ n_1=n_3}} + \frac{1}{6}\sum_{\substack{(n_1,n_2,n_3)\in\mathcal{C}_3\\ n_2=n_3}}\right)\right.$$

$$\left. + \frac{1}{3}\sum_{\substack{(n_1,n_2,n_3)\in\mathcal{C}_3\\ n_1=n_2=n_3}}\right]\frac{1}{(n_1n_2n_3)^s}$$

$$= \left[\frac{1}{6}\left(\sum_{\text{all } n_i \text{ unequal}} + \sum_{n_1=n_2\neq n_3} + \sum_{n_1=n_3\neq n_2} + \sum_{n_1\neq n_2=n_3} + \sum_{n_1=n_2=n_3}\right)\right.$$

$$+ \frac{1}{6}\left(\sum_{n_1=n_2\neq n_3} + \sum_{n_1=n_2=n_3}\right) + \frac{1}{6}\left(\sum_{n_1=n_3\neq n_2} + \sum_{n_1=n_2=n_3}\right)$$

$$\left. + \frac{1}{6}\left(\sum_{n_1\neq n_2=n_3} + \sum_{n_1=n_2=n_3}\right) + \frac{1}{3}\sum_{n_1=n_2=n_3}\right]\frac{1}{(n_1n_2n_3)^s}$$

$$= \left[\frac{1}{6}\left(\sum_{\text{all } n_i \text{ unequal}}\right) + \frac{1}{3}\left(\sum_{n_1=n_2\neq n_3} + \sum_{n_1=n_3\neq n_2} + \sum_{n_1\neq n_2=n_3}\right)\right.$$

$$\left. + 1\sum_{n_1=n_2=n_3}\right]\frac{1}{(n_1n_2n_3)^s}$$

$$= \left[\frac{1}{6}\sum_{\text{all } n_i \text{ unequal}} + \frac{1}{3}\sum_{\substack{n_1=n_2\lesseqgtr n_3,\text{or}\\ n_1=n_3\lesseqgtr n_2,\text{or}\\ n_1\lesseqgtr n_2=n_3}} + \sum_{n_1=n_2=n_3}\right]\frac{1}{(n_1n_2n_3)^s}$$

$$= \left[\sum_{n_1>n_2>n_3\geq 1} + \left(\sum_{n_1>n_2=n_3\geq 1} + \sum_{n_1=n_2>n_3\geq 1}\right) + \sum_{n_1=n_2=n_3}\right]\frac{1}{(n_1n_2n_3)^s}$$

$$= \sum_{\substack{\pi\in\mathcal{P}\\ \ell(\pi)=3}} \frac{1}{N(\pi)^s} = \zeta_{\mathcal{P}}(\{s\}^3),$$

where "$a \lessgtr b$" means $a \neq b$, but we mean to emphasize that only two of the six cases where exactly two of the n_i are equal, namely "$n_1 > n_2 = n_3$" and "$n_1 = n_2 > n_3$," yield a partition.

As the length k increases, analogous combinatorial principles of splitting up, sorting compositions and regrouping are at play in the indices of summation. For the corresponding proof of the general k case, the reader is referred to [10]; while they are too lengthy to conveniently reproduce here, the exact steps of the proof of MacMahon's partial fraction formula given in [10] also prove Theorem 2.4 under certain substitutions, which we detail in Sec. 4.

4. Correspondence with MacMahon Partial Fractions

The combinatorial identities in Sec. 3 mimic the second author's proof [10] of MacMahon's partial fraction decomposition of the generating function for partitions of length $\leq k$ (see [3, 4]). Let us recall MacMahon's result.

Theorem 4.1 (MacMahon). *For $|q| < 1$, we have*

$$\prod_{j=1}^{k} \frac{1}{1-q^j} = \sum_{\lambda \vdash k} \frac{1}{N(\lambda)\; m_1!\; m_2!\; \cdots m_k!\; (1-q)^{m_1}(1-q^2)^{m_2}\cdots(1-q^k)^{m_k}}. \tag{4.1}$$

Multiplying both sides of (4.1) by q^k and noting for every partition of k that $k = m_1 + 2m_2 + 3m_3 + \cdots + km_k$, gives the generating function for partitions of length *exactly* k

$$\frac{q^k}{\prod_{j=1}^{k}(1-q^j)} = \sum_{\lambda \vdash k} \frac{q^{m_1} q^{2m_2} q^{3m_3} \cdots q^{km_k}}{N(\lambda)\; m_1!\; m_2! \cdots m_k!\; (1-q)^{m_1}(1-q^2)^{m_2}\cdots(1-q^k)^{m_k}}. \tag{4.2}$$

We point out that there is a simple bijection between partitions generated in (4.2) and those in (4.1): for a partition $\lambda = (\lambda_1, \lambda_2, \ldots, \lambda_j)$ of length $j \leq k$, identify λ with the k-tuple $(\lambda_1, \lambda_2, \ldots, \lambda_j, 0, 0, \ldots, 0)$ and then add 1 to each component of the k-tuple to obtain a partition of length exactly k.

Then comparing Eq. (4.2) to Theorem 2.4, one sees an apparent correspondence between q-series generating functions and the respective zeta function components in Theorem 2.4.[b] We summarize these observations in the table below;

[b]Similar correspondences between q-series and zeta functions have been studied by Schneider and his collaborators (see [5, 6, 8, 9]).

corresponding entries evidently encode the same partition-theoretic information:

Generating function component	Analogous zeta function component
$q^{\|\lambda\|}$	$N(\lambda)^{-s}$
$\dfrac{q^j}{1-q^j}$	$\zeta(js)$
$\dfrac{q^k}{\prod_{j=1}^{k}(1-q^j)}$	$\zeta_{\mathcal{P}}(\{s\}^k)$

The analogy between the left-hand sides of (4.2) and Theorem 2.4 is clear when we compare the following summation representations:

$$\frac{q^k}{\prod_{j=1}^{k}(1-q^j)} = \sum_{\ell(\lambda)=k} q^{|\lambda|} \longleftrightarrow \zeta_{\mathcal{P}}(\{s\}^k) = \sum_{\ell(\lambda)=k} N(\lambda)^{-s}. \tag{4.3}$$

There is a one-to-one correspondence between the terms in each summation, i.e. between the partitions being generated. We note that multiplication of terms of either shape q^n or n^{-s} generates partitions in exactly the same way, viz. for partition $\lambda = (\lambda_1, \lambda_2, \lambda_3, \ldots, \lambda_r)$

$$q^{\lambda_1}q^{\lambda_2}q^{\lambda_3}\cdots q^{\lambda_r} = q^{|\lambda|} \longleftrightarrow \lambda_1^{-s}\lambda_2^{-s}\lambda_3^{-s}\cdots\lambda_r^{-s} = N(\lambda)^{-s}. \tag{4.4}$$

Therefore, taken in finite combinations, geometric series and zeta functions behave identically as partition generating functions.[c] Whereas the size $|\lambda|$ is the partition-encoding statistic in the former scheme, the norm $N(\lambda)$ encodes partitions in the latter.

Moreover, the summand q^{jn} in the geometric series $\sum_{n=1}^{\infty} q^{jn}$ and the respective term n^{-js} of $\zeta(js)$ can both be viewed as encoding the partition $(n)^j := (n, n, \ldots, n)$ consisting of j repetitions of the part n, viz.

$$\frac{q^j}{1-q^j} = \sum_{n=1}^{\infty} q^{|(n)^j|} \longleftrightarrow \zeta(js) = \sum_{n=1}^{\infty} N((n)^j)^{-s}. \tag{4.5}$$

The nth terms of the summations in (4.5) are in one-to-one correspondence.

Now, since the multiplicities $m_1, m_2, \ldots, m_k$ associated to any length-k partition must add up to k, these nonzero multiplicities themselves represent a permutation of some partition of size k. This observation brings compositions into the picture, and also provides a link between partitions of length k and those of size k.

[c]For infinite combinations, a different analogy holds, between product generating functions like $\prod_{j=1}^{\infty}(1-q^j)^{-1}$ and corresponding Euler products (see [8]).

To prove MacMahon's partial fraction formula in [10], roughly speaking, one replaces the term q^j with a multivariate product $x_1x_2x_3\ldots x_j$, $|x_i|<1$, then permutes the x_i's. More exactly, the geometric factors in the right-hand summands of (4.1) will be rewritten

$$\frac{1}{(1-x_1)^{m_1}(1-x_2x_3)^{m_2}\ (1-x_4x_5x_6)^{m_3}\cdots}. \tag{4.6}$$

One then permutes the x_i's of nonempty geometric factors, effectively giving rise to compositions, while keeping track of the permutations in the indices of summation as in the examples in Sec. 3, using properties of the symmetric group to enumerate multiply-counted terms. In the end one sets each dummy variable $x_i = q$ to produce (4.1).

The counting arguments of [10] are unaltered in the case of partitions of length exactly k. One can rewrite the geometric aspect of each right-hand summand of (4.2) as

$$\frac{x_1^{m_1}(x_2x_3)^{m_2}(x_4x_5x_6)^{m_3}\cdots}{(1-x_1)^{m_1}(1-x_2x_3)^{m_2}(1-x_4x_5x_6)^{m_3}\cdots}, \tag{4.7}$$

and by precisely the same steps taken in [10], Eq. (4.2) is proved. Going a step further in this direction: as $|x_i|<1$, one can rewrite the jth geometric factor of (4.7) above in series form

$$\frac{x_{i_1}x_{i_2}\cdots x_{i_j}}{1-x_{i_1}x_{i_2}\cdots x_{i_j}} = \sum_{n=1}^{\infty} x_{i_1}^n x_{i_2}^n\cdots x_{i_j}^n, \tag{4.8}$$

noting $i_{m+1} = i_m + 1$ in the indices, then in the sum on the right side of (4.8) make the change

$$x_i^n \mapsto n^{-x_i}, \tag{4.9}$$

with the x_i's now needing to satisfy $\mathrm{Re}(x_{i_1}+x_{i_2}+\cdots+x_{i_j})>1$ for convergence of the series. This mapping produces a one-to-one termwise correspondence

$$\sum_{n=1}^{\infty} x_{i_1}^n x_{i_2}^n\cdots x_{i_j}^n \longleftrightarrow \sum_{n=1}^{\infty} n^{-x_{i_1}}n^{-x_{i_2}}\cdots n^{-x_{i_j}}. \tag{4.10}$$

Moreover, the terms of both these series are in one-to-one correspondence with the summands of the jth geometric factor of (4.6) when it is expanded as a series

$$\frac{1}{1-x_{i_1}x_{i_2}\cdots x_{i_j}} = \sum_{n=1}^{\infty} x_{i_1}^{n-1}x_{i_2}^{n-1}\cdots x_{i_j}^{n-1}, \tag{4.11}$$

and the symmetric group acts on the x_i's identically in finite combinations of series of any one of these types.

Then one can replace the jth geometric factor of (4.6), $j=1,2,3,\ldots,k$, with the right-hand multivariate zeta series in (4.10), permute the x_i's and enumerate permutations following exactly the steps in [10]; finally, one sets each $x_i = s$ to arrive at Theorem 2.4.

Acknowledgments

The authors are grateful to the organizers of Analytic and Combinatorial Number Theory: The Legacy of Ramanujan, a Conference in Honor of Bruce C. Berndt's 80th Birthday, where we conceived this work. Moreover, we are thankful to Ken Ono for comments that strengthened the paper, and to the anonymous referee for useful suggestions.

References

[1] G. E. Andrews, *The Theory of Partitions* (Addison–Wesley, Reading, MA, 1976). Reissued: (Cambridge University Press, 1998).

[2] B. C. Berndt, *Number Theory in the Spirit of Ramanujan* (American Mathematical Society, 2006).

[3] P. A. MacMahon, *Combinatory Analysis*, Vol. II (Cambridge University Press, 1916).

[4] P. A. MacMahon, Seventh memoir on the partition of numbers: A detailed study of the enumeration of the partitions of multipartite numbers, *Phil. Trans. R. Soc. London A* **CCXVII** (1917) 81–113.

[5] K. Ono, L. Rolen and R. Schneider, Explorations in the theory of partition zeta functions, in *Exploring the Riemann Zeta Function, 190 Years from Riemann's Birth*, eds. H. Montgomery, A. Nikeghbali and M. Rassias (Springer, 2017), pp. 223–264.

[6] K. Ono, R. Schneider and I. Wagner, Partition-theoretic formulas for arithmetic densities, in *Analytic Number Theory, Modular Forms and q-Hypergeometric Series: in Honor of Krishna Alladi's 60th Birthday, University of Florida, Gainesville, March 2016*, eds. G. E. Andrews and F. Garvan (Springer, 2018), pp. 611–624.

[7] B. Riemann, Über die Anzahl der Primzahlen unter einer gegebenen Grösse, *Ges. Math. Werke Wiss. Nachlass* **2** (1859) 145–155.

[8] R. Schneider, Partition zeta functions, *Res. Number Theory* **2** (2016), Article id: 9, 17 pp.

[9] R. Schneider, Eulerian series, zeta functions and the arithmetic of partitions, Ph.D. thesis, Emory University (2018).

[10] A. V. Sills, On MacMahon's partial fractions, *Ann. Comb.* **23** (2019) 1073–1086.

© 2024 World Scientific Publishing Company
https://doi.org/10.1142/9789811277375_0038

Automatic discovery of irrationality proofs and irrationality measures*

Doron Zeilberger

Department of Mathematics, Rutgers University (New Brunswick)
Hill Center-Busch Campus, 110 Frelinghuysen Rd.
Piscataway, NJ 08854-8019, USA
doronzeil@gmail.com

Wadim Zudilin†

Department of Mathematics, IMAPP
Radboud University, PO Box 9010
6500 GL Nijmegen, Netherlands
w.zudilin@math.ru.nl

Dedicated to Bruce Berndt (b. March 13, 1939), on his $80\frac{3}{4}$*th birthday*

We illustrate the power of *Experimental Mathematics* and *Symbolic Computation* to suggest irrationality proofs of natural constants, and the determination of their irrationality measures. Sometimes such proofs can be fully automated, but sometimes there is still need for a human touch.

Keywords: Irrationality measure; experimental mathematics; Almkvist–Zeilberger algorithm; Wilf–Zeilberger algorithmic proof theory.

Mathematics Subject Classification 2020: 11J71, 11J82, 11Y60, 33F10

The Maple packages. This paper is accompanied by *four* Maple packages:

- `ALLADI.txt`: A Maple package, inspired by the Alladi–Robinson [1]. It does an automated redux of Theorem 1 in their paper, and extends their results to proving irrationality, as well as finding irrationality measures, of constants of the form $\int_0^1 \frac{dx}{P(x)}$, where $P(x)$ is a quadratic polynomial with integer coefficients.

*This article originally appeared in *International Journal of Number Theory*, **17**(3), pp. 815–825
© 2021 World Scientific Publishing Company. DOI: 10.1142/S1793042120400230
†Corresponding author.

- `GAT.txt`: A Maple package that includes the former case, but generalizes it to integrals of the form $\int_0^1 \frac{dx}{a+x^k}$, where a and k are positive integers.
- `BEUKERS.txt`: A Maple package inspired by Beukers' [5] for getting $\mathbb{Z}$-linear combinations of 1, $\mathrm{dilog}(a/(a-1)))$, and $\log(a/(a-1))$ that are very small, for integers $a \geq 2$. (*NB*: We follow Maple's convention $\mathrm{dilog}(1-x) = \sum_{n=1}^{\infty} \frac{x^n}{n^2}$ for the dilogarithmic function.)
- `SALIKHOV.txt`: A Maple package that generalizes Salikhov's method [11], to discover and prove linear independence measures of $1, \log(a/(a+1)), \log(b/(b+1))$ for many pairs of integers $b > a \geq 2$, in particular, for $1, \log(a/(a+1))$, $\log((a+1)/(a+2))$ for all $a \geq 1$.

They are available, along with numerous output files, in the form of computer-generated papers, from the following url:
`http://www.math.rutgers.edu/~zeilberg/mamarim/mamarimhtml/gat.html`,
and its mirrors,
`http://wain.mi.ras.ru/gat/` and `http://www.math.ru.nl/~wzudilin/gat/`.
The packages are reusable, as long as Maple exists and at least one of the webpages survives.

Preface: Apéry's astounding proof (and Berndt's seminar talk) and the Almkvist–Zeilberger algorithm. In 1978, 64-year-old Roger Apéry announced, and sketched an amazing proof that $\zeta(3) = \sum_{n=1}^{\infty} \frac{1}{n^3}$ is an irrational number. Some of the details were filled in by Henri Cohen and Don Zagier, and the completed proof was the subject of Alf van der Poorten's classic [12].

One of us (DZ) first learned about this proof from an excellent talk by Bruce Berndt, delivered at the University of Illinois, way back in Fall 1979 (when and where DZ had his third postdoc). It is therefore quite appropriate that we dedicate this paper to Bruce Berndt, since it deals with irrationality of constants inspired by Apéry's seminal proof, exposited so lucidly by Berndt.

Another *leitmotif* of this paper is the Almkvist–Zeilberger algorithm. Gert Almkvist and Bruce Berndt co-authored a classic expository paper [2] that won a prestigious *MAA Lester Ford award* in 1988. Since Almkvist (1934–2018) was also a good friend, and long-time collaborator, of the second-named author (WZ), and both authors are good friends and admirers of Berndt, it is more than fitting to dedicate this paper to Bruce Berndt.

Apéry's proof of the irrationality of $\zeta(3)$. Apéry (see [12]) *pulled out of a hat* two *explicit* sequences of rational numbers. The first sequence consisted only of integers

$$b_n := \sum_{k=0}^{n} \binom{n}{k}^2 \binom{n+k}{k}^2$$

now known as the Apéry numbers, while the second one was a sequence of rational numbers

$$a_n := \sum_{k=0}^{n} \binom{n}{k}^2 \binom{n+k}{k}^2 \left(\sum_{m=1}^{n} \frac{1}{m^3} + \sum_{m=1}^{k} \frac{(-1)^{m-1}}{2m^3 \binom{n}{m}\binom{n+m}{m}} \right).$$

It is reasonably easy to check that $p_n := \mathrm{lcm}(1,\ldots,n)^3 a_n$ are integers and, of course, $q_n := \mathrm{lcm}(1,\ldots,n)^3 b_n$ are integers. (Thanks to the Prime Number Theorem $\mathrm{lcm}(1,\ldots,n)$ is $\Omega(e^n)$, where we use the convenient notation $F(n) = \Omega(c^n)$ meaning $\lim_{n\to\infty}(\log F(n))/n = c$.) Then he came up with a real number

$$\delta = \frac{4\log(1+\sqrt{2}) - 3}{4\log(1+\sqrt{2}) + 3} = 0.080259\ldots > 0$$

such that, for some constant C (independent of n),

$$\left| \zeta(3) - \frac{p_n}{q_n} \right| \leq \frac{C}{q_n^{1+\delta}}.$$

Once all the claims are verified, it follows that $\zeta(3)$ is irrational. (Indeed, if $\zeta(3)$ would have been rational with denominator c, the left-hand side would have been bounded below by $1/(cq_n)$.) It also follows from the latter estimate that the *irrationality measure* of $\zeta(3)$ (see [12]) is bounded above by $1 + 1/\delta = 12.417820\ldots$.

Beukers' version. Months later, Frits Beukers [5] gave a much simpler rendition of Apéry's construction by introducing a certain explicit triple integral

$$I(n) := \int_0^1\int_0^1\int_0^1 \left(\frac{x(1-x)y(1-y)z(1-z)}{1-(1-xy)z} \right)^n \frac{\mathrm{d}x\,\mathrm{d}y\,\mathrm{d}z}{1-(1-xy)z},$$

and pointing out that

(i) $I(n)$ is small and can be explicitly bounded: $I(n) = O((\sqrt{2}-1)^{4n})$;
(ii) $I(n) = A(n) - B(n)\zeta(3)$ for certain sequences of rational numbers $A(n)$, $B(n)$, that can be explicitly bounded; and
(iii) $\mathrm{lcm}(1,\ldots,n)^3 A(n)$ and $\mathrm{lcm}(1,\ldots,n)^3 B(n)$ are integers.

The final argument is as in Apéry's proof, and both the irrationality of $\zeta(3)$ and the estimate $12.417820\ldots$ for its irrationality measure follow.

Shortly after, Krishna Alladi and Robinson [1] used one-dimensional analogs to reprove the irrationality of $\log 2$, and established an upper bound for its irrationality measure of 4.63 (subsequently improved, see [13]) by considering the simple integral

$$I(n) := \int_0^1 \left(\frac{x(1-x)}{1+x} \right)^n \frac{\mathrm{d}x}{1+x}.$$

Man muss immer umkehren. Carl Gustav Jacob Jacobi told us that *one must always invert.* Of course, he meant that if you have a complicated looking function like $\int_0^x \frac{1}{\sqrt{1-t^2}}\,dt$, its inverse function — in this case $\sin x$ — can be much more user-friendly. We understand Jacobi's quip in a different way. Rather than start with a famous constant, say $\zeta(3)$ or $\log 2$, and wreck our brains trying to find Beukers-like or Alladi–Robinson-like integrals that would produce good diophantine approximations, start with a family of integrals, $I(n)$ say, and see

(i) whether $I(0)$ is a familiar constant, let us call it x;
(ii) whether $I(n)$, for integers $n > 0$, can be written as $A(n) - B(n)x$, with $\{A(n)\}$ and $\{B(n)\}$ sequences of rational numbers;
(iii) whether $I(n)$ has exponential decay, i.e. is 'small';

Since $A(n) - B(n)x$ is so close to 0, the rational $A(n)/B(n)$ is very close to x. Write $A(n)/B(n)$ as $A'(n)/B'(n)$, where now $A'(n)$ and $B'(n)$ are *integers*, and verify

(iv) how close is $A(n)/B(n) = A'(n)/B'(n)$ to x, from a *diophantine* (rather than numerical-analysis) point of view.

We are looking for what we identify as an 'empirical delta', let us call it $\delta(n)$, such that

$$\left|x - \frac{A'(n)}{B'(n)}\right| = \frac{1}{B'(n)^{1+\delta(n)}};$$

in other words, we define

$$\delta(n) := \frac{-\log|x - A'(n)/B'(n)|}{\log B'(n)} - 1.$$

If the values of $\delta(n)$, for $990 \le n \le 1000$ say, are all strictly positive, and safely not *too* close to 0, then we can be assured that there *exists* a proof of irrationality of x together with a corresponding rigorous upper bound for its irrationality measure.

However being sure that a proof *exists* is not the same as having one. It would be nice to

(v) have a fully rigorous proof.

First, try to find one yourself, and you are welcome to get help from your computer, that excels not only in *number-crunching*, but also in *symbol-crunching*, but is still not so good in *idea-crunching* (although it is getting better and better!).

If you are stuck, you can always email an expert number-theorist and offer her or him to collaborate with you on the paper

"Proof of the irrationality of x".

If x was not yet proved to be irrational and, furthermore, is sufficiently famous (e.g., Euler's constant, γ, or Catalan's constant, G), you and your collaborator would become famous too (that what happened to Apéry). If the constant in question is complicated and obscure, it is still publishable, at least in the `arXiv`. If x is already proved to be irrational, and there is currently a proved upper bound for the irrationality measure of x and the implied (rigorous) bound from your sequence $A(n)/B(n)$ is better (i.e. smaller) than the previous one, you (and your expert collaborator) can write a paper

"A new upper bound for the irrationality measure of x",

and you (and your expert collaborator) would be known as the current holders of the world record of the irrationality measure of x, until someone else, in turn, would break your record.

In this paper, we follow the methodology outlined and show how, using the amazing Almkvist–Zeilberger algorithm [3] that finds (and at the same time, proves!) a linear recurrence equation with polynomial coefficients for such integrals $I(n)$, one can accomplish the four steps (i)–(iv) (that we call *reconnaissance*) very fast and very efficiently, and sometimes (but not always!) the last — 'rigorous' — step (v) can also be automated.

Warm up: Computerized irrationality proof of log 2. Consider the sequence of definite integrals

$$I(n) := \int_0^1 \left(\frac{x(1-x)}{1+x} \right)^n \frac{dx}{1+x}.$$

The Almkvist–Zeilberger algorithm [3] produces a linear recurrence equation with polynomial coefficients satisfied by $I(n)$:

$$(n+2)I(n) + (-6\,n-9)I(n+1) + (n+1)I(n+2) = 0.$$

From this we can compute, very fast, many values, and find out the 'empirical deltas'. For example,

$$I(50) = -\frac{18270835389224940244881539949907869989471021543939584297773}{172169139124777594800} \\ + 15310086199495855930932559804210504653 \log 2.$$

This implies the rational approximation to $\log 2$ (by 'pretending' that $I(50)$ is zero) of

$$\frac{18270835389224940244881539949907869989471021543939584297773}{2635924360893339481850468164186010894239049450495548604400},$$

whose empirical delta is

$$0.33269846131126944438\ldots.$$

This is encouraging! But we cannot judge from just one data point. We next find that the 'empirical deltas' for $n = 51$ and $n = 53$ are $0.31992792581569268673\ldots$ and $0.30031107795443952791\ldots$, respectively. To get more confidence, we need to go to higher values of n. The lowest empirical delta between $n = 990$ and $n = 1000$ turns out to be

$$0.28193333613008344616\ldots,$$

that is not as good, but is way above 0. It leads to an upper estimate $1+1/\delta(n) \le 4.5469377751717949058$ for the irrationality measure of $\log 2$. This trend persists, so we can be convinced that the integral $I(n)$ is promising. But this is *not* yet a rigorous proof.

Writing $I(n) = A(n) + B(n)\log 2$, the next step is to (automatically!) find the rate of growth of $A(n)$ and $B(n)$. The original proof in [1] used partial fractions, and the saddle-point method, but thanks to the Poincaré lemma (see [12, 14] for related use and references) we can do it very fast and automatically.

Note that $A(n)$ and $B(n)$ satisfy the same recurrence. In other words,

$$(n+2)A(n) + (-6\,n-9)A(n+1) + (n+1)A(n+2) = 0,$$
$$(n+2)B(n) + (-6\,n-9)B(n+1) + (n+1)B(n+2) = 0.$$

The 'constant-coefficient approximation' of the above recurrence is (taking the leading coefficient in n, that happens to be $n^1 = n$)

$$\bar{A}(n) - 6\bar{A}(n+1) + \bar{A}(n+2) = 0,$$

where $\bar{A}(n)$ is an approximation to $A(n)$ that Poincaré proved has the property that

$$\lim_{n\to\infty} \frac{\log \bar{A}(n)}{\log A(n)} = 1,$$

and similarly for $B(n)$ and $\bar{B}(n)$. The *indicial* equation of this constant-coefficient linear recurrence is

$$1 - 6N + N^2 = 0,$$

whose roots are

$$a = 3 + 2\sqrt{2}, \quad b = 3 - 2\sqrt{2}.$$

It follows that $|A(n)|, |B(n)| = \Omega(a^n)$ and that $|I(n)| = \Omega(b^n)$, since $I(n)$ is obviously the sub-dominant solution, of exponential *decay*. There is only one problem: $A(n)$ is not an integer. The computer can easily check, empirically that $A'(n) := \operatorname{lcm}(1,\dots,n)A(n)$ is an integer for $1 \le n \le 1000$, and then try to prove it in general (or get a little help from a human friend). Then defining additionally

$$B'(n) := \operatorname{lcm}(1,\dots,n)B(n), \quad I'(n) := \operatorname{lcm}(1,\dots,n)I(n),$$

we have that $A'(n), B'(n)$ are integers. As before, $\operatorname{lcm}(1,\dots,n) = \Omega(e^n)$, hence

$$|A'(n)|, B'(n) = \Omega(a^n\, e^n), \quad |I'(n)| = \Omega(b^n\, e^n).$$

Since we want

$$|I'(n)| = \frac{1}{\Omega(B'(n)^\delta)},$$

we take

$$\delta = \frac{-\log(be)}{\log(ae)} = -\frac{\log b + 1}{\log a + 1} = \frac{\log(3+2\sqrt{2}) - 1}{\log(3+2\sqrt{2}) + 1},$$

leading to the Alladi–Robinson upper bound of $1 + 1/\delta$ that equals

$$2\,\frac{\log\,(3+2\,\sqrt{2})}{\log\,(3+2\,\sqrt{2})-1} = 4.62210083245423133342\ldots.$$

This has been improved several times [13], first by Ekaterina Rukhadze [10] (see also [15]); the current record of 3.57455391 is due to Raffaele Marcovecchio [7].

The novelty of our approach is that it can be taught to a computer, and everything, except possibly proving the 'divisibility lemma' — the right denominators of $A(n)$ (that in this case is extremely simple, but in other cases may be quite involved). Using this method, our computer, Shalosh Ekhad, proved *ab initio, all by itself* (except the simple divisibility lemma) [1, Theorem 1]. Note that nowhere did we mention *Legendre polynomials* (they turned, in hindsight, to be unnecessary). Furthermore, our approach is *streamlined*, and the formulation of the theorem is more explicit.

Theorem (Alladi–Robinson [1], but with a more explicit formulation). *Let a and b be positive integers such that $a > (b - e^{-1})^2/4$. Then* $\log(1 + b/a)$ *is an irrational number with the irrationality measure that is at most*

$$\frac{\log(2a+b-2\sqrt{a(a+b)}) - \log(2a+b+2\sqrt{a(a+b)})}{\log(2a+b-2\sqrt{a(a+b)})+1}.$$

Computer-generated proof. See
`http://sites.math.rutgers.edu/~zeilberg/tokhniot/oALLADI1.txt`. □

The above computer-generated paper was produced with the Maple package `ALLADI.txt`. We now mention other papers generated by this Maple package.

• If you want to see computer-generated proofs of irrationality of 89 different constants, and possibly new irrationality measures for each of them, you are welcome to read
`http://sites.math.rutgers.edu/~zeilberg/tokhniot/oALLADI2.txt`.
They were 'cherry-picked' from the 'candidate pool' of

$$\int_0^1 \frac{\mathrm{d}x}{a+bx+cx^2} \quad \text{with } 1 \le a,b,c \le 10, \quad \gcd(a,b,c)=1,$$

that consisted of 841 'applicants', and naturally we only listed our successes. Of course, all such constants are already proved to be irrational by heavy-artillery theorems (like the Lindemann–Weierstrass theorem), but these theorems do not give explicit bounds for the irrationality measure, and may be *ineffective*.

• Moving right along, the computer-generated paper
`http://sites.math.rutgers.edu/~zeilberg/tokhniot/oALLADI3.txt`

gives irrationality proofs, and irrationality measures, to 43 constants, for integrals of the form

$$\int_0^1 \frac{dx}{a+cx^2}$$

for relatively prime integer pairs a, c in the range $3 \le a, c \le 40$. These are probably subsumed in previous works of Salikhov and his students.

• Even more impressive is `http://sites.math.rutgers.edu/~zeilberg/tokhniot/oALLADI4.txt`, that like the above Alladi–Robinson theorem is true for 'infinitely many' constants, i.e. it is true for symbolic a (subject to a congruence condition).

This theorem may be not new, but the *novelty* is that it was completely *computer-generated.* Let us state the theorem proved in that paper.

Theorem. *Let* a *be a positive integer such that* $a \bmod 4$ *is* 3. *Then* $\arctan(\sqrt{a})/\sqrt{a}$ *is an irrational number, with the irrationality measure at most*

$$\frac{\log(-a+\sqrt{a(a+1)}) - \log(a+\sqrt{a(a+1)})}{\log(-a+\sqrt{a(a+1)}) - \log\sqrt{a}+1}.$$

The Maple packages `GAT.txt` and `BEUKERS.txt`. The Maple package `GAT.txt` did not produce (so far) anything exciting, so we do not talk about it here, but the readers are welcome to explore it on their own.

Up to this point, all our integrals were one-dimensional. The Maple package `BEUKERS.txt` attempts to tweak Beukers' elegant proof [5] of the irrationality of $\zeta(2)$, by trying to see what happens when you look at the double integral

$$E(n,a) := \int_0^1\int_0^1 \left(\frac{x(1-x)y(1-y)}{1-xy/a}\right)^n \frac{dx\,dy}{1-xy/a}.$$

The original case of $a=1$ gave an irrationality proof (and measure) for $\zeta(2) = \pi^2/6$, and indeed $E(0,1) = \zeta(2)$. The recurrence for $E(n,1)$ is second order (the same one with which Apéry proved the irrationality of $\zeta(2)$). However, things get more complicated for higher a.

Since $E(0,a) = a\,\mathrm{dilog}((a-1)/a)$, one could hope that considering the above double integral would yield irrationality proofs for them. Alas, $E(n,a)$ gets 'contaminated' with $\log((a-1)/a)$, and it turns out that

$$E(n,a) = A(n,a) + B(n,a)\,\mathrm{dilog}((a-1)/a) + C(n,a)\log((a-1)/a),$$

for all n, for *three* sequences of rational numbers $\{A(n,a)\}$, $\{B(n,a)\}$, $\{C(n,a)\}$. This can be proved directly, but it follows immediately from the fact that it is true for $n = 0,1,2$, and that $E(n,a)$, and hence $A(n,a), B(n,a), C(n,a)$, all satisfy the third-order linear recurrence equation

$$\begin{aligned}
&-a^3(n+1)^2(a-1)(32na+76a-27n-66)E(n,a)\\
&\quad +a^2(512a^3n^3+2752a^3n^2-1072a^2n^3+4800a^3n-5792a^2n^2\\
&\qquad +636an^3+2736a^3-10140a^2n+3456an^2-81n^3-5796a^2\\
&\qquad +6068na-441n^2+3472a-768n-432)\,E(n+1,a)\\
&\quad +a(256a^2n^3+1632a^2n^2-120an^3+3376a^2n-780an^2-81n^3\\
&\qquad +2232a^2-1670na-522n^2-1170a-1086n-717)E(n+2,a)\\
&\quad +(32na+44a-27n-39)(3+n)^2E(n+3,a)=0.
\end{aligned}$$

This complicated recurrence was obtained using the Apagodu–Zeilberger *multivariable Almkvist–Zeilberger algorithm* [4].

Using this recurrence we prove that, for every integer $a \geq 2$, there exists a *positive* $\delta = \delta(a)$ and three sequences of integers $C_1(n,a)$, $C_2(n,a)$, $C_3(n,a)$ such that

$$\begin{aligned}
&|C_1(n,a)+C_2(n,a)\,\mathrm{dilog}((a-1)/a)+C_3(n,a)\log((a-1)/a)|\\
&\quad \leq \frac{\mathrm{CONSTANT}}{\max(|C_1(n,a)|,|C_2(n,a)|,|C_3(n,a)|)^{\delta(a)}}.
\end{aligned}$$

The explicit expression for $\delta(a)$ is a bit involved and we refer the reader to the computer-generated paper
`http://sites.math.rutgers.edu/~zeilberg/tokhniot/oBEUKERS1.txt`,
that contains a fully rigorous proof of this theorem.

If we did not know that $\log((a-1)/a)$ was irrational, this theorem would have implied that $\mathrm{dilog}((a-1)/a)$ and $\log((a-1)/a))$ cannot be both rational. It is not enough, by itself, to prove the linear independence of the three numbers $1, \log((a-1)/a), \mathrm{dilog}((a-1)/a)$ over the rationals, but some human modification of it makes the things work well — see the latest achievements in this direction, together with historical notes, in the wonderful paper [9] of Georges Rhin and Carlo Viola.

The Maple package `SALIKHOV.txt`. We are most proud of this last Maple package, since it generated a *new* theorem that should be of interest to 'mainstream', human mathematicians. It was obtained by generalizing Vladislav Salikhov's proof [11] of the linear independence of $1, \log 2, \log 3$. Our computer proved the following result.

Theorem. *For any integer $a > 0$, three numbers 1, $\log(a/(a+1))$, and $\log((a+1)/(a+2))$ are linearly independent over the rationals. Moreover, there exists a positive number $\nu(a)$ such that if q, p_1, p_2 are integers and $Q = \max(|q|, |p_1|, |p_2|) \geq Q_0$, where Q_0 is a sufficiently large number, then*

$$|q + p_1\log(a/(a+1)) + p_2\log((a+1)/(a+2))| > Q^{-\nu(a)}.$$

The full proof is in the following paper:
`http://sites.math.rutgers.edu/~zeilberg/tokhniot/oSALIKHOV2.txt`,
where an exact expression for $\nu(a)$ can be found (see also below). The theorem was previously known to be true for $a \geq 53$ by Masayoshi Hata [6] (recently improved, though somewhat implicitly, by Volodya Lysov to $a \geq 32$).

Because of the novelty, we choose this result to feature some human-generated highlights of the proof. The integrals

$$E_1(n) = \int_0^{2a+1} \left(\frac{x^2(x^2-(2a+1)^2)(x^2-(2a+3)^2)}{(x^2-(2a+1)^2(2a+3)^2)^2} \right)^n \frac{dx}{x^2-(2a+1)^2(2a+3)^2}$$

and

$$E_2(n) = \int_0^{2a+3} \left(\frac{x^2(x^2-(2a+1)^2)(x^2-(2a+3)^2)}{(x^2-(2a+1)^2(2a+3)^2)^2} \right)^n \frac{dx}{x^2-(2a+1)^2(2a+3)^2}$$

are generalizations of integrals in Salikhov's [11], and we have

$$E_1(n) = A_1(n) + B(n) \log \frac{a+1}{a+2} \quad \text{and} \quad E_2(n) = A_2(n) + B(n) \log \frac{a}{a+1},$$

where all $E_1(n)$, $E_2(n)$, $A_1(n)$, $A_2(n)$, $B(n)$ satisfy the *same* third-order linear recurrence equation with polynomial coefficients, the indicial polynomial of whose 'constant-coefficients recurrence approximation' is

$$1 + (4a^4 + 16a^3 - 11a^2 - 54a - 34)N$$
$$- (108a^6 + 648a^5 + 1440a^4 + 1440a^3 + 614a^2 + 76a - 1)N^2 + a(a+2)N^3.$$

This polynomial has three real roots $C_1(a)$, $C_2(a)$, $C_3(a)$ for $a \geq 1$ located as follows:

$$-\frac{1}{4a^2(a+2)^2} < C_3(a) < 0 < C_2(a) < \frac{1}{27a(a+2)} < 108a^2(a+1)^2 < C_1(a);$$

also $C_2(a) > \frac{1}{4a^2(a+2)^2} > |C_3(a)|$ for $a \geq 2$. Furthermore, choosing $K(a) = a(a+2)$ if a is odd, and $K(a) = (a/2)(a/2+1)$ if a is even, we get numbers $K(a)^n A_1(n)$, $K(a)^n A_2(n)$, and $K(a)^n B(n)$ integral. Then

$$\nu(a) \leq -\frac{\log C_1(a) + \log K(a) + 2}{\log C_2(a) + \log K(a) + 2}$$

for $a \geq 2$, and the same formula for $a = 1$ with $\log C_2(a)$ replaced with $\log |C_3(a)|$. The upper bound for $\nu(a)$ is asymptotically $\frac{3 \log a(a+2)}{\log 27 - 2}$ if a is odd and $\frac{3 \log a(a+2)}{\log 108 - 2}$ if a is even, as $a \to \infty$.

Conclusion. Humans, no matter how smart, can only go so far. Machines, no matter how fast, can also only go so far. The future of mathematics depends on a fruitful *symbiosis* of the strong points of both species, as we hope we demonstrated in this modest tribute to Bruce Berndt.

References

[1] K. Alladi and M. L. Robinson, Legendre polynomials and irrationality, *J. Reine Angew. Math.* **318** (1980) 137–155.

[2] G. Almkvist and B. Berndt, Gauss, Landen, Ramanujan, the arithmetic-geometric mean, ellipses, pi and the Ladies Diary, *Amer. Math. Mon.* **95** (1988) 585–608.

[3] G. Almkvist and D. Zeilberger, The method of differentiating under the integral sign, *J. Symb. Comput.* **10** (1990) 571–591;
`http://www.math.rutgers.edu/~zeilberg/mamarimY/duis.pdf`.

[4] M. Apagodu and D. Zeilberger, Multi-variable Zeilberger and Almkvist–Zeilberger algorithms and the sharpening of Wilf–Zeilberger theory, *Adv. Appl. Math.* **37** (2006) 139–152;
`http://sites.math.rutgers.edu/~zeilberg/mamarim/mamarimhtml/multiZ.html`.

[5] F. Beukers, A note on the irrationality of $\zeta(2)$ and $\zeta(3)$, *Bull. London Math. Soc.* **11**(3) (1979) 268–272.

[6] M. Hata, The irrationality of $\log(1+1/q)\log(1-1/q)$, *Trans. Amer. Math. Soc.* **350** (1998) 2311–2327.

[7] R. Marcovecchio, The Rhin–Viola method for $\log 2$, *Acta Arith.* **139** (2009) 147–184.

[8] M. Petkovsek, H. S. Wilf and D. Zeilberger, $A = B$ (A. K. Peters, 1996);
`https://www.math.upenn.edu/~wilf/AeqB.html`.

[9] G. Rhin and C. Viola, Linear independence of 1, Li_1 and Li_2, *Moscow J. Combin. Number Theory* **8**(1) (2019) 81–96.

[10] E. A. Rukhadze, A lower bound for the approximation of $\ln 2$ by rational numbers, *Vestnik Moskov. Univ. Ser. I Mat. Mekh.* [*Moscow Univ. Math. Bull.*] **42**(6) (1987) 25–29 (Russian).

[11] V. Kh. Salikhov, On the irrationality measure of $\ln 3$, *Doklady Math.* **76** (2007) 955–957.

[12] A. van der Poorten, A proof that Euler missed Apéry's proof of the irrationality of $\zeta(3)$, An informal report, *Math. Intell.* **1** (1979) 195–203;
`http://www.ega-math.narod.ru/Apery1.htm`.

[13] E. W. Weisstein, *Irrationality Measure*, from MathWorld — A Wolfram Web Resource;
`http://mathworld.wolfram.com/IrrationalityMeasure.html`.

[14] W. Zudilin, Difference equations and the irrationality measure of numbers, *Proc. Steklov Inst. Math.* **218** (1997) 160–174.

[15] W. Zudilin, An essay on irrationality measures of π and other logarithms, *Chebyshevskii Sb.* **5** (2004) 49–65 (Russian); English version:
`http://arxiv.org/abs/math/0404523`.

www.ingramcontent.com/pod-product-compliance
Lightning Source LLC
Chambersburg PA
CBHW082046171225
36884CB00006B/781